Cherry Kearton
University of Durham
21st February 1991

Egbert Brieskorn
Horst Knörrer

Plane Algebraic Curves

Translated from the German by John Stillwell

1986 Birkhäuser Verlag
Basel · Boston · Stuttgart

Authors:

Egbert Brieskorn
Horst Knörrer
Mathematisches Institut
Universität Bonn
Wegelerstrasse 10
D-5300 Bonn

First published 1981 by Birkhäuser Boston under the title "Ebene algebraische Kurven"

ISBN 0-8176-3030-9

Library of Congress Cataloging-in-Publication Data

Brieskorn, Egbert.
Plane algebraic curves.

Translation of: Ebene algebraische Kurven.
Bibliography: p.
Includes index.
1. Curves, Algebraic. 2. Curves, Plane.
3. Geometry, Projective. I. Knörrer, Horst. II. Title.
QA565.B8413 1986 516.3'52 86-17019
ISBN 3-7643-1769-8

CIP-Kurztitelaufnahme der Deutschen Bibliothek

Brieskorn, Egbert:
Plane algebraic curves / Egbert Brieskorn ; Horst Knörrer. Transl. from the German by John Stillwell. – Basel ; Boston ; Stuttgart : Birkhäuser, 1986.
Dt. Ausg. u.d.T.: Brieskorn, Egbert: Ebene algebraische Kurven
ISBN 3-7643-1769-8

NE: Knörrer, Horst:

Printed in Germany
ISBN 3-7643-1769-8
ISBN 0-8176-3326-X

Foreword to the English edition

It is natural that authors should be delighted when their works are translated into other languages, and even more so when the translation is done very nicely – as it is the case with the present translation of our notes on plane algebraic curves by John Stillwell. On the other hand, it is also true that as time goes by one gets more aware of the defects of one's work. One of our friends has criticized us for writing a heavy volume on such an elementary subject, and we have to admit that this criticism is not totally unjustified. However, we would like to say in our defence that a number of people who are now doing research on singularities found the book quite useful as a first introduction, and so it is our hope that readers of the English edition will have the same experience.

We would like to point out that there is now a more effective approach to iterated torus knots than the one presented in this book. This is developed in the beautiful new book by David Eisenbud and Walter Neumann, "Three-dimensional link theory and invariants of plane curve singularities", and in the forthcoming work of Françoise Michel and Claude Weber.

We would like to thank John Stillwell for all his work. We realize that in some instances translating such a book must have been a really difficult task. We feel that he has succeeded very well.

Egbert Brieskorn and Horst Knörrer Bonn, April 1986

"Es ist die Freude an der Gestalt in einem höheren Sinne, die den Geometer ausmacht."
(Clebsch, in memory of Julius Plücker, Göttinger Abh. Bd. 15).

Foreword

In the summer of 1976 and winter 1975/76 I gave an introductory course on plane algebraic curves to undergraduate students. I wrote a manuscript of the course for them. Since I took some trouble over it, and some colleagues have shown interest in this manuscript, I have now allowed it to be reproduced, in the hope that others may find it useful.

In this foreword I should like to explain what I wanted to achieve with this course. I wanted – above all – to show, by means of beautiful, simple and concrete examples of curves in the complex projective plane, the interplay between algebraic, analytic and topological methods in the investigation of these geometric objects. I did not succeed in developing the theory of algebraic curves as far as is possible – I must even say that in some places the course stops where the theory is just beginning. Rather, I aimed to allow the listeners to develop as much familiarity as possible with the new objects, and the best possible intuition. For this reason I almost always used the most elementary and concrete methods. Also for this reason, I have taken the trouble to make a great number of drawings. I once read a remark of Felix Klein to the effect that what a geometer values in his science is that he sees what he thinks.

Another principle which I have tried to put into effect in this course is that of breaking through the formal lecture style – the style which replaces the development of ideas by a staccato of definitions, theorems and proofs. Thus heuristic, historical and methodological considerations took up a substantial portion of the course, which they likewise occupy in the manuscript. I am well aware of the disadvantages of this method for the reader, and the resulting lack of formal precision, conciseness, clarity and elegance is an annoyance to me too. However, I have accepted this annoyance in order to be able to develop the ideas in a natural way and to promote understanding and thought.

I have developed no new scientific ideas in this course, but have drawn much from other sources. Thus in Chappter II I have depended heavily on notes of a course by R. Remmert on algebraic curves which brought me into contact with algebraic geometry for the first time as a student, and on the book by Walker. I have also used the introduction to algebraic geometry by van der Waerden. In the historical remarks I have relied a lot on the corresponding Enzyclopädie articles and the books of Smith and Struik. My aim was not historical refinement but to give students a picture of the beginnings from which the theory has developed. The whole later history – from the second half of the 19th century onwards – was not so important for this pedagogical purpose, and for it I refer to the new book of Dieudonné or the beginning of the book by Shafarevich. For the local investigations of topology and resolution of singularities I have depended on lecture notes of F. Pham and H. Hironaka as well as original work of A'Campo. Finally, I have used many other sources, which I perhaps have not always acknowledged. I hope the authors will forgive me.

What is lacking in the course? It lacks a chapter on the deformation of singularities, in which I would have liked to introduce the beautiful results of A'Campo and Gusein-Zade on the computation of the monodromy groups of plane curves. For this I refer to the report of A'Campo to the International Congress of Mathematicians in Vancouver in 1974. I have tried to admit the deformation viewpoint at least implicitly in sections 8.5, 9.2 and 9.3. What is not lacking? There is no lack of introductions to modern algebraic geometry. This course is not intended to be such an introduction. For this purpose there are now the beautiful books of D. Mumford, I.R. Shafarevich, P.A. Griffiths and R. Hartshorne. The purpose of the course is to familiarise students, in a natural, intuitive and concrete way, with the various methods for the investigation of singularities, and to lead in this way to my own field of work. I believe that it has reached this goal. This is shown by a series of beautiful Diplomarbeiten which have come into being in the meantime. If the notes presented here can also serve others similarly, then they have fulfilled their purpose, even if only as a source of suggestions or as a collection of material.

In conclusion, I should like to thank all of those who have helped in the production of this extensive piece of work : diploma mathematician Mr. Ebeling and above all Dr. Knörrer for working out some lectures in Chapter III, and for a critical inspection and proof-reading of the manuscript, and help with assembling the references, the three secretaries Mrs. Schmirler, Mrs. Weiss and Mrs. Eligehausen for the laborious and alienating work of typing the text, and the printers at the Mathematische Institut, Mr. Vogt and Mr. Popp, for printing the almost one thousand pages. But above all I should like to thank two students, Mr. Koch and Mr. Scholz, for taking on the enormous job of producing the manuscript, proof-reading, making the index and redrawing more nicely many of the figures. Without them the manuscript would never have been completed. Once again : many thanks!

Bonn, Spring 1978 E. Brieskorn

TABLE OF CONTENTS

I	**History of algebraic curves**	1
1.	Origin and generation of curves	2
1.1	The circle and the straight line	2
1.2	The classical problems of antiquity	3
1.3	The conic sections	4
1.4	The cissoid of Diocles	9
1.5	The conchoid of Nicomedes	13
1.6	The spiric sections of Perseus	16
1.7	From the epicycles of Hipparchos to the Wankel motor	19
1.8	Caustics and contour curves in optics and perspective	41
1.9	Further examples of curves from science and technology	58
2.	Synthetic and analytic geometry	66
2.1	Coordinates	66
2.2	The development of analytic geometry	69
2.3	Equations for curves	71
2.4	Examples of the application of analytic methods	79
2.5	Newton's investigation of cubic curves	87
3.	The development of projective geometry	102
3.1	Descriptive geometry and projective geometry	102
3.2	The development of analytic projective geometry	106
3.3	The projective plane as a manifold	118
3.4	Complex projective geometry	136
II	**Investigation of curves by elementary algebraic methods**	172
4.	Polynomials	172
4.1	Decomposition into prime factors	173
4.2	Divisibility properties of polynomials	174
4.3	Zeroes of polynomials	190
4.4	Homogeneous and inhomogeneous polynomials	193

5. Definition and elementary properties of plane algebraic curves 202

5.1 Decomposition into irreducible components 202

5.2 Intersection of a curve by a line 208

5.3 Singular points of plane curves 211

6. The intersection of plane curves 227

6.1 Bézout's theorem 227

6.2 Applications of Bézout's theorem 239

6.3 The intersection ring of $P_2(\mathbb{C})$ 260

7. Some simple types of curves 278

7.1 Quadrics 278

7.2 Linear systems of cubics 281

7.3 Inflection point figures and normal forms of cubics 283

7.4 Cubics, elliptic curves and abelian varieties 306

III **Investigation of curves by resolution of singularities** 323

8. Local investigations 325

8.1 Localisation-local rings 325

8.2 Singularities as analytic set germs 337

8.3 Newton polygons and Puiseux expansions 370

8.4 Resolution of singularities by quadratic transformations 455

8.5 Topology of singularities 535

9. Global investigations 576

9.1 The Plücker formulae 576

9.2 The formulae of Clebsch and Noether 601

9.3 Differential forms on Riemann surfaces and their periods 627

Bibliography 694

Index 702

I. HISTORY OF ALGEBRAIC CURVES

The plane algebraic curves have a history of more than 2000 years. At the end of this development stands a definition of algebraic curves, an understanding of the main problems of the theory, and methods for handling them, to the extent that almost all problems about algebraic curves admit methodical treatment and solution. These questions and methods are part of a general field, algebraic geometry, whose development proceeded mainly during the past and present century and still continues. The theory of algebraic curves is a part of this general theory which has the character of a paradigm, and thereby serves as a good introduction to it for the beginner. Today this is the main point of view in introductory treatments of this field, e.g. the book of Fulton [Fl].

This viewpoint will also play a role in our course, but it will not be the only one. Our emphasis will be not so much on the development of the algebraic conceptual apparatus and the algebraic methods, as on the geometric viewpoint, where possible, in working out some of the many connections between this field and analysis and topology. Moreover, I will not, as is almost always done today, completely ignore the two-thousand-year history of algebraic curves.

In the first lectures of this course I shall give an introduction to this history, naturally quite sketchy, in which I omit all proofs to save time. Of course I shall later formulate and prove all assertions with the usual rigour.

The purpose of this historical introduction is to give provisional answers to the following questions :

1. What is the origin of the objects of our investigation, the algebraic curves? Why were mathematicians concerned with them originally?
2. From which viewpoints have they been investigated? What has led

to the viewpoints and methods which predominate today?

3. What were or are the main objects of investigation?

It will be understood that the answers to these questions, as long as theory has not been given, will be incomplete. For those who are interested in more details, I recommend looking at older books on algebraic curves. I particularly recommend the following literature: the survey articles of Berzolari [B2] and Kohn-Loria [K3] in Band III, 2.1 of the Encyclopädie der Mathematischen Wissenschaften, as well as the books of Loria [L4] and Gomes Teixeira [T2].

1. Origin and generation of curves

First I should like to tell you something of the reasons why the first interesting curves were already considered in antiquity. We shall then see how these were taken up again in the Renaissance, after which there was a permanent enrichment of the theory by new content and methods. We shall see that there were many causes for the origin and generation of algebraic curves: the development of historically important mathematical problems, playful mathematical constructions and the joy of solving problems, but also, and very important in this field, numerous applications of mathematics in other fields: perspective, optics, astronomy, architecture, kinematics, mechanics and technology.

1.1 The circle and the straight line

The simplest curves are the circle and the line. Their origins lie in prehistoric times. Knowledge of them was necessary for the solution of numerous practical problems, such as land measurement (= geometry in the original meaning of the word) and building construction. Straight lines and circles are treated intuitively in the earliest mathematical texts, and the first attempt at strict definition and proof in mathematics, about 600 B.C. (Thales), concerns these objects. The attempt of the Greek mathematicians, such as Euclid (300 B.C.), to define a line was inadequate from a logical standpoint, even though the Greeks of course already knew almost all essential properties of lines in the plane. From our present-day standpoint, the question of the definition of lines depends on the conceptual framework : one can make the line an implicitly defined basic axiomatic concept as, e.g., in Hilbert's axiomatisation of euclidean geometry, or take it to be a derived concept as, e.g., in analytic geometry or linear algebra, where it is a 1-dimensional affine subspace.

The definition of the circle causes no difficulties – and did not cause any for Euclid – when one assumes the euclidean plane as known : it is the locus of all points in the plane which have a given distance from a given point. This type of definition of curves as loci was typical of the way the Greeks handled curves. They were defined as the loci of points having certain distance relationships (specific for each curve) to given points, lines and circles.

To construct, i.e. draw, a circle one uses a simple mechanism, the pair of compasses. To draw a straight line one mostly uses a somewhat problematic instrument, the ruler, and hence a template.

It is very clear that numerous simple practical and mathematical problems can be solved by construction with compasses and ruler. From the analytic-algebraic standpoint, the construction of the intersection of two lines is the solution of two linear equations, the construction of the intersection of a line and a circle, or of two circles, is the solution of a quadratic equation. By iterating such constructions one can therefore solve all problems of the form : from given segments of length $a_1,\ldots,a_n$, construct a segment of length a , where a results from $a_1,\ldots,a_n$ by repeated rational operations and extractions of square roots. Criteria for this to be the case are obtained from Galois theory. Of course these conditions for the solvability of problems by ruler and compass construction were unknown to the Greeks, so it is understandable why they tried to solve the great problems of antiquity by such constructions.

1.2 The classical problems of antiquity

The classical problems of antiquity were:

1. The trisection of an arbitrary angle.
2. Squaring the circle (before 1500 B.C.?).
3. Duplication of the cube. (Delian problem) (5th century B.C.?)

Squaring the circle means determining the number π. Lindemann (1882) showed π is transcendental, i.e. not the root of any algebraic equation

$$x^n + a_1x^{n-1} + \ldots + a_n = 0$$

with rational coefficients a_i. This means, in particular, that π cannot be constructed with compasses and ruler.

Duplication of the cube leads to solving the equation $x^3 - 2 = 0$ which, by Galois theory, likewise cannot be done by ruler and compass construction.

As for the trisection of an arbitrary angle α : the addition theorem for trigonometric functions immediately yields

$$\sin 3\beta = 3 \sin \beta - 4 \sin^3 \beta,$$

hence, setting $\beta = \alpha/3$, $x = \sin \beta$ and $c = \sin \alpha$,

$$4x^3 - 3x + c = 0,$$

and it again follows from Galois theory that the solution of this equation for arbitrary c cannot be found by ruler and compass construction.

Thus the classical problems cannot be solved by ruler and compass constructions. It is true that the Greeks had no proof of this, but they saw the futility of their attempts – and found solutions with the help of curves less simple than the circle and the line. I shall say something about them in what follows.

1.3 The conic sections

The discovery of the conic sections is attributed to Menaechmus (c. 350 B.C.). They were intensively investigated by Apollonius of Perga (c. 225 B.C.). These mathematicians generated the conic sections as intersections of cones with planes. Apollonius and, to some extent, Menaechmus could also characterise the conic sections by area application properties, which are expressed in today's notation by

$$y^2 = px + qx^2.$$

In the terminology of Apollonius we obtain for

$q = 0$ the parabola, i.e. "application"

$q > 0$ the hyperbola, i.e. "application with excess"

$q < 0$ the ellipse, i.e. "application with defect".

Of course the Greeks also knew the definitions of these curves as loci.

A parabola is the locus of all points having equal distances from a given point P and a given line g.

An ellipse (hyperbola) is the locus of all points for which the sum (difference) of distances from two given points P, Q has a fixed value.

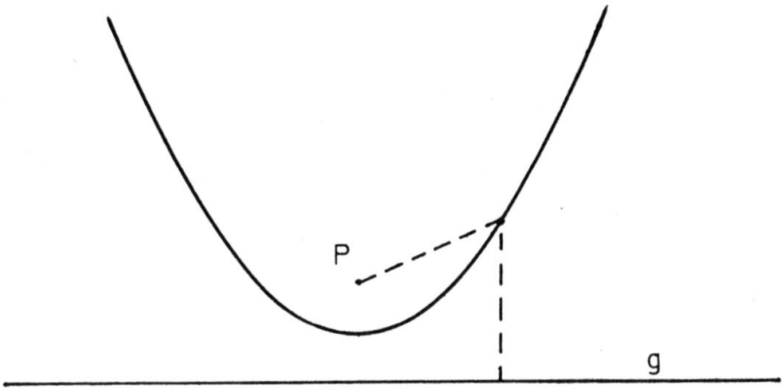

Parabola

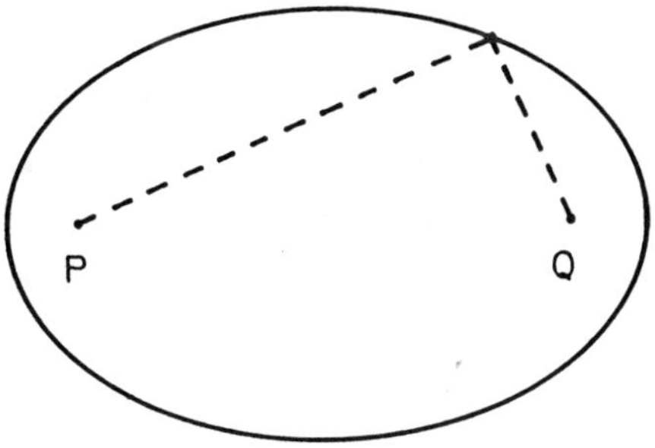

Ellipse

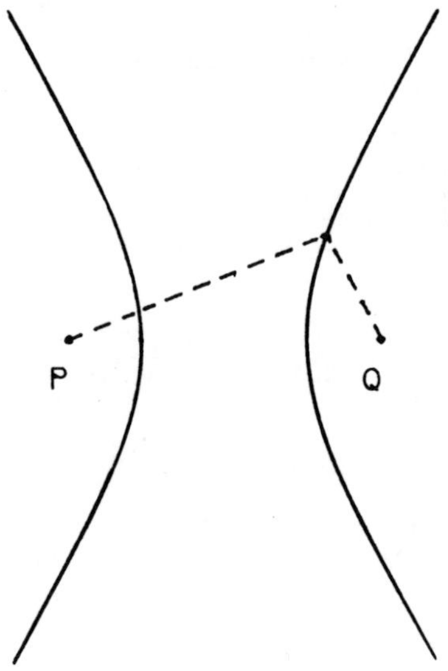

Hyperbola

Menaechmus had already seen that one could use conic sections to solve the Delian problem. Determination of the intersection of the parabolas with equations

$$y^2 = 2x$$
$$x^2 = y$$

leads to the solution of the equation

$$2x = x^4$$

and hence to $x = 0$ and $x = \sqrt[3]{2}$, whence one has a method to solve the Delian problem when one has a method to construct parabolas which yields not just particular points but the whole curve. It is not known which constructions the Greeks used to draw the conic sections. Later, many such constructions — so-called "organic generations" - were found for conic sections. The best known is undoubtedly the generation of an ellipse by a point A on a line which moves in the plane so that two other points B, C on the line move on a pair of axes.

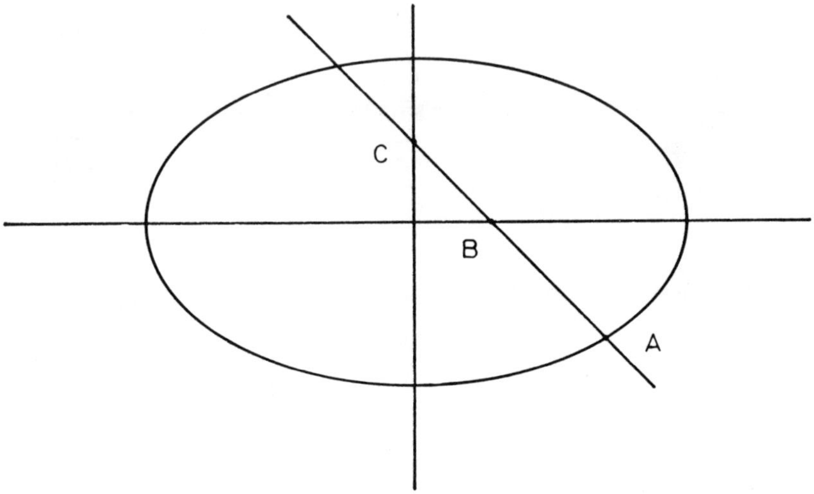

By using this manner of generation one can construct an apparatus convenient for drawing ellipses. The next page shows such an ellipsograph. There are similar parabolagraphs and hyperbolagraphs.

Like the Delian problem, the angle trisection problem can also be solved with the help of conic sections, but I shall not go further into this. Of course there is no such solution for squaring the circle, but here the Greeks also found a way : they used certain other curves, such as the quadratrix of Hippias (c. 425 B.C.).

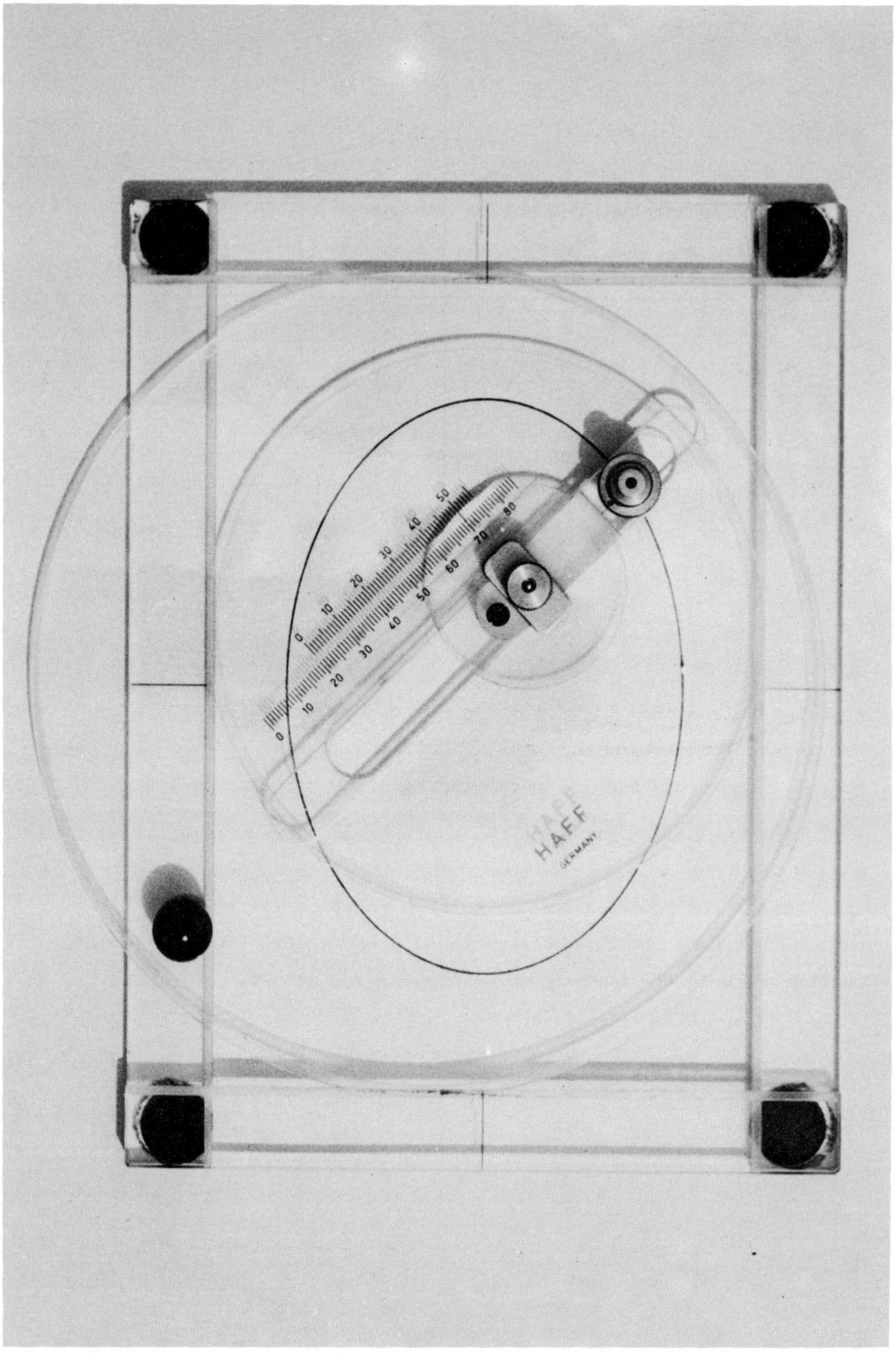

Ellipsograph

1.4 The cissoid of Diocles

Apart from solutions of the problems of doubling the cube and trisecting the angle with the help of conic sections, the Greeks found still other solutions with the help of other curves. The oldest and most important of these curves were the cissoid of Diocles and the conchoid of Nicomedes.

The cissoid is constructed pointwise as follows. We draw a circle about the intersection 0 of two perpendicular lines $\overline{AA'}$ and $\overline{A''A'''}$. Parallel to $\overline{A''A'''}$ we draw two lines equidistant from it, meeting $\overline{AA'}$ in B and B' respectively. Let their intersections with the circle be C, C', C'', C''' as in the figure. The line through one of these points, say C', and A cuts the other parallel, $\overline{CC''}$, in a point P. The totality of points constructed in this way is the cissoid of Diocles.

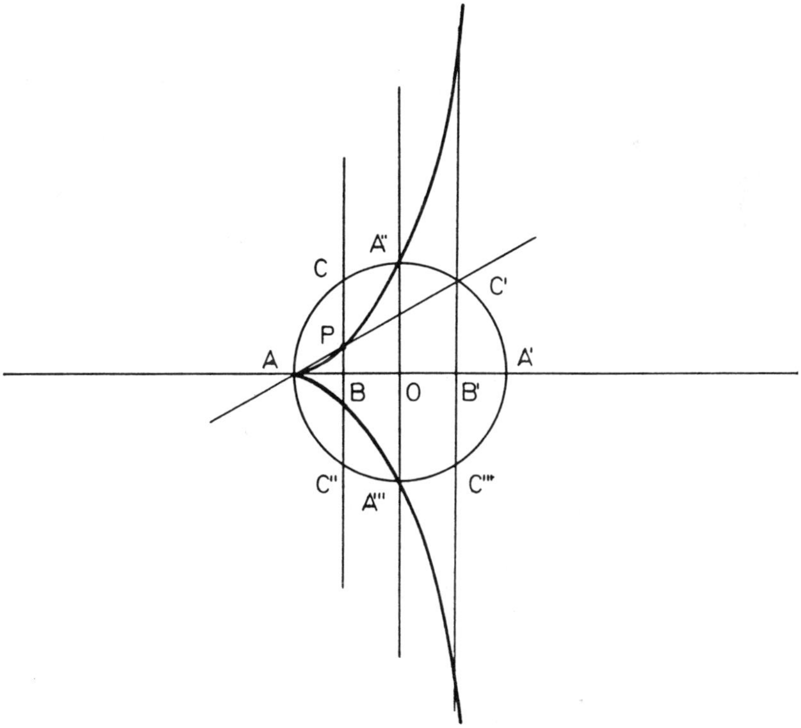

The name "cissoid" was used for this curve from the early seventeenth century on. The Greek word κισσοειδής means "ivy-shaped". We know from Proclus and Pappus that there were curves called κισσοειδής (e.g. Pappus, Collection III, 20 and IV, 58). We do not know which curves were so called, but probably the cissoid of Diocles was not one of them ([T3], p. 24). We may assume that the word κισσοειδής refers to the way the curve comes to a singular point, "making an angle with itself", as the Greeks put it, reminding us in a way of the edge of an ivy leaf.

The curve obviously goes through the points A, A'', A''' of the circle and has the tangent to the circle at A' as asymptote. The cissoid has a cusp or, as one says, a return point. Such singular points will be carefully studied in this course, and the cissoid is perhaps the oldest example (c. 180 B.C.) of a curve with such a singularity.

We now suppose that the cissoid is already drawn, as described above. Then Diocles solves the Delian problem by the following construction :

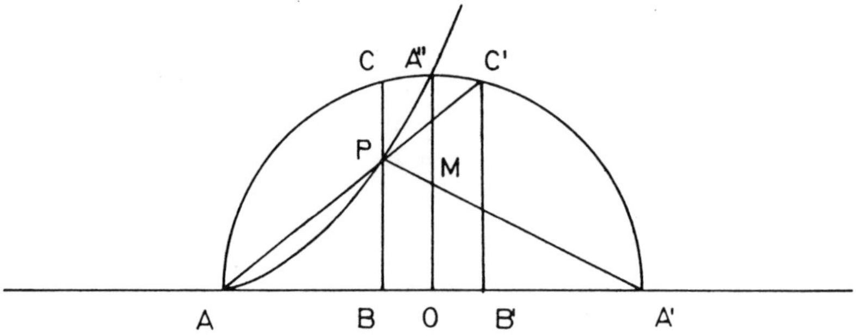

Let M be the midpoint of $\overline{OA''}$ and let P be the intersection of $\overline{A'M}$ with the cissoid, with $\overline{CB}$ and $\overline{C'B'}$ as above. Let x,y,z be the segments $x = A'B$, $y = BC$, $z = AB$. Then obviously

$$\frac{A'B}{PB} = \frac{A'0}{M0} = 2$$

$$\frac{A'B}{BC} = \frac{CB}{AB} = \frac{C'B'}{B'A'} = \frac{AB'}{B'C'} = \frac{AB}{BP}$$

hence

$$\frac{x}{y} = \frac{y}{z} = \frac{z}{\frac{1}{2}x},$$

which immediately implies $x^3 = 2y^3$, and hence the Delian problem is solved.

For this to really solve the problem it is of course necessary to have not just a pointwise construction of the cissoid, but a process which continuously draws a whole piece of the curve, and hence an organic generation. The following organic generation of the cissoid was given by Newton.

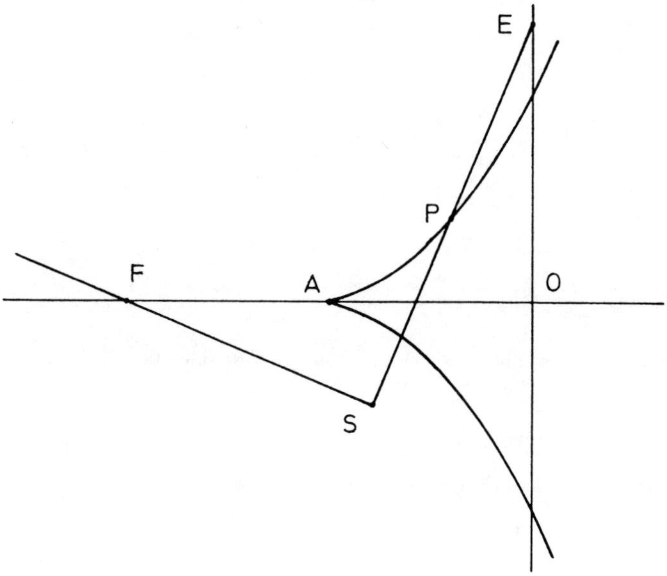

A right angle with an arm of fixed length 2r is moved in the plane in such a way that its endpoint E moves on a line and the other arm always goes through a fixed point F at distance 2r from the line. The midpoint P of the arm $\overline{SE}$ then describes the cissoid. It is interesting, incidentally, to use other points P', P'' on the arm $\overline{SE}$ in place of P. These describe curves like the following:

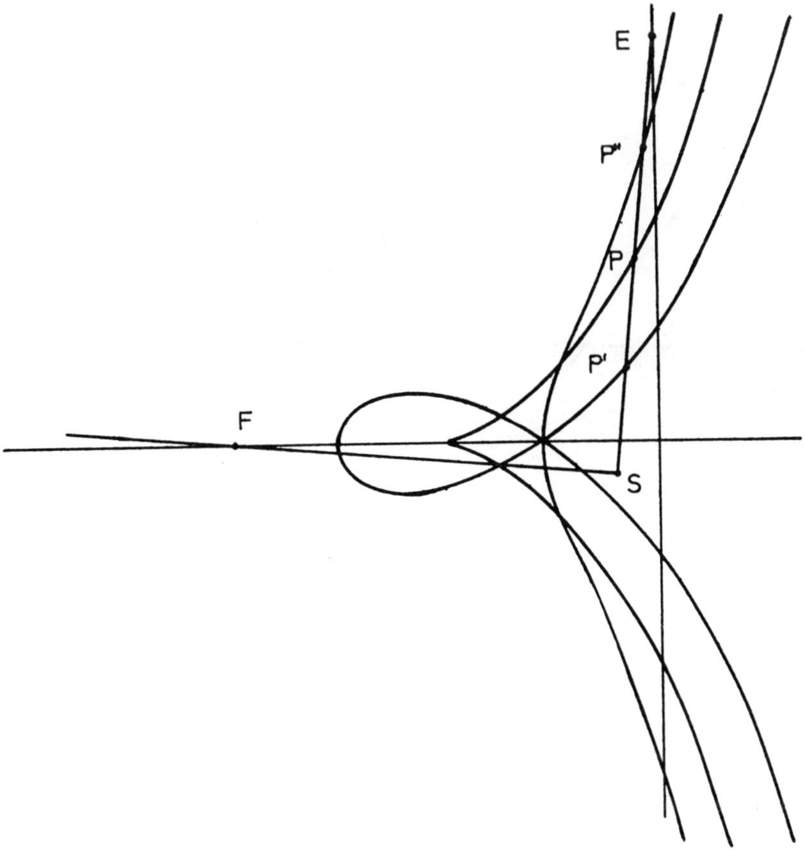

In this way one obtains a whole 1-parameter family of curves which are in a sense deformations of the cissoid.

1.5 The conchoid of Nicomedes

The conchoids of Nicomedes are curves, found c. 180 B.C., which solve the Delian problem and also the problem of trisecting the angle.

They are constructed as follows. Given a line ℓ, a point 0 at distance d from ℓ, and a segment k, let A be an arbitrary point on ℓ and P, P' the points on the line $\overline{0A}$ at distance k from A. The locus of all these points P, P' is a conchoid. Its form depends on the relation between d and k as follows:

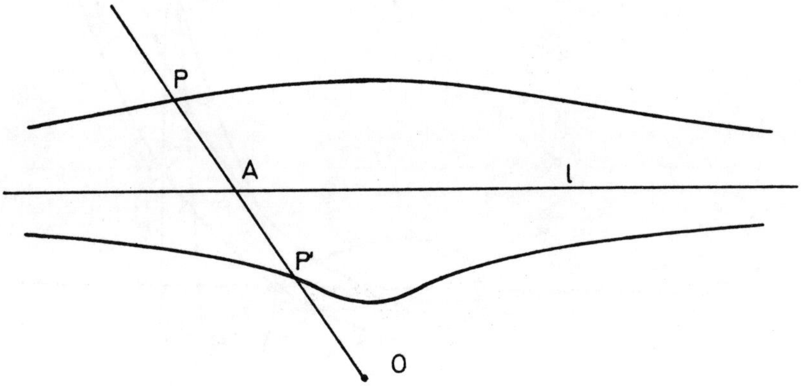

k < d

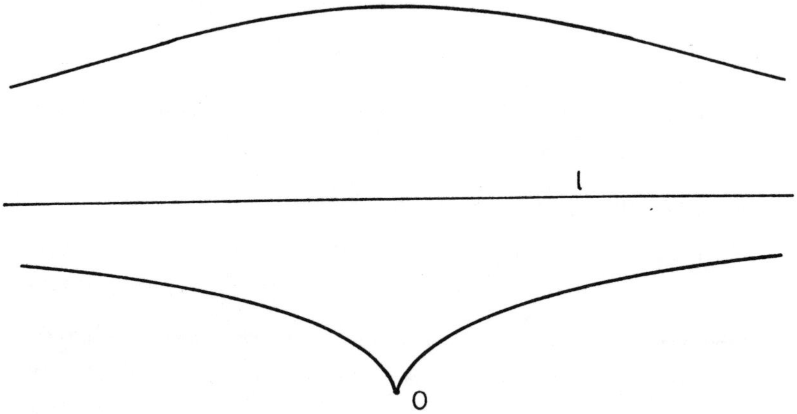

k = d

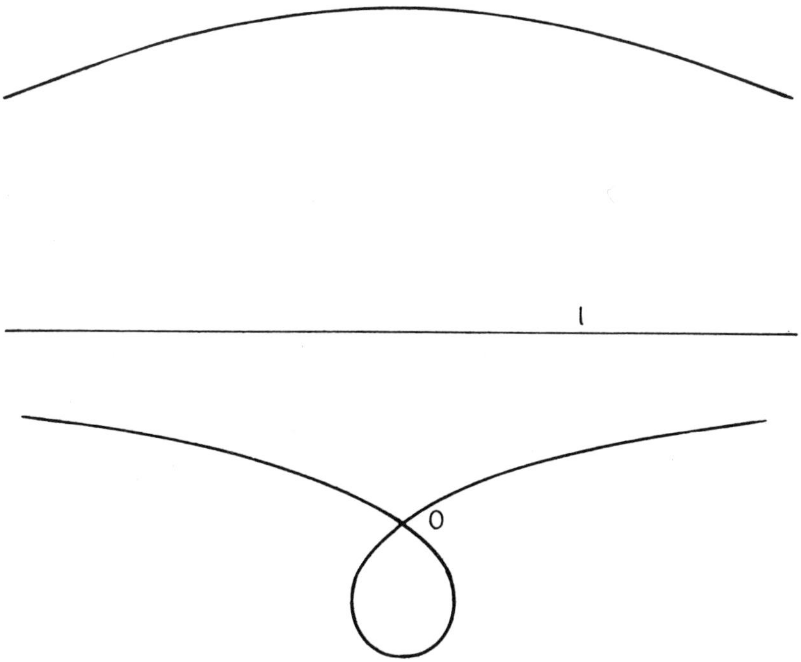

k > d

Thus conchoids have two branches, one of which can have a cusp or double point. The name stems from the shell-like shape (κόγχη = concha = seashell).

In antiquity the conchoid was also used for the construction of vertical sections of columns.

We remark that the construction just described yields an organic generation. From it we obtain the following process of Nicomedes for the trisection of an arbitrary angle α.

Let $\sphericalangle AOB$ be the given acute angle α, where B is the foot of the perpendicular ℓ from A onto $\overline{OB}$. One draws the conchoid for ℓ and 0 with $k = 2\overline{OA}$. The parallel to $\overline{OB}$ through A cuts the conchoid on the side away from 0 at C. Let $\gamma = \sphericalangle BOC$. Then $\gamma = \frac{\alpha}{3}$ and hence the problem is solved.

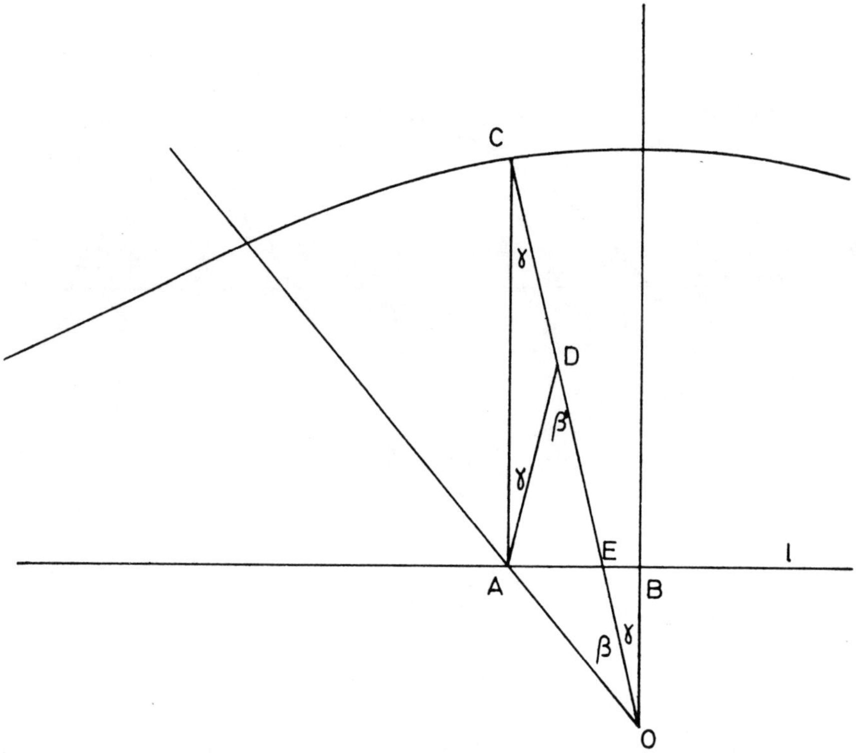

Proof : Let E be the intersection of AB and CO and let D be the midpoint of the segment $\overline{CE}$. Then $\overline{DA} = \overline{AO}$ by construction of the conchoid. Then the base angles of the isosceles triangle ODA satisfy $\beta = \beta'$. Thus it follows from $\beta + \gamma = \alpha$ and $\beta' = 2\gamma$ that $\alpha = 3\gamma$.

1.6 The spiric sections of Perseus

Following Menaechmus' construction of the ellipse, hyperbola and parabola by cutting a cone by a plane, around 150 B.C. the Greek mathematician Perseus had the idea of cutting a torus by a plane parallel to the axis of rotation, and thereby obtained interesting curves. Because the Greeks called the torus the spira (σπεῖρα), these curves are called the spiric sections of Perseus. We shall not construct these curves exactly here, but only give a rough picture of their form by a drawing.

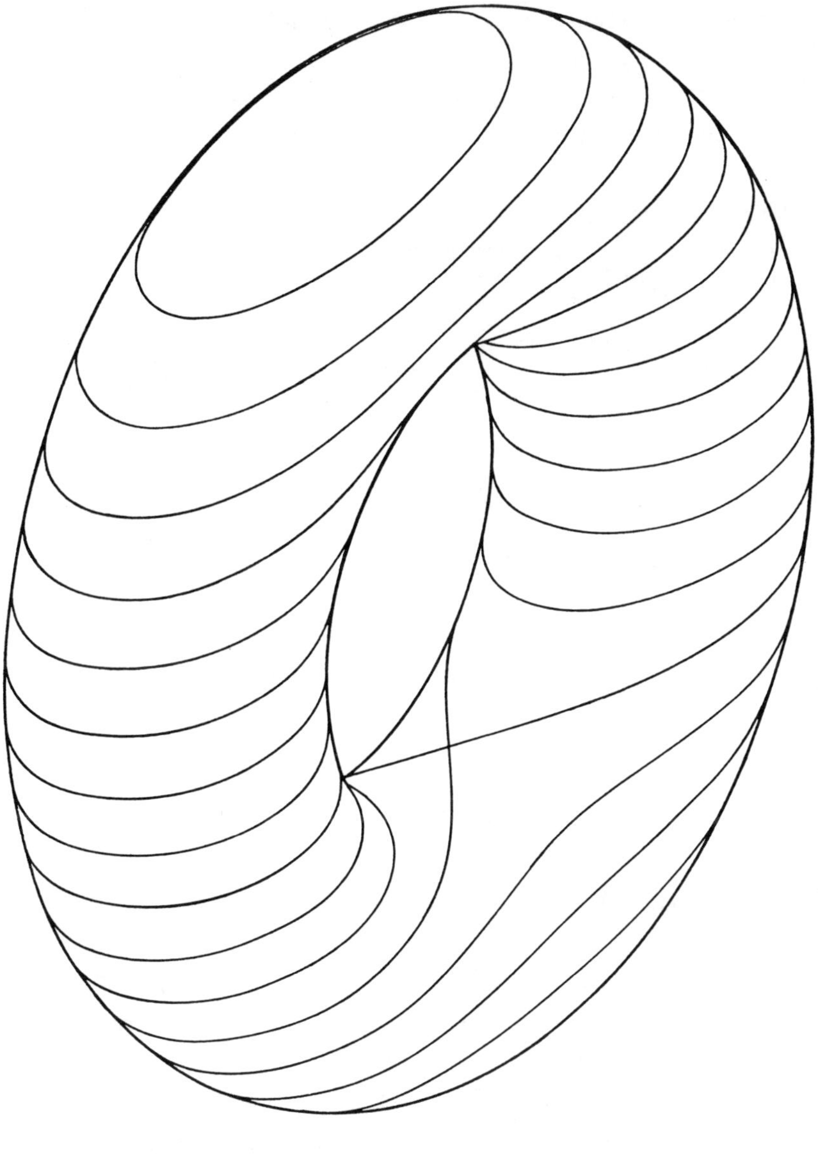

Perhaps this look at the spiric sections of Perseus will remind some of you of the Cassini curves. Rightly so!

The Cassini curves are defined as follows : one is given two points A, B distance 2a apart, and a positive number c. Then the associated Cassini curve is the locus of all points P such that $\overline{PA} \cdot \overline{PB} = c^2$. For fixed A, B and different values of c these curves look like the following. These curves were found by the astronomer Giovanni Domenico Cassini (c. 1650-1700).

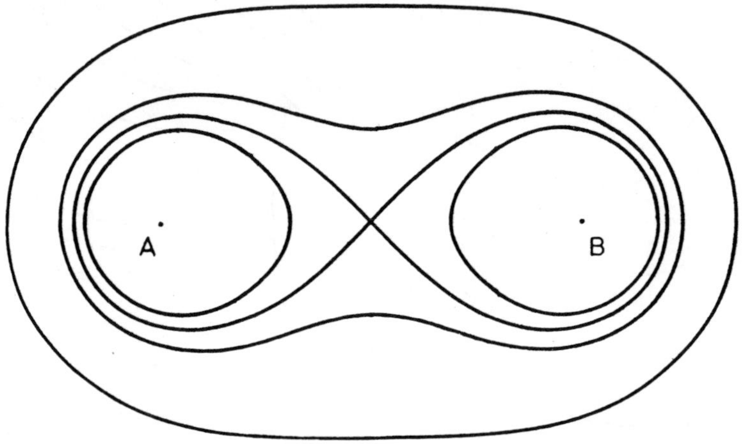

He believed that the sun travelled around the earth on one such convex curve, with the earth at a focus.

For $a = c$ the curve has an ordinary double point and forms a double loop. This is the lemniscate of Jacob Bernoulli (1694). (λημνίσκος = loop in the form of an 8). This curve played an important role in the development of the theory of elliptic functions. When the torus on which we consider the spiric sections of Perseus results from rotation of a circle of radius r about an axis at distance R from the centre of the circle, and when the plane of intersection is distance d from the axis, then the spiric sections of Perseus are Cassini curves precisely when $d = r$. Then $a = R$ and $2rR = c^2$. Thus the Cassini curves are special cases of the spiric sections of Perseus.

1.7 From the epicycles of Hipparchos to the Wankel motor

Mathematics is indebted to astronomy for the knowledge of a series of interesting special curves. We have met one example already : the Cassini curves. A more important example was the epicycle, used by the ancient astronomers Hipparchos (c. 180-125 B.C.) and Ptolemy (c. 150 A.D.) to describe the paths of the planets. Such a curve is the path of a point on a circle which turns with constant angular velocity about its centre, while the centre at the same time travels with constant angular velocity on another circle.

During the Renaissance another interesting class of curves, the wheel curves, was found, and it was noticed only later that these wheel curves were included among the classical epicyclic curves.

A wheel curve is the path of a point on a circle K which rolls on a fixed circle K' without slipping. One distinguishes between epicycloids, hypocycloids and pericycloids according to the position of the circles. If K' degenerates to a line, then the resulting curve is called a cycloid. If K degenerates to a line, then the curve is called a circle involute.

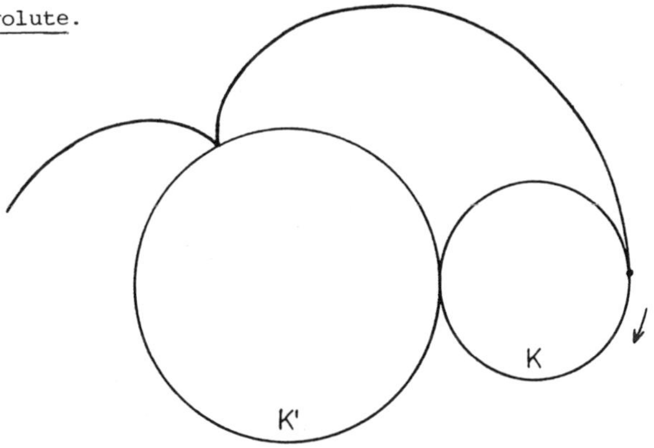

Epicycloid

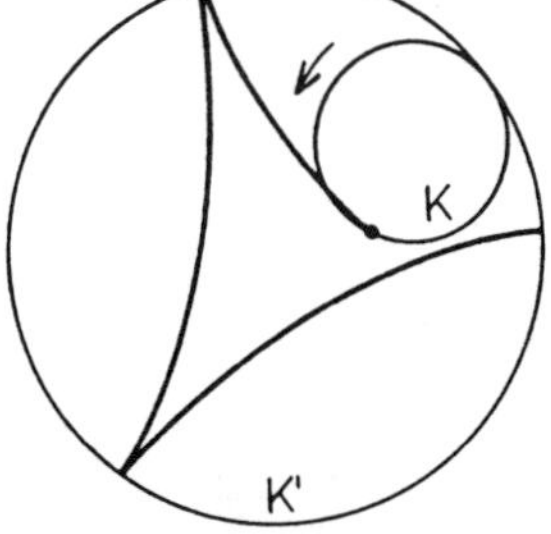

Hypocycloid

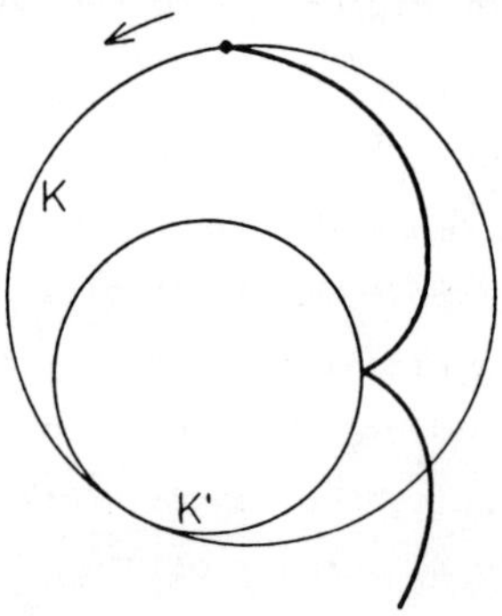

Pericycloid

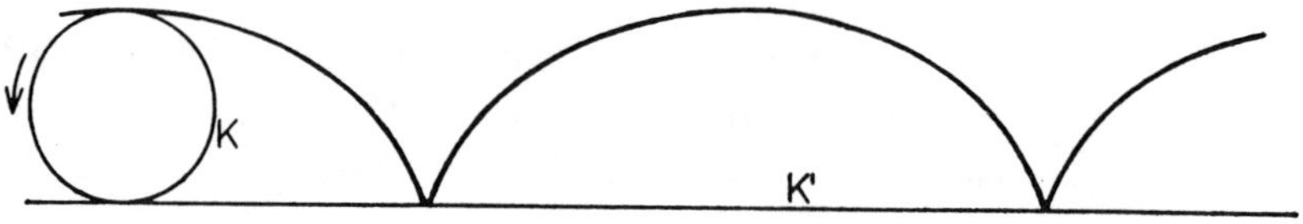

Cycloid

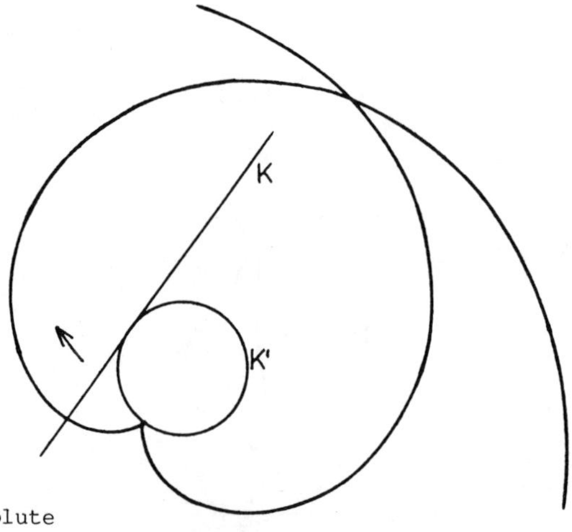

Circle involute

Epi-, hypo- and pericycloids are closed curves or not according as the ratio of the radii r and r' of K and K' is rational or not. They have cusps and, when the situation demands it, ordinary double points.

One can – and this is of practical importance – generalise the definition of these curves still further, by considering the path, not just of a point on the rolling circle K, but of any point in the plane of K which moves with K as K rolls on K'. These paths are called lengthened or shortened epi-, hypo- and pericycles, or simply trochoids.

The following is a toy with which one can draw such trochoids, and some of them are shown. (One sees easily that the peritrochoids are epitrochoids and that the class of epicyclic curves of antiquity is identical with the class of trochoids.)

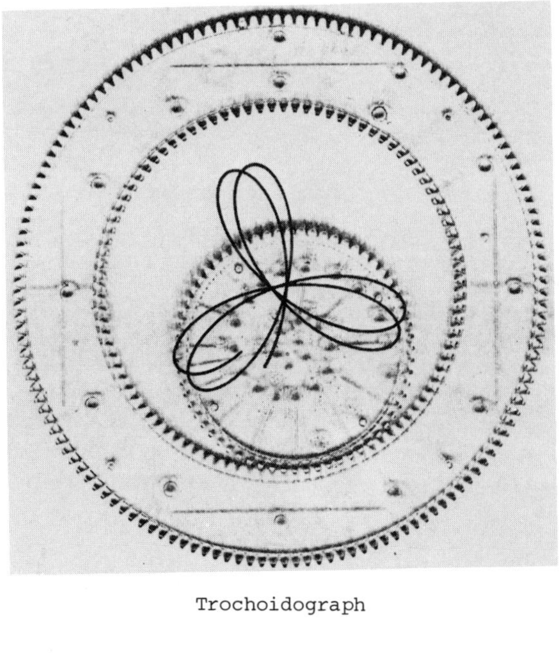

Trochoidograph

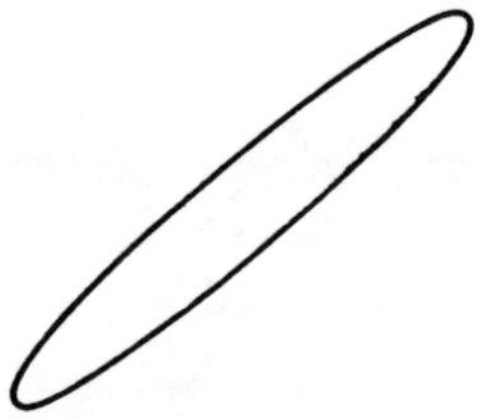

r/r' = 1/2

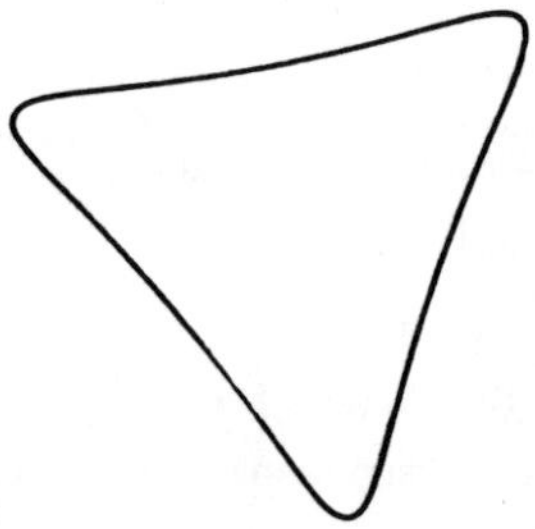

r/r' = 1/3

r/r' = 1/4

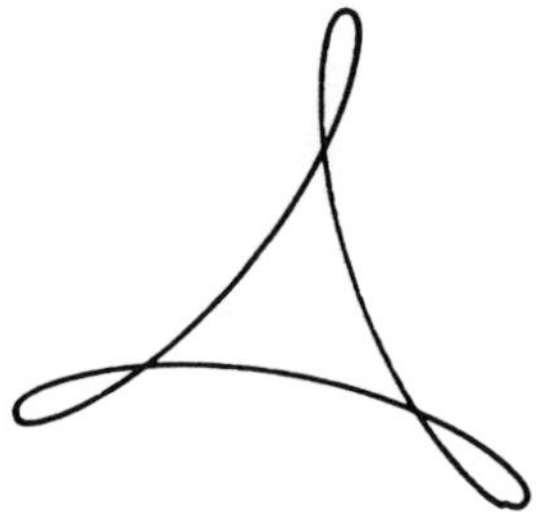

r/r' = 2/3

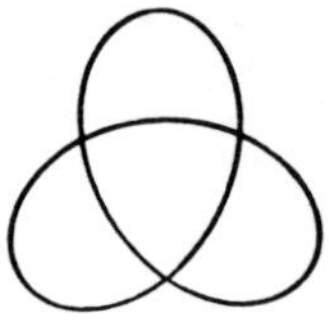

r/r' = 2/3

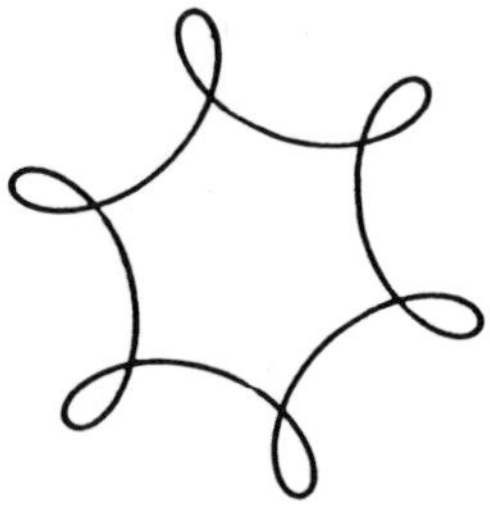

r/r' = 5/6

Hypotrochoids

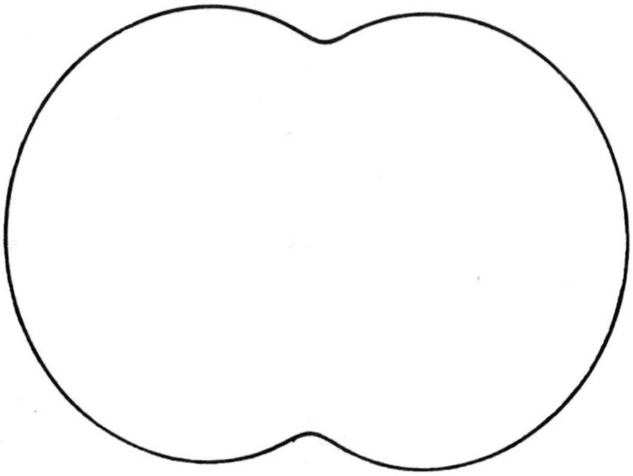

r/r' = 1/2

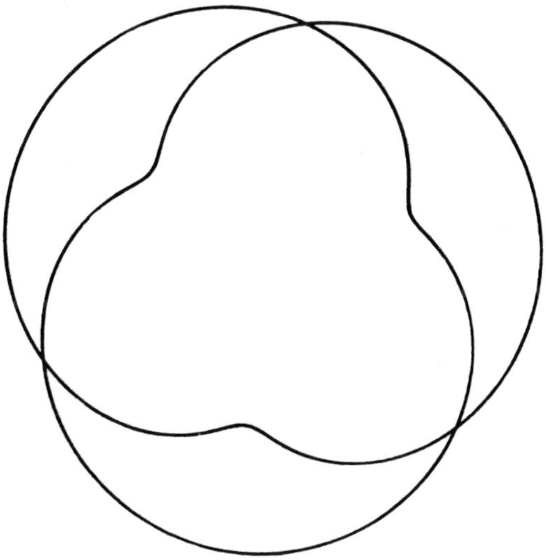

r/r' = 2/3

Epitrochoids

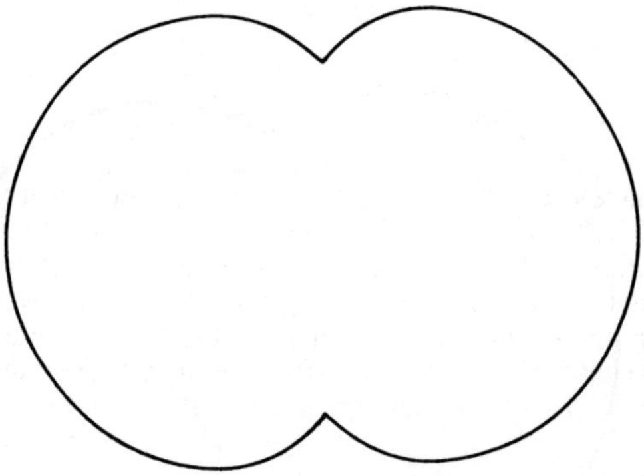

r/r' = 3/2

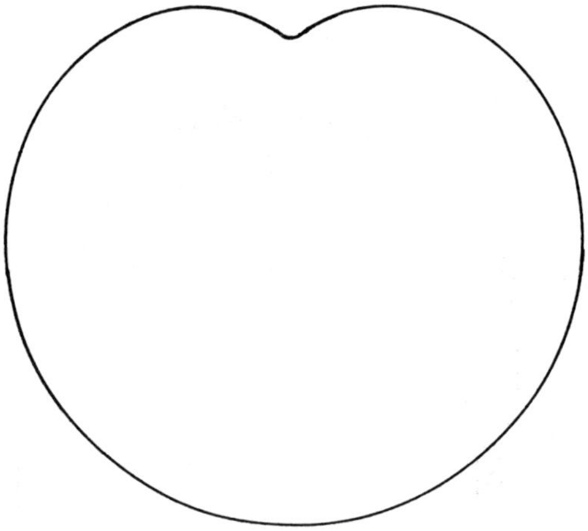

r/r' = 2

Peritrochoids

An epitrochoid with r' = r is found in Albrecht Dürer's "Unterweisung der Messung mit dem Zirkel und Richtscheit" (Instruction in measurement with compasses and straight edge), 1525, where these curves are called spider lines because of the spider-like configuration of Dürer's construction lines, as the accompanying reproduction shows.

ABer ein andre lini/die sey genant ein spiñen lini/ darum dz sie im aufreissen/dardurch mans macht scheir einer spinnen enlich ist/die mach jch durch ein zwifache bewegung alsô/Jch reis eyn aufrechte lini.a.b.daran setz jch ein andre lini der end sey.c.vñ die lini.a.b.laß jch im end a stet bleiben/Aber das end.b.für jch in zirckels weis herumb/wie jch dañ der end im vmblauf vberal mit b.vertzeichent hab/ Darnach soll im end.b.die ander daran gestossen lini.c. mit jrem hyndern ende im punckten.b.auch stett bleiben/aber das förder end.c.soll in zirckels weiß herum gefürt werden/ So dañ die erst lini vmgefürt/vnd die ander anstosset auch sonderlich herum gefürt wirdet/so zeichent das end c.ein sonderliche lini/damit aber dise lini gewyß gefürt werd/so setz jch eyn zirckel mit dem ein fuß in dē punctē.a.vñ reiß mit dem andern fuß eyn zirckellini vnder dem/b/die gradir jch auch in theyl mit ziffern/dardurch die lini.a.b.von punckt zů punckt gewyß gee/Des gleichē thů jch jm auch im punctē.b. vnnd so offt jch mit der lini.a.b .eyn grad gee / so oft gee jch auch ein grad im zirckel.b.mit der lini.c.so zeichnet das end.c.die punckten zwischen den jr lini zůsamen soll getzogen werden/die jch vberall mit.c verzeichent hab/wie das nachfolgett aufgeryssen ist.

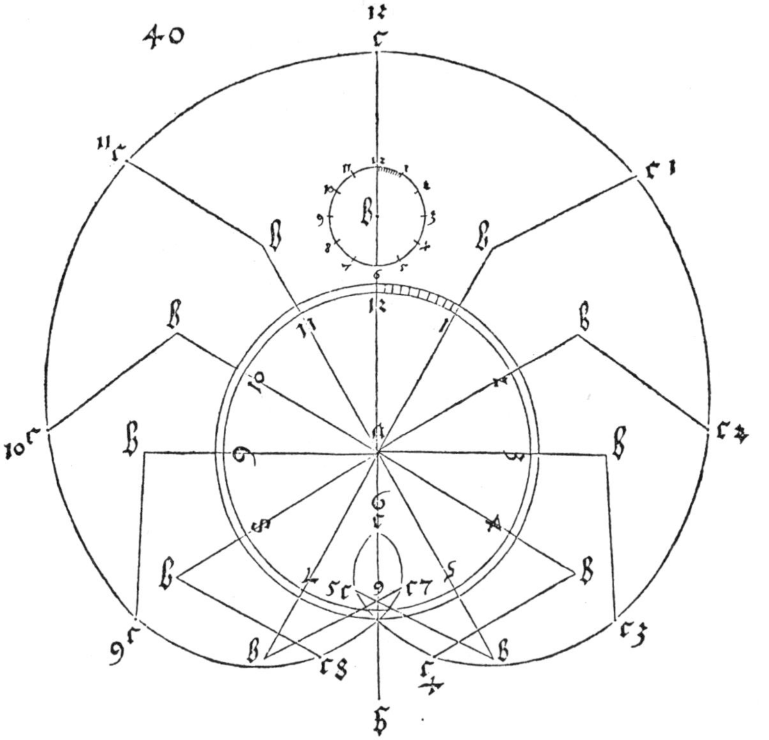

NAchfolget will ich ein Jnstrument machen/damit man an vil end/hoch/nyder/ zůn seytten/ fürsich oder hyndersich/eyn schlangen lini deüten vñ reissen mag/Solchs instrument wirdt an stangen gebogen gewendt vnd vmbgeryben / vnnd in glidern der stangen söllen scheiben seyn/Jn der Centrum söllen die büg sein darin es vmb geet/ein teil mag fürsich das ander hyndersich oder wo man hyn will gebogen werden/oder alle mit eynander fürsich oder hyndersich/ vñ in welchen

In the course of the next three hundred years a series of important mathematical works were written on these curves. We give just a few of the best known names. Those concerned with epi- and hypocycloids and the cycloids included :

Epi- and hypocycloids

Daniel Bernoulli	1725
Jacob Bernoulli	(1692-1699)
Johann Bernoulli	1695
Desargues	(1593-1662)?
Dürer	1525
Euler	(1745, 1781)
de la Hire	1694
L'Hospital	(1661-1704)
Huygens	1679
Newton	1686

Cycloids

Jacob Bernoulli	
Johann Bernoulli	
Charles Bouvelles	1501
Nicolaus von Cues	1454?
Descartes	(1596-1650)
Fermat	(1601-1665)
Galilei	1590
L'Hospital	
Leibniz	
Mersenne	1615
Newton	
Pascal	(1623-1662)
Roberval	(1602-1675)
Wallis	(1616-1703)
Guido Grandi	1728 :

"Flores geometrici ex rhodonearum et claeliarum curvarum descriptione resultantes" (rose curves = special trochoids).

The <u>cycloid</u> has the following characteristic property :

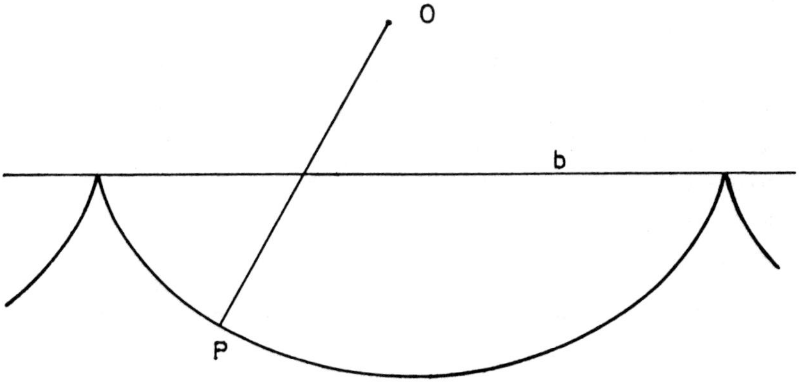

For each point P of the cycloid, the radius of curvature OP is halved by the base b.

In the Acta eruditorum of 1696, Johann Bernoulli posed the following problem, whose solution he already knew, and published in 1697 : given two points A and B in a vertical plane, find a curve from A to B so that a point P which slides from A to B along the curve under gravity takes the shortest possible time.

He called the curve which solves this problem the <u>brachistochrone</u> (βράχιστος χρόνος = shortest time).

The problem was solved by several mathematicians, some immediately: Newton, Leibniz, Jacob Bernoulli and L'Hospital. The solution is the following.

Let b be the horizontal line through the higher point, A. Then the cycloid through A and B with basis b is the brachistochrone, provided it has no minimum between a and b.

<u>Proof</u>

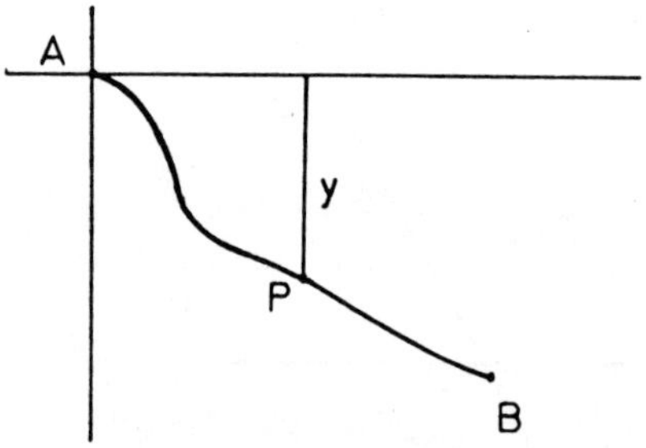

Whichever curve P slides along, energy conservation implies that

$$mgy = \frac{m}{2} v^2,$$

where m is the mass, v the velocity, and y the height fallen by P, and hence

$$v = \sqrt{2gy}.$$

Now we compare the cycloid through A and B with another curve K from A to B, and show by infinitesimal considerations that when the time of fall along K is minimal, K must be the cycloid. We illustrate our infinitesimal reasoning by the following figure.

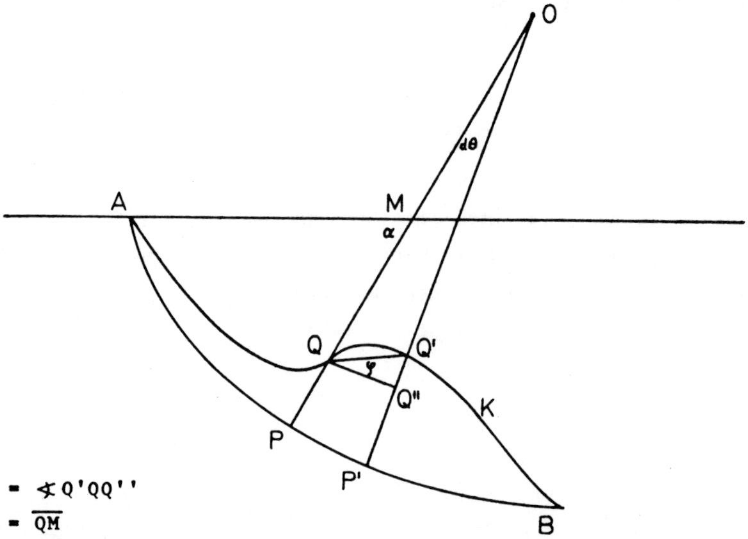

$\varphi = \sphericalangle Q'QQ''$

$q = \overline{QM}$

$p = \overline{PM} = \overline{MO}$

Let $\overline{OP}$ be the radius of curvature at the point P on the cycloid, and let Q be the intersection of $\overline{OP}$ with K. Let P' resp. Q' be infinitesimally neighbouring points on the two curves. We compare the infinitesimal times dt for the path from P to P' and $d\tau$ for the path from Q to Q'. The velocity at P is $\sqrt{2gp \sin \alpha}$ and at Q it is $\sqrt{2gq \sin \alpha}$. The infinitesimal distances to be traversed satisfy :

$$\overline{PP'} = 2p d\theta \quad \text{and} \quad \overline{QQ''} = \overline{QQ'} \cos \phi = (q+p) d\theta .$$

Hence

$$dt = \frac{2p\, d\theta}{\sqrt{2gp \sin \alpha}} , \quad d\tau = \frac{(q+p) d\theta}{\sqrt{2gq \sin \alpha} . \cos \phi}$$

$$\frac{q+p}{\sqrt{q}} = \frac{(\sqrt{q}-\sqrt{p})^2}{\sqrt{q}} + 2\sqrt{p} \geq 2\sqrt{p}$$

Thus $d\tau \geq dt$ and equality holds only when $\sqrt{q} - \sqrt{p} = 0$ and $q = p$, i.e. $P = Q$. Q.E.D.

I have investigated this example in detail because the brachistochrone problem was one of the earliest variational problems, and its investigation was the starting point in the development of the calculus of variations. Thus we again see a special curve, like the lemniscate of Bernoulli, as the starting point in the development of a whole mathematical discipline.

Another important property of the cycloid had been discovered earlier by Huygens 1673 (Horologium oscillatorium) : it is the tautochrone (ταὖτος χρόνος = same time), i.e. the curve with the property that a heavy point traversing it always reaches the bottom in the same time, regardless of the starting point of its motion. This leads to the construction of the cycloidal pendulum, for which the duration of swing is independent of the amplitude, obtained by rolling the thread of the pendulum on a cycloid to form its involute, which is itself a cycloid (congruent to the original).

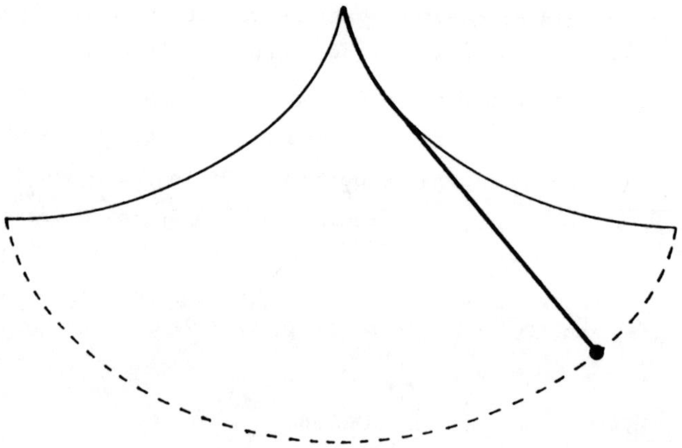

Cycloidal pendulum

Following the cycloid, we now introduce a few epi- and hypocycloids which have been investigated particularly frequently : the cardioid, the nephroid, the astroid and the three-cusped hypocycloid.

The cardioid (heart curve) is the epicycloid with $r/r' = 1$.

The nephroid (kidney curve) is the epicycloid with $r/r' = 1/2$.

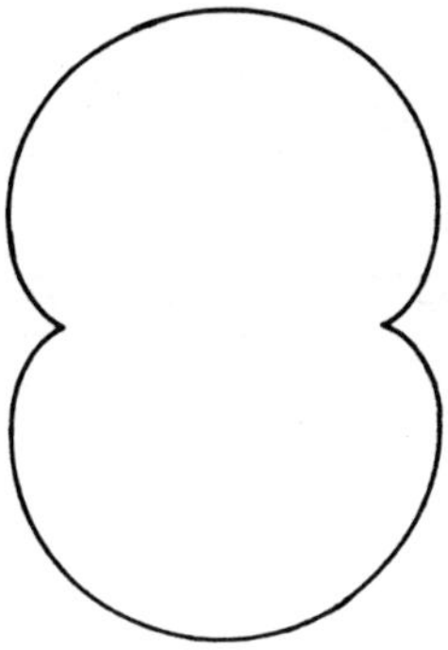

This curve comes up as the solution to a classical optical problem* : it is the catacaustic of parallel light rays falling on a circle. The catacaustic is the curve enveloping the family of reflected rays, their envelope. It had already been observed in antiquity that the intensity of reflected light was higher along this curve, whence the Greek name. A rigorous explanation was not supplied by ray optics, but first came through wave optics (the first such investigation was by Airy 1838 in his explanation of the rainbow by means of caustics**.) We shall return to the interesting curves which result from caustics again in section 1.8.

* Huygens : Traité de la lumière, 1690.

** Such results which were correct qualitatively, and to some extent quantitatively, already appear in Descartes and Newton.

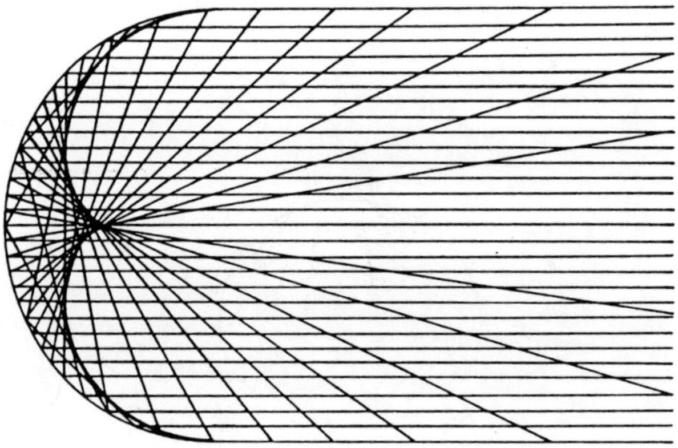

Caustic of the circle with parallel incoming light

The general caustic of the circle by reflection of light from a point source at arbitrary position was first determined by Cayley 1858.

The three-cusped hypocycloid is the hypocycloid with $r/r' = 1/3$. It was investigated by Euler 1745 in connection with an optical problem, and later studied by Steiner (1857), which is why it is also called the Steiner hypocycloid. It has the interesting property that the segments of its tangents lying inside it have constant length. Thus one can move a segment around the interior so that it always touches the cycloid and has its endpoints on the cycloid.

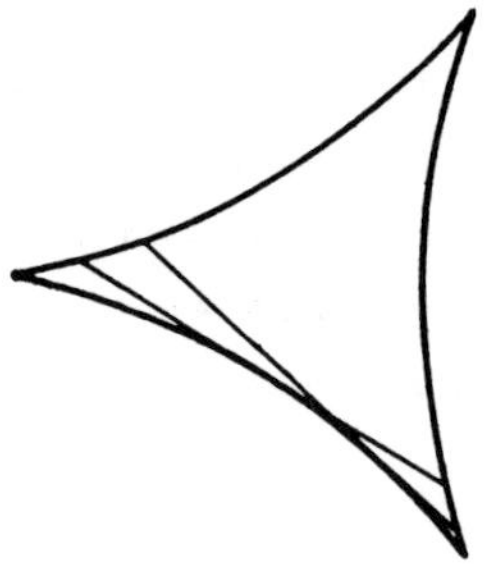

Three-cusped hypocycloid

The astroid (star curve) is the hypocycloid with $r/r' = 1/4$. It was already known in the time of Leibniz. An interesting way to generate it is the following : slide a segment of fixed length along a pair of perpendicular axes, with one endpoint on each. The envelope of these segments is the astroid.

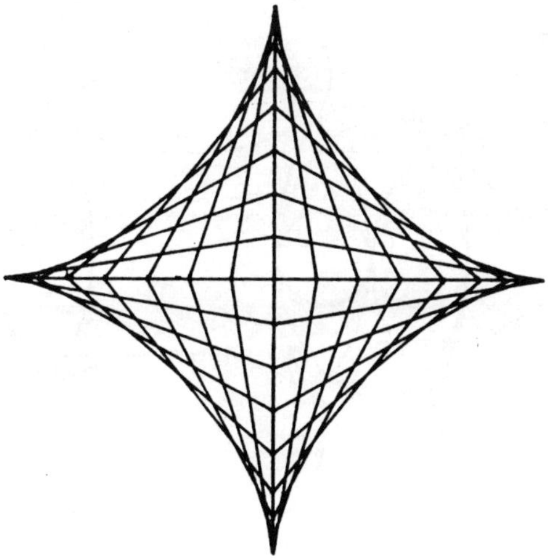

To conclude this short survey of the cycloids we report on two technically important applications : the cycloidal gear and the trochoidal-rotation-reciprocator.

It appears that the cycloidal gear was already known to Desargues (1593-1662). One arrives at this construction as follows. A circle K rolls on the outside of a fixed circle K'. Inside K a circle K'' rolls, and in such a way that it simultaneously rolls on the outside of K', i.e. at each moment it passes through the point where K meets K'. Then a point on K'' describes a hypocycloid H in a plane fixed to K, while it describes an epicycloid E in a plane fixed to K'.

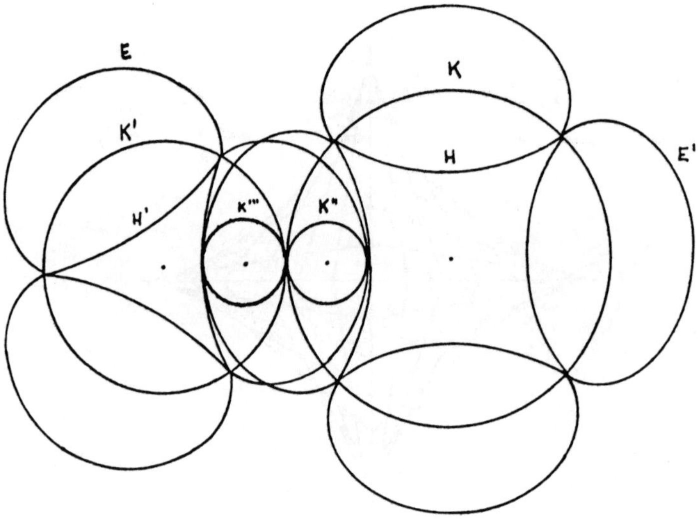

Similarly, a point on a circle K''' rolling in K' describes a hypocycloid H' in the plane of K', and an epicycloid E' in the plane of K. It is clear that with the rolling of K on K' the curves E and H always have the same tangent, so that they slide over each other. The same holds for E' and H'.

This fact gives rise to the cycloidal gear. The cogwheels of such a gear have teeth consisting of pieces of the epi- and hypocycloids, while the remaining, inessential parts of the boundary are circular arcs concentric with the axis. For the teeth of the cogwheel corresponding to K one uses pieces of H and E' and the cycloids resulting from them by a rotation, and corresponding to K' one uses pieces of H' and E. The pieces of E slide on those of H, those of H' slide on those of E', and the following drawing shows a simple example, in which the radii of K'' and K''' have been taken equal for the sake of simplicity, and the gaps between the teeth have also been taken equal to the width of the teeth. (In practice they are not allowed to be quite equal, so that the teeth of the opposing wheel have some play.) Moreover, the number of teeth in the drawing has been chosen smaller than is usual in practice. Further details, and descriptions of other gears and models may be found, e.g., in the work of Schilling [S1]. Here we reproduce some pictures of the models described there. (See also Wunderlich [W5].)

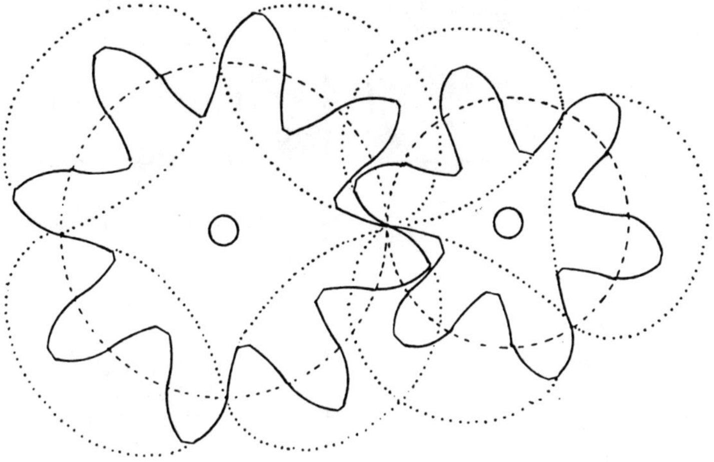

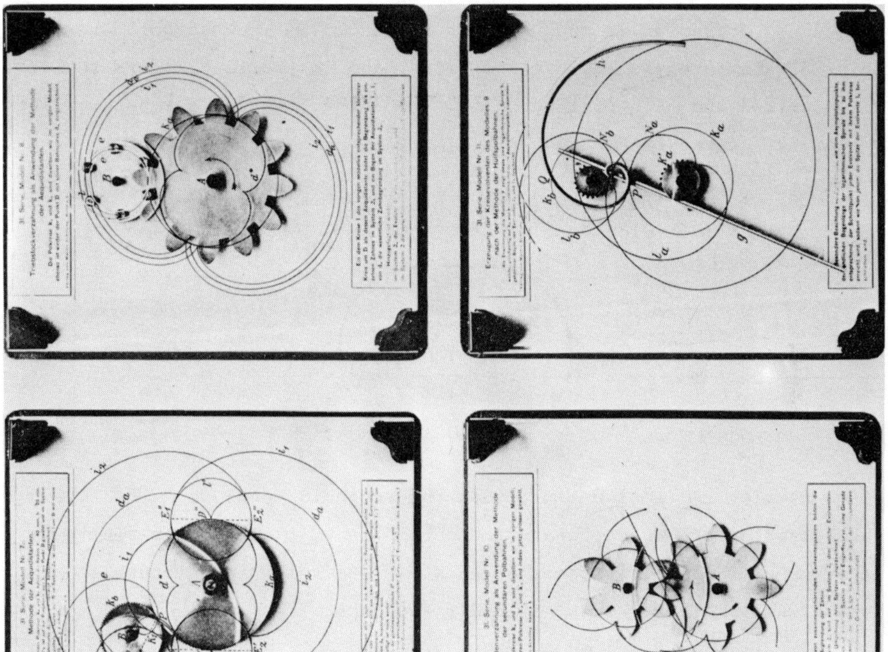

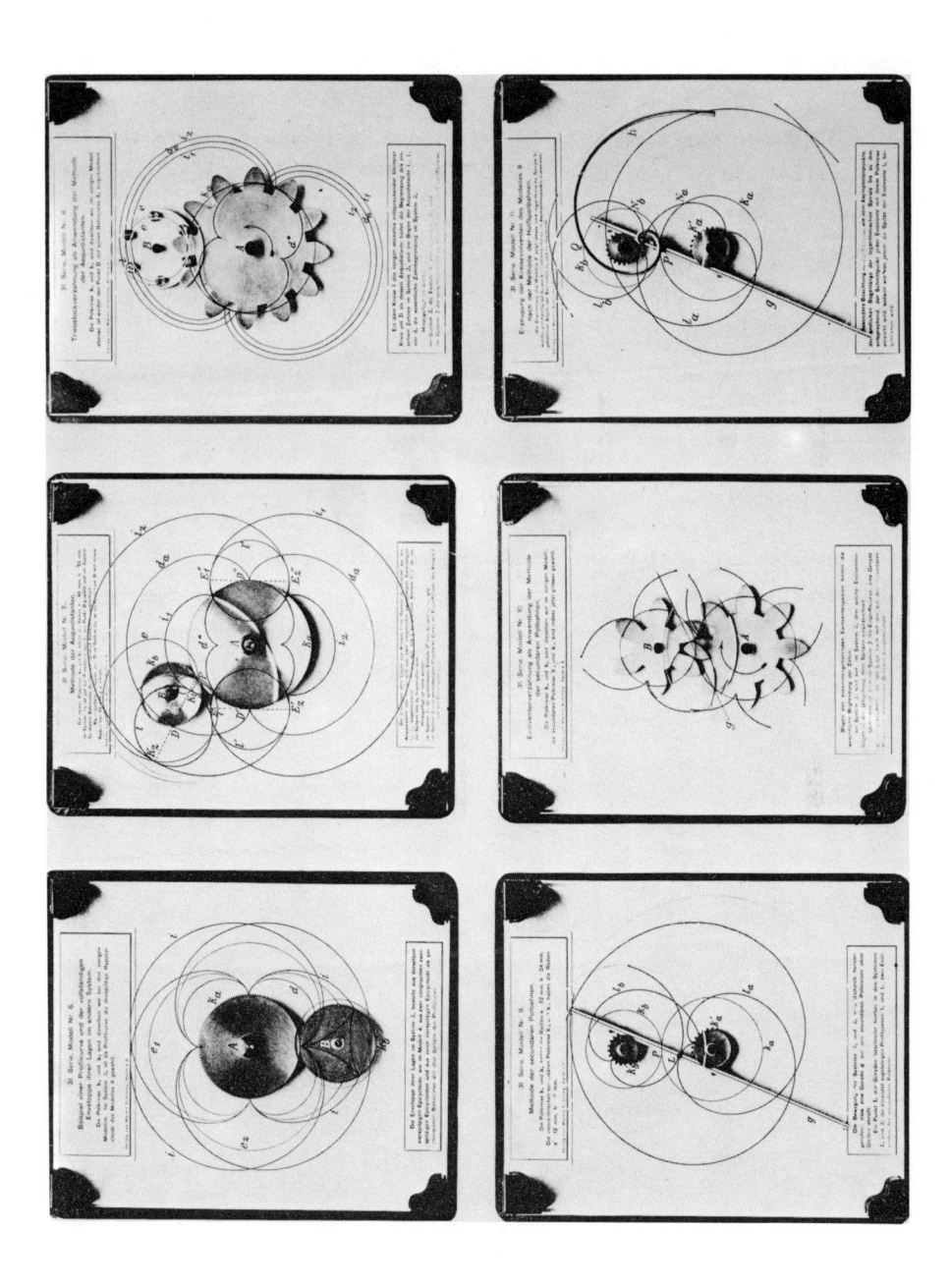

The mathematical starting point for the construction of the rotation reciprocator is similar to the one above. We again consider two circles K and K' which roll on each other. To fix ideas, we choose K of radius 2 and K' of radius 3. In what follows we think alternatively of K' being fixed, and K rolling in its interior, or K fixed and K' rolling on it. We also consider again two planes V and V' fixed to K and K'.

When K' rolls on K, a point outside K' describes a trochoid in V of the following form :

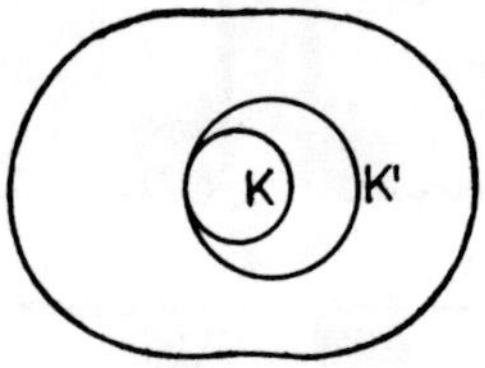

If we now let K roll on K', then the trochoid moves with it. At each position of K we obtain a trochoid in V', altogether forming a 1-parameter family of trochoids in K'. The following picture shows such a family of trochoids and their envelope.

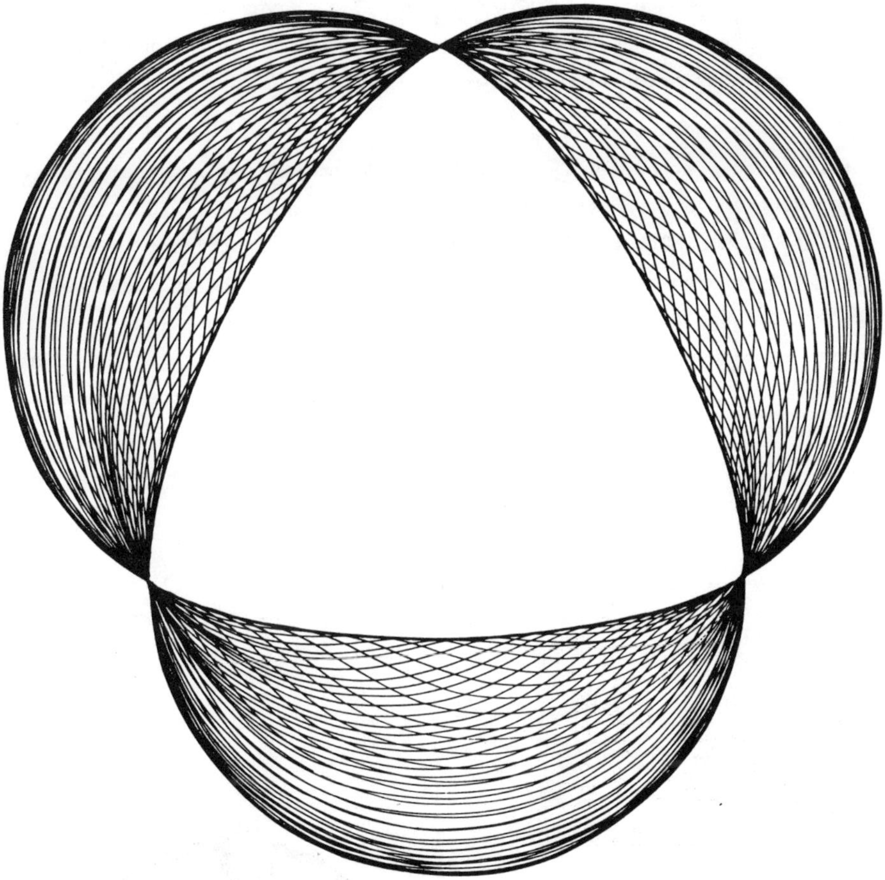

The envelope, which looks similar to a trefoil knot, has three double points, and all trochoids of the family go through these three double points.

Now we consider the curvilinear triangle Δ with these three double points as vertices, and the three inner arcs of the envelope as edges. When we again let K' roll on K, Δ moves inside the trochoid, and in such a way that its vertices always run on the trochoid, while its edges slide along it.

This is the geometric basis for the construction of the Wankel motor. The piston of this motor is (in principle) cylindrical and has cross-section Δ, and the cylinder of the motor has the trochoid as cross-section. The circles K and K' form an inner gear which gives guidance to the piston in addition to that of the cylinder wall. The axis of the piston moves on a circle around the centre of K; this motion is transmitted by means of an eccentric cylinder to an axle turning in a central bearing, and is thus transformed into a rotation of this axle. Details may be found in an essay of H.R. Müller [M3], from which the accompanying pictures are reproduced, or e.g. in the book of Wunderlich [W5] on plane kinematics.

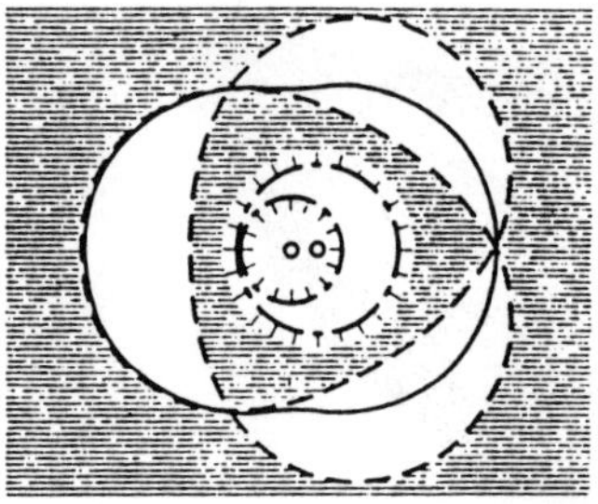

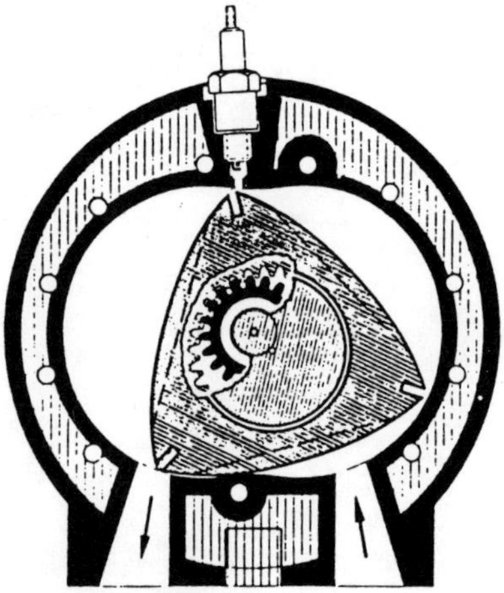

1.8 Caustics and contour curves in optics and perspective

In the perspective representation of curved surfaces, interesting curves with singularities appear as outlines. The following pictures show, as an example, one and the same torus under parallel projection in different positions. The curves appearing as outlines were investigated by Cauchy and called toroids.

They can be constructed as parallel curves of the ellipse which is the projection of the spine of the torus. The parallel curves of a given curve are defined as the envelopes of systems of circles with centres on the curve and fixed radius.

One sees easily that the toroids are in fact outlines of the torus under orthogonal projection by thinking of the torus as the envelope of spheres with centres on the spine. The images of the spheres under orthogonal projection are circles with centres on the ellipse which is the image of the spine. Their envelope is the toroid.

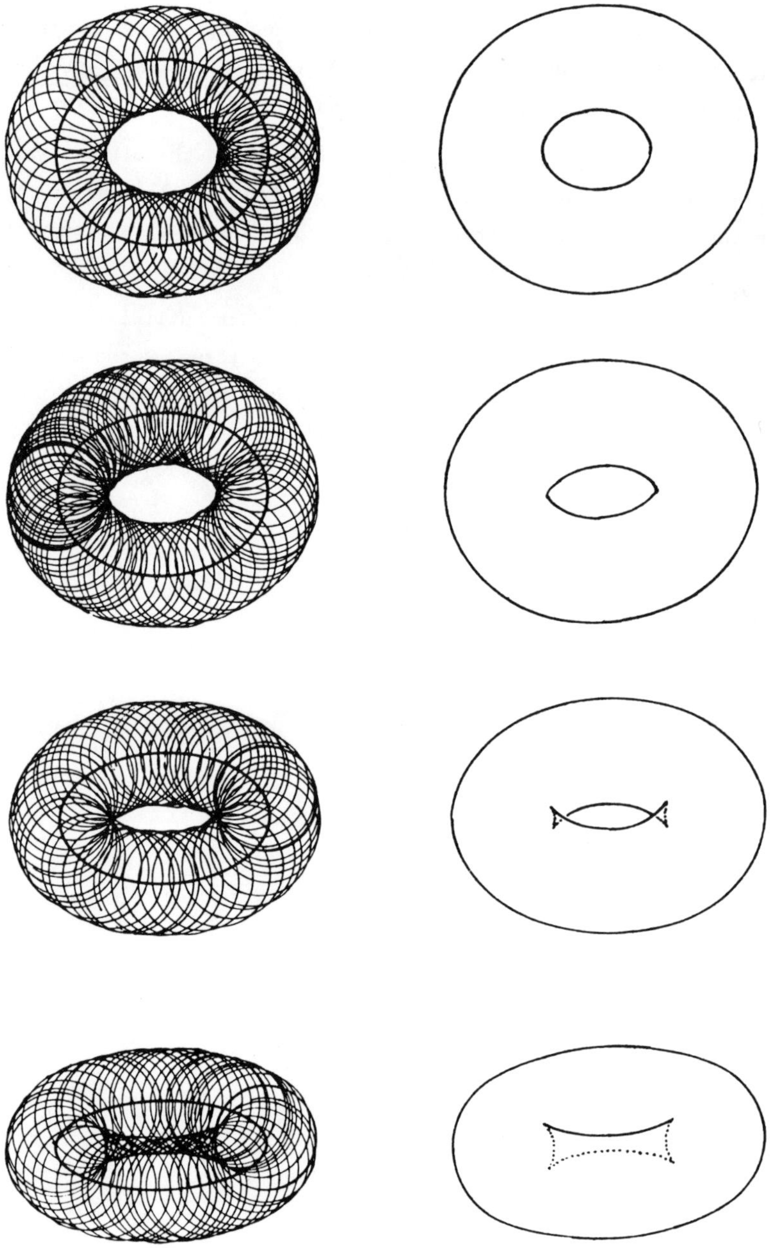

The problem of constructing outlines had already been encountered by the artists of the Renaissance – Uccello, Piero della Francesca, Alberti, Leonardo da Vinci and Dürer. Perspective representations of the torus may be found, e.g., in drawings and paintings of Uccello and Leonardo. The following reproductions give an impression of them.

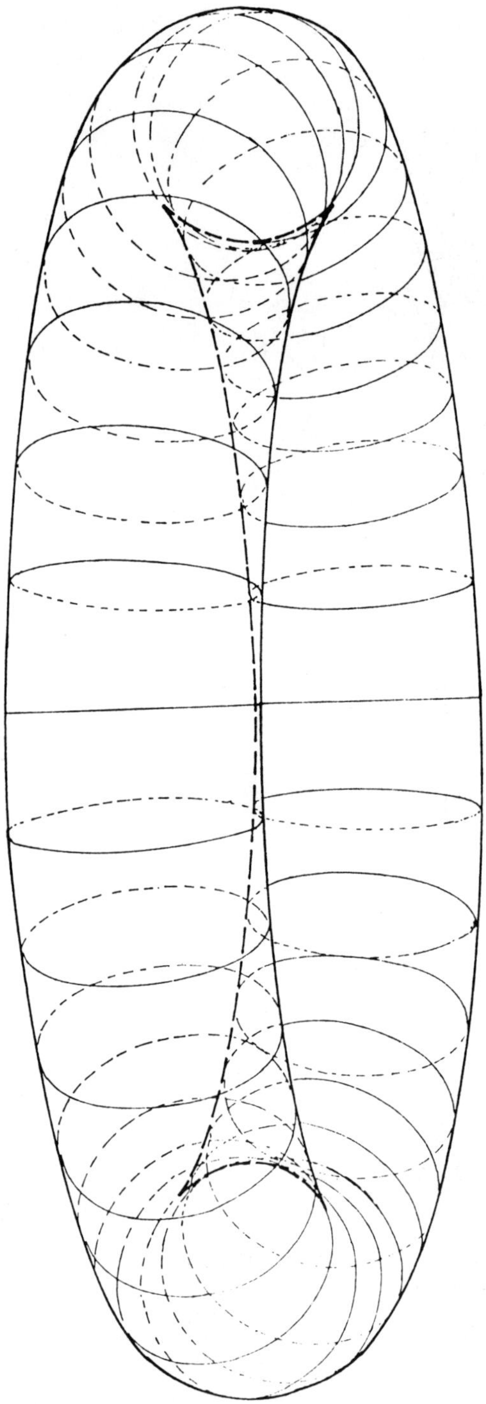

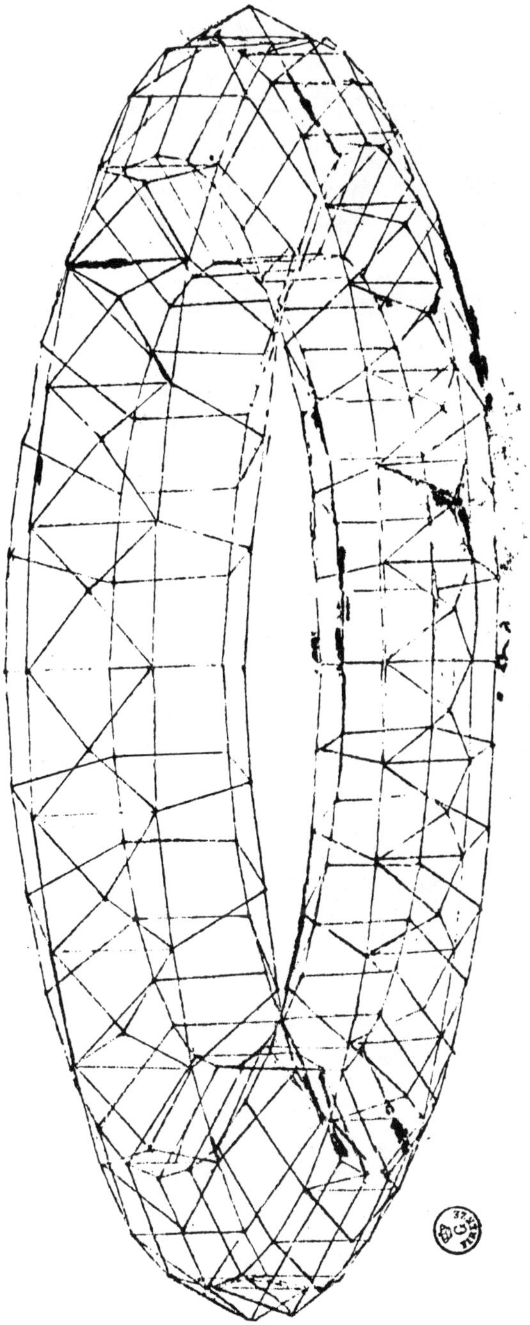

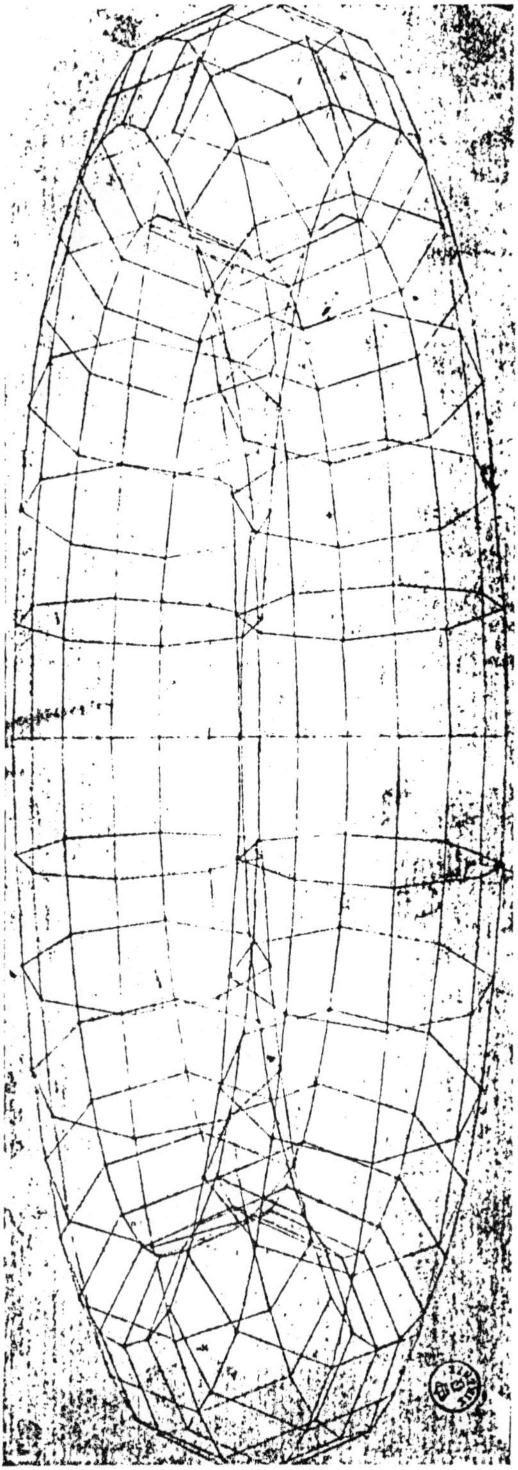

PAOLO UCCELLO, PERSPEKTIVISCHE STUDIEN

FLORENZ, HANDZEICHNUNGSKABINETT DER UFFIZIEN

Incidentally, it was his concern with perspective that led Dürer to the discovery of new algebraic curves which today are known as Dürer's shell curves. . The following are the relevant pages of Dürer's book on measurement with compasses and straight edge [D6], in which a definition and organic generation of the curves is given. Dürer's definition is not quite complete, inasmuch as the complete algebraic curve consists of two branches, only one of which is drawn by Dürer. We reproduce pictures of these curves from the book of Loria [L4].

seyten abgeschnyden sind worden/vnd drag sie zů der aufrechten lini/f/vnd kum mit zal auf zal/vnd puncktir die breyten zů beyden seyten der aufrechten/f/ neben dem kegel von der zal. 1. herab byß auf/g/h/ Darnach zeuch die gabellini Hiperbole/vonn punckt zů punckt/wie jch das hie vnden hab aufgeryssen/so eygentlich/ob schon keyn schryft dabey wer/vermeint jch/diß solt alles durch sehen käntlich seyn.

37

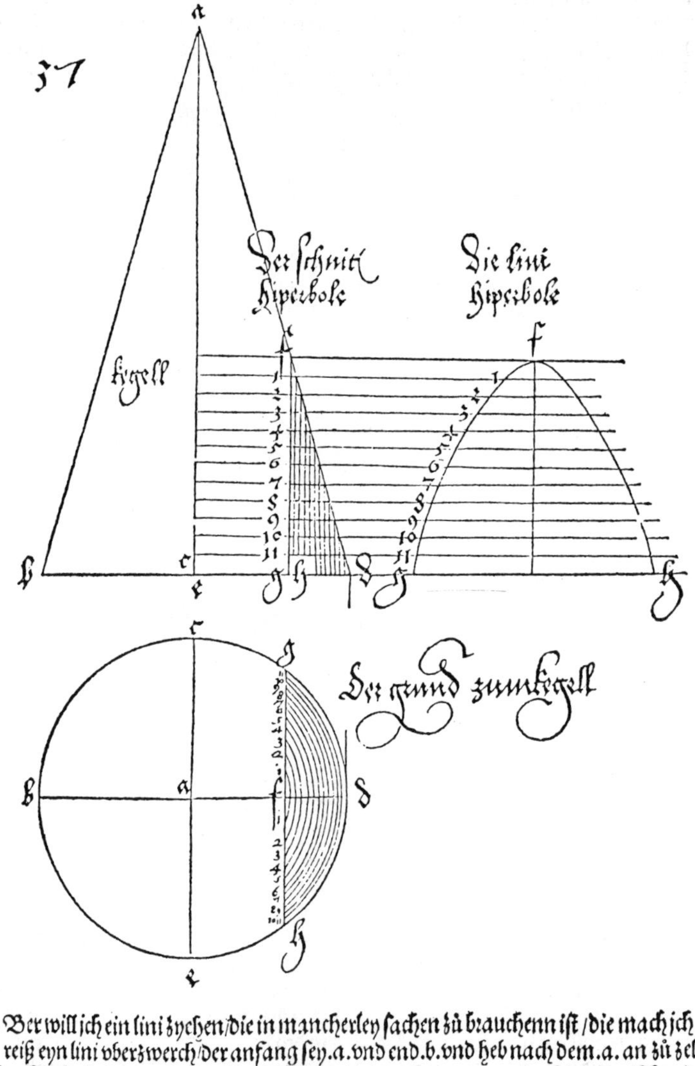

ABer will jch ein lini zyehen/die in mancherley sachen zů brauchenn ist/die mach jch also/Jch reiß eyn lini vberzwerch/der anfang sey.a.vnd end.b.vnd heb nach dem.a. an zů zelen/vn̄ setz auf diser lini.16.punckten in gleicher weiten nach eynander/ doch das zwischen dem end.b.

vnd dē punckten 16 ein trum vngeferlich vber bleib/Darnach setz jch ein aufrechte lini/auf die zwerch lini.a.b.in den punckten.13.so lang die zwerch lini.a.16.ist/ vnd puncktir sie auch mit disen zalen / vnd zel von vndē jch vbersich/Darnach nym ein richtscheydt/vñ stich darauf die leng.a/b.vnd setz das mit dem ein end/auf die zwerch lini.a.b.in den punckten .1. vnd leg es in der aufrechten in den punckten.1. vnnd wo das ander end des richtscheitz hyn trift/da setz jch auch ein punckten.1.Darnach leg jch das richtscheid mit dem ein end auf der zwerch lini.a.b.in den punckten.2.vnd erhebs an der aufrechten lini in den punckten.2.wo dann das ander end des richtscheitz hintrift/ dahyn setz jch auch ein punckten 2. Also thů jch im durch die gantzē zal der zwerch vnd aufrechten lini / biß das jch im durchschliefen zů 16 zaln kom/Darnach zeůch jch dise muschellini vonn punckt zů punckt / wie jch dann hie vnden hab aufgeryssen/dise lini ist in mancherley weis zů verkeren.

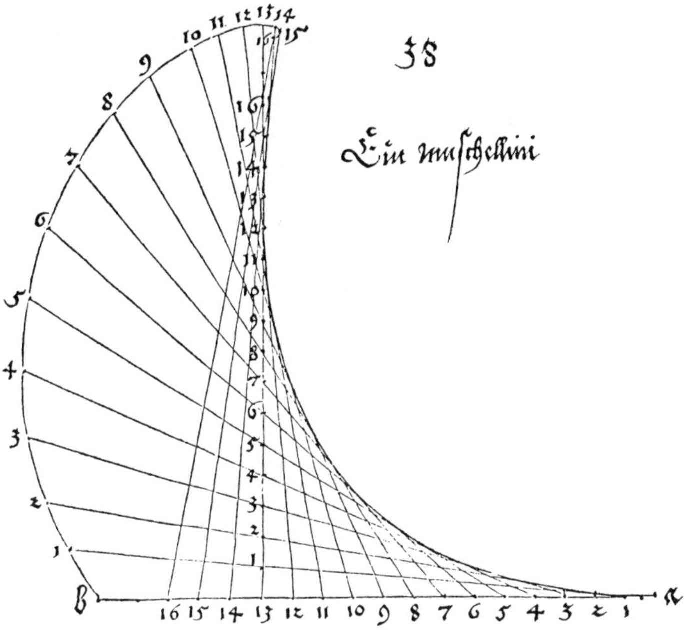

ZV diser egemachten lini mag man ein werckzeůg zů richten/ damit man sie machen kan leichtiglich/ nemlich also / Mach ein vierecket holtz so lang du seyn bedarfst vberzwerch / des anfang sey vorn.a.hinden.b.darein stoß oben ein tiefe niet/ das etwas darinn hyn vnd her gefürt můg werden/vnd theyl das holtz/mit punckt vnd zalen in so vil teyl du wilt / vnd heb die zal bey dem.a. an/Darnach mach in der mitt an ytlicher seyten diser zwerch laden/zwey aufrechte dünne richtscheytlein/so lang das zwerch holtz oder ladē ist/vnd das sie eng bey einander sten/vnd puncktir sie gleich mit der zal/als das zwerch hoĺtz oder die ladē puncktirt ist/vñ heb die zal vnden an/Darnach mach ein feins lentzlein/so lang oder kurtz du das habē wilt/vñ mach jm zůhynderst ein vmblaufetz redlein/ das in die mitt/der niet/in der zwerch laden.a.b.gerecht sey/darines hin vnd her geen můg/Darnach scheůb das lentzlein zwischen die zwey richtscheitlein hinauß gegen dem teyl.b.vñ setz das hynden oder vnden mit dem redlein in die mitt des zwerch holtz/gegē dem.a.in den erstē punckten.1.vñ leg das lentzlein zwischē den holtzern auch nider in den vnderstē puncktē . 1. vnd wañ du dañ mit dem redlein vnden gegen dem richtscheyt ferst/ vñ so weyt du hinein ferst/ so weyt far allweg mit dem lentzlein zwischē dein richtscheitlein vbersich/biß das du vnden mit dem redlein durch die zwey richtscheytlein vnd gar ans end kumst /

wirdet dir das lentzlein vorn mit der spitz dise lini furen/wie sie werdenn soll/ Dise mein meinung hab ich nachfolget aufgeryssen.

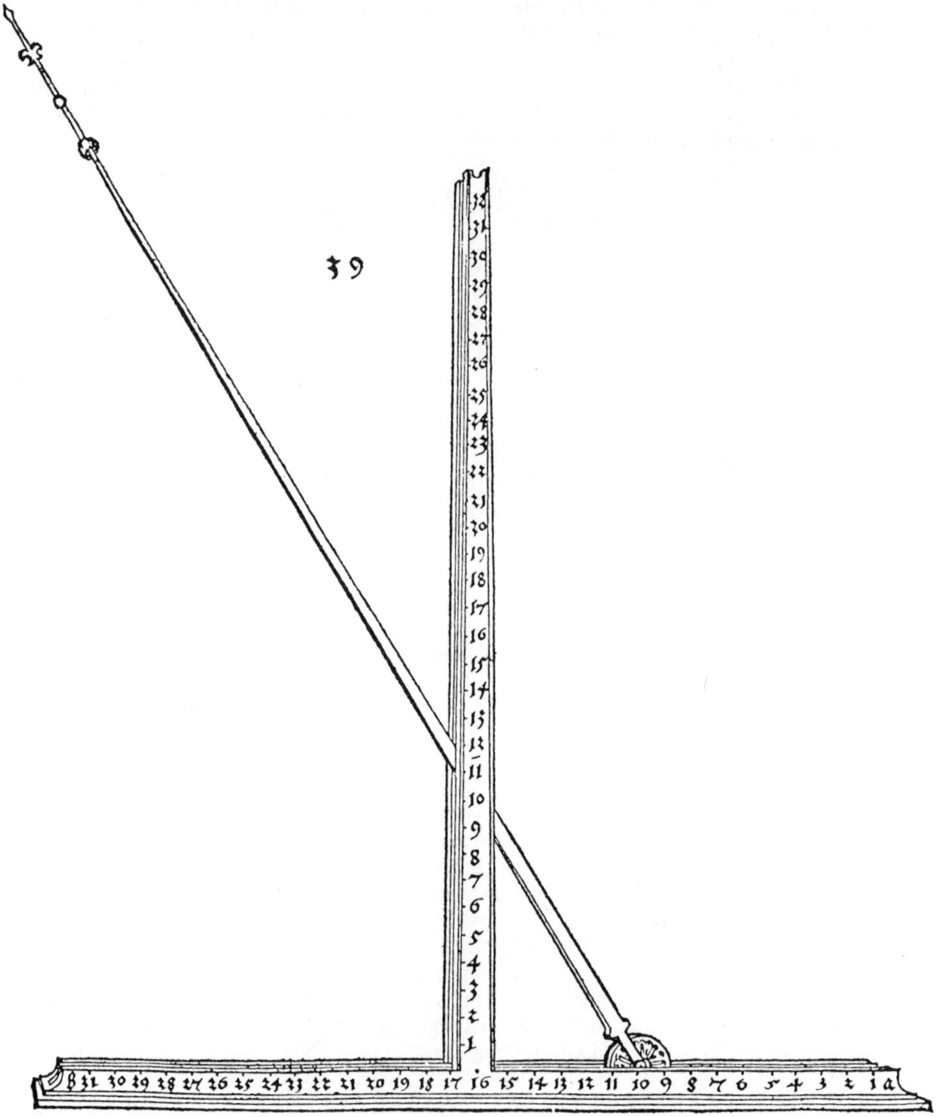

39

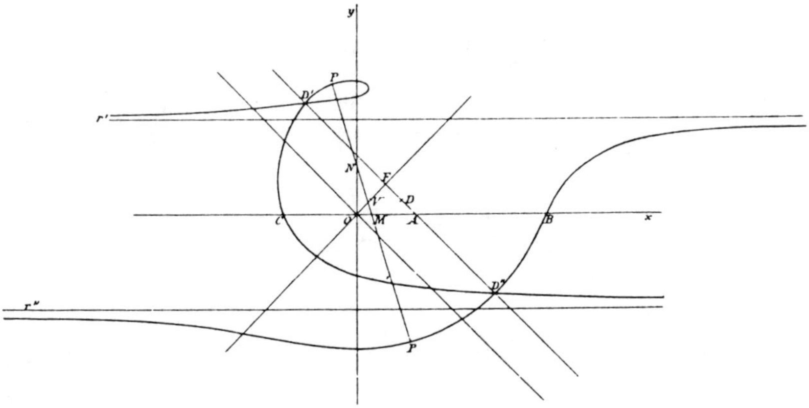
y
P
D'
r'
N
F
V
D
C
O
M
A
B
x
D''
r''
P

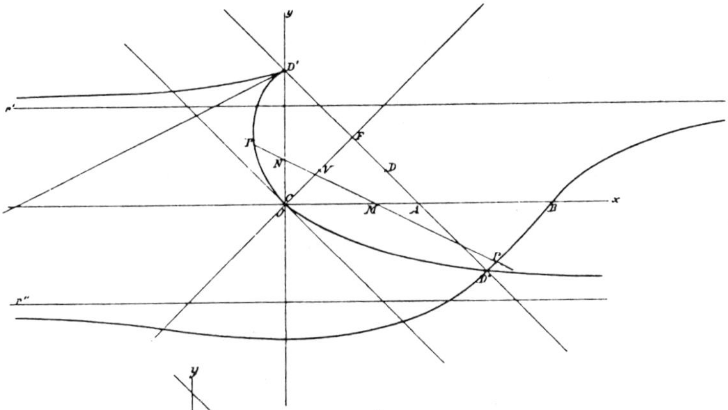
y
D'
r'
F
P
N
V
D
O
M
A
B
x
D''
r''

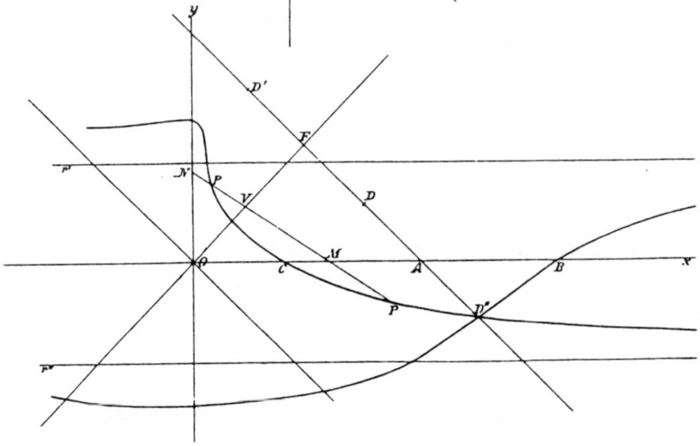
y
D'
F
r'
N
P
V
D
M
O
C
A
B
x
P
D''
r''

We now return from this digression to the outlines occurring in perspective representation, and clarify the appearance of singular points on outlines by two particularly simple examples.

Here is the first example : we consider a plane E in space and a smooth surface F lying over it as in the following picture.

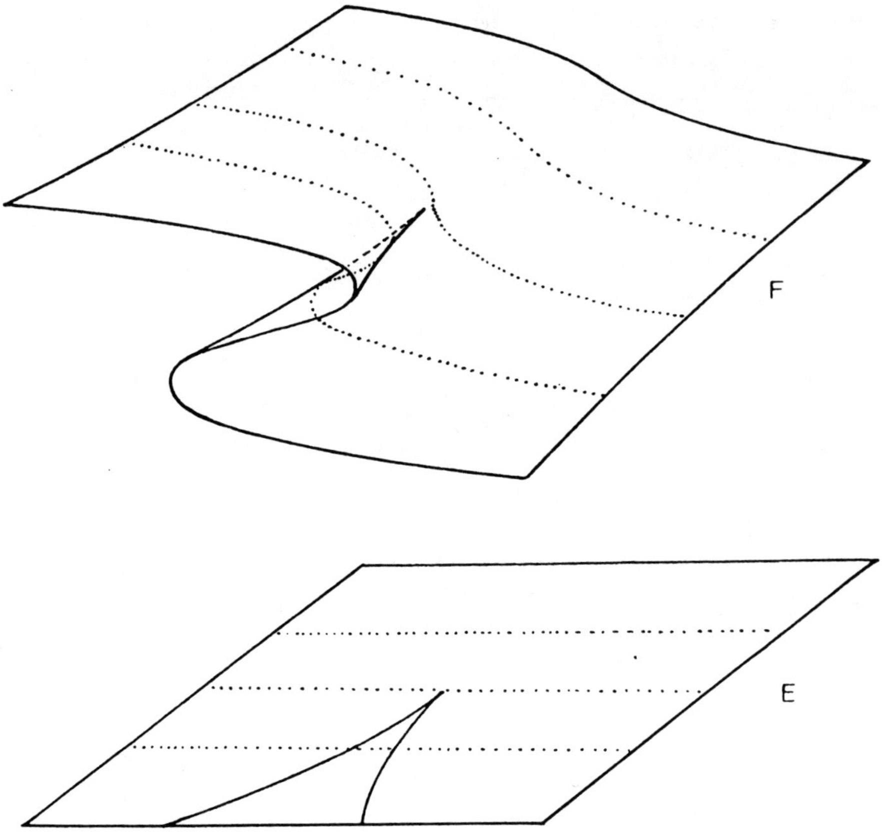

We hope it is clear from the picture what the surface looks like. In any case, the exact description of the surface does not enter. (One could easily give a particularly simple surface of this type : the planes perpendicular to E cut F in the dotted lines, which have the form of the curves $y = z^3 + xz$, where z is a coordinate perpendicular to E and x, y are coordinates in E. Thus the equation $y = z^3 + xz$ describes a surface of the type considered.) The surface F is a smooth surface, but orthogonal parallel projection onto E yields an outline with a cusp. (In the case of the surface with the equation above, one calculates quite easily that the outline has the equation : $27y^2 + 4x^3 = 0$.) Now an essential fact is that the generation of such an outline is a stable phenomenon : any other surface lying sufficiently close to F generates a similar outline with one cusp. One can also express this as follows : each small alteration in the projection map leads to essentially the same map, i.e. the mapping is stable. For this reason, the investigation of such singularities plays an important role in Mather's theory of stable mappings (see e.g. [Gl]).

This theory of stable mappings is connected with another important interpretation of the mapping of F onto E : we suppose that the coordinates x, y in the plane E are parameters which control the state z of a physical system. The points of the surface F are the possible equilibrium states of the system. For parameter values "outside" the outline there is exactly one equilibrium state, "inside" there are three. However, the "middle" one of these three is unstable. When one continuously varies the control parameters along a curve in E, it is obvious that the system may suddenly jump from one of the two stable states to the other on crossing the outline. Thus a quantitatively small change generates a qualitatively large jump. Such a jump is also known as a "catastrophe", about which there is a whole theory, the catastrophe theory of Thom, which appears to have interesting applications in widely different fields ([Tl], [Zl]).

We shall give another example of this kind, somewhat more complicated, but not offering anything new in principle. The surface F now has two "folds" instead of one. (An equation for such a surface would be, say, $y = z^4 + cz^2 + xz$).

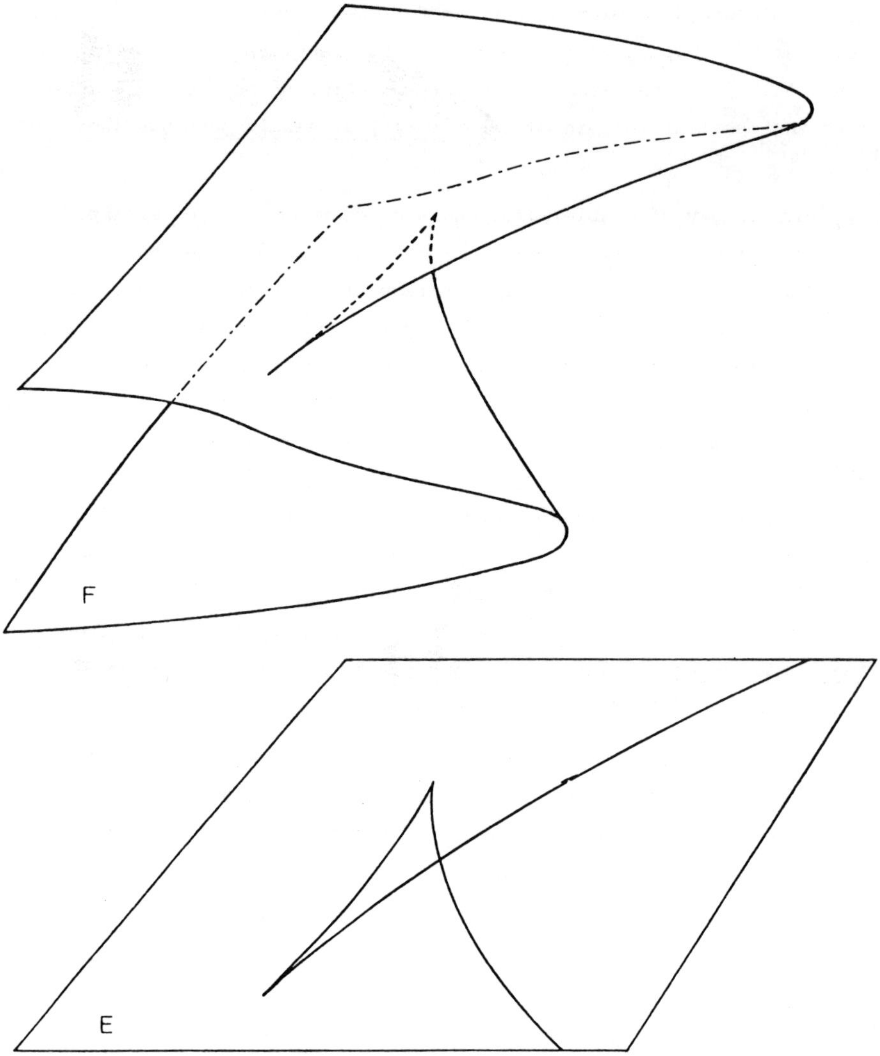
F
E

We see that the outline of F under orthogonal projection onto E has two cusps and an ordinary double point, and looks quite like a part of a toroid. In fact, what we see here is the paradigm for the outline of the projection of a smooth surface on a plane. Each such stable projection (i.e. one for which a small disturbance causes no essential alteration) generates outlines whose only singularities are ordinary double points and cusps (see also 4.2).

It is interesting to vary the parameter c in the surface equation considered above. This corresponds to varying the distance of parallel curves in our previous consideration of the ellipse. Of course, we can consider parallels to any other curve in place of the ellipse, e.g. the parabola. The following picture shows the family of curves parallel to a parabola. We have also shown the normals to the parabola. These normals are perpendicular to the parallel curves. The envelope of the normals is a curve K with a cusp, and the cusps of the parallel curves lie precisely on this curve K.

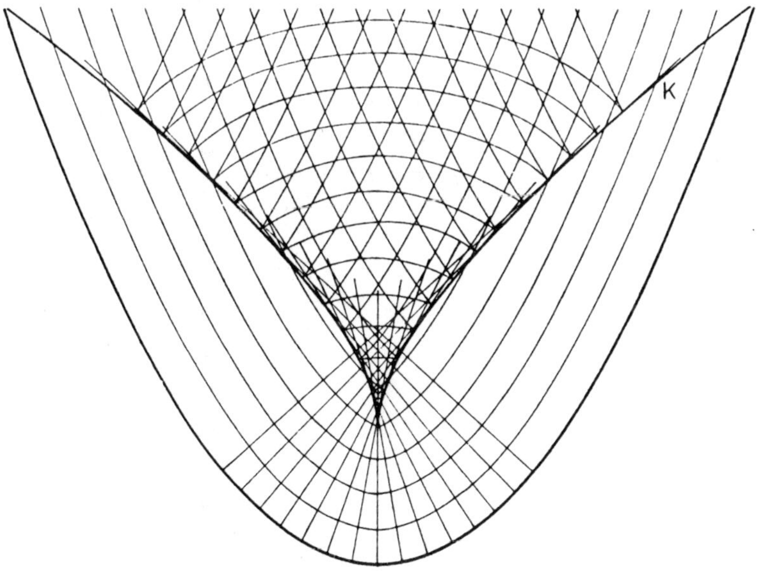

This mathematical situation is the model for an important class of physical problems. We think of the parabola as a wave front, of light waves, say. As the wave front continues to move, it describes precisely the parallel curves. In the optical interpretation, the normals to the parallel curves are precisely the light rays which one considers in ray optics, and their envelope K is the caustic. Thus one sees that the caustic is exactly the locus of singularities of the wave front.

Naturally, similar considerations apply in space instead of the plane. The caustics are then surfaces in space with certain singularities. Just as stable plane caustics can have only ordinary singularities and double points as singularities, one can prove that there are only three kinds of singularities of stable caustics in space. Here are pictures of them (see [Al]).

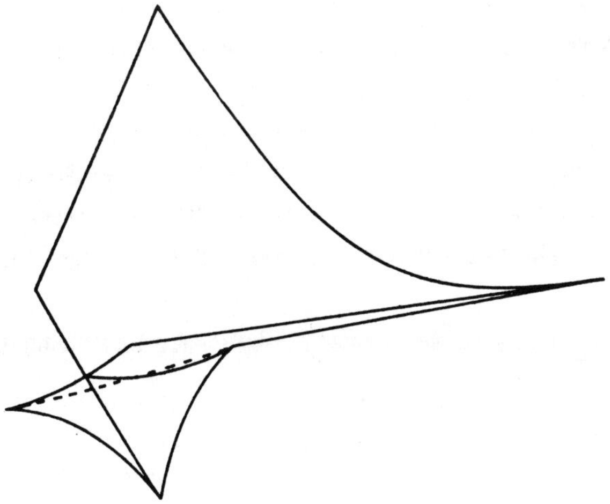

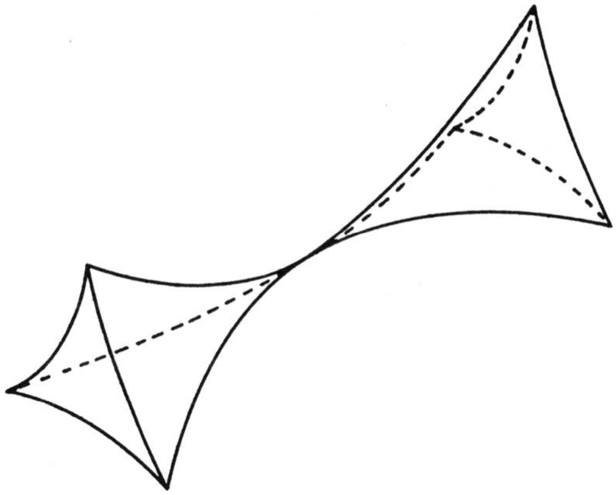

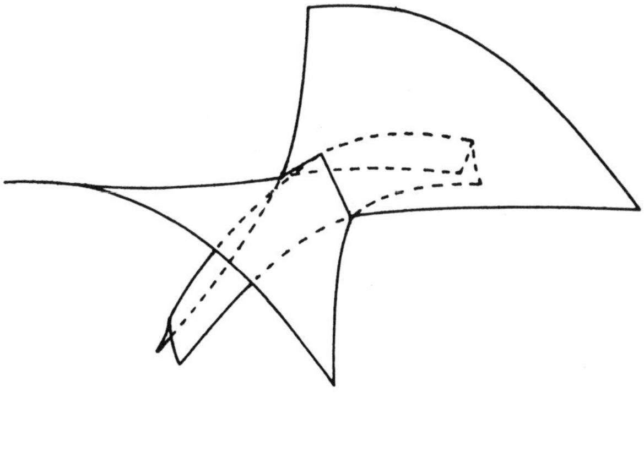

The systematic investigation of caustics extends, however, far beyond the starting point in geometric optics and wave optics chosen here, into the general theory of asymptotic solutions of partial differential equations by rapidly oscillating integrals (see e.g. the survey articles of Duistermaat [D5] and Arnold [A1]). But this leads far beyond the scope of this introduction, and we therefore leave optics and perspective and consider in conclusion a few examples of curves whose study had its origin in the science and technology of the last century.

1.9 Further examples of curves from science and technology

We have already seen some examples of technological applications of curves in section 1.7 : the cycloidal gear and the reciprocating motor. Here is another example of curves whose investigation stemmed from an important technological discovery : the Watt curves. These curves result from the discovery of J. Watt, which he made in 1784 for the purpose of guiding the piston rod of a steam engine, namely the Watt parallelogram. The Watt curves are generated as follows : we consider a quadrilateral ABCD, movable in the plane. The points A and D are fixed, while B and C can move on circles around A and D respectively. Now we choose a point M on the line BC. When B and C move, M describes a curve. The resulting curves are called Watt curves.

Their form depends on the relative sizes of the segments $\overline{AB}$, $\overline{BC}$, $\overline{CD}$, $\overline{AD}$. The following drawings show a few such curves.

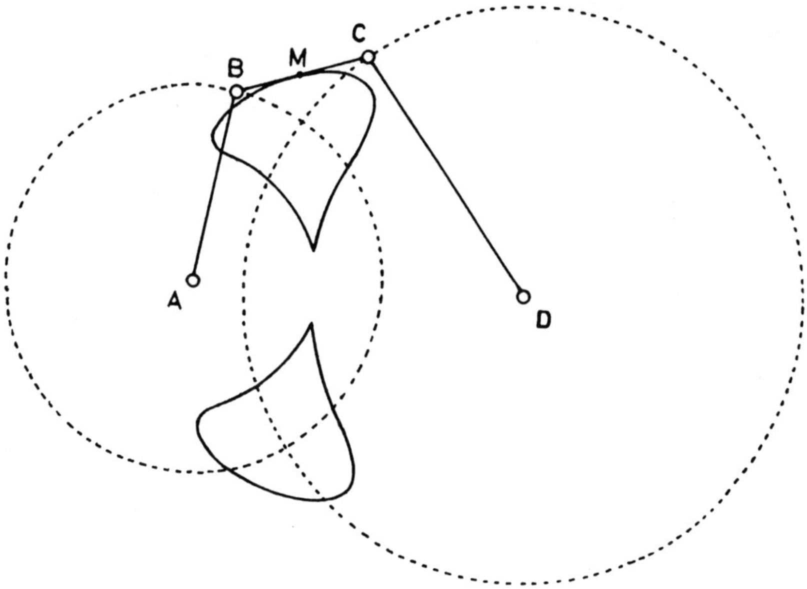
A
B
M
C
D

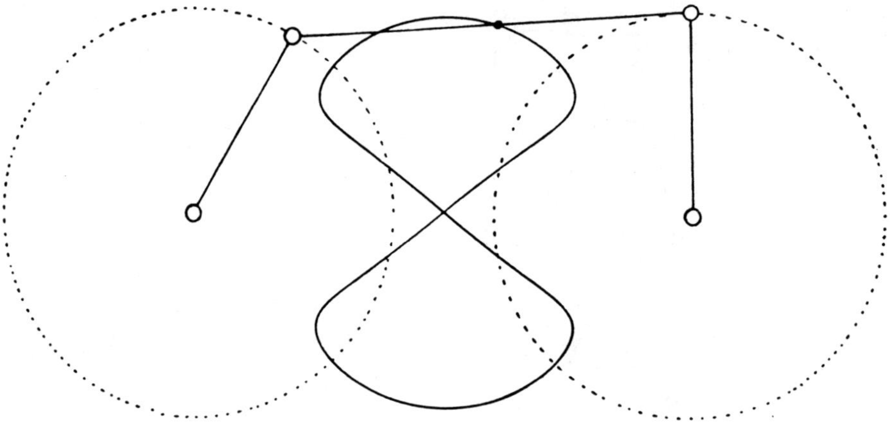

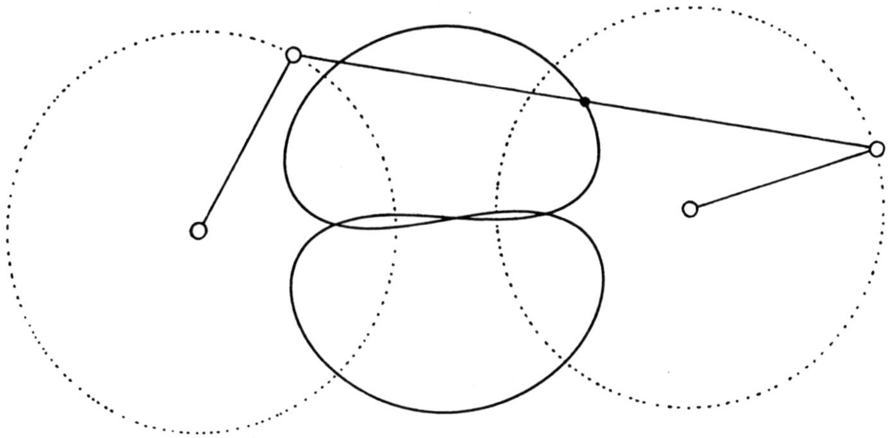

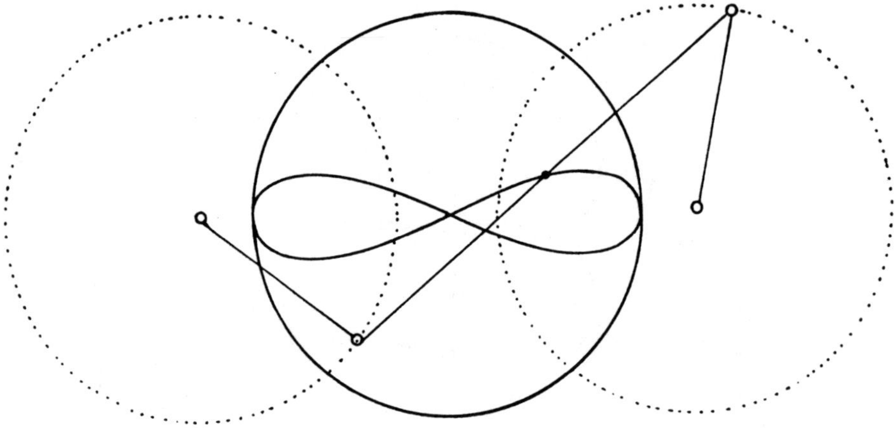

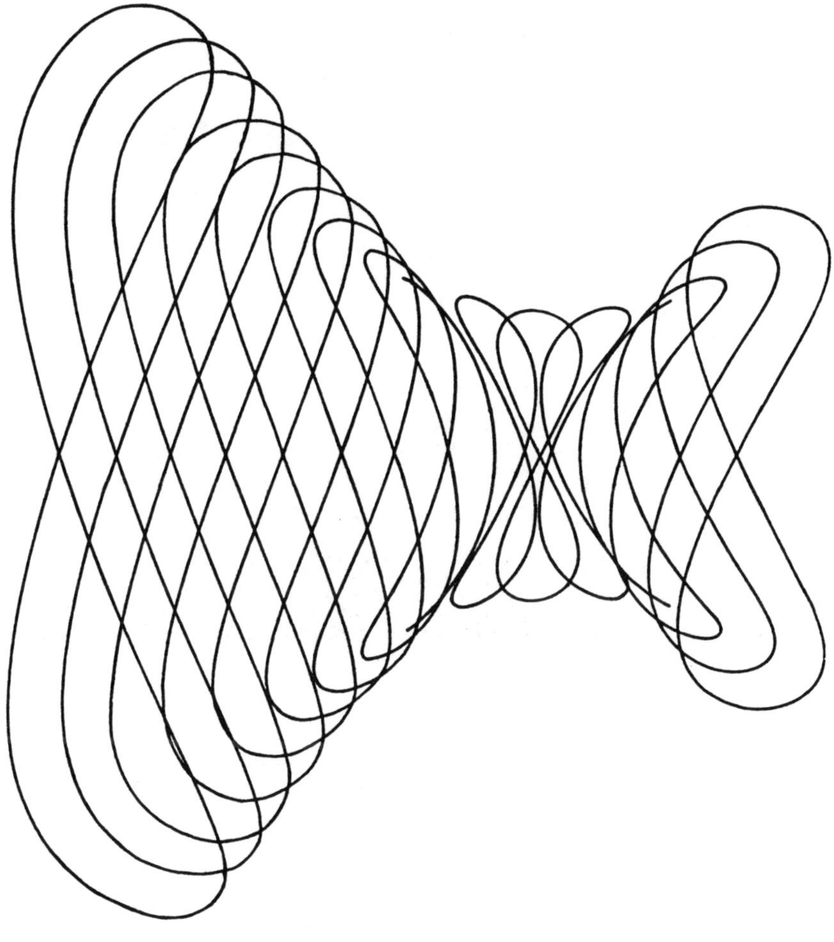

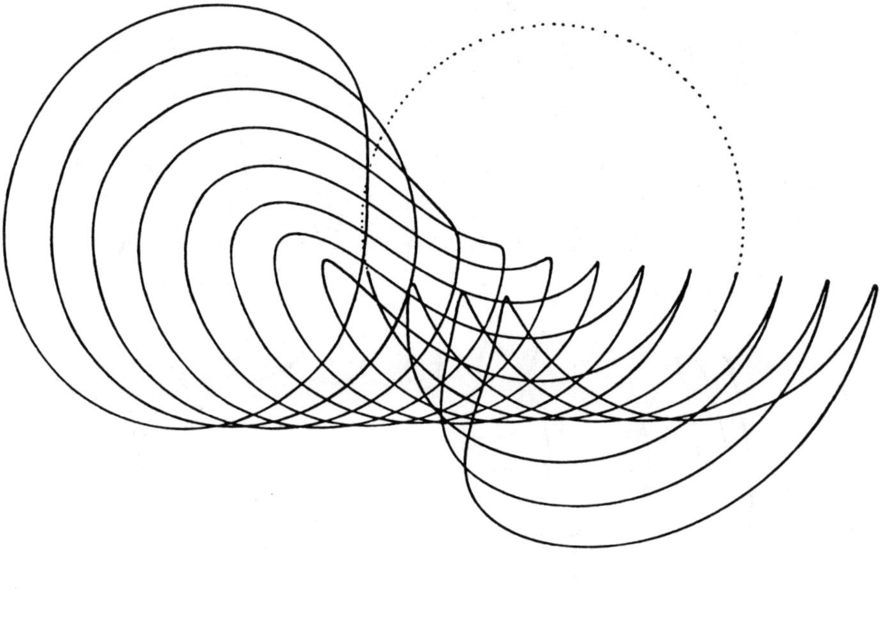

Thus one sees that with the help of a quadrilateral linkage one can already generate quite interesting curves. If one takes linkages with more rods, one can generate still more complicated curves. All these curves are algebraic curves or, more precisely, pieces of such, and conversely, each finite piece of an algebraic curve can be generated in this way, as was proved by Kempe. Admittedly, these linkages are mostly too complicated in practice. However, a few of them are of practical significance, e.g. in the construction of approximate linear motion. (Wunderlich [W5])

To conclude we shall briefly mention a class of curves which are usually known as Lissajou curves because (in Europe) they were first investigated by Lissajou 1850 in connection with vibration problems. They had already (1815) been investigated by the American mathematician and astronomer Bowditch in connection with the motion of a double pendulum. Nowadays one can make them visible in a particularly beautiful way by means of the cathode ray oscilloscope, because these curves are parametrised by

$$x = a \sin \omega t$$
$$y = b \sin \omega'(t+\tau).$$

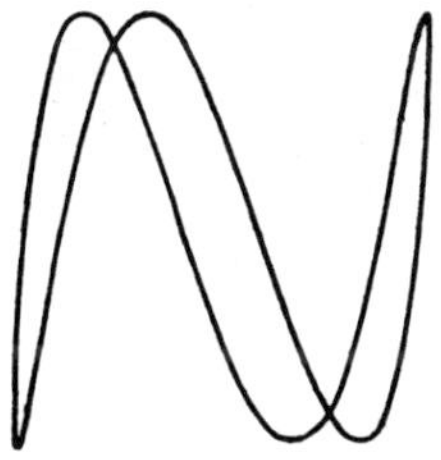

$\omega:\omega' = 1:3$

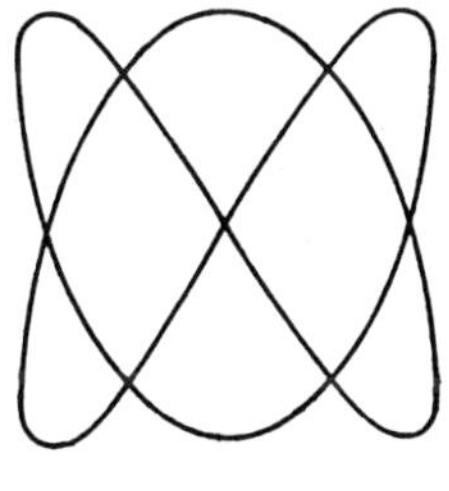

$\omega:\omega' = 2:3$

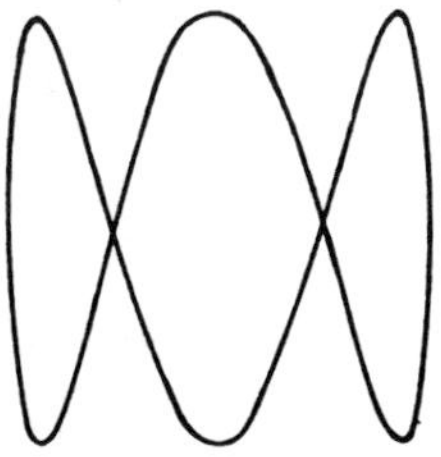

$\omega:\omega' = 1:3$

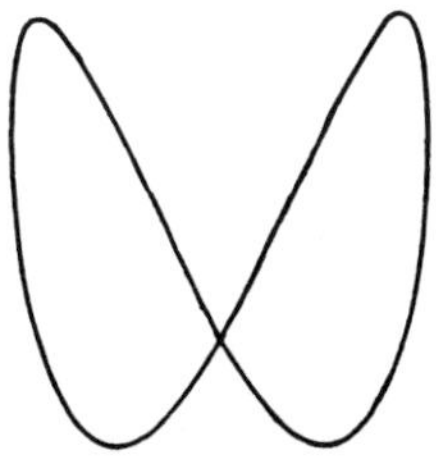

$\omega:\omega' = 1:2$

2. Synthetic and analytic geometry

Human activity and thought are extremely complex historical processes, in which many conflicting tendencies unfold. This also holds for mathematics, in which the unfolding of these conflicts is an important element pushing development forwards. In the process of mathematical research, as well as in the mathematical method itself, dialectical conflicts are of fundamental significance : analytic-synthetic, axiomatic-constructive, exact-intuitive, abstract-concrete, special-general, simple-complex, finite-infinite, regular-singular, algebraic-geometric, qualitative-quantitative. All these conflicts have had a marked influence on many fields of mathematics. I cannot go further into details here, so I shall refer to my essay on dialectic in mathematics [B5], as well as to the article by Alexandroff in the same book.

In what follows we shall confine ourselves to showing the unfolding of the conflict between analytic and synthetic in the development of the theory of plane curves.

2.1 Coordinates

The introduction of coordinates into geometry and the development of analytic geometry are usually attributed to Descartes and Fermat. This is, in the main, correct. However, it is important to emphasize that this brilliant achievement, which decisively affected the further development of mathematics, was made possible by earlier works which prepared the way for it – as indeed is the case with all brilliant achievements in human history. This is not to deny the novelty, the jump in evolution, but merely to conceive it as part of a historical process.

The idea of using coordinates in land measurement and city planning appears to have been the basis of Egyptian and Roman land surveying (for this and what follows see, e.g., Smith, History of Mathematics [S6], II, p. 316). The Greek geographers and astronomers, e.g. Hipparchus and Ptolemy, used degrees of longitude and latitude to describe points on the surface of the earth and in the sky.

The Greek geometers, in discussing curves, used relations between segments appearing in the construction which amount, from a present-day standpoint, to equations for the curves in cartesian coordinates. However, this analytic element was not fully developed : the basic idea of each analytic method is the reduction of a system to a few basic elements. The advantage of simplification gained in this way may possibly be opposed by the disadvantage of complexity in the reconstruction of the system from the basic elements. In the case of analytic geometry the reduction consists in choosing two perpendicular lines in the plane and determining the position of points, on a curve for example, by their distances x, y from these lines.

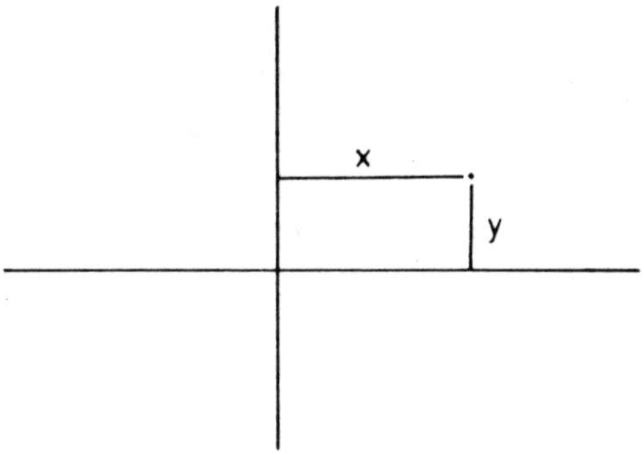

More generally, one can take any two intersecting lines as coordinate axes.

The simplicity of the reduction of the description to a position relative to only two lines is opposed by the complexity of the equation $f(x,y) = 0$ for the coordinates x, y of the points of the curve. The Greeks, whose thinking was perhaps more synthetic than analytic, preferred to use many auxiliary lines, in order to attain simple relationships.

The use of coordinates developed further in the Middle Ages, particularly in the work of Oresme (c. 1360), where there are already attempts to represent functions by graphs. Kepler and Galilei were influenced by this.

The decisive advantage of the analytic method, the reduction of qualitative geometric relations to complex quantitative relations – equations between coordinates – could only come to light after methods for handling such quantitative relations, i.e. algebra, had been developed. This development was due mainly to the eastern mathematicians of the Middle Ages, Arabs, Persians and Indians.

With the decline of the Roman empire came a period of socio-economic and cultural degeneration and poorer communication with the Orient, then trade began to develop between the blossoming Italian merchant cities and the Arab world, European scholars and traders such as Leonardo of Pisa (known as Fibonacci) (c. 1200-1220) studied the eastern culture, among other reasons "to put to use in their mercantile civilisation, which already in the twelfth and thirteenth century had seen the growth of banking and the beginnings of a capitalistic form of industry" (for this and what follows cf. D. Struik [S8], pp. 86-129*). Fibonacci cited the Arab algebraists and contributed to the gradual spread of Arabic numerical calculation in the account books of European merchants and bankers. This development strengthened in the next 300 years in the mercantile towns, growing under the direct influence of trade, navigation, astronomy and surveying. The townspeople were interested in numbers, arithmetic and calculation. "The fall of Constantinople in 1453, which ended the Byzantine empire, led many Greek scholars to the western cities. Interest in the original Greek texts increased and it became easier to satisfy this interest. University professors joined with cultured laymen in studying the texts, ambitious reckon masters listened and tried to understand the new knowledge in their own way" (quotation from Struik). The first objective was to pick up the old knowledge of the Greeks and Arabs. But the Renaissance was also a new age : "Characteristic of the new age was the desire not only to absorb classical information but also to create new things, to penetrate beyond the boundaries set by the classics". In mathematics, algebra was developed far beyond that of the Arabs.

At the beginning of the 16th century, Scipio del Ferro at the University of Bologna solved the cubic equation

$$x^3 + px = q$$

for non-negative p, q by

$$x = \sqrt[3]{\sqrt{\frac{p^3}{27} + \frac{q^2}{4}} + \frac{q}{2}} - \sqrt[3]{\sqrt{\frac{p^3}{27} + \frac{q^2}{4}} - \frac{q}{2}} .$$

---

*In the English edition, pp. 98-123 (Translator's note).

Ferrari reduced the equation of degree 4 to a cubic, Bombelli introduced a theory of pure imaginary numbers in his "Algebra" of 1572, in connection with the investigation of cubic equations, and Vieta (1540-1603) perfected the theory of equations, and was one of the first to denote the constants and unknowns in equations by letters. This brought algebra to a level of development at which it could be applied to geometry.

2.2 The development of analytic geometry

In Europe, the value of algebra for the solution of geometric problems was already known to Leonardo of Pisa in his "Practica geometriae" (1220), following the Arab mathematicians who had earlier known and used the relation between algebraic and geometric problems. Conversely, geometric methods were used for the solution of equations, e.g. by Cardano (1545). The relation between algebra and geometry later became common knowledge, e.g. with Vieta and Ghetaldi (1630).

But it was Fermat and Descartes who first arrived at a programme, a general method for the treatment of geometric problems by algebraic-analytic methods, i.e. at analytic geometry.

Fermat had the idea of analytic geometry in 1629. The details were first published posthumously. He had the rectangular coordinate axes as well as the equations $y = mx$ for lines and $x^2 \pm a^2y^2 = b^2$ for the conic sections. He knew that the extrema of a function $y = f(x)$ lie where the tangents to the corresponding curve in the (x,y)-plane are parallel to the x-axis.

Descartes published his geometry in 1637, but had worked on it earlier, perhaps since 1619. This is not the place to thoroughly assess Descartes' achievement within the general social development of his time. I content myself with a few quotations from the book of Struik already cited : Descartes' "Geometry" "brought the whole field of classical geometry within the scope of the algebraists". The book was originally published as an appendix to the "Discours de la Méthode"*, the "discourse on reason in which the author explained his rationalistic approach to the study of nature". ([S8])

In accordance with many other great thinkers of the seventeenth

---

* Complete title : Discours de la méthode pour bien conduire sa raison et chercher la vérité dans les sciences.

century, Descartes searched for a general method of thinking in order to be able to facilitate inventions and to find the truth in the sciences. Since the only known natural science with some degree of systematic coherence was mechanics, and the key to understanding of mechanics was mathematics, mathematics became the most important means for the understanding of the universe. Moreover, mathematics with its convincing statements was itself a brilliant example that truth could be found in science "... Cartesians, believing in reason ... found in mathematics the queen of the sciences".

His "Geometry" actually contains little analytic geometry in the modern sense, no "cartesian" axes and no derivations of the equations of conic sections as in Fermat. Nevertheless, the influence of this work on the development of geometry has not been overestimated. Descartes' programme and merit is the "consistent application of the well-developed algebra of the early seventeenth century to the geometrical analysis of the ancients, and by this, an enormous widening of its applicability".

A further merit of Descartes is the following : the Greeks and subsequent mathematicians had indeed considered not only lengths x, y of segments but also their products, such as x^2, x^3, xy etc., but the latter were not regarded as numbers of the same type, i.e. segments. Descartes abolished this distinction : "An algebraic equation became a relation between numbers, a new advance in mathematical abstraction necessary for the general treatment of algebraic curves, which one can regard as the final adoption of the algorithmic-algebraic tradition of the east by the west".

Thus Descartes' achievement lay in a unification of the numerous attempts already existing and in the conception of a general, quasi-mechanical method for the solution of geometric problems. Unlike the mathematicians before him he does not want to investigate just individual curves, he wants a general method for the investigation and classification of all curves. This is clearly expressed in the following quotation from him :

"I could give here several other ways of tracing and conceiving curved lines, of ever-increasing complexity ; but in order to comprehend all those which occur in nature and to separate them by order into certain genera, I know of no better way than to say that, for those we may call "geometric", that is those which are determined by some precise and exact measure, their points must bear a certain relation to the points of a straight line which can be expressed by a single uniform equation."

Thus Descartes sees the equation of a curve as the starting point of this general method. The method itself consists in the application of algebra to this equation.

The execution of this programme by Descartes, Fermat and their contemporaries, by Wallis, Pascal, Newton and Leibniz, was however not carried out by algebraic methods alone, but also by those which led to infinitesimal calculus – infinitesimal calculations in which limit processes complemented the algebraic methods.

The development of infinitesimal calculus was made possible by numerous and very varied earlier developments, which we cannot describe adequately here. These preparatory developments included investigations of curves such as the calculation of arc lengths (rectification), areas and volumes, as well as centres of mass, which go back to ancient traditions and from a modern viewpoint amount to calculation of integrals.

To these were added the investigations of Fermat and others on problems such as the tangent problem, which from our viewpoint amount to calculation of derivatives. Descartes' method consists in transforming these problems, which had previously been handled by geometric methods and more or less strict limit arguments, into problems which could be solved by algebraic calculations and, if necessary, by limit arguments. In the case of limits, each separate case was handled by a new and different argument, until Leibniz and Newton, with the differential and integral calculus, found a uniform method for handling all these problems. Leibniz conceived infinitesimal calculus to be, among other things, a method for unifying the different investigations of curves. Struik : "The search for a universal method by which he could obtain knowledge, make inventions, and understand the essential unity of the universe was the mainspring of his life."

Thus it was that the geometry of plane curves became analytic geometry, the investigation of the equations defining curves by algebraic and analytic methods. This was the starting point for two hundred years of development, after which the tendency to synthesis again came to the fore.

2.3 Equations for curves

We shall present equations for some of the curves considered earlier.

The equation for a <u>line</u> is of course

$$ax + by + c = 0.$$

The equation for a circle with centre (x_0, y_0) and radius r is

$$(x-x_0)^2 + (y-y_0)^2 = r^2 ,$$

as follows immediately from Pythagoras' theorem.

The <u>parabola</u> is the locus of all points equidistant from a line g and a point P. We choose the line through P and perpendicular to g as the y-axis and the perpendicular bisector of the perpendicular from P to g as the x-axis.

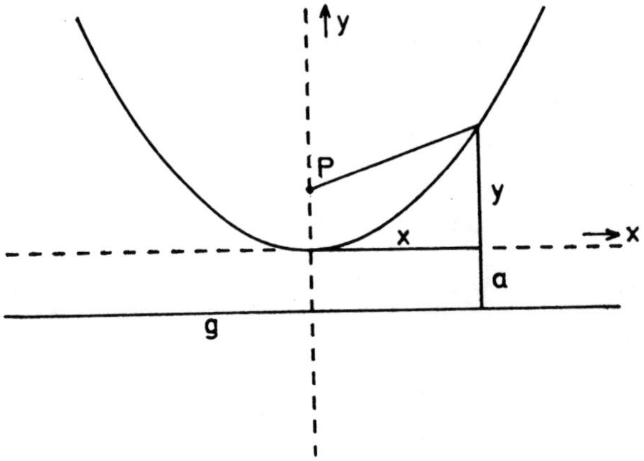

When the distance from P to g equals $2a$, the points of the parabola satisfy

$$(y+a)^2 = (y-a)^2 + x^2$$

by Pythagoras' theorem, and we obtain the equation of the parabola as:

$$x^2 - 4ay = 0.$$

For the <u>ellipse</u> and <u>hyperbola</u> we choose the x-axis to be the line through the foci, and the y-axis as their perpendicular bisector. Let 2e be the distance between the foci, and let c be the sum resp. difference of the distances from the foci.

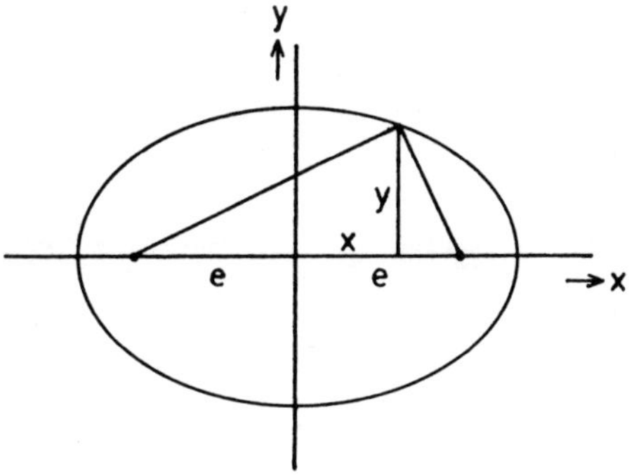

It follows immediately from Pythagoras' theorem and the curve definition that:

$$\sqrt{(x+e)^2+y^2} \pm \sqrt{(e-x)^2+y^2} = c.$$

For the ellipse one has the plus sign and $c > 2e$, for the hyperbola the minus sign and $c < 2e$. Squaring both sides one obtains

$$(x+e)^2 + y^2 + (x-e)^2 + y^2 - c^2 = \mp 2\sqrt{(x^2+y^2+e^2)^2-4x^2e^2} ,$$

and squaring again :

$$(4c^2-16e^2)x^2 + 4c^2y^2 = c^2(c^2-4e^2).$$

If one sets $c = 2a$ and $4c^2 - 16e^2 = \pm 16b^2$, one obtains the equation for the ellipse resp. hyperbola

$$\frac{x^2}{a^2} \pm \frac{y^2}{b^2} = 1.$$

If one does not choose the axes in the special way described above, but arbitrarily, and if one also admits degenerate conic sections, i.e. line pairs or double lines, then one obtains the general <u>equation of a</u>

<u>conic section</u> as

$$ax^2 + by^2 + cxy + dx + ey + f = 0,$$

i.e. an equation $f(x,y) = 0$ where $f(x,y)$ is a polynomial of degree 2 in the variables x and y.

To present the equation of the <u>cissoid</u> we recall the figure appearing in the definition, to which we have added coordinate axes.

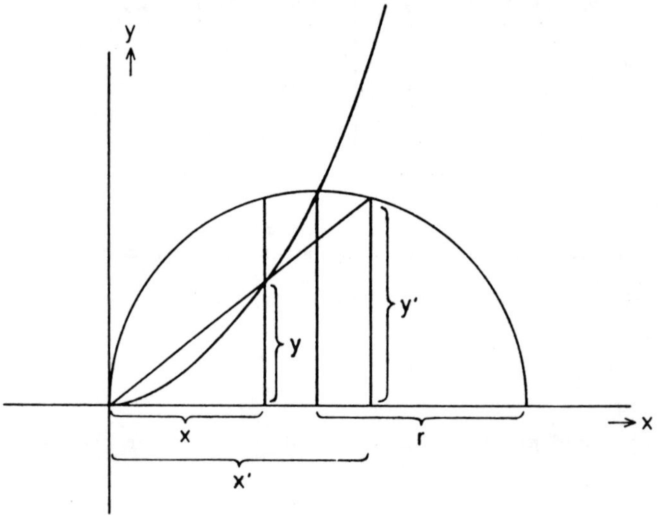

With the notation shown on the figure, the coordinates x, y of a point on the cissoid satisfy :

$$\frac{y}{x} = \frac{y'}{x'}$$

$$x' = 2r-x$$

$$y' = \sqrt{r^2-(r-x)^2} .$$

Squaring both sides of the first equation and substituting the other two yields the equation of the cissoid :

$$y^2(2r-x) - x^3 = 0,$$

and thus an equation $f(x,y) = 0$, where $f(x,y)$ is a polynomial of degree 3 in x and y.

We establish the equation of the conchoid in quite an analogous way. We choose coordinate axes and notations as in the following figure.

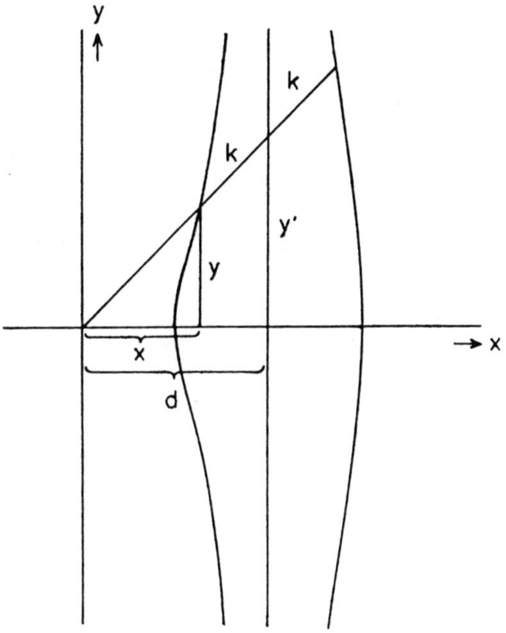

Then :

$$\frac{y}{x} = \frac{y'}{d}$$

$$(y'-y)^2 + (d-x)^2 = k^2.$$

An easy calculation gives the equation of the conchoid :

$$(y^2+x^2)(d-x)^2 - k^2x^2 = 0.$$

Thus it is an equation $f(x,y) = 0$ where f is a polynomial of degree 4 in x and y.

The spiric sections of Perseus result from cutting a torus by a plane. We therefore begin by setting up the equation of a torus relative to cartesian space coordinates x, y, z. The torus results from rotation of a circle about the z-axis. The circle lies in a plane which contains this axis, and has radius r. Its centre lies in the (x,y)-plane and has distance R from the z-axis.

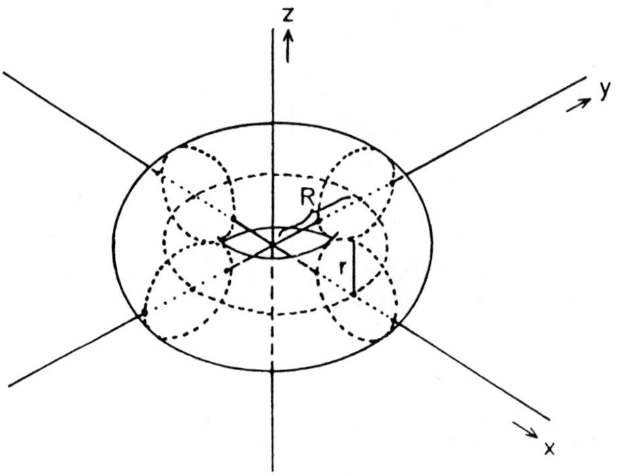

Then we obviously have

$$(R-\sqrt{x^2+y^2})^2 + z^2 = r^2 ,$$

hence

$$R^2 + x^2 + y^2 + z^2 - r^2 = 2R\sqrt{x^2+y^2} .$$

Thus one obtains the equation

$$(R^2-r^2+x^2+y^2+z^2)^2 - 4R^2(x^2+y^2) = 0$$

for the torus. If one now sets y equal to a constant c, then one obtains the equation of the spiric sections :

$$(R^2-r^2+c^2+x^2+z^2)^2 - 4R^2(x^2+c^2) = 0.$$

This is an equation $f(x,y) = 0$ where $f(x,y)$ is a polynomial of degree 4.

It is likewise easy to set up the equation of the Cassini curves on the basis of the definition. When one does this one sees that the equation coincides with the one above for $c = r$. In this way one easily sees that the Cassini curves are special cases of the spiric sections and with this we have the first example of the effectiveness of the analytic method.

The epicyclic curves are most simply described by a parametric representation

$$x = R \cos m\phi \pm r \cos n\phi$$
$$y = R \sin m\phi \pm r \sin n\phi.$$

When $\frac{m}{n}$ is rational one can easily derive an equation for the curve from this. We can assume without loss of generality that m, n are non-negative integers, because if not there are integers m', n', q with $m = \frac{m'}{q}$, $n = \frac{n'}{q}$, and when we use $\phi' = \phi/q$ as parameter we obtain a parametric representation with integral m', n'.

It now follows easily by iteration of the addition theorems for trigonometric functions that $\cos m\phi$ and $\sin m\phi$ are polynomial functions of $\cos \phi$, $\sin \phi$. For example

$$\cos 2\phi = \cos^2\phi - \sin^2\phi$$
$$\sin 2\phi = 2 \sin \phi \cos \phi.$$

If one then sets $t = \tan \frac{\phi}{2}$,

$$\sin \phi = \frac{2t}{1+t^2}, \quad \cos \phi = \frac{1-t^2}{1+t^2}.$$

Using the parameter t in place of ϕ, one now obtains a parametrisation for the epicyclic curve :

$$x = R_1(t)$$
$$y = R_2(t)$$

where the $R_i(t)$ are rational functions of t, i.e. quotients $P_i(t)/Q_i(t)$ of polynomials. The points of our curve are therefore the points (x,y) for which there is a t which is a common zero of the polynomial equations

$$P_1(t) - xQ_1(t) = 0$$
$$P_2(t) - yQ_2(t) = 0.$$

But it is well known that this holds precisely for the (x,y) for which the resultant of these two polynomials, which is of course a polynomial $f(x,y)$ in x and y, vanishes. Thus $f(x,y) = 0$ is the desired equation for our epicyclic curve for rational m/n.

For irrational m/n and arbitrary coordinates ξ, η in the plane there can be no continuous equation $f(\xi,\eta) = 0$ whose zero set is the epicyclic curve, because the points of this curve lie densely in an annulus. But when one works with polar coordinates, which are essentially coordinates in the universal covering of the original punctured plane, then the curve has an equation of the form

$$\phi^2 - a + b \cos \phi = 0.$$

Thus one sees that these curves, despite having the same type of generation and trigonometric parametrisation, are very different from the point of view of equations. One satisfies a polynomial equation and has a very nice, namely rational, parametrisation, the other has no reasonable equation at all in the plane, and only a transcendental equation in the covering.

Similar remarks apply to the Lissajou curves. For a rational frequency ratio the two vibrations satisfy an algebraic equation, which one can derive by the method above, for an irrational ratio they do not.

Finally we mention the equation of the cycloid in cartesian coordinates:

$$x = r \text{ arc cos } \frac{r-y}{r} - \sqrt{2ry-y^2}$$

for a circle of radius r which rolls on the x-axis.

Thus we see that many interesting curves can be defined as the zero sets of equations $f(x,y) = 0$, where $f(x,y)$ is a polynomial in x and y. Such curves were called algebraic curves by Leibniz.

Some curves have an analytic equation $f(x,y) = 0$ in cartesian coordinates, where $f(x,y)$ is not a polynomial, and hence a transcendental, function. Leibniz called such curves transcendental curves. (Example : the cycloid.)

Other curves satisfy, e.g., a transcendental equation relative to polar coordinates. Often these are also called transcendental curves.

According to Loria [L4], Descartes considered a somewhat different class of curves, which he called geometric curves. In our language they are defined as those which satisfy a differential equation $f(x,y,y') = 0$, where $f(x,y,y')$ is a polynomial in all three variables. This class includes the algebraic curves and many special transcendental ones. I do not know whether Loria's assertion is historically correct.

It is clear that algebraic methods will apply most successfully when $f(x,y)$ is a polynomial. Only for this class, the algebraic curves, is there a general and self-contained algebraic theory. In what follows we will therefore confine ourselves to the investigation of algebraic curves.

2.4 Examples of the application of analytic methods

The tangent problem consists in constructing the tangent to a given plane curve at a given point.

The meaning of "constructing" remains open at first. The mathematicians of the 17th century gave solutions to this problem in many individual cases. The unification of these results was a very important element in the development of differential calculus. After the completion of this development the solution of the tangent problem could be presented as follows.

The curve in the plane is given by an equation $f(x,y) = 0$ in cartesian coordinates x, y. If (x_0,y_0) is a given point on the curve, then $f(x_0,y_0) = 0$. It may be that the curve does not have a uniquely defined tangent at this point, e.g. in the case of an ordinary double point, as we have seen in many examples. One hypothesis which guarantees the existence and uniqueness of the tangent at a point, is that the curve be smooth, or regular, at this point. Analytically, this is equivalent to the hypothesis of the implicit function theorem, i.e. that the partial derivatives of f do not both vanish at (x_0,y_0).

Thus we can assume without loss of generality that $\frac{\partial f}{\partial y}(x_0,y_0) \neq 0$. Then, by the implicit function theorem, there is a unique solution $y = y(x)$ with $y(x_0) = y_0$ and $f(x,y(x)) \equiv 0$ in an interval containing x_0, and $\frac{dy}{dx} = -\frac{\partial f}{\partial x} / \frac{\partial f}{\partial y}$.

This function $y(x)$ therefore gives a local parametrisation $x \mapsto (x,y(x))$ for our curve in the neighbourhood of the given point.

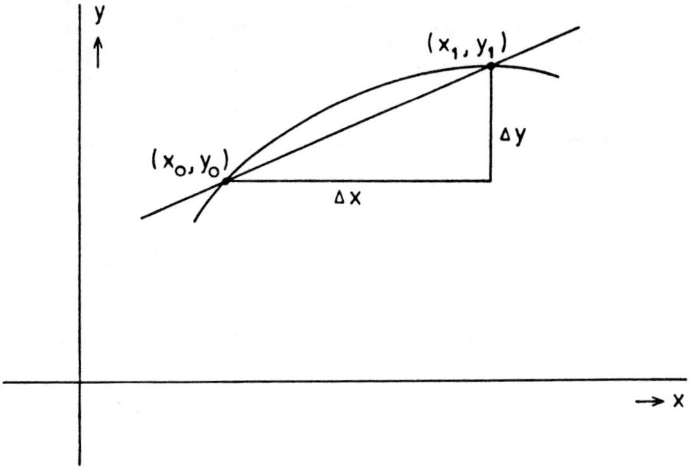

If one considers a secant of the curve through a neighbouring point (x_1,y_1) on the curve, then this has the equation

$$(y-y_0) = \frac{y_1-y_0}{x_1-x_0}(x-x_0) = \frac{\Delta y}{\Delta x}(x-x_0).$$

Here one sees the advantage of the application of algebra to geometry made possible by Descartes : the description of the secant is reduced to the simplest algebraic operations on the differences between the coordinates of the points. Now comes the infinitesimal part of the analysis : if one lets (x_1,y_1) tend toward (x_0,y_0), then the secant becomes the desired tangent, and correspondingly the difference quotient $\frac{\Delta y}{\Delta x}$ becomes the differential quotient $\frac{dy}{dx}$. Thus the computation above and this differential quotient finally give us the <u>equation of the tangent</u> at the point (x_0,y_0) :

$$\frac{\partial f}{\partial x}(x_0,y_0)(x-x_0) + \frac{\partial f}{\partial y}(x_0,y_0)(y-y_0) = 0.$$

This determines the tangent analytically. Now when our curve is algebraic, $\frac{\partial f}{\partial x}(x_0,y_0)$ and $\frac{\partial f}{\partial y}(x_0,y_0)$ may be computed from x_0,y_0 and the coefficients of the polynomial $f(x,y)$ by rational operations, i.e. the corresponding segments may be constructed by ruler and compass and one has thereby – in principle – a construction of the tangent in an entirely classical sense.

We have just defined regular points. A non-regular point is a singular point. Thus it is a point (x_0,y_0) with $f(x_0,y_0) = 0$, $\frac{\partial f}{\partial x}(x_0,y_0) = 0$, $\frac{\partial f}{\partial y}(x_0,y_0) = 0$. Examples of such points that we have seen previously (in non-analytic description) are the cusps and ordinary double points.

We give an example of the way in which the analytic description of singular points enables us to determine them by algebraic operations : we determine the singular points of the astroid.

With respect to suitable cartesian coordinates the astroid has the equation

$$(x^2+y^2-1)^3 + 27x^2y^2 = 0,$$

as we shall see later. Thus the partial derivatives, which we shall denote by f_x and f_y instead of $\frac{\partial f}{\partial x}$ and $\frac{\partial f}{\partial y}$, satisfy

$$3(x^2+y^2-1)^2 2x + 54xy^2 = 0$$

$$3(x^2+y^2-1)^2 2y + 54x^2y = 0$$

at the singular points. The three equations imply

either (a) $x = 0$ and $y = \pm 1$

or (b) $y = 0$ and $x = \pm 1$

or (c) $(x^2+y^2-1)^2+9y^2 = 0$ and $(x^2+y^2-1)^2+9x^2 = 0$.

From (c) it follows that $x^2 = y^2$ and hence

$$(2x^2-1)^2 + 9x^2 = 0.$$

If one sets $\xi = x^2$, this gives the equation

$$4\xi^2 + 5\xi + 1 = 0.$$

The solutions are $\xi = -1$ and $\xi = -\frac{1}{4}$.

Thus one obtains $(x,y) = (\pm i, \pm i)$ resp. $(x,y) = (\pm \frac{i}{2}, \pm \frac{i}{2})$ as solutions of (c). The first of these solutions are also solutions of the astroid equation, the latter are not.

Thus we have obtained exactly 8 points as solutions $f = 0$, $f_x = 0$, $f_y = 0$:

$(0, \pm 1)$

$(\pm 1, 0)$

$(\pm i, \pm i)$.

The 4 points $(0, \pm 1)$ and $(\pm 1, 0)$ are the 4 cusps of the astroid. These are precisely the singular points of these real curves. What is the meaning of the other four points? This cannot be understood until one replaces the real curve $f(x,y) = 0$ by a complex curve, in which points (x,y) with complex coordinates are also admitted as solutions of $f(x,y) = 0$, and therefore we shall do this later.

The singular points of the astroid are thus determined. The real ones are simple cusps. But curves can have much more complicated "higher" singularities, and since Cramer, around 1750, the investigation of such singularities has been a permanently important theme of this theory.

In the pair of opposites "<u>regular-singular</u>" we have before us another dialectical pair of opposites which plays a role in all fields of mathematics and relates closely to other pairs of opposites such as "special-general" and "finite-infinite". A <u>singularity</u> within a totality is a place of uniqueness, of speciality, of degeneration, of indeterminacy or infinity. All these basic meanings are closely connected. I cannot go into more detail here and refer to my lecture on singularities [B7].

In various examples we have seen that the construction of the <u>envelope</u>, the curve enveloping a given family of curves, can lead to interesting and important new curves. We shall now show in an example how one can conceive this process, or construction, analytically.

A one-parameter family of curves is given by a one-parameter family of equations $f_a(x,y) = 0$, where a is the parameter and x,y are cartesian coordinates in the plane. One can best express the fact that this family depends on the parameter a in a reasonable way by saying $f_a(x,y) = f(x,y,a)$, where the function $f(x,y,z)$ depends on x,y,z in the desired way, e.g. differentiably or analytically. Now how does one describe resp. define the envelope? Intuitively speaking, it should be a curve which touches all curves of the family and which is touched at each of its points by some curve of the family. One possible way to make this precise is the following:

We consider the surface F in 3-dimensional space, with coordinates x,y,z given by $f(x,y,z) = 0$. When we cut F with the plane $z = a$ and project the curve of intersection onto the (x,y)-plane by $(x,y,z) \mapsto (x,y)$ we obtain precisely the curve $f_a(x,y) = 0$. This leads us to consider the mapping $F \to \mathbb{R}^2$ defined by projection.

We look at an example. Let the family of curves be the family of circles

$$(x-a)^2 + (y-a)^2 = a^2$$

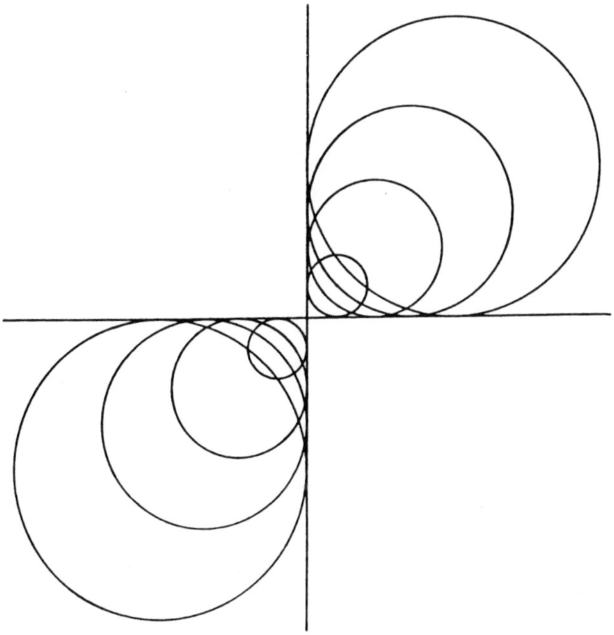

The enveloping curve is obviously just the pair of axes, the surface F is an oblique cone, and the axis pair is obviously its outline under projection onto the (x,y)-plane.

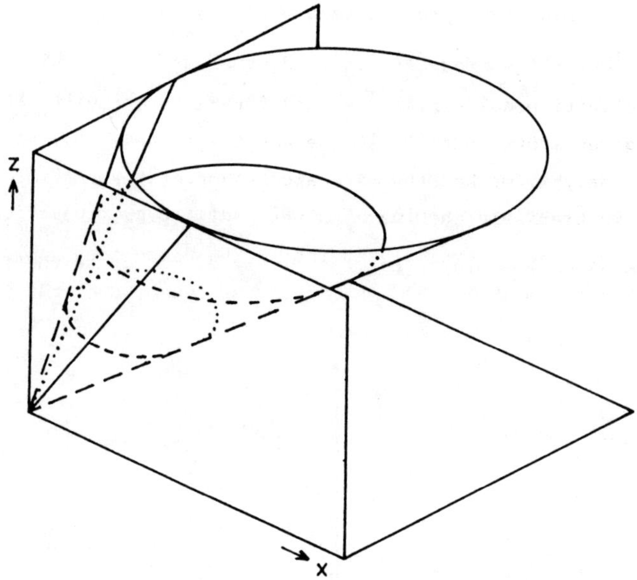
z
x

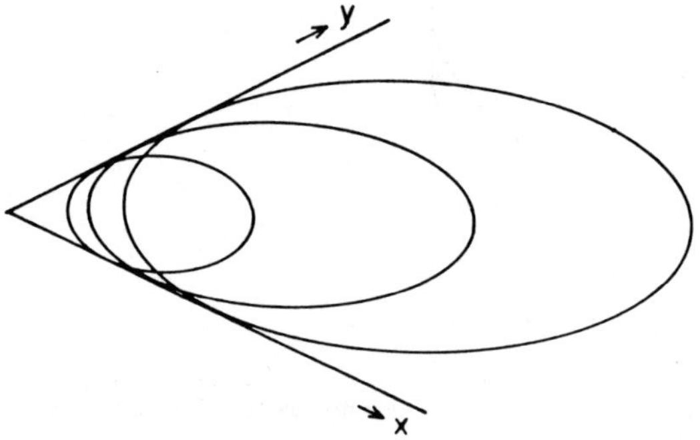
y
x

This brings us to the following definition :

The envelope of the family of curves given by $f_a(x,y) = 0$ is the outline of the surface with equation $f(x,y,z) = 0$ under projection onto the (x,y)-plane (where $f_a(x,y) = f(x,y,a)$.)

This definition does not correspond entirely to the intuitive definition given above, but it has the advantage that it immediately gives us an analytic method for the calculation of the envelope. For the outline is precisely the image in the (x,y)-plane of all points of F at which there is a vertical tangent. These are precisely the points at which $\frac{\partial f}{\partial z} = 0$. Thus we obtain the equation of the envelope by elimination between the equations

$$f(x,y,z) = 0$$

$$\frac{\partial f}{\partial z}(x,y,z) = 0.$$

We remark that not every family of curves has a curve as envelope. For example, one sees immediately that for the family of all lines through the origin the envelope in the above sense consists only of the origin itself.

The families of curves which appear most frequently in our context are not parametrised by a real parameter a, but by the points (a,b) of a curve which itself is given by an equation $g(a,b) = 0$. Thus one has a family of equations $f(x,y,a,b) = 0$ for the family of curves. By carrying over the considerations above one now sees easily that :

One obtains the equations of the envelope by elimination of a,b from the equations

$$f(x,y,a,b) = 0 \qquad (1)$$

$$g(a,b) = 0 \qquad (2)$$

$$f_a g_b - f_b g_a = 0 \qquad (3)$$

As an example, we shall compute the equation of the astroid. The astroid was the envelope of the family of lines resulting from the motion of a line with two points constrained to slide on the axes. We choose the distance between the points equal to one. When a,b are the intercepts of the line on the axes, the equations then read:

(1) $\frac{x}{a} + \frac{y}{b} - 1 = 0$

(2) $a^2 + b^2 - 1 = 0$

(3) $\frac{b}{a^2}x - \frac{a}{b^2}y = 0.$

Multiplication of (1) by b^3 yields

$$\frac{b^3}{a} x + b^2 y - b^3 = 0.$$

Substitution of (3) in this yields

$$(a^2+b^2)y - b^3 = 0.$$

Hence, by (2)

$$y = b^3$$

and similarly $x = a^3$.

Substitution in (1) gives the natural equation of the astroid

$$x^{2/3} + y^{2/3} = 1.$$

When one raises both sides to the power 3, carries roots to the right-hand side and again raises to the power 3, one finally obtains the equation of the astroid

$$(x^2+y^2-1)^3 + 27x^2y^2 = 0.$$

We remark that any curve can be regarded as the envelope of the family of its tangents. This fact plays an important role in Plücker's idea of regarding the lines as elements of a new manifold, and taking the coefficients of their equations as coordinates in this manifold. The totality of tangents to a curve then constitutes a new curve in this manifold of lines, the dual curve of the original.

After these examples of the application of differential calculus to the analytic geometry of curves, we shall now discuss in detail one example among the generally known applications of integral calculus (such as computation of area), namely the problem of rectifying a curve, i.e. calculating arc length.

Suppose a curve in cartesian coordinates is given by a parametrisation

$$x = x(t)$$
$$y = y(t).$$

We know that such a parametrisation always exists locally, in the neighbourhood of a regular point. Then the arc length of the curve from $(x(t_0), y(t_0))$ to $(x(t_1), y(t_1))$ is

$$\int_{t_0}^{t_1} \sqrt{\dot{x}(t)^2 + \dot{y}(t)^2}\, dt.$$

Even for simple curves, e.g. the ellipse, such an integral is in general not elementarily computable, i.e. expressible in terms of elementary functions, and this was a factor leading to the theory of elliptic integrals and the development of function theory.

We shall look at quite a simple example. We consider a generalised parabola

$$x^p - y^q = 0.$$

Incidentally, one sees here again how the analytic base allows interesting curves to be defined through mere generalisation of the form of equations. A global parametrisation of this curve (for p, q relatively prime) is

$$x = t^q$$
$$y = t^p.$$

However, we prefer to choose the parametrisation by x

$$y = x^{p/q}.$$

We then find the arc length to be

$$\int_0^x \sqrt{1 + \frac{p^2}{q^2} x^{2(p/q - 1)}}\, dx.$$

This cannot be elementarily evaluated for general p, q. However, for the special case $p = 3, q = 2$, i.e. for Neil's parabola (or semicubical parabola)

$$x^3 - y^2 = 0$$

we obtain

$$\int_0^x \sqrt{1 + \frac{9}{4} x}\, dx = \frac{8}{27}\left(1 + \frac{9}{4} x\right)^{3/2} \Big|_0^x .$$

And hence in this case we can construct a segment of the same length as a given arc of the curve by ruler and compasses. Neil's parabola was one of the first curves to be rectified, by Neil himself in 1657.

2.5 Newton's investigation of cubic curves

The analytic method of Descartes is in a certain sense the most general method for the generation of curves, because it allows the concept of plane algebraic curve to be defined in full generality, as the zero set of an equation $f(x,y) = 0$, where $f(x,y)$ is an arbitrary polynomial in x and y. Each particular choice of equation defines a curve, whose particular properties can be investigated, e.g. the generalised parabola $x^p - y^q = 0$ or the folium of Descartes $x^3 + y^3 + axy = 0$.

The generality underlying the definition of algebraic curves makes for a corresponding generality in the resulting problem of bringing order into the totality of all these curves, of classifying them in a suitable sense.

A first approach to a classification was already found by Newton. It comes directly from the form of the equation : obviously the equation $f(x,y) = 0$, and hence the curve itself, can be more complex as the degree of the polynomial increases. This number, the order of the curve, is therefore a certain quantitative measure for the geometric, qualitative complexity of curves. It is easy to see that, under a change of cartesian coordinates, the equation is transformed into another of the same order, and thus in the order we have found a measure invariant under such transformations, an invariant for the qualitative geometric properties of the curve.

It is easy to capture the geometric meaning of this invariant more precisely. Let $f(x,y) = 0$ be the equation of a given curve of order d, and let $ax + by + c = 0$ be the equation of an arbitrary line. We can eliminate one of the variables from the line equation, say $y = \alpha x + \beta$ for $b \neq 0$. If we substitute this in the polynomial $f(x,y)$, then in $g(x) = f(x,\alpha x+\beta)$ we obtain a polynomial in x of degree $\leq d$, and for all but at most finitely many lines this polynomial is exactly of degree d.

Now the zeroes of $g(x)$ are obviously just the abscissas of the intersections of the line with the curve, and hence we obtain : a curve of order d is cut by an arbitrary line in at most d points. Is it perhaps exactly d ? We consider an example, say the semicubical parabola $y^2 - x^3 = 0$.

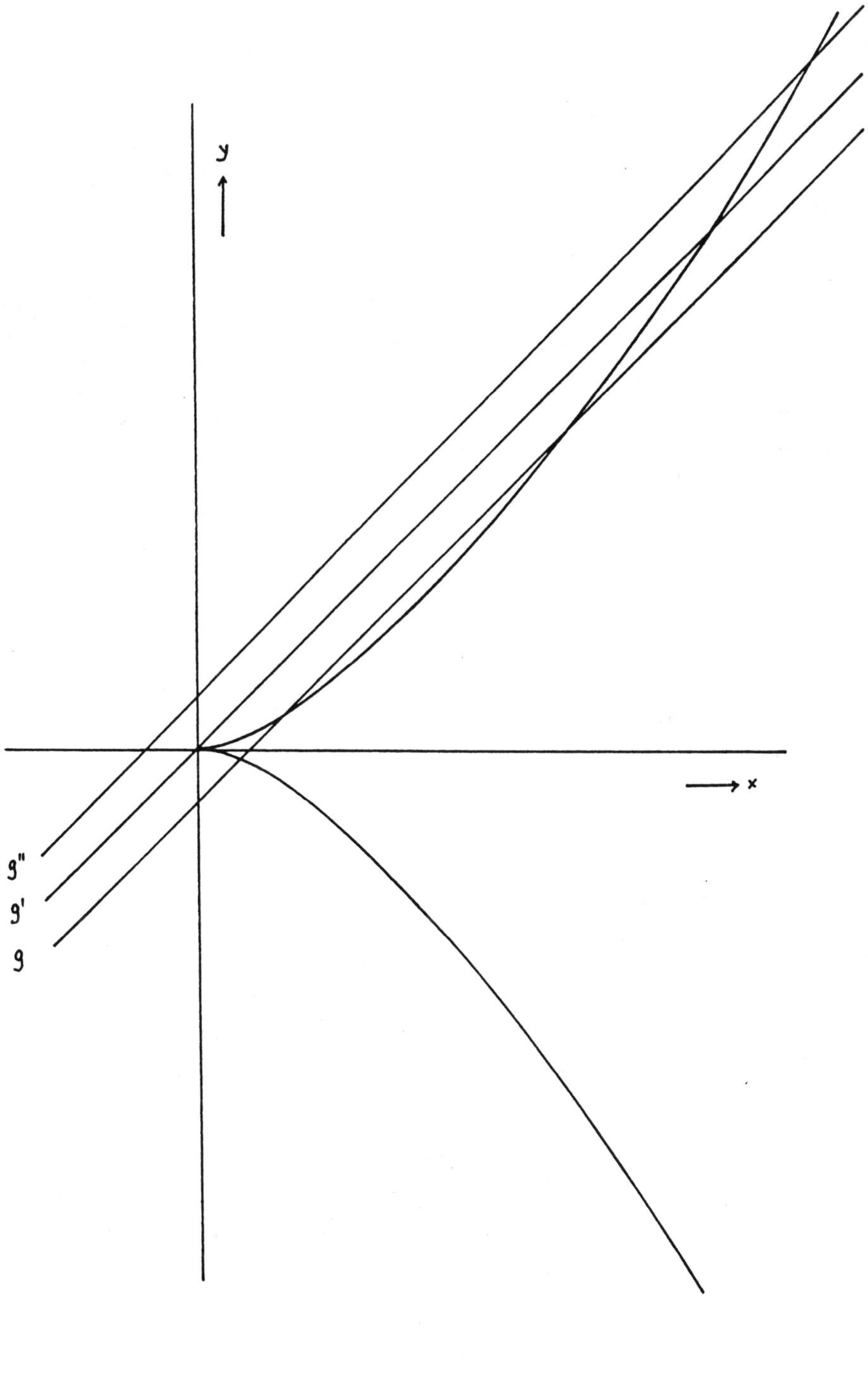
y
x
g"
g'
g

The figure above shows the semicubical parabola and some of the lines intersecting it. The line g cuts the curve at three points, as we would expect on the basis of the preceding considerations, and each line which results from g by a sufficiently small displacement has the same property. But when we displace g so far that the line goes through the origin, say at position g', two of the three points of intersection coincide. If one assigns this intersection a multiplicity 2, then g' still cuts in three points. Algebraically, this means that in the polynomial equation for the intersections of the curve with the line $y = ax$, namely $a^2x^2 - x^3 = 0$, the zero $x = 0$ has multiplicity 2. If we displace our line still further, say to the position g'', then the two intersections at the origin vanish! Algebraically, this corresponds to the fact that for $y = ax + b$ with $a \cdot b > 0$ the intersection equation $x^3 - (ax+b)^2 = 0$ has two non-real roots. By the fundamental theorem of algebra, each polynomial of degree d has d zeros — with multiplicities counted — but of course these need not all be real. We see here that it will be useful to also admit complex solutions for the equation $f(x,y) = 0$ of our curve.

Let us consider again the intersection of the semicubical parabola with all lines through the origin. All of these lines, apart from the two coordinate axes, cut the curve in three points : the origin with multiplicity two, and a further point P with multiplicity one. When the line moves to the position of the x-axis, P also moves to the origin, the single intersection of the curve with the x-axis. For this reason, one should now count this intersection with multiplicity 3 ! We see from this that it will become necessary to develop a theory of the multiplicities with which curves are cut at their singular points.

If our line now moves to the other distinguished position, that of the y-axis, then P moves to infinity, so that the only intersection with the curve which remains is the origin, counted with multiplicity two.

We see from this that it will become necessary to develop a systematic theory of infinitely distant points, in order to capture the infinitely distant intersections as well. This is a factor which led to the development of projective geometry.

We shall in fact develop a satisfactory theory of multiplicity and intersections in complex projective geometry.

We have seen that with the order d of curves we have at our disposal the first invariant which can be applied to the classification of curves. It is now quite natural to proceed with the classification problem by obtaining an overview of all curves of lower order, say by considering the cases $d = 1,2,3,4$.

The case of curves of order 1 is trivial : these are just all lines $ax + by + c = 0$.

The curves of order 2 are, when one admits degenerate cases such as line pairs and double lines, just the conic sections

$$ax^2 + bxy + cy^2 + dx + ey + f = 0.$$

One also calls these curves of order 2 quadrics.

The first interesting case is that of the cubics, i.e. the curves of order 3.* This case was first investigated systematically by Newton in his article on curves, [N2]. We shall give a brief survey of the results which Newton obtained in it.

In his investigation of cubic curves, Newton systematically applies the analytic method : he investigates the equation $f(x,y) = 0$, where $f(x,y)$ is a polynomial of degree 3. He starts with the asymptotic behaviour of curves. For "large" x,y this behaviour is of course determined by the terms of highest, i.e. third, degree in f, and hence by a homogeneous polynomial of third degree, $\tilde{f}(x,y) = ax^3 + bx^2y + cxy^2 + dy^3$. Now there are obviously two possible cases. Case A : $\tilde{f}$ is the third power of a linear form. By a change of cartesian coordinates (not necessarily rectangular) one then has $\tilde{f}(x,y) = ax^3$ with $a \neq 0$. Case B : $\tilde{f}$ is not a third power. Then one sees that with a suitable change of coordinates $\tilde{f}$ can be brought into the form $\tilde{f}(x,y) = xy^2 - ax^3$. In the remaining terms of f, i.e. in $f - \tilde{f}$, the monomials y^2, xy and y can appear, in addition to powers of x. In case B one can make the y^2 and xy terms vanish by displacement of the origin, and in case A one can make xy and y vanish, by a new change of coordinates, when y^2 appears. If y^2 does not appear, but xy does, one can make y vanish. In summary, one has : a cubic curve which contains no lines has an equation

---

*Particular examples of cubics had already long been known, e.g. the cissoid of Diocles, the folium of Descartes and Neil's parabola.

of one of the 4 normal forms which follow, relative to suitable cartesian coordinates :

I. $xy^2 + ey = ax^3 + bx^2 + cx + d$.

II. $xy = ax^3 + bx^2 + cx + d$.

III. $y^2 = ax^3 + bx^2 + cx + d$.

IV. $y = ax^3 + bx^2 + cx + d$.

Newton now discusses these equations, having to make case distinctions according to the roots of $ax^3 + bx^2 + cx + d = 0$. In this way he obtains 72 cases, and there are six cases that he overlooked. The discussion of cases II and IV is of course trivial, case I is the most complicated. We shall not go into this discussion in detail here, but in order to give an impression of it we reproduce the first pages of Newton's work and the section in which he discusses case III. In this case he obtains 5 types of "diverging parabolas" : there are two which are non-singular, when all three roots of $ax^3 + bx^2 + cx + d = 0$ are different, with two curved paths in the case of three real zeros, and one in the case of one. If two of the three roots coincide one obtains a curve with an isolated singular point or an ordinary double point. If all three roots coincide one of course obtains the semicubical parabola. Newton's manuscript also contains pictures of all these curves, so that we can simply refer to it.

CURVES. *The incomparable Sir* Iſaac Newton *gives this following Ennumeration of Geometrical Lines of the Third or Cubick Order; in which you have an admirable account of many Species of* Curves *which exceed the* Conick-Sections, *for they go no higher than the Quadratick or Se-Second Order.*

The Orders of Geometrick Lines.

1: GEOMETRICK-LINES, are beſt diſtinguiſh'd into *Claſſes*, *Genders*, or *Orders*, according to the Number of the Dimenſions of an Equation, expreſſing the relation between the *Ordinates* and the *Abſciſſæ*; or which is much at one, according to the Number of Points in which they may be cut by a Right Line. Wherefore, a Line of the *Firſt Order* will be only a *Right Line*: Theſe of the *Second* or *Quadratick Order*, will be the *Circle* and the *Conick-Sections*; and theſe of the *Third* or *Cubick Order*, will be the *Cubical* and *Nelian Parabola's*, the *Ciſſoid* of the Antients, and the reſt as belew ennumerated. But a *Curve* of the *Firſt Gender* (becauſe a Right Line can't be reckoned among the *Curves*) is the ſame with a Line of the *Second Order*, and a *Curve* of the *Second Gender*; the ſame with a Line of the *Third Order*, and a Line of an *Infiniteſimal Order*, is that which a Right Line may cut in infinite Points, as the *Spiral*, *Cycloid*, the *Quadratrix*, and every Line generated by the Infinite Revolutions of a *Radius* or *Rota*.

2. The chief Properties of the *Conick-Sections* are every where treated of by Geometers; and of the ſame Nature are the Properties of the *Curves* of the *Second Gender*, and of the reſt, as from the following Ennumeration of their Principle Properties will appear.

3. For if any right and parallel Lines be drawn and terminated on both Sides by one and the ſame *Conick-Section*; and a Right Line biſecting any two of them, ſhall biſect all the reſt; and therefore ſuch a Line is called the Diameter of the Figure; and all the Right Lines ſo biſected, are called *Ordinate Applicates* to that Diameter, and the Point of Concourſe to all the Diameters is called the *Center* of the *Figure*; as the Interſection of the *Curve* and of the Diameter, is called the *Vertex*, and that Diameter the *Axis* to which the Ordinates are *Normally* applied. And ſo in *Curves* of the *Second Gender*, if any two right and parallel Lines are drawn occurring to the Curve in Three Points; a right Line which ſhall ſo cut thoſe Parallels, that the Sum of Two Parts terminated at the Curve on one Side of the Interſecting Line ſhall be equal to the third Part terminated at the Curve on the other ſide, this Line ſhall cut, after the ſame manner, all others parallel to theſe, and occurring to the Curve in Three-Points; that is, ſhall ſo cut them that the Sum of the Two Parts on one Side of it, ſhall be equal to the Third Part on the other.

And therefore theſe Three Parts one of which is thus every where equal to the Sum of the other two, may be called *Ordinate Applicates* alſo: And the Interſecting Line to which the Ordinates are applied, may be called the *Diameter*; the Interſection of the Diameter and the Curve, may be called the *Vertex*, and the Point of Concourſe of any two Diameters, the *Center*.

And if the Diameter be Normal to the Ordinates, it may be called the *Axis*; and that Point where *all* the Diameters terminate, the *General Centre*.

Aſſymptotes and their Properties.

4. The Hyperbola of the Firſt Gender has Two *Aſſymptotes*, that of the Second, Three; that of the Third, Four, and it can have no more, and ſo of the reſt. And as the Parts of any Right Line lying between the *Conical Hyperbola* and its Two Aſſymptotes are every where equal; ſo in the Hyperbola's of the Second Gender, if any Right Line be drawn, cutting both the Curve and its Three Aſſymptotes, in Three Points; the Sum of the Two Parts of that Right Line, being drawn the ſame way from any Two Aſſymptotes to Two Points of the Curve, will be equal to the Third Part drawn a contrary way from the Third Aſſymptote to a Third Point of the Curve.

Latera Tranſverſa and Recta.

5. And as in *Non Parabolick Conick Sections*, the Square of the Ordinate Applicate, that is, the Rectangle under the Ordinates, drawn at contrary Sides of the Diameter, is to the Rectangle of the Parts of the Diameter, which are terminated at the Vertex's of the *Ellipsis* or *Hyperbola*; as a certain Given Line which is called the *Latus Rectum*, is to that Part of the Diameter which lies between the Vertex's, and is called the *Latus Transversum*: so in *Non-Parabolick Curves* of the Second Gender, a Parallelopiped under the Three Ordinate Applicates, is to a Parallelopiped, under the Parts of the Diameter terminated at the Ordinates, and the Three Vertex's of the Figure in a certain Given Ratio; in which Ratio, if you take Three Right Lines to the Three Parts of a Diameter situated between the Vertex's of the Figure, one answering to another, then these Three Right Lines may be called the *Latera Recta* of the Figure, and the Parts of the Diameter between the *Vertices*, the *Latera Transversa*. And as in the Conick Parabola having to one and the same Diameter but one only Vertex, the Rectangle under the Ordinates is equal to that under the Part of the Diameter cut off between the Ordinates and the Vertex, and a certain Line called the *Latus Rectum*: So in the Curves of the Second Gender, which have but two Vertex's to the same Diameter; the Parallelopiped under Three Ordinates, is equal to the Parallelopiped under the Two Parts of the Diameter cut off between the Ordinates and those Two Vertexes, and a given Right Line, which therefore may be called the *Latus Rectum*.

The Ratio of the Rectangles under the Segments of Parallels.

Lastly, As in the Conick-Sections when two parallels terminated on each side at the Curve, are cut by two other Parallels terminated on each side by the Curve, the First being cut by the Third, and the Second by the Fourth; as here the Rectangle under the Parts of the First, is to the Rectangle under the Parts of the Third; as the Rectangle under the Parts of the Second is to that under the Parts of the Fourth: So when Four such Right Lines occur to a Curve of the Second Gender, each one in Three Points, then shall the Parallelopiped under the Parts of the First right Line be to that under the Parts of the Third, and as the Parallelopiped under the Parts of the Second Line into that under the Pares of the Fourth.

Hyperbolick and Parabolick Legs.

All the Legs of Curves of the second and higher Genders, as well as of the first, infinitely drawn out, will be of the *Hyperbolick* or *Parabolick* Gender; and I call that an *Hyperbolick* Leg, which infinitely approaches to some *Asymptote*; and that a *Parabolick* one, which hath no *Asymptote*. And these Legs are best known from the Tangents: For if the Point of Contact be at an infinite Distance, the Tangent of an *Hyperbolick Leg* will coincide with the *Asymptote*, and the Tangent of a Parabolick Leg will recede *in infinitum*, will vanish and no where be found. Wherefore the Asymptote of any Leg is found, by seeking the Tangent to that Leg at a Point infinitely distant: And the *Course*, *Place* or *Way* of an infinite Leg, is found by seeking the Position of any Right Line, which is parallel to the Tangent where the Point of Contact goes off *in infinitum*: For this Right Line is directed towards the same way with the infinite Leg.

The Reduction of all Curves of the Second Gender, to Four Cases of Equations.

CASE I

All Lines of the First, Third, Fifth and Seventh Order, and so of any one, proceeding in the Order of the odd Numbers, have at least two Legs or Sides proceeding *ad infinitum*, and towards contrary ways. And all Lines of the Third Order have two such Legs or Sides running out contrary ways, and towards which no other of their infinite Legs (except in the *Cartesian Parabola*) do tend. If the Legs are of the Hyperbolick Gender, let *C A S* be their Asymptote; and to it let the Parallel *C Bc* be drawn, terminated (if possible) at both Ends at the Curve. Let this Parallel be bissected in *X*; and then will the Place of that Point *X*

Fig. 1.

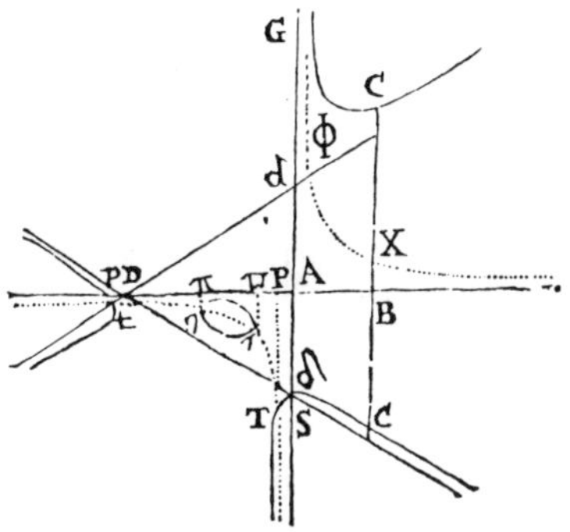

be the *Conical Hyperbola* $X\Phi$, one of whose Asymptotes is AS: Let its other Asymptote be AB; then the *Equation* by which the Relation between the Ordinate BC and the Absciſſa AB is determined, if AB be put $=x$ and $BC=y$, will always be in this Form, $xyy+ey=ax^3+bxx+cx+d$, where the Terms e, a, b, c and d denote given Quantities, affected with their Signs $+$ and $-$; of which any one may be wanting, ſo the Figure, through their Defect, don't turn into a Conick-Section. And this Conical Hyperbola may coincide with its Aſymptotes, that is, the Point X may come to be in the Line AB, and then the Term $+ey$ will be wanting.

CASE II.

9. But if the Right Line CBc cannot be terminated both ways at the Curve, but will occur to the Curve only in one Point; then draw any Line in a given Poſition, which ſhall cut the Aſymptote AS in A; as alſo any other Right Line, as BC, parallel to the Aſymptote, and meeting the Curve in the Point C: And then the Equation by which the Relation between the Ordinate BC and the Abſciſſa AB is determined, will always put on this Form, $xy=ax^3+bxx+cx+d$.

CASE III.

10. But if the oppoſite Legs are of the Parabolick Gender, draw the Right Line CBc, terminated at both Ends, if it's poſſible, at the Curve; and running according to the Courſe of the Legs; which biſſect in B: Then ſhall the Place of B be a Right Line. Let that Right Line be AB, terminated at any given Point, as A; and then the Equation by which the Relation between the Ordinate BC and the Abſciſſa AB is determined, will always be in this Form, $yy=ax^3+bxx+cx+d$.

CASE IV.

11. But if the Right Line CBc meet the Curve but in one Point, and therefore can't be terminated at the Curve at both Ends; let the Point where it occurs to the Curve be C; and let that Right Line at the Point B, fall on any other Right Line given in Poſition, as AB, and terminated at any given Point, as A: Then will the Equation, by which the Relation between the Ordinate BC and the Abſciſſa AB is determined, always be in this Form, $yy=ax^3+bxx+cx+d$.

The Names of the Forms.

12. In the Enumeration of Curves of theſe Caſes, we call that an *Inſcribed Hyperbola*, which lies entirely within the Angle of the Aſymptotes, like the *Conical Hyperbola*; and that a *Circumſcribed* one, which cuts the Aſymptotes, and contains the Parts cut off within its own proper Space; and that an *Ambigenal* one, which hath one of its infinite Legs inſcribing it, and the other circumſcribing it. I call that a *Converging* one, whoſe Concave Legs bend inwards towards one another, and run both the ſame way; but that I call a *Diverging* one, whoſe Legs turn their Convexities towards each other, and tend towards quite contrary ways. I call that *Hyperbola contrary leg'd*, whoſe Legs are Convex towards contrary Parts, and run infinitely on towards contrary ways; and that a *Conchoidal one*, which is applied to its Aſymptote with its Concave Vertex and diverging Legs; and that an *Anguineal or Eel-like* one, which cuts its Aſymptote with contrary Flexions, and is produced both ways into contrary Legs. I call that a *Cruciform or Croſs-like one*, which cuts its Conjugate croſs wiſe; and that *Nodate*, which, by returning round into, decuſſates it ſelf. I call that *Cuſpidate*, whoſe two Parts meet and terminate in the Angle of Contact; and that *Punctate*, whoſe Oval *Conjugate* is infinitely ſmall, or a Point: And that Hyperbola I

gles of the Asymptotes, but in the contiguous or adjoining ones, and that on each Side the Abscissa

Fig. 68.

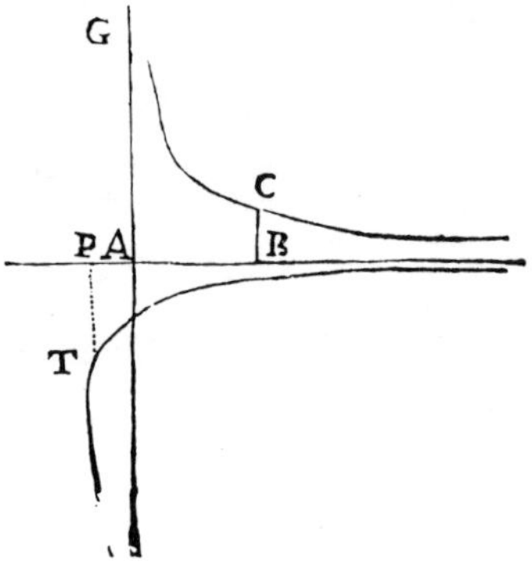

Fig. 69.

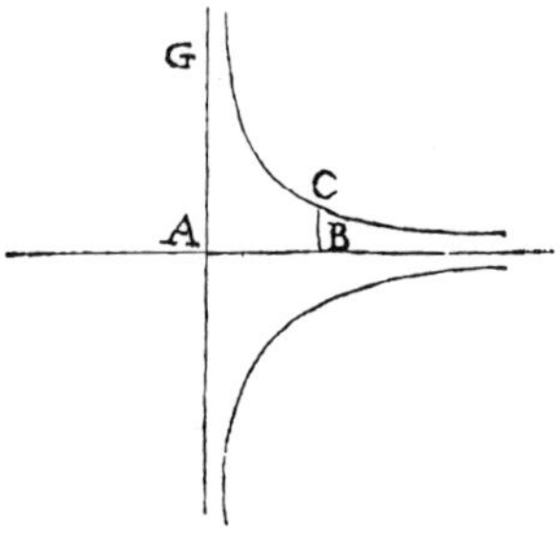

AB; and even without any Diameter, if the Term *ey* be there, but with one if that be wanting. Which two *Species* are the *Sixty fourth* and *Sixty fifth*.

A Trident.

26. In the second Case of the Equations there is $xy = ax^3 + bxx + cx + d$: And the Figure in this Case will have four infinite Legs, of which

Fig. 76.

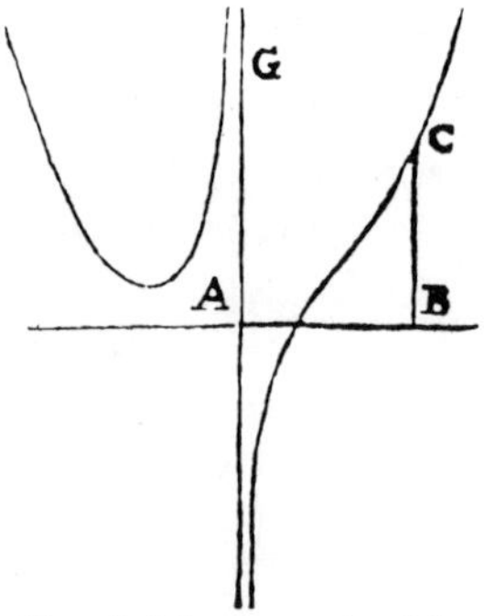

two are Hyperbola's about the Asymptote *AG* tending towards contrary Parts; and two converging Parabola's, and, with the Former, making as it were the Figure of a *Trident*. And this Figure is that Parabola by which *D. Cartes* constructed Equations of six Dimensions. This therefore is the *Sixty sixth Species*.

Five Diverging Parabola's.

27. In the third Case the Equation was $yy = ax^3 + bxx + cx + d$; and designs a Parabola, whose Legs diverge from one another, and run out infinitely contrary ways. The Abscissa *AB* is its Diameter, and its five Species are these:

If, of the Equation $ax^3 + bxx + cx + d = 0$, all the Roots *Aτ*, *AT*, *At*, are real and unequal; then the Figure is a diverging Parabola

Fig. 70.

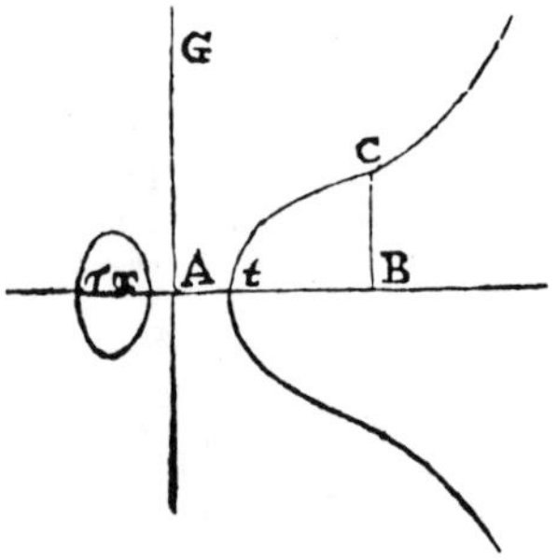

Fig. 71.

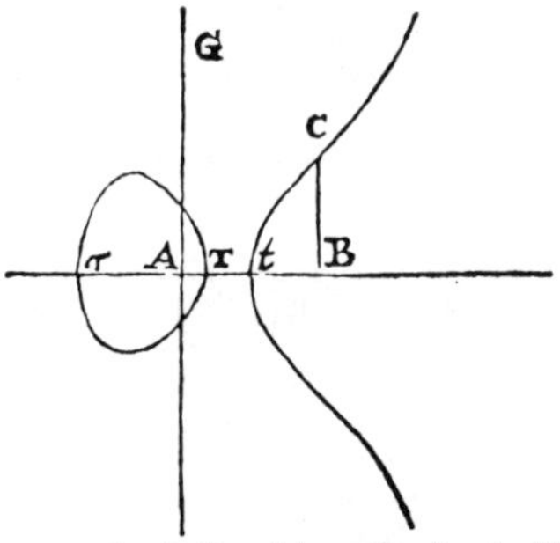

of the Form of a Bell, with an Oval at its Vertex. And this makes a *Sixty seventh Species.*

If two of the Roots are equal, a Parabola will be formed, either *Nodated* by touching an Oval,

Fig. 72.

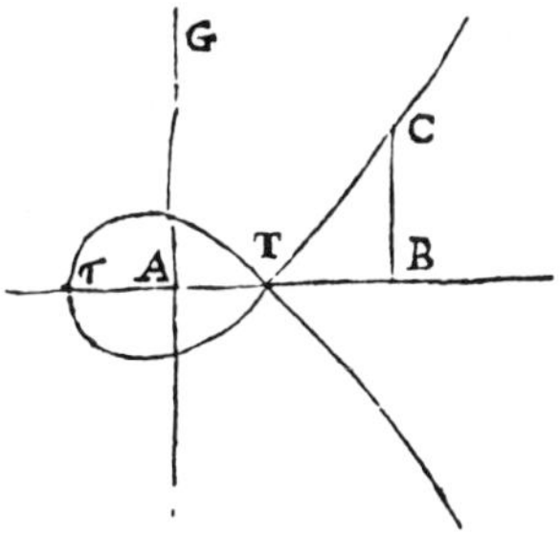

Fig. 73.

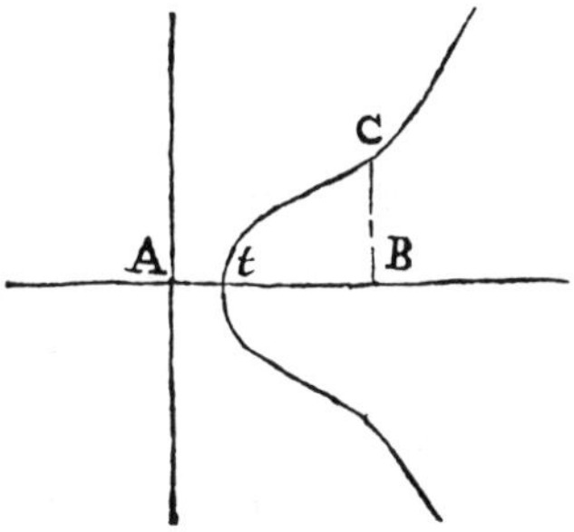

or *Punctate,* by having the Oval infinitely small. Which two *Species* are the *Sixty eighth* and *Sixty ninth.*

If three of the Roots are equal, the Parabola will be *Cuspidate* at the Vertex. And this is the

Fig. 75.

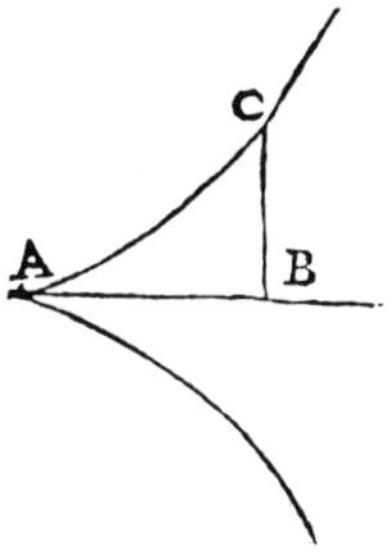

Neilian Parabola, commonly called Semi-cubical. Which makes the *Seventieth Species.*

If two of the Roots are impossible, there will (See *Fig.* 73.)

Fig. 73.

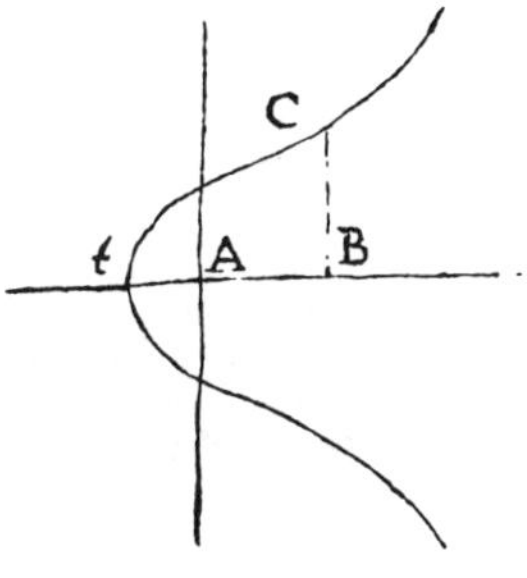

be a *Pure* Parabola of a Bell-like Form. And this makes the *Seventy first Species.*

The Cubical Parabola.

28. in the Fourth Cafe, let the Equation be $y = ax + bxx + cx + d$; then will it denote

Fig. 77.

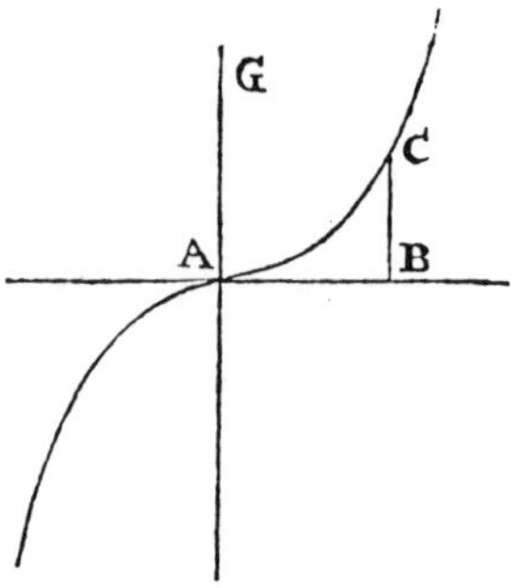

the *Cubical Parabola* with contrary turn'd Legs. And this makes up, or compleats, the Number of the Species of thefe Curves to be in all *Seventy two.*

Of the Genefis of Curves by Shadows.

29. If the Shadows of Figures are projected on an infinite Plane illuminated from a lucid Point, the Shadows of the Conick-Sections will always be Conick-Sections; thofe of the Curves of the Second Gender, will always be Curves of the Second Gender; and the Shadows of Curves of the Third Gender, will themfelves be of the fame Gender, and foon *in infinitum.* And as a Circle, by the Projection of its Shade, generates all the Conick-Sections; fo will the five diverging Parabola's fpoken of in *ch.* 28. by their Shadows generate and exhibit all Curves of the Second Gender; and fo fome more fimple Curves of other Genders may be found, which, by the Projection of their Shadows from a lucid Point upon a Plane, fhall from all other Curves of the fame kinds.

Of the double Points of Curves.

30. I faid above, that Curves of the Second Gender might be cut by a Right Line in three Points; but two of thofe Points are fometimes co-incident. As when the Right Line paffes by an Oval infinitely fmall, or by the Concourfe of two Parts of a Curve mutually interfecting each other, or running together into a *Cufpis.* And if at any time all the Right Lines tending the fame way with the infinite Leg of any Curve, do cut it in one only Point, (as happens in the Ordinates of the *Cartefian,* and in the *Cubical Parabola*, and in the Right Lines which are parallel to the Abfciffa of the Hyperbolifms of Hyperbola's and Parabola's;) then you are to conceive that thofe Right Lines pafs through two other Points of the Curve (as I may fay) placed at at an infinite Diftance; and thefe two co-incident Interfections, whether they be at a finite or an infinite Diftance, I call the *Double Point.* And fuch Curves as have this *Double Point,* may be defcribed by the following Theorems.

Theorems for the Organical Defcription of Curves.

31. *Theor.* I. If two Angles, as *PAD* and *PBD,* whofe Magnitude is given, be turned round the Poles *A* and *B,* given alfo in Pofition; and their Legs *AP, BP,* by their Point of Concourfe

Fig. 78.

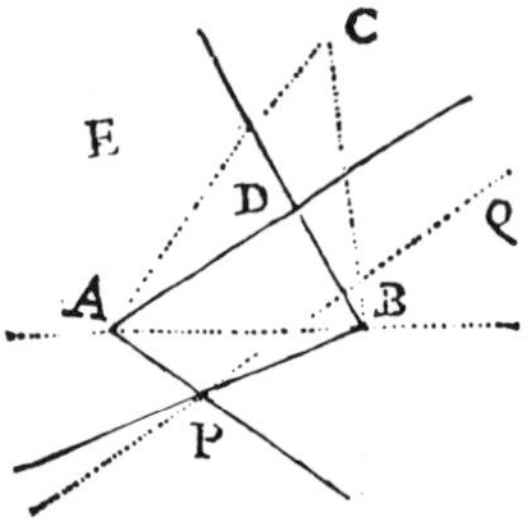

P, defcribe a Conick-Section paffing thro' their Poles *A* and *B*; except when that Right Line happens to pafs through either of the Poles *A* or *B*; or when the Angles *BAD* and *ABD* vanifh together into nothing; for in fuch Cafes the Point will defcribe a Right Line.

II. If the firft Legs *AP, BP,* by their Point of Concourfe *P,* do defcribe a Conick-Section paf-

Newton's description of all the cubics was later criticised by some mathematicians, e.g. Euler, because this classification lacks a general principle. Plücker later gave a more refined classification, in which he arrives at 219 types.

In the face of this multiplicity of types, the question of a unifying principle naturally arises. And such a principle already has its basis in the work of Newton, namely the short section (29) : Of the Genesis of Curves by Shadows, reproduced on the previous page.

The following two pictures illustrate these ideas of Newton. The first shows how projection of a circle in a plane E from a point P onto the planes E', E'', E''' yields ellipse, parabola and hyperbola. The second picture shows the projection of a cubical parabola onto a semicubical parabola, and conversely.

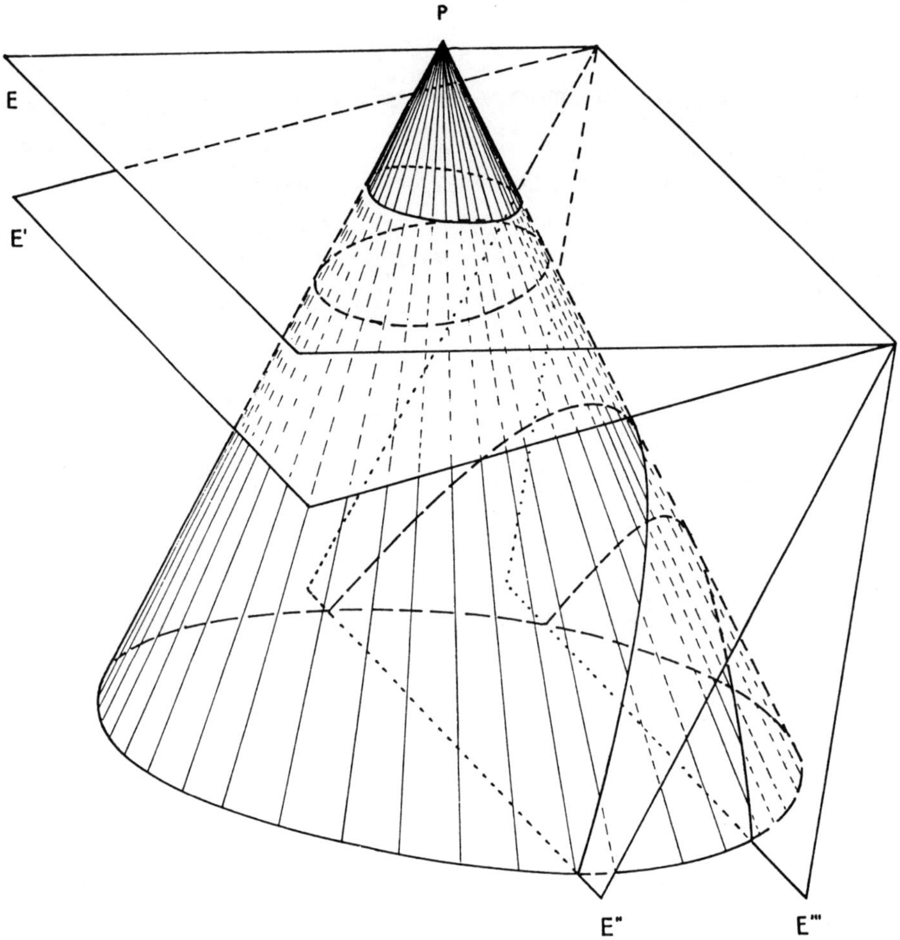
P
E
E'
E"
E'"

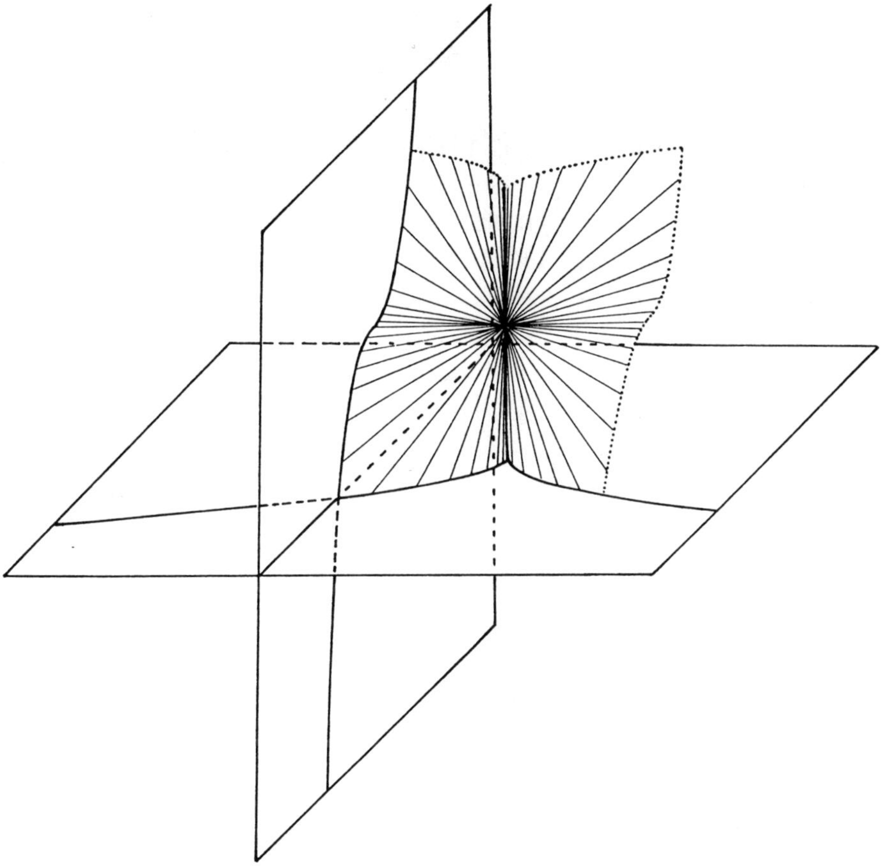

Newton's theorem contains an important idea : that the classification of the great multiplicity of cubic curves is essentially simplified when one considers two curves to belong to the same class when one results from the other by projection from a centre. The unfolding of this idea led to the development of projective geometry, which will be discussed in the next section.

3. The development of projective geometry

We have seen how the methods of perspective representation of spatial objects, which had their beginnings in antiquity, were developed into a fine art during the Renaissance in the work of artists, architects and inventors. In the 16th and 17th century this art became a geometric theory, which at the end of the 18th century unfolded into a mature system, descriptive geometry. This theory is concerned with generating planar pictures of spatial or planar objects by projection. We have already seen how Newton used this projection method to unify the classification of cubic curves. As descriptive geometry itself unfolded, geometers realised that the consistent introduction of the "projective" viewpoint made possible a far-reaching unification and generalisation of geometric methods. This led to the development of projective geometry, in which the synthetic and analytic methods developed alternately. In what follows we shall look more closely at some essential steps in this development.

3.1 Descriptive geometry and projective geometry

The beginnings of perspective representation, of buildings for instance, already appear in antiquity. In the Renaissance these were developed into an art of drawing, with elements of a geometric theory, by Alberti, Franceschi, Viator, Brunelleschi, Ghiberti, Leonardo da Vinci and Albrecht Dürer. With G. Ubaldo del Monte (1545-1607) there appears the basic idea of central projection, in which the perspective of an object appears as a section of the "visual ray pyramid" by the picture plane, and a geometric theory for the solution of problems of perspective representation begins to develop.

One of the main problems of descriptive geometry is to draw planar pictures of spatial objects. This is done by central projection in the following way : one connects the points P of the object with a fixed point Z, the centre of projection, cuts the projection rays ZP

by a fixed plane E' , the "picture plane", and draws the point P' of intersection as the image of the point P.

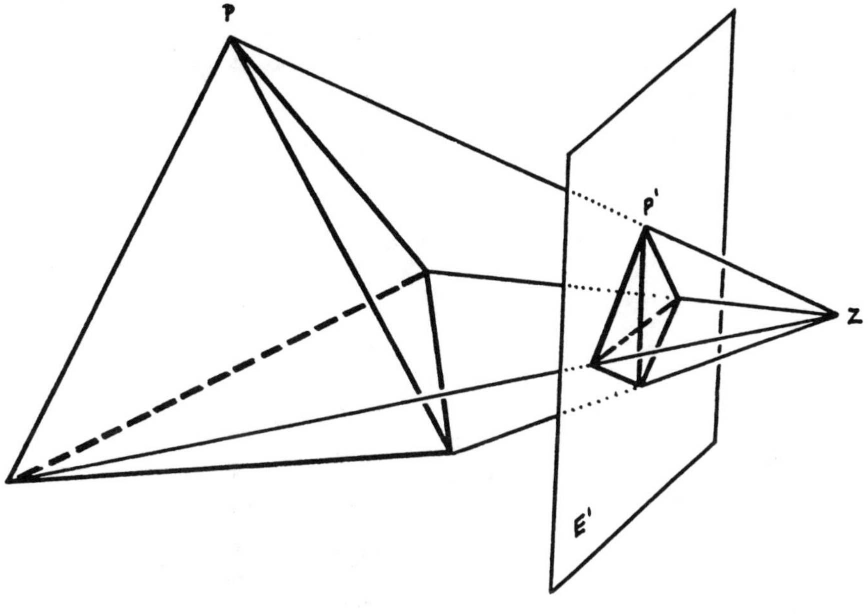

The centre of projection Z can be taken to be the eye or a light source, the projection rays as visual rays or light rays. When Z is an infinitely distant point, the projection rays are parallel, and the result is the orthogonal or oblique <u>parallel projection</u>, according as the plane E' is orthogonal to the direction of projection or not.

The image surfaces of the visible and invisible boundary surfaces of the object being mapped will be separated by true and apparent <u>outlines</u>. Such outlines can be interesting curves – we have seen, e.g., how a toroid appears in the case of orthogonal parallel projection of a torus.

We shall now confine ourselves to the special case in which the projected figures are planar, i.e. we investigate the projection of a plane E onto a plane E' from a centre Z , as indicated by the following picture.

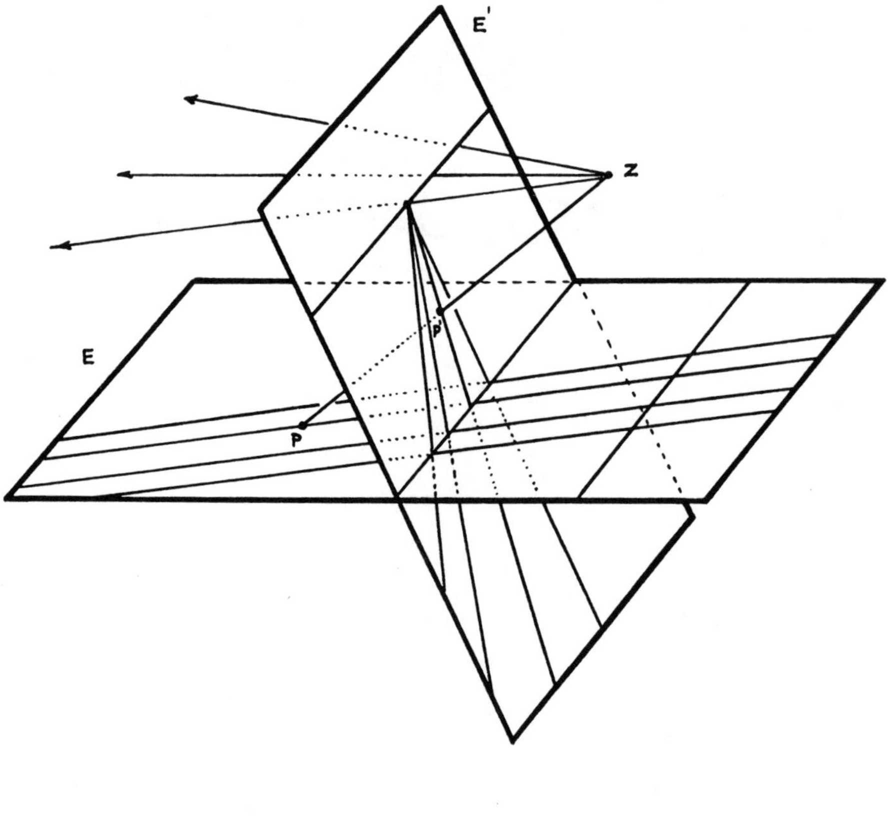

This projection associates each point P of the plane E with a unique point P' of the plane E' as image. An exception occurs with the points on the disappearing line, i.e. the line in which the plane through Z parallel to E' cuts the plane E. The points on the disappearing line have no image points in E', their images disappear at infinity, so to speak. A line in E has an image in E' which is also a line. Parallel lines in E have images in E' which obviously meet at a point, the vanishing point. The collection of all vanishing points constitutes the vanishing line, which is the intersection of E' with the plane through Z parallel to E. The points on the vanishing line are the only points in E' which are not images of points in E. They are in a way the images of the "points at infinity" of E.

The fact that the images of parallel lines under central projection meet at a vanishing point was already seen by Ubaldo around 1600. But the view of this point as the image of a point at infinity is due to the master builder and engineer Desargues (around 1640). With this introduction of the points at infinity and the line at infinity into the framework of his theoretical work on perspective, and the application of projective methods in the proof of geometrical theorems (Desargues' theorem, cf. 6.2), Desargues was a precursor of projective geometry.

The fruitful influence which the development of descriptive geometry had on that of projective geometry strengthened at the time of the French revolution, when G. Monge, director of the École Polytechnique, founded in 1794 and the prototype of all engineering and military schools in the early 19th century, published his "Géométrie descriptive", a thorough system of descriptive geometry he had worked out in connection with his lectures on fortification, and which he published in order to make French industry, with these tools in the hands of its engineers and architects, self-sufficient.

One student of Monge was Poncelet. Poncelet addressed himself to the purely synthetic content of Monge's geometry and thereby arrived at a way of thinking already hinted at two hundred years earlier by Desargues. Poncelet became the founder of projective geometry.

The "Traité des propriétés projectives des figures" (Treatise on the projective properties of figures) of Poncelet appeared in 1822. "This heavy volume contains all the essential concepts underlying the new form of geometry, such as cross ratio, perspectivity, projectivity, involution, and even the circular points at infinity" (quoted from Struik). The starting point of Poncelet was to ask which properties of figures are retained under arbitrary projections. These are the projective properties. They can involve quantitative relations such as the cross ratio of four points z_1, z_2, z_3, z_4 on a number line, $\frac{z_1-z_3}{z_1-z_4} : \frac{z_2-z_3}{z_2-z_4}$. However, most of the relations concerned are qualitative, positional relations, which is why this geometry is often known as the geometry of position.

In the systematic application of projections one had a synthetic method of great generality for the generation and investigation of geometric figures, and particularly curves. It was only natural for immediate attempts to be made to give this method an analytic form. This occurred in the work of Möbius and Plücker.

3.2 The development of analytic projective geometry

The first step in the application of analytic methods to plane geometry was the introduction of cartesian coordinates. In this one chooses two systems of parallel lines, and the position of a point is determined by the two lines on which it lies. The two lines are given by two numbers (x,y), which are just the cartesian coordinates.

Parallel projection takes parallels to parallels, hence the corresponding mappings can be properly described in cartesian coordinates. However, parallelism is not preserved under arbitrary central projection. One would like to know how to introduce coordinates which are adequate for use with arbitrary projections. Under arbitrary central projection the two systems of parallels of a cartesian coordinate system go to two pencils of lines through the corresponding vanishing points.

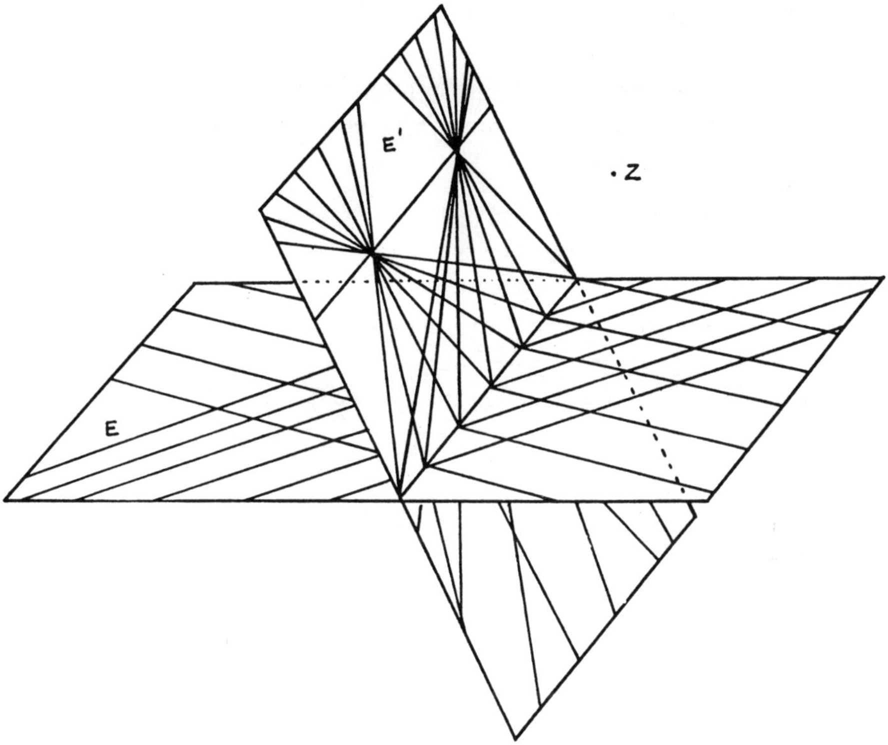
E'
Z
E

Since the position of a point P in the plane E is determined by the two lines on which it lies, the position of the image point P' in E' is determined by the two image lines. Thus when we have described these two lines by suitable coordinates, we shall have coordinates which allow a quantitative description of projection. In order to arrive at the right idea, we consider a special case. We take E and E' orthogonal to each other, and the centre of projection at distance 1 from E and E'. In E we use a system of rectangular cartesian axes, whose origin lies on the intersection line L of E and E' and whose axes make an angle of 45° with L. The following figure illustrates the situation. It shows half planes of E and E' folded into the same plane.

Let u,v,w be the distances of P' from line QQ' and the images of the coordinate axes respectively. Here, distances in the interior of the triangle OQQ' are taken to be positive. The position of P' is of course uniquely determined by the triple (u,v,w) ; (u,v,w) are called the <u>triangular coordinates</u> of P'. The line QP' is obviously given by $\frac{v}{u} = \frac{x}{1}$, and the line Q'P' by $\frac{w}{u} = \frac{y}{1}$.

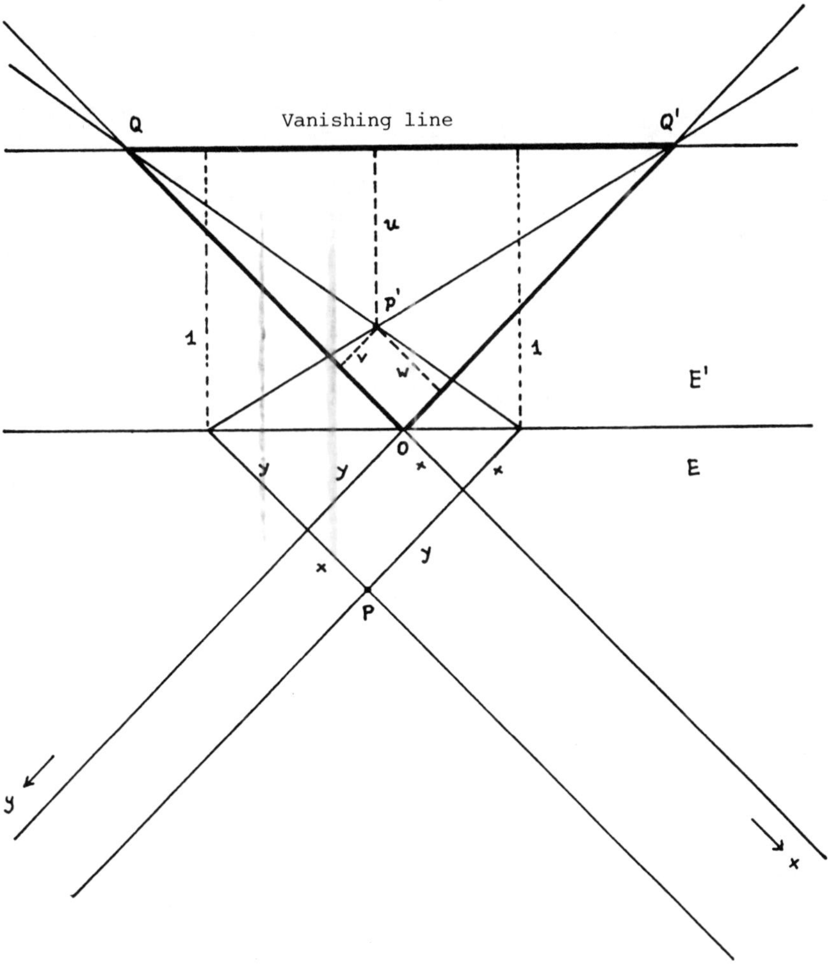

Vanishing line
Q
Q'
u
P'
1
1
v
w
E'
E
O
y
y
x
x
x
y
P
y
x

Thus to determine the position of a point in the plane E', as the intersection of the lines QP' and Q'P', one does not need the triangular coordinates (u,v,w) themselves, but only the ratios $x = \frac{v}{u}$ and $y = \frac{w}{u}$. If one were to multiply the triangular coordinates (u,v,w) by a common factor $\lambda \neq 0$, constructing the triple $(x_0,x_1,x_2) = (\lambda u,\lambda v,\lambda w)$, then the same ratios

$$x = \frac{x_1}{x_0}$$

$$y = \frac{x_2}{x_0}$$

would result. This suggests that we admit not only the triangular coordinates (u,v,w) as coordinates of a point in the plane E', but also – as a generalisation of the coordinate concept – all triples $(x_0,x_1,x_2) = (\lambda u,\lambda v,\lambda w)$. Thus a point no longer corresponds to one number triple, but to infinitely many. Nevertheless, one says each of these triples are <u>the</u> coordinates of the point, because all the different coordinates differ only by a factor. One calls these coordinates <u>homogeneous</u> (because all the coordinates x_0,x_1,x_2 are multiplied uniformly by the same factor λ). Thus we see how one proceeds quite naturally from the triangular coordinates, which come up in the analytic treatment of central projection and which were introduced by Plücker in 1830, to homogeneous coordinates. Plücker carried out the step to homogeneous coordinates in 1834.

We remark that the homogeneous coordinates (x_0,x_1,x_2) of a point can never satisfy $x_0 = x_1 = x_2 = 0$, because $u = v = w = 0$ would mean that the point lay on the three lines QQ', QQ and OQ', whereas they have no point in common.

One more historical remark : homogeneous coordinates had already come up earlier – 1827 – in Möbius' <u>barycentric calculus</u>. Möbius fixed three points p_1,p_2,p_3 in the plane, gave these points weights, and then considered the centre of mass of this system of three weighted points. It is clear immediately that two such systems with weights (x_0,x_1,x_2) and (x_0',x_1',x_2') have the same centre of mass when there is a $\lambda \neq 0$ with $(x_0',x_1',x_2') = (\lambda x_0,\lambda x_1,\lambda x_2)$. Thus one also arrives at the homogeneous extension of the coordinate concept from this starting point.

Möbius realised that homogeneous coordinates were suitable for the description of central projections. If E and E' are two planes in space and if one introduces homogeneous coordinates (x_0, x_1, x_2) in E and (x'_0, x'_1, x'_2) in E', then each central projection of E onto E' with a projection centre on neither E nor E' is described in terms of these coordinates by

$$\begin{aligned} x'_0 &= a_{00}x_0 + a_{01}x_1 + a_{02}x_2 \\ x'_1 &= a_{10}x_0 + a_{11}x_1 + a_{12}x_2 \qquad (*) \\ x'_2 &= a_{20}x_0 + a_{21}x_1 + a_{22}x_2 \end{aligned}$$

where this transformation is invertible, i.e. its matrix is invertible. The converse does not quite hold : if E and E', homogeneous coordinates, and the transformation (*) are given, then one cannot unconditionally regard the latter as a central projection of E onto E' at their given positions in space. However, one can prove that each such linear transformation can be realised as the composite of a chain of central projections $E = E_1 \to E_2 \to \ldots \to E_k = E'$ with a series of auxiliary planes. It is therefore natural to consider all invertible linear transformations of the form (*). Möbius called these transformations collineations because they preserve the collinearity of points, i.e. they map lines to lines.

Figures which go into each other under collineations are called projectively related. This projectivity is an equivalence relation, like congruence, similarity and affinity of figures, whereas the perspectivity of figures, defined by central projection, is not an equivalence relation – it is not transitive, as one sees from the following figure. (In this figure the lines E_1 and E_2, as well as the lines E_2 and E_3, are perspectively related. The relation between E_1 and E_3 resulting from composition is, however, not a perspectivity, because the lines connecting the corresponding points p_1, p_3 resp. q_1, q_3 resp. r_1, r_3 no longer go through the same point. The relation between E_1 and E_3 is only a projective relation.)

Thus we see that in the course of the development of geometry different, and ever more general, equivalence relations and corresponding geometries have arisen for the investigation of planar and spatial figures :

congruence
similarity
affinity
projectivity.

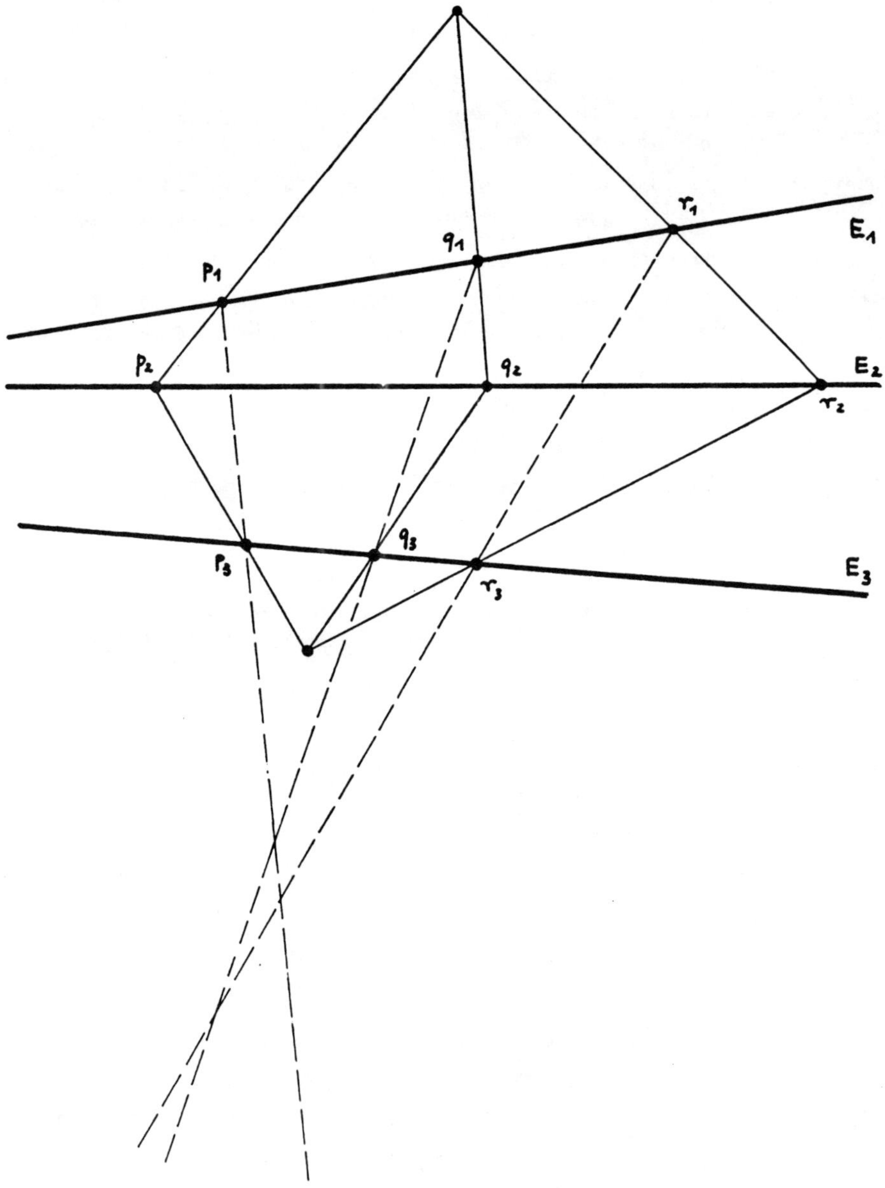

To each of these equivalence relations belongs a class of transformations which relate figures which are equivalent under the respective equivalence relations. In fact we have the following correspondence between classes of transformations and equivalence relations :

congruence	-	motions of the euclidean plane
similarity	-	central projection of parallel planes
affinity	-	composition of parallel projections
projectivity	-	collineations, i.e. compositions of arbitrary central projections.

Motions have obviously been considered since antiquity, affine transformations appear in Euler and have been familiar to mathematicians since 1800. Collineations were introduced in full generality by Möbius, and he realised that homogeneous coordinates were the means of handling them analytically, since collineations are simply the invertible linear transformations in homogeneous coordinates.

Möbius characterises the real collineations as the continuous mappings of one plane onto another which preserve collinearity, and hence map lines to lines. In order to have bijective correspondences between the lines of the two planes and the points of the two planes it is necessary to complete the affine planes by lines at infinity.

This is already clear when one considers central projection of a plane E onto a plane E' (see the figure in 3.1.). The vanishing line in E' corresponds to a line at infinity in E, and the disappearing line in E corresponds to the line at infinity in E'. In intuitive form, such ideas are already present in Desargues, and Poncelet later used them in synthetic form, without rigorous justification, as a basis for the construction of projective geometry.

The introduction of homogeneous coordinates gives a rigorous analytic setting for this process of <u>completion of an affine plane to a projective plane</u>.

Let us recall once again how we came to the triangular coordinates (u,v,w) and from there to the homogeneous coordinates $(x_0,x_1,x_2) = (\lambda u,\lambda v,\lambda w)$ by consideration of central projection of a plane E onto a plane E'. The homogeneous coordinates, which correspond to points of the affine plane E, are all the (x_0,x_1,x_2) with $x_0 \neq 0$, and indeed the homogeneous coordinates (x_0,x_1,x_2) define the point in the affine plane with cartesian coordinates

$$x = \frac{x_1}{x_0}$$

$$y = \frac{x_2}{x_0} .$$

The points with $x_0 = 0$ are just the points on the vanishing line, which corresponds to the line at infinity of the plane E, and which we want to introduce with the help of homogeneous coordinates.

I think it is now clear what we have to do. We simply admit all triples (x_0,x_1,x_2) as homogeneous coordinates, naturally with the restriction that not all three of $x_0,x_1,x_2 = 0$. And we declare that (x_0,x_1,x_2) and (x_0',x_1',x_2') define the same point of the projective plane when there is a real number $\lambda \neq 0$ such that

$$(x_0',x_1',x_2') = (\lambda x_0,\lambda x_1,\lambda x_2).$$

When this holds we shall also write

$$(x_0,x_1,x_2) \sim (x_0',x_1',x_2').$$

This defines an equivalence relation $\sim$ on the set of $(x_0,x_1,x_2) \in \mathbb{R}^3 - \{0\}$, and what we have just said means that the points of the projective plane are defined as the equivalence classes of the (x_0,x_1,x_2). Thus :

Definition : The <u>real projective plane</u> P_2 is the set of equivalence classes

$$P_2 = \{(x_0,x_1,x_2) \mid x_i \in \mathbb{R},\ (x_0,x_1,x_2) \neq (0,0,0)\} /\sim.$$

The class of a triple (x_0,x_1,x_2) is called the point of P_2 with the <u>homogeneous coordinates</u> (x_0,x_1,x_2). The subset $P_1 = \{(x_0,x_1,x_2) \in P_2 \mid x_0 = 0\}$ is called the <u>line at infinity</u> of P_2.

We remark that the correspondence $(x_0,x_1,x_2) \mapsto (x,y)$ with

$$x = \frac{x_1}{x_0}$$

$$y = \frac{x_2}{x_0}$$

defines a bijective mapping of $P_2 - P_1$ onto the affine plane E with cartesian coordinates (x,y). The inverse mapping is the correspondence

$$(x,y) \mapsto (1,x,y).$$

When we identify E with $P_2 - P_1$ by means of this mapping, we can regard P_2 as the completion of E by the line P_1 at infinity. We can also interpret this completion as follows :

We consider lines in the affine plane E with cartesian coordinates (x,y), and in fact such lines through the origin. Each such line has a parametric representation

$$x = at$$
$$y = bt.$$

All $(a,b) \neq (0,0)$ appear, and (a,b) defines the same line as (a',b') just in case $(a',b') = (\lambda a, \lambda b)$ for some $\lambda \neq 0$. Thus we obtain a bijection between all lines in the plane through the origin and the points of the line at infinity when we associate the line with the parametric representation above to the point $(0,a,b) \in P_1$. This association has a geometric meaning : to the point $(x,y) = (at,bt)$ corresponds the point in P_2 with homogeneous coordinates (1,at,bt). For $t \neq 0$ this is the same point as (t^{-1},a,b). If one now lets t go to infinity, then this point goes to (0,a,b). Thus what this says is : we have completed the affine plane E to the projective plane by adjoining a line P_1 at infinity, each point of which has been added to a line G in E through the origin and which is the intersection of P_1 with the thus completed line.

In this way we see how the introduction of homogeneous coordinates allows the introduction of <u>"ideal" elements</u> at infinity to be simply described, as is necessary for a consistent development of projective geometry from Poncelet's starting point.

The passage to the projective plane P_2 also permits the invertible homogeneous linear transformations

$$x' = Ax$$

to be interpreted as bijective mappings of the projective plane P_2 onto itself : the inverse mapping is

$$x = A^{-1}x'.$$

Two matrices define the same mapping just in case they are the same up to a scalar factor. The totality of all these mappings of P_2 onto itself obviously constitute a group, the group of all collineations of P_2 , and by what we have just said this group is canonically isomorphic to the <u>projective linear group</u>

$$PGL(3, \mathbb{R}) = GL(3, \mathbb{R})/\{\lambda \cdot Id\}.$$

(GL(n,K) denotes the general linear group of all invertible $n \times n$ matrices with entries from the field K.)

We can of course regard the linear transformation $x' = Ax$ as a coordinate change of P_2 rather than a self-mapping, i.e. we assign to a point with the homogeneous coordinates (x_0,x_1,x_2) the new homogeneous coordinates (x_0',x_1',x_2'), which are computed from the old by $x' = Ax$. Naturally there is no great difference between the two : projective geometry is the investigation of the properties of geometric figures which are preserved under arbitrary collineations – projective geometry is the investigation of the properties of geometric figures which are invariant under arbitrary homogeneous coordinate transformations.

Now which geometric figures shall we investigate in projective geometry? In the first place, curves, of course. We consider some quite simple examples :

(1) A line G in the affine plane E has the equation

$$bx + cy + a = 0$$

in cartesian coordinates (x,y). If we view E as $P_2 - P_1$, and hence set $x = \frac{x_1}{x_0}$, $y = \frac{x_2}{x_0}$, then

$$b\frac{x_1}{x_0} + c\frac{x_2}{x_0} + a = 0$$

or equivalently, after multiplication by x_0,

$$ax_0 + bx_1 + cx_2 = 0.$$

If we now consider this equation not only in E, but in the whole projective plane P_2, then the points (x_0,x_1,x_2) in P_2 which satisfy this equation are just the points of G plus the point $(0,c,-b)$, which is precisely the point on the line at infinity which completes the affine line G in P_2. We therefore make the definition :

A projective line in P_2 is the zero set of an equation

$$ax_0 + bx_1 + cx_2 = 0.$$

The set of these projective lines consists precisely of the completed affine lines and the line P_1 at infinity.

Two completed lines of the affine plane obviously intersect in the same point of the line P_1 at infinity when they are parallel in the affine plane. It is clear that there is exactly one line through two points of P_2 and that two lines intersect in exactly one point. In a similar way to the axiomatic construction of euclidean geometry, one can also base the geometry of the projective plane on the basic concepts of point and line, as Hilbert has done in his classical book "Grundlagen der Geometrie". For this one needs axioms of incidence, order and continuity. We shall not go further into this here, but instead use from the outset the analytic definition of the projective plane, and the lines in it, which we have just developed.

As soon as one has defined the lines in P_2 and as soon as one has defined the topology on P_2 (naturally the quotient topology on $P_2 = (\mathbb{R}^3-\{0\})/\sim)$, one can, following Möbius, define the collineations of P_2 as the continuous bijective mappings of P_2 onto itself which send lines to lines. As we have already said, Möbius showed that the collineations are exactly the mappings described by invertible homogeneous linear transformations in homogeneous coordinates.

(2) We have seen in (1) that the passage from an affine line to the corresponding projective line corresponds to the passage from the affine equation to the homogeneous equation $ax_0 + bx_1 + cx_2 = 0$. An equation $F(x_0,x_1,x_2) = 0$ is called homogeneous of degree n when the function F satisfies $F(\lambda x_0,\lambda x_1,\lambda x_2) = \lambda^n F(x_0,x_1,x_2)$ for all $\lambda \neq 0$. The same process of homogenisation now describes the general passage from an algebraic curve in the affine plane to the corresponding curve in the projective plane.

Consider for example the affine equations of the circle, hyperbola and parabola :

(i) $x^2 + y^2 - 1 = 0$

(ii) $x^2 - y^2 - 1 = 0$

(iii) $x^2 - y = 0.$

Homogenisation gives

(i) $-x_0^2 + x_1^2 + x_2^2 = 0$

(ii) $-x_0^2 + x_1^2 - x_2^2 = 0$

(iii) $-x_0x_2 + x_1^2 = 0.$

The homogeneous equation (ii) goes to (i) by the transformation $(x_0,x_1,x_2) \mapsto (x_1,x_0,x_2)$, and (iii) goes to (i) by the transformation $(x_0,x_1,x_2) \mapsto (x_0+x_2,x_1,x_0-x_2)$. Thus we see, by quite simple homogeneous coordinate transformations, that circle, hyperbola and parabola are projectively equivalent, i.e. the associated projective curves are the same – they transform into each other by collineations.

In 2.5 we saw this simply and directly by application of central projection and the conic section definition of these curves. Here we see that the analytic proof by calculation with homogeneous equations is at least as simple.

(3) As the last example we choose the two cubics

(i) $y - x^3 = 0$

(ii) $y^2 - x^3 = 0$

which we found in 2.5, following Newton, to result from each other by central projection.

Homogenisation gives

(i) $x_0^2 x_2 - x_1^3 = 0$

(ii) $x_0 x_2^2 - x_1^3 = 0$

and these two equations are interchanged by the transformation $(x_0, x_1, x_2) \mapsto (x_2, x_1, x_0)$. In this case the analytic argument is simpler than the synthetic.

These examples should suffice to show that with the introduction of the projective plane, homogeneous coordinates, the description of collineations by homogeneous linear transformations and the description of curves by homogeneous equations one has the correct analytic framework for the investigation of curves from the standpoint of projective geometry.

3.3 The projective plane as a manifold

An n-dimensional manifold M is a topological space which looks locally like the n-dimensional affine space $\mathbb{R}^n$, i.e. the space admits a covering by open sets $U_i \subset M$ for which there are homeomorphisms $\phi_i : U_i \to V_i$ onto open sets $V_i \subset \mathbb{R}^n$. (In addition, one usually assumes that M is Hausdorff and paracompact.) More precisely, one calls such a topological space an n-dimensional topological manifold.

When one is able to choose the system of U_i and ϕ_i in such a way that these homeomorphisms are compatible with each other in a suitable sense, then one can give the manifold additional structure by choice of the U_i and ϕ_i. E.g. when all the $\phi_i \circ \phi_j^{-1}$ are differentiable, they define a differentiable structure on M. It is then meaningful to speak of differentiable functions f on M : f is differentiable when all the $f \circ \phi_i^{-1}$ are differentiable in $V_i \subset \mathbb{R}^n$. One can then carry out real analysis on M. A manifold with a differentiable structure is called a differentiable manifold. The U_i are called coordinate neighbourhoods and the $\phi_i : U_i \to \mathbb{R}^n$ are called charts. With their help one can use the cartesian coordinates in $\mathbb{R}^n$ as local coordinates in $U_i \subset M$. If the $\phi_i \circ \phi_j^{-1}$ are also real analytic, then M has the structure of an n-dimensional real analytic manifold. If one replaces $\mathbb{R}^n$ by $\mathbb{C}^n$ in this definition and demands that the $\phi_i \circ \phi_j^{-1}$ be complex analytic, then M has the structure of an n-dimensional complex manifold, and one can speak of holomorphic functions on M and carry out complex analysis. For $n = 1$ one obtains in this way just the Riemann surfaces and the function theory of a complex variable.

One can give an n-dimensional differentiable manifold M yet another structure by giving a euclidean metric on the tangent space of each point $p \in M$, depending differentiably on p. Such a metric is called a Riemannian metric, and when we have chosen such a metric on M, as is always possible (in many different ways), M becomes a Riemannian manifold, and one can carry out differential geometry on M, Riemannian geometry.

Finally, analogously to the definition of differentiable real and complex-analytic manifolds, one can also define abstract n-dimensional real algebraic resp. complex algebraic manifolds. Here the open sets $U_i \subset M$ are Zariski-open sets, and the mappings $\phi_i : U_i \to V_i$ are bijective mappings onto n-dimensional affine algebraic subsets V_i of $\mathbb{R}^n$ resp. $\mathbb{C}^n$, such that the $\phi_i \circ \phi_j^{-1}$ are regular, i.e. rational functions with non-vanishing denominator (cf. 4.3 for affine algebraic sets).

For each class of manifolds defined above one can define a corresponding class of mappings : continuous resp. differentiable mappings, real resp. complex analytic mappings, regular mappings, and thus one obtains the categories of topological manifolds, differentiable manifolds, real analytic manifolds, complex manifolds, abstract real algebraic manifolds and complex algebraic manifolds. Of course, one can make even finer distinctions. For example, one calls a differentiable manifold a C^r-manifold when the coordinate exchange mappings $\phi_i \circ \phi_j^{-1}$ are r times continuously differentiable. In speaking of a differentiable manifold without further qualification one means generally, as we shall in this course, a C^{∞}-manifold. An example where this distinction is important will appear later (in 5.3).

In what follows we shall assume the elementary facts about differentiable manifolds to be known – whoever does not know these things can easily learn them from the introduction to differential topology by Bröcker and Jänich [B8]. Here we want only to recall the basic concept of "manifold" in order to make clear what we are talking about in what follows. The concept of manifold is quite central in mathematics, and similar in unifying power to the concept of group, hence it is worthwhile to say at least a few words on the origin of this concept. We must content ourselves with a brief suggestion ; for details we refer to the book of E. Scholz [S2].

When one wants to understand, today, the decisive progress implicit in the introduction of the manifold concept, one must first understand the position originally held by geometry in mathematics, and the paramount position of euclidean geometry up to the time of Gauss. Until this time, geometry was, in a quite naive way, the science of real physical space, and euclidean geometry was the description of the properties of this space. An internal mathematical development, the discovery of non-euclidean geometries by Gauss, Bolyai and Lobachevsky, put the paramount position of euclidean geometry into question and led to fundamental scientific and philosophical discussion on the nature of space and the position of geometry, in relation to which the idealistic philosophies of Kant and Hegel were partly a hindrance and partly helpful.

The result of this discussion, which continued for decades, was, to the extent that one can speak of "a" result in view of the continued divergence of idealistic and materialistic positions in the philosophy of science, a new assessment of the position of geometry in mathematics. The purpose of geometry is no longer the direct description of real physical space and its laws – that is a problem of physics. The geometer's problem is, independently of immediate ties with reality, to develop mathematical concepts and structures which seem suitable for the use of physicists on their problem. In particular, this leads to an unfolding of the concept of space, i.e. an investigation of mathematical objects and structures which can serve as possible models for spatial reality, or which are in some way related to such models. This is the position reached by Riemann at the end of his famous Habilitation lecture of 1854, in which he developed the foundations of Riemannian geometry starting from Gauss' investigations on the differential geometry of surfaces.

The historical development has shown that Riemann's conceptual preparation achieved exactly what he wanted, because his geometry became the mathematical foundation of general relativity theory. It was precisely the release of geometry from the task of directly comprehending spatial reality which made possible the degree of conceptual abstraction needed for later deeper penetration into the nature of space.

The central concept of geometry became the concept of manifold in its numerous variations. From the present standpoint, manifolds are space-like mathematical objects, possible models for physical space, they are "spaces" – but mathematical spaces now, no longer naively identified with "the" physical space as euclidean space was previously. The mathematicians of the last hundred years have hesitated to regard the manifolds as "spaces". They have rightly and intentionally chosen the concept "manifold", i.e. a collection of different objects of the same kind, in order to stress the abstract character of the concept. In fact, one had already arrived, in various ways, at certain collections of objects which we call manifolds today, and treated them in a geometric way. Thus the work of Lagrange and Cauchy in theoretical physics had shown that it was desirable to treat multidimensional manifolds in a geometric way, because a geometric treatment made the analytic and algebraic arguments more comprehensible.

Thus it was – in a dialectic development – precisely the application of analytic and algebraic methods made possible by the introduction of the coordinate concept which led to an extension of the domain of geometric concepts and methods. This appears particularly clearly in the work of the English mathematician Cayley, who developed the theory of homogeneous polynomials, or more precisely, algebraic invariant theory, and interpreted it geometrically at the same time. A collection of algebraic objects, e.g. the set of homogeneous polynomials of given degree and given number of variables, was regarded as a manifold. Before this, manifolds of geometric objects had been considered by Plücker, but Plücker had not been prepared to regard these objects as "points" of a space. We have already seen quite a simple example of this in 3.2 : the collection of all lines through a point of the plane can be identified with a manifold P_1 , the projective line. There will be more on such manifolds, whose elements are curves, later (in the discussion of linear systems and the discussion of duality in projective geometry, 6.2).

A very important contribution to the development of the manifold concept was made by H. Grassmann in his work "Die lineare Ausdehnungslehre", published in 1844, in which he spoke of n-tuply extended manifolds for the first time and developed, among other things, modern n-dimensional analytic geometry and linear algebra, in which the mathematical structure is worked out in a coordinate-free way, allowing the simplest treatment of problems in n-dimensional geometry, and in other fields as well.

To summarise : the essential elements in the development of the concept of manifold were : the fall of euclidean space as the privileged object of geometric investigation and the related reassessment of the position of geometry, and an extension of the domain of geometric objects, which could now be arbitrary manifolds of other mathematical objects. In connection with this, geometry was freed from the restriction to three-dimensionality. The coordinate concept was extended to cope with these wider requirements : indeed, the definition of a manifold which we have given — though it is a more precise statement, which first appeared later — expresses the intuitive idea that a manifold is a space with a covering by coordinate neighbourhoods which are compatible with each other.

After these, admittedly incomplete, remarks on this decisive phase in the development of geometry we shall now concern ourselves quite concretely with a particular manifold, the projective plane, and somewhat more generally with projective spaces.

<u>Definition</u>

(i) The <u>n-dimensional real projective space</u> $P_n(\mathbb{R})$ is the quotient

$$P_n(\mathbb{R}) = (\mathbb{R}^{n+1} - \{0\})/\sim$$

by the equivalence relation $\sim$ which is defined as follows:

$(x'_0, \ldots, x'_n) \sim (x_0, \ldots, x_n)$ if and only if
$(x'_0, \ldots, x'_n) = (\lambda x_0, \ldots, \lambda x_n)$

with $\lambda \in \mathbb{R}$, $\lambda \neq 0$.

(ii) We denote the equivalence class of $(x_0,\ldots,x_n)$ by $(x_0,\ldots,x_n)$ itself, and call the $(n+1)$-tuple of real numbers $(x_0,\ldots,x_n)$ the homogeneous coordinates of the point $(x_0,\ldots,x_n)$.

(iii) The real projective linear group $PGL(n+1,\mathbb{R})$ is the group

$$PGL(n+1,\mathbb{R}) = GL(n+1,\mathbb{R})/\{\lambda\cdot Id\}.$$

It acts on $P_n(\mathbb{R})$ by linear transformations

$$x' = Ax.$$

These bijective mappings of $P_n(\mathbb{R})$ onto itself are called collineations.

One can also view the transformation $x' = Ax$ as a coordinate change. Each coordinate system x' defined in this way will also be called a homogeneous coordinate system.

(iv) A projective linear subspace of $P_n(\mathbb{R})$ is a subset of $P_n(\mathbb{R})$ described by a system of linear homogeneous equations :

$$a_{i0}x_0 + \ldots + a_{in}x_n = 0, \quad i = 1,\ldots,m.$$

When this system has rank $n-k$, one can obviously introduce homogeneous coordinates $y_0,\ldots,y_n$ so that the projective subspace is given by $y_{k+1} = \ldots = y_n = 0$. One then has a natural bijection between this projective linear subspace P and $P_k(\mathbb{R})$, and hence P is a k-dimensional projective subspace. The collineations map any k-dimensional projective subspace onto another and preserve incidences. The $(n-1)$-dimensional projective subspaces are called hyperplanes.

Around 1870 the concept of n-dimensional projective space was universally known , and an n-dimensional projective geometry began to be systematically developed in these spaces. We cannot do this here and we shall only concern ourselves briefly with the projective spaces as manifolds.

The topology on the projective space is by definition the quotient topology on $\mathbb{R}^{n+1} - \{0\}/\sim$. With this topology, $P_n(\mathbb{R})$ is a topological manifold.

Proof : Let $U_i \subset P_n(\mathbb{R})$ be the complement of the hyperplane $x_i = 0$, thus

$$U_i = \{(x_0,\ldots,x_n) \in P_n(\mathbb{R}) \mid x_i \neq 0\}.$$

This is obviously an open set (by definition of the quotient topology). Let

$$\phi_i : U_i \to \mathbb{R}^n$$

be the mapping

$$\phi_i(x_0,\ldots,x_n) = \left(\frac{x_0}{x_i},\ldots,\frac{\hat{x}_i}{x_i},\ldots,\frac{x_n}{x_i} \right).$$

This mapping is continuous and bijective and has the continuous inverse

$$\phi_i^{-1}(x_1,\ldots,x_n) = (x_1,\ldots,x_i,1,x_{i+1},\ldots,x_n),$$

hence ϕ_i is a homeomorphism. Moreover, $P_n(\mathbb{R})$ is a compact Hausdorff space, as we see immediately and hence it is proved that $P_n(\mathbb{R})$ is a topological manifold. But we have proved much more. The chart mappings $\phi_i : U_i \to \mathbb{R}^n$ above satisfy : $\phi_i \circ \phi_j^{-1}$ on $\phi_j(U_i \cap U_j)$ is described by rational functions with non-vanishing denominator, and hence is a regular mapping. Thus we obtain, by this system of charts on $P_n(\mathbb{R})$, the structure of an abstract real algebraic manifold. In particular, we thereby obtain the structure of a differentiable manifold on $P_n(\mathbb{R})$, and we shall always provide $P_n(\mathbb{R})$ with this differentiable structure in future. (Warning : there are also "exotic" differentiable structures on the topological manifold $P_n(\mathbb{R})$.)

We shall now attempt to get an intuitive picture of this differentiable manifold, at least for $n = 2$. We first note the following :

The equivalence classes of $\mathbb{R}^{n+1} - \{0\}/\sim$ are exactly the lines in $\mathbb{R}^{n+1}$ through the origin, with the origin itself removed from all lines. Thus we have a canonical bijection of $P_n(\mathbb{R})$ onto the set of all lines through the origin in $\mathbb{R}^{n+1}$, and we can also define $P_n(\mathbb{R})$ as the manifold of these lines. Here we see an example of the way in which geometric objects, in this case lines, can be regarded as points of a new space, in this case the manifold $P_n(\mathbb{R})$.

Now we consider the standard sphere S^n in $\mathbb{R}^{n+1}$, namely

$$S^n = \{(x_0,\ldots,x_n) \in \mathbb{R}^{n+1} \mid x_0^2 + \ldots + x_n^2 = 1\}.$$

Each line through the origin meets S^n in exactly two points, which are antipodal, i.e. when one point is $(x_0,\ldots,x_n)$, the other is $(-x_0,\ldots,-x_n)$. Thus we let $\mathbb{Z}_2$, the cyclic group of order 2, act on S^n by the antipodal map

$$(x_0,\ldots,x_n) \mapsto (-x_0,\ldots,-x_n)$$

and construct the orbit space $S^n/\mathbb{Z}_2$, i.e. we identify each point of S^n with its antipodal point.

We can make $S^n/\mathbb{Z}_2$ into a differentiable manifold. To do this we first choose a suitable covering of the manifold S^n by coordinate neighbourhoods. Indeed, let

$$U_i^+ = \{(x_0,\dots,x_n) \in S^n | x_i > 0\}$$

$$U_i^- = \{(x_0,\dots,x_n) \in S^n | x_i < 0\}.$$

These are the two open hemispheres into which S^n is divided by the hyperplane $x_i = 0$. Now we map $U_i^{\pm}$ onto the hyperplanes $x_i = \pm 1$ by projection from the centre.

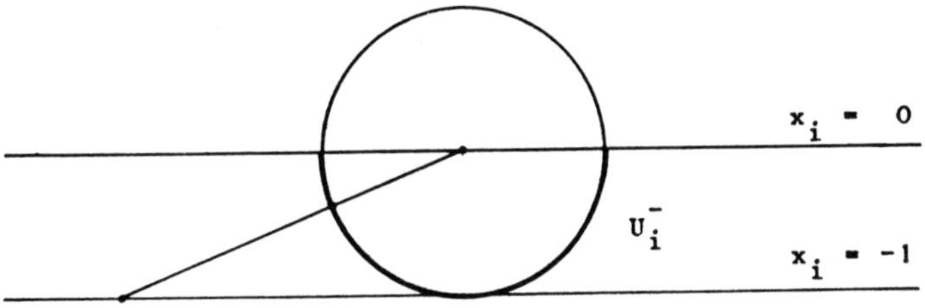

If we use $(x_0,\dots,\hat{x}_i,\dots,x_n)$ as cartesian coordinates in these hyperplanes, then these projections become described by the mappings $\phi_i^{\pm} : U_i^{\pm} \to \mathbb{R}^n$ with

$$\phi_i^{\pm}(x_0,\dots,x_n) = \left(\frac{x_0}{x_i}, \dots, \frac{\hat{x}_i}{x_i}, \dots, \frac{x_n}{x_i} \right).$$

These $\phi_i^{\pm}$ are homeomorphisms, and by means of these charts S^n becomes a differentiable manifold.

Now this whole construction of charts has been chosen, obviously, to be compatible with the quotient mapping $S^n \to S^n/\mathbb{Z}_2$. The homeomorphic images of U_i^+ and U_i^- are the same open set $\tilde{U}_i$ in $S^n/\mathbb{Z}_2$, and $\phi^{\pm}$ both induce the same homeomorphism

$$\tilde{\phi}_i : \tilde{U}_i \to \mathbb{R}^n ,$$

namely the mapping

$$\tilde{\phi}_i(x_0,\ldots,x_n) = \left(\frac{x_0}{x_i},\ldots,\frac{\hat{x}_i}{x_i},\ldots,\frac{x_n}{x_i}\right).$$

In this way, $S^n/\mathbb{Z}_2$ also becomes a differentiable manifold.

Now we have already implicitly established – this was the starting point of our investigation of $S^n/\mathbb{Z}_2$ – that the inclusion $S^n \subset \mathbb{R}^{n+1}$ defines a bijective mapping

$$S^n/\mathbb{Z}_2 \to \mathbb{R}^{n+1} - \{0\}/\sim \; = P_n(\mathbb{R})$$

by passing to the ~-equivalence classes. The above description of the coordinate neighbourhoods and charts for the differentiable manifolds $S^n/\mathbb{Z}_2$ and $P_n(\mathbb{R})$ shows immediately that this bijective mapping is differentiable and has a differentiable inverse. One calls such a mapping of differentiable manifolds a diffeomorphism. Thus we have shown :

Proposition 1

The canonical mapping

$$S^n/\mathbb{Z}_2 \to P_n(\mathbb{R})$$

is a diffeomorphism.

It follows, in particular, that $P_n(\mathbb{R})$ is compact, connected and Hausdorff, since these assertions clearly hold for $S^n/\mathbb{Z}_2$.

With Proposition 1 we have learned another description of projective space, but this still does not give us a clear intuitive picture of the manifold. We shall therefore make a further modification of this description. First of all, the following is clear : the inclusion of the closed hemisphere $\overline{U_0^-} \subset S^n$ induces a homeomorphism

$$\overline{U_0^-}/\sim \; \to S^n/\mathbb{Z}_2 \, .$$

What happens to the hemisphere $\overline{U_0^-}$ on passing to $\overline{U_0^-}/\sim$? The antipodal points on the boundary of the hemisphere become identified. Thus we obtain

Proposition 2

As a topological space, $P_n(\mathbb{R})$ results from a closed n-dimensional hemisphere S^n_+ by identifying antipodal points on the boundary.

One obtains immediately, as a consequence, that $P_1(\mathbb{R})$ results from a semicircle, in other words an interval, by identification of the endpoints, and hence is again homeomorphic to a circle. Indeed, it is easy to convince oneself that

<u>Corollary 3</u>

$P_1(\mathbb{R})$ is diffeomorphic to S^1.

This settles the nature of the projective line. Now for the projective plane $P_2(\mathbb{R})$. One can interpret Proposition 2 as saying that $P_2(\mathbb{R})$ results from a 2-dimensional disc by identification of diametrically opposite boundary points. From this one already obtains a first intuitive impression of the behaviour of curves in $P_2(\mathbb{R})$: one represents $P_2(\mathbb{R})$ diagrammatically by a disc. The boundary represents the line at infinity, and the behaviour of the curve at the boundary of the disc represents the behaviour at infinity of the actual curve in $P_2(\mathbb{R})$. Of course, one must think of antipodal points on the boundary as being identified. The following schematic pictures of this kind already make e.g. the projective equivalence of hyperbola, circle and parabola, and of the cubics $y = x^3$ and $y^2 = x^3$, intuitively clear, as we have proved analytically in 3.2.

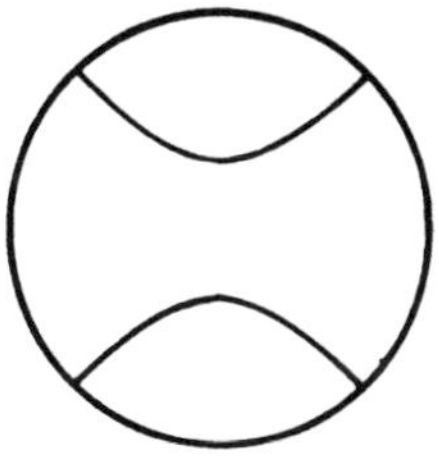

Hyperbola

$y^2 - x^2 = 1$

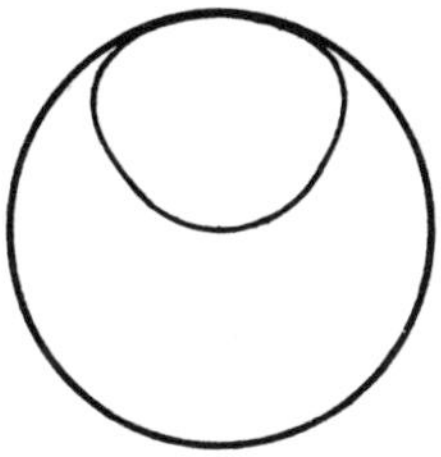

Parabola

$y = x^2$

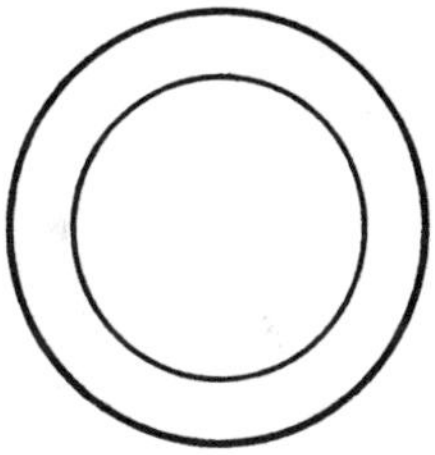

Circle

$x^2 + y^2 = 1$

$y = x^3$

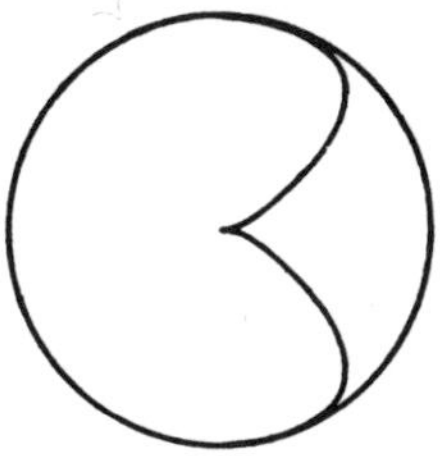

$y^2 = x^3$

Such representations have also been used to advantage in the investigation of the behaviour at infinity of the phase portrait of ordinary differential equations in the plane (cf. e.g. Lefschetz [L2]).

While this schematic representation is certainly useful, we want to obtain an even more intuitive representation of the surface $P_2(\mathbb{R})$. We therefore return to Proposition 2, but deform the hemisphere gradually, as shown in the following pictures.

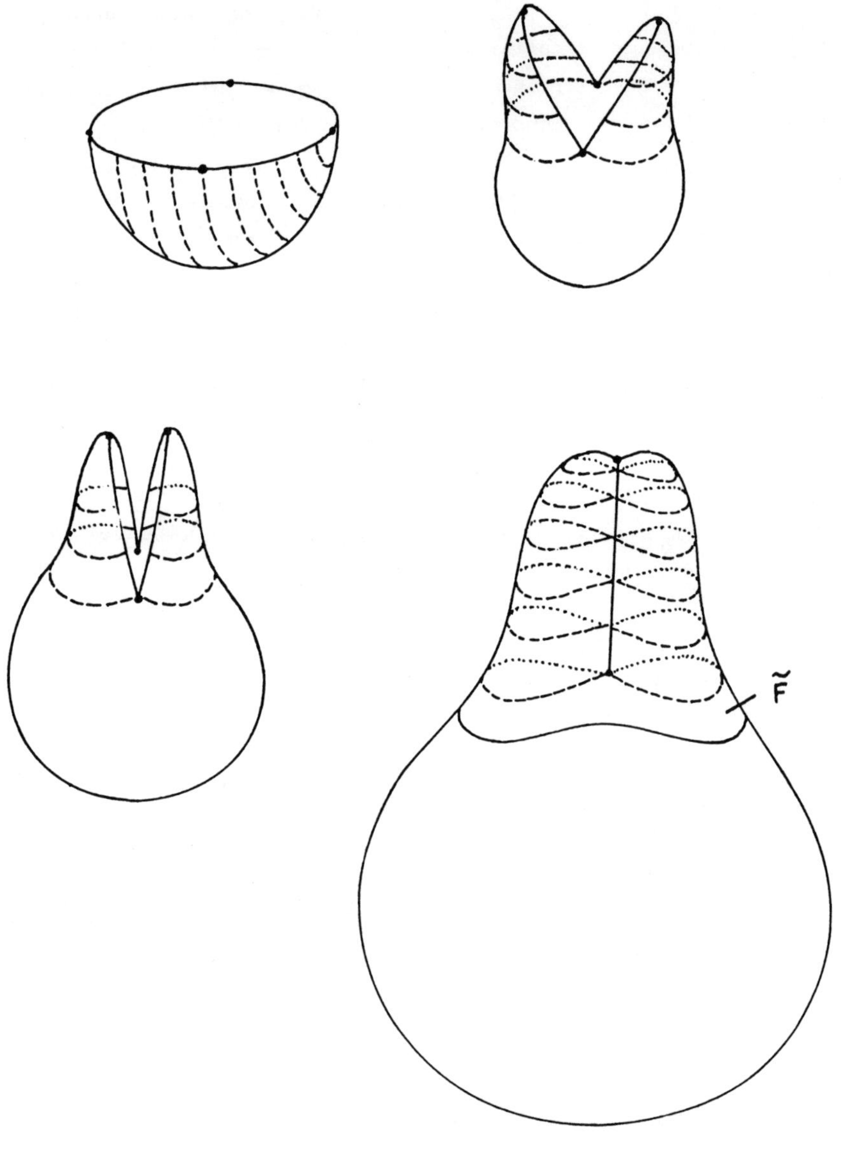

The last picture in this series shows the model of the projective plane obtained by this process. This model is a compact surface F lying in $\mathbb{R}^3$, with a self-intersection along a line segment and two complicated singular points at the ends of this segment. This surface results from the hemisphere by identifying not only the diametrically opposite points on the boundary circle, but also their mirror image points in one coordinate plane. This means : the singular surface F results from the projective plane $P_2(\mathbb{R})$ by identifying pairs of points on the line P_1 at infinity in a certain way. The identification corresponds to that in which (x,y) is identified with $(x,-y)$ on the circle $x^2 + y^2 = 1$ homeomorphic to P_1. Thus an interval results from P_1 by identification, and this is the interval along which F intersects itself.

If one views the picture of F as if the self-intersection did not really exist, then one obtains an intuitive representation of the projective plane. The surface F can be put together from two pieces : the piece $\tilde{F}$ with the self-intersection, and a perforated sphere. One calls $\tilde{F}$ a crosscap. Thus our model F of the projective plane results from cutting a disc from a sphere and replacing it by a crosscap. In a certain sense our model is not particularly nice, because of the two singular points. There are, though we shall not go into this further here, models of $P_2(\mathbb{R})$ in $\mathbb{R}^3$ – more precisely, immersions of $P_2(\mathbb{R})$ in $\mathbb{R}^3$ – whose singularities at worst appear like the intersections of two or three coordinate planes in $\mathbb{R}^3$. (See e.g. Hilbert-Cohn-Vossen [Hl].) There is nothing better than this : one can prove that there is no embedding of $P_2(\mathbb{R})$ as a submanifold of $\mathbb{R}^3$, and hence without singularities.

However, our model with the crosscap is good enough to establish an interesting topological property of $P_2(\mathbb{R})$, namely that $P_2(\mathbb{R})$ is not orientable. Intuitively, this means that the surface F is one-sided in $\mathbb{R}^3$; when a fly crawls along the curve shown (it ignores the self-intersection!) it ends up on the "other" side, i.e. there is only one side.

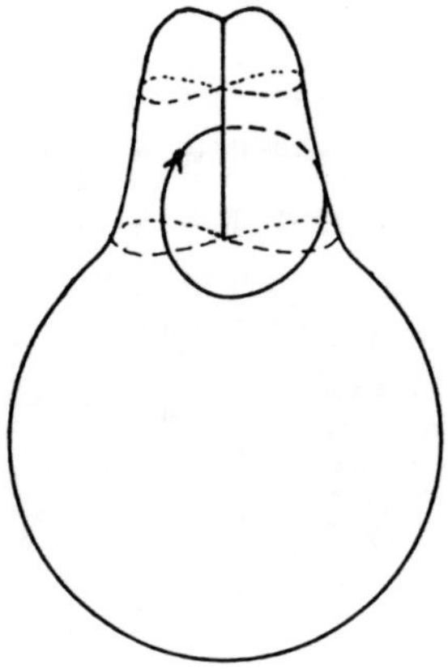

The one-sidedness is a property of the surface in $\mathbb{R}^3$, non-orientability is an intrinsic property of the surface. Non-orientability means : if one continuously extends an orientation of the surface, chosen at a point, along a suitable closed curve, then one returns with the opposite orientation. On F, for example, this is the case for the curve already considered. Thus we see:

Proposition 4

$P_2(\mathbb{R})$ is a non-orientable surface.

The first non-orientable surface was discovered in 1858 by Möbius (in the context of an investigation of polyhedra!) and in the same year, independently, by Listing, a student of Gauss. This first non-orientable surface, which caused great astonishment at the time, is the Möbius band, pictures of which are familiar nowadays. It results from the product $[0,1] \times [-1,1]$ when one identifies $(0,t)$ with $(1,-t)$. The identification of the endpoints 0 and 1 of $[0,1]$ results in a circle S^1, and the Möbius band is therefore a kind of twisted product of S^1 with an interval. Over the neighbourhood U of each point in S^1 it looks like the product $U \times [-1,1]$, but globally it is twisted. One calls such a thing a locally trivial fibre bundle with basis S^1 and fibre $[-1,1]$.

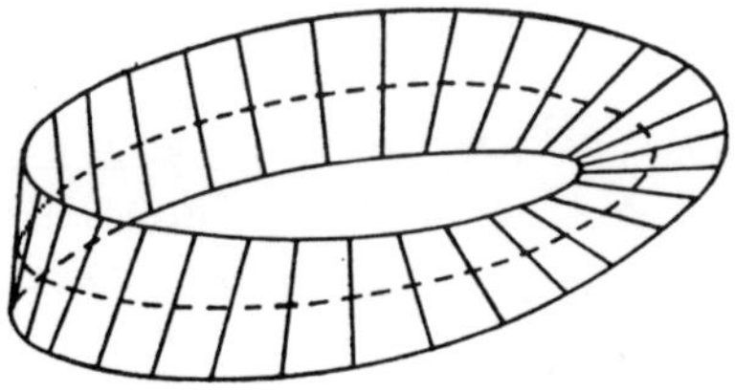

The non-orientability of the long "known" projective plane $P_2(\mathbb{R})$ was first discovered in 1873 by Felix Klein and Schläfli.

In fact there is a close connection between the non-orientability of these two surfaces. Namely, we shall show that the Möbius band can be regarded as a part of the surface $P_2(\mathbb{R})$, and it will then be clear that $P_2(\mathbb{R})$ is non-orientable along with the Möbius band. We show

Proposition 5

The projective plane results from identification of a disc and a Möbius band along their boundaries.

Proof : Just for fun we give a very intuitive topological proof. The Möbius band M and the disc D both have a circle as boundary. We paste them together along their common boundary and we want to see why the resulting surface

$$F = M \cup D$$

is the projective plane. To do this we cut F along the centre line of M and obtain a surface F' whose boundary $\partial F'$ is a circle S^1. Now everyone who has played with a Möbius band M knows that cutting it along its centre line gives an annulus M' (M' has a twisted embedding in $\mathbb{R}^3$ but that does not matter, we are interested only in the abstract surface M', which is an annulus). Thus

$$F' = D \cup M'$$

is a disc $\tilde{D}$ with boundary S^1. The surface F now results from F' by identification of the points of $\partial F'$ which correspond to the same point on the centre line of the Möbius band. That is, F results from $\tilde{D}$ by identification of antipodal points on the boundary S^1 of $\tilde{D}$. Thus F is the projective plane by Proposition 2. Q.E.D.

The description of $P_2(\mathbb{R})$ in Proposition 5 nicely shows, in particular, what the projective plane P_2 looks like in the neighbourhood of the line P_1 at infinity : P_1 is the centre line of the Möbius band, and the Möbius band M is a neighbourhood of P_1 in P_2, in fact a particularly nice neighbourhood U, a tubular neighbourhood, as the differential topologists say. This neighbourhood $U = M$ of P_1 is simply the complement of a "large" disc D around the origin of the affine plane $E \subset P_2$, and the intervals into which M is fibred are the intersections of $U = M$ with the projective lines in P_2 through the origin of E.

To conclude, we shall give another intuitive description of $P_2(\mathbb{R})$ as the union of a Möbius band M and a disc D. To do this we again consider a surface $\tilde{F}$ in $\mathbb{R}^3$ with self-intersection along a line segment and two complicated singularities at the endpoints, namely the surface shown in the following figures.

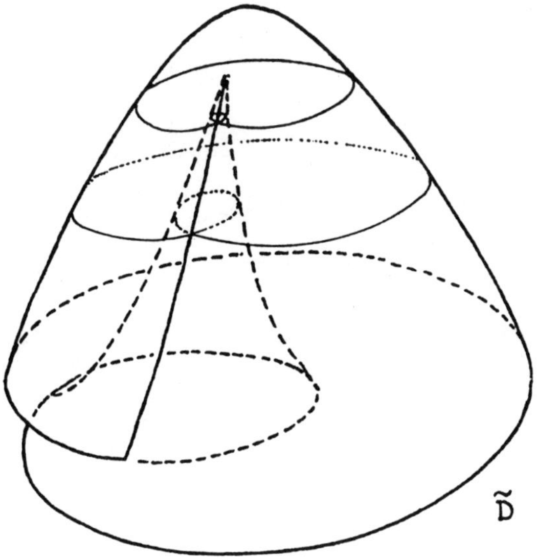

D̃

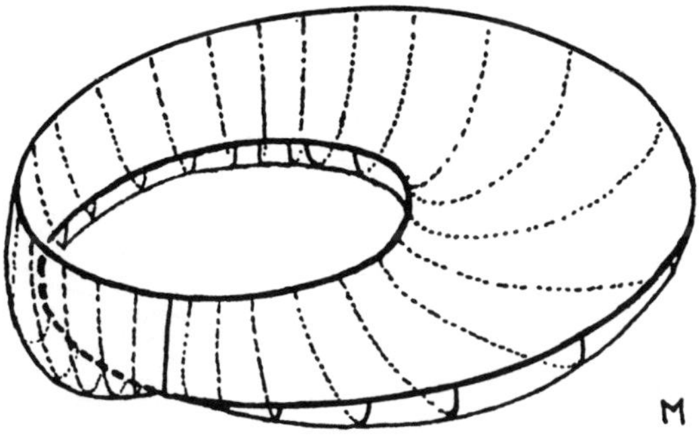

M

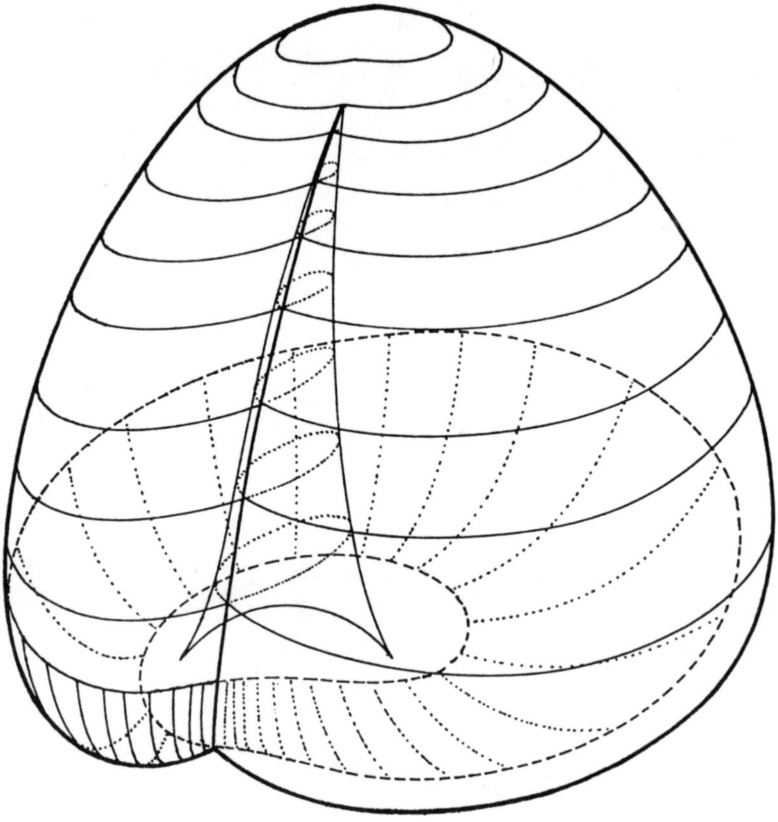

Roughly speaking, this surface is constructed as follows: a Möbius band M is embedded in $\mathbb{R}^3$ in the usual way. (The fibres of the Möbius band are shown as dotted lines in the picture.) Then a point p is chosen in the complement of M and connected by curved rays to all the boundary points of M. When one does this correctly, the union of these rays becomes a disc $\tilde{D}$ with a self-intersection, as in the first picture.

Attachment of the disc $\tilde{D}$ to the Möbius band M then results in the singular surface $\tilde{F}$

$\tilde{F} = \tilde{D} \cup M.$

For clarification we give two more pictures, showing the topological type of the singularities of $\tilde{F}$ at the endpoints of the line of self-intersection.

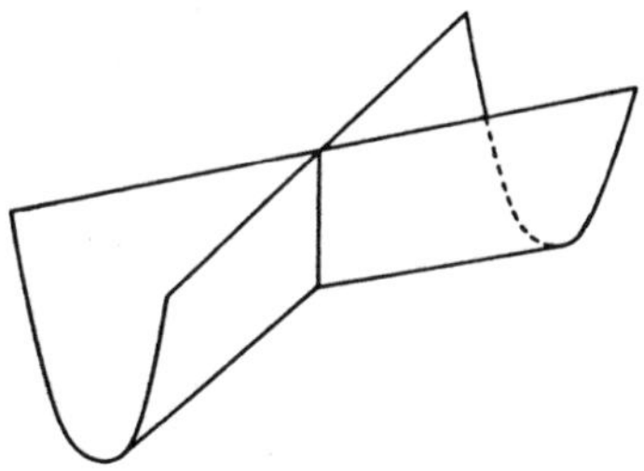

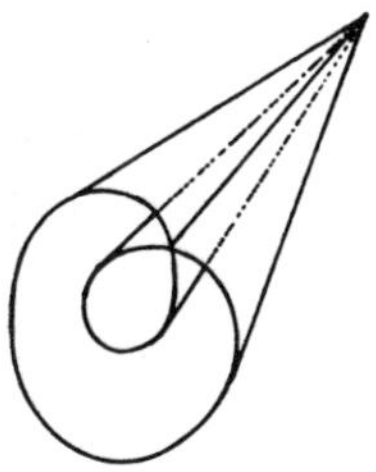

With this we shall leave the topology of the real projective plane and turn to the complex projective plane and the curves in it.

3.4 Complex projective geometry

We have already seen, in 2.4 and 2.5, that the application of analytic methods to problems such as the determination of the coordinates of singular points of a curve, or the intersection of two curves, leads to equations whose solutions are complex numbers. In the synthetic treatment of these problems imaginary elements also appeared; the first time this occurred to a significant extent was in Poncelet, with the application of his "continuity principle". This principle, whose mathematical content is today reduced to the identity theorem for analytic functions and the fundamental theorem of algebra, says that "a property known of a figure in sufficient generality also holds for all other figures obtainable from it by continuous variation of position" (F. Klein [Kl]), p. 81-82). For example, it follows from the principle that two conics always intersect in four points. When this does not happen in the real domain, the missing points of intersection are imaginary. A strict treatment of such imaginary elements was not carried out by Poncelet, and it is not easy to do on a synthetic basis.

In an analytic treatment, the introduction of imaginary elements, which was done systematically by Plücker, offers no difficulties in principle: one has simply to admit complex numbers wherever real numbers were used previously. The coordinates of points, the coefficients of polynomials, the solutions of equations etc. – all these quantities can now be complex.

In particular, one now considers the n-dimensional complex projective space $P_n(\mathbb{C})$. It is defined quite analogously to the real projective space :

$$P_n(\mathbb{C}) = \mathbb{C}^{n+1} - \{0\}/\sim$$

where $(x'_0,\dots,x'_n) \sim (x_0,\dots,x_n)$ exactly when there is a $\lambda \in \mathbb{C} - \{0\}$ with

$$(x'_0,\dots,x'_n) = (\lambda x_0,\dots,\lambda x_n).$$

A whole series of earlier definitions carry over immediately to the complex case. Thus one defines the coordinate neighbourhoods U_i by

$$U_i = \{(x_0,\dots,x_n) \in P_n(\mathbb{C}) \mid x_i \neq 0\}$$

and affine coordinates on U_i by

$$(x_0,\dots,x_n) \mapsto \left(\frac{x_0}{x_i},\dots,\frac{\hat{x}_i}{x_i},\dots,\frac{x_n}{x_i}\right).$$

The coordinate exchanges are then described by regular functions, and $P_n(\mathbb{C})$ thereby receives, in particular, the structure of an abstract complex-algebraic manifold. One is then free to view $P_n(\mathbb{C})$ as a complex manifold, a differentiable manifold or a topological manifold.

As in the real case, one can also define the complex d-dimensional projective linear subspaces of $P_n(\mathbb{C})$, and again these are mapped onto other such subspaces by the collineations, i.e. the operations of the complex projective linear group

$$PGL(n+1,\mathbb{C}) = GL(n+1,\mathbb{C})/\{\lambda\cdot Id\}.$$

However, one can no longer characterise the collineations as the continuous incidence-preserving bijections of $P_n(\mathbb{C})$ onto itself which carry projective linear subspaces to projective linear subspaces, because conjugation of $P_n(\mathbb{C})$ also has these properties. Conjugation κ is the mapping

$$(x_0,\dots,x_n) \mapsto (\bar{x}_0,\dots,\bar{x}_n)$$

of $P_n(\mathbb{C})$ onto itself. Of course, $\kappa^2 = Id$, i.e. κ is an involution. One can show that each homeomorphism of $P_n(\mathbb{C})$ which carries linear subspaces to linear subspaces is either a collineation or the product of a collineation with the conjugation κ. The fixed point set $P_n(\mathbb{C})^{\kappa}$, i.e. the set of $x \in P_n(\mathbb{C})$ with $\kappa(x) = x$ is obviously just the real projective space

$$P_n(\mathbb{R}) = P_n(\mathbb{C})^{\kappa} \subset P_n(\mathbb{C}).$$

In 3.3.1 we described the real projective space as $S^n/\mathbb{Z}_2$. Corresponding to this, we have the following description of the complex projective space. We consider the $(2n+1)$-sphere

$$S^{2n+1} = \{(x_0,\dots,x_n) \in \mathbb{C}^{n+1} \mid \Sigma x_i\bar{x}_i = 1\}.$$

The multiplicative group S^1 of complex numbers λ with $|\lambda| = 1$ acts on S^{2n+1} by

$$(x_0,\dots,x_n) \mapsto (\lambda x_0,\dots,\lambda x_n).$$

One can make the orbit space S^{2n+1}/S^1 into a differentiable manifold in a natural way, and then the canonical mapping

$$S^{2n+1}/S^1 \to P_n(\mathbb{C})$$

is a diffeomorphism.* The coset mapping

$$S^{2n+1} \to P_n(\mathbb{C})$$

is a differentiable locally trivial fibre bundle with basis $P_n(\mathbb{C})$, total space S^{2n+1} and fibre S^1. It is also known as the Hopf fibre bundle or the Hopf fibration. When $n = 1$, then $P_n(\mathbb{C})$ is homeomorphic to the 2-sphere S^2, as we will soon see, and the Hopf fibration is therefore a fibration

$$S^3 \to S^2$$

of the 3-sphere over the 2-sphere with fibre S^1. One can visualise this fibration as follows. One divides S^2 by the equator into two discs, D^+, D^-. The preimages of these discs, considered as abstract spaces, are simply products $S^1 \times D^+$, $S^1 \times D^-$, and hence solid tori. Let us see how these solid tori, fibred by circles S^1, lie in S^3. We view S^3 as the compactification of $\mathbb{R}^3$ by a point at infinity, and so we want to see how such a solid torus, with its fibration, is embedded in $\mathbb{R}^3$. To do this we take an ordinary solid torus with its ordinary fibration by circles parallel to the centre line, cut it along a disc perpendicular to the centre line, twist through 360° and paste back together. This does not change the solid torus as an abstract space, nor its fibration into circles, but we have altered the embedding of the fibration in $\mathbb{R}^3$. Any two fibres are now linked together.

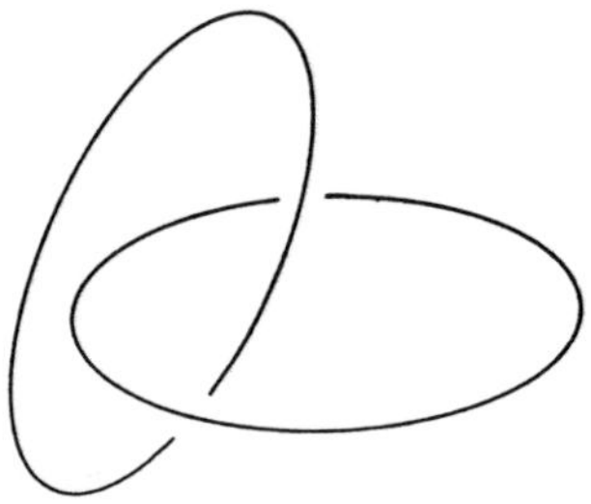

The Hopf fibration results from pasting together two solid tori fibred in this way. (The decomposition of S^3 into two tori will be useful to us later, e.g. in 5.3 in the study of singularities of the

---

*It follows, in particular, that $P_n(\mathbb{C})$ is compact and connected.

complex curves.) The next picture shows the fibration of the solid torus just constructed, embedded in $\mathbb{R}^3$.

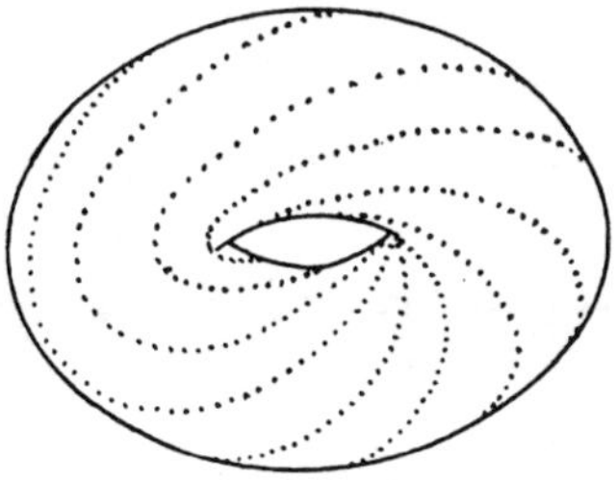

In general, i.e. for $n > 1$, it is difficult to obtain an intuitive picture of the topology of $P_n(\mathbb{C})$, because $P_n(\mathbb{C})$ has real dimension $2n$ as a differentiable manifold. Thus the complex plane is already a real 4-dimensional manifold, and for that reason I cannot give an intuitive picture of it.

In what follows we therefore confine ourselves to picturing the figures within the complex projective plane we want to study – and these are the curves.

From what we have said in 3.3 about curves in the real projective plane, and from the remarks at the beginning of this section on the passage from real to complex projective geometry, it is already clear how we define curves in the <u>complex projective plane</u> : as subsets of $P_2(\mathbb{C})$ which are the zero sets of homogeneous polynomials.

$$F(x_0,x_1,x_2) = \sum_{i+j+k=m} a_{ijk}x_0^i x_1^j x_2^k$$

with complex coefficients a_{ijk}. The number m is the "order" of the curve. Now when we want to study these curves, we can view them from several aspects : we can regard such a curve C as a <u>projective algebraic subset of</u> $P_2(\mathbb{C})$, i.e. as simply the zero set of a homogeneous polynomial, or as a <u>complex analytic subset</u> of the complex manifold $P_2(\mathbb{C})$, i.e. as a subset which in a suitable coordinate neighbourhood of each point is the zero set of a complex analytic function, or finally as simply a <u>topological subset</u> of the manifold $P_2(\mathbb{C})$.

The first viewpoint is the viewpoint of algebraic geometry, and it is the one we shall mostly adopt later in this course (in Chapter II), though not always consistently. The second viewpoint is that of function theory. In particular, each non-singular curve C in $P_2(\mathbb{C})$ is a 1-dimensional complex manifold, by the implicit function theorem, and hence, disregarding the embedding, a one-dimensional compact complex manifold. When the curve has singular points one can associate it, as we shall see later (in 9.1), with a unique one-dimensional compact complex manifold, its "normalisation", by "resolution" of its singular points. But the one-dimensional compact complex manifolds are precisely the abstract <u>Riemann surfaces</u> which one studies in complex variable theory, and thus one sees that from the second viewpoint, the study of curves in the complex projective plane is really a part of complex function theory.

There is really little difference between the algebraic-geometric and function theoretic starting points. The important <u>theorem of Chow</u> shows that : each closed complex analytic subset of $P_n(\mathbb{C})$ is a projective algebraic subset.

The third, topological, viewpoint is essentially qualitative in comparison with the other two. For that reason it is perhaps especially suitable for obtaining a first impression of the nature of the geometric figures we wish to investigate. In what follows we shall therefore investigate a few simple examples of curves mainly from the topological viewpoint, and they will be curves of orders 1, 2 and 3 in turn, in other words lines, quadrics and cubics.

We first consider a <u>complex projective line</u> $L \subset P_2(\mathbb{C})$, i.e. the zero set of an equation

$$a_0x_0 + a_1x_1 + a_2x_2 = 0.$$

After a suitable change of coordinates the equation has the form simply $x_2 = 0$. This shows that our line L is isomorphic to the 1-dimensional complex projective space $P_1(\mathbb{C})$. Thus it suffices to find out what the manifold $P_1(\mathbb{C})$ looks like.With the homogeneous coordinates (x_0,x_1), $P_1(\mathbb{C})$ becomes covered by two coordinate neighbourhoods U_0 and U_1

$$U_i = \{(x_0,x_1) \in P_1(\mathbb{C}) \mid x_i \neq 0\}.$$

U_0 and U_1 are isomorphic to $\mathbb{C}$. One obtains a coordinate z_i in U_i by

$$z_0 = \frac{x_1}{x_0} \quad \text{in } U_0$$

$$z_1 = \frac{x_0}{x_1} \quad \text{in } U_1.$$

Thus the coordinate change in $U_0 \cap U_1$ is described by

$$z_1 = \frac{1}{z_0} .$$

$P_1(\mathbb{C})$ results from two copies of $\mathbb{C}$ when one identifies the complement of the origin in one copy with the complement of the origin in the other copy by means of the formula $z_1 = z_0^{-1}$. One can also view this as identifying, say, U_0 with $\mathbb{C}$ and then adding another point where $x_0 = 0$, i.e. $z_0 = \infty$. Thus the one-dimensional complex projective space is none other than the <u>Riemann number sphere</u>. As a differentiable manifold it is diffeomorphic to the standard sphere S^2.

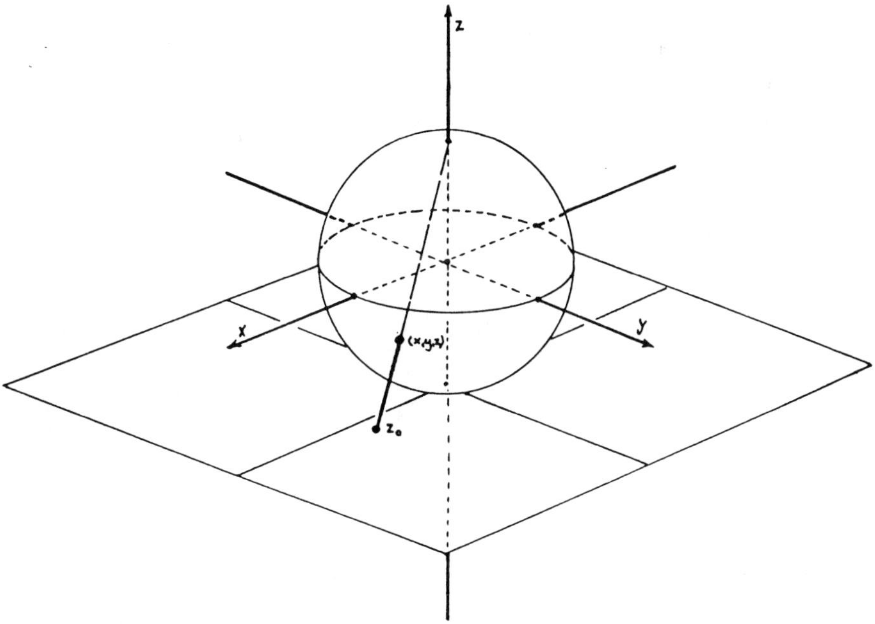

One obtains a diffeomorphism of the standard sphere $S^2 \subset \mathbb{R}^3$ with equation $x^2 + y^2 + z^2 = 1$ onto $P_1(\mathbb{C})$ with affine coordinates z_0 in U_0 and z_1 in U_1 by <u>stereographic projection</u> :

$$(x,y,z) \mapsto z_0 = \frac{2(x+iy)}{1-z} \quad \text{for} \quad z \neq 1$$

$$(x,y,z) \mapsto z_1 = \frac{x-iy}{2(1+z)} \quad \text{for} \quad z \neq -1.$$

This makes the real projective line $P_1(\mathbb{R}) \subset P_2(\mathbb{R})$ correspond to the great circle $y = 0$ on S^2.

So much for projective lines. We now consider the <u>quadrics</u>. It is easy to prove that, up to equivalence under collineations, there are only two distinct complex projective quadrics, namely the reducible quadric, which consists simply of two projective lines, and the irreducible quadric. The latter is a non-singular curve (proof in 7.1).

Reducibility of a quadric, i.e. decomposition into two components – both lines – means that the homogeneous polynomial of degree 2 in its equation $F(x_0,x_1,x_2) = 0$ decomposes into two linear factors. Mind you, the decomposition into linear factors is in general only possible when we admit linear factors with complex coefficients. F may or may not decompose as a real polynomial. Example

(i) $x_1 \cdot x_2$ real decomposition,

(ii) $x_1^2 + x_2^2 = (x_1+ix_2)(x_1-ix_2)$ non-real decomposition.

In case (i) the real zero set in $P_2(\mathbb{R})$ is the union of two distinct real projective lines.

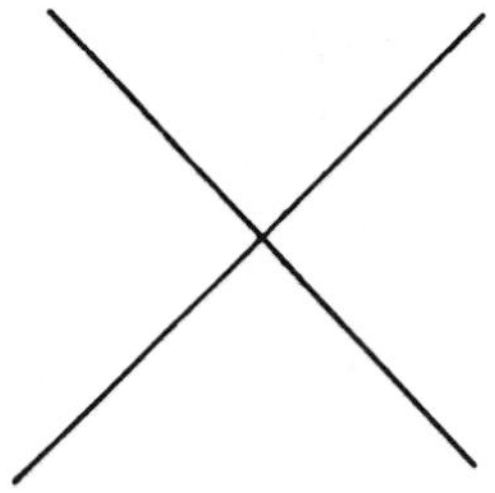

Thus the real picture leads to an adequate view of the nature of the complex curve. But in case (ii) the real zero set in $P_2(\mathbb{R})$ is just a single point, namely $(1,0,0)$, and this seemingly wayward curve gives no adequate picture of the associated complex curve.

As we have already said, the complex curve consists of two complex projective lines which meet transversely. It is not possible to draw an adequate picture of this situation, because the lines are real 2-dimensional and lie in the real 4-dimensional manifold $P_2(\mathbb{C})$. The transverse intersection looks exactly like that between the two complex coordinate axes $x_1 = 0$ and $x_2 = 0$ in $\mathbb{C}^2$. Thus it is not possible to make a good picture of the reducible quadric Q embedded in $P_2(\mathbb{C})$. However, one can easily visualise the topological space itself, apart from the embedding : Q is the one point union of two projective lines. Each projective line is homeomorphic to S^2. Thus Q results from two copies of S^2 when one identifies a point on one copy with a point on the other. Thus Q is – in the language of topologists – a <u>bouquet</u> $S^2 \vee S^2$ of two 2-spheres.

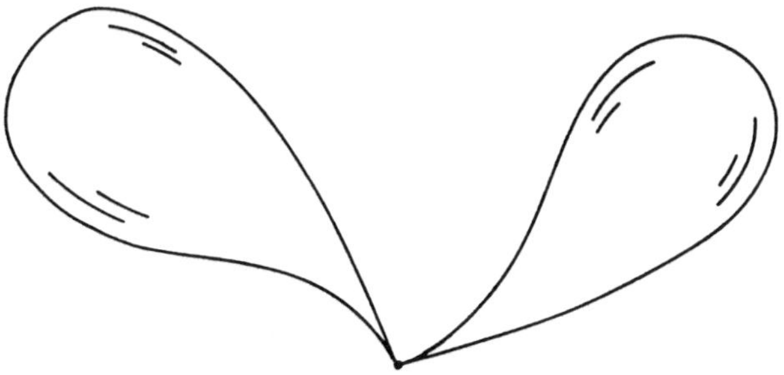

To repeat : this picture does not adequately represent the embedding $Q \subset P_2(\mathbb{C})$, because there the two spheres are of course embedded as smooth submanifolds. But when one draws them as smooth and without penetration of each other, the result is a picture of the following type, which indeed is also topologically correct, but psychologically counter to the idea of a transverse intersection.

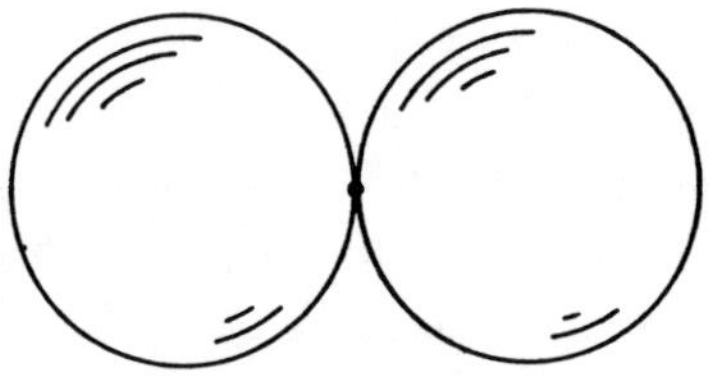

The following picture is likewise dangerous, though useful, because it may give the impression that the singular point of the 1-dimensional quadric Q looks like a 2-dimensional cone vertex.

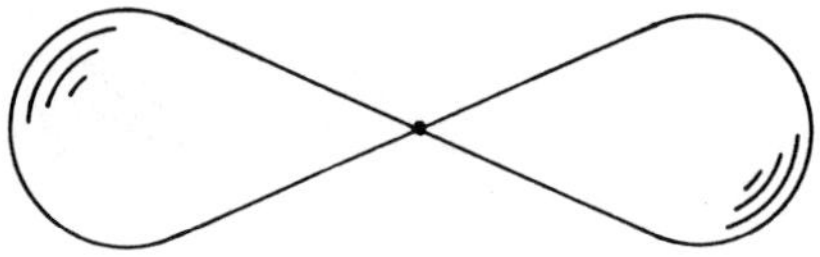

A picture of a real cone vertex, such as the following, is more adequate as a representation of the real surface and — by analogy — also the complex surface in three-dimensional space with the equation

$$x^2 + y^2 - z^2 = 0$$

I explain these different figures in such detail because in my experience it always leads to confusion when one does not say what one means by them.

So much for the reducible quadric, the union of two lines. Now to the irreducible quadric. In 3.3 we have already established that the circle, hyperbola and parabola, which are distinct from the affine viewpoint, are real projectively equivalent. The same is just as true in complex projective geometry : by suitable coordinate transformations one can bring the equations of all irreducible quadrics into the same normal form (cf. 7.1). One can choose the normal form to be, e.g. :

$$x_0^2 + x_1^2 + x_2^2 = 0.$$

In spite of this we shall begin our investigation of the irreducible complex quadric with the real affine normal forms :

(i) $y = x^2$ parabola

(ii) $xy = 1$ hyperbola

(iii) $x^2 + y^2 = 1$ circle.

We do this because we want to read off some characteristics of the complex curves from the pictures of the corresponding real curves, and because we want to demonstrate different methods of investigation. The corresponding complex projective curves are projectively equivalent — this is shown by a proof corresponding to the real case given in 3.3. In the investigation we mostly use the affine complex coordinates x, y. For the time being we choose the homogeneous complex coordinates (x_0, x_1, x_2) so that

$$x = \frac{x_1}{x_0}$$

$$y = \frac{x_2}{x_0}.$$

In order to investigate a curve like our irreducible quadric, or more generally any curve C, we can make use of the method of projection. To do this we choose a point p of the projective plane, which can lie on C but need not, a projective line L with $p \notin L$, and project C onto L. We shall see examples shortly.

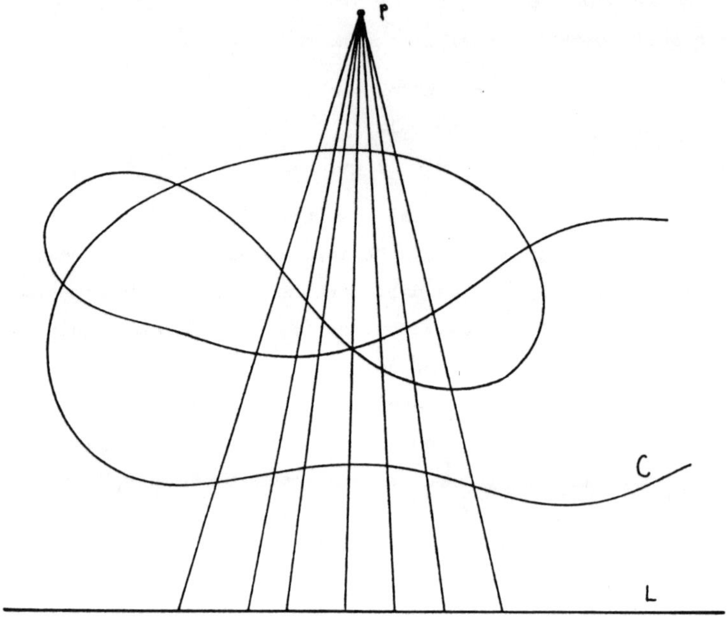
P
C
L

In the complexes, each point of L has in general the same number of pre-image points on C. When p does not lie on C their number equals the order of C (see 5.2). But for particular points of L some of the pre-image points may coincide. The mapping

$$C \to L$$

which one obtains by projecting from a point p outside C is what the topologists and function theorists call a branched covering. The points where several pre-image points coincide in C are called branch points. We shall see in examples that the description of C as a branched covering of L permits, e.g., conclusions to be drawn about the topology of C.

This general method of investigation by projection will now be applied in a variety of ways to quadrics.

We begin with the affine equation of the parabola :

$$y - x^2 = 0.$$

The homogeneous equation of the corresponding quadric C reads :

$$x_0x_2 - x_1^2 = 0.$$

The line at infinity, $x_0 = 0$, contains a point of C, namely $p = (0,0,1)$. We project C from this point onto the line $x_2 = 0$. In affine coordinates (x,y), this is projection in the direction of the y-axis onto the x-axis. Over each point of the x-axis lies one point of the quadric, namely $(x,y) = (x,x^2)$. The real affine picture of this is the following :

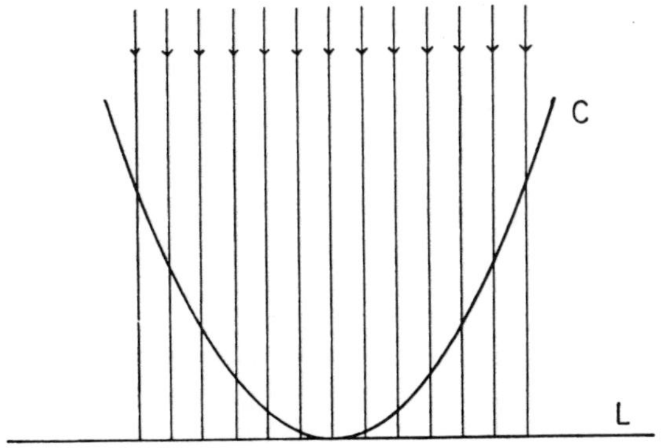

The real projective picture is schematically something like this :

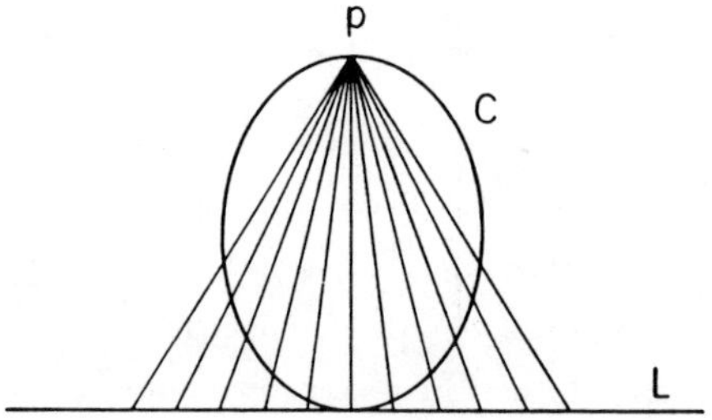

Thus projection yields a bijective mapping onto $L = P_1(\mathbb{C})$

$$C \to P_1(\mathbb{C}),$$

where the point p is mapped onto the point at infinity of $P_1(\mathbb{C})$. The inverse mapping

$$P_1(\mathbb{C}) \to C$$

is given, in the homogeneous coordinates (z_0, z_1) of $P_1(\mathbb{C})$ and (x_0, x_1, x_2) of $P_2(\mathbb{C})$, by

$$(z_0, z_1) \mapsto (z_0^2, z_0 z_1, z_1^2) = (x_0, x_1, x_2).$$

This is a bijective regular mapping with regular inverse. Thus the projective line and the irreducible quadric C are isomorphic as abstract curves, even though their embeddings in $P_2(\mathbb{C})$ are completely different and there is no collineation which carries a line to a quadric! In particular, the two curves are also the same as topological spaces resp. differentiable manifolds, i.e. the quadric is a 2-sphere. The real quadric corresponds, under the above isomorphism $P_1(\mathbb{C}) \to C$, to the real projective space $P_1(\mathbb{R}) \subset P_1(\mathbb{C})$, and hence to a great circle on the 2-sphere. This explains the nature of this curve.

Nevertheless, we shall go on to see what happens when we project from a point outside C, say from $p = (0,1,0)$ onto $x_1 = 0$. Affinely, this is projection in the direction of the x-axis onto the y-axis. Here again are the corresponding real-affine and real-projective pictures (schematic analogue pictures : we show only the real curves).

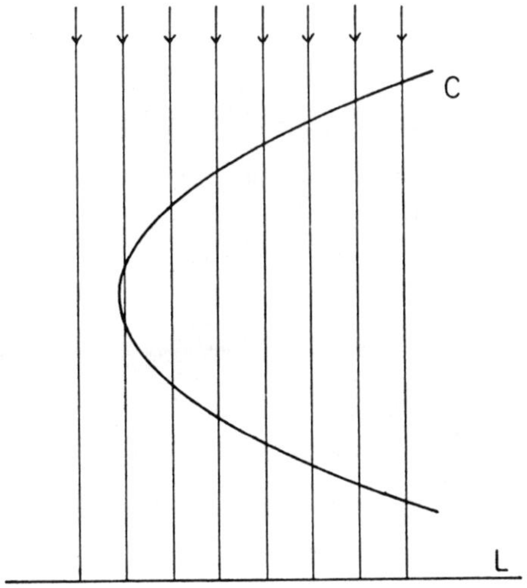
C
L

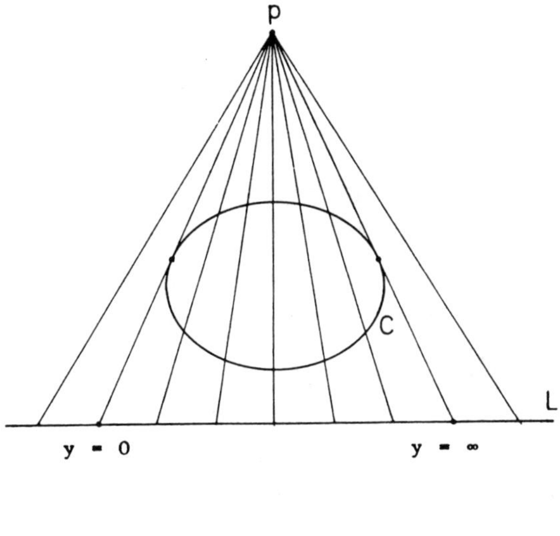
p
C
L
y = 0
y = ∞

If we use x as affine coordinate on C and y as affine coordinate on the y-axis L, then the projection is described by

$$y = x^2.$$

If we identify, as before, C with $P_1(\mathbb{C})$ in the homogeneous coordinates (z_0,z_1) and if we use the homogeneous coordinates (x_0,x_2) on the projective line L with equation $x_1 = 0$, then the mapping $C \to L$ induced by the projection becomes identified with the mapping

$$P_1(\mathbb{C}) \to P_1(\mathbb{C})$$

with

$$(z_0,z_1) \mapsto (x_0,x_2) = (z_0^2,z_1^2).$$

One sees immediately that this is a well-defined, regular and hence certainly complex analytic mapping of the Riemann sphere onto itself. Apart from the two points $(x_0,x_2) = (1,0)$ and $(x_0,x_2) = (0,1)$, each point of the image space has exactly two preimages, while $(1,0)$ and $(0,1)$ each have one. The latter are the branch points. In the affine coordinates $x = \frac{z_1}{z_0}$ resp. $y = \frac{x_2}{x_0}$ the mapping, as we have already said, is described by

$$y = x^2 ,$$

and in the affine coordinates $x' = \frac{z_0}{z_1}$ resp. $y' = \frac{x_0}{x_2}$ it is correspondingly described by

$$y' = x'^2.$$

As one says in function theory, the mapping $P_1(\mathbb{C}) \to P_1(\mathbb{C})$ is a two-sheeted branched covering of the Riemann sphere, branched over the points 0 and ∞.

We shall represent this mapping as intuitively as possible. To do this we first consider the complement of the point ∞. Thus we consider the mapping

$$\mathbb{C} \to \mathbb{C}$$

described by $y = x^2$. One can obtain a good intuitive impression of this mapping with the help of the following picture. The advantage of this picture is that one sees immediately how the preimage space $\mathbb{C}$

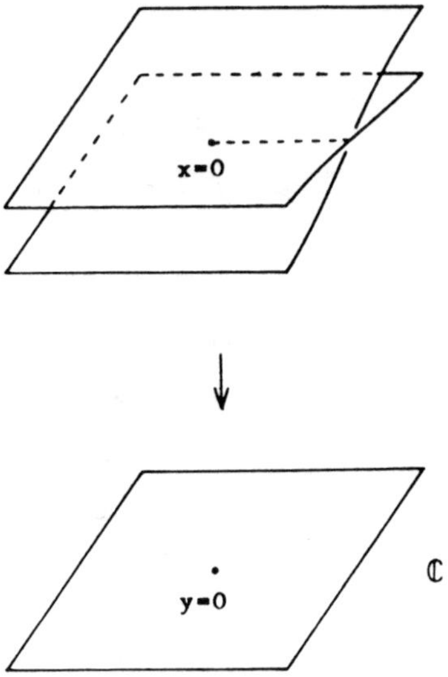

"lies over the image space $\mathbb{C}$ ", how the preimage space consists of two "sheets", and how one comes from one sheet to the other by travelling around the branch point $y = 0$. When the point x traverses the circle $x = e^{2\pi i\phi}$, $0 \leq \phi \leq 1$, once, then the image point $y = x^2 = e^{2\pi i 2\phi}$ traverses the circle around $y = 0$ twice, and indeed with twice the speed.

A certain disadvantage of the above picture is that in passing from one sheet to the other across the dotted line a self-intersection seems to occur. Of course this is only because we are trying to represent the graph of the mapping $\mathbb{C} \to \mathbb{C}$ with $y = x^2$ as a subset of 3-dimensional space, where in reality this graph, and indeed the curve C' with affine equation $y - x^2 = 0$ is in the real 4-dimensional space $\mathbb{C}^2$. Such self-intersections cannot be avoided with models in $\mathbb{R}^3$.

The "sheets" we spoke of above can be explicitly described quite simply, namely as the upper half plane H^+ and the lower half plane H^- of $\mathbb{C}$. The line separating them is the real axis, which is mapped by $y = x^2$ onto the positive real semi-axis. When we cut the image space along this positive real semi-axis and the preimage space along the real axis, the preimage space falls into two separate sheets, and the graph of our mapping looks as follows:

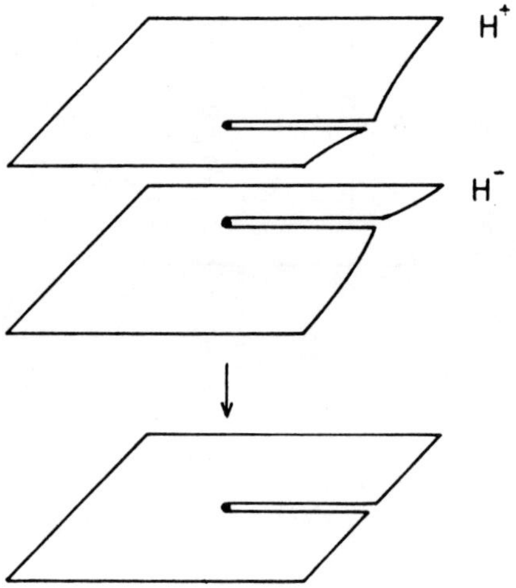

Conversely, we can reconstruct our double covering of $\mathbb{C}$ as follows : we cut the image space $\mathbb{C}$ along the positive real semi-axis. The resulting space is homeomorphic to a half plane, and the mapping $y = x^2$ yields a homeomorphism of precisely H^+ or H^- onto the cut plane. The boundary of the cut plane is a line, which is divided by the origin into two half lines, the "banks". We now take two copies of the cut plane, call them H^+ and H^-, and denote their banks by a resp. b as in the next figure.

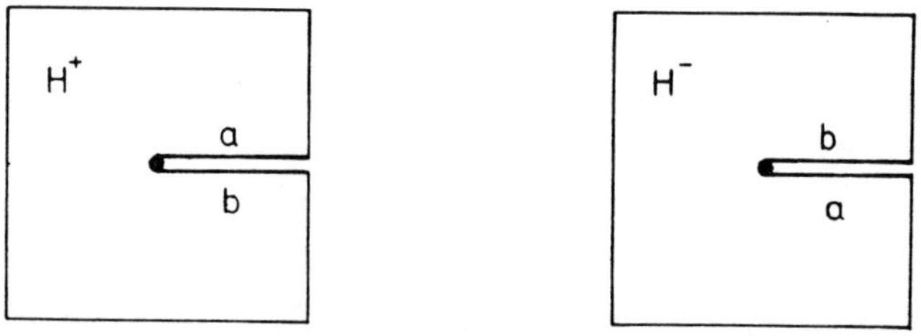

Then we paste bank a of H^+ to bank a of H^-, and bank b of H^- to bank b of H^+. The resulting space is of course the plane $\mathbb{C}$.

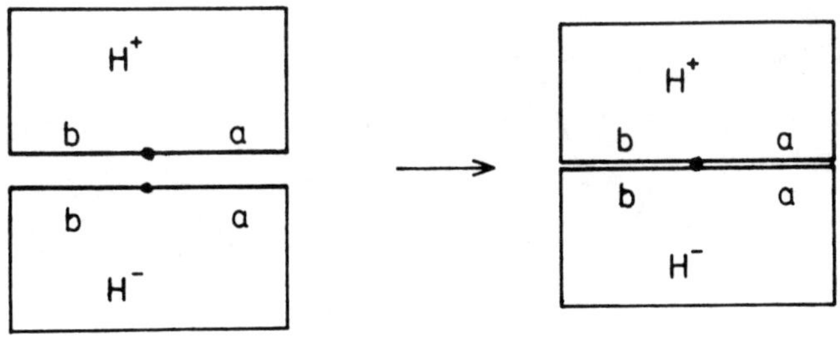

But H^+ and H^- were copies of the cut plane $\mathbb{C}$ and they therefore have canonical mappings onto $\mathbb{C}$. Thus we obtain a mapping of $\mathbb{C} = H^+ \cup H^-$ onto $\mathbb{C}$, and it is obviously the required double covering.

I hope that nobody is bewildered by the naive sounding terminology which refers to topological operations as "cutting" and "pasting". Naturally one can formulate all this in the rigorous language of set theoretic terminology – for example, the "pasting together" of two spaces is the passage to a suitable quotient space of the disjoint sum, with the corresponding quotient topology. However, the terminology of cutting and pasting – introduced by Riemann long before the invention of set theoretic topology – is so intuitive that I believe it is the most suitable for giving the correct impression of our branched coverings.

Now that we have thoroughly investigated the branched covering $\mathbb{C} \to \mathbb{C}$, there is no further difficulty in analysing the complete two-sheeted covering

$$P_1(\mathbb{C}) \to P_1(\mathbb{C})$$

of the Riemann sphere in the same way. In the image space we make a branch cut from the south pole 0 to the north pole ∞, in fact along the half of the great circle through the poles which corresponds to the positive real axis. If we cut the sphere along this semicircle, then the result is a space homeomorphic to a hemisphere.

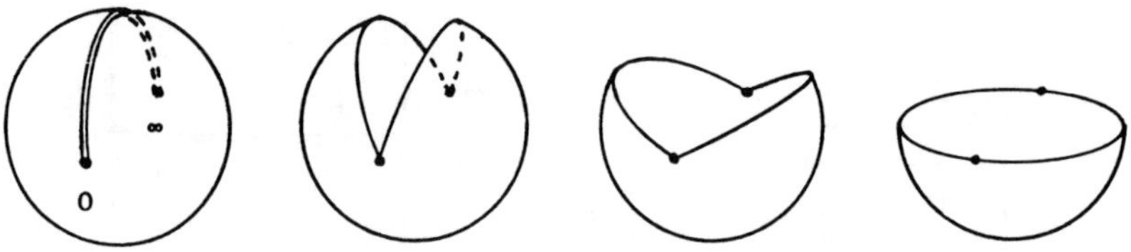

This cutting separates the preimage space into two sheets, namely the hemisphere into which the sphere $P_1(\mathbb{C})$ is divided by the great circle $P_1(\mathbb{R})$. Conversely, when one knows that the preimage space is a double-branched covering of $P_1(\mathbb{C})$ with two branch points over 0 and ∞, then one can reconstruct the covering space by pasting together two hemispheres, and conclude that the covering space must be a 2-sphere. In this way one can topologically analyse more general covering spaces, e.g. a curve C which is a branched covering of a projective line under projection. The following picture gives an intuitive impression of the two-sheeted branched covering

$$S^2 \to S^2$$

just considered.

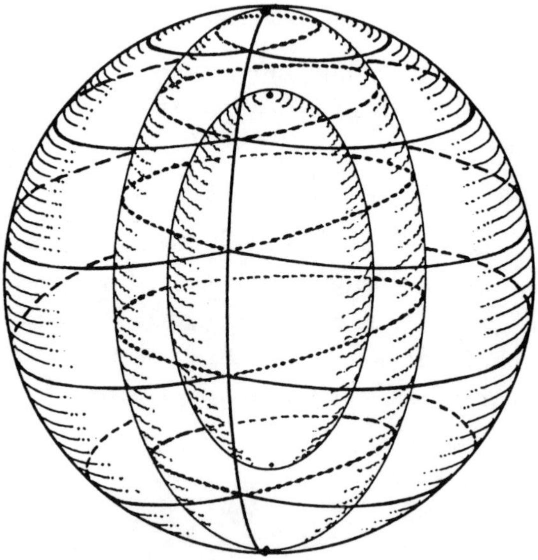

We shall now briefly explain why coverings such as the two-sheeted branched covering above are interesting not only topologically, but also function theoretically – we already know that they appear in algebraic geometry in a natural way, namely with the projection of curves onto lines.

We begin by considering again the affine curve $C' \subset \mathbb{C}^2$ with the equation $y - x^2 = 0$, and the double covering

$$C' \to \mathbb{C}$$

obtained by projection onto the y-axis. When we restrict any entire rational function on $\mathbb{C}^2$, e.g. the function x, to the 1-dimensional complex manifold C', then we obtain a well-defined holomorphic function x on C'. (Indeed we have seen before that one can take x as a coordinate on C'.) We now consider a point $y \in \mathbb{C}$ and the two points of C' lying over y. Then at these two points the function x has precisely the values

$$x = \pm \sqrt{y}.$$

Thus we see the following : on the y-plane a function $\sqrt{y}$ cannot, as is well known, be defined properly, i.e. as a single-valued analytic function of y, because when one travels once around the origin and analytically continues one of the two power series expansions, then one returns with the expansion with the opposite sign. But when we go to the double covering $C' \to \mathbb{C}$, $x = \sqrt{y}$ becomes a well-defined complex analytic function on C'. If one also admits the value ∞ for y, then one must likewise do so for x, i.e. one comes to precisely the branched covering

$$P_1(\mathbb{C}) \to P_1(\mathbb{C})$$

that we have analysed above, and on the covering space $x = \sqrt{y}$ becomes a well-defined meromorphic function. One calls this branched covering the Riemann surface of the many-valued "function" $\sqrt{y}$. In general one can consider a many-valued algebraic "function" $y(x)$ defined by an implicit equation

$$y^m + a_1(x)y^{m-1} + \ldots + a_m(x) = 0$$

where the $a_i(x)$ are polynomials in x. (In our case the equation was simply $y^2 - x = 0$.) One interprets the $a_i(x)$ as meromorphic functions on $P_1(\mathbb{C})$ and y as a many-valued meromorphic "function" on

$P_1(\mathbb{C})$. One now seeks a 1-dimensional compact complex manifold $\tilde{C}$ and a holomorphic mapping

$$\tilde{C} \to P_1(\mathbb{C})$$

as a branched covering so that y can be interpreted as a well-defined meromorphic function on $\tilde{C}$. (One hopes that there is such a thing, because when one takes an m-sheeted covering one has in general m points over each point of $P_1(\mathbb{C})$, at which the function y can take its m values.) Such a $\tilde{C}$, for which the sheet number of the covering is chosen as small as possible, always exists and is essentially unique. One calls this branched covering $\tilde{C} \to P_1(\mathbb{C})$ the Riemann surface of the many-valued "function" y.

Now this many-valued Riemann surface has a lot to do with algebraic curves in $P_2(\mathbb{C})$. Namely, the affine equation

$$y^m + a_1(x)y^{m-1} + \ldots + a_m(x) = 0$$

also defines a projective algebraic curve C in $P_2(\mathbb{C})$, and projection in the direction of the x-axis onto the y-axis defines a mapping

$$C \to P_1(\mathbb{C}).$$

When C is non-singular, as in the previous case of the quadric, then $C \to P_1(\mathbb{C})$ is already the Riemann surface of y. When C is singular, one must go to a non-singular normalisation $\tilde{C} \to C$, and the composition

$$\tilde{C} \to P_1(\mathbb{C})$$

is then the desired Riemann surface.

Summary : The connection between the algebraic geometric and function theoretic viewpoints in the investigation of algebraic curves is the following : the concrete Riemann surfaces $\tilde{C} \to P_1(\mathbb{C})$ of many-valued algebraic functions are the branched coverings of $P_1(\mathbb{C})$ which result from composition of a normalisation $\tilde{C} \to C$ and projection $C \to P_1(\mathbb{C})$ of an algebraic curve C. In addition, one can show that each abstract Riemann surface $\tilde{C}$, i.e. each compact 1-dimensional complex manifold, can be viewed as a branched covering $\tilde{C} \to P_1(\mathbb{C})$, i.e. as the concrete Riemann surface of a many-valued algebraic "function".

Of course we cannot here go precisely into all these things from function theory, and we must refer to textbooks on function theory such as those of Behnke and Sommer [B1] or Siegel [S5]. But perhaps it was useful to at least explain the connection between function theory and the theory of algebraic curves here.

We shall now study the quadrics from another viewpoint, and to do so we begin with the second affine normal form, the equation of the hyperbola

$$xy = 1.$$

More generally, we consider the whole family of quadrics defined by the affine equations

$$xy = c,$$

where c is a complex parameter. For $c \neq 0$ these quadrics are irreducible, while for $c = 0$ we have a reducible quadric, which decomposes into the two lines $x = 0$ and $y = 0$. The real images of these curves already suggest what happens in this family of quadrics as $c \to 0$. The irreducible quadrics are drawn towards the reducible quadric

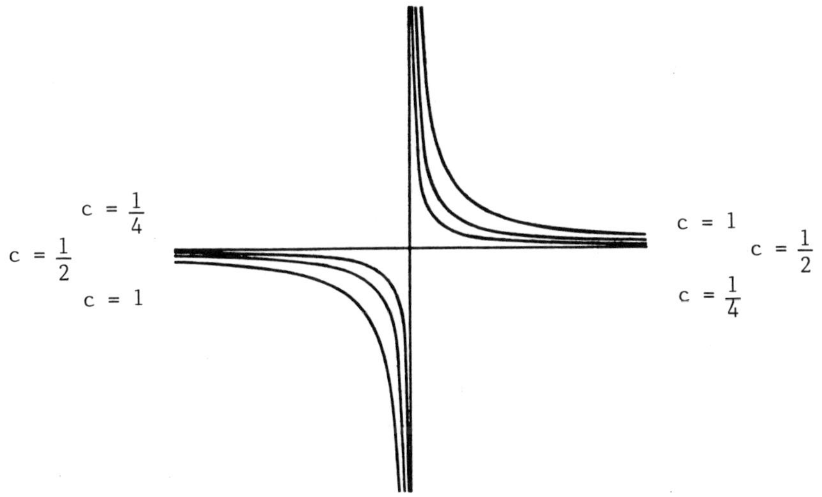

as $c \to 0$. In the process, the topological type changes : we already know that an irreducible complex projective quadric is homeomorphic to S^2, whereas a reducible one is homeomorphic to $S^2 \vee S^2$.

In order to see more precisely what is happening, we now use the third real normal form, that of the circle, and consider the family of complex projective quadrics with the affine equations

$$x^2 + y^2 = c^2,$$

$c \geq 0$ real.

We could again analyse this family of quadrics by viewing them as branched coverings of the y-axis with the branch points $y = \pm c$.

Instead of that we choose another method here, which brings out the relation between real and complex quadrics particularly clearly. In order to be able to denote real and imaginary parts of coordinates in the usual way by x and y, we change the notation for affine complex coordinates : we now call them z_1 and z_2 with

$$z_1 = x_1 + iy_1$$

$$z_2 = x_2 + iy_2 .$$

Thus we consider the family of affine quadrics

$$z_1^2 + z_2^2 = c^2.$$

This equation is a complex equation, equivalent to the following two real equations

$$x_1^2 + x_2^2 - y_1^2 - y_2^2 = c^2 \qquad (1)$$

$$x_1y_1 + x_2y_2 = 0. \qquad (2)$$

Let $Q_c \subset \mathbb{C}^2$ be the affine quadric defined by these two equations. What does this real 2-dimensional manifold look like ? In order to see, we project $\mathbb{C}^2$ onto $\mathbb{R}^2$, i.e. we consider the mapping

$$\mathbb{C}^2 \to \mathbb{R}^2$$

with

$$(x_1,y_1,x_2,y_2) \mapsto (x_1,x_2).$$

Because of (1), Q_c is mapped onto the region $x_1^2 + x_2^2 \geq c$ outside the circle with radius c which is in fact the associated real quadric. In order to analyse the mapping of Q_c onto this region B more precisely, we partition this region into the rays which are its intersections with lines through the origin.

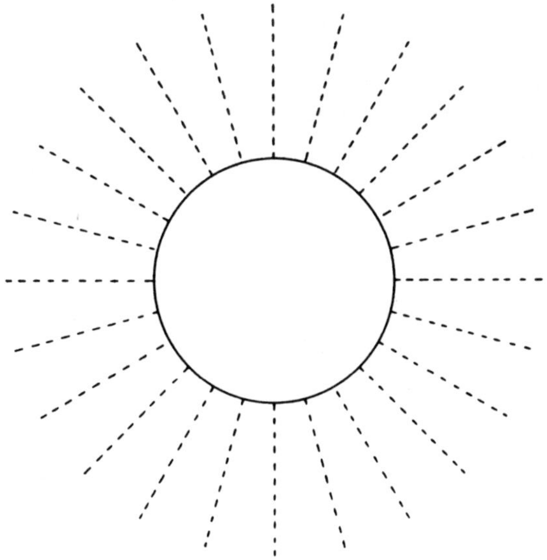

What is the appearance of the part of Q_c which is mapped by $Q_c \to B$ onto such a ray ?

Equation (2) says : the y coordinates of the preimage points of the ray through (x_1, x_2) lie on the line $x_1 y_1 + x_2 y_2 = 0$ in the (y_1, y_2)-plane. When we map this line into the (x_1, x_2)-plane by $y_1 \mapsto x_1$, $y_2 \mapsto x_2$ it becomes precisely the line through the origin perpendicular to the ray in question, and hence parallel to the tangent to the circle $x_1^2 + x_2^2 = c^2$ at the point where the ray meets the circle.

Equation (1) determines the distance ρ of the two points of Q_c on this tangent from the origin in terms of the distance r of (x_1, x_2) from the origin :

$$\rho^2 = y_1^2 + y_2^2$$

$$r^2 = x_1^2 + x_2^2 .$$

Then equation (1) says :

$$r^2 - \rho^2 = c^2.$$

This is the equation of a real hyperbola in the (r, ρ)-plane. The following picture suggests the situation. It shows the intersection of Q_c with the 3-dimensional space $y_1 = 0$, and consists of the real circle $x_1^2 + x_2^2 = c^2$ and the two hyperbola branches which lie over the positive and negative x_1-axes.

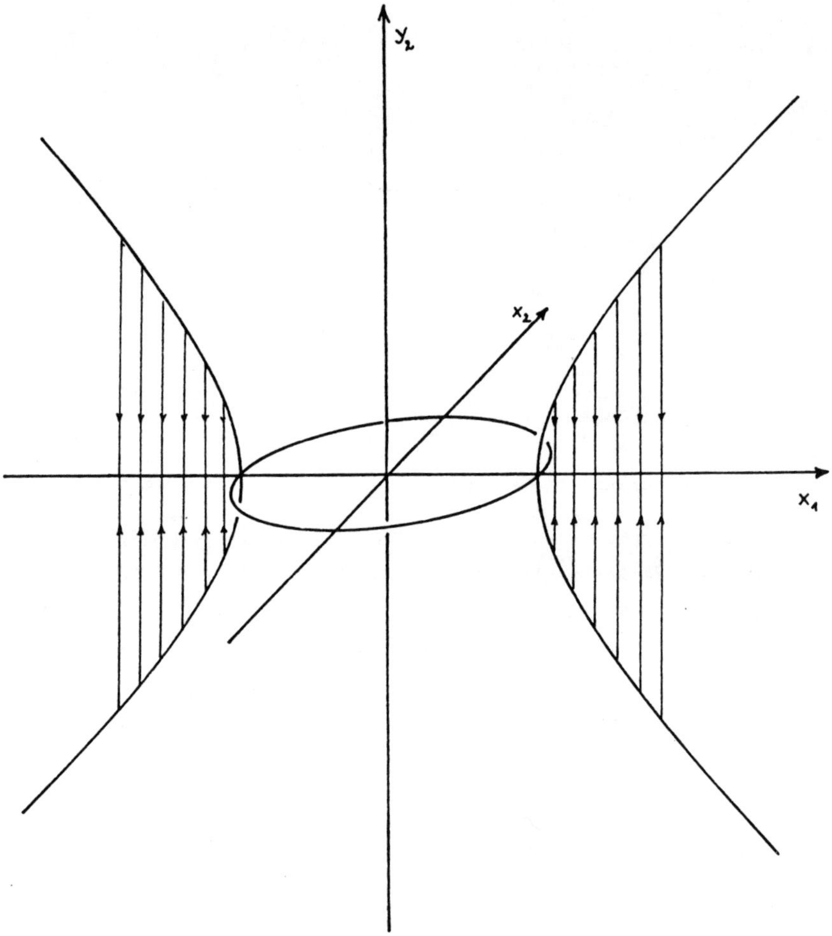

When we let the ray $x_2 = 0$, $x_1 > c$ just considered rotate in the (x_1, x_2)-plane it sweeps out the region B and traverses all the rays into which we have decomposed B. Over each line lies a real hyperbola branch in the affine quadric Q_c, as we have seen above. This brings us to the idea that our surface Q_c could be viewed topologically as a one-sheeted hyperboloid of revolution H_c, i.e. as the surface in $\mathbb{R}^3$ with the equation

$$u^2 + v^2 - w^2 = c^2.$$

And in fact one easily sees from our previous considerations how to explicitly define a homeomorphism of the complex affine quadric Q_c onto the hyperboloid H_c. The mapping

$$\phi_c : Q_c \to H_c$$

is defined by

$$u = x_1$$
$$v = x_2$$
$$w = \sqrt{y_1^2 + y_2^2}\ .$$

The sign of the root is determined as follows : the circle $x_1^2 + x_2^2 = c^2$ divides Q_c into two regions, namely the regions with $x_1y_2 - x_2y_1 > 0$ and $x_1y_2 - x_2y_1 < 0$. (This condition describes the decomposition of tangents to the circle into half tangents to left and right of the point of contact.) In one region we choose the positive root, in the other, the negative root. Then ϕ_c is well defined, and one sees easily that it is a homeomorphism. Since we can carry out the construction for each c, and since it depends continuously on c, we obtain a homeomorphism of the family of affine complex quadrics Q_c onto the family of one-sheeted hyperboloids of revolution H_c, so that we can describe the degeneration of the quadrics as $c \to 0$ topologically in terms of the corresponding degeneration of the family H_c.

But one sees immediately, for the family of hyperboloids of revolution, what happens as $c \to 0$: the "waist", i.e. the circle $x_1^2 + x_2^2 = c^2$, contracts to a point.

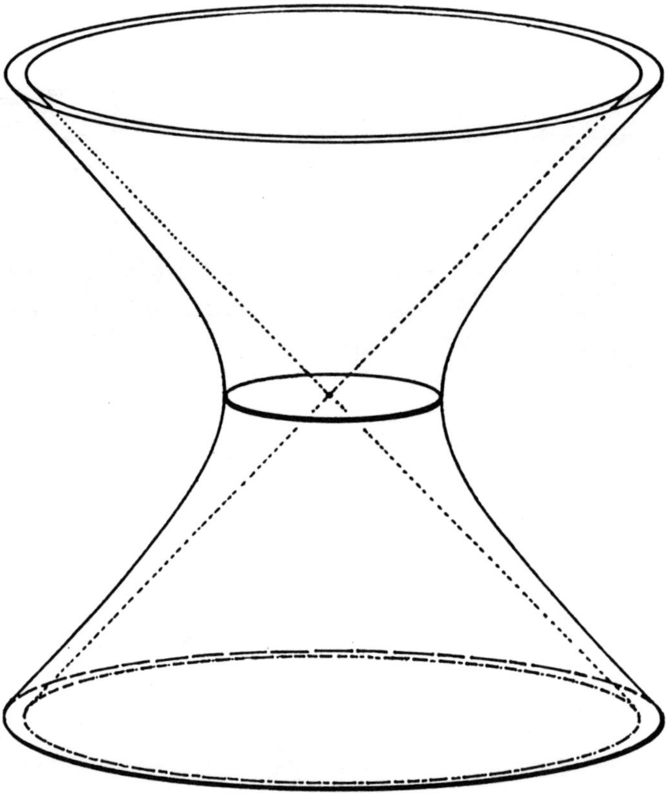

The circle, which has radius c and hence contracts to a point as $c \to 0$, is called a vanishing cycle.

We have now come to a better understanding of the degeneration, not only of the affine quadrics, but also of the complex projective quadrics. When one lets c tend to 0 in the family of complex projective quadrics

$$z_1^2 + z_2^2 - c^2 z_0^2 = 0$$

then the quadrics degenerate to a reducible quadric. In the process, the vanishing cycle contracts to a point. The non-degenerate quadrics are topological 2-spheres, with the vanishing cycle, the circle in the real affine plane

$$x_1^2 + x_2^2 = c^2,$$

a great circle on this sphere. As $c \to 0$ the equator of this sphere constricts and one obtains a bouquet $S^2 \vee S^2$. We have already established that a reducible quadric is topologically a bouquet.

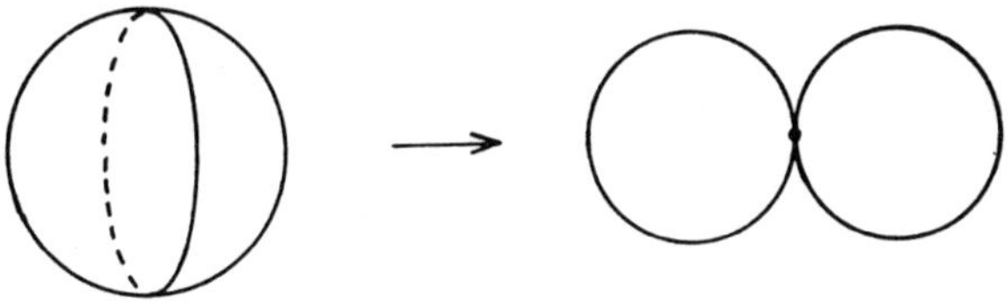

We now want to make an investigation similar to the one we have just made for a family of quadrics, but now for a family of cubics, namely the family of diverging cubic parabolas with the affine equation

$$z_2^2 - z_1^3 + z_1 - t = 0. \qquad (1)$$

Here, t is any complex number. We denote the cubic associated with t in the complex projective plane by C_t. By equating partial derivatives to zero one finds two values of t for which the curve C_t is singular, namely

$$t = \pm \frac{2\sqrt{3}}{9} .$$

In order to obtain a first impression of the course of the curves C_t, we again divide into real and imaginary parts $z_1 = x_1 + iy_1$ and $z_2 = x_2 + iy_2$. Equation (1) is then equivalent to

$$x_2^2 - y_2^2 - x_1^3 + 3x_1y_1^2 + x_1 = t \qquad (2)$$

$$2x_2y_2 - 3x_1^2y_1 + y_1^3 + y_1 = 0. \qquad (3)$$

If we now intersect the affine curve with the equation (1) in $\mathbb{C}^2$ with the real three-dimensional linear subspace $y_1 = 0$, then (2) and (3) give the equations

$$y_2 = 0 \quad \text{and} \quad x_2^2 - x_1^3 + x_1 = t$$

$$\text{or} \quad x_2 = 0 \quad \text{and} \quad -y_2^2 - x_1^3 + x_1 = t.$$

The following picture, which I received from the American mathematician Richard Bassein, shows these curves for various values of t. Curve branches which correspond to the same value of the parameter t are denoted by the same letters a, b, c, d, e. We have

$t > \frac{2\sqrt{3}}{9}$ for a

$t = \frac{2\sqrt{3}}{9}$ for b

$t = 0$ for c

$t = -\frac{2\sqrt{3}}{9}$ for d

$t < -\frac{2\sqrt{3}}{9}$ for e.

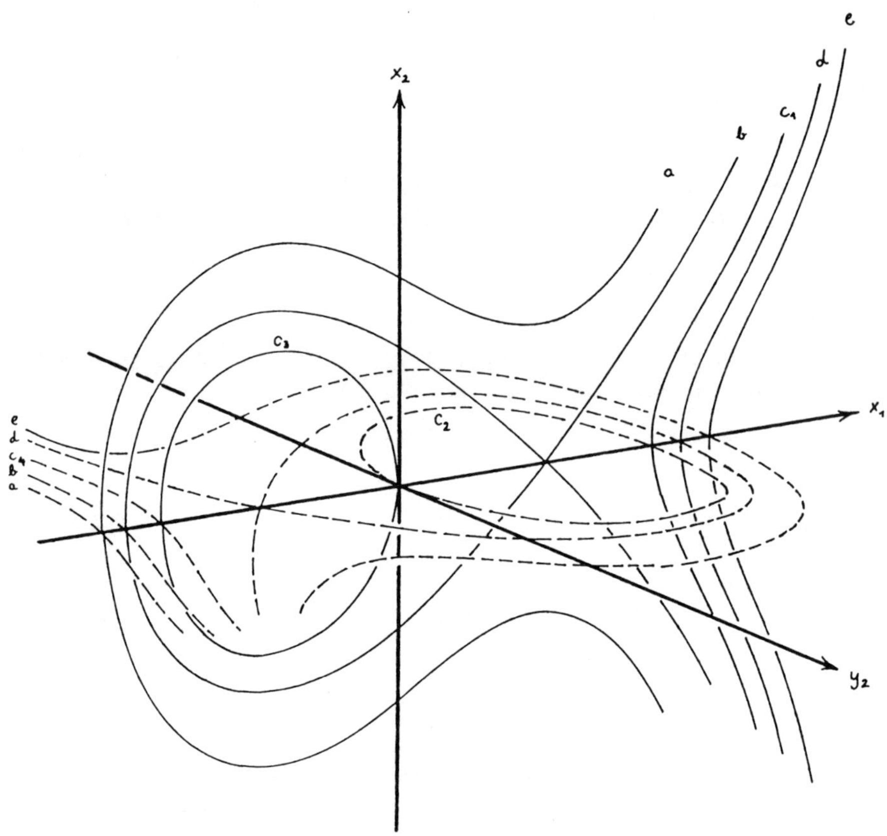

We shall now investigate the curves C_t topologically, again by suitable projection onto a projective line.

We first investigate one of the two singular curves, say the curve $C = C_t$ for $t = \frac{2\sqrt{3}}{9}$. The investigation of the other singular curve proceeds analogously, with analogous result. The picture of the real affine curve already suggests how to project most conveniently : from the singular point $p = (z_1, z_2)$ with $z_1 = \frac{1}{3}\sqrt{3}$, $z_2 = 0$ onto the y-axis $z_1 = 0$.

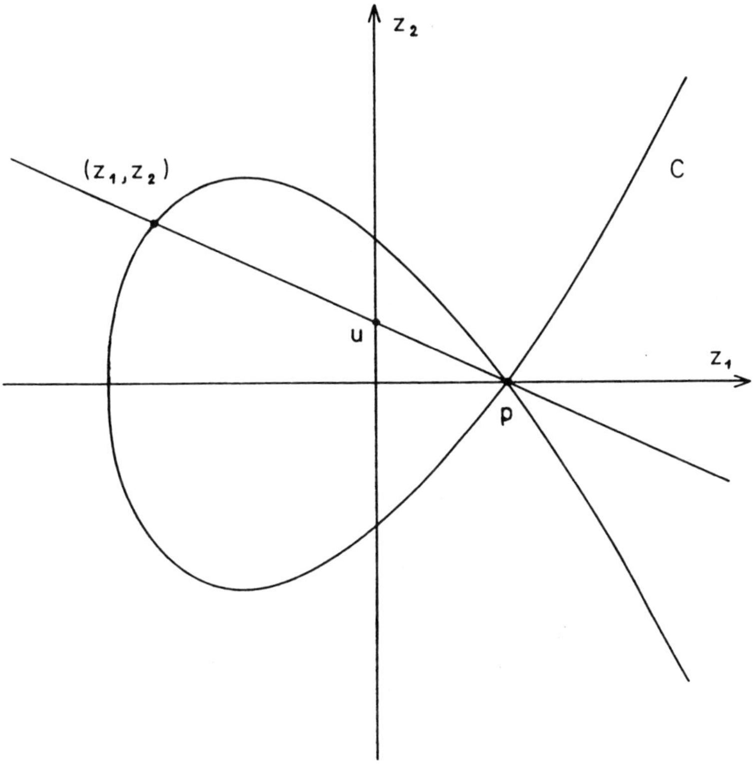

One immediately computes that the line through p and the point $(0,u)$ on the z_2-axis cuts the curve C in exactly one point other than p, namely the point (z_1,z_2) with

$$z_1 = 3u^2 - \frac{2}{3}\sqrt{3}$$

$$z_2 = -3\sqrt{3}\,u^3 + 3u.$$

Conversely, $u \mapsto (z_1,z_2)$ defines a holomorphic mapping of $\mathbb{C}$ onto the affine cubic. This admits an immediate extension to a holomorphic mapping

$$P_1(\mathbb{C}) \to C,$$

namely, in homogeneous coordinates :

$$(u_0,u_1) \mapsto (u_0^3, 3u_0u_1^2 - \tfrac{2}{3}\sqrt{3}\,u_0^3, -3\sqrt{3}\,u_1^3 + 3u_0^2u_1).$$

One easily sees that this mapping is almost bijective — more precisely, the following holds : if one removes the singular point p from C and the two points $p_{\pm} = (1, \pm 1/\sqrt{\sqrt{3}})$ from $P_1(\mathbb{C})$ then the mapping

$$P_1(\mathbb{C}) - \{p_+,p_-\} \to C - \{p\}$$

is bijective. The two points p_+ and p_- are the two preimages of p in $P_1(\mathbb{C})$. They are the two points at which the two tangents to C at p meet the z_2-axis. Thus we obtain :

<u>**Result**</u> : The singular cubic C with an ordinary double point p results from the Riemann sphere by identification of two points p_+ and p_- .

We can see this topological space intuitively in various ways, e.g. by one of the following pictures. The first picture suggests that one can also obtain this space, which results from S^2 by identification of

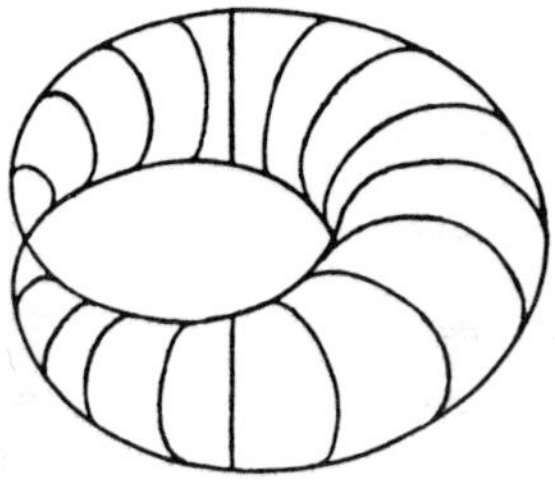

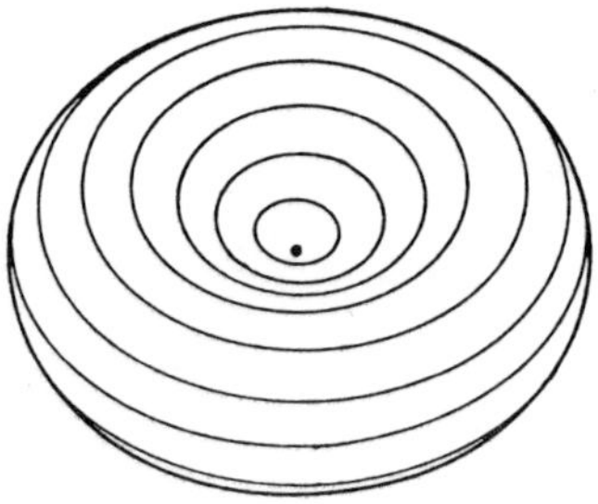

two points, by contracting a meridian on a torus to a point.

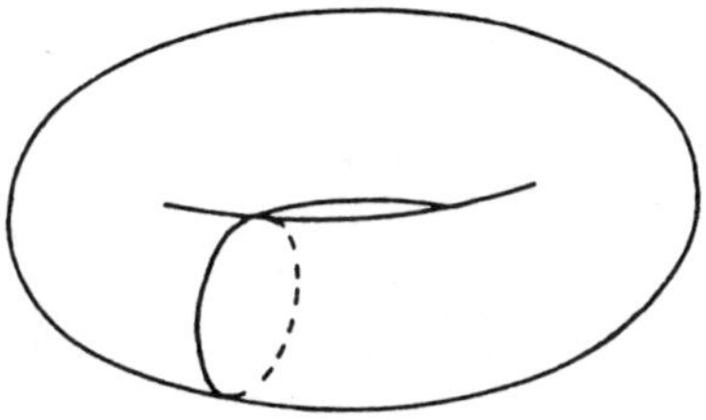

Since one can, of course, exchange meridian and latitude circles on a torus by exchange of the factors in $S^1 \times S^1$, one can equally well contract a latitude circle, say the inner equator.

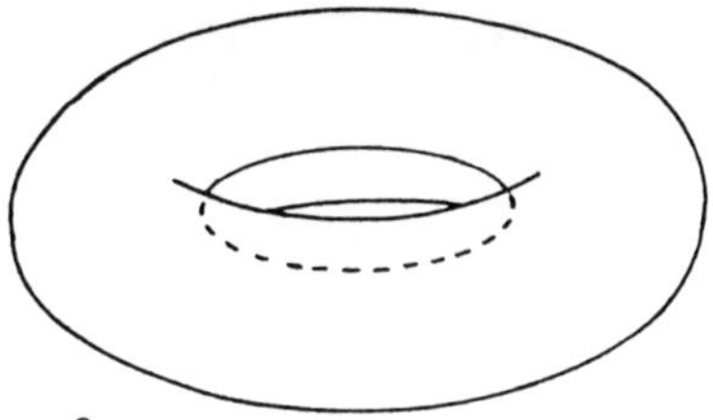

When one does this in $\mathbb{R}^3$, the result is the second of the above figures, which one can obtain by rotation of a figure 8.

We make these remarks at this stage because we shall see later that this contraction of meridian resp. latitude circles describes precisely the topological degeneration which occurs in our family of cubics C_t when the parameter t tends to the two singular values.

We shall now topologically analyse the non-singular cubics of our family. To do this we project C_t in the direction of the z_2-axis onto the z_1-axis. In coordinates this is the mapping

$$(z_1, z_2) \mapsto z_1.$$

This affine mapping may be extended to a holomorphic mapping

$$C_t \to P_1(\mathbb{C})$$

in which the point $(0,0,1)$ is sent to the point $(0,1)$: C_t then becomes a two-sheeted branched covering of the 2-sphere. The branch points (in affine coordinates) are obviously the three solutions $z_1 = e_i$ of the equation

$$z_1^3 - z_1 + t = 0,$$

together with the point $z_1 = \infty$, as one easily convinces oneself. The three branch points can be seen very nicely, by the way, on the curve denoted by c in Richard Bassein's picture.

Now we analyse the branched covering $C_t \to P_1(\mathbb{C})$ by the same method that we applied previously to quadrics. The position of the three points e_1, e_2, e_3 on the sphere is unimportant for the topological analysis. We can therefore identify the 2-sphere S^2 with the surface of a tetrahedron with vertices e_1, e_2, e_3 and ∞. Then we cut the tetrahedron along the three edges from e_i to ∞ and obtain a triangle.

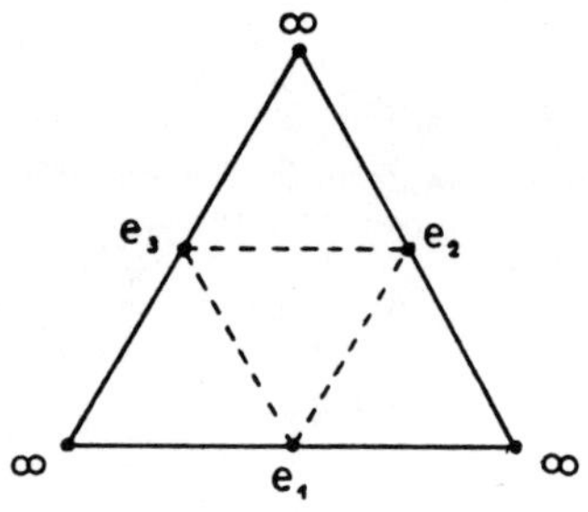

If we also cut the surface C_t doubly covering the 2-sphere along the preimages of the cut edges, then it divides into two pieces, two copies of the triangle above. (This is a general fact from topology : an m-sheeted unbranched covering of a simply connected region G decomposes simply into m copies of G , and we can apply this here, because our triangle of course is simply connected.)

Conversely, we can reconstruct the surface C_t by suitably pasting two copies of the triangle. Naturally the sides on which the corresponding points e_i lie must be identified. First we paste the sides with e_2 together. Then a parallelogram results.

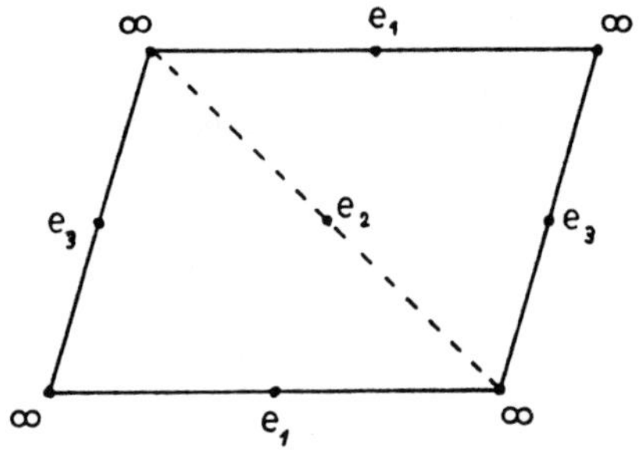

Then we identify the sides with e_1 – with the correct orientation of course – and the result is a cylinder.

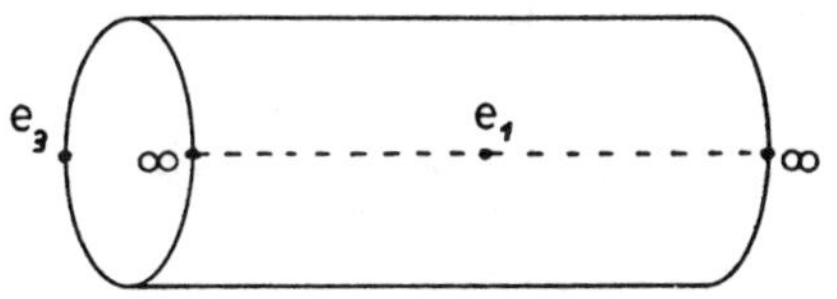

Finally, we identify the sides with e_3 , and the result is a torus.

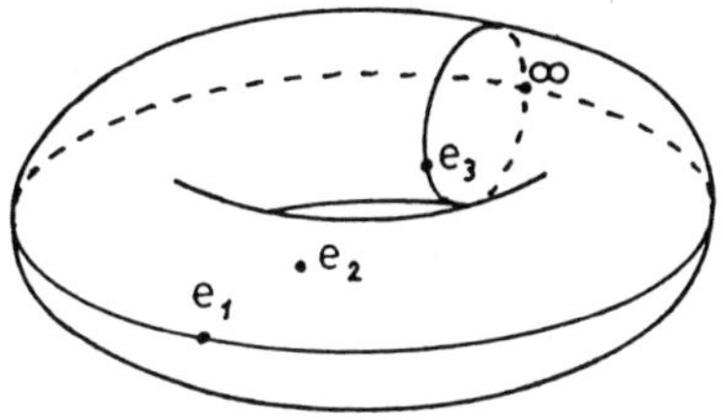

Result : The non-singular cubic curves C_t in $P_2(\mathbb{C})$ are homeomorphic to a torus.

It is natural to ask where the curves we see in Richard Bassein's picture lie on the torus. This question only acquires a precise meaning when we specify precisely how the curve C_t is to be identified with the torus, for example, by saying how the edges of the tetrahedron lie on the sphere. We shall do this for the parameter value $t = 0$. The curve C_0 cuts the real x-axis in the three points

$e_1 = -1$
$e_2 = 0$
$e_3 = +1$

and at ∞. We choose the segments of the real axis from ∞ to e_1, e_1 to e_2 , e_2 to e_3 and e_3 to ∞ as edges of our tetrahedron. The remaining edges do not interest us. When we identify C_0 with the torus in $\mathbb{R}^3$ as previously prescribed, then we obtain the following picture for the arrangement on the torus of the curves shown in Bassein's picture :

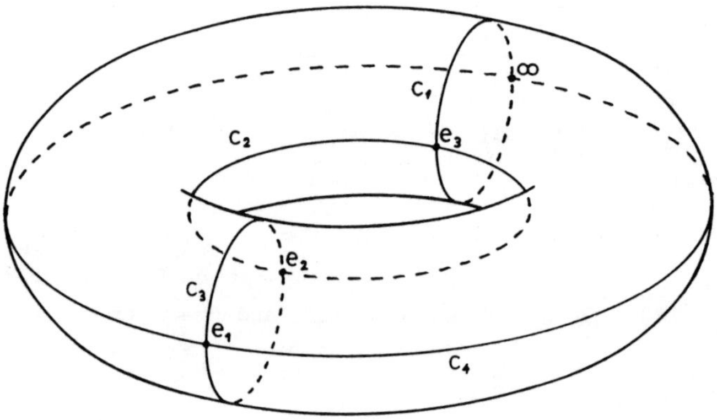

Result : If t runs from 0 to the critical value $+\frac{2\sqrt{3}}{9}$, then the latitude circle c_2 contracts to a point, and if t runs from 0 to the critical value $-\frac{2\sqrt{3}}{9}$, then the meridian circle c_3 contracts to a point.

In this way we have obtained a deeper understanding of what happens topologically when a family of non-singular cubics degenerates to an irreducible cubic with an ordinary double point. One can similarly investigate, e.g., how non-singular cubics degenerate to three lines in general position. In this case three meridians of a torus contract to points.

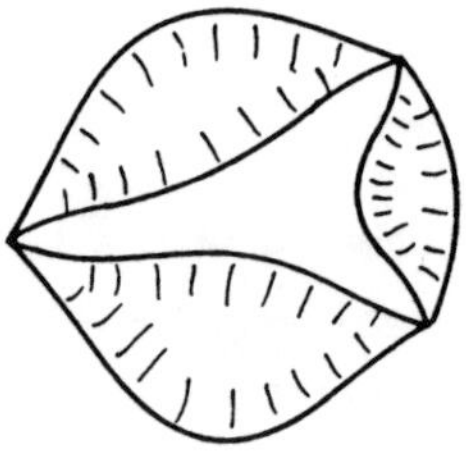

Another example would be the degeneration of a non-singular cubic into the semicubical parabola. In this case a meridian and latitude circle on the torus both contract. The result is a sphere, and this harmonises with the fact that the mapping

$$(u_0,u_1) \mapsto (u_0^3,u_1^3,u_0u_1^2)$$

is a homeomorphism of $P_1(\mathbb{C})$ onto the semicubical parabola $z_2^3-z_0z_1^2 = 0$.

Summary : We summarise the long discussion in this section on the geometry of curves in the complex projective plane.

One can treat curves algebraically – this will happen in Chapter II. One can treat them function-theoretically as abstract or concrete Riemann surfaces – a viewpoint to which we shall occasionally return (e.g. in 7.4). Finally, we have investigated the examples of quadrics and cubics topologically. In these examples we have seen that curves in the complex projective plane, viewed as topological spaces, result from compact orientable surfaces, possibly with the identification of finitely many points of the surfaces. The genus of these surfaces is the most fundamental invariant. Later we shall derive formulae, the Plücker formulae, for the computation of genus.

We have also seen examples of the way in which the topological type – the genus – of a family of curves alters when a curve acquires a singularity. The Plücker formulae cover this phenomenon.

We have seen in addition that one can interpret this alteration of topological type as the contraction of vanishing cycles. This viewpoint plays an important role in algebraic geometry today, in the topological investigation of algebraic manifolds (Lefschetz [L1], Griffiths [G2], Deligne [D1], Milnor [M1], Brieskorn [B6].

With this prospect we conclude this introductory chapter.

II. INVESTIGATION OF CURVES BY ELEMENTARY ALGEBRAIC METHODS

In Chapter I of this course we have seen how the historical development of the theory of plane curves led to the investigation of curves in the complex projective plane which are the zero sets of homogeneous polynomials. In this second chapter we want to begin such an investigation – from an analytic geometric standpoint – using essentially only elementary algebraic methods, namely simple facts about computation with polynomials.

In the later chapters we shall then carry out deeper and more extensive investigations – likewise from an analytic geometric standpoint and mostly by algebraic methods – guided strongly, however, by qualitative geometric viewpoints.

4. Polynomials

In this section we collect a few simple facts about computation with polynomials which we shall use regularly in what follows. We confine ourselves only to essentials, and for details refer to algebra textbooks, e.g. [Wl] or [Rl].

An affine algebraic curve C in the complex affine plane $\mathbb{C}^2$ is the zero set of an equation $f(x,y) = 0$, where $f(x,y) = \sum_{i+j\leq n} a_{ij}x^iy^j$ is a polynomial in the two indeterminates x and y. If C' is another curve with equation $g(x,y) = 0$, where g is likewise a polynomial, then we can consider the polynomial $f\cdot g$, and the zero set of $f\cdot g(x,y)$ is obviously the curve $C \cup C'$. The decomposition of polynomials into factors corresponds to the decomposition of curves into components. For the algebraic treatment of this decomposition into components it is therefore necessary to investigate divisibility and factorisation properties of polynomials. The investigation of polynomials in several indeterminates reduces largely to the investigation of polynomials in one indeterminate, because, e.g., one can regard the polynomial f with two indeterminates x, y and complex coefficients

a_{ij} as a polynomial in the single indeterminate x

$$a_0(y)x^n + a_1(y)x^{n-1} + \ldots + a_n(y),$$

where the coefficients $a_i(y)$ are in turn polynomials in the indeterminate y with complex coefficients.

4.1 Decomposition into prime factors

Let R be a commutative integral domain with unit (e.g. a polynomial ring over a field).

Definition: Let $p \in R$, $p \neq 0$ be a non-unit.

(i) Then p is called irreducible when $p = a \cdot b$ implies either a or b is a unit in R. (Instead of irreducible one also says indecomposable.)

(ii) p is called a prime element if, whenever p divides a product $a \cdot b$, p divides one of the factors a or b.

Remark : Each prime element is irreducible.

Proof : When $p = a \cdot b$, p divides a factor, say $a = \alpha p$. Cancellation of p gives $\alpha \cdot b = 1$, so that b is a unit. Q.E.D.

The converse does not hold in general. For example, in $\mathbb{C}\{x,y,z\}/(z^2-xy)$ the class of z is irreducible but not a prime element, because z divides xy, but neither x nor y. This, together with the natural question of existence and uniqueness of decomposition into indecomposable factors, leads us to consider the following factorisation properties for the ring R (cf. Scheja [R1], §14).

Definition : Let R be a commutative integral domain with 1 and let $a \in R$ be any non-zero non-unit. R has property

- F, when each a is a product of finitely many prime elements;
- F_0, when each a is a product of finitely many irreducible elements;
- F_1, when F_0 holds and the decomposition into irreducible factors is unique up to order and multiplication by units;
- F_2, when each irreducible element is prime.

Proposition 1

$$F \Leftrightarrow F_1 \Leftrightarrow F_0 \text{ and } F_2.$$

The proof is simple (cf. e.g. Scheja [R1], Theorem 160).

Definition : A ring with these properties is called factorial (also : UFD, unique factorisation domain).

Example : In $R = \mathbb{C}\{x,y,z\}/(z^2-xy)$, z^2 has two essentially different decompositions into irreducible elements : $z^2 = z\cdot z = x\cdot y$. Thus R is not a UFD. However, in the following sections we shall see that the rings which interest us, namely polynomial rings over fields, are UFD's.

4.2 Divisibility properties of polynomials

Let A be a commutative integral domain with 1. We consider polynomials in an indeterminate x with coefficients from A

$$a_0x^n + a_1x^{n-1} + \dots + a_n \;,\; a_i \in A,\; a_0 \neq 0 \;;$$

n is called the degree of the polynomial. The collection of all such polynomials forms a ring, the polynomial ring A[x], which is obviously also an integral domain. We shall investigate divisibility and factorisation properties in this ring.

We first assume that A is a field so that one can divide by all non-zero $b \in A$. Then the most important fact concerning A[x] is the existence of a division algorithm. If

$$f = a_0x^n + a_1x^{n-1} + \dots + a_n \quad \text{and} \quad g = b_0x^m + \dots + b_m \neq 0$$

are polynomials of degrees n, m respectively (with $n \geq m$) then there are unique polynomials q and r with

$$f = q\cdot g + r$$

where degree r < degree g or r = 0. The polynomial r is the "remainder" on division of f by g. It is clear how the algorithm for carrying out this division with remainder proceeds : the term of highest degree in q is $\frac{a_0}{b_0}x^{n-m}$. One subtracts $\frac{a_0}{b_0}x^{n-m}\cdot g$ from f and obtains a polynomial of lower degree, then divides by g again, etc.

Rings which admit such a division with remainder are called euclidean rings.

We now use the division algorithm in order to determine the greatest common divisor of two polynomials $a, a' \in A[x]$ (euclidean algorithm) :

We iterate division with remainder :

$$a = qa' + a''$$
$$a' = q'a'' + a'''$$
$$\cdots$$
$$a^{(k-2)} = q^{(k-2)}a^{(k-1)} + a^{(k)}$$

until the remainder $a^{(k)}$ becomes a constant in A. Now if b divides both a and a', then it also divides a'', and hence also a''', etc. Thus it follows that if $a^{(k)} \neq 0$ then a, a' have no proper (i.e. non-constant) common divisor, they are "relatively prime". If $a^{(k)} = 0$, then $a^{(k-1)}$ is the greatest common divisor of a and a'.

It follows immediately from the above algorithm that a'' is a linear combination of a, a' with coefficients in $A[x]$, hence so too is a''', etc. Thus : if c is the greatest common divisor of a and a' there are polynomials u and v with

$$ua + va' = c.$$

In particular, for relatively prime a, a' there are polynomials u, v with

$$ua + va' = 1.$$

We can use this to show :

Proposition 1

A polynomial ring over a field is factorial.

Proof : F_0 obviously holds. We prove F_2. Let p be irreducible and suppose $p|a \cdot b$ but $p \nmid a$. By $p \nmid a$ and irreducibility of p, a and p are relatively prime, and hence there are u, v with

$$ua + vp = 1$$

hence

$$uab + vpb = b.$$

Then since $p|a \cdot b$ we also have $p|b$. Q.E.D.

Example : Let $f \in \mathbb{C}[x]$ be a polynomial of nth degree, $f = x^n + \ldots$ and let $a_1, \ldots, a_n \in \mathbb{C}$ be its zeroes. Then $f = (x-a_1)(x-a_2)\ldots(x-a_n)$ is the decomposition of f into irreducible factors.

Remark : One sees from this example that there is no general algorithm for computing the prime factor decomposition.

We now give up our restriction to polynomial rings over a field and consider the general problem of deciding when $A[x]$ is factorial for an integral domain A. Of course this cannot be the case unless A is itself factorial. However, this is also sufficient, as one proves by reduction to $Q[x]$, where Q is the quotient field of A : in order to decompose a polynomial in $A[x]$ into prime factors, one first divides it by the g.c.d. of its coefficients and obtains a "primitive" polynomial, i.e. one with relatively prime coefficients. Then one separately decomposes the g.c.d. into prime factors in A and it remains to find a prime factorisation for primitive polynomials. But for a primitive polynomial $f \in A[x]$ one can show without difficulty that f is irreducible resp. prime in $A[x]$ if and only if it is irreducible resp. prime in $Q[x]$. (See [R1], Theorem 180.) Since it is easy to see that each primitive polynomial f in $A[x]$ is a product of finitely many primitive irreducible polynomials f_i , we thereby also obtain a prime factor decomposition of f, because the f_i are irreducible in $A[x]$, hence in $Q[x]$, hence prime in $Q[x]$, hence prime in $A[x]$. Thus one has the following theorem of Gauss.

Theorem 2

A polynomial ring $A[x]$ is factorial if and only if A is factorial.

Corollary : A factorial, $x_1,\dots,x_n$ indeterminates $\Rightarrow$ $A[x_1,\dots,x_n]$ factorial. In particular, polynomial rings over fields are factorial.

The theoretically proven possibility of decomposing each polynomial in $A[x_1,\dots,x_n]$ into primes of course says nothing about how one may find the decomposition for a given polynomial f, in other words, how one finds the prime divisors of f. Certainly, one can use the division algorithm to establish whether a given polynomial g divides f, but in general one must try infinitely many g to find the divisors of f, e.g. for $f \in \mathbb{C}[x]$, all $g(x) = x-a$ where a is an arbitrary complex number. When A is a UFD in which prime factor decomposition can be carried out in finitely many steps, and which has only finitely many units, then the same is true for $A[x]$, as one easily shows by an argument of Kronecker (see e.g. van der Waerden [W1], §30). Thus, e.g., the prime factor decomposition in $\mathbb{Z}[x_1,\dots,x_n]$, and hence also in $\mathbb{Q}[x_1,\dots,x_n]$, can be carried out in finitely many steps. But for the case in which we are particularly interested, namely $\mathbb{C}[x_1,\dots,x_n]$, this unfortunately does not hold (cf. also Scheja [R1]).

We now consider the question of when two polynomials $f, g \in A[x]$ have a common divisor, under the assumption that A is a UFD.

Let

$$f = a_0x^m + a_1x^{m-1} + \dots + a_m$$

$$g = b_0x^n + b_1x^{n-1} + \dots + b_n .$$

We admit the possibility that either a_0 or b_0 vanish. In order to establish whether f and g have a non-constant common divisor, one can of course apply the euclidean algorithm when A is a field, and it will indeed yield the greatest common divisor. However, if we do not want the g.c.d. itself, but only to know whether it is non-constant, then it suffices to establish whether the constant $a^{(k)}$ appearing last in the algorithm vanishes or not. It is clear from the description of the algorithm that this constant $a^{(k)}$ is a uniquely defined rational expression in the coefficients a_i and b_j of the two polynomials which must vanish when there is a non-constant g.c.d. . It is clear from this that there must be a polynomial $R(a,b)$ in the coefficients a_i and b_j such that $R(a,b) = 0$ if and only if f and g have a common divisor.

In order to compute this polynomial explicitly, we could analyse the euclidean algorithm. However, that would be troublesome. Instead, we proceed as follows.

Suppose that f and g have a non-constant common divisor h. Thus we have

$$f = u \cdot h$$

$$g = v \cdot h$$

where u, v are polynomials with degree $u < m$, degree $v < n$, say

$$u = c_0x^{m-1} + c_1x^{m-2} + \dots + c_{m-1}$$

$$v = d_0x^{n-1} + d_1x^{n-2} + \dots + d_{n-1} ,$$

and

$$vf - ug = 0 \quad (u, v \neq 0) .$$

The existence of this equation is therefore necessary for the existence of a non-constant common divisor. However, it is also sufficient, because the prime divisors of f must then all appear in $u \cdot g$ as well, but they cannot all appear in u, since degree $u <$ degree f, when

$a_0 \neq 0$. Thus a prime divisor of f divides the polynomial g. Multiplying out the above equation and comparing coefficients yields the following linear equations for the coefficients c_i and d_j :

$$\begin{array}{llllllll}
a_0d_0 & & & & -b_0c_0 & & & = 0 \\
a_1d_0 + a_0d_1 & & & & -b_1c_0 & -b_0c_1 & & = 0 \\
\cdots\ \cdots\ \cdots & & & & \cdots & \cdots & & \\
\cdots\ \cdots\ \cdots & & & & -b_{m-1}c_0 & \cdots & -b_0c_{m-1} & = 0 \\
a_md_0\ \cdots\ \cdots + a_0d_m & & & & -b_mc_0 & \cdots & -b_1c_{m-1} & = 0 \\
& + a_md_1\ \cdots\ \cdots + a_0d_{m+1} & & & \cdots & \cdots & \cdots & \cdots \\
& \cdots\ \cdots\ \cdots & & & \cdots & \cdots & \cdots & \cdots \\
& & \cdots\ \cdots & + a_0d_{n-1} & \cdots & \cdots & \cdots & \cdots \\
& & \cdots & + a_1d_{n-1} & -b_nc_0 & \cdots & \cdots & \cdots \\
& & & \cdots & & \cdots & \cdots & \cdots \\
& & & + a_md_{n-1} & & & -b_nc_{m-1} & = 0
\end{array}$$

For these equations to have a non-trivial solution it is necessary and sufficient that the determinant of the coefficients vanish. By multiplying columns by -1 and transposition one obtains the determinant in the form :

$$R_{f,g} = \left|\begin{array}{ccccccccc}
a_0 & a_1 & a_2 & \cdots & a_m & & & & \\
& a_0 & a_1 & a_2 & \cdots & a_m & & & \\
& & a_0 & a_1 & a_2 & \cdots & a_m & & \\
& & & \cdots & \cdots & & & & \\
& & & & \cdots & \cdots & & & \\
& & & & a_0 & a_1 & a_2 & \cdots & a_m \\
b_0 & b_1 & \cdots & b_{n-1} & b_n & & & & \\
& b_0 & b_1 & \cdots & & b_n & & & \\
& & b_0 & b_1 & \cdots & & b_n & & \\
& & & \cdots & & \cdots & & & \\
& & & & b_0 & b_1 & \cdots & & b_n
\end{array}\right| \begin{array}{l} \left.\vphantom{\begin{array}{c}1\\1\\1\\1\\1\\1\end{array}}\right\} n \\ \left.\vphantom{\begin{array}{c}1\\1\\1\\1\\1\end{array}}\right\} m \end{array}$$

__Definition__ : $R_{f,g}$ is called the resultant of $f = a_0x^m + \ldots + a_m$ and $g = b_0x^n + \ldots + b_n$.

We have proved :

__Theorem 3__

Let A be a UFD and let

$$f = a_0x^m + \ldots + a_m$$

$$g = b_0x^n + \ldots + b_n$$

be polynomials from $A[x]$ with $a_0 \neq 0$ or $b_0 \neq 0$. Then f and g have a non-constant common divisor if and only if the resultant R of f and g vanishes.

Remark : The resultant R also arises as follows : Suppose that f and g have a common zero x_0 , so that

$$a_0x_0^m + a_1x_0^{m-1} + \ldots + a_m = 0$$

$$b_0x_0^n + b_1x_0^{n-1} + \ldots + b_n = 0.$$

By multiplying these equations by suitable powers of x_0 and polynomial expressions in a_i, b_j and subtracting, one can successively eliminate all powers of x_0 until what remains is a polynomial in the a_i and b_j which must vanish, namely the resultant R. This means : the equation $R = 0$ results from the two equations above by elimination. It is a necessary condition for the existence of a common zero $x_0 \in A$ (because in that case $x - x_0$ is a common divisor).

One can also carry out an analogous elimination process for the equations

$$a_0x^m + a_1x^{m-1} + \ldots + a_m = f$$

$$b_0x^n + b_1x^{n-1} + \ldots + b_n = g.$$

Then one obtains, instead of the equation $R = 0$, an equation $R = Pf + Qg$ with polynomials P and Q. In this way one then proves (exercise, see van der Waerden [W1], §34) the following :

Proposition 4

Let A be a UFD, and let $f,g \in A[x]$ with $f,g \notin A$. Then for $f = a_0x^m + \ldots + a_m$ and $g = b_0x^n + \ldots + b_n$ with resultant R there are polynomials P, Q with degree $P < n$ and degree $Q < m$ such that

$$R = Pf + Qg.$$

The coefficients of P and Q are integral polynomials in the a_i and b_j which depend only on the numbers m, n. R is an integral homogeneous polynomial in the a_i and b_j of degree $m+n$, of degree n in the a_i and degree m in the b_j.

Finally, we mention yet another description of the resultant for the case when f and g decompose into linear factors. (For the simple proof we refer to van der Waerden [W1], §35.)

Proposition 5

Let $f = a_0(x-x_1) \ \ldots \ (x-x_m)$

$g = b_0(x-y_1) \ \ldots \ (x-y_n)$

and let R be the resultant of f and g. Then

(i) $$R = a_0^n b_0^m \prod_{i,j} (x_i - y_j)$$

(ii) $$R = a_0^n \prod_i g(x_i)$$

(iii) $$R = (-1)^{m \cdot n} b_0^m \prod_j f(y_j).$$

Remarks : In (i), R is symmetric in the x_i and y_i hence – as we expect – a polynomial in the elementary symmetric functions of the x_i and y_j respectively, hence in the coefficients of f and g.

Formula (ii) also holds when g does not decompose into linear factors, and (iii) holds when f does not decompose.

Corollary 6

Two polynomials $f, g \in \mathbb{C}[x]$ have a common zero if and only if their resultant vanishes.

In 6.1 we shall see the usefulness of the resultant of two polynomials in establishing the existence of common components of two curves and in calculating the number of intersection points of two curves with the equations $f = 0$ and $g = 0$. In order to see an application of the resultant immediately, we show how one can use it to find by elimination the equation of a rational curve given by a parametrisation

$$x = \frac{P_1(t)}{Q_1(t)}$$

$$y = \frac{P_2(t)}{Q_2(t)} .$$

(We have already seen this in connection with the presentation of equations for epicyclic curves in 2.3.)

Let f and g be the following polynomials in the indeterminate t with coefficients in $\mathbb{C}[x,y]$.

$$f = P_1(t) - xQ_1(t)$$

$$g = P_2(t) - yQ_2(t),$$

and let R be their resultant.

For (x_0,y_0) to be an image point of $t \in \mathbb{C}$ it is necessary that

$$f = 0$$
$$g = 0.$$

For this it is necessary that $R(x_0,y_0) = 0$. Thus

$$R(x,y) = 0$$

is the desired equation of the curve.

Example :

$$x = t^2$$
$$y = t^3 - t.$$

Of course in this case one can eliminate t immediately:

$$t = x^{\frac{1}{2}}$$
$$y = x^{\frac{1}{2}}(x-1)$$
$$y^2 = x(x-1)^2.$$

Hence the desired equation is

$$y^2 - x^3 + 2x^2 - x = 0.$$

However, we can also construct the resultant

$$R = \begin{vmatrix} 1 & 0 & -x & 0 & 0 \\ 0 & 1 & 0 & -x & 0 \\ 0 & 0 & 1 & 0 & -x \\ 1 & 0 & -1 & -y & 0 \\ 0 & 1 & 0 & -1 & -y \end{vmatrix} = y^2 - x^3 + 2x^2 - x$$

and with $R = 0$ we obtain the same equation.

To conclude, we briefly consider the important special case of the resultant of f and g when g is the formal derivative f' of the polynomial f, thus

$$f = a_0x^m + a_1x^{m-1} + \dots + a_m$$
$$f' = ma_0x^{m-1} + (m-1)a_1x^{m-2} + \dots + a_{m-1}.$$

Definition : $R_{f,f'}$ is the discriminant D_f of f.

Since f and f' have a proper common factor if and only if f has a multiple factor, it follows from Theorem 3 that

Corollary 7

A polynomial f has a multiple factor if and only if its

discriminant vanishes.

It is important to understand clearly the geometric meaning of the discriminant. We shall achieve this by considering the discriminant of polynomials of the third and fourth degree.

We first consider polynomials of degree 3 with real resp. complex coefficients, $x^3 + a_1x^2 + a_2x + a_3$. By a coordinate transformation $x = x' - a_1/3$ (a "Tschirnhaus-transformation") one can always arrange that the coefficient of the second highest power vanishes. Thus without loss of generality we assume that our polynomial f is of the following form :

$$f = x^3 + bx + a.$$

Eliminating x between

$$x^3 + bx + a = 0$$
$$3x^2 + b \quad = 0$$

yields the discriminant equation $R_{f,f'} = 0$

$$27a^2 + 4b^3 = 0.$$

This is the equation of a semicubical parabola in the (a,b)-plane (i.e. in $\mathbb{R}^2$ when we are discussing polynomials with real coefficients, and in $\mathbb{C}^2$ in the complex case). For the sake of clearness we carry out the discussion which follows for the real case.

Now it is important to analyse not just a single polynomial $x^3 + bx + a$, but to allow the parameter (a,b) to vary in the plane $\mathbb{R}^2$ and to investigate how the behaviour of solutions of $x^3 + bx + a = 0$ depends on a, b. We could do this in the present case with the help of the Cardano formula, see 2.1, e.g. for $b \geq 0$, $a \leq 0$ the (real) solution is given by

$$x = \sqrt[3]{\sqrt{\frac{b^3}{27} + \frac{a^2}{4}} - \frac{a}{2}} - \sqrt[3]{\sqrt{\frac{b^3}{27} + \frac{a^2}{4}} + \frac{a}{2}}$$

in which the square root term is just the discriminant (up to a factor). Instead, we shall be content with a qualitative consideration.

We consider the graphs of the function $y = x^3 + bx$. There are three qualitatively different cases : $b < 0$, $b = 0$ and $b > 0$.

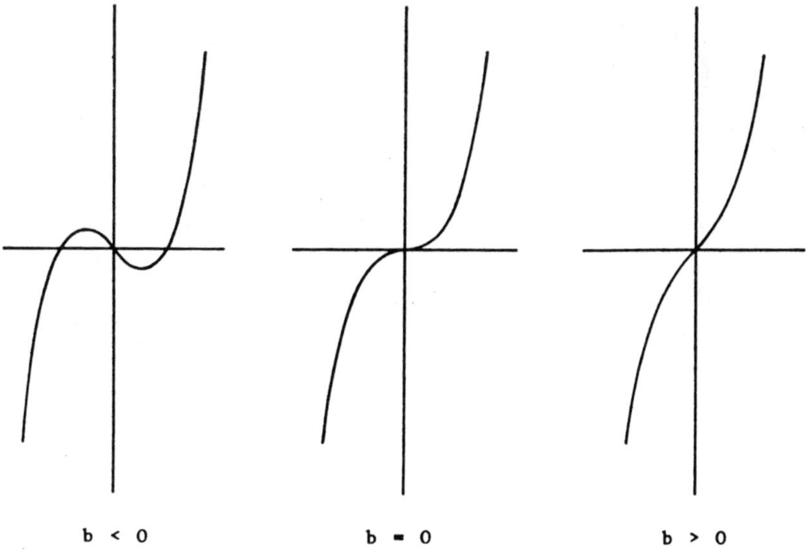

Now we consider the zeroes of $x^3 + bx + a$. In the case $b > 0$ there is obviously exactly one real zero. In the case $b = 0$ there is exactly one real zero when $a \neq 0$, and a triple zero when $a = 0$. The most interesting case is when $b < 0$. As long as a lies between the minimum and maximum of $y = x^3 + bx$ there are obviously three real zeroes, otherwise one. If a becomes equal to one of the extrema, two of the zeroes become a double zero. The condition that a lie between the extrema may be expressed algebraically immediately. It is the condition

$$27a^2 + 4b^3 < 0.$$

Thus we see : the discriminant $27a^2 + 4b^3 = 0$ divides the plane into two regions G resp. G' where $27a^2 + 4b^3 < 0$ resp. $27a^2 + 4b^3 > 0$. For $(a,b) \in G$, $x^3 + bx + a$ has three real zeroes, for $(a,b) \in G'$ it has one. Along the discriminant curve D there are multiple zeroes.

We can visualise this situation geometrically as follows :

In $\mathbb{R}^3$ with the coordinates (x,y,b) we consider the collection of all cubical parabolas $y = x^3 + bx$, i.e. we consider the smooth surface F in $\mathbb{R}^3$ with equation $x^3 + bx - y = 0$. If we set $y = -a$, then for constant a, b the points (x_i, a, b) on the surface are exactly the points whose first coordinate x_i is a zero of $x^3 + bx + a$. Thus when we project $\mathbb{R}^3$ onto the $\mathbb{R}^2$ with the coordinates (a,b) by $(x,y,b) \mapsto (-y,b)$ and restrict to the surface F we obtain a mapping

$$F \to \mathbb{R}^2$$

with the following property :

The preimage in F of each point $(a,b) \in \mathbb{R}^2$ consists of the points (x_i, a, b), $1 \leq i \leq 3$, where x_i is a zero of $x^3 + bx + a$. The following picture illustrates this mapping $F \to \mathbb{R}^2$. (It shows F in a parallel projection in which the planes b = constant in $\mathbb{R}^3$, and the cubical parabolas $y = x^3 + bx$ therein, are parallel to the plane of projection.)

One sees how projection in the direction of the x-axis sends three points of F to each point of G, and one to each point of G'. Over each point of the discriminant curve D there is one simple point and a doubly counted point where two preimage points "run together". Over the cusp of the discriminant curve there is a triply counted point.

One also sees that the discriminant curve is precisely the outline of the projection $F \to \mathbb{R}^2$, as we have already observed in section 1.8.

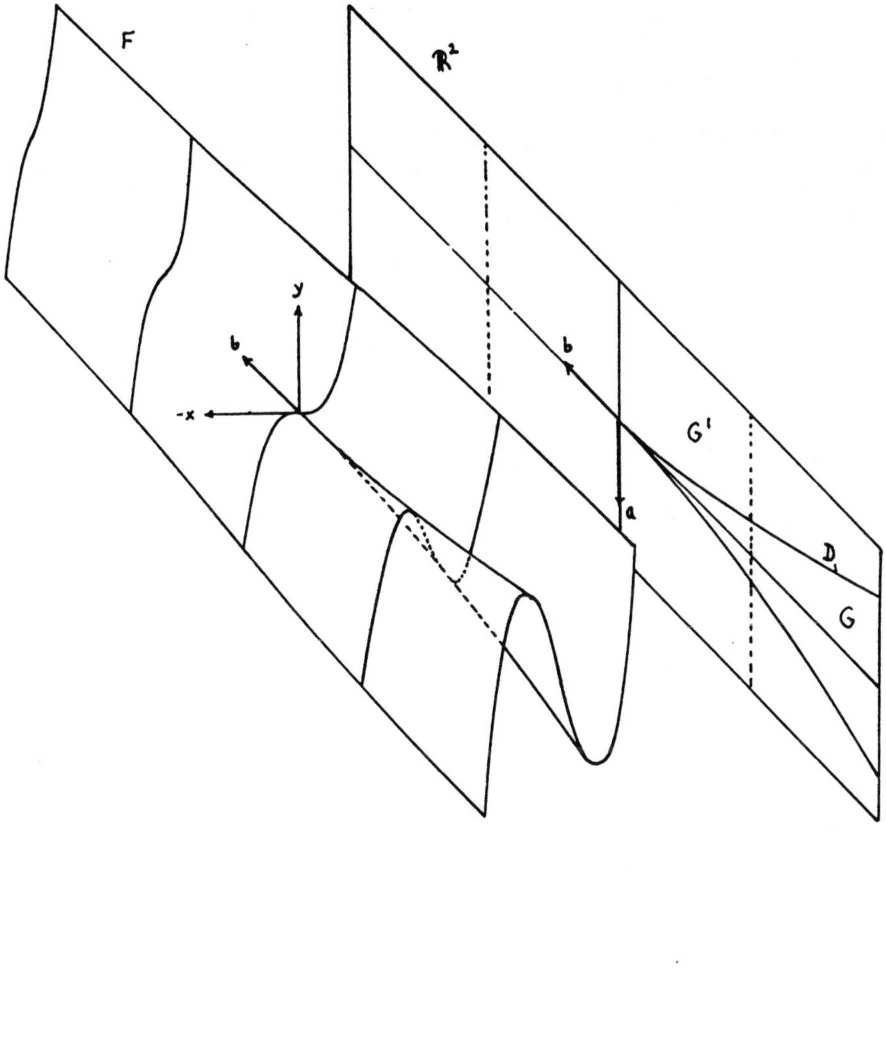
F
$\mathbb{R}^2$
y
b
-x
b
a
G'
D_1
G

Now we consider polynomials of fourth degree. By a Tschirnhaus transformation we can assume without loss of generality that the polynomials are of the form

$$f = x^4 + cx^2 + bx + a.$$

We shall again regard a, b, c as variables in an $\mathbb{R}^3$, and we shall investigate the vanishing behaviour of f in relation to (a,b,c).

The discriminant of f is D_f, where

$$27D_f = 4(c^2+12a)^3 - (2c^3-72ac+27b^2)^2.$$

Thus it is a polynomial of degree 5 in a, b, c and it is not quite easy to see what the discriminant surface described by the equation $D_f = 0$ looks like in $\mathbb{R}^3$.

We shall determine the form of this surface by considering the curves in which it cuts the planes $c =$ constant. Then $x^4 + cx^2 + bx + a$ still depends on the two parameters a and b. Now, exactly as in the case of the cubical parabolas in $\mathbb{R}^3$, we can consider the family of biquadratic parabolas $y = x^4 + cx^2 + bx$. They form a surface F_c in $\mathbb{R}^3$. The following figure shows a particular parallel projection of such a surface F_c for a $c < 0$.

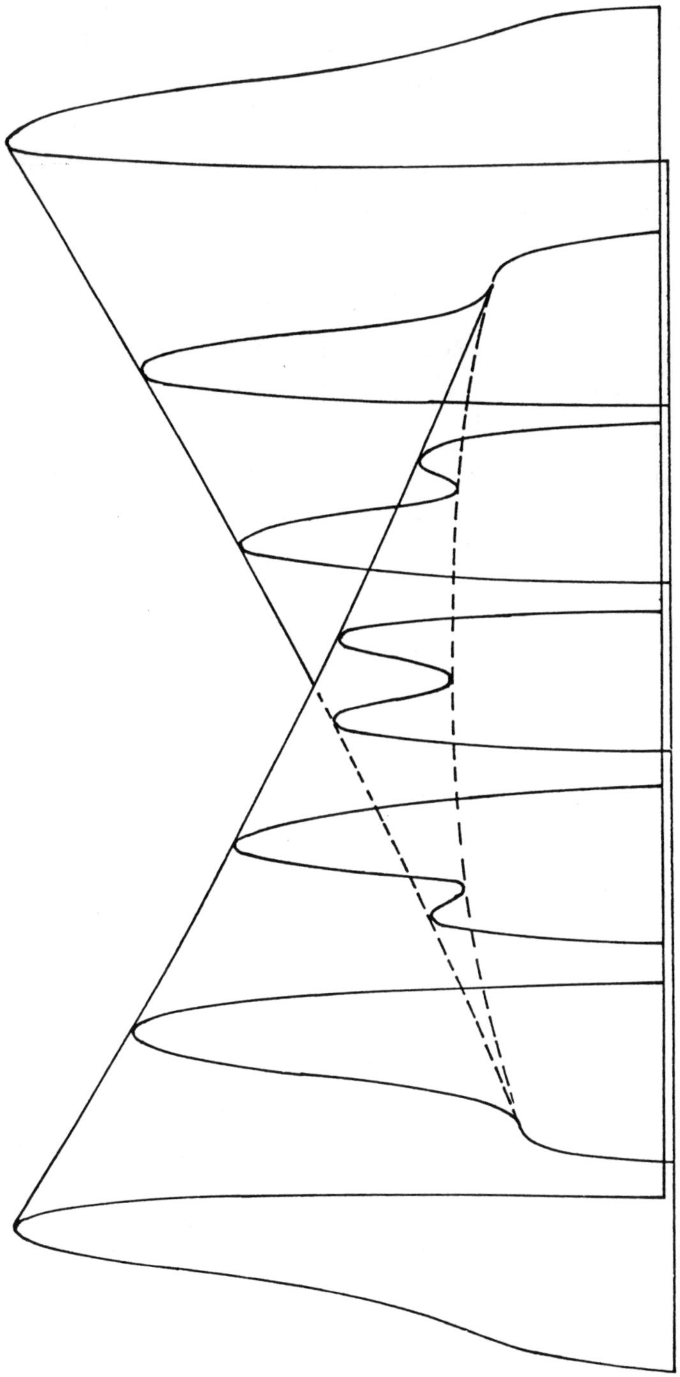

If one maps F_c onto the (a,b)-plane by the projection $(x,y,b) \mapsto (a,b)$ with $a = -y$, then the outline obtained is precisely the intersection D_c of the discriminant surface with the plane c = constant. One sees from the following picture that for $c < 0$ this is a curve of the following form, with two cusps and an ordinary double point (cf. section 1.8).

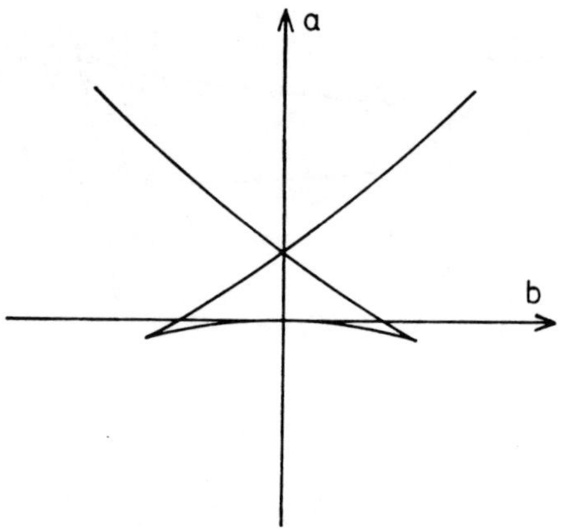

When we want to see the whole discriminant surface, we must vary c. As $c \to 0$, the triangle enclosed by the curve contracts to a singular point. For positive c, D_c consists of a non-singular curve and an isolated singular point. Altogether, the discriminant surface has the form shown in the following picture :

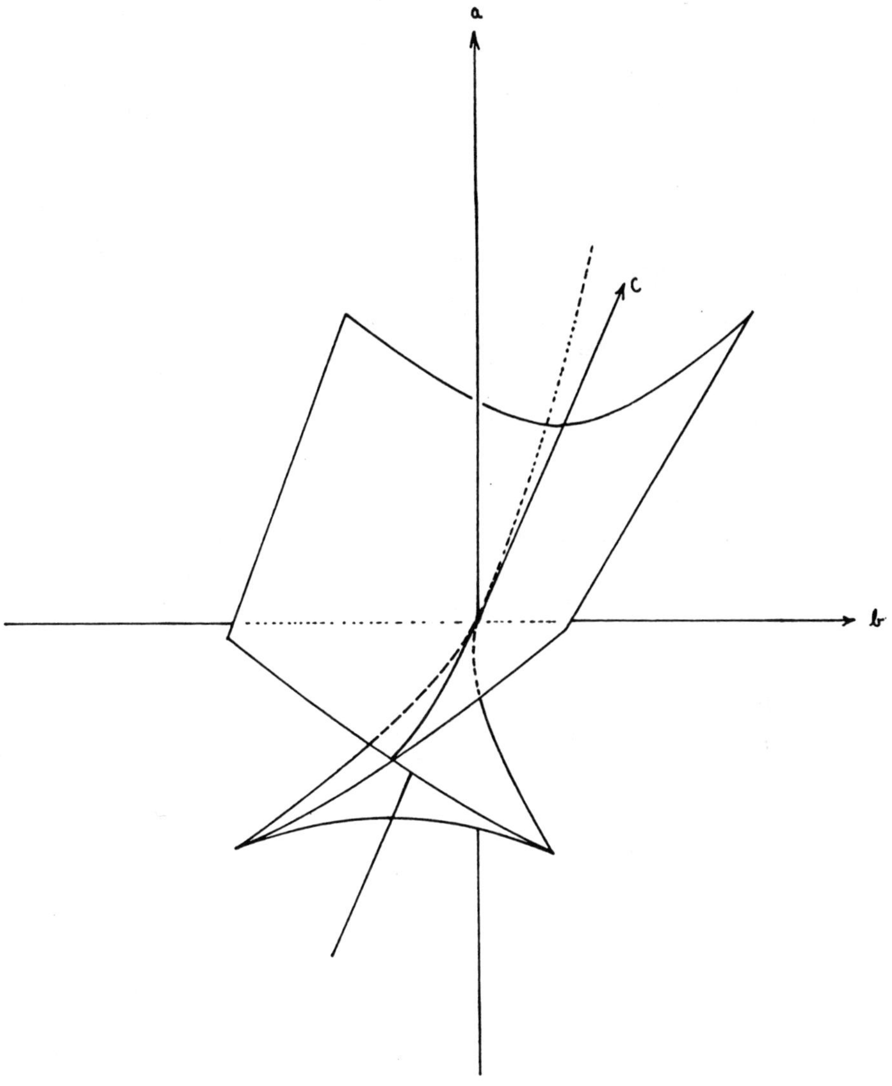
a
c
b

The singular loci of the discriminant surface are the fold edge with the equations

$$c^2 + 12a = 0$$
$$2c^3 - 8ac + 9b^2 = 0$$

and the double line with the equations

$$b = 0$$
$$c^2 - 4a = 0.$$

Along the fold edge the surface looks locally like the product of a cusp with a line (except of course at $(a,b,c) = (0,0,0)$). These edge points correspond to polynomials with a triple zero. The origin corresponds to the polynomial x^4 with a quadruple zero. Along the double line the surface intersects itself (for $c < 0$). These points correspond to polynomials with two double zeroes. The remaining points on the discriminant surface correspond to polynomials with only one double zero. The points outside the discriminant surface correspond to polynomials without multiple zeroes. They fall into three regions. In the region "enclosed" by the discriminant surface there are 4 real zeroes, in that next to the enclosing part of the surface there are two real zeroes, and in the third region there are none.

Thus we see that an understanding of the geometry of the discriminant surface is very useful in understanding equations of degree 4. The discriminant surface and its significance for the discussion of the biquadratic equation were first indicated by Kronecker (Monatsbericht der Berliner Akademie, 14.2.1878). In good old textbooks of algebra one therefore finds pictures of this surface (e.g. Weber's Lehrbuch der Algebra, Band 1, §84, p. 279, [W4]). Unfortunately, with the coming of "modern" algebra these pictures vanished from the algebra texts, and they only surfaced again when it was realised how these discriminants played a role in various places, e.g. as catastrophe sets in Thom's catastrophe theory or as caustics in optics (see [B9], [D5], [T1], [Z1]), as we have already seen in section 1.8.

After this digression on the geometric meaning of discriminants of polynomials, we now turn to the investigation of zero sets of polynomials.

4.3 Zeroes of polynomials

Let A be a commutative ring with 1 and let $f(x_1,\ldots,x_k)$ be a polynomial with coefficients in A. Then a k-tuple of constants $(c_1,\dots,c_k)$ with $c_i \in A$ and $f(c_1,\dots,c_k) = 0$ is called a <u>zero</u> of f in A^k. The most important example is the case where A is a field K. The <u>zero set</u> $V(f)$ of f is then a subset of the affine space K^k:

$$V(f) = \{(c_1,\dots,c_k) \in K^k \mid f(c_1,\dots,c_k) = 0\}.$$

Such a zero set is also called an <u>affine algebraic hypersurface</u>. When $k = 3$, $V(f)$ is a surface in the space K^3, when $k = 2$ it is a plane <u>affine algebraic curve</u>. Naturally, one can also consider common zero sets of finitely many functions $f_1,\dots,f_r$. These sets $V(f_1,\dots,f_r)$ of common zeroes of $f_1,\dots,f_r$ are called <u>affine algebraic sets</u>, and their investigation is an important problem of algebraic geometry. However, in accordance with the theme of this course, we shall confine ourselves mostly to curves here.

As long as the coefficient ring A remains an arbitrary ring, the relation between a polynomial $f \in A[x]$ and its zeroes seems to be uncontrollable. This is already shown by the following two examples.

<u>Example 1</u>. In $\mathbb{Z}/9\mathbb{Z}$, $f(x) = x^2$ has three zeroes, namely $0, 3, 6$, whereas e.g. the fundamental theorem of algebra leads us to expect that a polynomial of degree 2 will have at most 2 zeroes.

<u>Example 2</u>. Consider the polynomial $f(x) = x^2 + x$ in $\mathbb{Z}/2\mathbb{Z}[x]$. This is a polynomial different from zero, but for all $c \in \mathbb{Z}/2\mathbb{Z}$ (namely $c = 0$ and $c = 1$) $f(c) = 0$, so that f identically vanishes as a function on $K = \mathbb{Z}/2\mathbb{Z}$, and f and 0 have the same zero set.

Under suitable assumptions about the domain of coefficients, such things cannot happen.

<u>Proposition 1</u>

Let A be a commutative integral domain. Then

(i) If $a \in A$ is a zero of $f \in A[x]$, then $(x-a)$ is a divisor of f.

(ii) $f \in A[x]$ of degree n has at most n zeroes.

(iii) If A has infinitely many elements and if $f \in A[x_1,\dots,x_k]$ vanishes identically as a function on A^k, then $f = 0$ in $A[x_1,\dots,x_k]$.

<u>**Proof**</u> : (i) We have $f = q(x-a)+r$ with $r \in A$ by division with remainder*. If one sets $x = a$, then $r = 0$ follows because $f(a) = 0$.
(ii) This assertion follows from (i), since f can have at most n linear factors.
(iii) Suppose $f \neq 0$. By the finiteness of the number of roots of $f(x_1,\ldots,x_{k-1},x_k) \in A[x_1,\ldots,x_{k-1}][x_k]$, from (ii), and the infinitude of A, there is a $c_k \in A$ with $f(x_1,\ldots,x_{k-1},c_k) \neq 0$ in $A[x_1,\ldots,x_{k-1}]$. Induction yields the existence of $(c_1,\ldots,c_k) \in A^k$ with $f(c_1,\ldots,c_k) \neq 0$. Contradiction!

<u>**Remark**</u> : A always has infinitely many elements, e.g., when A is an algebraically closed field. (Proof : if A were finite, $\prod_{a \in A}(x-a) + 1$ would be a polynomial without zeroes in A, contrary to algebraic closure.)

The following theorem now shows that for an algebraically closed field like, e.g., the field $\mathbb{C}$, there is a reasonable connection between polynomials with coefficients in the field and their zero sets. This theorem is a special case of the theorem which is basic to algebraic geometry, <u>Hilbert's Nullstellensatz</u>.

<u>**Theorem 2**</u> (Study's lemma)

Let K be an algebraically closed field and let $f,g \in K[x_1,\ldots,x_k]$ be polynomials with zero sets $V(f)$ and $V(g)$ respectively. Then : when f is irreducible and $V(f) \subset V(g)$, f is a divisor of g.

<u>**Proof**</u> : When $g = 0$, f is trivially a divisor. We therefore assume that $g \neq 0$. Then $f \neq 0$ also by (iii) in Proposition 1. If f is constant, then it is trivially a divisor. Hence we assume that $f \notin K$. Then without loss of generality we can assume $f \notin K[x_1,\ldots,x_{k-1}]$. It then follows, as we shall show later, that $g \notin K[x_1,\ldots,x_{k-1}]$ also. Thus by Proposition 4.2.4 we have a representation

$$R = Pf + Qg$$

for the resultant R of $f,g \in K[x_1,\ldots,x_{k-1}][x_k]$. R is a polynomial in $K[x_1,\ldots,x_{k-1}]$.

---

*Although A need not be factorial, this causes no problems for division with remainder, because the coefficient of x in $(x-a)$ equals 1.

Let the coefficient of the highest power of f be $a_0(x_1,\dots,x_{k-1})$. Because of the algebraic closure of K there is a zero of f for each $(c_1,\dots,c_{k-1})$ with $a_0(c_1,\dots,c_{k-1}) \neq 0$, and by hypothesis this is also a zero of g. Thus substituting these $(c_1,\dots,c_{k-1})$ in the above equation always gives $R(c_1,\dots,c_{k-1}) = 0$. Consequently, the function $a_0(x_1,\dots,x_{k-1})R(x_1,\dots,x_{k-1})$ vanishes identically on K^{k-1}. Hence by 1(iii), $a_0 R = 0$ in $K[x_1,\dots,x_{k-1}]$. Since $a_0 \neq 0$, it follows that $R = 0$. Thus, by Theorem 4.2.3, f and g have a non-constant common divisor, and by the irreducibility of f it must be f itself. Thus f is a divisor of g, as was to be shown.

We still have to carry out the proof that $g \notin K[x_1,\dots,x_{k-1}]$. We give a proof by contradiction.

Suppose $g \in K[x_1,\dots,x_{k-1}]$. Then by Proposition 1(iii) there is a $(c_1,\dots,c_{k-1})$ which is not a zero of ga_0 as a function on K^{k-1}. Since g, viewed as a function on K^k, does not depend on the last component, it follows that $g(c_1,\dots,c_{k-1},c_k) \neq 0$ for all $c_k \in K$. But in contradiction to this, there is a c_k with $f(c_1,\dots,c_{k-1},c_k) = 0$ since K is algebraically closed, $f \notin K[x_1,\dots,x_{k-1}]$ and $a_0(c_1,\dots,c_{k-1}) \neq 0$. This proves the theorem.

Thus for algebraically closed fields K, such as $\mathbb{C}$, there is a reasonable connection between polynomials and their zero sets. E.g., if f and g are irreducible polynomials with the same zero set $V(f) = V(g)$, then $f = a\cdot g$ for some $a \in K - \{0\}$. For non-algebraically closed fields K such as, e.g., the field of real numbers $\mathbb{R}$ this is not at all true.

Example : In $\mathbb{R}$, $f(x) = 1 + x^2$ has a zero set which is empty, and hence contained in any other zero set. Also, $1 + x^2$ is irreducible. However, it certainly does not divide every other polynomial, e.g. it does not divide 1.

4.4 Homogeneous and inhomogeneous polynomials

We have already seen in sections 3.3 and 3.4 how one comes to define curves in the real or complex projective plane as zero sets of homogeneous polynomials. Since the homogeneous coordinates $(x_0,\dots,x_n)$ of a projective space are defined only up to multiplication by a scalar factor $t \neq 0$, one can only call such a point a "zero" of a polynomial f when f satisfies the following:

For all $(x_0,\dots,x_n) \neq (0,\dots,0)$, $f(x_0,\dots,x_n) = 0$ if and only if $f(tx_0,\dots,tx_n) = 0$ for all $t \neq 0$.

The homogeneous polynomials have this property, and for polynomials with coefficients in an algebraically closed field homogeneity also follows from this property by 4.3.1(iii).

This makes it clear why we must be concerned with homogeneous polynomials in what follows : because curves in the projective plane are precisely the zero sets of homogeneous polynomials.

Definition : A non-zero polynomial $f \in A[x_1,\dots,x_k]$ is <u>homogeneous of degree</u> n just in case

$$f(tx_1,\dots,tx_k) = t^n f(x_1,\dots,x_k)$$

holds in $A[x_1,\dots,x_k][t]$.

First we shall give various other characterisations of homogeneity. The simplest is the following :

Proposition 1

$0 \neq f \in A[x_1,\dots,x_k]$ is homogeneous of degree n when f is of the following form :

$$f = \sum_{i_1+\dots+i_k=n} a_{i_1\dots i_k} x_1^{i_1}\dots x_k^{i_k}.$$

Proof : Let $f = \sum_{\nu=0}^{N} \sum_{i_1+\dots+i_k=\nu} a_{i_1\dots i_k} x_1^{i_1}\dots x_k^{i_k}$. Then

$f(tx_1,\dots,tx_k) = \sum_{\nu=0}^{N} t^{\nu} \sum_{i_1+\dots+i_k=\nu} a_{i_1\dots i_k} x_1^{i_1}\dots x_k^{i_k}$. Comparing coefficients with $t^n f(x_1,\dots,x_k)$ yields the assertion.

Corollary 2

Each polynomial $f \neq 0$ is uniquely expressible as a sum of non-vanishing homogeneous terms, $f = f_{n_1} + \dots + f_{n_r}$, where f_{n_i} is homogeneous of degree n_i (and $n_1 < \dots < n_r$).

Remark : (i) We could also have tried to define homogeneous polynomials somewhat differently by viewing a polynomial $f(x_1,\dots,x_k) \in A[x_1,\dots,x_k]$ as a function on A^k. In this sense one would call f a <u>homogeneous function</u> of degree n on A^k when, for all $(c_1,\dots,c_k) \in A^k$ and all $t \in A$:

$$f(tc_1,\dots,tc_k) = t^n f(c_1,\dots,c_k).$$

In general this is a weaker concept (cf. 4.3, example 2). But when A contains infinitely many elements, it follows from 4.3.1(iii) that :

$f \neq 0$ is a homogeneous polynomial if and only if the corresponding function is homogeneous.

(ii) When A is a field K with infinitely many elements we can express the homogeneity property somewhat differently : let K* be the multiplicative group of non-zero elements of K. Then K* acts on K^k by $(c_1,\ldots,c_k) \overset{t}{\mapsto} (tc_1,\ldots,tc_k)$. Correspondingly, one obtains an action on polynomials by $f(x_1,\ldots,x_k) \overset{t}{\mapsto} f(tx_1,\ldots,tx_k)$.

The formula

$$f(tc_1,\ldots,tc_k) = t^n f(c_1,\ldots,c_k)$$

then describes the transformation behaviour of homogeneous polynomials under this action of K*.

Now it is natural to consider, in addition to the simplest conceivable K*-action $(c_1,\ldots,c_k) \mapsto (tc_1,\ldots,tc_k)$, other simple actions, e.g.

$$(c_1,\ldots,c_k) \mapsto (t^{w_1}c_1,\ldots,t^{w_k}c_k)$$

where $w_1,\ldots,w_k$ are particular natural members. Then one again has a corresponding K* action on polynomials and one can consider the polynomials which have the following transformation behaviour with respect to this action :

$$f(t^{w_1}x_1,\ldots,t^{w_k}x_k) = t^n f(x_1,\ldots,x_k).$$

One easily sees, exactly as in the homogeneous case, that these are precisely the polynomials of the following form :

$$f = \sum_{w_1 i_1+\ldots+w_k i_k=n} a_{i_1\ldots i_k} x_1^{i_1}\ldots x_k^{i_k}.$$

These polynomials are called <u>weighted-homogeneous</u> with the <u>weights</u> $w_1,\ldots,w_k$ and <u>degree</u> n (other names for them are <u>quasi-homogeneous</u> or <u>isobaric</u> (where w_i is called the weight of x_i and n is called the weight of f)).

Examples :

(1) $f(x_1,\ldots,x_k) = x_1^{a_1} + \ldots + x_k^{a_k}$

is weighted homogeneous with weights $w_i = a_1 \ldots \hat{a}_i \ldots a_k$ and degree $n = a_1 \ldots a_k$. These polynomials have played an important rôle in establishing connections between algebraic geometry and differential topology : [B4], [M1], [M2].

(2) Let f, g be polynomials of degrees m, n

$$f = a_0x^m + a_1x^{m-1} + \dots + a_m$$

$$g = b_0x^n + b_1x^{n-1} + \dots + b_n.$$

We already know that the resultant $R(a,b)$ is a homogeneous polynomial in the a_i, b_j of degree $m+n$, and hence weighted homogeneous with weights $w_i = 1$ and degree $m+n$. But when we set $a_0 = 1$, $R(a,b)$ is no longer homogeneous in $a_1,\dots,a_m$. We can, however, view it as weighted homogeneous in the following way. Suppose f and g split into linear factors :

$$f = (x-x_1) \dots (x-x_m)$$

$$g = (y-y_1) \dots (y-y_n).$$

Then we know (4.2.5) that

$$R = \prod_{i,j} (x_i-y_j).$$

Thus R is homogeneous in the $x_1,\dots,x_m$, $y_1,\dots,y_n$ of degree $n\cdot m$. Now a_i is, up to sign, the ith elementary symmetric function in $x_1,\dots,x_m$; likewise b_j is, up to sign, the jth elementary symmetric function in $y_1,\dots,y_n$. Since the $x_1,\dots,x_m$, $y_1,\dots,y_n$ have weight 1, this leads us to give a_i weight i and b_j weight j. Then, by the homogeneity of R in the x_i, y_j, we obviously have :

The resultant of $x^m + a_1x^{m-1} + \dots + a_m$ and $x^n + b_1x^{n-1} + \dots + b_n$ is weighted homogeneous of degree $m\cdot n$ when a_i receives weight i and b_j receives weight j.

It follows in particular that :

The discriminant of $x^n + a_1x^{n-1} + \dots + a_n$ is weighted homogeneous of degree $n(n-1)$, with the weight i for a_i.

Examples : (i) $f = x^3 + a_2x + a_3$

$$D_f = 27a_3^2 + 4a_2^3$$

(ii) $f = x^4 + a_2x^2 + a_3x + a_4$

$$27D_f = 4(a_2^2 + 12a_4)^3 - (2a_2^3 - 72a_2a_4 + 27a_3^2)^2.$$

The corresponding zero surfaces, which we have considered previously, are therefore invariant under the corresponding K^*-action.

We now come to another important characterization of homogeneous, and more generally quasi-homogeneous, polynomials.

Theorem 3

Let K be a field of characteristic 0. A non-vanishing polynomial $f \in K[x_1,\ldots,x_k]$ is quasi-homogeneous with weights $w_1,\ldots,w_k$ and degree n if and only if it satisfies the following partial differential equation of Euler :

$$\sum_{j=1}^{k} w_j x_j \frac{\partial f}{\partial x_j} = nf.$$

Proof : When f is quasi-homogeneous it is of the form

$$f = \sum_{w_1 i_1 + \ldots + w_k i_k = n} a_{i_1 \ldots i_k} x_1^{i_1} \ldots x_k^{i_k}.$$

Hence

$$\sum_{j=1}^{k} w_j x_j \frac{\partial f}{\partial x_j} = \sum_{j=1}^{k} \sum_{w_1 i_1 + \ldots + w_k i_k = n} w_j i_j x_j a_{i_1 \ldots i_k} x_1^{i_1} \ldots x_j^{i_j - 1} \ldots x_k^{i_k}$$

$$= \sum_{w_1 i_1 + \ldots + w_k i_k = n} (w_1 i_1 + \ldots + w_k i_k) a_{i_1 \ldots i_k} x_1^{i_1} \ldots x_k^{i_k}$$

$$= nf.$$

To prove the converse we write a given polynomial f as $f = f_{n_1} + \ldots + f_{n_r}$, where the f_{n_s} are quasi-homogeneous of degree n_s with weights $w_1,\ldots,w_k$. Then, as we have just proved,

$$\sum_{j=1}^{k} w_j x_j \frac{\partial f}{\partial x_j} = n_1 f_{n_1} + \ldots + n_r f_{n_r}.$$

Since f now satisfies the Euler differential equation, it follows that $n_1 f_{n_1} + \ldots + n_r f_{n_r} = nf_{n_1} + \ldots + nf_{n_r}$, and hence by comparing coefficients (since $\operatorname{char}(K) = 0$), $r = 1$ and $n = n_1$, so that f is quasi-homogeneous.

Remark : When D denotes the partial differential operator $D = \sum_{j=1}^{k} w_j x_j \frac{\partial}{\partial x_j}$, the preceding theorem reads

$$Df = nf$$

for quasi-homogeneous f. Repeated application of D of course gives

$$D^r f = n^r f.$$

D^r is a partial differential operator of rth order. Its principal part, i.e. the sum of the highest order terms, is

$$D_0^r := \sum_{j_1,\dots,j_r=1}^{k} w_{j_1} \cdot \dots \cdot w_{j_r} x_{j_1} \dots x_{j_r} \frac{\partial^r}{\partial x_{j_1} \dots \partial x_{j_r}} .$$

In the special case where all $w_j = 1$, the homogeneous case, D_0^r also satisfies a differential equation of the form $D_0^r f = \text{constant} \cdot f$.

Corollary 4

When $f(x_1,\dots,x_k)$ is homogeneous of degree n,

$$\sum_{j_1,\dots,j_r=1}^{k} x_{j_1} \dots x_{j_r} \frac{\partial^r f}{\partial x_{j_1} \dots \partial x_{j_r}} = n(n-1)\dots(n-r+1)f.$$

Proof : The proof is by complete induction. Theorem 2 is the base step because $D_0^1 = D$.

$$Df = n \cdot f . \tag{1}$$

Now suppose we have proved

$$D_0^r f = n(n-1)\dots(n-r+1)f \tag{2}$$

Application of D to (2) gives

$$DD_0^r f = n(n-1)\dots(n-r+1)nf \tag{3}$$

because of (1). By means of the product rule for differentiation, one shows easily that

$$DD_0^r f = D_0^{r+1} f + rD_0^r f. \tag{4}$$

It follows immediately from (3) and (4) that :

$$D_0^{r+1} f = n(n-1)\dots(n-r+1)(n-r)f,$$

which completes the induction step from r to $r+1$.

Now that we have learned a series of characterisations of homogeneous polynomials, we shall concern ourselves with the relations between homogeneous and inhomogeneous equations for curves in the projective and affine plane respectively. More generally, we consider the case of hypersurfaces in projective space.

We recall the connection between the projective space $P_k(\mathbb{C})$ and the affine space $\mathbb{C}^k$. (Cf. sections 3.3, 3.4.) $P_k(\mathbb{C})$ is covered by $k+1$ coordinate neighbourhoods U_i, where

$$U_i = \{(x_0,\dots,x_k) \in P_k(\mathbb{C}) \mid x_i \neq 0\}.$$

One has a homeomorphism $U_i \to \mathbb{C}^k$ defined by

$(x_0,\ldots,x_k) \mapsto \left(\frac{x_0}{x_i},\ldots,\frac{\hat{x}_i}{x_i},\ldots,\frac{x_k}{x_i}\right)$. In particular, one has a homeomorphism

$$U_0 \to \mathbb{C}^k$$

with

$$(x_0,\ldots,x_k) \mapsto \left(\frac{x_1}{x_0},\ldots,\frac{x_k}{x_0}\right)$$

and the inverse mapping

$$(x_1,\ldots,x_k) \mapsto (1,x_1,\ldots,x_k).$$

This is how we always identify the affine space $\mathbb{C}^k$ with U_0. $\mathbb{C}^k$ is then the complement in $P_k(\mathbb{C})$ of the "hyperplane at infinity"

$$P_{k-1}(\mathbb{C}) = \{(x_0,\ldots,x_k) \in P_k(\mathbb{C}) \mid x_0 = 0\}.$$

Now when $F(x_0,\ldots,x_k)$ is a non-constant homogeneous polynomial the equation $F(x_0,\ldots,x_k) = 0$ defines a <u>projective algebraic hypersurface</u> $V(F)$ in $P_k(\mathbb{C})$:

$$V(F) = \{(x_0,\ldots,x_k) \in P_k(\mathbb{C}) \mid F(x_0,\ldots,x_k) = 0\}.$$

Then the part of $V(F)$ lying in the affine space $U_0 \subset P_k(\mathbb{C})$, namely $V(F) \cap U_0$, is obviously just

$$V(F) \cap U_0 = \{(1,x_1,\ldots,x_k) \in P_k(\mathbb{C}) \mid F(1,x_1,\ldots,x_k) = 0\}.$$

Thus if we set

$$f(x_1,\ldots,x_k) := F(1,x_1,\ldots,x_k)$$

the zero set $V(f) \subset \mathbb{C}^k$ of f satisfies, by the identification of $\mathbb{C}^k$ with U_0

$$V(F) \cap U_0 = V(f).$$

The passage from F to f therefore describes the passage from the projective to the affine hypersurface. One obtains the converse passage by <u>homogenisation</u> of the polynomial f. When f is a polynomial of degree n in the indeterminates $x_1,\ldots,x_k$, the homogenisation of f is the homogeneous polynomial F of degree n in the variables $x_0,\ldots,x_k$ defined by

$$F(x_0,\ldots,x_k) = x_0^n f\left(\frac{x_1}{x_0},\ldots,\frac{x_k}{x_0}\right).$$

This F obviously satisfies $F(1,x_1,\dots,x_k) = f(x_1,\dots,x_k)$ so that if one homogenises f and then returns to the affine situation, f is recovered. The converse does not quite hold, because in passing from $F(x_0,\dots,x_k)$ to $F(1,x_1,\dots,x_k)$ a possible factor x_0^r in F becomes 1. However, when no factor x_0^r, $r \geq 1$, is present we in fact have : the homogenisation of $F(1,x_1,\dots,x_k)$ is $F(x_0,\dots,x_k)$, so that one has a bijection between the polynomials $f(x_1,\dots,x_k)$ of degree n and the homogeneous polynomials $F(x_0,\dots,x_k)$ of degree n which are not divisible by x_0. We shall describe the polynomials which correspond in this way as "associated".

In what follows we shall use the divisibility properties of polynomials, obtained in section 4.2, in order to prove the corresponding divisibility properties of the associated homogeneous polynomials. We shall always use f, g, h etc. to denote polynomials in the indeterminates $x_1,\dots,x_k$, and mostly use F, G, H etc. to denote homogeneous polynomials in the indeterminates $x_0,\dots,x_k$. For the sake of simplicity we let the coefficient ring be a field K.

Lemma 5

Each factor of a homogeneous polynomial is homogeneous.

Proof : Suppose $F = G \cdot H$, where F is homogeneous of degree m. Suppose also that H is homogeneous, say of degree $n \leq m$. Then G must also be homogeneous, because it follows from $F(tx) = G(tx) \cdot H(tx) = G(tx) \cdot t^n H(x) = t^m F(x) = t^m G(x) \cdot H(x)$ that $G(tx) = t^{m-n} G(x)$. Thus it suffices to show that G and H are not both inhomogeneous. If this were so, then there would be representations of $G(tx)$, $H(tx)$ in $A[x,t]$ of the form

$$G(tx) = t^r (P(x) + tR(x,t))$$
$$H(tx) = t^s (Q(x) + tS(x,t))$$

with $P, Q, R, S \neq 0$. Then we would have

$$F(tx) = G(tx)H(tx) = t^{r+s}(PQ+t(RQ+PS)+t^2 RS).$$

Here we would have $t(RQ+PS) + t^2 RS \neq 0$ because the degrees in t satisfy degree $(RQ+PS) \leq$ degree RS. But this means $F(tx) \neq t^n F(x)$, contrary to the homogeneity of F.

(**Alternative proof** : Let $G = G_{m_1} + \dots + G_{m_r}$ and $H = H_{n_1} + \dots + H_{n_s}$ be the decompositions into homogeneous components. If $r = s = 1$

were not the case, $F = G\cdot H$ would have homogeneous components of different degrees, namely the term $G_{m_1}H_{n_1}$ of lowest degree and the term $G_{m_r}H_{n_s}$ of highest degree.)

Proposition 6

Let $f \neq 0$ be a polynomial with the (essentially unique) decomposition into irreducible factors $f = f_1 \cdot \ldots \cdot f_r$, and let F resp. F_i be the homogeneous polynomials associated with f resp. f_i. Then $F = F_1 \cdot \ldots \cdot F_r$ is the (essentially unique) decomposition of F into irreducible factors.

Proof : When g resp. h are associated with the homogeneous polynomials G resp. H, then $f = g\cdot h$ is obviously associated with $F = G\cdot H$. Therefore, $F = F_1 \cdot \ldots \cdot F_r$. Moreover, it is obvious from Lemma 5 that a polynomial f is irreducible if and only if the associated homogeneous polynomial F is irreducible. Thus the F_i are irreducible.

Remark : The geometric meaning of this (for $k = 2$), as we shall see later, is that for a curve in the projective plane which does not contain the line at infinity the decomposition into irreducible components corresponds precisely to the decomposition of the corresponding curve in the affine plane into irreducible components.

The decomposition into irreducible factors is particularly simple and useful for homogeneous polynomials in two variables with coefficients in an algebraically closed field.

Proposition 7

Let K be algebraically closed and let $F \in K[x_0,x_1]$, $F \neq 0$ be homogeneous of degree n. Then there are $(a_i,b_i) \in K^2$, $i = 1,\ldots,n$, with $(a_i,b_i) \neq (0,0)$, such that

$$F(x_0,x_1) = a \prod_{i=1}^{n} (b_i x_0 - a_i x_1)$$

where $a \in K$ and the a_i, b_i are determined up to a factor from K.

Proof : This is a corollary to Proposition 6, since over an algebraically closed K each polynomial in one variable decomposes into linear factors.

Remark : For $K = \mathbb{C}$ the geometric meaning of this is that, for each nth degree homogeneous polynomial $F(x_0,x_1)$ the zero set in $P_1(\mathbb{C})$ decomposes into n points (counting multiplicities).

To conclude our investigation of the divisibility properties of homogeneous polynomials, we consider the resultant of two homogeneous polynomials $F, G \in K[x_0,\dots,x_k]$, which we view as a polynomial in the indeterminate x_k with coefficients in $K[x_0,\dots,x_{k-1}]$.

Proposition 8

Let
$$F = A_0x_k^m + A_1x_k^{m-1} + \dots + A_m$$
$$G = B_0x_k^n + B_1x_k^{n-1} + \dots + B_n$$

where A_i , $B_j \in K[x_0,\dots,x_{k-1}]$ are homogeneous polynomials of degrees i, j respectively. Then if $R_{F,G} \in K[x_0,\dots,x_{k-1}]$ is the resultant of F and G we have $R_{F,G} = 0$ or $R_{F,G}$ is a homogeneous polynomial of degree $m\cdot n$.

Proof : This follows from the assumption degree $A_i = i$ resp. degree $B_j = j$, and the earlier remark that the resultant is weighted homogeneous in A_i , B_j with degree $m\cdot n$ when one gives A_i , B_j the weights i, j respectively. Of course, one can also prove the equation

$$R_{F,G}(tx_0,\dots,tx_{k-1}) = t^{m\cdot n}R_{F,G}(x_0,\dots,x_{k-1})$$

directly by suitable determinant manipulations.

With these theorems on polynomials our algebraic preparations are concluded, and we now apply ourselves to the algebraic investigation of curves in the projective plane.

5. Definition and elementary properties of plane algebraic curves

5.1 Decomposition into irreducible components

After the foregoing historical and methodological discussions, the following definition of a plane algebraic curve should – I hope – be sufficiently motivated.

Definition

A plane, complex projective-algebraic curve is the zero set of a non-constant homogeneous polynomial in the complex projective plane $P_2(\mathbb{C})$.

Remarks : We use the terminology "plane curve" because the curve is embedded in the plane – one could also consider curves which are embedded in other manifolds, or one could define a concept of algebraic curve quite independently of these embeddings. We shall not do this here, but we shall make remarks later on such "abstract curves" defined independently of embeddings. Curves embedded in other spaces will also appear naturally later, with the resolution of singularities of curves in §9. We say "algebraic curve" because the curve has an algebraic description, as the zero set of a polynomial, and also because the abstract curve can be given an algebraic structure. (One gives such a structure by giving a system of open sets – the Zariski topology – and saying what are the admissible functions on these open sets : the everywhere defined rational functions.) We shall not go into this further here since, in this paragraph and the next, we shall consider curves quite concretely as curves embedded in $P_2(\mathbb{C})$.

We say "projective-algebraic curve" because we define a curve as the zero set of a homogeneous polynomial in the projective plane. The other possibility would be to consider plane affine-algebraic curves, which are the zero sets of non-constant polynomials in the affine plane. We shall say some more about the connection between the two concepts shortly. Generally, we shall use affine algebraic curves only as a means of handling projective algebraic curves, and we shall adhere to the projective standpoint. Considering projective curves means, in particular, that we consider complete curves, i.e. curves which do not lack points "at infinity" etc. Topologically (in the sense of ordinary topology) this means that the curves are compact, as closed subsets of the compact plane $P_2(\mathbb{C})$. We have already seen, in the report on Newton's classification of cubics, that the projective standpoint leads to a far-reaching simplification in the problem of classifying curves (2.5). This simplification brings out the essence of the theory of plane curves more strongly.

Finally, we say "complex-projective" curves because we shall develop our whole theory for curves defined over the field of complex numbers. One could consider other fields, say the field $\mathbb{R}$ of real numbers, which perhaps suggests itself in view of the prehistory of algebraic curves. Then one would arrive at real-projective algebraic curves in $P_2(\mathbb{R})$, resp. real-affine algebraic curves in $\mathbb{R}^2$. In general, algebraic geometers consider curves in $P_2(k)$ resp. k^2, where k is an arbitrary field. E.g., when $F(x_0,x_1,x_2)$ is a homogeneous (resp. arbitrary) polynomial with integer coefficients, it defines a curve in $P_2(\mathbb{F}_q)$ resp. $\mathbb{F}_q^2$ for almost every finite field $\mathbb{F}_q$. The consideration of such curves has a lot to do with number theory.

Since we are more interested in the connections of the theory of algebraic curves with function theory, geometry and topology, we shall confine ourselves entirely to the case $k = \mathbb{C}$ in this course, with occasional remarks on the case $k = \mathbb{R}$. Thus we are considering plane complex projective algebraic curves. When it is clear what is meant, we shall speak simply of "curves" or "plane curves".

A final comment on the definition of plane projective algebraic curves is the following. We have defined such curves $C \subset P_2(\mathbb{C})$ as zero sets of homogeneous polynomials $F(x_0,x_1,x_2)$. This assumes that homogeneous coordinates are given in $P_2(\mathbb{C})$. Now we shall frequently have occasion to go from a given homogeneous coordinate system (x_0,x_1,x_2) to another (x_0',x_1',x_2') by a linear transformation $x_i = \Sigma a_{ij}x_j'$. This transformation carries $F(x_0,x_1,x_2)$ to a homogeneous polynomial $G(x_0',x_1',x_2')$, and, relative to the new coordinates, the curve C is the zero set of G. Thus one sees that the property of being a projective algebraic curve does not depend on the choice of homogeneous coordinates. The coordinates, and the curve equations relative to them, are for us only an algebraic means of describing the curves.

An aspect of the interplay between analysis and synthesis in the mathematical investigation of possible kinds of objects is the attempt to decompose the objects under investigation into simpler ones which do not decompose further, investigating the indecomposable objects first and then building up all the others. This going back to simplest elements is a reduction process, and hence one often calls the indecomposable objects "<u>irreducible</u>". We now proceed along these general lines for curves.

<u>Definition</u>

A plane curve is <u>irreducible</u> if it is not the union of two distinct plane curves.

<u>Proposition 1</u>

Let C be a plane curve and let F be a homogeneous polynomial with zero set C. Then C is irreducible just in case F is a power of an irreducible polynomial.

<u>Proof</u> : Suppose that C is the union $C' \cup C''$ of two different curves C', C''. Let C', C'' be the zero sets of G, H respectively. Since $C' \neq C''$, there must be at least two distinct irreducible poly-

nomials F_1, F_2 which divide either G or H. But then F_1, F_2 must also divide each polynomial F with zero set C, because the zero sets of the F_i are contained in that of F, and hence $F_i | F$ follows by 4.4.6 and Study's lemma 4.3.2. Thus F is not a power of an irreducible polynomial, because the decomposition into irreducible polynomials is unique.

Conversely, let C be the zero set of a polynomial F which is not a power of an irreducible polynomial. Then F has a factorisation $F = G \cdot H$ into relatively prime homogeneous polynomials G and H (this follows from 4.4.6). Let C', C'' be the zero sets of G, H respectively. Obviously $C = C' \cup C''$. We have to show that the curves C' and C'' are different. If $C' \subset C''$, then the zero sets C_i' of the prime factors G_i of G are contained in C'', and by 4.3.2 and 4.4.6 the G_i divide the polynomial H, contrary to the assumption that G and H are relatively prime. This proves the proposition.

Proposition 2

Let C be any plane algebraic curve, the zero set of the homogeneous polynomial F with prime factors $F_1, \ldots, F_r$, so that

$$F = F_1^{k_1} \ldots F_r^{k_r}.$$

Then

(i) C has a unique decomposition into irreducible curves,

$$C = C_1 \cup \ldots \cup C_r.$$

(ii) With suitable numbering, the irreducible curve C_i is the zero set of F_i. The polynomials F_i are determined, up to a constant factor, by C_i (and hence by C).

Proof : Assertion (ii) is an immediate consequence of Proposition 1. If an irreducible curve C' were the zero set of two distinct (i.e. differing by more than a constant factor) irreducible homogeneous polynomials G and H, then C' would also be the zero set of $G \cdot H$, in contradiction to Proposition 1.

Proof of (i). The existence of a decomposition into irreducible curves C_i follows immediately from Proposition 1 and the existence of the decomposition $F = F_1^{k_1} \ldots F_r^{k_r}$ proved in 4.4.6 when one defines C_i to be the zero set of F_i. One sees the uniqueness as follows. If C' is an irreducible curve with $C' \subset C$ and if C' is the zero set of the irreducible polynomial G, then it again follows from Study's lemma and the unique factorisation of F into irreducibles that G must be one of the F_i, up to a constant factor, and hence $C' = C_i$. Any other decomposition of C into irreducible curves must therefore consist of curves C_i, though perhaps not all of them. However, the latter cannot happen, because if a C_{i_0} were missing in the decomposition, F_{i_0} would divide the product $\prod_{i \neq i_0} F_i$ by Study's lemma, contrary to the uniqueness of prime factorisation.

This proves Proposition 2.

Definition

The irreducible curves appearing in the unique decomposition of a plane curve into irreducibles are called the irreducible components.

Proposition 2 allows us to associate each plane curve C with a homogeneous polynomial which is unique up to a constant factor : if C is irreducible, then it is the zero set of an essentially unique irreducible polynomial. If C decomposes into irreducible components, $C = C_1 \cup \ldots \cup C_r$, then each C_i is associated with an essentially unique irreducible polynomial F_i, and we associate C with the polynomial $F = F_1 \cdot \ldots \cdot F_r$.

Definition

If F is the unique (up to a constant factor) homogeneous polynomial associated with the curve C as above, then we call

$$F(x_0, x_1, x_2) = 0$$

the equation of C.

In this way we have chosen a particular equation from among all the possible equations for a curve C (in fixed homogeneous coordinates), namely the one for which the homogeneous polynomial $F = F_1 \cdot \ldots \cdot F_r$ has no multiple factors. By Proposition 2, all other equations for C are of the form $F_1^{k_1} \cdot \ldots \cdot F_r^{k_r} = 0$. Thus we have a bijection between the plane curves on the one hand and the equivalence classes of homogeneous polynomials without multiple factors on the other, where two homogeneous polynomials are considered to be equivalent when they differ only by a constant factor. Polynomials with multiple factors should properly be associated with "curves" having multiple components - cf. 6.2 for further explanation.

Definition

The order of a plane curve C is the degree of the polynomial associated with C. One frequently says degree instead of order.

The degree of a polynomial does not change under linear transformations, which represent coordinate changes resp. collineations of $P_2(\mathbb{C})$. With this number, the order of a curve, we have therefore defined the first invariant for curves. (Cf. the earlier explanation in 2.5.) We shall become acquainted with the geometric meaning of this invariant in the next section, 5.2. Perhaps it is not superfluous to mention that order is an invariant of curves embedded in the plane, but not of abstract curves. For example, the non-singular quadrics, i.e. curves of order 2, are isomorphic, as abstract curves, to lines, i.e. curves of order 1, but naturally there is no collineation carrying a quadric to a line.

Definition

A plane curve of order 1, 2, 3, 4, 5, 6 is called a line, quadric, cubic, quartic, quintic, sextic respectively.

To conclude this section we shall explain further the relation between plane projective-algebraic and affine-algebraic curves, and between homogeneous and affine equations. A plane projective-algebraic curve C has only finitely many irreducible components, and certainly only finitely many lines are irreducible components of C. If L is any line which is not an irreducible component of C, then L cuts the curve C in only finitely many points (see section 5.2). We choose such a line L and choose homogeneous coordinates (x_0,x_1,x_2) so that L has equation $x_0 = 0$. Let $F(x_0,x_1,x_2) = 0$ be the equation of C relative to these coordinates. Now let $f(x,y) = F(1,x,y)$, so that f and F are associated polynomials in the sense of 4.4. Let $\mathbb{C}^2$ be the affine plane $P_2(\mathbb{C}) - L$ with the affine coordinates $x = \frac{x_1}{x_0}$, $y = \frac{x_2}{x_0}$.

Obviously, $C \cap \mathbb{C}^2$ is the affine-algebraic curve C' with equation $f(x,y) = 0$. The projective-algebraic curve C results from the affine-algebraic C' by completing C' with the addition of the finitely many points of $C \cap L$ on the "line at infinity" L. Since $f(x,y)$ uniquely determines the associated homogeneous polynomial $F(x_0,x_1,x_2)$, we can describe C just as well by means of the affine equation $f(x,y) = 0$, and with suitable choice of coordinates this may be clearer and show special features of the curve more quickly. Hence we shall often make use of this description by an affine equation.

5.2 Intersection of a curve by a line

In Chapter I we have seen that curves have been studied since antiquity for the reason that certain problems reduce to constructing or finding the intersection of certain curves. Accordingly, investigating the intersection of two curves is an important task, which we shall undertake in detail in §6. By way of preparation, we shall now concern ourselves with the particularly simple case in which one of the curves is a line L. When L is a component of C, $L \cap C = L$ of course consists of infinitely many points. We therefore assume that L is not a component of C.

Proposition 1

Let C be a plane complex projective-algebraic curve and let L be a line which is not a component of C. Then the number of points of intersection of L and C, counting multiplicities, is equal to the order of C. (Multiplicity is defined in the proof.)

Proof : Without loss of generality, we can assume that L has the equation $x_2 = 0$. Then (x_0,x_1) are homogeneous coordinates in the one-dimensional projective space L. If $F(x_0,x_1,x_2) = 0$ is the equation of C, then $F(x_0,x_1,0) = 0$ is the equation of the intersection of C and L. Since L is not a component of C, x_2 does not divide F, and hence $F(x_0,x_1,0)$ is a non-vanishing homogeneous polynomial of the same degree, m, as $F(x_0,x_1,x_2)$. Hence by 4.4.7 it has exactly m zeroes in L when these are counted with the right multiplicity, i.e., as often as the corresponding linear factors appear in $F(x_0,x_1,0)$. This definition of multiplicity is an "ad hoc" definition. In 6.1 we shall be concerned with a general definition of multiplicity which in the end will reduce to the present one, and we shall prove in particular that the definition is independent of coordinates. At any rate, with the present definition, the theorem is proved.

Proposition 1 shows the meaning of the order m of a curve C : almost all lines cut C in m points, when one counts the multiplicities of the points. In fact, most lines cut C in m distinct points, hence with multiplicity one. We have :

Proposition 2

Let C be a plane curve of order m and let p be a point in the complement of C. Then, with at most $m(m-1)$ exceptions L_i, $i = 1,\dots,m(m-1)$, a line L through p cuts the curve C in exactly m distinct points.

Proof : We choose homogeneous coordinates (x_0,x_1,x_2) so that $p = (0,0,1)$. Let $F(x_0,x_1,x_2) = 0$ be the equation of C. We obtain a bijection between the family $\mathcal{L}$ of all lines through p and the projective line G with equation $x_2 = 0$ by associating each $L \in \mathcal{L}$ with its intersection $\lambda = (\lambda_0,\lambda_1,0)$ with G. We denote the line through λ and p by L_λ. Obviously

$$L_\lambda - \{p\} = \{(\lambda_0,\lambda_1,t) \in P_2(\mathbb{C}) \,|\, t \in \mathbb{C}\}.$$

Thus the intersections of L_λ with C are given by the equation $F(\lambda_0,\lambda_1,t) = 0$. The polynomial $F(\lambda_0,\lambda_1,t)$ does not vanish identically, because L_λ is not a component of C. Since $p \notin C$ it has degree precisely m. Then by 4.2.7, this polynomial has a multiple zero just in case its discriminant vanishes. By 4.4.8, this discriminant

$$R(\lambda_0,\lambda_1) = R_{F(\lambda_0,\lambda_1,t),\ \frac{\partial F}{\partial t}(\lambda_0,\lambda_1,t)}$$

is a homogeneous polynomial in λ_0, λ_1 of degree $m(m-1)$, hence it has, counting multiplicities, $m(m-1)$ zeroes $\lambda^{(i)}$. The preceding considerations show that at most the lines $L_i = L_{\lambda^{(i)}}$ in $\mathcal{L}$ have intersections of multiplicity > 1 with the curve C.

The special case $m = 2$ is of course well known : there are at most two tangents to a quadric through a given point p. The next intersecting case is that of cubics. In this case, Proposition 2 leads us to expect at most $3.2 = 6$ tangents. The following figure illustrates this situation, where C and p are chosen so that all 6 tangents are real.

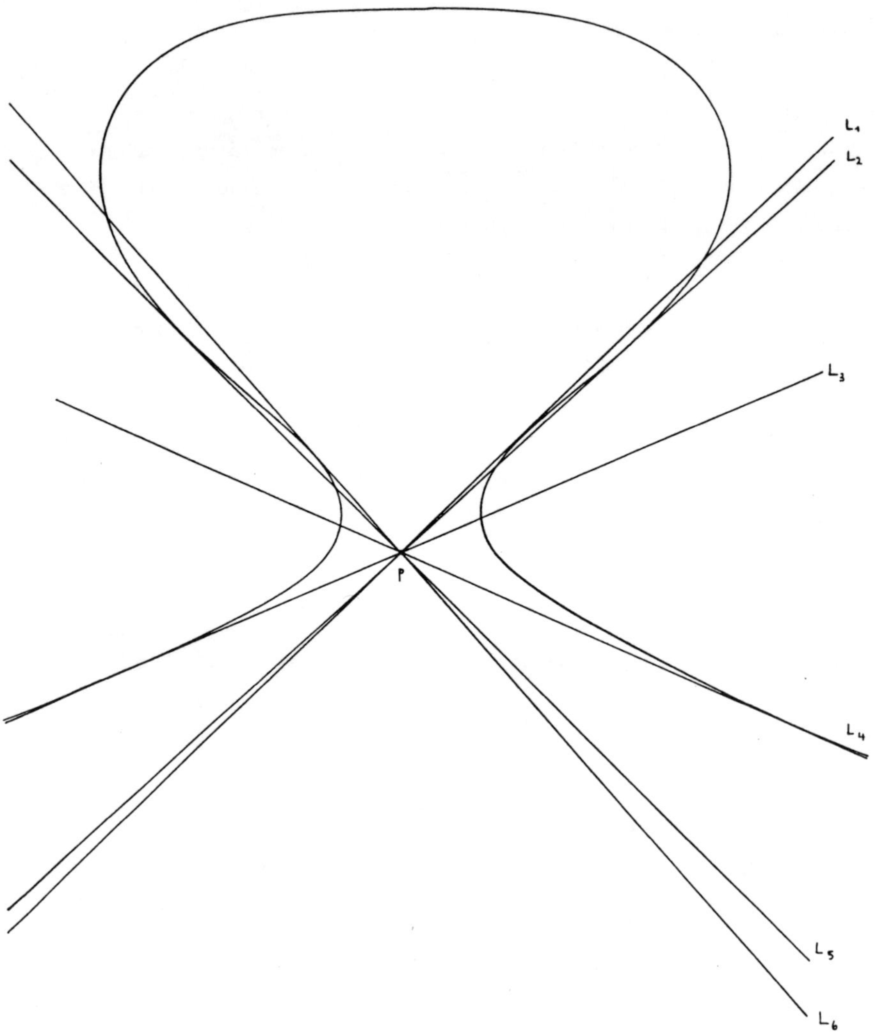

L_1
L_2
L_3
L_4
L_5
L_6
P

5.3 Singular points of plane curves

Right at the beginning of this course we saw many examples of curves on which some points looked different from others. There were cusps, and double points where the curves intersected themselves, on the cissoid of Diocles, the conchoid of Nicomedes, on the epicyclic curves of Ptolemy, and on many other curves. Then in 2.4 we gave a preliminary indication of the way in which these singular points could be handled, for real affine-algebraic curves, by analytic methods. The non-singular, or regular, points then appear as those points at which the tangent problem has a simple, unique solution.

The same idea underlies the following treatment of singular points of plane complex projective-algebraic resp. affine-algebraic curves. For the moment we wish only to define these points and characterise them in a simple way. A detailed analysis of singular points will be carried out later - in §8.

We want to define the singular points of a curve C to be those points p at which the curve has more than one tangent – counting multiplicity. In order to make this idea precise, we first consider an arbitrary line L through p and investigate the intersection of L and C at p.

The clearest way to carry out these investigations is with the help of the affine equation $f(x,y) = 0$ for C. Naturally we choose coordinates so that the point p being investigated does not lie on the line at infinity, and so that the latter is not a component of C. Let the point p have the affine coordinates (a,b). The lines through p then have parametric representation

$$\begin{aligned} x &= a + \lambda t \\ & \qquad\qquad t \in \mathbb{C} \\ y &= b + \mu t. \end{aligned}$$

Here, the $(\lambda,\mu) \neq (0,0)$ is determined by the line up to a constant factor. Thus the lines through p constitute a 1-dimensional projective space $\mathcal{L}$ with the homogeneous coordinates (λ,μ). The multiplicity of the intersection of C with the lines $L_{(\lambda,\mu)}$ defined in the proof of 5.1.1, with the above parametric representation, is precisely the multiplicity of the zero of the polynomial

$$f(a+\lambda t, b+\mu t)$$

in the indeterminate t at the position $t = 0$. As is well known, this multiplicity is described by the vanishing of the derivatives with respect to t at the position $t = 0$. More precisely : if the derivatives of order less than r vanish at $t = 0$, then the polynomial has an r-tuple zero at $t = 0$. One can also express this as follows.

If $\sum \frac{f^{(k)}(a,b)t^k}{k!}$ is the Taylor expansion of $f(a+\lambda t,b+\mu t)$ at $t = 0$, and if this expansion begins with the term of order r, then $f(a+\lambda t,b+\mu t)$ has an r-tuple zero at $t = 0$. This Taylor expansion results from the Taylor expansion of f at (a,b), namely

$$f(x,y) = \sum_{k=0}^{\infty} \frac{1}{k!} \sum_{i=0}^{k} \binom{k}{i} \frac{\partial^k f}{\partial x^i \partial y^{k-i}}(a,b)(x-a)^i(y-b)^{k-i} \tag{1}$$

where all terms of the same order have been collected into a sum, by setting $x = a + \lambda t$, $y = b + \mu t$. Thus the coefficient of the term of order k in the Taylor expansion of $f(a+\lambda t,b+\mu t)$ is, up to a factor $\frac{1}{k!}$, equal to

$$\sum_{i=0}^{k} \binom{k}{i} \frac{\partial^k f}{\partial x^i \partial y^{k-i}}(a,b)\lambda^i \mu^{k-i}. \tag{2}$$

This is a homogeneous polynomial of degree k in the indeterminates λ, μ. If one replaces these indeterminates in (2) by $x-a$, $y-b$ then one obtains exactly the term of order k in the Taylor expansion (1) of f at the point (a,b).

In this way we have obtained the following result : Suppose that the terms in the Taylor expansion (1) of f vanish for orders $< r$, while the term of order r does not vanish. Then all lines $L_{(\lambda,\mu)}$ through $p = (a,b)$ cut the curve C with multiplicity $\geq r$ at p, because (2) also vanishes for $k < r$ and hence $f(a+\lambda t,b+\mu t)$ has at least an r-tuple zero. Counting multiplicity, there are exactly r lines $L_{(\lambda_j,\mu_j)}$, $j = 1,\dots,r$ which cut C with multiplicity $> r$ at p, namely those in $\mathcal{L}$ corresponding to the r zeroes (λ_j,μ_j) of

$$\sum_{i=0}^{r} \binom{r}{i} \frac{\partial^r f}{\partial x^i \partial y^{r-i}}(a,b)\lambda^i \mu^{r-i} = 0.$$

These results motivate the following definition :

<u>Definition</u>

Let C be a plane complex projective-algebraic curve and let p be a point of C. Let (x,y) be affine coordinates such that p has coordinates (a,b) and C has the affine equation $f(x,y) = 0$.

(i) The <u>multiplicity</u> $\nu_p(C)$ of C at p is the order of the lowest non-vanishing term in the Taylor expansion of f at p.

(ii) The <u>tangents</u> to C at p are the lines through p which cut C with multiplicity $> \nu_p(C)$ at p. Counting multiplicity, C has exactly $\nu_p(C)$ tangents at p.

(iii) p is a <u>regular point</u> of C when $\nu_p(C) = 1$.
p is a <u>singular point</u> of C when $\nu_p(C) > 1$.
p is a <u>double point</u> of C when $\nu_p(C) = 2$.
p is a <u>triple point</u> of C when $\nu_p(C) = 3$.

Remarks : (i) In the definition of multiplicity we have used affine coordinates. However, the definition is obviously independent of the choice of coordinates.

(ii) It is clear from the foregoing discussion how one computes the $\nu = \nu_p(C)$ tangents to C at p. The affine equation of the curve they make up (with multiple components when tangents have to be multiply counted) reads :

$$\sum_{i=0}^{\nu} \binom{\nu}{i} \frac{\partial^{\nu} f}{\partial x^i \partial y^{\nu-i}} (a,b)(x-a)^i (y-b)^{\nu-i} = 0,$$

and therefore results from equating to zero the lowest non-vanishing term in the Taylor expansion of f at p.

(iii) We shall soon see (in §6.2) that at most finitely many points of a plane curve can be singular, while all others are regular. This justifies the choice of the words regular and singular : the singularities are the exception, regular points are the rule.

(iv) At this stage we should list a few examples of singular points. The simplest examples are the double points and simple cusps, which we have already met with several curves in §1 :

The typical equation for an ordinary double point is $xy = 0$; the typical equation for a simple cusp is $x^2 - y^3 = 0$, the equation of the semicubical parabola. As a common generalisation of both, we have already met, in 2.4, the generalised parabolas

$$x^p - y^q = 0,$$

which have a singularity at $x = y = 0$ for $p > 1, q > 1$. The different pairs (p,q) yield different singularities – the pair (p,q) is a characteristic invariant of the singularity. In the case of the generalised parabolas this pair suffices to describe the type of the singularity. In general, however, the singularities of plane curves are much more complicated, and even when the curve consists of a single piece in the neighbourhood of the singular point one needs several number pairs, the so-called Puiseux pairs in order to describe the type of the singularity – we shall do this later (in 8.3). The complexity of singularities is not adequately reflected, as we shall see in an example shortly, by the real pictures of curves we have used intuitively most of the time until now. As an example, the generalised parabolas $x^p - y^q = 0$ yield only three qualitatively different pictures for relatively prime p and q, namely those represented by $x^2 - y^3 = 0$, $x^3 - y^4 = 0$ and $x^3 - y^5 = 0$.

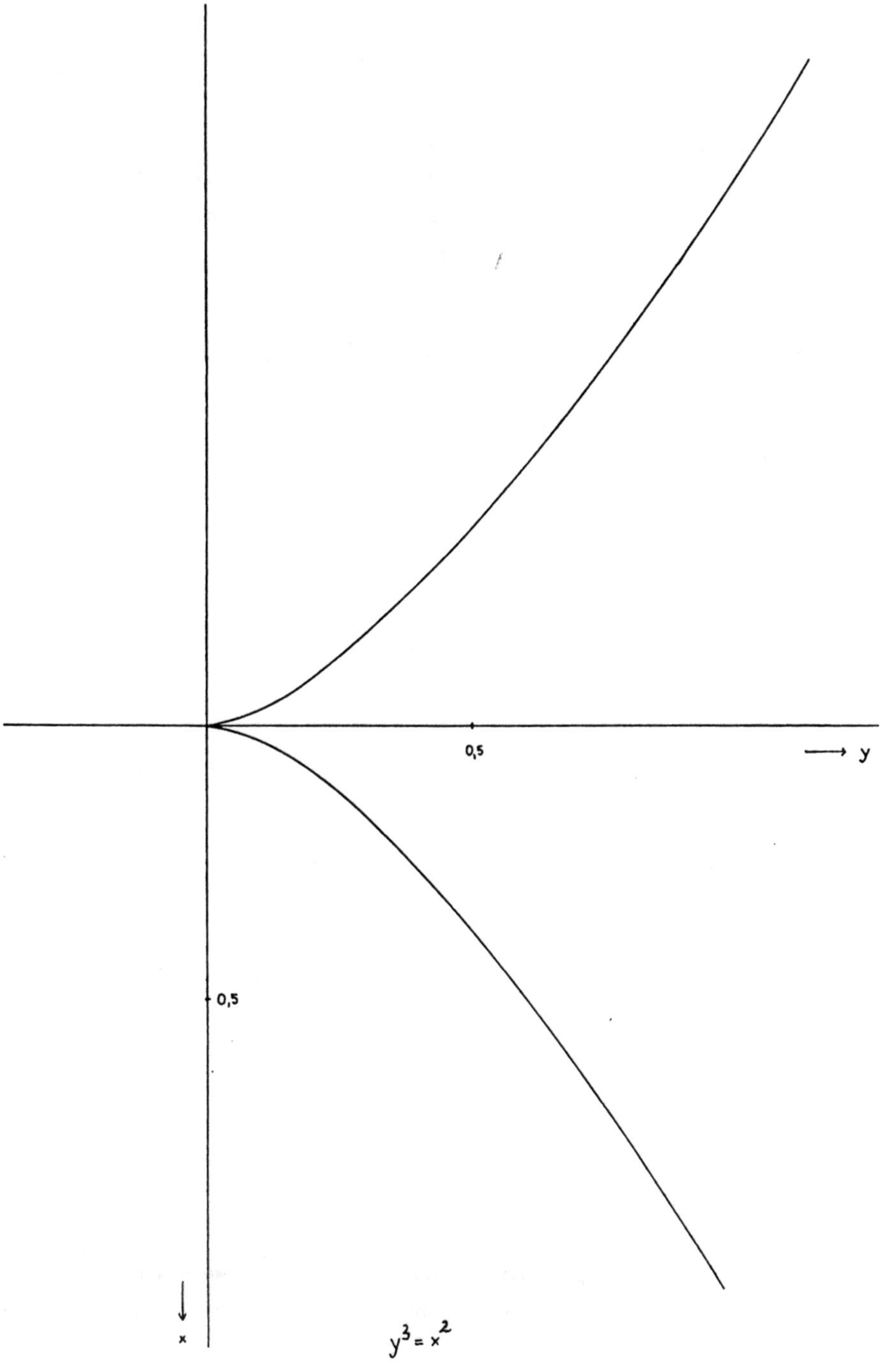

$y^3 = x^2$

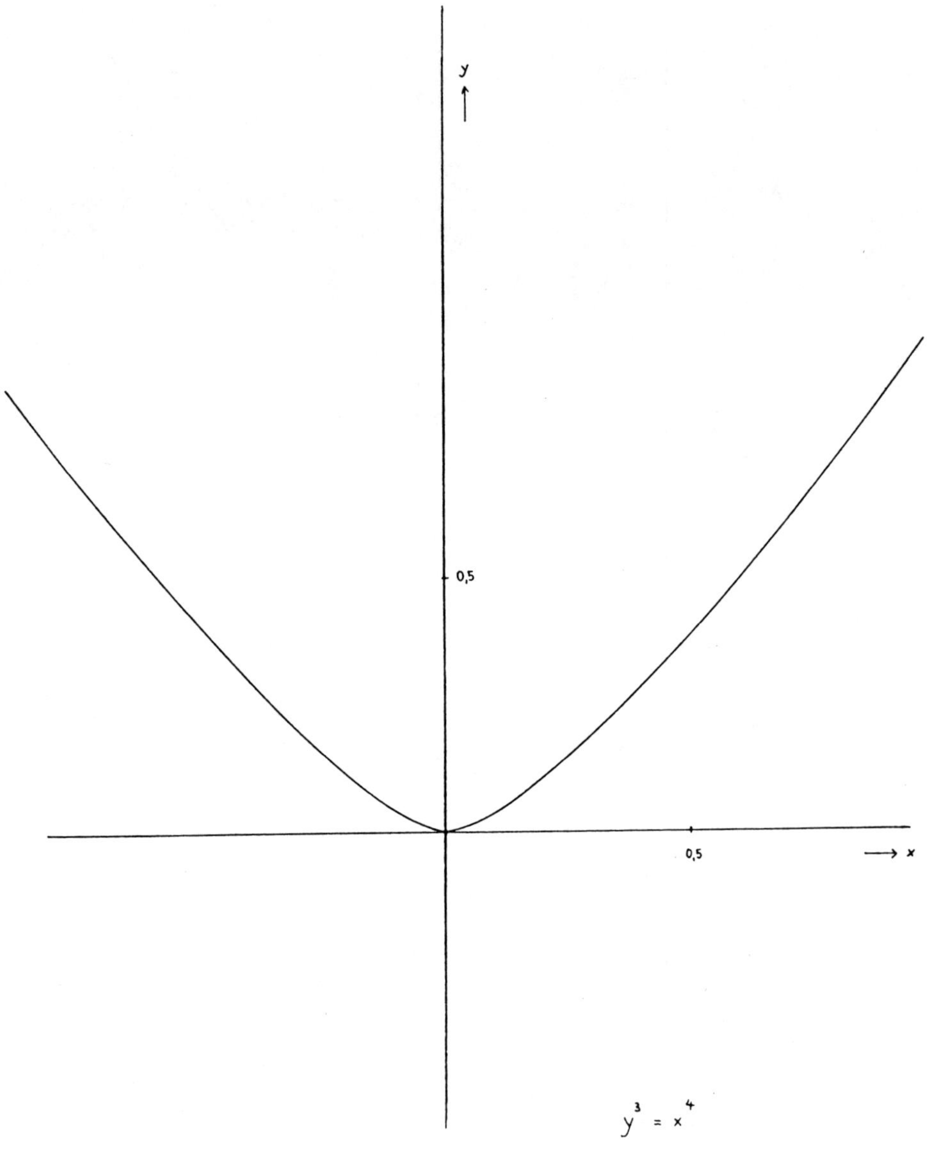

$y^3 = x^4$

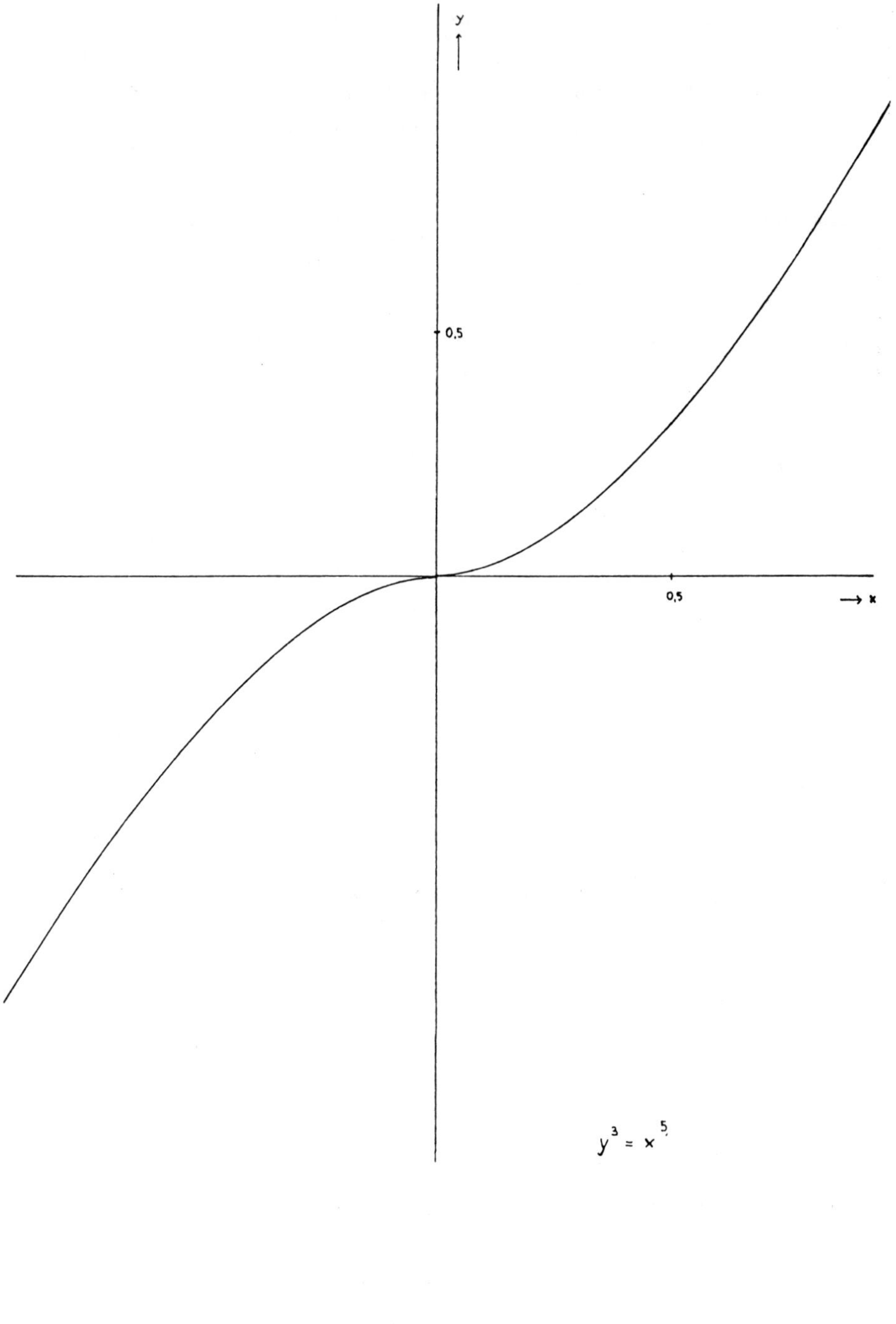

$y^3 = x^5$

It may well be surprising to learn that the curves $x^3 - y^4 = 0$ and $x^3 - y^5 = 0$ have the origin $x = 0$, $y = 0$ as a singular point, because the real curves appear to be smooth and have a well-defined tangent. On the other hand, the two partial derivatives $3x^2$ and $4y^3$ vanish at this point, so there is a singularity there according to our definition. It seems a contradiction occurs between our definition and our intention that the definition of regular point should describe the point where the curve is smooth.

We shall show that the apparent contradiction is resolved when we consider the complex curve instead of the real one.

First we analyse the real affine-algebraic curve with the equation $x_1^3 - x_2^4 = 0$. We begin by showing that this subset of $\mathbb{R}^2$ is a 1-dimensional, once continuously differentiable submanifold (C^1-manifold) of $\mathbb{R}^2$. To do this it suffices to give a once continuously differentiable mapping $t \mapsto (x_1(t), x_2(t))$ of an interval, which maps the interval bijectively onto the affine algebraic curve in such a way that we always have $(x_1'(t), x_2'(t)) \neq (0,0)$. We can easily give such a mapping :

$$x_1(t) = t^{4/3}$$
$$x_2(t) = t$$

has the desired properties. This mapping is continuously differentiable once, but not twice. There is in fact no twice continuously differentiable mapping $t \mapsto (x_1(t), x_2(t))$ at all from an interval onto our curve such that $(x_1'(t), x_2'(t)) \neq (0,0)$.

<u>**Proof**</u> : If we had such a mapping, then the two functions $x_1 = x_1(t)$, $x_2 = x_2(t)$ would satisfy

$$x_1^3 = x_2^4 \; . \qquad (1)$$

Differentiation of (1) gives

$$3x_1^2 x_1' = 4x_2^3 x_2' \; . \qquad (2)$$

Dividing by x_1^2 and combining with (1) gives

$$3x_1' = 4x_2^{1/3} x_2' \; . \qquad (3)$$

Differentiation of (3) gives

$$3x_1'' = 4/3 \; x_2^{-2/3} (x_2')^2 + 4x_2^{1/3} x_2'' \; . \qquad (4)$$

Multiplication of (4) by $x_2^{2/3}$ gives

$$4/3(x_2')^2 = 3x_2^{2/3}x_1'' - 4x_2x_2'' . \qquad (5)$$

But the relation (5) contains a contradiction : because when t_0 is the point where $(x_1(t_0), x_2(t_0)) = (0,0)$. we have $x_1'(t_0) = 0$ by (3) and hence $x_2'(t_0) \neq 0$. Thus the left-hand side of (5) is non-zero, while the right-hand side is zero, and that is the contradiction. Hence there are no twice continuously differentiable functions with the desired properties.

We have therefore obtained the following result : the real affine-algebraic curve with the affine equation $x_1^3 - x_2^4 = 0$ is a C^1-submanifold of $\mathbb{R}^2$, but not a C^r-submanifold for $r > 1$. It is indeed "smooth", but not especially smooth. This situation is typical. A real affine-algebraic curve can indeed be a C^r-submanifold of $\mathbb{R}^2$ at a singular point, for low r. But there is always an r_0 such that the curve is not a C^r-submanifold at this point for $r > r_0$. At a regular point, on the other hand, the curve is always a C^∞-submanifold of $\mathbb{R}^2$ and in fact even a real analytic submanifold – this follows immediately from the Implicit Function Theorem.

We have therefore established that the singular points on a real algebraic curve can be recognised as places where the curve is less smooth, in the sense that the curve is not at that point a submanifold of $\mathbb{R}^2$ or, if it is, it is not differentiable as often there as at regular points. However, one cannot necessarily see this with the naked eye on the picture of the real curve.

The situation is much less equivocal for complex curves in the complex plane. One can prove : at a singular point p, a complex affine-algebraic curve C in $\mathbb{C}^2$ is never a submanifold of $\mathbb{C}^2$, not even a topological submanifold. This means : there is no neighbourhood U of p on $\mathbb{C}^2$ such that the pair $(U, U \cap C)$ is homeomorphic to $(\mathbb{R}^4, \mathbb{R}^2)$. We shall not prove this fact here – it is not altogether easy to demonstrate. Instead we shall clarify our example $z_1^3 - z_2^4 = 0$ when the associated affine-algebraic curve C is embedded in $\mathbb{C}^2$. It then quickly becomes clear that C cannot be a topological submanifold of $\mathbb{C}^2$.

We now consider the mapping $\phi : \mathbb{C} \to \mathbb{C}^2$ defined by

$$z_1(t) = t^4$$
$$z_2(t) = t^3.$$

One sees immediately that the image of ϕ is precisely the curve C with equation $z_1^3 - z_2^4 = 0$, and that the mapping $\phi : \mathbb{C} \to C$ is a homeomorphism. As a topological space, therefore, the curve C is a 2-dimensional topological manifold. But it is not a topological submanifold of $\mathbb{C}^2$! We now want to convince ourselves of this.

To do so we consider the circle S_c^1 , $|t| = c$, in $\mathbb{C}$. Its image $K_c = \phi(S_c^1)$ under ϕ is contained in a 2-dimensional torus $S^1 \times S^1$, namely the torus consisting of the $(z_1, z_2) \in \mathbb{C}^2$ with

$$|z_1| = c^4$$
$$|z_2| = c^3.$$

For its part, this torus is contained in a 3-sphere enclosing the origin of $\mathbb{C}^2$, namely the 3-sphere S_c^3

$$|z_1|^2 + |z_2|^2 = c^8 + c^6.$$

If one now allows c to run from zero to infinity, then $c^6 + c^8$ increases strictly monotonically from zero to infinity. The corresponding spheres fill $\mathbb{C}^2 - \{0\}$, and different values of c give different spheres. For this reason, we obviously have

$$K_c = S_c^3 \cap C.$$

Now how does the circle K_c lie in the 3-sphere S_c^3 ? K_c lies on the torus described above. Thus one must first understand how this torus lies in the 3-sphere.

This is now a standard situation, which has nothing to do with our curve. We consider the 3-sphere S_1^3 , and hence the points in $\mathbb{C}^2$ with

$$|z_1|^2 + |z_2|^2 = 2,$$

and the torus therein, of points of $\mathbb{C}^2$ with

$$|z_1| = 1, \ |z_2| = 1.$$

This torus divides the 3-sphere into 2 pieces, namely

$$T^+ = \{(z_1, z_2) \in S_1^3 \mid |z_1| \le 1\}$$
$$T^- = \{(z_1, z_2) \in S_1^3 \mid |z_2| \le 1\}.$$

Obviously

$$S_1^3 = T^+ \cup T^- .$$

We claim that T^+ and T^- are 3-dimensional solid tori. The proof, say for T^+, is simple. If T^0 is the 3-dimensional solid torus

$$T^0 = \{(z_1,z_2) \in \mathbb{C}^2 \mid |z_1| \leq 1, |z_2| = 1\},$$

then one obtains a homeomorphism $T^+ \to T^0$ by

$$(z_1,z_2) \mapsto (z_1,z_2/|z_2|).$$

Thus one finds that the representation $S_1^3 = T^+ \cup T^-$ of the 3-sphere S_1^3 results from pasting together two copies of the solid torus T^0 along the boundary $S^1 \times S^1$, with the two factors of the boundary $S^1 \times S^1$ being exchanged by the pasting. One can visualise this situation well by omitting a point of S_1^3, so that one obtains the 3-dimensional euclidean space $\mathbb{R}^3$. Then one can describe the decomposition as follows : one solid torus is simply embedded in the usual way. The other is its complement. One easily sees that the complement in $\mathbb{R}^3$ is likewise a torus, more precisely a solid torus, from which a point has been removed. The following picture may help.

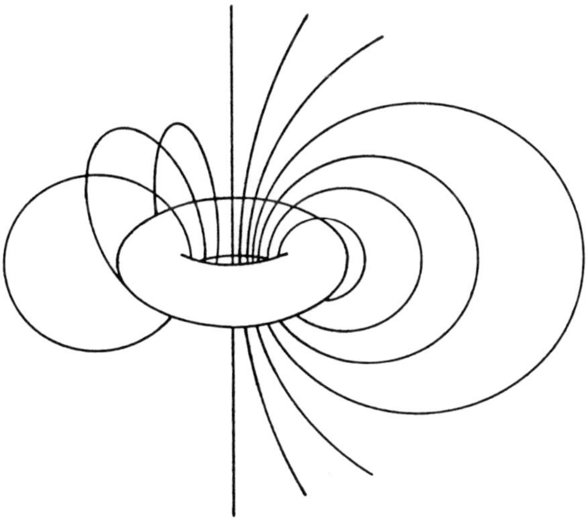

We now return to the topological investigation of the curve C. The intersection

$$K_1 = S_1^3 \cap C$$

is now seen to lie in the torus T with $|z_1| = 1$, $|z_2| = 1$, and the latter torus lies in the sphere in the way described above, hence it is embedded in $\mathbb{R}^3 = S_1^3 - \{p\}$ in the usual way. Now how does the circle K_1 lie in the torus T ? K_1 is the image of S_1^1 in $T = S_1^1 \times S_1^1$ under the mapping

$$t \mapsto (t^4, t^3).$$

This is a circle which winds around the torus, in such a way that it turns four times in one direction while it turns three times in the other. The following picture presents this intuitively.

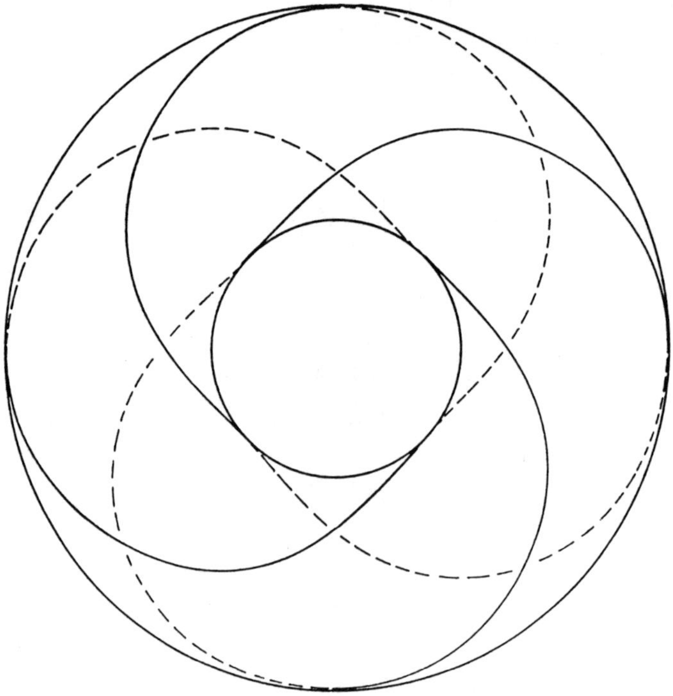

The curve K_c in S^3 resp. $\mathbb{R}^3$ is thus a circle looped around itself in a certain way or, as topologists say, a <u>knot</u>. Such knots were first investigated by Listing, a student of Gauss, in the middle of the 19th century, after Gauss' own reflections on electrodynamics had led him to consider several linked circles, and to define a linking number for them. The knots defined by mappings $S^1 \to S^1 \times S^1$ of the form $t \mapsto (t^q, t^p)$ with relatively prime p, q are called <u>torus knots</u>. They are precisely the result of intersecting the generalised parabolas $z_1^p - z_2^q = 0$ with the standard 3-sphere in $\mathbb{C}^2$. It was first observed by Brauner in 1928 that the singularities of plane curves have something to do with knots. The simplest knot occurring in this way is that which results from the semicubical parabola $z_1^2 - z_2^3 = 0$. It is the trefoil knot, an ancient pattern, examples of which occur in the history of art from the earliest times.

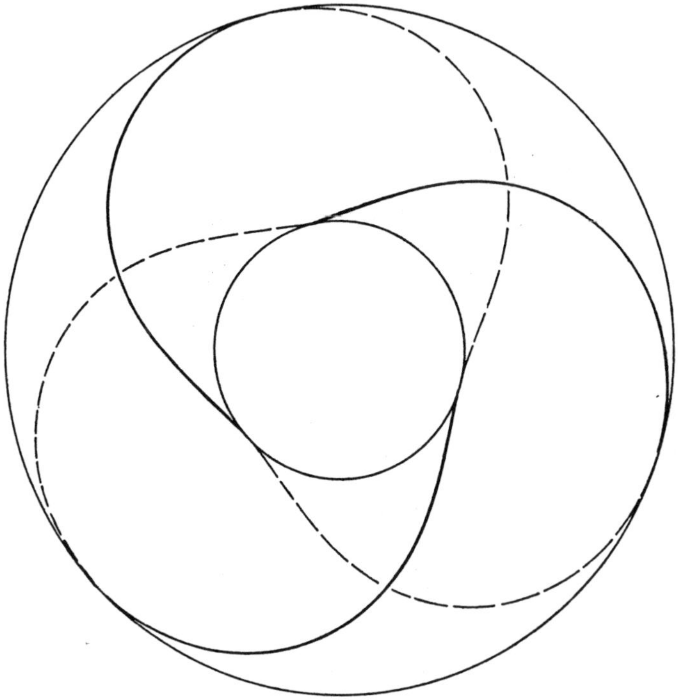

For curves with more complicated singularities, intersection with a small sphere round the origin yields more complicated knots, the so-called iterated torus knots. When the curve divides into several pieces in a neighbourhood of the singular point, intersection with a sphere yields a link of several iterated torus knots. We shall go into details later (in 8.4).

We now return to the topological investigation of our curve C with the equation $z_1^3 - z_2^4 = 0$. We have investigated the intersection

$$K_c = S_c^3 \cap C$$

for a particular value $c = 1$ and found that K_c is a particular torus knot in the 3-sphere. The situation for arbitrary c is exactly the same. In fact we can now precisely describe the topological type of the pair $(\mathbb{C}^2, C)$. We view the sphere S_1^3 and the knot K_1 in it as a subset of $\mathbb{C}^2$ and construct the cone $\tilde{C}$ over K_1, i.e.

$$\tilde{C} = \{tz \in \mathbb{C}^2 \mid t \geq 0,\ z \in K_1\}.$$

Of course, we can view $\mathbb{C}^2$ itself as the cone over S_1^3, i.e. write each point of $\mathbb{C}^2$ as tz with $z \in S_1^3$ and $t \geq 0$. Then

$$(tz_1, tz_2) \mapsto (t^4 z_1\ ,\ t^3 z_2)$$

defines a homeomorphism $\mathbb{C}^2 \to \mathbb{C}^2$ which maps the cone $\tilde{C}$ onto the affine algebraic curve C. Thus we have proved :

Proposition 1

Let $C \subset \mathbb{C}^2$ be the curve with equation $z_1^p - z_2^q = 0$, where p, q are relatively prime. Then the pair $(\mathbb{C}^2, C)$ is homeomorphic to the pair $(\mathbb{C}^2, \tilde{C})$, where $\tilde{C}$ is the cone over a torus knot of type (p,q).

What we do not want to prove here is the purely topological fact that the pair $(\mathbb{C}^2, \tilde{C})$ is not homeomorphic to the pair $(\mathbb{R}^4, \mathbb{R}^2)$. That this is so is, nevertheless, very plausible, because it is intuitively immediate that the torus knot of type (3,4) cannot be unknotted. Thus we already seem to have obtained a far-reaching insight into the topological nature of the singular points of plane curves. When these curves do not divide into several pieces in the neighbourhood of such a point, they are indeed 2-dimensional topological manifolds. However, they are not embedded flatly in the real 4-dimensional $\mathbb{C}^2$, as topological submanifolds, but in the tricky way described above : they are locally cones over iterated torus knots.

When the curve divides locally into several pieces, the situation is still more complicated. Then the curve is locally the cone over a link of iterated torus knots. The simplest situation is that in which the individual pieces are non-singular and meet transversely.

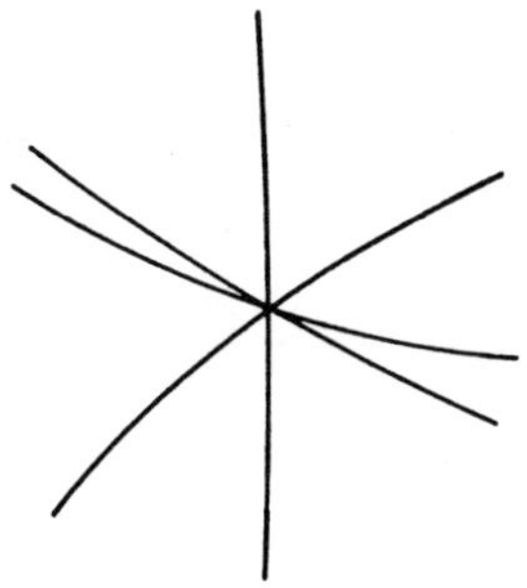

One calls such a singular point an ordinary n-tuple point when the curve divides locally into n components. Since we shall first treat the local decomposition into components later, we shall give the following definition here instead of the intuitive one :

Definition

A singular point p of a plane curve C is called an ordinary n-tuple point of C when C has exactly $n = \nu_p(C)$ distinct tangents at C.

To conclude this section we want to describe the singular points of a curve with the help of the homogeneous equation, instead of the affine equation we have used till now.

Proposition 2

Let C be a plane complex projective algebraic curve with homogeneous equation $F(x_0,x_1,x_2) = 0$, and let p be a point of C. Then $\nu_p(C) = r$ exactly when all partial derivatives of F up to, but not including, order r vanish at p.

Proof : Without loss of generality, we can choose coordinates as we please, since the statement about the vanishing of the derivatives of F at p is obviously coordinate-invariant. We choose the homogeneous coordinates (x_0,x_1,x_2) so that x_0 does not divide F and p does not lie on the line $x_0 = 0$, say $p = (1,a,b)$. Then $f(x,y) = F(1,x,y)$, hence $f(x,y) = 0$ is the affine equation of C and

$$\frac{\partial f^{i+j}}{\partial x^i \partial y^j}(a,b) = \frac{\partial F^{i+j}}{\partial x_1^i \partial x_2^j}(1,a,b).$$

It follows that : if all the partial derivatives of F of order $\leq r-1$ vanish at p, then the same is true for the partial derivatives of f ($r < n := \text{degree}(F)$)). Then if not all the rth derivatives of F vanish at p, at least one of the derivatives $\frac{\partial^r F}{\partial x_1^i \partial x_2^{r-i}}$ does not vanish at p, and hence an rth derivative of f also does not vanish. Otherwise $\frac{\partial^r F}{\partial x_0^r}$ would be the single non-vanishing derivative and the Euler formula 4.4.3 would give

$$\frac{\partial^{r-1} F}{\partial x_0^{r-1}}(1,a,b) = \frac{1}{n-r+1} x_0 \frac{\partial^r F}{\partial x_0^r}(1,a,b)$$

contrary to the fact that the left side vanishes and the right does not. This proves one direction of the theorem. The converse direction is proved similarly : if all the partial derivatives of f of order $\leq r-1$ vanish at p, then the corresponding derivatives of F also vanish. With the help of the Euler formula, one then proves by induction that the derivatives $\frac{\partial^i F}{\partial x_0^i}$, for $i \leq r-1$, also vanish at p. This proves the theorem.

The multiplicity $\nu_p(C)$ of a point p on a curve of order m naturally satisfies $\nu_p(C) \leq m$; this follows immediately from the definition. Thus to find the singular points of C we need only seek points with $\nu_p(C) = r \leq m$. The following corollary shows that the points with $\nu_p(C) \geq r$ can be found with the help of the $(r-1)^{th}$ derivatives alone.

Corollary 3

Let $C \subset P_2(\mathbb{C})$ be a plane curve of order m with homogeneous equation $F(x_0,x_1,x_2) = 0$. Then the points p of C with $\nu_p(C) \geq r$ (where $r \leq m$) are precisely the points of $P_2(\mathbb{C})$ where all the $(r-1)^{th}$ partial derivatives vanish. In particular, the singular points of C are the points of $P_2(\mathbb{C})$ where all the first partial derivatives vanish.

Proof : By Proposition 2, the vanishing derivative condition is trivially necessary, and by the Euler formula 4.4.4 it is also sufficient.

To conclude, we shall now give the equation of the tangent to a curve at a regular point in homogeneous coordinates.

Proposition 4

The equation of the tangent to a curve C with homogeneous equation $F(x_0,x_1,x_2) = 0$ at a regular point p reads

$$\sum_{i=0}^{2} x_i \frac{\partial F}{\partial x_i}(p) = 0.$$

Proof : One can choose the coordinates for the proof arbitrarily, since the given equation transforms correctly under coordinate transformations. We therefore choose the coordinates (x_0,x_1,x_2) so that $p = (1,a,b)$ and so that x_0 does not divide F.

Let $x = \frac{x_1}{x_0}$, $y = \frac{x_2}{x_0}$ and let $f(x,y) = F(1,x,y)$. By remark (ii) after the definition of tangent, the equation of the tangent to the affine curve $f(x,y) = 0$ is

$$\frac{\partial f}{\partial x}(a,b)(x-a) + \frac{\partial f}{\partial y}(a,b)(y-b) = 0 \tag{1}$$

hence

$$\frac{\partial F}{\partial x_1}(1,a,b)\left(\frac{x_1}{x_0} - a\right) + \frac{\partial F}{\partial x_2}(1,a,b)\left(\frac{x_2}{x_0} - b\right) = 0. \tag{2}$$

Since $F(1,a,b) = 0$, it follows from the Euler formula that

$$-a\frac{\partial F}{\partial x_1}(1,a,b) - b\frac{\partial F}{\partial x_2}(1,a,b) = \frac{\partial F}{\partial x_0}(1,a,b). \tag{3}$$

After multiplication by x_0, (2) and (3) yield

$$x_0 \cdot \frac{\partial F}{\partial x_0}(1,a,b) + x_1 \cdot \frac{\partial F}{\partial x_1}(1,a,b) + x_2 \cdot \frac{\partial F}{\partial x_2}(1,a,b) = 0$$

and this is the equation of the tangent which we wished to derive.

6. The intersection of plane curves

6.1 Bézout's theorem

In this section we shall prove the main result in the intersection theory of plane algebraic curves, Bézout's theorem (1779). This theorem gives the precise number of intersection points of two plane curves C, C' in the complex projective plane $P_2(\mathbb{C})$. To be sure, the points of intersection must be chosen with suitable multiplicity, and defining this intersection multiplicity is a problem in itself. Here we shall choose a definition which makes the proof of Bézout's theorem particularly simple.

We shall first concern ourselves with a simpler qualitative

question, namely, whether the number of points in $C \cap C'$ is finite or infinite. Since each curve in $P_2(\mathbb{C})$ naturally has infinitely many points, $C \cap C'$ consists of infinitely many points when C and C' have a component in common. In what follows we shall therefore assume that C and C' have no common component. First we shall show that, under this assumption, $C \cap C'$ consists of only finitely many points. In fact, we shall give a bound on the number of intersection points, given the orders $\operatorname{ord}(C) = m$ and $\operatorname{ord}(C') = n$ of the curves.

The basic idea which we use to bound the number of these intersection points and in Bézout's theorem is quite simple : as in the following figure we choose a point q in the complement of C and C', and a line L which does not go through q, and project $P_2(\mathbb{C}) - \{q\}$ from q onto L. This projection π maps the line L_c through q and a point $c \in L$ onto c. On each line L_c there are obviously only finitely many points of $C \cap C'$ (by 5.2). Thus to show that there are only finitely many intersection points, it suffices to show intersection points lie on L_{c^i} for only finitely many $c^i \in L$. A refinement of this argument also yields the desired bound and a determination of the number of intersection points.

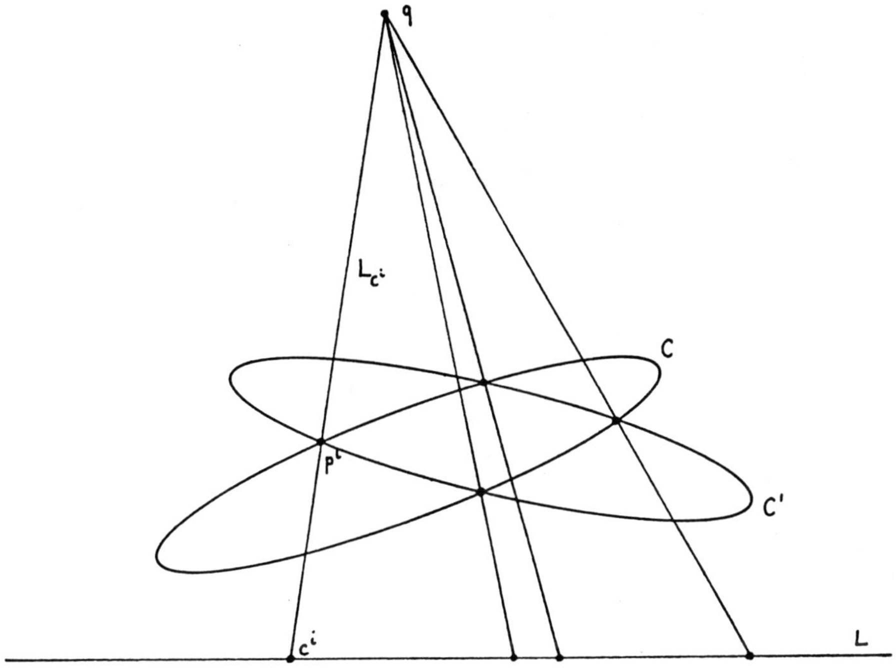
q
L_{c^i}
C
p^i
C'
c^i
L

In order to determine the L_{c^i}, we proceed as follows : we choose homogeneous coordinates x_0, x_1, x_2 in $P_2(\mathbb{C})$ so that $q = (0,0,1)$ and L has the equation $x_2 = 0$, hence x_0, x_1 are homogeneous coordinates in the projective line L and the projection $\pi : P_2(\mathbb{C}) - \{q\} \to L$ is given by $\pi(x_0,x_1,x_2) = (x_0,x_1)$. Relative to these coordinates, let C, C' have the equations $F(x_0,x_1,x_2) = 0$, $G(x_0,x_1,x_2) = 0$ respectively, where F and G are homogeneous polynomials of degree m and n respectively. Because of the above description of the projection, it is convenient to view these polynomials as polynomials in x_2 with coefficients in $\mathbb{C}[x_0,x_1]$, thus

$$F = A_0x_2^m + A_1x_2^{m-1} + \dots + A_m$$

$$G = B_0x_2^n + B_1x_2^{n-1} + \dots + B_n \, .$$

Here A_i, $B_j \in \mathbb{C}[x_0,x_1]$ are homogeneous polynomials of degrees i resp. j, and since $q \notin C, C'$ and $q = (0,0,1)$ we obviously have $A_0 \neq 0$, $B_0 \neq 0$. Therefore, since C and C' have no common component, and hence F and G have no non-constant common factor, it follows from 4.2.3 and 4.4.8 that $R_{F,G}(x_0,x_1)$ is a non-zero homogeneous polynomial of degree $m \cdot n$.

Now let $p = (c_0,c_1,c_2)$ be an intersection point of C and C'. Then $F(c_0,c_1,c_2) = 0$ and $G(c_0,c_1,c_2) = 0$, i.e. c_2 is a common solution of the two equations

$$A_0(c_0,c_1)x_2^m + A_1(c_0,c_1)x_2^{m-1} + \dots + A_m(c_0,c_1) = 0$$

$$B_0(c_0,c_1)x_2^n + B_1(c_0,c_1)x_2^{n-1} + \dots + B_n(c_0,c_1) = 0.$$

Hence it follows by 4.2.6 that $R_{F,G}(c_0,c_1) = 0$, and conversely, this condition is sufficient for the existence of a solution. Thus we obtain :

The zeroes of $R_{F,G}$ are precisely the $c \in L$ for which the lines L_c contain intersection points of C and C'. This determines the lines L_{c^i}, and yields a method for explicitly finding the intersections of C and C' by computing the zeroes of a certain polynomial.

After these preparations we now obtain the desired bound on the number of intersection points very quickly.

<u>**Proposition 1**</u>

Two curves C and C', of orders m resp. n, without common component in the complex projective plane have at most $m\cdot n$ intersection points.

<u>**Proof**</u> : Suppose there are at least $m\cdot n+1$ intersection points $p^1,\ldots,p^{m\cdot n+1}$. Then we choose the point q and the line L as above, and in addition arrange that q lies on none of the lines L_{ij} connecting p^i to p^j, where $i \neq j$. Then p^i and p^j have different images c^i and c^j in L under the projection π. In this way we obtain at least $m\cdot n+1$ different points $c^1,\ldots,c^{m\cdot n+1}$ in L which are zeroes of the resultant $R_{F,G}$ of the equations of C and C'. But this contradicts the fact that $R_{F,G}$ is homogeneous of degree $m\cdot n$. Thus there are at most $m\cdot n$ points of intersection.

Since we now know that the two curves C and C' have only finitely many points p^i in common, we can choose our centre of projection q so that it lies on no line L_{ij} connecting two intersection points p^i and p^j. Then the projection π yields a bijection between the intersection points p^i and their images $c^i = \pi(p^i)$ in L. This suggests the following :

<u>**Definition**</u>

Let C and C' be curves in the complex projective plane without common components, and let p^i be the intersection points of C and C'. Let (x_0,x_1,x_2) be coordinates such that the point $(0,0,1)$ lies on no line connecting intersection points, and let $F(x_0,x_1,x_2) = 0$ resp. $G(x_0,x_1,x_2) = 0$ be the equation of C resp. C' with respect to these coordinates. Let $R_{F,G}(x_0,x_1)$ be the resultant of the polynomials F and G, viewed as polynomials in x_2 with coefficients in $\mathbb{C}[x_0,x_1]$. Let $p^i = (c_0^i,c_1^i,c_2^i)$ and let $c^i = (c_0^i,c_1^i)$. Then the <u>intersection multiplicity</u> $\nu_{p^i}(C,C')$ of C and C' at p^i is defined to be the multiplicity of the zero c^i of $R_{F,G}$.

In this way we finally have the definition of intersection multiplicity reduced to quite a simple multiplicity definition, namely that of the multiplicity of zeroes of a polynomial in one variable (4.4.7). Since $R_{F,G}$, being a homogeneous polynomial of degree $m\cdot n$, has exactly $m\cdot n$ zeroes when their multiplicities are counted, it obviously follows that the sum of intersection multiplicities $\nu_{p^i}(C,C')$ is exactly $m\cdot n$.

This is the desired determination of the number of intersection points.

Theorem 2 (Bézout's Theorem)

The sum of intersection multiplicities of two curves C, C' of orders m, n respectively in the complex projective plane satisfies

$$\sum_{p \in C \cap C'} \nu_p(C,C') = m \cdot n.$$

This theorem, which originated in remarks of Newton and MacLaurin, was already proved by Euler 1748 and Cramer 1750. In the next section we shall see some applications of this theorem, and in the section after that we shall investigate it from a different point of view. But first let us consider more closely the concept of intersection multiplicity just introduced.

The problem of defining intersection multiplicity can be approached in different ways : on the one hand, one can choose an axiomatic method. This means presenting a series of postulates for intersection multiplicity, and then proving that there is at most one intersection theory satisfying these conditions. An example of this method is the presentation in the book of Fulton [Fl]. In order to show that there really is an intersection theory satisfying these postulates, one must still define intersection multiplicity in some explicit way (explicit as opposed to implicit definition by axioms). Our definition above is an example of such an explicit definition. There is a series of other, equivalent definitions, some of which we shall meet later.

Some of these are better, and others worse, for generalising to higher-dimensional situations. An example of such a higher- dimensional situation is the following : let V and W be projective-algebraic varieties in $P_n(\mathbb{C})$ of co-dimensions a resp. b. When the intersection $V \cap W$ is pure-dimensional of co-dimension $a+b$, one can again associate intersection multiplicities with the components of $V \cap W$. If in particular $a + b = n$, then $V \cap W$ consists of finitely many points, and we again have Bézout's theorem : the sum of the intersection multiplicities is the product of the orders of V and W.

Our definition of intersection multiplicity above has the advantage that Bézout's theorem follows immediately from known results about resultants. But it has the disadvantage of being a priori dependent on the choice of coordinates. We first want to convince ourselves that it is really independent of this choice.

Lemma 3

The intersection multiplicity $\nu_p(C,C')$ defined above does not depend on the choice of coordinate system in the projective plane.

Proof : Let (x_0,x_1,x_2) be a homogeneous coordinate system which satisfies the conditions for the definition of multiplicity, so that $(0,0,1) \notin C \cup C' \cup \bigcup_{i,j} L_{ij}$, where L_{ij} is the line connecting the intersection points p^i and p^j of C with C' (an "admissible" coordinate system for short). We obtain any other admissible coordinate system (x'_0,x'_1,x'_2) by a linear transformation $x_i = \Sigma a_{ij}x'_j$, where $A = (a_{ij})$ is an invertible 3×3 matrix, i.e. an element of $GL(3,\mathbb{C})$. The matrices which yield admissible coordinate systems are precisely those for which the point with x'-coordinates $(0,0,1)$ again does not lie in $C \cup C' \cup \bigcup_{i,j} L_{ij}$. In the x-coordinate system this point is (a_{02},a_{12},a_{22}). If $H(x_0,x_1,x_2) = 0$ is the equation of $C \cup C' \cup \bigcup_{i,j} L_{ij}$ in the (x_0,x_1,x_2)-coordinate system, then the set X of all matrices A which yield admissible coordinate systems is described by

$$X = \{A \in GL(3,\mathbb{C}) \mid H(a_{02},a_{12},a_{22}) \neq 0\}.$$

One can also describe this set of matrices as follows : let $M(3,\mathbb{C})$ be the set of all 3×3 matrices $A = (a_{ij})$ with $a_{ij} \in \mathbb{C}$. This is a $\mathbb{C}^9$ with the coordinates a_{ij}. The determinant is a homogeneous polynomial $D(a_{ij})$ of third degree in the a_{ij}.

Let $P(a_{ij}) = D(a_{ij}) \cdot H(a_{02},a_{12},a_{22})$. Then :

$$X = \{(a_{ij}) \in M(3,\mathbb{C}) \mid P(a_{ij}) \neq 0\}.$$

Thus X is the complement of an affine hypersurface in the 9-dimensional affine space $\mathbb{C}^9 = M(3,\mathbb{C})$. X is a topological space as a subspace of $\mathbb{C}^9$. We claim that X is path-connected. This results from the following general fact : if V is a connected algebraic manifold over the field of complex numbers and $W \subset V$ is an algebraic subvariety, then $X = V - W$ is also connected. (One can prove this general assertion by induction on $\dim W$: by the induction hypothesis one can consider $V\text{-Sing}(W)$ instead of V, where $\text{Sing}(W)$ is the singularity set of W.

Hence it suffices to prove the theorem when W is non-singular. The proof then follows from the fact that a curve connecting two points of $V-W$ in V can be made transverse to W by an arbitrarily small perturbation. But since W has real co-dimension in V of at least 2, transversality means that the curve does not meet W.) In our case $V = \mathbb{C}^9$ and W is the hypersurface with equation $P = 0$. In this case one can see the connectedness of $X = V - W$ directly quite easily : if $A,B \in X$ and L is the complex-affine line in $\mathbb{C}^9$ through A and B, then $L' = L \cap X$ results from L by removal of finitely many points, namely the zeroes of the polynomial which results from P by substitution of an affine coordinate for L. $L' \subset X$ is of course connected and it contains A, B. Then it is clear that X is path-connected.

We now come to the proof of the coordinate independence of intersection multiplicity. We shall prove that the intersection multiplicities $\nu_p(C,C')$ and ν'_p defined for two coordinate systems (x_0,x_1,x_2) and (x'_0,x'_1,x'_2) are the same. Let $x = Ax'$. By the connectedness of X, we can find a continuous curve A_t, $0 < t < 1$, in X with $A_0 = 1$, $A_1 = A$. An admissible coordinate system $x_t = (x_{t,0},x_{t,1},x_{t,2})$ is defined for each t by $A_t x_t$. If one substitutes the latter in the equations $F(x) = 0$ resp. $G(x) = 0$ of C resp. C', then one obtains the equations $F_t(x_t) = 0$ resp. $G_t(x_t) = 0$ of C resp. C' relative to the coordinates x_t. As above, we consider the resultant $R_{F_t,G_t}(x_{t,0},x_{t,1})$. By construction, $R_t(x_0,x_1) = R_{F_t,G_t}(x_0,x_1)$ is a one-parameter family of homogeneous polynomials of degree $m \cdot n$ with the following property : the number of distinct zeroes $c^i(t)$ of R_t (not counting multiplicity) is constant (namely, equal to the number of distinct intersection points p^i of C and C', to which they correspond under projection). But it follows from this that the multiplicity of zeroes $c^i(t)$ of R_t, which depends continuously on t, is also constant. This is immediate from the following well-known fact about the roots of polynomials in one indeterminate :

Let $f \in \mathbb{C}[x]$ be a polynomial of degree d and let $c \in \mathbb{C}$ be a zero of f of multiplicity ν. Let $\varepsilon > 0$ be so small that all other zeroes have distance $> \varepsilon$ from x. Then there is a $\delta > 0$ for ε such that each polynomial whose coefficients differ from those of f by less than δ has exactly ν zeroes in the circle of radius ε around c, counting multiplicities.

This proves that the multiplicity of the zeroes $c^i(t)$, and with it the intersection multiplicity of C and C' at p^i with respect to the coordinate system x_t, is independent of t. Thus the intersection multiplicity is coordinate-independent.

Q.e.d.

The computation of intersection multiplicity from the definition in terms of the multiplicity of zeroes of the resultant is naturally very troublesome. For this reason, simpler formulae for the intersection multiplicity are useful. The most important one is given by the following theorem.

Proposition 3

Let C and C' be curves without common components and let p be an intersection point of C and C'. Then

(i) $\nu_p(C,C') \geq \nu_p(C) \cdot \nu_p(C')$

(ii) Equality, $\nu_p(C,C') = \nu_p(C) \cdot \nu_p(C')$, holds if and only if tangents to C at p are pairwise distinct from the tangents to C' at p.

Proof : Let $r = \nu_p(C)$ and $s = \nu_p(C')$. Without loss of generality we can choose homogeneous coordinates (x_0,x_1,x_2), relative to which we have equations $F(x_0,x_1,x_2) = 0$ resp. $G(x_0,x_1,x_2) = 0$ for C and C' and their resultant $R_{F,G}$ (with respect to x_2), so that $q = (0,0,1)$ is not on C or C' or on a line connecting intersection points of C and C', or on a tangent to C or C' at p, and so that $p = (1,0,0)$.

Let $f(x,y) = F(1,x,y)$ and $g(x,y) = G(1,x,y)$ be the associated polynomials and let $R_{f,g}(x)$ be their resultant, where f, g are viewed as polynomials in y with coefficients in $\mathbb{C}[x]$. Obviously : the order of the zero $x = 0$ of $R_{f,g}(x)$ equals the order of the zero $(1,0)$ of $R_{F,G}$, hence it equals $\nu_p(C,C')$.

Thus we investigate $R_{f,g}(x)$.

It follows from $\nu_p(C) = r$, $\nu_p(C') = s$ and $p = (1,0,0)$ that :

$$f(x,y) = f_0y^m + f_1y^{m-1} + \dots + f_{m-r}y^r + f_{m-r+1}xy^{r-1} + \dots + f_mx^ry^0$$

$$g(x,y) = g_0y^n + g_1y^{n-1} + \dots + g_{n-s}y^s + g_{n-s+1}xy^{s-1} + \dots + g_nx^sy^0$$

with f_i, $g_j \in \mathbb{C}[x]$.

The equations of the tangents to C resp. C' at $p = (1,0,0)$ are

$$f_{m-r}(0)y^r + f_{m-r+1}(0)xy^{r-1} + \ldots + f_m(0)x^r = 0$$
$$g_{n-s}(0)y^s + g_{n-s+1}(0)xy^{s-1} + \ldots + g_n(0)x^s = 0.$$

The tangents to C are pairwise distinct from those to C' if and only if these equations have no non-trivial common solution, i.e. when

$$D_1 = \begin{vmatrix} f_{m-r}(0) & \cdots & f_m(0) & & \\ & \ddots & & \ddots & \\ & & f_{m-r}(0) & \cdots & f_m(0) \\ g_{n-s}(0) & \cdots & g_n(0) & & \\ & \ddots & & \ddots & \\ & & g_{n-s}(0) & \cdots & g_n(0) \end{vmatrix} \neq 0$$

Moreover, the assumption that no further intersection points lie on the line $x = 0$ from p to q has the consequence that the equations $f(0,y) = 0$ and $g(0,y) = 0$ have only $y = 0$ as common solution, i.e.

$$f_0(0)y^{m-r} + f_1(0)y^{m-r-1} + \ldots + f_{m-r}(0) = 0$$
$$g_0(0)y^{n-s} + g_1(0)y^{n-s-1} + \ldots + g_{n-s}(0) = 0$$

have no common solution, because $f_{m-r}(0) \neq 0$ and $g_{n-s}(0) \neq 0$ since $x = 0$ is not a tangent to C resp. C' at p. This means that the following resultant does not vanish :

$$D_2 = \begin{vmatrix} f_0(0) & \cdots & f_{m-r}(0) & & \\ & \ddots & & \ddots & \\ & & f_0(0) & \cdots & f_{m-r}(0) \\ g_0(0) & \cdots & g_{n-s}(0) & & \\ & \ddots & & \ddots & \\ & & g_0(0) & \cdots & g_{n-s}(0) \end{vmatrix} \neq 0$$

Now we consider the determinant of the $(m+n)$-rowed matrix which defines $R_{f,g}$. We multiply the $(n-s+1)^{th}$ row by x, the next by x^2 etc., down to the nth, which we multiply by x^s. Similarly, we multiply the $(n+m-r+1)^{th}$ row by x, the next by x^2 etc., down to the $(n+m)^{th}$, which we multiply by x^r. As a result, $R_{f,g}$ is multiplied by $x^{1+2+...+r+1+2+...+s}$. Now we divide the last column by x^{r+s}, the second last by x^{r+s-1} etc., back to the $(n+m-r-s+1)^{th}$, which is divided by x. As a result, the determinant is divided by

$$x^{1+2+...+r+(r+1)+(r+2)+...+(r+s)} = x^{1+2+...+r+1+2+...+s+r.s} .$$

Altogether we have

$$R_{f,g}(x) = x^{r.s}D(x). \qquad (1)$$

$D(x)$ is the determinant of a matrix whose entries are polynomials in x. The form of this matrix is indicated in the following schema. Expanding by the first $n+m-r-s$ columns, one sees that

$$D(0) = D_1 \cdot D_2. \qquad (2)$$

Then assertion (i) follows from (1) and assertion (ii) follows from (1) and (2) and $D_2 \neq 0$, together with the fact that $D_1 \neq 0$ just in case the transversality condition for the tangents is satisfied.

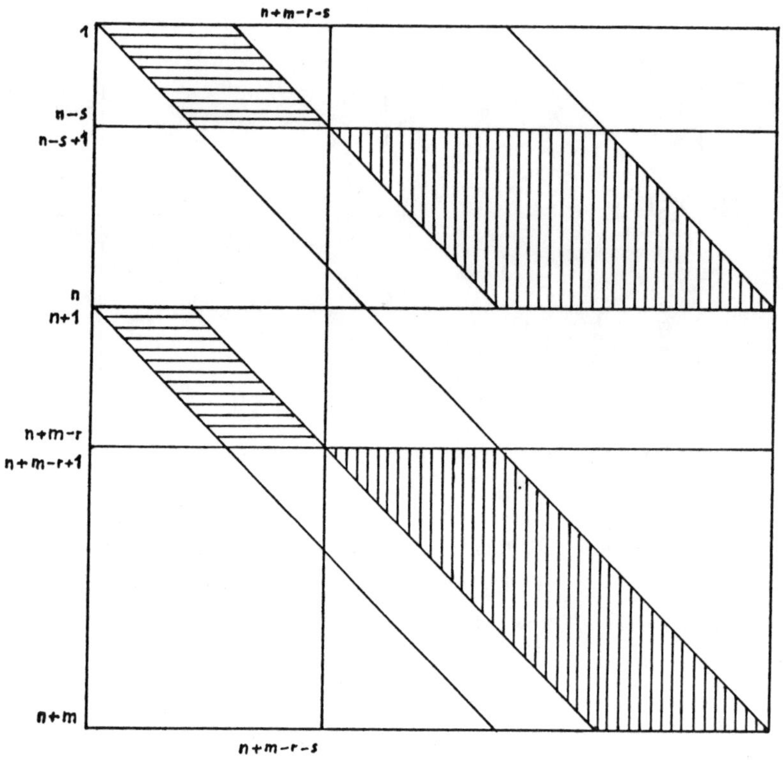

To conclude, we can use the local intersection numbers to define a global <u>intersection number</u> $C \cdot C'$ <u>of two curves</u> without common components :

$$C \cdot C' = \sum_{p \in C \cap C'} \nu_p(C,C').$$

6.2 Applications of Bézout's theorem

In the section "Origin and generation of curves" at the beginning of this course we saw that an important reason for the study of curves is the fact that the solution of many problems - e.g. the classical problems of antiquity - reduces to finding the intersection of suitable curves. This viewpoint was also stressed by Newton. Bézout's theorem solves the problem of determining the number of intersection points of two curves C and C', and results in many applications to geometric problems, as we shall see in this paragraph and the next (on cubics).

One can also invert the question in a certain sense : given finitely many points in the projective plane, one can ask which curves go through these points. In this way one obtains a whole system of curves such that the given points lie in the intersection of all curves in the system. Before we start on applications of Bézout's theorem, we make a series of remarks on such systems of curves.

We shall describe our system of curves with the help of the corresponding system of equations. To do so we must extend the earlier definition of curves in 5.1, in order to have a completely satisfactory relation between curves and equations. In 5.1 we defined a curve as the zero set of a homogeneous polynomial in the projective plane, and hence as a subset of $P_2(\mathbb{C})$. We then described each curve by an equation $F(x_0,x_1,x_2) = 0$, where F is a homogeneous polynomial, determined by the curve up to a constant factor. The decomposition of the curve into irreducible components then corresponds to the decomposition of F into irreducible factors. For this reason, the polynomials appearing in equations of curves are precisely those with no multiple factors. When we want all polynomials to appear, we must correspondingly admit "curves" with "multiple" components. Thus by a curve with multiple components we mean a formal linear combination

$$C = n_1C_1 + \ldots + n_rC_r$$

where the C_i are irreducible curves (in the old sense) and the n_i are natural numbers. The number n_i gives the multiplicity with which we count the component C_i. If $F_i = 0$ is the equation of C_i, then the equation of C is naturally

$$F_1^{n_1} \cdot \ldots \cdot F_r^{n_r} = 0$$

and the order of C is the degree of this homogeneous polynomial. If two such curves $C = \Sigma m_i C_i$ and $C' = \Sigma n_j C'_j$ have no common components, then one naturally defines the multiplicity of a point of intersection by

$$\nu_p(C,C') = \Sigma m_i n_j \nu_p(C_i, C'_j)$$

and the ordinary Bézout theorem trivially yields a Bézout's theorem for these curves with multiple components. When we mean such curves with multiple components in what follows we shall place the word "curve" in quotation marks.

(If one also admits negative n_i in the linear combinations $D = \Sigma n_i C_i$, then one calls such a D a <u>divisor</u> on the projective plane. In general, one understands a divisor on an algebraic manifold W to be an integral formal linear combination of subvarieties of co-dimension 1.)

The set of all homogeneous polynomials

$$\sum_{i_0+\ldots+i_k=m} a_{i_0\ldots i_k} x_0^{i_0} \ldots x_k^{i_k}$$

of degree m in the indeterminates $x_0,\ldots,x_k$ obviously forms a complex vector space of dimension

$$N = \binom{m+k}{k},$$

because this is the number of combinations of m elements (with repetitions) which can be chosen from a set of $k+1$ elements. If one regards two homogeneous polynomials as equivalent when they differ only by a constant factor, i.e. when they define the same "curve", then the set of these equivalence classes forms a complex projective space P_{N-1}, and the coefficients $a_{i_0\ldots i_k}$ of the homogeneous polynomial are homogeneous coordinates in this P_{N-1}. Thus we obtain : <u>the set of all "curves" of order</u> m <u>in the projective plane forms a complex projective space of dimension</u> $\frac{m(m+3)}{2}$.

One can now consider projective subspaces of this projective space. One calls such a linear subspace P_d of curves of a particular degree m a <u>linear subsystem</u>, and the dimension d of P_d is called the <u>dimension of the linear system</u>. If $d = 1$, a case which one considers particularly often, then instead of speaking of a 1-dimensional linear system one also speaks of a linear <u>pencil</u> of curves. For example, the lines through a fixed point in the projective plane form a pencil.

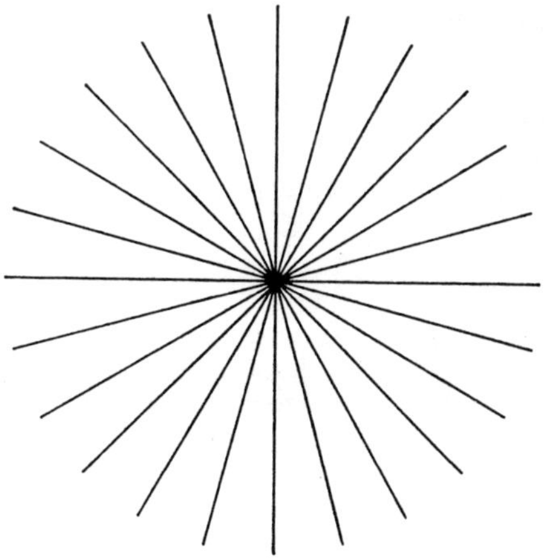

One meets linear systems in the treatment of various problems in algebraic geometry, e.g. – and this is a particularly important application – in the construction of rational mappings (see also [M8], §6A).

One can describe linear systems in various ways, e.g. by giving a basis $F_0, \ldots, F_d$ for the $(d+1)$-dimensional vector space of homogeneous polynomials of degree m corresponding to the d-dimensional linear system P_d. The linear system then consists of all "curves" with equations

$$\lambda_0 F_0 + \ldots + \lambda_d F_d = 0.$$

Often one considers linear systems of "curves" of a particular degree m which are defined by conditions of the following kind. One gives a certain finite set of points in the plane and demands that the "curves" go through these points, or that they have multiplicities exceeding particular values at these points. The "curves" which satisfy these conditions then form a linear system.

We shall now study such conditions in somewhat more detail. We consider "curves" of degree m, given by equations

$$F(x_0, x_1, x_2) = 0,$$

where

$$F(x_0, x_1, x_2) = \sum_{i+j+k=m} a_{i,j,k} x_0^i x_1^j x_2^k .$$

Since we first consider the vector space of all such polynomials, we regard the $a_{i,j,k}$ as variables, as coordinates in this vector space. Now we consider a particular point $c = (c_0, c_1, c_2)$ in $P_2(\mathbb{C})$. The condition for the "curve" to go through this point is that $F(c_0, c_1, c_2) = 0$, thus

$$\sum_{i+j+k=m} a_{i,j,k}\, c_0^i c_1^j c_2^k = 0.$$

This is a linear equation for the indeterminates $a_{i,j,k}$, with the complex numbers $c_0^i c_1^j c_2^k$ as coefficients.

Thus the condition for a "curve" of degree m to go through a point is a <u>linear condition</u>, and the "curves" which satisfy it form a projective subspace of codimension 1 in the projective space of all "curves" of degree m.

More generally, one sees that the condition for a "curve" of degree m to have multiplicity at least s at a particular point p is equivalent to $s(s+1)/2$ linearly independent conditions.
Proof : by 5.3.3 the multiplicity condition for the curve is equivalent to the vanishing at p of the $(s-1)^{th}$ partial derivatives of the associated homogeneous polynomial F. A homogeneous polynomial in three variables has exactly $s(s+1)/2$ partial derivatives of order $s-1$ (namely, as many as the number of terms up to order $s-1$ in the Taylor expansion of the affine equation at p). Hence the multiplicity condition is equivalent to $s(s+1)/2$ linear conditions. These conditions are linearly independent. To prove this one chooses the coordinates, without loss of generality, so that $p = (1,0,0)$. Then, since $F = \sum_{i+j+k=m} a_{ijk} x_0^i x_1^j x_2^k$, the vanishing of

$$\frac{\partial^{s-1} F}{\partial x_0^{s-1-j-k} \partial x_1^j \partial x_2^k}(1,0,0)$$ means precisely that $a_{i,j,k} = 0$, where $j+k \leq s-1$, and these are $s(s+1)/2$ linearly independent conditions.

Now one can, as already indicated at the beginning of this section, choose several points $p_1, \dots, p_r$ in $P_2(\mathbb{C})$, and natural numbers $s_1, \dots, s_r$, and consider the set of "curves" of mth order which have multiplicity at least s_ρ at p_ρ. These are purely linear conditions, and the "curves" which satisfy them therefore form a linear system $\mathcal{L}$ in the $(m(m+3)/2)$-dimensional projective space of all "curves" of mth order. We have proved :

Proposition 1

The "curves" of mth order which have multiplicity at least s_ρ at p_ρ form a linear system $\mathcal{L}$ of dimension d, where

$$d \geq \frac{m(m+3)}{2} - \sum_\rho \frac{s_\rho(s_\rho+1)}{2} .$$

Whether equality or inequality holds here depends on whether the conditions for the different points are linearly independent or not. Both can happen. One sees this already in a trivial example : let $m = 1$, $r = 3$, $s_1 = s_2 = s_3 = 1$. Thus we consider the condition for a line to go through three given points p_1 , p_2 , p_3. If the three points are in "general position", i.e. not collinear, then the three conditions are linearly independent, and $d = \frac{1.4}{2} - 3 = -1$. I.e. P_d is empty and there is no line which goes through p_1 , p_2 , p_3. But if the points are collinear, and hence in "special position", then $d > -1$ and there is one line, so that $d = 0$, when the points do not all coincide. In the latter case $d = 1$, and the lines form a pencil.

This situation is also typical for "curves" of higher order and systems of conditions for the multiplicities of the "curves" at given points p_ρ. In what follows we confine ourselves to the case $s_\rho = 1$.

Suppose we consider the case of quadrics, so that $m = 2$. They form a linear system of dimension $2(2+3)/2 = 5$. Thus at least one quadric goes through five different points. When exactly one goes through them, we shall say that the 5 points are in general position (relative to quadrics). When 5 points $p_1,\ldots,p_5$ are in general position and one chooses a point p_6 which does not lie on the quadric through $p_1,\ldots,p_5$, then there is no quadric through $p_1,\ldots,p_6$. This is the normal case, and hence we shall say that the six points are in general position (relative to quadrics) when no quadric goes through them, while they are in special position when they lie on a quadric.

Definition

The r different points $p_1,\ldots,p_r \in P_2(\mathbb{C})$ are in general position relative to "curves" of mth order when the linear system $\mathcal{L}$ of "curves" of mth order through $p_1,\ldots,p_r$ has the smallest possible dimension d, i.e.

$$d = m(m+3)/2 - r$$

when $m(m+3)/2 \geq r$ and $\mathcal{L}$ is empty when $m(m+3)/2 < r$.

In order to justify this definition, we must show that general position is really "general", i.e. that almost all r-tuples of points have general position. To do this we must first explain what "almost all" is to mean. We can view the r-tuple $(p_1,\dots,p_r)$ as a point in the m-fold cartesian product $P_2(\mathbb{C}) \times \dots \times P_2(\mathbb{C})$. The condition of general position is a condition on the rank of a system of linear equations, and it may therefore be expressed by saying that certain determinants do not vanish. These determinants are polynomials in the homogeneous coordinates of the points $p_1,\dots,p_r$. The set of $(p_1,\dots,p_r)$ in general position is therefore the complement $P_2(\mathbb{C}) \times \dots \times P_2(\mathbb{C}) - A$ of a certain algebraic set A, namely the zero set of these determinants. Two cases are now conceivable. Either the exceptional set A equals $P_2(\mathbb{C}) \times \dots \times P_2(\mathbb{C})$ – then there is no point $(p_1,\dots,p_r)$ in general position. Or else A is a proper algebraic subset. Then the points $(p_1,\dots,p_r)$ in general position in $P_2(\mathbb{C}) \times \dots \times P_2(\mathbb{C})$ form a non-empty so-called Zariski-open subset. In the sense of the usual topology (not the Zariski topology), such a set is then open and dense. When this is the case one is entitled to say that r distinct points <u>in general</u> are in general position, and we shall use the expression in this sense.

<u>Proposition 2</u>

In the projective plane, r distinct points in general are in general position relative to "curves" of mth order. In particular, there is in general exactly one "curve" of mth order through $m(m+3)/2$ points.

Proof : From what we have already said it is clear that it suffices to show : for each m there are $m(m+3)/2$ points through which exactly one "curve" of mth order passes. We show this by induction on m. For $m = 1$ the assertion is clear : there is exactly one line through two distinct points. To make the induction step from $m-1$ to m, we observe that

$$m(m+3)/2 = (m-1)(m-1+3)/2 + m + 1.$$

We first choose $m+1$ distinct points $p_1,\dots,p_{m+1}$ on a line L, and then $(m-1)(m-1+3)/2$ further points $p_{m+2},\dots,p_{m(m+3)/2}$ which do not lie on L and which are in general position. This is possible by the induction hypothesis. Claim : only one "curve" of mth order goes through $p_1,\dots,p_{m(m+3)/2}$.

Proof : by Bézout's theorem (5.2.1), L must be a component of each "curve" C of mth order through $p_1,\ldots,p_{m+1}$, hence $C = L \cup C'$, where C' is a "curve" of $(m-1)^{th}$ order. But the "curve" C' must go through the remaining points, hence it is uniquely determined, by the choice of these points. This proves the theorem.

Example :

Through 5 points there is in general exactly one quadric. Through 9 points there is in general exactly one cubic.

Now, as a first application of Bézout's theorem, we shall prove a theorem on special positions of intersection points of two "curves". As special cases, we obtain two theorems about particularly beautiful symmetric configurations of points and lines, namely the theorems of Pascal and Brianchon.

Theorem 3

Let C, C' be "curves" of order n which meet in exactly n^2 distinct points. When exactly $m \cdot n$ of these points lie on an irreducible "curve" C'' of order m, then the remaining $n(n-m)$ intersection points lie on a "curve" C''' of order $n-m$.

Proof : Let $F(x_0,x_1,x_2) = 0$ be the equation of C and let $G(x_0,x_1,x_2) = 0$ be the equation of C'. Let $\mathcal{L}$ be the pencil of "curves" of nth order with the equations

$$\lambda F + \mu G = 0.$$

All "curves" in $\mathcal{L}$ go through the n^2 intersection points of C and C'. Since $\mathcal{L}$ is 1-dimensional, we can find a "curve" in $\mathcal{L}$ to satisfy a given linear condition. To set up such a condition, we choose a point $q \in C''$ different from the $n \cdot m$ points in $C \cap C''$. Let $\tilde{C} \in \mathcal{L}$ be a "curve" with $q \in \tilde{C}$ – we have just established that $\tilde{C}$ exists. Then $\tilde{C}$ and C'' meet in at least $nm+1$ points. Hence they must have a common component, by Bézout's theorem, and since C'' is irreducible by hypothesis, this can only be C''. Thus

$$\tilde{C} = C'' \cup C''',$$

with C''' a "curve" of order $n-m$. The remaining $n(n-m)$ intersection points of C and C', which do not lie in C'', lie in $\tilde{C}$, and hence in C'''. This proves the theorem.

This theorem is a theorem about special position of points. The hypothesis says that $m \cdot n$ of the intersection points have special position, because they lie on a "curve" of order m, whereas for $n > m \geq 1$ we always have $m \cdot n \geq m(m+3)/2$, and hence $m \cdot n$ points in general position do not lie on a "curve" of order m (cf. Proposition 2). Likewise, of course, the hypothesis that n^2 points lie, not only on C, but also on another "curve" C' of order n, is a hypothesis about the special position of these points, because for $n > 2$, n^2 points in general determine a "curve" C of order n uniquely. Finally, the conclusion of the theorem is an assertion about the special position of points, because for $n > m \geq 2$, $n(n-m)$ points in general position do not lie on a "curve" of order $n-m$, by Proposition 2.

As an application of Theorem 6.2.3, we now prove Pascal's theorem on the configuration corresponding to a hexagon inscribed in an irreducible quadric. A <u>hexagon</u> is a system of six distinct points $p_1, \ldots, p_6$ and six lines $L_1, \ldots, L_6$ in the projective plane such that : L_i is the line through p_i and p_{i+1} for $i = 1, \ldots, 6$, where we set $p_7 = p_1$.

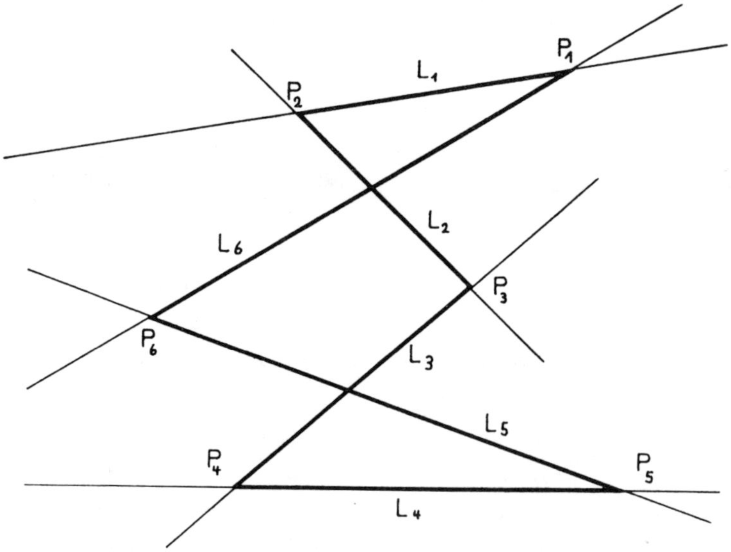

We call $p_1, \ldots, p_6$ the <u>vertices</u> of the hexagon and L_i , L_{i+3} the <u>opposite sides</u> (where indices are reduced modulo 6). We shall say that a hexagon is <u>inscribed in a quadric</u> when its six vertices lie on the quadric. A hexagon <u>circumscribes</u> an irreducible quadric when its sides are tangents to the quadric.

Corollary 4 (Pascal's theorem)

For a hexagon inscribed in an irreducible quadric the intersection points of the three pairs of opposite sides lie on a line.

Proof : Let $p_1,\ldots,p_6$ be the vertices and let $L_1,\ldots,L_6$ be the sides of the hexagon inscribed in the quadric Q. We apply Theorem 6.2.3 to $C = L_1 \cup L_3 \cup L_5$, $C' = L_2 \cup L_4 \cup L_6$ and $C'' = Q$. We check the hypotheses of 6.2.3. It is clear first of all that the L_i are pairwise distinct, because if two sides L_i , L_j were to coincide, at least three vertices would lie on L_i , hence on $L_i \cap Q$, contrary to Bézout's theorem. Now we claim that C and C' meet in exactly 9 points. Proof : it is clear from the hypotheses that $C \cap C'$ consists of the 6 vertices and the three intersection points q_1,q_2,q_3 of the pairs (L_1,L_4), (L_2,L_5), (L_3,L_6) of opposite sides. Also, the q_i are distinct from the p_j , since each side L_j meets Q_i in two vertices p_j , p_{j+1} , and hence in no other point by Bézout. But the q_i are also different from each other, for if two of them did coincide, so too would the corresponding side pairs, whereas all the sides are distinct.

Thus $C \cap C'$ consists of the 9 distinct points $p_1,\ldots,p_6,q_1,q_2,q_3$. Of these, $p_1,\ldots,p_6$ lie on the irreducible quadric C''. Hence by 6.2.3 the three points q_1 , q_2 , q_3 lie on a line C''', as was to be proved. This proof of Pascal's theorem is due to Plücker.

Remarks :

(i) The proof of Pascal's theorem which we have just given is not exactly the same as Pascal's. Pascal first proves his theorem – in the real case – for the circle. Since every irreducible conic section is projectively equivalent to a circle, the theorem then follows in general, since the incidences are preserved by collineations.

(ii) We have proved Pascal's theorem only for a hexagon inscribed in an irreducible quadric. However – and this was already known to the Greek mathematician Pappus – it also holds for the reducible quadric consisting of two distinct lines. This follows from a refinement of Theorem 6.2.3 where C and C' are cubics and C'' is a reducible quadric. We shall prove this refinement later (in 7.2). In the reducible case the Pascal theorem yields a beautiful configuration of 9 lines and 9 points. The 9 points are the 6 vertices of the hexagon and the 3 intersection points of the pairs of opposite sides. The 9 lines are the 6 sides of the hexagon, the lines of the reducible quadric, on which the 6 vertices lie, and the line through the 3 intersection points of the opposite sides. Through each of the 9 points go three lines, and on each of the 9 lines lie three points. This configuration, the Pascal configuration, was extensively investigated in the 19th century. It belongs, together with the Desargues configuration and the inflection point configuration of the plane cubic which we shall meet later (in 7.3), among the configurations of classical projective geometry.

The Desargues configuration consists of 10 points and 10 lines. The points are the 6 vertices of two triangles in perspective position, the 3 intersection points of corresponding sides, which are collinear by the Desargues theorem, and the centre of perspectivity. The 10 lines are the 6 sides of the triangles, the 3 lines connecting the vertices to the centre of perspectivity, and the line through the 3 intersection points of corresponding sides. Through each of the 10 points go 3 lines, and on each of the 10 lines lie 3 points. The following pictures illustrate Pascal's theorem and the configurations of Pascal and Desargues.

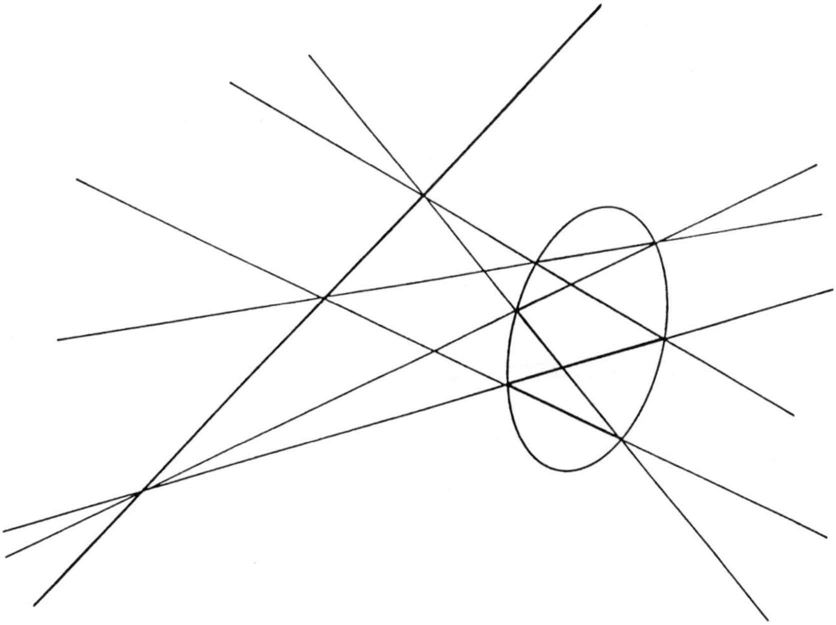

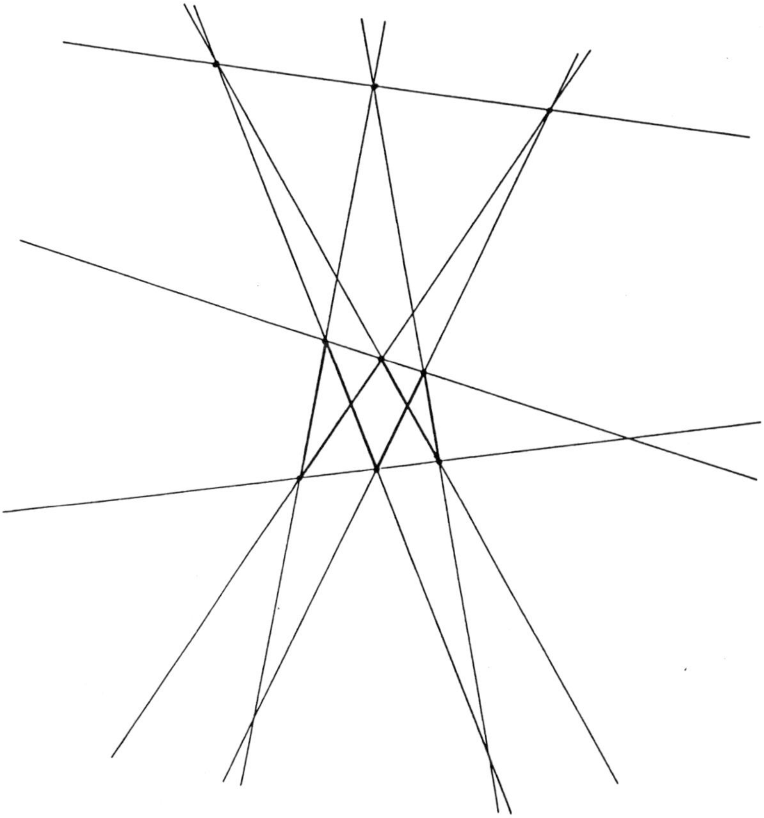

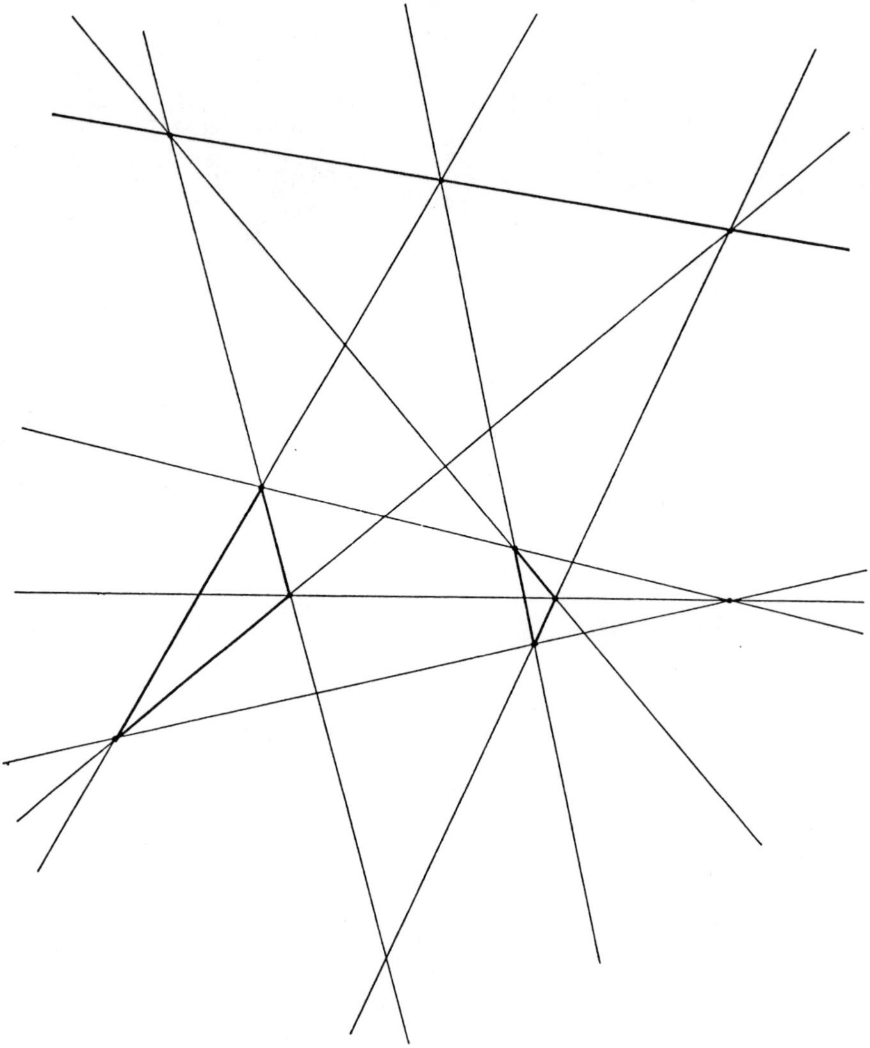

We shall now dualise Pascal's theorem to obtain Brianchon's theorem. We have not previously mentioned the duality principle of projective geometry in this course, because a suitable time and opportunity were lacking. Now we can say a few words about it.

Let $P_2(\mathbb{C})$ be the projective plane. To abbreviate we set $P = P_2(\mathbb{C})$. In P we have homogeneous coordinates (x_0,x_1,x_2). If L is any line in P, then this line is given by an equation

$$a_0x_0 + a_1x_1 + a_2x_2 = 0$$

where (a_0,a_1,a_2) is determined up to a scalar factor. If we regard (a_0,a_1,a_2) as a point of $P_2(\mathbb{C})$, then $L \mapsto (a_0,a_1,a_2)$ yields a bijection between the set P^* of all lines in P and $P_2(\mathbb{C})$. Thus we can again view the set P^* of all lines in P as a projective plane P^*, the projective plane dual to P.

In 6.2 we have viewed many more general linear systems of "curves" of higher order as higher-dimensional projective spaces.

A pencil of lines through a point $p \in P$ is a line in P^*, and each line L^* in P^* is a pencil of lines through a point $p \in P$, namely $p = \bigcap_{L \in L^*} L$. Thus the lines of P correspond to the points of P^*, and the points of P correspond to the lines of P^*, bijectively in fact. Thus we have identified the dual space $(P^*)^*$ with P again.

We denote the line L^* in P^* dual to a point $p \in P$ by p', and the point p^* in P^* dual to a line L in P by L'. Similarly, we denote the line L in P dual to a point $p^* \in P^*$ by $(p^*)'$, and the point p in P dual to a line L^* in P^* by $(L^*)'$. Then we have

$$p'' = p$$
$$L'' = L$$

and similarly

$$(p^*)'' = p^*$$
$$(L^*)'' = L^*.$$

Moreover, the following obviously holds for the incidence of points and lines : if $L^* = p'$ and $p^* = L'$ then

$$p \in L \iff L^* \ni p^*. \qquad (*)$$

For this reason, one can translate statements about the incidence of points and lines in P into equivalent statements about the incidence of points and lines in P*.

For example, the intersection point of two lines in P corresponds to the connecting line of the corresponding points in P* and, dually, the connecting line of two points in P corresponds to the intersection point of the corresponding lines in P*. Or : three points in P are collinear when the three corresponding lines in P* meet at a point, etc. Under this type of translation, true statements about the incidence of points and lines obviously go over to true statements by (*). This is the duality principle, which has played a great rôle in the development of projective geometry.

<u>Duality Principle</u>:

If, in a theorem on the incidence of points and lines, one everywhere replaces the word "point" by the word "line", and the word "line" by the word "point", and if one replaces each assertion of the form "point p lies on the line L " by the dual assertion "line L* goes through the point p* ", then one obtains a valid new theorem on incidences of points and lines, dual to the original theorem.

Of course, the principle in this form is not a theorem of projective geometry, but rather a statement on a higher level, a statement about theorems of projective geometry. From the modern standpoint the principle is an aside, since we now know how projective geometry can be derived from incidence axioms. But from the historical standpoint the discovery of the duality principle and the clarification of the nature of duality was a decisive event in developing the conception of projective geometry as a geometry based on axioms of incidence.

In the above formulation, the duality principle applies only to theorems in which (in principle) only the concepts "point" and "line" appear, together with the incidence relation $p \in L$ only. For example, one can apply it to Desargues' theorem, but not to Pascal's theorem, because the concept of "quadric" appears in the latter. However, one can extend the duality principle by defining "dual curves" for curves of higher order.

The definition of the <u>dual curve</u> C* of C is very simple. C* is the set of tangents of C in the sense of 5.3. This is a subset of the dual plane P*. When C is a line, C* consists of only one point, hence it is not a true curve. For this reason we assume, in the above definition, that C contains no lines as components. We then still have to show that C* is in fact an algebraic curve.

Proposition 5

Let C be a curve in the complex projective plane P, without lines as components, and let C* in the dual plane P* be the set of tangents of C. Then C* is a complex projective-algebraic curve in the plane P*, the dual curve to C.

Proof : We just sketch the proof. Let (x_0, x_1, x_2) be homogeneous coordinates in P and let (y_0, y_1, y_2) be the dual homogeneous coordinates in P*. Let $F(x_0, x_1, x_2) = 0$ be the equation of C. We want to find an equation $F^*(y_0, y_1, y_2) = 0$ with zero set C*. If (x_0, x_1, x_2) is a regular point of C, then by 5.3.4 the tangent to C at (x_0, x_1, x_2) is the point in P* with the homogeneous coordinates (y_0, y_1, y_2), where

$$y_i = \frac{\partial F}{\partial x_i}(x_0, x_1, x_2) \qquad i = 0,1,2. \tag{1}$$

For a regular point with $\frac{\partial F}{\partial x_0}(x_0, x_1, x_2) \neq 0$ these three equations are obviously equivalent to the following two :

$$\begin{aligned} y_0 \frac{\partial F}{\partial x_1}(x_0, x_1, x_2) - y_1 \frac{\partial F}{\partial x_0}(x_0, x_1, x_2) &= 0 \\ y_0 \frac{\partial F}{\partial x_2}(x_0, x_1, x_2) - y_2 \frac{\partial F}{\partial x_0}(x_0, x_1, x_2) &= 0. \end{aligned} \tag{2}$$

Moreover, a point (x_0, x_1, x_2) of C satisfies

$$\sum_{i=0}^{2} x_i \frac{\partial F}{\partial x_i} = mF = 0 \tag{3}$$

by the Euler formula 4.4.3. Because of (1), this equation is equivalent to

$$\sum_{i=0}^{2} x_i y_i = 0 \tag{4}$$

for a regular point. The set of points (y_0, y_1, y_2) in P* for which the three equations (2) and (4) have a solution obviously consists of just the points of C* together with finitely many lines in P*, which correspond to the pencils of lines in P through the singular points of C. (We shall see later that C has only finitely many singular points, and it also follows from 5.2.2 together with 6.1.3.)

Thus it suffices to show that the set $\tilde{C}$ of (y_0,y_1,y_2) for which the three equations (2) and (4) have a solution (x_0,x_1,x_2) is the zero set of a homogeneous polynomial $R(y_0,y_1,y_2)$. One finds such a polynomial by elimination of x_0,x_1,x_2 from the three equations (2) and (4). One can carry out this elimination systematically with the help of a kind of resultant. One has the following general result from elimination theory, which can be proved with the help of theorems from elimination theory and the resultant from 4.2 (van der Waerden, Algebra II [Wl], 3rd edition, §88).

Lemma 6

Let $\Phi_0,\ldots,\Phi_n$ be homogeneous polynomials with undetermined coefficients in the variables $(x_0,\ldots,x_n)$. Then there is a polynomial R in the undetermined coefficients of the Φ_i such that, for given values a_{ij} of these undetermined coefficients, $R(a_{ij}) = 0$ is equivalent to the existence of a nontrivial solution of the corresponding equations

$$\Phi_i(x_0,\ldots,x_n) = 0, \quad i = 0,\ldots,n.$$

The polynomial R is homogeneous in the undetermined coefficients from the individual polynomials Φ_i.

Applying 6.2.6 to the three homogeneous polynomials on the left sides of equations (2) and (4), one immediately obtains the existence of a homogeneous polynomial $R(y_0,y_1,y_2)$ with zero set $\tilde{C}$. Thus $\tilde{C}$ is in fact an algebraic curve, and hence the same is true of C^*. The $F^*(y_0,y_1,y_2)$-equation for C^* results from R when one divides by the highest possible powers of the linear forms which correspond to the pencils through the singular points of C. This proves 6.2.5.

The order of the dual curve C^* is called the <u>class</u> of C. In 5.2.2 we have proved the following : let C be a curve of order $m > 1$, let p be a point in the complement of C, and let $L^* = p'$ be the pencil of lines through p. Then in L^* there are, counting multiplicity, exactly $m(m-1)$ lines L_i which meet C anywhere with multiplicity greater than one. These lines include, of course, all lines connecting p to singular points of C. The remaining L_i are precisely the tangents to C through p. But the number of these tangents, counting multiplicity, is just the class of C. Thus we obtain :

Proposition 7

A nonsingular curve C of order $m > 1$ has class $m(m-1)$. For singular curves C of order $m > 1$ the class is less than $m(m-1)$. Each singular point lowers the class.

There are formulae of Plücker which say precisely how much a singular point lowers the class. For example, one can prove that an ordinary double point lowers the class by two, and an ordinary cusp lowers it by three. Thus a nonsingular cubic has class 6, an irreducible cubic with an ordinary double point has class 4, and an irreducible cubic with an ordinary cusp has class 3. A nonsingular quadric C naturally has class 2, and C* is again nonsingular. Proof : one writes the equation of C in the normal form $x_0^2 + x_1^2 + x_2^2 = 0$ (see 7.1), then C* obviously has the equation $y_0^2 + y_1^2 + y_2^2 = 0$.

The interpretation of lines in the plane P as points in a dual projective plane P* goes back to Poncelet, Gergonne and Plücker, as does the consideration of curves of such points in P*.* Each curve in P* which contains no lines as components arises as a dual curve C* to a curve C. One has C** = C, and the composition of the canonical rational mappings $C \to C^*$ and $C^* \to C^{**}$ is the identity (cf. e.g. van der Waerden [W2], §19, Satz 2). This means :

If C* is the dual curve of C, p is a regular point of C and L is the tangent to C at p, and if also $p^* = L'$ is the point of C* corresponding to L and $L^* = p'$ is the line of P* corresponding to p, then

$p \in C \iff L^*$ is tangent to C* at p*.

If one combines this equivalence with the one considered previously,

$p \in L \iff L^* \ni p^*$,

then one obtains a rule for turning each theorem about incidences of points and lines, or points and curves, or contact of lines and curves into a corresponding dual theorem. The duality principle then acquires an extended meaning in which each theorem has a dual.

We now want to use this principle to derive Brianchon's theorem from Pascal's theorem.

---

* The germ of this idea was already present in the description of curves as envelopes of their tangents, which we met in 2.4.

Proposition 8 (Brianchon's theorem)

The lines connecting opposite vertices of a hexagon which circumscribes an irreducible quadric meet at a point.

Proof : As in Pascal's theorem, we consider a hexagon with vertices $p_1,\dots,p_6$ and sides $L_1,\dots,L_6$, inscribed in a quadric Q. As before, let q_1,q_2,q_3 be the intersection points of the three pairs of opposite sides. Now we consider the dual quadric Q^* in the dual plane P^*, with the points $p^*_{i+1} = L'_i$, $i = 1,\dots,6$ (modulo 6) and lines $L^*_i = p'_i$, $i = 1,\dots,6$, together with $G^*_j = q'_j$, $j = 1,2,3$. The dual of the assertion that $p_1,\dots,p_6$ and $L_1,\dots,L_6$ form a hexagon H is that $p^*_1,\dots,p^*_6$ and $L^*_1,\dots,L^*_6$ form a hexagon H^*. The dual of the assertion that H is inscribed in the quadric Q is that H^* circumscribes the quadric Q^*. The dual of the assertion that the q_j are intersection points of opposite sides is that the G^*_j are connecting lines of opposite vertices. And finally, the dual of the assertion that the q_j are collinear is that the G^*_j meet at a point. This proves that Brianchon's theorem is just the dual of Pascal's theorem.

The following picture illustrates Brianchon's theorem.

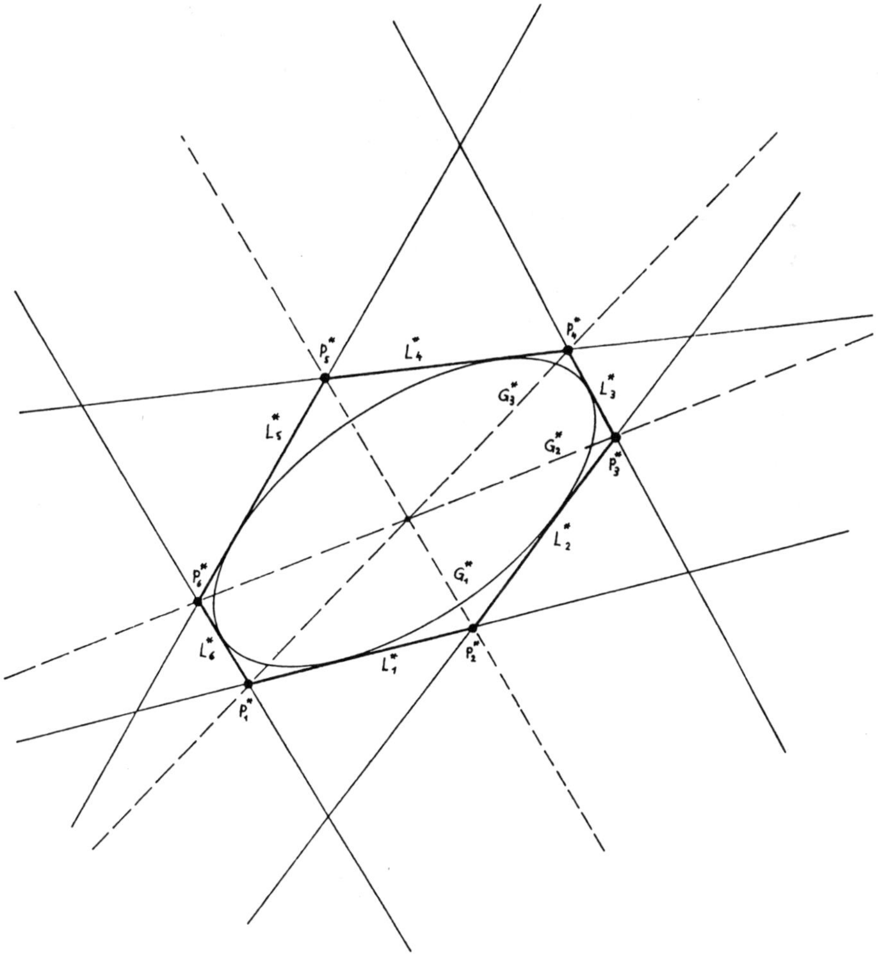
P_4^*
P_5^*
L_4^*
L_3^*
G_3^*
L_5^*
G_2^*
P_3^*
L_2^*
P_6^*
G_1^*
L_6^*
L_1^*
P_2^*
P_1^*

Our first application of Bézout's theorem was Theorem 3 on the special position of intersection points of two curves, and the consequent theorems of Pascal and Brianchon.

As a second application of Bézout's theorem, we shall now derive bounds for the number of singular points of a curve. The basic idea is the following. Given a curve C, one constructs a curve C' such that all singular points of C lie in the intersection of C and C'. One then bounds the number of intersection points of C and C' using Bézout's theorem.

Proposition 9

Let C be a curve of order n without multiple components. Then the number of singular points of C is bounded as follows.

$$\sum_{p\in C} \nu_p(C) \cdot (\nu_p(C)-1) \leq n(n-1).$$

We first prove the following

Lemma : Let $f(x)$ and $g(x)$ be polynomials with coefficients in $\mathbb{C}[x_1,\ldots,x_n]$ such that $g(x)$ is irreducible and a divisor of both f and its derivative f'. Then g^2 also divides the polynomial f.

Proof of the lemma : Since g is a divisor of f, $f = g\cdot h$, hence $f' = g'\cdot h + g\cdot h'$. Thus g also divides $g'\cdot h$. Since g is irreducible, and hence prime by 4.1.1 and 4.2.2, and since g does not divide g', g necessarily divides h in dividing $g'h$. Thus g^2 is a divisor of $f = g\cdot h$.

Proof of the proposition : Let $F(x_0,x_1,x_2) = 0$ be the equation of C, where coordinates are chosen so that $(1,0,0)$ does not lie on C and $x_0 = 0$ is not a component of C. Then it follows that x_0 actually appears in each factor of the polynomial $F(x_0,x_1,x_2)$, so that each nonconstant factor of $F(x_0,x_1,x_2)$ is also a nonconstant factor of F, viewed as a polynomial in x_0 with coefficients in $\mathbb{C}[x_1,x_2]$. Therefore $F(x_0,x_1,x_2)$ and $\frac{\partial F}{\partial x_0}(x_0,x_1,x_2)$ have no nonconstant common factor, otherwise F would have a multiple factor by the lemma. Hence one can apply Bézout's theorem to the curve C and the curve C' with the equation

$$\frac{\partial F}{\partial x_0}(x_0,x_1,x_2) = 0,$$

and obtain precisely $n(n-1)$ as the sum of the intersection numbers of C and C'.

Now if p is a singular point of C, then $p \in C'$, and hence by 5.3.3

$$\nu_p(C') \geq \nu_p(C) - 1.$$

Thus by 6.1.3 and 6.1.1 we obtain

$$\sum_{p \in C} \nu_p(C) \cdot (\nu_p(C)-1) \leq \sum_{p \in C \cap C'} \nu_p(C) \cdot \nu_p(C') \leq n(n-1),$$

which is the required bound.

The bound in 6.2.9 is optimal under the assumptions made there, because if, e.g., C consists of n different lines through a point p, then C has only one singular point p and $\nu_p(C) = n$, so that equality holds in 6.2.9.

Under additional assumptions on the number of components of C the bound may be improved. For example, we have

Proposition 10

The number of singular points of an irreducible curve C of order n satisfies

$$\sum_{p \in C} \nu_p(C) \cdot (\nu_p(C) - 1) \leq (n-1)(n-2).$$

Proof : By 6.2.9, C has only finitely many singular points, say $p_1, \ldots, p_k$. For brevity we set $r_i = \nu_{p_i}(C)$. Let $\mathcal{L}$ be the linear system of "curves" of $(n-1)^{th}$ order which have multiplicity $\geq r_i - 1$ at p_i. By the remark early in this section 6.2, $\mathcal{L}$ has dimension

$$d \geq \frac{(n-1)(n+2)}{2} - \sum_{i=1}^{k} \frac{r_i(r_i-1)}{2} .$$

It follows immediately from the bound 6.2.9 for $\Sigma r_i(r_i-1)$ that $d > 0$ for $n > 1$. Thus if we choose d points on C apart from the p_i, then there is a $C' \in \mathcal{L}$ which meets C at these d points.

By Bézout's theorem and 6.1.3, the intersection number of C and C' then satisfies

$$n(n-1) \geq \sum_{p \in C \cap C'} \nu_p(C) \nu_p(C') \geq d + \sum_{i=1}^{k} r_i(r_i-1).$$

The asserted bound follows immediately from this.

Corollary 11

(i) An irreducible quadric has no singular points.

(ii) An irreducible cubic has at most one singular point, and this is a double point.

(iii) An irreducible curve of order n with an $(n-1)$-tuple point has no other singular points.

Proof : Trivial.

The example (ii) of an irreducible cubic with one double point (e.g. the semicubical parabola) shows, incidentally, that the bound 10 is sharp.

These few applications of Bézout's theorem will suffice us for the time being.

6.3 The intersection ring of $P_2(\mathbb{C})$

In this last section of this paragraph on Bézout's theorem we want to interpret Bézout's theorem from a topological standpoint. Naturally, the relationship with topology can only be properly understood when one already has the basic concepts of algebraic topology, from a two-semester course, say. However, to provide at least a reasonable understanding for those without this previous knowledge, I shall briefly define a few of the concepts used.

Let M be a compact orientable differentiable manifold without boundary. The example we shall later investigate in detail is $M = P_2(\mathbb{C})$, but first we develop the theory of the intersection ring for an arbitrary M.

Let $H_p(M)$ be the pth homology group of M. For those not familiar with homology groups, we briefly indicate the definition. $H_p(M)$ is an abelian group defined as the quotient of two other groups:

$$H_p(M) = Z_p(M)/B_p(M).$$

Here $Z_p(M)$ is the group of p-cycles of M, and $B_p(M)$ is the group of bounding p-cycles of M.

There are several ways to define p-cycles. We first describe a very intuitive definition. To do this we choose a triangulation T of the manifold M, i.e. we describe M as the union of simplexes of various dimensions, subject to certain incidence conditions which we shall not go into here (they say that M is a simplicial complex with this triangulation). A simplex of dimension n (n-simplex for short) is the n-dimensional analogue of a tetrahedron. A 0-simplex is a point, a 1-simplex is an interval, a 2-simplex is a triangle, a 3-simplex is a tetrahedron, and the standard n-simplex Δ_n is the convex hull of the standard basis of $\mathbb{R}^{n+1}$. The simplexes of the triangulation of M are topological images of standard simplexes. One can prove that a differentiable manifold M always has a triangulation. We shall denote the manifold M, together with the chosen triangulation, by M_T.

The decomposition of the manifold M_T into the simplest building blocks, the simplexes, or the construction of M_T from these building blocks, allows us to carry out a topological analysis of M_T by algebraic methods. To do this we consider finite integral formal linear combinations

$$z = \Sigma n_i \sigma_i ,$$

where the σ_i are p-simplexes. Such linear combinations of p-simplexes are called p-chains. Intuitively one can view, e.g., a 1-chain as a sequence, a "chain", of 1-simplexes, where the chain of course may branch or fall into several pieces, and have multiply counted edges.

Of particular interest now are the closed chains, the cycles.

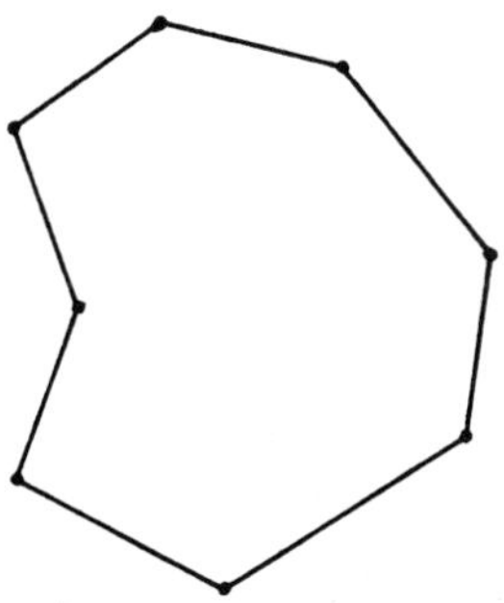

In general one can define closed chains as those chains which have no boundary. How does one define the boundary? A p-simplex has p+1 sides, e.g. a triangle with the vertices 0, 1, 2 has the sides with endpoints (0,1), (0,2), (1,2). The sides are oriented when our triangle's vertices are traversed from the lowest to highest number.

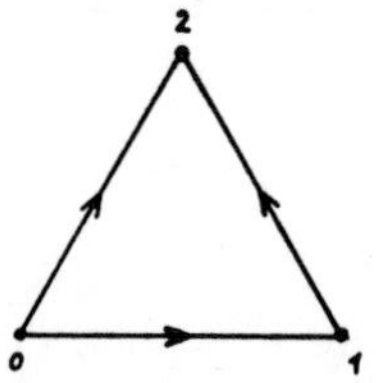

One naturally regards the three sides as the boundary of the triangle, but one also traverses them so that the boundary as a whole is traversed in the same direction, from 0 to 1 to 2 to 0 :

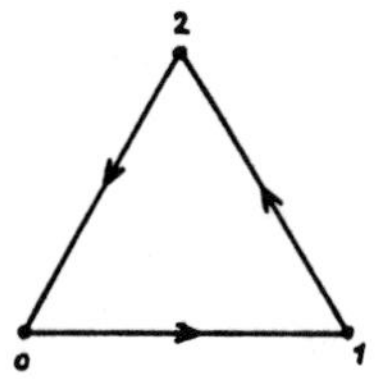

In general, when Δ_i is the ith side of the simplex Δ, i.e. the side opposite the ith vertex, one defines the boundary of Δ to be

$$\partial\Delta = \Sigma(-1)^i \Delta_i \ .$$

When $z = \Sigma n_i \sigma_i$ is any p-chain of M_T, one defines the <u>boundary of</u> z by

$$\partial z = \Sigma n_i \partial\sigma_i \ .$$

Then one can define :

z is a p-<u>cycle</u> just in case $\partial z = 0$.

z is a p-<u>boundary</u> just in case $z = \partial y$ for a (p+1)-chain y.

And finally one can define :

$Z_p(M_T)$ is the group of p-cycles of M_T .

$B_p(M_T)$ is the group of p-boundaries of M_T .

$H_p(M_T) = Z_p(M_T)/B_p(M_T)$ is the pth homology group of M.

The classes [z] in $H_q(M_T)$ generated by the cycles z are called <u>homology classes</u>. This definition is very intuitive and concrete.

However, it has the disadvantage of inflexibility, brought about by the choice of triangulation. For example, one is frequently forced to go from one triangulation T to a finer triangulation T', a refinement, and it is then annoying that our definition of homology groups – at least a priori – depends on the choice of triangulation.

For this reason one nowadays usually employs a variant of the construction above, the "singular homology theory". In it one has singular simplexes of M in place of simplexes. These are no longer topological simplexes, but continuous images, or more precisely, mappings of standard simplexes into M.

Then one defines, completely analogously to the foregoing,

$Z_p(M)$ the group of singular p-simplexes
$B_p(M)$ the group of singular p-boundaries
$H_p(M) = Z_p(M)/B_p(M)$ the pth singular homology group.

It is clear that one can carry out this construction for each topological space M, and that a continuous mapping $f : M \to N$ induces a homomorphism

$$f_q : H_q(M) \to H_q(N).$$

In this way one obtains, for each q, a functor H_q from the category of topological spaces into the category of abelian groups with certain typical properties, the well-known Eilenberg-Steenrod axioms for homology theory, which we shall not go into here. This homology theory is a means of translating geometric problems into algebraic problems, and the theory of the intersection ring, which we want to sketch here, is an example of it.

The connection between the homology theory described first, using the simplicial complexes M_T, and the singular homology of M is simply that the canonical homomorphism

$$H_q(M_T) \to H_q(M)$$

is an isomorphism.

The geometric problem that we want to translate into an algebraic problem by means of homology theory is the description of the intersection of two cycles. The simplest example would be two 1-cycles on a surface M, which intersect in finitely many points. The problem would then be the determination of the intersection number. We have met another problem of intersection number determination with Bézout's theorem. In this case the manifold M is the complex projective plane $P_2(\mathbb{C})$, hence of real dimension 4, and we want to intersect complex curves C, C' in M. In 3.4 we have seen examples of what such curves look like topologically : they result from compact orientable surfaces by possible identifications of some points. Since one can triangulate these surfaces without boundary by decomposition into triangles, i.e. 2-simplexes, it is plausible that one can associate the curves C, C' with cycles resp. homology classes $[C]$, $[C']$, and thereby turn the problem of determining the intersection number of C and C', solved by Bézout's theorem, into a problem about the homology of M. In what follows we want to explore these ideas somewhat further.

Our program is the following :

(1) Definition of the intersection class $[z]\cdot[z']$ of two homology classes of M, and with it the definition of the intersection ring $H_*(M)$ of M.

(2) Definition of the homology class $[C]$ corresponding to a curve, the so-called fundamental class.

(3) Comparison of topological and algebraic definitions of intersection.

(4) Computation of the intersection ring of the projective plane.

(5) Topological "proof" of Bézout's theorem.

(1) On the definition of intersection class :

In order to define the intersection class of two homology classes, one must find cycles z_1, z_2 in these homology classes which meet as reasonably as possible, i.e. transversely in a cycle z. One then defines the intersection class $[z_1]\cdot[z_2]$ to be the homology class $[z]$. Thus the problem first is to find such z_1 and z_2 which meet reasonably.

There are various methods for this. E.g. one can push one of the two cycles a little in the manifold M, so that it is in general position relative to the other, and then obtain the desired intersection cycle. Here we want to describe another method, which is particularly intuitive, evocative of the essence of the situation, and which has played an important rôle historically. The essential property of manifolds that it involves is the existence of dual cell decompositions.

We first describe dual cell decompositions for 2-dimensional manifolds, because one can form an intuitive picture of them particularly easily. Let M be a 2-dimensional compact oriented manifold and let T be a triangulation of M. Let T' be the barycentric subdivision. It is constructed as follows : first one subdivides each 1-simplex by a point. Then one chooses a point in the interior of each 2-simplex and connects it to the vertices of the 2-simplex and the three subdivision points of the edges. In this way each 2-simplex is subdivided into 6 smaller 2-simplexes. One defines the barycentric subdivision for triangulated higher-dimensional manifolds correspondingly by inductively subdividing first the 1-simplexes, then the 2-simplexes, then the 3-simplexes, etc. The following picture illustrates this subdivision process.

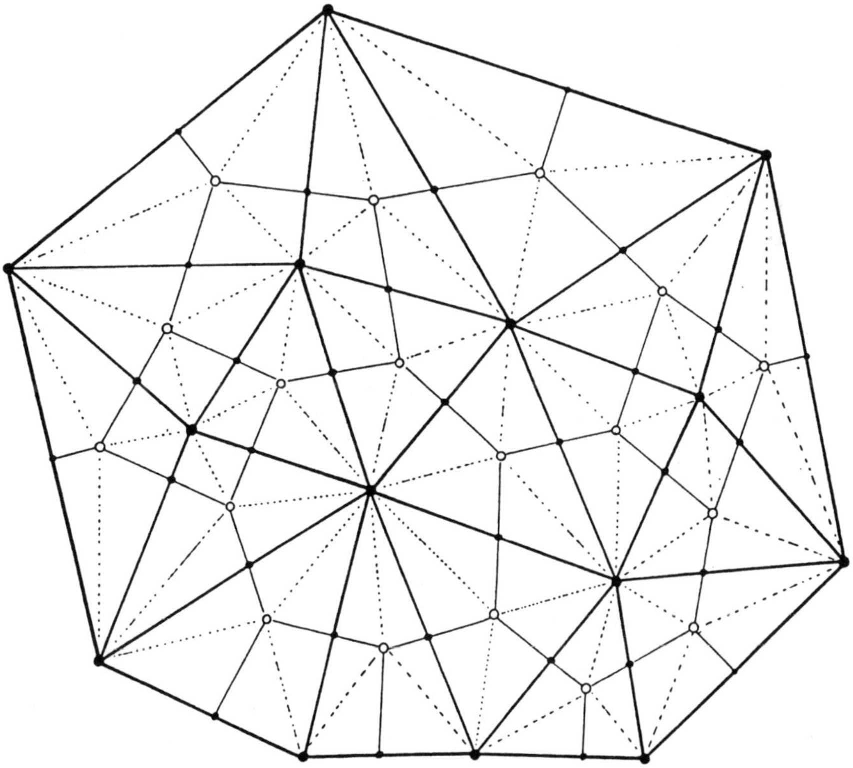

Each 2-simplex σ of the old triangulation T is the union of all 2-simplexes in the new triangulation T' which have a newly chosen subdivision point p in the interior of σ as vertex. These form, as one says, the star of the vertex p. Thus we can view the old triangulation T as the decomposition of M into the stars of the new vertices which have been chosen in the simplexes of highest dimension.

Now we consider the stars of the old vertices, i.e. the unions of simplexes of T' which have an old vertex in common. These new stars are no longer triangles, but polygons. They are topologically the same as the previous stars, however, in that each star is a cell, i.e. homeomorphic to a circular disc. And one sees immediately that these stars again cover M completely. We have constructed two cell decompositions of M, and they are dual to each other. The vertices, i.e. 0-cells, of one cell decomposition correspond to the 2-cells of the other, and the 1-cells of one decomposition correspond to the 1-cells of the other. All this generalises to manifolds of dimension n : by barycentric subdivision of a triangulation and collecting simplexes in stars one obtains two cell decompositions, dual to each other, such that a p-cell of one decomposition meets a q-cell of the other decomposition transversely in a chain of T', if at all.

A beautiful example of dual cell decompositions occurs with the five Platonic solids, the regular polyhedra. The tetrahedron, octahedron and icosahedron are triangulations of the 2-sphere with 4, 8, 20 triangles respectively. The dual cell decomposition for the tetrahedron is again a tetrahedron, for the octahedron it is a cube with its 8 vertices, and for the icosahedron it is the dodecahedron with its 12 pentagons and 20 vertices.

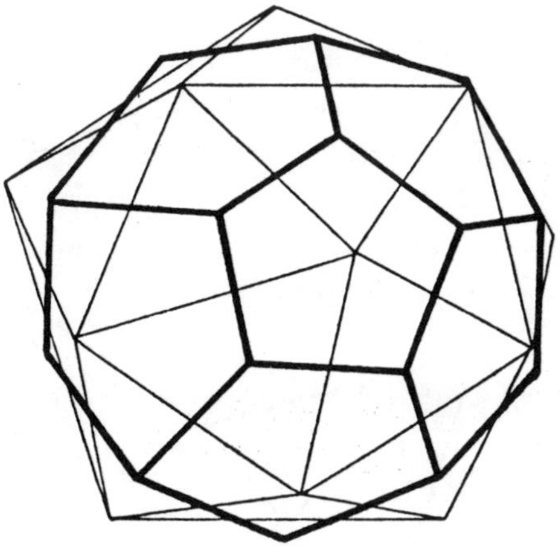

The p-cells of one cell decomposition are simply the p-simplexes of the triangulation T. Thus p-chains of the triangulation T are certain special p-chains of T', which we call cell chains. Correspondingly, one can also consider the p-chains of T' which are defined by linear combinations of cells of the dual cell decomposition. We call these the cell chains of the dual cell decomposition. We call either type of cell chain closed resp. bounding when it lies in $Z_p(M_{T'})$ resp. $B_p(M_{T'})$. By dividing each group of closed p-cell chains by the corresponding group of bounding p-cell chains, we obtain two new homology groups for M. In one case this is simply the old homology group $H_p(M_T)$, and we already know that the natural homomorphism $H_p(M_T) \to H_p(M_{T'})$ is an isomorphism. One can show that it is likewise for the dual cell decomposition.

In particular, one can not only represent each homology class in $H_p(M_{T'})$ by a cycle z_1 which is a closed p-cell chain in one cell decomposition, but one can also represent each homology class in $H_q(M_{T'})$ by a q-cycle z_2 which is a closed q-cell chain in the dual cell decomposition. The p-cells of z_1 meet the q-cells of z_2 transversely in a (p+q-n)-chain of $M_{T'}$, if at all. By recalling the orientation of M one can give these intersection chains a sign ± 1, and one then sums all these chains. It then turns out that this sum is a cycle, the intersection cycle $z_1 \cdot z_2$. The next picture illustrates the intersection of two dual 1-cycles in a surface.

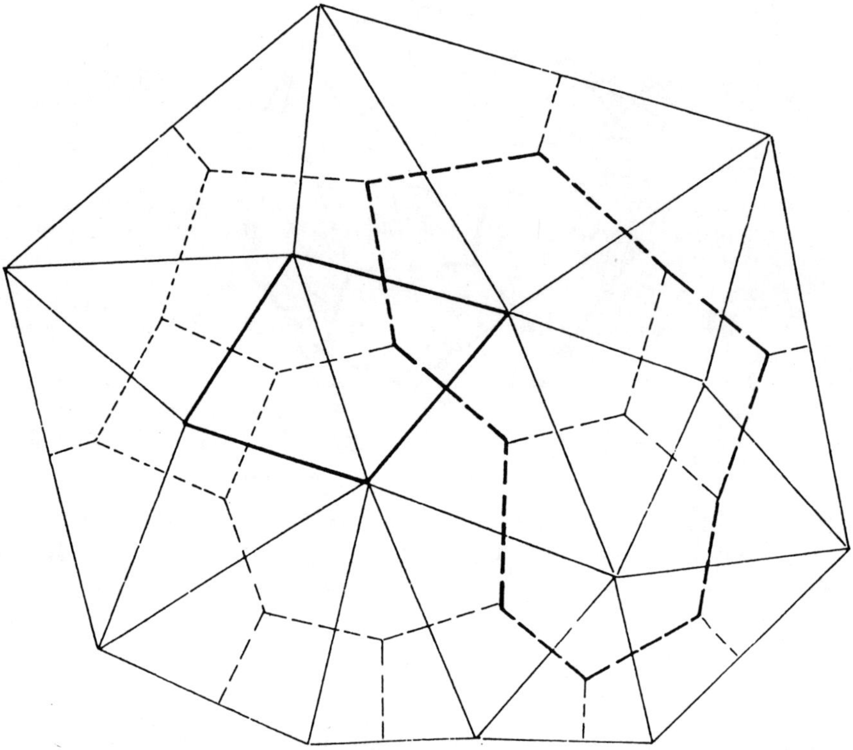

One can show that the homology class $[z_1 \cdot z_2]$ depends only on the homology classes $[z_1]$ and $[z_2]$, so that one can define the <u>intersection class</u> $[z_1]\cdot[z_2]$ by

$$[z_1]\cdot[z_2] = [z_1 \cdot z_2].$$

For $[z_1] \in H_p(M)$, $[z_2] \in H_q(M)$ we have $[z_1]\cdot[z_2] \in H_{p+q-n}(M)$.

This settles the first point of our program, the definition of the intersection class. However, we want to note a few more properties of this intersection product.

When M_T is a connected compact oriented manifold, there are two distinguished homology classes. One is the 0-dimensional homology class $[m]$ generated by an arbitrary 0-simplex, i.e. a point $m \in M$.

Each $m \in M$ represents the same homology class, and one easily sees that $H_0(M)$ is the infinite cyclic group generated by $[m]$, thus

$$H_0(M) = \mathbb{Z} \cdot [m].$$

For $a \in H_p(M)$ and $b \in H_q(M)$ with $p+q = n = \dim M$ one therefore has a definition of the <u>intersection number</u> $\langle a,b\rangle$, namely

$$\langle a,b\rangle \cdot [m] = a \cdot b.$$

If one represents a and b as above by cycles z_1, z_2 of the dual cell decompositions, then $\langle a,b\rangle$ is simply the number of intersection points of z_1 and z_2, where each intersection point is taken with the sign ± 1 given by the orientation of the simplexes of z_1, z_2 which meet in it.

For cycles z_1, z_2 of complementary dimensions which meet only at isolated points one can similarly define a local <u>intersection multiplicity</u> $\langle z_1, z_2\rangle_p$ for each intersection point p, and we then have

$$[z_1]\cdot[z_2] = [\sum_{p \in z_1 \cap z_2} \langle z_1, z_2\rangle_p [p]].$$

Thus in particular the intersection numbers satisfy

$$\langle [z_1], [z_2]\rangle = \sum_{p \in z_1 \cap z_2} \langle z_1, z_2\rangle_p .$$

The other distinguished homology class associated with M_T is the <u>fundamental class</u> $[M] \in H_n(M_T)$. This is the homology class represented by the <u>fundamental cycle</u>. The fundamental cycle is the sum of all n-simplexes of M_T, where each n-simplex is oriented in the same way as M. This is obviously a cycle, and one can show that $H_n(M)$ is the infinite cyclic group generated by $[M]$:

$$H_n(M) = \mathbb{Z}[M].$$

Obviously $[M]$ has the following property : if z is a closed cell chain of the dual cell decomposition, then $z \cdot M_T = z$, hence

$$[z] \cdot [M] = [z].$$

Thus $[M]$ behaves like an identity element for the intersection product. All this suggests that we introduce a multiplication on

$$H_*(M) = \bigoplus_p H_p(M)$$

by means of the intersection product. The result is :

Theorem 1

For a compact oriented manifold M, $H_*(M)$ with the intersection product is a graded ring with unit element $[M]$. Multiplication is commutative in the following sense : for $a \in H_p(M)$, $b \in H_q(M)$,

$$a \cdot b = (-1)^{p \cdot q} b \cdot a.$$

Definition : $H_*(M)$ is called the intersection ring of M.

The intersection ring of a manifold was introduced by S. Lefschetz around 1926/27. Lefschetz made very important contributions to the topology of algebraic manifolds, [L1]. A treatment of the intersection ring in the style of this course may be found, e.g., in the classical topology book of Seifert-Threlfall [S3], Chap. 10 ; a modern definition of intersection numbers and intersection classes may be found in the book of Dold [D4] (VII.4 and VIII.13).

What does the intersection ring $H_*(M)$ look like ? First of all, one can say that the individual groups $H_p(M_T)$ are obviously finitely generated abelian groups, since the triangulation T of the compact manifold M has only finitely many simplexes. Moreover, $H_p(M) = 0$ for $p > n$, $H_0(M) \cong \mathbb{Z}$ and $H_n(M) \cong \mathbb{Z}$. In general, $H_p(M)$, like any finitely generated abelian group, has the form

$$H_p(M) = F_p \oplus T_p.$$

Here T_p is the finite abelian group of torsion elements of $H_p(M)$, and F_p is a finitely generated free abelian group, i.e. F_p is a finite direct sum

$$F_p = \mathbb{Z} \oplus \ldots \oplus \mathbb{Z}.$$

The number b_p of summands is called the pth Betti number of M.

These Betti numbers cannot be completely arbitrary. They satisfy the important Poincaré duality theorem.

Theorem 2 (Poincaré duality theorem)

The Betti numbers of an n-dimensional manifold satisfy

$$b_p = b_{n-p} .$$

In fact, one can prove a better version of this theorem which also says something about the ring structure of $H_*(M)$. We consider the bilinear form

$$H_p(M) \times H_{n-p}(M) \to \mathbb{Z}$$

given by the intersection product $(a,b) \mapsto \langle a,b\rangle$. With the help of dual cell decompositions one can prove that this bilinear form is non-degenerate on the free components F_p, F_{n-p} and that it can be described by a matrix with determinant ± 1 for given bases. One can also express this as follows : the canonical homomorphism of F_{n-p} into the dual group $F_p^* = \mathrm{Hom}(F_p, \mathbb{Z})$, defined by the bilinear form, is an isomorphism

$$F_{n-p} \cong F_p^*. \tag{1}$$

Naturally, it follows in particular that $b_p = b_{n-p}$. In addition, one can show

$$T_{n-p-1} \cong T_p. \tag{2}$$

Assertions (1) and (2) together form the full content of the Poincaré duality theorem in its classical formulation.

Nowadays one usually formulates the theorem somewhat differently, namely, as a theorem on the relation between homology and cohomology. The cohomology groups $H^q(M)$ of a compact manifold are also finitely generated abelian groups. We shall not go into the definition, but only say that they are defined dually to the homology groups, as it were, with the help of linear forms on the chains. Because of this one has a natural surjective homomorphism $H^q(M) \to \mathrm{Hom}(H_q(M), \mathbb{Z}) = F_q^*$ which maps the free component of $H^q(M)$ isomorphically onto F_q^*. It is not difficult to show that the kernel is isomorphic to T_{q-1}. The Poincaré duality theorem therefore gives us an isomorphism

$$H^q(M) \cong H_{n-q}(M). \tag{3}$$

The modern version of the Poincaré duality theorem just says that there is such an isomorphism, and that it is defined in a particular, natural way (namely by the cap product with the fundamental class $[M]$).

Now one shows in algebraic topology that one can always provide the direct sum $H^*(M)$ of the cohomology groups with a ring structure, i.e. for any topological space. One can then prove (Dold [D4]) that when M is a compact oriented manifold then this ring structure on $H^*(M)$ goes via the <u>Poincaré isomorphism</u> $H^*(M) \cong H_*(M)$ into the ring structure on the intersection ring. If one already had the Poincaré isomorphism, then one could define the intersection ring with the help of the cup product from cohomology. However, that would be much less intuitive, even if perhaps more elegant technically. Also, it is the case historically that one first found the intersection ring $H_*(M)$ for manifolds, and discovered only later that $H^*(M)$ had a ring structure for any space. So much for the intersection ring.

(2) <u>**The homology class of a submanifold**</u>

(3) <u>**Comparison of topological and algebraic intersection theory**</u>

Let M be a compact oriented differentiable manifold, and let $V \subset M$ be a differentiable submanifold which is likewise compact and oriented. Further, let $i : V \to M$ be the inclusion mapping, and let $i_* : H_*(V) \to H_*(M)$ be the mapping on homology induced by i. (Of course, this is not in general a ring homomorphism, but only a homomorphism of abelian groups.) In $H_*(V)$ one has a distinguished element, the fundamental class $[V]$. We recall that, for a given triangulation of V, $[V]$ is represented by a fundamental cycle, the sum of properly oriented simplexes of highest dimension. Naturally one can also view this cycle as a cycle in M, a k-cycle when V is k-dimensional. It then represents a homology class in $H_k(M)$, obviously just $i_*([V])$. For the sake of simplicity we denote the latter by $[V]$ also.

$$[V] \in H_*(M)$$

is the <u>homology class of the submanifold</u> $V \subset M$.

If V and W are oriented, compact submanifolds of M which meet transversely, and if one orients the intersection manifold $V \cap W$ by a suitable rule which takes account of the orientations of V, W and M, then one can show :

$$[V \cap W] = [V] \cdot [W].$$

This means : the geometric intersection of submanifolds corresponds exactly to the algebraic product in the intersection ring.

Naturally one can apply all this to the case where M is a non-singular complex projective-algebraic manifold of complex dimension n, and V, W are non-singular complex submanifolds of complex dimensions p, q respectively. As real differentiable manifolds, M, V and W then have real dimensions $2n$, $2p$ and $2q$ respectively. If, for example, $M = P_2(\mathbb{C})$ and C_1, C_2 are non-singular plane curves, then M is real four-dimensional and $[C_1], [C_2] \in H_2(P_2(\mathbb{C}))$. If these curves meet transversely, then the topological intersection number

$<[C_1],[C_2]>$ precisely equals the number of intersection points, because the rules for the orientation of the complex manifolds $C_1, C_2, P_2(\mathbb{C})$ are such that the local intersection multiplicities are always positive. On the other hand, in this simple case the number of intersection points is also the intersection number $C_1 \cdot C_2$ in the sense of algebraic geometry, as defined in 6.1. Thus the topological and algebraic geometric intersection numbers coincide :

$$C_1 \cdot C_2 = <[C_1],[C_2]>.$$

But 6.1 allows the algebraic geometry intersection number to be defined for quite arbitrary curves, including ones which are singular or which have non-transverse intersection. Hence there should be a corresponding more general topological theory, and this is in fact the case. In order to develop it, one must first define the homology class

$$[V] \in H_{2k}(M)$$

associated with a projective-algebraic subset V of a non-singular projective-algebraic manifold M, where V has pure complex dimension k and possible singularities.

There are several possibilities for such a definition. The one most convenient for us again uses triangulations. We have already seen in examples in 3.4, and we shall prove generally by resolution of singularities in §9, that topologically each plane curve C results from a compact orientable surface $\tilde{C}$ when certain points of $\tilde{C}$ are identified. It is clear what $\tilde{C}$ looks like : each connected component results from a 2-sphere by attaching handles. In particular, one can triangulate $\tilde{C}$. If one takes the triangulation to be so fine that the points to be identified are vertices, then one obtains a triangulation of C, and this triangulation defines a 2-cycle in $P_2(\mathbb{C})$, because the boundaries of any two adjacent 2-simplexes cancel out. This 2-cycle defines a homology class

$$[C] \in H_2(P_2(\mathbb{C})) \ ;$$

the homology class $[C]$ is just the image of the fundamental class $[\tilde{C}]$ under the resolution mapping $\tilde{C} \to C \subset P_2(\mathbb{C})$, and for that reason it is independent of the choice of triangulation. When C is a "curve" with multiple components, $C = \Sigma n_i C_i$ with irreducible C_i, then naturally one defines

$[C] = \Sigma n_i [C_i]$.

One can now prove, though we do not wish to do so here, that

Proposition 3

For "curves" C, C' in the complex projective plane without common components, the topological and algebraic intersection numbers coincide, i.e.

$C \cdot C' = \langle [C],[C'] \rangle$.

All this can be considerably generalised. One has the following theorem, already asserted in the thirties [W2], but first proved later by Lojasiewicz and Giesecke [L3], in a form even more general than we state here.

Theorem 4

Let M be a complex projective algebraic variety* and let $V_1,\dots,V_k$ be subvarieties. Then there is a triangulation of M as a finite simplicial complex for which $V_1,\dots,V_k$ are subcomplexes.

Recently, Hironaka has shown how one can use resolution of singularities to prove such results.

With the help of this theorem one can associate a homology class $[V] \in H_*(M)$ with any pure-dimensional subvariety V of a non-singular M, and one can show that the algebraic geometrically defined intersection cycle $V \cdot W$ satisfies

$[V \cdot W] = [V] \cdot [W]$.

Another version of this is found in Borel-Haefliger [B3]. Naturally we cannot go into the algebraic geometric intersection theory in any such generality here, but one gets roughly the right idea of it when one knows that for two curves C, C' without common component the intersection cycle is defined by

$$C \circ C' = \sum_{p \in C \cap C'} \nu_p(C,C') \cdot p.$$

For algebraic cycles, i.e. integral formal linear combinations of subvarieties V_i of a non-singular projective-algebraic manifold M, one can introduce a suitable equivalence relation (rational equivalence), and with the definition of intersection cycles these equivalence

*I.e. an irreducible algebraic subset of a $P_n(\mathbb{C})$, see [M8].

classes of cycles become a ring, the Chow ring $A_*(M)$. $V \mapsto [V]$ then defines a homomorphism from this purely algebraically defined Chow ring to the topologically defined intersection ring $H_*(M)$, and in many cases, e.g. for $M = P_n(\mathbb{C})$, this is an isomorphism. (More on this is in Grothendieck [G3].) The problem of deciding which homology classes of $H_*(M)$ lie in the image of $A_*(M)$ is in general extraordinarily difficult and has been a much-investigated problem of recent algebraic geometry. However, we are again looking far beyond the framework of our course, and we now return to deal with the next point of our program.

(4) Computation of the intersection ring of $P_n(\mathbb{C})$

In computing $H_*(P_n(\mathbb{C}))$ it is important to have a sequence of projective subspaces

$$P_0(\mathbb{C}) \subset P_1(\mathbb{C}) \subset P_2(\mathbb{C}) \subset \ldots \subset P_{n-1}(\mathbb{C}) \subset P_n(\mathbb{C})$$

such that $P_i(\mathbb{C}) - P_{i-1}(\mathbb{C})$ is isomorphic to $\mathbb{C}^i$. One can define such a sequence, for example, by

$$P_i(\mathbb{C}) = \{(x_0,\ldots,x_n) \in P_n(\mathbb{C}) \mid x_{i+1} = \ldots = x_n = 0\}.$$

To abbreviate we set $P_i = P_i(\mathbb{C})$. By (2), P_i corresponds to a homology class

$$[P_i] \in H_{2i}(P_n(\mathbb{C})).$$

It is clear that any other i-dimensional projective subspace P_i' of P_n defines the same homology class. (Proof : There is a collineation g of P_n with $g(P_i') = P_i$. Because of the connectedness of $GL(n+1,\mathbb{C})$ there is then a continuous family of collineations g_t, $0 \leq t \leq 1$, with $g_0 = 1$, $g_1 = g$. Thus g is homotopic to the identity and it follows that the mapping g_* on homology induced by g is the identity (by the axioms of homology theory). Thus

$$[P_i] = [gP_i'] = g_*[P_i'] = [P_i'].)$$

We can use the classes $[P_i]$ to describe the additive structure of $H_*(P_n(\mathbb{C}))$. We show :

Proposition 5

$$H_i(P_n(\mathbb{C})) \cong \begin{cases} \mathbb{Z} & \text{for } i \text{ even, } 0 \leq i \leq 2n \\ 0 & \text{otherwise.} \end{cases}$$

$H_{2j}(P_n(\mathbb{C}))$ is generated by $[P_j]$.

<u>**Proof**</u> : The proof is by induction on n. For $n = 0$ the theorem is trivial. We assume it holds for $n-1$, and prove it for n. $H_{2n}(P_n(\mathbb{C}))$ is generated by the fundamental class $[P_n]$ and is infinite cyclic, as a special case of the corresponding general fact about compact oriented manifolds which we ascertained earlier in (1) and use here without proof.

Thus we consider $H_p(P_n(\mathbb{C}))$, $p < 2n$. Let z be a p-cycle in $P_n(\mathbb{C})$, $p < 2n-1$. Then there is a point which lies in no simplex of z. By the earlier remarks we can assume, without loss of generality, that this point has coordinates $(0,\ldots,0,1)$. Now we consider the family of singular cycles z_i , $0 \leq t \leq 1$, which are images of z under the mappings $(x_0,\ldots,x_{n-1},x_n) \overset{\phi_t}{\longmapsto} (x_0,\ldots,x_{n-1},tx_n)$. Since the mappings ϕ_0 and ϕ_1 are homotopic, z_0 and z_1 are homologous. (Intuitively speaking : the "union" of all z_t , $0 \leq t \leq 1$, yields a $(p+1)$-cycle with boundary z_0-z_1.) But the cycle z_0 is a cycle in P_{n-1} and hence homologous to 0 in P_{n-1} when p is odd, and to $a[P_k]$ for some $a \in \mathbb{Z}$ when $p = 2k$. Therefore the corresponding homologies hold just as well in P_n and we have proved : $H_p(P_n(\mathbb{C})) = 0$ for odd p, and $H_{2k}(P_n(\mathbb{C}))$ is generated by $[P_k]$.

Now it only remains to show that if a cycle y in P_{n-1} is a boundary as a cycle in $P_n(\mathbb{C})$, say $y = \partial z$ in P_n , then y is also a boundary in P_{n-1}. This goes as follows. We associate z with the cycle z_0 in P_{n-1} as above, and then $y = \partial z_0$. The required result then follows from the induction hypothesis that the $H_{2k}(P_{n-1})$ are infinite cyclic, together with the preceding arguments for the corresponding assertion about $H_{2k}(P_n)$, and this completes the proof of the proposition.

We now determine the multiplicative structure of the intersection ring $H_*(P_n(\mathbb{C}))$.

<u>**Proposition 6**</u>

The multiplicative structure of the intersection ring $H_*(P_n(\mathbb{C}))$ is given by the following formula for the product of generators $[P_i]$:

$$[P_i]\cdot[P_j] = [P_{i+j-n}].$$

<u>**Proof**</u> : Because of what was said earlier, the proof is now trivial, since we can assume by the above remarks that P_i , P_j are any subspaces of dimension i, j respectively, e.g. let P_i be the one with

$x_{i+1} = \ldots = x_n = 0$ and let P_j be $x_0 = \ldots = x_{n-j-1} = 0$. These P_i and P_j meet transversely in a P_{i+j-n}. The assertion $[P_{i+j-n}] = [P_i]\cdot[P_j]$ then follows from the general formula $[V \cap W] = [V]\cdot[W]$ in the intersection ring from (1).

Remark : Let $H \subset P_n(\mathbb{C})$ be a hyperplane. It follows from Proposition 6 that $[H]^i = [P_{n-i}]$ for $i \leq n$, and of course, by Proposition 5, $[H]^i = 0$ for $i > n$. Hence we can also express the description of the intersection ring in Propositions 5 and 6 as follows : if x is an indeterminate over $\mathbb{Z}$, then $[H] \mapsto x$ defines an isomorphism

$$H_*(P_n(C)) \cong \mathbb{Z}[x]/(x^{n+1})$$

from the intersection ring $H_*(P_n(\mathbb{C}))$ to the quotient of the polynomial ring $\mathbb{Z}[x]$ by the ideal generated by x^{n+1}. For many important manifolds one can also describe the intersection ring in a similar way.

(5) "Proof" of Bézout's theorem

Just as in the algebraic-geometric case (as in 5.2) we first treat the intersection of a "curve" with a line. If C is a "curve" of order m, then it meets a general line L in exactly m points, by 5.2.2, hence

$$L \cdot C = m.$$

Hence by Proposition 3 we also have

$$<[L],[C]> = m.$$

By Proposition 5, $[C] = a[L]$ for a certain whole number a. But then by Proposition 6,

$$<[L],[C]> = <[L],a[L]> = a<[L],[L]> = a.$$

Thus it follows that $a = m$. We have proved :

Proposition 7 :

The homology class $C \in H_2(P_2(\mathbb{C}))$ of a plane "curve" of order m is $m\cdot[L]$, where $[L] \in H_2(P_2(\mathbb{C}))$ is the generating homology class of a line L.

Now we can quite easily give a topological "proof" of Bézout's theorem. Proof in quotes, because in this section we have used so many facts proved only in part. We should therefore regard this as an interpretation of Bézout's theorem rather than a proof. It shows that the global invariant, the "intersection number" $C\cdot C'$, which is the sum of the local intersection numbers, can be computed quite easily for plane curves by means of other global invariants, namely the orders of C and C', because the intersection ring of $P_2(\mathbb{C})$ is so simple. Thus :

Bézout's theorem

The intersection number $C \cdot C'$ of two plane curves C and C' of orders m and m' is given by

$$C \cdot C' = m \cdot m' .$$

Proof : By Propositions 3, 6 and 7 we have :

$$C \cdot C' = \langle [C],[C'] \rangle = \langle m[L],m'[L] \rangle = m \cdot m' \langle [L],[L] \rangle = mm'.$$

7. Some simple types of curves

7.1 Quadrics

Next to lines, quadrics are the simplest plane curves. From the complex-projective standpoint they are the analogues of the conic sections of antiquity. If one also admits curves with multiple components, and thus understands the quadrics to include all "curves" with equations of degree 2, then a quadric is just a curve with a homogeneous equation

$$\sum_{i,j=0}^{2} a_{ij}x_i x_j = 0,$$

where one can assume without loss of generality that $a_{ij} = a_{ji}$. The polynomial $\Sigma a_{ij}x_i x_j$ is a form of degree 2, a quadratic form. If A is the matrix (a_{ij}), then one can write this form in matrix fashion as follows :

$$x^t Ax.$$

Under a linear coordinate transformation $x = By$, this goes over to

$$y^t B^t ABy.$$

Under linear coordinate transformations, a quadric transforms like a quadratic form, and we can therefore bring the equation into a normal form by bringing the matrix A into a normal form through a transformation $B^t AB$. Now it is known from linear algebra that, over the complex numbers, a symmetric matrix can always be brought into diagonal form by such a transformation, with 1's and 0's on the diagonal. With real symmetric matrices one can arrive by real transformations at a diagonal form in which the diagonal entries are 0 or ± 1. The projective classification of quadrics follows immediately from this.

Theorem 1

Relative to suitable homogeneous coordinates, the equation of each complex projective plane quadric Q has exactly one of the following three normal forms

(i) $x_0^2 + x_1^2 + x_2^2 = 0$

(ii) $x_0^2 + x_1^2 = 0$

(iii) $x_0^2 = 0.$

In case (i) Q is an irreducible, non-singular quadric, in case (ii) Q decomposes into two distinct lines, and in case (iii) Q is a double line.

Thus the complex-projective classification of quadrics is very simple : there are only three different kinds. The most interesting are, of course, the non-singular curves.

What else can one say about the non-singular quadrics? Let us go from the normal form $x_0^2 + x_1^2 + x_2^2 = 0$ to another normal form by introducing new coordinates $y_0 = x_0$, $y_1 = ix_1 + x_2$, $y_2 = ix_1 - x_2$. Then one obtains the equation

$$y_0^2 - y_1 y_2 = 0.$$

This equation has the advantage, among others, that it immediately shows the non-singular quadric Q to be isomorphic, as an abstract curve, to the 1-dimensional complex projective space $P_1(\mathbb{C})$, i.e. to the Riemann number sphere. Namely, we choose homogeneous coordinates (z_0,z_1) in $P_1(\mathbb{C})$ and consider the mapping

$$(z_0,z_1) \mapsto (z_0 z_1, z_0^2, z_1^2).$$

One sees immediately that this is an everywhere defined, rational, bijective mapping

$$P_1(\mathbb{C}) \to Q$$

and that the inverse mapping is likewise rational and everywhere defined. This mapping is therefore an isomorphism of $P_1(\mathbb{C})$ onto Q, when one considers them as abstract curves. In particular, the two curves are homeomorphic as topological spaces, namely, both are homeomorphic to the sphere S^2. But more than that : they are not just the same as topological spaces, but also as abstract curves.

This fact, which we have proved quite directly here, is essentially a special case of a much more difficult theorem of function theory, the main theorem of conformal mapping (cf. Behnke and Sommer [B5], V, §5, Satz 28). It follows from this theorem that : each non-singular projective-algebraic curve which is homeomorphic to S^2 as a topological space is isomorphic to the projective line $P_1(\mathbb{C})$ as an abstract algebraic curve. Thus there is only one curve, up to isomorphism, which looks topologically like $P_1(\mathbb{C})$, namely $P_1(\mathbb{C})$ itself. The non-singular quadric is abstractly the same as this curve – but of course, as a curve embedded in $P_2(\mathbb{C})$, it is different from $P_1(\mathbb{C}) \subset P_2(\mathbb{C})$. It even has a different homology class, by 6.3.7.

We shall see later (in §9) that none of the non-singular curves of order greater than 2 in $P_2(\mathbb{C})$ is homeomorphic to $P_1(\mathbb{C})$. In contrast to this, singular curves of higher order can very well be homeomorphic to $P_1(\mathbb{C})$. An example of this is the semicubical parabola C :

$$y_0 y_1^2 - y_2^3 = 0.$$

Here we obtain a homeomorphism of $P_1(\mathbb{C})$ onto C by

$$(z_0, z_1) \mapsto (z_0^3, z_1^3, z_0 z_1^2).$$

This is still a bijective, everywhere defined mapping, but the inverse is no longer regular at the singular point $(1,0,0)$ of C.

Also of interest is the example of the irreducible cubic C with a double point, which we have already treated in 3.4. In this case we have a regular mapping

$$P_1(\mathbb{C}) \to C$$

which is admittedly no longer bijective, but it is bijective and has a regular inverse when we remove some points from $P_1(\mathbb{C})$ and C. One can regard such a mapping as a rational parametrisation of C, because when one introduces an affine coordinate t in $P_1(\mathbb{C})$ and affine coordinates (x,y) in the plane of the curve then the essentially bijective mapping is described by two rational functions

$$x = x(t)$$
$$y = y(t).$$

Curves which have such a parametrisation are therefore called rational curves. The lines and the non-singular quadrics are, as we have already said, the only non-singular rational curves. The rational curves are the simplest curves of all.

In the following sections we shall see that the non-singular curves of the next highest order, the cubics, are essentially more interesting. We shall see that as topological spaces they are all homeomorphic to a torus $S^1 \times S^1$, but that as curves in the plane and also as abstract curves they are of infinitely many different kinds.

7.2 Linear systems of cubics

Recall our remarks on linear systems of "curves" at the beginning of section 6.2. We shall now be particularly interested in linear systems of cubics. We understand a cubic to be a curve of order 3, which can have multiple components. Naturally the really interesting cubics are the irreducible cubics, because the components of the others are only lines or quadrics. But when we consider linear systems we must also admit reducible "curves", since, e.g., often in a pencil of curves all but finitely many are irreducible, the exceptions having several, and also multiple, components. We shall see an example of this shortly (7.3.5).

The linear system of all cubics is 9-dimensional, and through 9 points there goes in general exactly one cubic, as we have seen in 6.2.2. Of course, there are systems of 9 points in special position, through which more than one cubic passes. For example, passing through the 9 intersection points of two cubics C and C' we have at least all the cubics of the pencil generated by C and C'. Thus the 9 intersection points of two cubics are in special position relative to cubics. When these 9 intersection points are all different, it is of interest whether any 8 of them are in general position relative to cubics. The following theorem says they are.

Proposition 1

Let C and C' be two cubics which meet in exactly 9 different intersection points, and let $\mathcal{L}$ be the pencil of cubics generated by C and C'. Then if $p_1,\ldots,p_8$ are any eight of the nine intersection points and if $\hat{\mathcal{L}}$ is the linear system of cubics through $p_1,\ldots,p_8$, then $\hat{\mathcal{L}} = \mathcal{L}$.

In other words : any eight of the nine intersection points are in general position relative to cubics.

Proof : We shall prove five assertions about the general position of subsystems of the nine intersection points $p_1,\dots,p_9$ relative to lines and quadrics, and from them derive the theorem in a sixth step.

(1) 4 points from $p_1,\dots,p_9$ are never collinear.

Proof : Otherwise the line through these points would be a common component of C and C' by Bézout's theorem.

(2) 7 points from $p_1,\dots,p_9$ never lie on a quadric.

Proof : An irreducible quadric which had 7 intersection points in common with C and C' would be a common component of C and C'. Of seven intersection points on a reducible quadric, at least four must lie on one of the components, and hence on a line – in contradiction to (1).

(3) Through any 5 points from $p_1,\dots,p_9$ there is exactly one quadric.

Proof : Suppose two quadrics Q, Q' went through $p_1,\dots,p_5$. By Bézout's theorem, both must be reducible, otherwise they could meet each other in only 4 intersection points. For the same reason, they must have a line L as common component, namely one on which 3 of the 5 points lie. More than three points cannot lie on L by (1). Thus the other two points must lie on the other component and thereby determine it uniquely. Hence $Q = Q'$.

(4) If $\hat{\mathcal{L}} \neq \mathcal{L}$, then no 3 points from $p_1,\dots,p_8$ are collinear.

Proof : Since $\mathcal{L} \subset \hat{\mathcal{L}}$, the assumption $\hat{\mathcal{L}} \neq \mathcal{L}$ means that $\hat{\mathcal{L}}$ is at least 2-dimensional, so that we can place two more linear conditions on the cubics from $\hat{\mathcal{L}}$. We do this as follows:

Suppose that p_1,p_2,p_3 lie on a line L. Then p_4,p_5,p_6,p_7,p_8 lie on exactly one quadric Q, by (3). We choose a $p \neq p_1,p_2,p_3$ in L and a $q \notin L \cup Q$. Since $\hat{\mathcal{L}}$ is at least 2-dimensional, there is a $\hat{C} \in \hat{\mathcal{L}}$ through p and q. Since $\hat{C} \cap L$ contains four points, L is a component of $\hat{C}$ by Bézout's theorem, hence $\hat{C} = L \cup Q'$ for some quadric Q'.

Now, by (1), $p_4,\dots,p_8$ do not lie on L, hence they lie on Q', and so $Q = Q'$, which means $\hat{C} = L \cup Q$ in contradiction to $q \in \hat{C}$, $q \notin L \cup Q$.

(5) If $\hat{\mathcal{L}} \neq \mathcal{L}$, then no six points from $p_1,\dots,p_9$ lie on a quadric.

Proof : If $p_1,\dots,p_6$ lay on a quadric Q, then Q would be irreducible by (4). Let L be the line through p_7, p_8. We choose a $p \neq p_1,\dots,p_6$ on Q and a $q \notin L \cup Q$. If $\hat{\mathcal{L}} \neq \mathcal{L}$, then there would again be a $\hat{C} \in \hat{\mathcal{L}}$ through p and q. Since $\hat{C}$ has seven points in common with Q, it follows by Bézout's theorem that $\hat{C} = Q \cup L'$ for some line L'. By (2), p_7, p_8 do not lie on Q, hence they lie on L'. Thus $L' = L$ and $\hat{C} = Q \cup L$ in contradiction to $q \in \hat{C}$, $q \notin L \cup Q$.

(6) The assumption $\hat{\mathcal{L}} \neq \mathcal{L}$ leads to a contradiction.

Proof : Let L be the line through p_1 and p_2 and let Q be the unique (by (3)) quadric through p_3, p_4, p_5, p_6, p_7. We choose two distinct points $q_1,q_2 \neq p_1,p_2$ on L. If $\hat{\mathcal{L}} \neq \mathcal{L}$ there is a $\hat{C} \in \hat{\mathcal{L}}$ through q_1, q_2. Since $p_1,p_2,q_1,q_2 \in C \cap L$, L is a component of $\hat{C}$ by Bézout's theorem, hence $\hat{C} = L \cup Q'$. Since $p_3,\dots,p_7$ do not lie on L, by (4), they lie on Q'. Thus $Q' = Q$ by (3), i.e. $\hat{C} = L \cup Q$. Now $p_8 \in \hat{C}$ because $\hat{C} \in \hat{\mathcal{L}}$, but $p_8 \notin L$ by (4) and $p_8 \notin Q$ by (5). This is a contradiction! Hence the theorem is proved.

As a corollary, we obtain the sharpening of 6.2.3 for cubics, mentioned in the remarks on Pascal's theorem, 6.2.4, and needed for the proof of Pappus' theorem.

Corollary 2

When two cubics C, C' meet in exactly 9 distinct points, 6 of which lie on a quadric, then the other three lie on a line.

Proof : Let L be the line through two of the three remaining intersection points. Then by Proposition 1, the cubic $Q \cup L$ contains all 9 intersection points, and L therefore goes through all three intersection points which do not lie on Q.

Q.E.D.

7.3 Inflection point figures and normal forms of cubics

In this section we shall present normal forms for the equations of plane cubic curves, and use them to obtain results about the inflection points of these curves. It will turn out that these inflection points form an interesting configuration. Conversely, results about the inflection points of cubics are useful in obtaining normal forms.

The investigation of inflection points of plane cubics was already recognised by Newton to be an interesting problem. When we look at the pictures in Newton's enumeration of the (real) plane curves of third order, we find that these real cubics have up to three (real) inflection points. We reproduce some of the pictures here [N2].

Fig. 8.

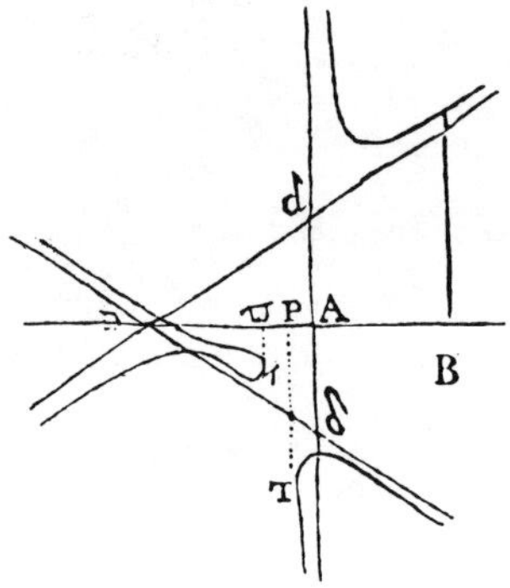

Fig. 11.

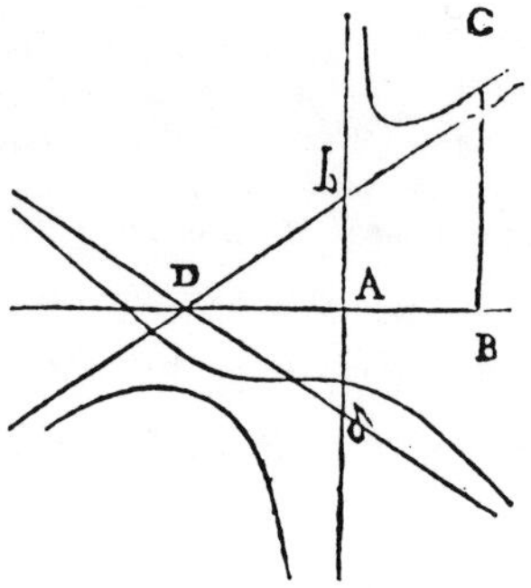

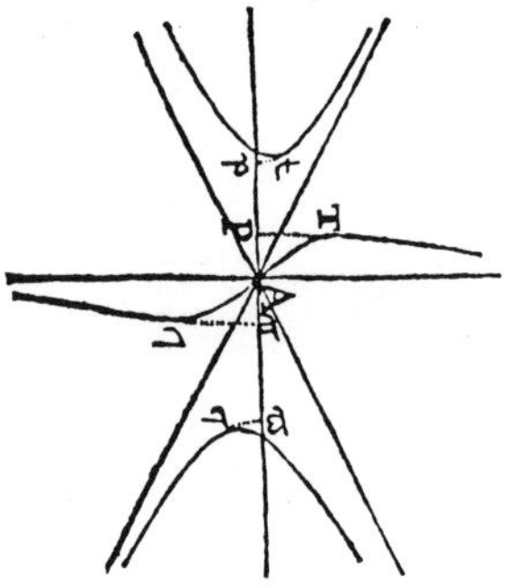

Fig. 33.

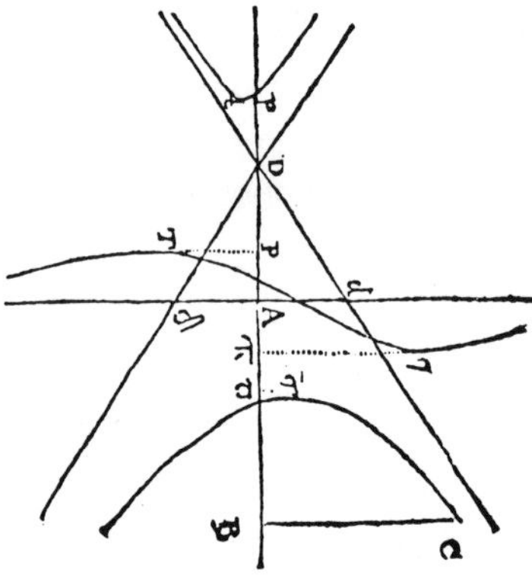

Fig. 12.

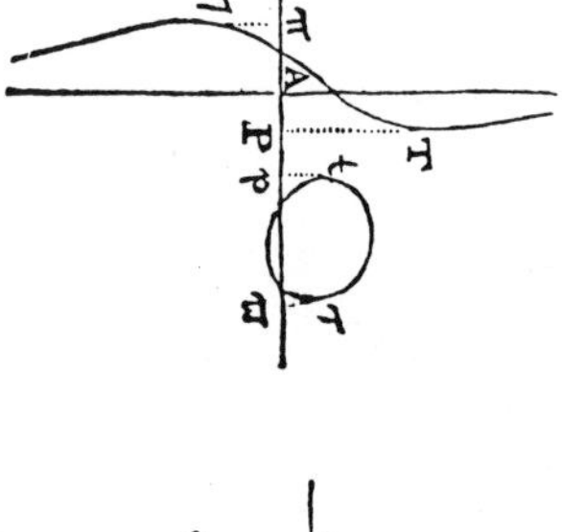

Fig. 39.

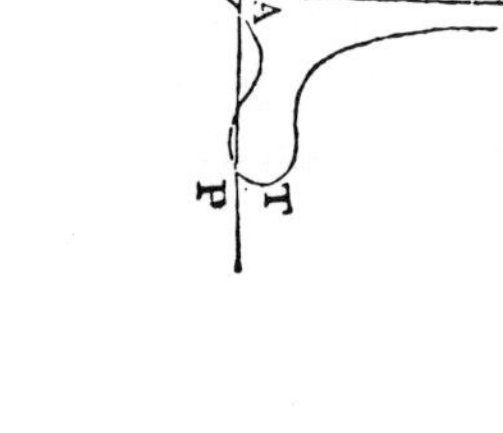

Fig. 42.

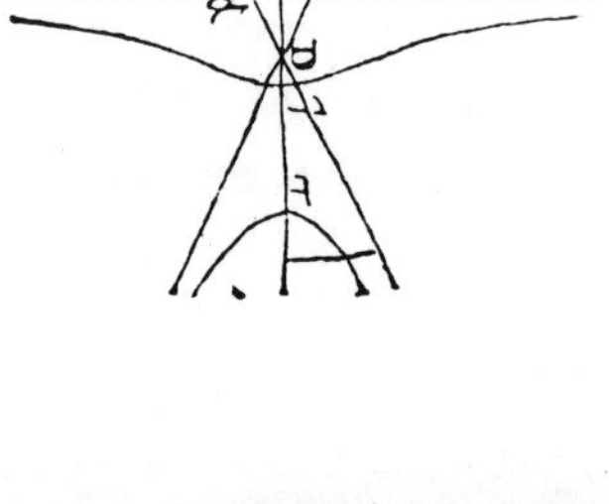

Fig. 27.

Fig. 43.

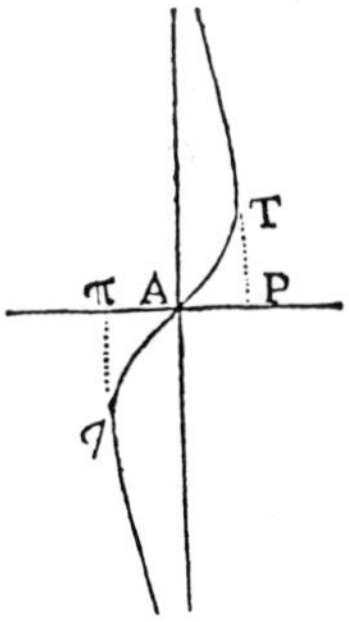

Fig. 44.

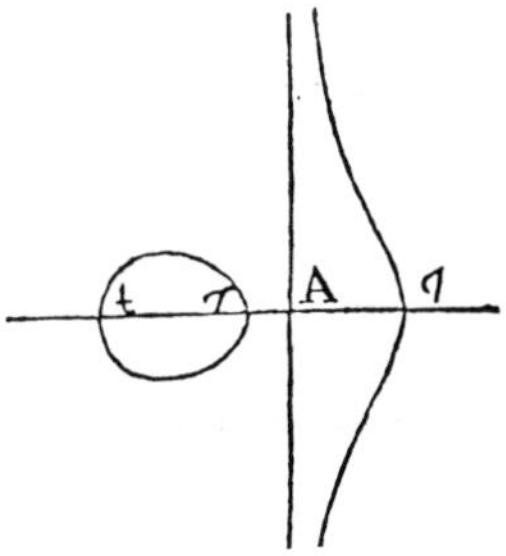

Fig. 45.

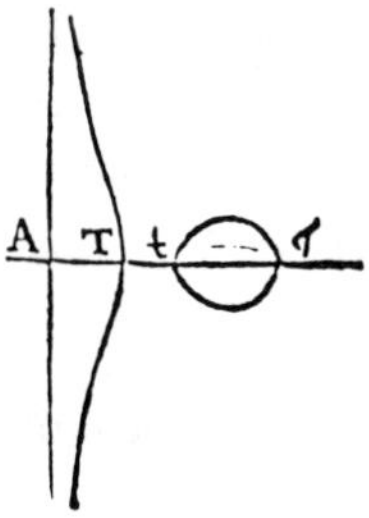

Fig. 56.

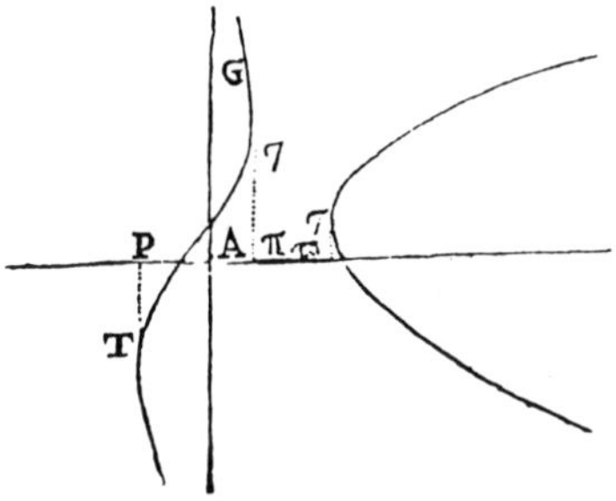

Fig. 63.

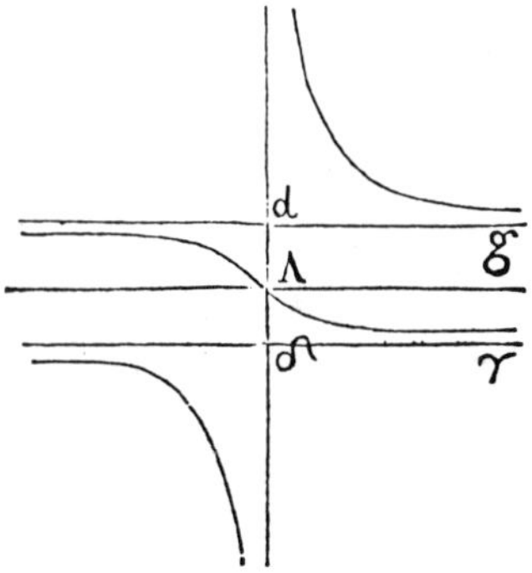

Fig. 53.

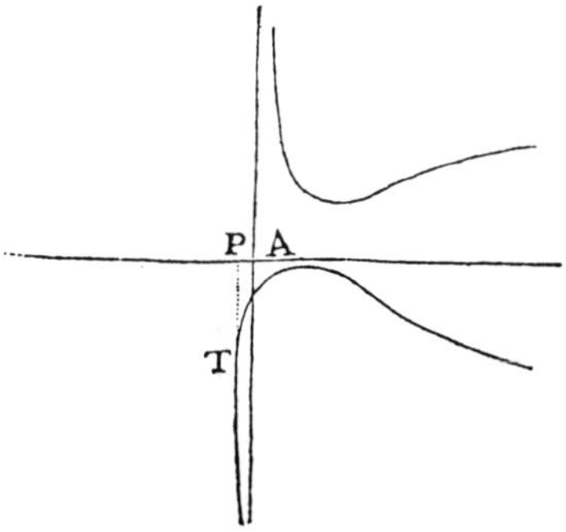

Fig. 54.

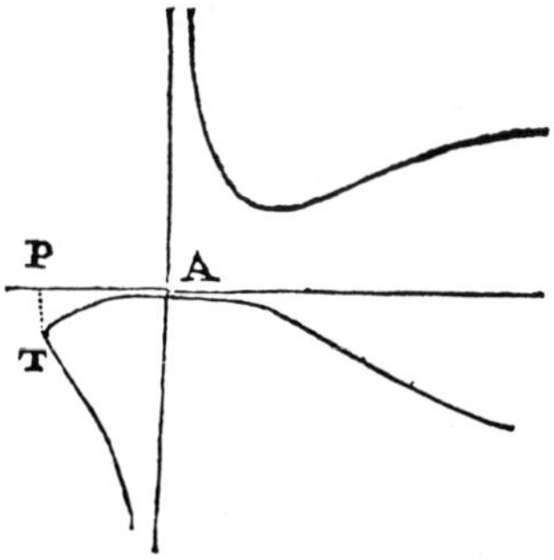

How shall we define an inflection point? It is sometimes proposed to define an inflection point of a curve as a point at which its curvature changes sign. (Intuitively : at an inflection point the curve changes form a "left-hand bend" to a "right-hand bend" or conversely.) This has the differential geometric consequence that the radius of curvature vanishes at the inflection point and changes sign there. This definition is in good agreement with intuition, however, as the following discussion shows, it is unsatisfactory in other respects :

We consider the curve with the affine equation $y-x^m = 0$. When $m > 2$ is odd, then this curve has an inflection point at the origin, in the intuitive sense above. But when n is even it has no inflection point in the above sense. Now we perturb the equation a little, i.e. we consider the equation

$$y = (x-\varepsilon_1)(x-\varepsilon_2) \ldots (x-\varepsilon_m),$$

where the ε_i are m different real numbers. Then the function $y(x)$ defined thereby has exactly m real zeroes and hence (by Rolle's theorem), $m-1$ extrema, at which y' must change from positive to negative. Thus y'' has exactly $m-2$ zeroes, which are simple zeroes, in fact each lies between two extrema. This means that the affine curve with the equation

$$y - (x-\varepsilon_1)(x-\varepsilon_2) \ldots (x-\varepsilon_m) = 0$$

has exactly $m-2$ inflection points in the intuitive sense above. Now we regard the $\varepsilon_1,\ldots,\varepsilon_m$ as variable and let them tend to zero. Then the $m-2$ inflection points tend to the origin, and for this reason one should define the origin to be an $(m-2)$-tuple inflection point, at least in algebraic geometry, where one wants a specialisation of the type just described (in which the parameters ε_i take the special value 0) to preserve the number of inflection points.

The following pictures of $y = (x^2-a^2)(x^2-4a^2)$ as $a \to 0$ show how the two inflection points merge at the origin, and the two inflection tangents tend toward the tangent to the curve $y = x^4$ at the origin.

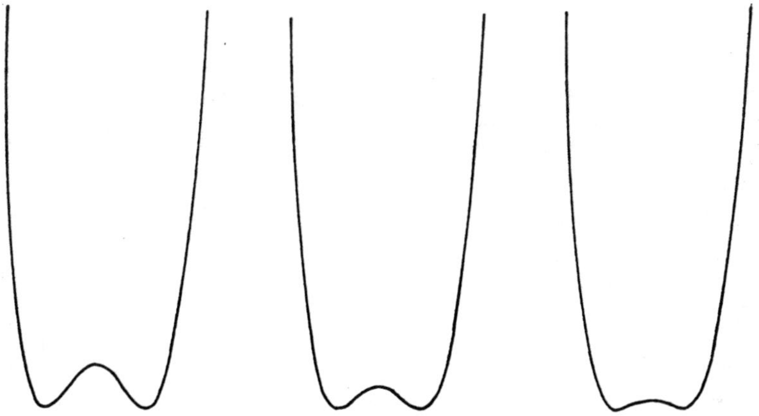

a = 0,6 a = 0,5 a = 0,4

In section 5.3 we have defined the tangents to a curve C at a point p of multiplicity $\nu_p(C)$ to be those lines through p which meet C at p with multiplicity greater than $\nu_p(C)$. If one counts tangents "with multiplicity", then there are exactly $\nu_p(C)$ tangents. In particular, if p is a regular point of C, i.e. if $\nu_p(C) = 1$, then C has a well-defined tangent T_p at p, and it meets C with multiplicity $\nu_p(C,T_p) > 1$ at p. Now what is the exact multiplicity with which C meets the tangent T_p at p ? This is a measure of how closely the tangent clings to the curve. Consider for example a plane affine curve with equation

$$y - f(x) = 0,$$

which is not a line.

Its tangent at the point $(x_0,f(x_0))$ has the equation

$$y = f(x_0) + f'(x_0)(x-x_0).$$

By the definition in 5.2, the intersection multiplicity of curve and tangent at this point is the order of vanishing of the restriction of $y-f(x) = 0$ to the tangent at x_0 , and hence the order of the zero x_0 of

$$f(x) - (f(x_0) + f'(x_0)(x-x_0)).$$

If one introduces the Taylor expansion for $f(x)$ at x_0, then one sees that the intersection multiplicity is r, where $f^{(r)}(x_0)$ is the first non-vanishing derivative at x_0. Thus for almost all x_0 the intersection multiplicity is equal to 2, and it is greater than 2 only at the finitely many points x_0 where $f''(x_0) = 0$. At these finitely many points the tangent clings to the curve better than at the other points, and these are also the points at which we expect inflections, after the analysis above. All this motivates the following :

Definition

A regular point p of a curve C in the complex projective plane is an inflection point of C when the tangent T to C at p is not a component of C and meets C at p with multiplicity at least 3.

More precisely, p is called an r-tuple inflection point when the intersection multiplicity satisfies $\nu_p(C,T) = r+2$.

If the tangent T is a component of C, then p is called an improper inflection point.

In what follows we shall sometimes use "inflection point" to include improper inflection points. It should be clear from the context what is meant. When it is not, we shall refer to inflection points in the sense of the above definition as "proper inflection points" for the sake of clarity.

We now want to develop a method for computing the inflection points of a given curve. Naturally one hopes to express the coordinates of the inflection points as solutions of suitable equations, i.e. as intersections of the given curve C with other associated curves. This idea goes back to de Gua (1740) and was carried out by Plücker in his "System der analytischen Geometrie", 1835, though Plücker cut C by a curve which was not "covariant", i.e. its definition was coordinate dependent. A covariant curve which achieved the same thing was then found by Hesse 1844. Hesse found his curve by a systematic discussion. To save time here we simply give his curve and content ourselves with the heuristic remark that, in the examples above, the inflection points of C were found as zeroes of equations defined with the help of the second derivatives of the equation for C. Thus we also expect a definition of this form for the Hessian curve.

Definition

Let C be a curve in the complex projective plane which does not decompose completely into lines. Let C be given by the equation $F(x_0,x_1,x_2) = 0$. Then the Hessian curve H_C of C is the curve with the equation

$$\det\left(\frac{\partial^2 F}{\partial x_i \partial x_j}\right) = 0.$$

Remarks : (i) Hesse showed that $\det\left(\frac{\partial^2 F}{\partial x_i \partial x_j}\right)$ vanishes identically just in case C decomposes into lines. (ii) One computes trivially that the curve H_C is independent of the choice of coordinates, since the Hessian matrix $\left(\frac{\partial^2 F}{\partial x_i \partial x_j}\right)$ transforms like a quadratic form under coordinate transformations. (iii) It is clear that the Hessian curve H_C of a curve of order m has order $3(m-2)$. In particular, the Hessian curve of a cubic is again a cubic. (iv) For a quadric C, H_C is empty. (v) It is clear that H_C goes through each singular point p of C. This is trivial when $\nu_p(C) > 2$, because then all $\frac{\partial^2 F}{\partial x_i \partial x_j}(p)$ vanish by definition. But it also holds when $\nu_p(C) = 2$, because the Euler formula 4.4.3 gives

$$0 = (m-1)\frac{\partial F}{\partial x_i}(p) = \sum_j \frac{\partial^2 F}{\partial x_i \partial x_j}(p)p_j$$

for $p = (p_0,p_1,p_2)$, and it follows that

$$\det\left(\frac{\partial^2 F}{\partial x_i \partial x_j}(p)\right) = 0.$$

The following theorem now shows that the Hessian curve H_C achieves exactly what is wanted : it picks out the inflection points on C.

Theorem 1

Let C be a plane curve without lines as components. Then a regular point p of C is an r-tuple inflection point just in case the Hessian curve H_C meets C at p with multiplicity r.

Proof : We choose homogeneous coordinates (x_0,x_1,x_2) so that the point p under investigation has coordinates $(1,0,0)$ and so that the tangent to C at p has equation $x_2 = 0$. If one introduces affine coordinates x, y by $x = \frac{x_1}{x_0}$, $y = \frac{x_2}{x_0}$, then C has an affine equation $f(x,y) = 0$, where we can write the polynomial f in the following form:

$$f(x,y) = y \cdot u(x,y) + x^{r+2} g(x).$$

Here $u(0,0) \neq 0$, $g(0) \neq 0$ and $r+2$ is the intersection multiplicity of C with its tangent $y = 0$ at p. The associated homogeneous polynomial F is then of the form

$$F(x_0,x_1,x_2) = x_2 U(x_0,x_1,x_2) + x_1^{r+2} G(x_0,x_1), \tag{1}$$

where U and G are the polynomials associated with u and g. It then follows from $g(0) \neq 0$ and $u(0,0) \neq 0$ that

$$\frac{\partial U}{\partial x_0}(1,0,0) \neq 0 \tag{2}$$

$$U(1,0,0) \neq 0 \tag{3}$$

$$G(1,0) \neq 0. \tag{4}$$

If one computes the second derivatives of F, then it follows from (1), (2) and (4) by simple calculation that the determinant D of the Hessian matrix of F is of the following form :

$$D = x_2 V(x_0,x_1,x_2) + x_1^r H(x_0,x_1) \tag{5}$$

where V and H are certain homogeneous polynomials and

$$H(1,0) \neq 0. \tag{6}$$

$D(x_0,x_1,x_2) = 0$ is by definition the equation of H_C. Thus the intersection points of C and H_C are the common zeroes of F and D. One easily concludes the following from (1), (3), (4), (5) and (6) :

When $r = 0$, $p \notin C \cap H_C$. When $r > 0$, p is an isolated intersection point of C and H_C. With this we have already proved : C and H_C have no common component and therefore meet in finitely many points, and these points are precisely the inflection points of C.

Thus it only remains to prove the statement $\nu_p(C,H_C) = r$ on the multiplicity of the inflection point. To do this we go back to the definition of intersection multiplicity as the multiplicity of the zero of a certain resultant in 6.1. We have to choose the coordinates suitably for this, namely, so that the point $(0,0,1)$ lies on none of the finitely many lines connecting intersection points of C and H_C. Then we construct the resultant $R_{F,D}$ of F and D, which we view as a polynomial in x_2 with coefficients in $\mathbb{C}[x_0,x_1]$. In this sense the constant terms of F resp. D are just $x_1^{r+2}G$ resp. $x_1^r H$. It then follows, by expanding the determinant form of $R_{F,D}$ by the last column that

$$R_{F,D} = x_1^r HP + x_1^{r+2} GQ \qquad (7)$$

for certain homogeneous polynomials $P, Q \in \mathbb{C}[x_0,x_1]$. We claim :

$$P(1,0) \neq 0. \qquad (8)$$

Proof : $P(1,0)$ is the resultant of $U(1,0,x_2)$ and $x_2 \cdot V(1,0,x_2)$. If it vanished, then these two polynomials would have a common zero x_2, and $(1,0,x_2)$ would be a common zero of F and D, hence a point on $C \cap H_C$ and the line through $(1,0,0)$ and $(0,0,1)$, hence equal to $(1,0,0)$, contrary to $U(1,0,0) \neq 0$.

It follows from (7) and (8) that the multiplicity of the zero $(1,0)$ of $R_{F,D}$ is equal to r. But this multiplicity is equal to $\nu_p(C,H_C)$ by definition, and hence the theorem is proved.

Corollary 2

A non-singular curve of order $m \geq 2$ in the complex projective plane has exactly $3m(m-2)$ inflection points, counting multiplicities.

Proof : The corollary follows immediately from Theorem 1, Bézout's theorem, and remark (iii) above.

It follows in particular from the corollary that a non-singular quadric has no inflection points, and a non-singular cubic has nine, counting multiplicity. But in the case of the cubic we can say even more. Since, by Bézout's theorem, the intersection multiplicity of an inflection tangent with a cubic at the inflection point is at most 3, while it is at least 3 by definition, it must exactly equal 3, and all inflection points of a cubic are simple. Thus we obtain the result already found by Plücker :

Corollary 3

A non-singular cubic has exactly 9 distinct inflection points, and these are all simple inflection points.

In what follows we shall see that we can determine the position of these inflection points exactly, and that they form a very beautiful symmetric configuration. We shall prove this with the help of a particularly symmetrical normal form for the equation of a non-singular cubic, found by Hesse in 1844. We now derive this normal form.

Let C be a non-singular cubic. We begin by using the existence of inflection points to choose a coordinate system for which as many terms as possible in the equation for C vanish, and then use further coordinate transformations to bring the equation into normal form.

First we choose homogeneous coordinates (z_0, z_1, z_2) so that $(0,0,1)$ and $(0,1,0)$ are inflection points of C and $z_1 = 0$ resp. $z_2 = 0$ are the corresponding inflection tangents.

Then, if one sets either $z_1 = 0$ or $z_2 = 0$ in the equation for C, all terms except z_0^3 vanish. Consequently, the equation contains only monomials divisible by $z_1 z_2$, apart from z_0^3, and hence it is of the form

$$z_1 z_2(\alpha z_0 + \beta z_1 + \gamma z_2) + \delta z_0^3 = 0.$$

Since C is non-singular, it follows by an easy calculation that $\beta, \gamma, \delta \neq 0$. Since we can also use the coordinates $(\delta^{1/3} z_0, \beta z_1, \gamma z_2)$ in place of (z_0, z_1, z_2), we can assume without loss of generality that the equation is of the following form :

$$z_1 z_2(\alpha z_0 + z_1 + z_2) + z_0^3 = 0. \qquad (1)$$

Then the non-singular nature of C is equivalent to the condition

$$\alpha^3 + 27 \neq 0, \qquad (2)$$

as one easily sees by computation of the common zeroes of the three partial derivatives of (1).

We now introduce new coordinates (y_0, y_1, y_2) by :

$$\begin{aligned} z_0 &= y_0 \\ 3z_1 &= -\alpha y_0 - y_1 + 2y_2 \\ 3z_2 &= -\alpha y_0 + 2y_1 - y_2 \end{aligned} \qquad (3)$$

One computes quite easily that this transformation converts equation (1) into the equation

$$(27+\alpha^3) y_0^3 - 2y_1^3 + 3y_1^2 y_2 + 3y_2^2 y_1 - 2y_2^3 - 3\alpha y_0(y_1^2 - y_1 y_2 + y_2^2) = 0. \qquad (4)$$

Now one recognises immediately that $y_1^2 - y_1 y_2 + y_2^2$ splits into the linear factors $(y_1 + \varepsilon y_2)(y_1 + \varepsilon^2 y_2)$, where ε and ε^2 are the two solutions $e^{\pm 2\pi i/3}$ of

$$\varepsilon^2 + \varepsilon + 1 = 0.$$

Then one sees that the sum of the homogeneous terms of order 3 in y_1, y_2

is, up to sign, the sum of the cubes of these two linear factors. We may therefore write (4) as :

$$(27+\alpha^3)y_0^3-(y_1+\varepsilon y_2)^3-(y_1+\varepsilon^2 y_2)^3-3\alpha y_0(y_1+\varepsilon y_2)(y_1+\varepsilon^2 y_2) = 0. \tag{5}$$

Thus if one introduces new coordinates (x_0,x_1,x_2) by

$$x_0 = y_0$$
$$x_1 = y_1 + \varepsilon y_2$$
$$x_2 = y_1 + \varepsilon^2 y_2 ,$$

then (5) becomes the equation

$$(27+\alpha^3)x_0^3 - x_1^3 - x_2^3 - 3\alpha x_0x_1x_2 = 0. \tag{6}$$

Since we can also replace the coordinate x_0 by $(27+\alpha^3)^{1/3}x_0$, by virtue of (2), we find in the end that for suitable coordinates (x_0,x_1,x_2) the equation for C is of the following form :

$$x_0^3 + x_1^3 + x_2^3 + \lambda x_0x_1x_2 = 0, \tag{7}$$

where λ can be any complex number apart from the three roots of $\lambda^3 + 27 = 0$. Thus we have proved :

Theorem 4

Relative to suitable homogeneous coordinates, each non-singular cubic has an equation of the form

$$x_0^3 + x_1^3 + x_2^3 + \lambda x_0x_1x_2 = 0.$$

(Such an equation describes a non-singular cubic just in case $\lambda^3 + 27 \neq 0$.)

Remark : Let C_λ be the cubic with the equation

$$x_0^3 + x_1^3 + x_2^3 + \lambda x_0x_1x_2 = 0.$$

We have proved that each non-singular cubic $C \subset P_2(\mathbb{C})$ can be carried into some curve C_λ by a collineation. In this sense, one can view the C_λ as normal forms of cubics. To be sure, this normal form is not uniquely determined by C. For example, if one multiplies the coordinates by (different) cube roots of unity, then λ is also multiplied by a cube root of unity, i.e. C_λ and $C_{\varepsilon\lambda}$ can be carried into each other by collineations. Later we shall investigate precisely when two curves C_λ, $C_{\lambda'}$ can be carried into each other.

We now compute the inflection points of C_λ. To do this we compute the Hessian determinant

$$D_\lambda = \begin{vmatrix} 6x_0 & \lambda x_2 & \lambda x_1 \\ \lambda x_2 & 6x_1 & \lambda x_0 \\ \lambda x_1 & \lambda x_0 & 6x_2 \end{vmatrix}$$

$$= -6\lambda^2(x_0^3+x_1^3+x_2^3) + (216+2\lambda^3)x_0x_1x_2.$$

Thus we see that, like C_λ, the Hessian corresponds to a 1-dimensional linear system $\mathcal{L}$ of cubics, defined by

$$\alpha(x_0^3+x_1^3+x_2^3) + \beta x_0x_1x_2 = 0.$$

The non-singular curves of this linear system are precisely the cubics C_λ, $\lambda = \frac{\beta}{\alpha}$, with $\lambda^3 + 27 \neq 0$. For $\alpha = 0$ one obtains the triangle C_∞ of the three coordinate axes, with equation $x_0x_1x_2 = 0$. For $\lambda^3 + 27 = 0$, C_λ likewise decomposes into three lines, because one easily computes that in this case the equation splits into linear factors as follows :

$$C_{-3} \qquad (x_0+x_1+x_2)(x_0+\varepsilon x_1+\varepsilon^2x_2)(x_0+\varepsilon^2x_1+\varepsilon x_2) = 0$$

$$C_{-3\varepsilon} \qquad (x_0+x_1+\varepsilon x_2)(x_0+\varepsilon x_1+x_2)(x_0+\varepsilon^2x_1+\varepsilon^2x_2) = 0$$

$$C_{-3\varepsilon^2} \qquad (x_0+x_1+\varepsilon^2x_2)(x_0+\varepsilon x_1+\varepsilon x_2)(x_0+\varepsilon^2x_1+x_2) = 0.$$

Altogether, $C_\infty \cup C_{-3} \cup C_{-3\varepsilon} \cup C_{-3\varepsilon^2}$ is a system of twelve different lines.

Moreover, it is clear that any two curves C_λ, $C_{\lambda'}$ of our linear system must meet in the same 9 points. In particular, the intersection points of C_λ with the Hessian curve H_{C_λ} are the same as those of C_0 with C_∞, and hence the common zeroes of

$$x_0x_1x_2 = 0$$

$$x_0^3 + x_1^3 + x_2^3 = 0.$$

But one can compute these immediately : they are the points

$$p_{00} = (0,-1,1) \quad p_{01} = (0,-\varepsilon,1) \quad p_{02} = (0,-\varepsilon^2,1)$$

$$p_{10} = (1,0,-1) \quad p_{11} = (1,0,-\varepsilon) \quad p_{12} = (1,0,-\varepsilon^2)$$

$$p_{20} = (-1,1,0) \quad p_{21} = (-\varepsilon,1,0) \quad p_{22} = (-\varepsilon^2,1,0).$$

These are therefore the inflection points of C_λ. It follows by Theorem 7.2.1 that the linear system of cubics through these 9 points is 1-dimensional, hence it coincides with our linear system $\mathcal{L}$. Thus we have proved the following :

Proposition 5

(i) The linear system of cubics through the 9 points p_{ij}, $i, j = 0,1,2,$ is the 1-dimensional system of cubics C_λ.

For finite λ, C_λ is the cubic with equation

$x_0^3 + x_1^3 + x_2^3 + \lambda x_0 x_1 x_2 = 0,$

and C_∞ is the triangle $x_0 x_1 x_2 = 0$.

(ii) C_∞, C_{-3}, $C_{-3\varepsilon}$, $C_{-3\varepsilon^2}$ are triangles, each of three different lines, making 12 lines altogether.

(iii) All other C_λ are non-singular and have inflections precisely at the nine points p_{ij}.

It is interesting and useful to study the configuration formed by the 9 inflection points and the 12 lines of the degenerate cubics somewhat more closely. All 9 inflection points lie on each of the 4 triangles. By comparing the equations of the 12 lines with the coordinates of the 9 points, one can easily determine which points lie on which lines. In this way one obtains the following result :

Proposition 6

The position of the 9 inflection points on the 12 lines of the degenerate cubics can be described in terms of the expansion scheme for 3×3 determinants, as follows :

If one lets p_{ij} correspond to the element in the ith row and j^{th} column, then 3 points p_{ij} lie on each of the 12 lines, namely, those whose corresponding elements are in the same row, or in the same column, or in the same element in the expansion of the determinant.

The following schema shows the incidences just described :

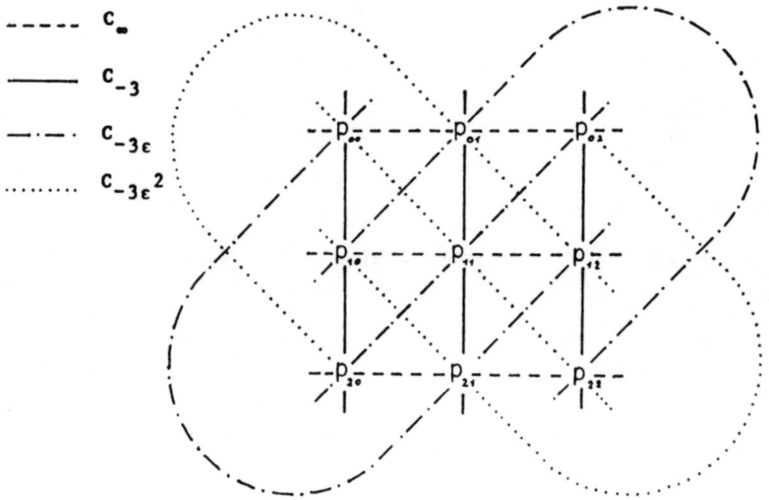

One sees that three of the 9 points lie on each line, and that 4 of the 12 lines pass through each point. Configurations of this type are also called "tactical configurations", and one indicates the numbers of incident points resp. lines by symbols such as $(9_4, 12_3)$ or $\Delta^{3,4}_{9,12}$. (Other examples are, e.g., the configurations one obtains from two triangles in perspective position, or the Pascal hexagon inscribed in a degenerate quadric, which we discussed previously. More on this may be found, e.g., in the Steinitz article "Konfigurationen der projektiven Geometrie" [S7], as well as in R.D. Carmichael's book "Introduction to the Theory of Groups of Finite Order" [C1].)

What is particularly nice is that one can also obtain the configuration $(9_4, 12_3)$ in quite a different way :

Proposition 7

The configuration $(9_4, 12_3)$ of the 9 inflection points and the 12 lines of the degenerate cubics of $\mathcal{L}$ is isomorphic to the configuration of 9 points and 12 lines in the affine plane over the prime field $\mathbb{F}_3$ of characteristic 3. The isomorphism is given by associating the point p_{ij} with the point with coordinates (i,j), $i,j = 0,1,2$, and associating the lines of one configuration with the lines through the corresponding points of the other configuration.

<u>Proof</u> : The statement is easily checked explicitly using the incidence schema from Proposition 6.

The advantage of the description of our configuration in Proposition 7 is that we can now easily give the symmetry group of this configuration. This is because the automorphisms of point and line configurations in the affine plane over $\mathbb{F}_3$ are known : they are, as one can easily convince oneself, just the affine automorphisms.

They form a group, the affine group

$A(2, \mathbb{F}_3)$.

Since there are nine translations in the plane $\mathbb{F}_3^2$, and the general linear group $GL(2, \mathbb{F}_3)$ has order 48, as one easily sees*, $A(2, \mathbb{F}_3)$ is a group of order $9.48 = 432$. In this group we have a normal subgroup of index 2, the special affine group $SA(2, \mathbb{F}_3)$, which is generated by the translations and the elements of $SL(2, \mathbb{F}_3)$.

$\mathrm{ord}(SA(2, \mathbb{F}_3)) = 216$.

The next question, of course, is whether we can induce the automorphisms of $(9_4, 12_3)$ by collineations in the projective plane. This is in fact the case, at least for $SA(2, \mathbb{F}_3)$, and it is achieved by the <u>Hessian group</u> G_{216}.

This is the group of 216 collineations generated by the 5 collineations A, B, C, D, E described by the following schema for $(x_0,x_1,x_2) \mapsto (x_0',x_1',x_2')$:

	A	B	C	D	E
x_0'	x_1	x_0	x_0	x_0	$x_0 + x_1 + x_2$
x_1'	x_2	x_2	εx_1	εx_1	$x_0 + \varepsilon x_1 + \varepsilon^2 x_2$
x_2'	x_0	x_1	$\varepsilon^2 x_2$	εx_2	$x_0 + \varepsilon^2 x_1 + \varepsilon x_2$

In fact, one easily shows that A and C are the translations in $\mathbb{F}_3^2$ by the vectors $(2,0)$ resp. $(0,2)$, and that B, D, E correspond to the special linear transformations

$$\begin{pmatrix} 2 & 0 \\ 0 & 2 \end{pmatrix}, \begin{pmatrix} 1 & 0 \\ 1 & 1 \end{pmatrix} \begin{pmatrix} 0 & 1 \\ 2 & 0 \end{pmatrix}$$

respectively, and that these 5 transformations generate $SA(2, \mathbb{F}_3)$.

---

*See L.E. Dickson : "Linear groups" [D2].

One can generate the remaining transformations of $A(2, \mathbb{F}_3)$ when one adds the conjugation transformation $(x_0,x_1,x_2) \mapsto (\overline{x_0},\overline{x_1},\overline{x_2})$, which one naturally cannot induce (as an automorphism of $(9_4, 12_3)$) by a projective linear transformation. Incidentally, it is of course clear that there is at most one projective linear transformation which induces a given automorphism of $(9_4, 12_3)$.

From these findings we can draw the following conclusion :

Proposition 8

G_{216} transforms $\mathcal{L}$ into itself. Two cubics C_λ, C_μ are then equivalent via a projective linear transformation A if and only if one is carried to the other by a transformation in the Hessian group G_{216}.

Proof : When A carries C_λ to C_μ, A induces an automorphism of the configuration $(9_4, 12_3)$, and hence A is an element of G_{216}. Conversely, since the elements of G_{216} are automorphisms of $(9_4, 12_3)$, they carry each C_λ to a cubic with the same inflection points and hence, by Proposition 5, to a C_μ.

With this, we can finally obtain a simple answer to the question following Theorem 4, asking when C_λ and C_μ can be carried into each other by collineations.

We consider the action of G_{216} on the linear system $\mathcal{L}$. In G_{216} we have a normal subgroup G_{18}, of order 18, which carries each C_λ into itself, and hence acts trivially on $\mathcal{L}$. It is the group generated by A, B, C. The quotient group $\mathcal{T} = G_{216}/G_{18}$ therefore acts on the 1-dimensional projective space $\mathcal{L}$, and it is clear from the foregoing that two cubics C_λ, $C_{\lambda'}$ are equivalent via a collineation just in case an operation in $\mathcal{T}$ carries one to the other.

We therefore study the action of $\mathcal{T}$ on $\mathcal{L}$ more closely. $\mathcal{T}$ is obviously a group of order 12, generated by the cosets $\bar{D}$ and $\bar{E}$ of D and E. We shall explicitly describe the action of these generators on $\mathcal{L}$. To do this it is convenient to use the affine coordinate $\mu = -\frac{\lambda}{3}$ instead of λ on the 1-dimensional projective space, and hence to write our curve equations in the form

$$x_0^3 + x_1^3 + x_2^3 - 3\mu x_0x_1x_2 = 0.$$

The parameter values for which these cubics degenerate are :

$$\mu = \infty$$
$$\mu = \varepsilon^i, \ i = 0,1,2.$$

It is now easy to compute the action of $\bar{D}$ resp. $\bar{E}$ by carrying out the substitutions corresponding to D resp. E in the above equation. One obtains :

$$\bar{D}(\mu) = \varepsilon^2 \mu$$
$$\bar{E}(\mu) = \frac{\mu+2}{\mu-1} .$$

In particular :

$$\bar{D}(\varepsilon^0) = \varepsilon^2 \quad \bar{D}(\varepsilon^1) = \varepsilon^0 \quad \bar{D}(\varepsilon^2) = \varepsilon^1 \quad \bar{D}(\infty) = \infty$$

$$\bar{E}(\varepsilon^0) = \infty \quad \bar{E}(\varepsilon^1) = \varepsilon^2 \quad \bar{E}(\varepsilon^2) = \varepsilon^1 \quad \bar{E}(\infty) = \varepsilon^0 .$$

But one can easily see from this how the group $\mathcal{T}$ acts on $\mathcal{L}$.

Proposition 9

The action of $\mathcal{T}$ on $\mathcal{L}$ is equivalent to the action of the tetrahedral group T on the Riemann number sphere S^2.

Proof : We inscribe a tetrahedron in the Riemann number sphere S^2. The tetrahedral group T is the group of rotations which carry the tetrahedron into itself. T consists of 12 elements : the identity, 8 elements of order 3 and 3 elements of order 2. The elements of order 3 are rotations of 120° about the 4 axes which connect a vertex to the centre of the opposite face. The 3 elements of order 2 are the rotations of 180° about the 3 axes which connect centres of opposite edges.

Now we choose an affine coordinate μ on the number sphere S^2 so that the 4 vertices of the inscribed tetrahedron are just $\mu = \infty, \varepsilon^0, \varepsilon^1, \varepsilon^2$. Each element of the tetrahedral group T then defines a certain permutation of the 4 vertices, and hence a certain projective transformation of the 1-dimensional projective space S^2. This is just the corresponding rotation of the sphere, and in this way the tetrahedral group acts on S^2. If we now identify S^2 and $\mathcal{L}$, then $\bar{D}$ and $\bar{E}$ are obviously operations in the tetrahedral group, and so $\mathcal{T}$ acts as a subgroup of T. Since both have the same order, it follows that $\mathcal{T} = T$.

We could now easily settle our problem of deciding when C_λ can be carried to $C_{\lambda'}$ by a collineation, as follows : the action of $\mathcal{T}$ on λ yields 12 numbers $\lambda_1,\ldots,\lambda_{12}$, and $C_{\lambda'}$ can be obtained from C_λ

just in case $\lambda' = \lambda_i$ for some i. However, we want to give this condition a still more elegant formulation by finding a function $\phi(\lambda)$ such that $\phi(\lambda)$ takes the same value precisely on $\lambda_1,\ldots,\lambda_{12}$. Then we can say : C_λ can be carried into $C_{\lambda'}$ just in case $\phi(\lambda) = \phi(\lambda')$.

Finding such a function is closely connected with describing the canonical mapping $S^2 \to S^2/T$ from S^2 onto the orbit space S^2/T.

It is clear that the quotient space S^2/T is again a 2-sphere, topologically. (Proof : if one divides the sphere into spherical triangles corresponding to the 4 triangles of the tetrahedron, and subdivides each of these 4 triangles into 3 smaller triangles by connecting their vertices to the centre, then one obtains a decomposition of the sphere into 12 fundamental triangles.

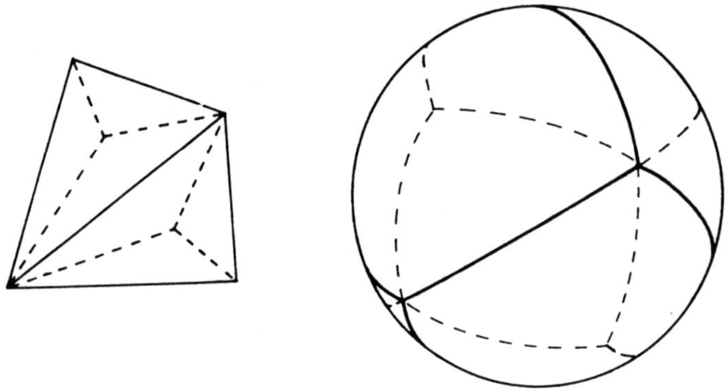

Now the quotient mapping obviously identifies the "inner" edges, while the two halves of the "outer" edge are identified by reflection in its midpoint. This of course gives a topological 2-sphere.)

One can now show that the quotient space can also be given the structure of an (abstract) algebraic curve in a natural way, that this curve is the Riemann sphere, and that the mapping $S^2 \to S^2/T$, when one views as a mapping $P_1(\mathbb{C}) \to P_1(\mathbb{C})$ in this sense, is given in homogeneous coordinates by $(u,v) \mapsto (P(u,v),Q(u,v))$, where $P(u,v)$, $Q(u,v)$ are homogeneous polynomials of degree 12. These polynomials must then be invariant under the action of T, in the sense that the action of each element of T multiplies them by the same constant. Conversely, when we find two such invariant polynomials which differ by more than a constant factor, then we can use them to define a mapping $P_1(\mathbb{C}) \to P_1(\mathbb{C})$ which factors over $P_1(\mathbb{C})/T$, so that we can identify $P_1(\mathbb{C})/T$ with $P_1(\mathbb{C})$.

Thus we now want to find such invariant polynomials. The zeroes of P resp. Q must, counting multiplicities, be 12 points which are permuted by the action of T. The most obvious thing is to take the vertices of the tetrahedron, namely the points with the inhomogeneous coordinates $\mu = \infty, \varepsilon^0, \varepsilon^1, \varepsilon^2$. We introduce homogeneous coordinates (u,v) with $\mu = v/u$. Then these 4 vertices are the zeroes of the homogeneous polynomial

$$u(u^3 - v^3).$$

In homogeneous coordinates one can describe the substitutions $\bar{D}$ and $\bar{E}$ by

$$(u,v) \mapsto (u,\varepsilon^2 v)$$
$$(u,v) \mapsto (v-u,\ v+2u).$$

The above polynomial is unaltered by the first substitution, and the second multiplies it by 9.

A further system of 4 points invariant under T are the centres of the triangles of the tetrahedron, i.e. the points $\mu = 0$ and $\mu = -2\varepsilon^i$, $i = 0,1,2$. They are the zeroes of the homogeneous polynomial

$$v(v^3 + 8u^3),$$

which $\bar{D}$ multiplies by ε^2, and $\bar{E}$ multiplies by 9. Therefore

$$P(u,v) = u^3(u^3-v^3)^3$$
$$Q(u,v) = v^3(v^3+8u^3)^3$$

are two invariant homogeneous polynomials of the desired kind, which describe the quotient mapping for us.

We shall now describe this mapping in affine coordinates. Thus we go to $\mu = v/u$ in the preimage space and to $-Q/P$ in the image space. Then we obtain the rational function

$$j(\mu) = \frac{\mu^3(\mu^3+8)^3}{(\mu^3-1)^3}.$$

This has triple poles at precisely the 4 values ∞, ε^i, $i = 0,1,2$, which correspond to degenerate cubics. For the remaining values we call the number $j(\mu)$ the <u>j-invariant</u> of the cubic curve with equation

$$x_0^3 + x_1^3 + x_2^3 - 3\mu x_0 x_1 x_2 = 0.$$

It is clear from the construction and the above remarks that for each $j \in \mathbb{C}$ there are, counting multiplicity, exactly 12 values $\mu_1, \ldots, \mu_{12} \in \mathbb{C} - \{\varepsilon^0, \varepsilon^1, \varepsilon^2\}$ for which $j(\mu_i) = j$, that these are permuted among themselves by the action of T, and that μ, μ' with $j(\mu) \neq j(\mu')$ are inequivalent under T. In view of the identification of the T-action with that of $\mathcal{T}$ of $\mathcal{L}$ given by Proposition 9, we finally have :

<u>Theorem 10</u>

Let $j(\mu) = \dfrac{\mu^3(\mu^3+8)^3}{(\mu^3-1)^3}$ be the j-invariant of the cubic curve $\tilde{C}_\mu$ with equation

$$x_0^3 + x_1^3 + x_2^3 - 3\mu x_0 x_1 x_2 = 0, \ \mu^3 \neq 1.$$

Then $\tilde{C}_\mu$ and $\tilde{C}_{\mu'}$ are isomorphic via a collineation if and only if $j(\mu) = j(\mu')$.

Since each nonsingular cubic curve is isomorphic, via a collineation, to a $\tilde{C}_\mu$, this gives a complete classification of the nonsingular cubic curves.

<u>Remark</u> : The j-invariant defined here agrees with the one most commonly used elsewhere only up to multiplication by a constant, which is unimportant in our context.

The normal form equation for the cubic which we have used until now is particularly suitable for the investigation of the inflection point configuration. For certain other purposes another normal form is natural, the <u>Weierstrass normal form</u>.

<u>Theorem 11</u>

Relative to suitable affine coordinates, each nonsingular cubic has an equation of the form

$$y^2 = 4x^3 - g_2 x - g_3 \ ,$$

where the polynomial $4x^3 - g_2 x - g_3$ has no multiple zeroes. Conversely, each such equation defines a nonsingular cubic.

Proof : Since we have already derived one normal form in Theorem 4, we shall skip a detailed derivation of the Weierstrass normal form and content ourselves with a few hints. One first chooses homogeneous coordinates so that $x_0 = 0$ is the inflection tangent at $(0,0,1)$. Then the cubic has an equation in affine coordinates of the form

$$y^2 + (ax+b)y + g(x) = 0$$

where $g(x)$ is a polynomial of degree 3. At this stage it is clear that one can reach the desired form of equation through further coordinate transformations.

Remarks :

(i) There is a reason for choosing the normal form in just this way, i.e. with the coefficient 4 for x^3, which we shall meet in 7.4 : one can define two functions $x(z)$, $y(z)$ in a very natural way and then show that they satisfy exactly this equation in Weierstrass normal form, and hence parametrise the curve. The function x is the Weierstrass $\wp$-function and y is its derivative. If these functions are taken as given from the beginning, then the equation comes automatically, as the differential equation of the Weierstrass $\wp$-function.

(ii) The normal form in Theorem 4 contains only one parameter λ, the Weierstrass form contains two, g_2 and g_3. With the Weierstrass normal form, each curve obviously has a whole one-parameter family of equations - along with $y^2 = 4x^3 - g_2x - g_3$ there are also all the equations $y^2 = 4x^3 - \alpha^2 g_2 x - \alpha^3 g_3$.

(iii) The condition that $4x^3 - g_2x - g_3$ have no multiple roots is equivalent to non-vanishing of the discriminant, and hence equivalent to the condition

$$g_2^3 - 27g_3^2 \neq 0.$$

(iv) One can of course compute the j-invariant of Theorem 10 as a function of the parameters g_2 and g_3 in the Weierstrass normal form. Up to a multiplicative constant it is

$$J = \frac{g_2^3}{g_2^3-27g_3^2} .$$

Thus two curves with equations in Weierstrass normal form are isomorphic just in case they have the same J invariant.

To conclude this section on inflection points and normal forms of plane cubics we briefly treat the case of singular cubics. The investigation of reducible cubics is not interesting from this viewpoint : they decompose into lines or into a nonsingular quadric and a line. Proper inflection points are not present.

We therefore confine ourselves to irreducible cubics. By Corollary 6.2.11, an irreducible cubic C can have at most one singular point p, and this point must have multiplicity 2. By 5.3, C either has two simple tangents at such a point, i.e. p is an ordinary double point, or C has a double tangent. The following two propositions now show that with this case distinction the projective classification of singular irreducible cubics is already settled. Since the proofs of the theorems are quite simple and along similar lines to the proofs of Theorem 4 and Proposition 6 (choice of suitable coordinates – coordinate transformation to normal form – computation of the Hessian curve) we omit them.

<u>Proposition 12</u>

An irreducible cubic with an ordinary double point has equation

$$y^2 = x^2(x+1),$$

relative to suitable affine coordinates. It has exactly 3 inflection points, and they are collinear.

<u>Proposition 13</u>

An irreducible cubic with one non-ordinary double point has equation

$$y^2 = x^3$$

relative to suitable affine coordinates. It has exactly one inflection point.

Theorem 4, Theorem 10, Proposition 12 and Proposition 13 completely settle the complex-projective classification of plane irreducible cubics which began in a certain sense with Newton's projective classification, which we discussed in 2.5.

Admittedly, Newton was of course concerned with real cubics. By way of conclusion, we want to carry over our results on complex cubics to real cubics, at least as far as inflection points are concerned. For this purpose we confine ourselves to nonsingular real cubics, i.e. to the zero sets in $P_2(\mathbb{R})$ of real homogeneous polynomials

$F(x_0,x_1,x_2)$ for which the cubic in $P_2(\mathbb{C})$ defined by $F(x_0,x_1,x_2) = 0$ is nonsingular.

Proposition 14

A nonsingular real cubic has exactly three real inflection points. These inflection points are collinear.

Proof : One easily sees that the Weierstrass normal form of a nonsingular complex cubic which is defined by a real equation can be chosen in such a way that the coefficients g_2, g_3 in the equation $y^2 = 4x^3 - g_2x - g_3$ are real. It follows easily that the real cubic has at least three real inflection points.

On the other hand, we claim that it has at most three real inflection points. Proof : conjugation $(x_0,x_1,x_2) \mapsto (\overline{x_0},\overline{x_1},\overline{x_2})$ of $P_2(\mathbb{C})$ induces an automorphism $\bar{\kappa}$ of the complex inflection point configuration $(9_4, 12_3)$, and hence an involution $\bar{\kappa}$ in $A(2, \mathbb{F}_3)$ by the remark at the end of Proposition 7. Now one easily proves that each involution in $A(2, \mathbb{F}_3)$, i.e. each $\kappa \in A(2, \mathbb{F}_3)$ such that $\kappa^2 = 1$, has either one, three or nine fixed points. But the fixed points of $\bar{\kappa}$ are just the real inflection points. Thus it suffices to show that one can never choose the coordinates in $P_2(\mathbb{C})$ so that all 9 inflection points become real. This follows from the computation of inflection points for the Hessian normal form at the end of Theorem 4. Assuming 9 real inflection points in some coordinate system, there would be a collineation A which carried the 9 inflection points of the Hessian normal form into 9 points in $P_2(\mathbb{R}) \subset P_2(\mathbb{C})$, and one could assume without loss of generality that $A(1,0,-1) = (1,0,0)$, $A(0,1,-1) = (0,1,0)$ and $A(0,1,-\varepsilon) = (0,0,1)$. This determines A up to a multiplicative constant and one computes immediately that, e.g., $A(0,1,-\varepsilon^2) = (0,1,\varepsilon)$, contrary to the assumption that A maps all inflection points to real points. Thus the real cubic has exactly three inflection points.

It is clear that they must also be collinear. Because the line through two of these inflection points meets the complex cubic in a third inflection point, and this third intersection point must also be real, and hence coincident with the third real inflection point. This proves Proposition 14. Of course one can also carry out the proof quite independently of the Weierstrass normal form and the associated Hessian curve.

7.4 Cubics, elliptic curves and abelian varieties

When C is a nonsingular plane cubic and p and q are points on C, then the line through p and q cuts the cubic C in just one further point r', and this property obviously holds only for cubics. Thus, for a cubic, $(p,q) \mapsto r'$ gives a mapping $C \times C \to C$, and this leads us to define a group structure on $C \times C$, which indeed must also be given by a mapping $C \times C \to C$ — if one exists. When there is one group structure on a set, there are of course many others, obtained from the given structure by conjugation with a translation. A translation sends the identity element to another element, which becomes the identity element of the new structure. Thus in order to pick out a particular structure among these conjugate ones, we must fix a particular element to serve as the identity. We choose a particular point $p_0 \in C$ for this purpose.

Now one easily sees that the product or sum of two elements $p, q \in C$ cannot simply be defined as the third intersection point, r', with C of the line through p and q, because then p_0 would obviously not be an identity element. One must therefore associate the point r' in a suitable way with some other point, which is then the sum of p and q. The simplest way to obtain this further point from r' is perhaps the following : one draws a line through r' and the given point p_0. This meets C in a further point r. When one defines this r to be $p+q$ it is certainly clear that $p + p_0 = p$, as we desire. This motivates the following definition, which will be justified further by what follows :

Definition

Let C be a nonsingular plane cubic and let $p_0 \in C$ be any particular point on C. Then the sum of $p, q \in C$ is defined as follows. Let r' be the third intersection point of C with the line through p and q, and let r be the third intersection point of C with the line through r' and p_0. Then

$$p + q = r.$$

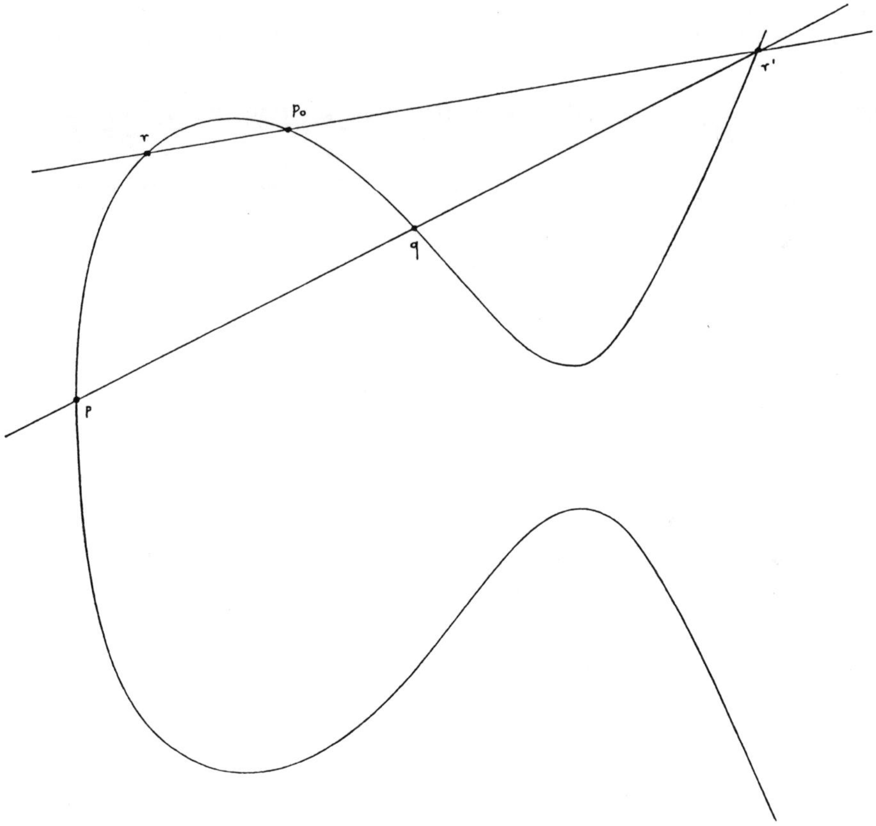
r
p_0
r'
q
p

In the above definition, the line through two coincident points of C is of course understood to be the tangent at the point in question.

Theorem 1

Let C be a nonsingular plane cubic, let p_0 be a point of C, and let + be the binary operation on elements of C just defined. Then (C,+) is an abelian group with p_0 as identity element.

Proof : As noted above, it is clear that p_0 is an identity element. It is likewise clear that the operation is commutative, because the line through p and q does not depend on the order of p and q. One constructs the inverse -p of p as follows : the tangent to C at p_0 cuts C at a further point p_1. Then the intersection point -p of C with the line through p and p_1 is obviously the inverse of p as the first of the two following figures illustrates.

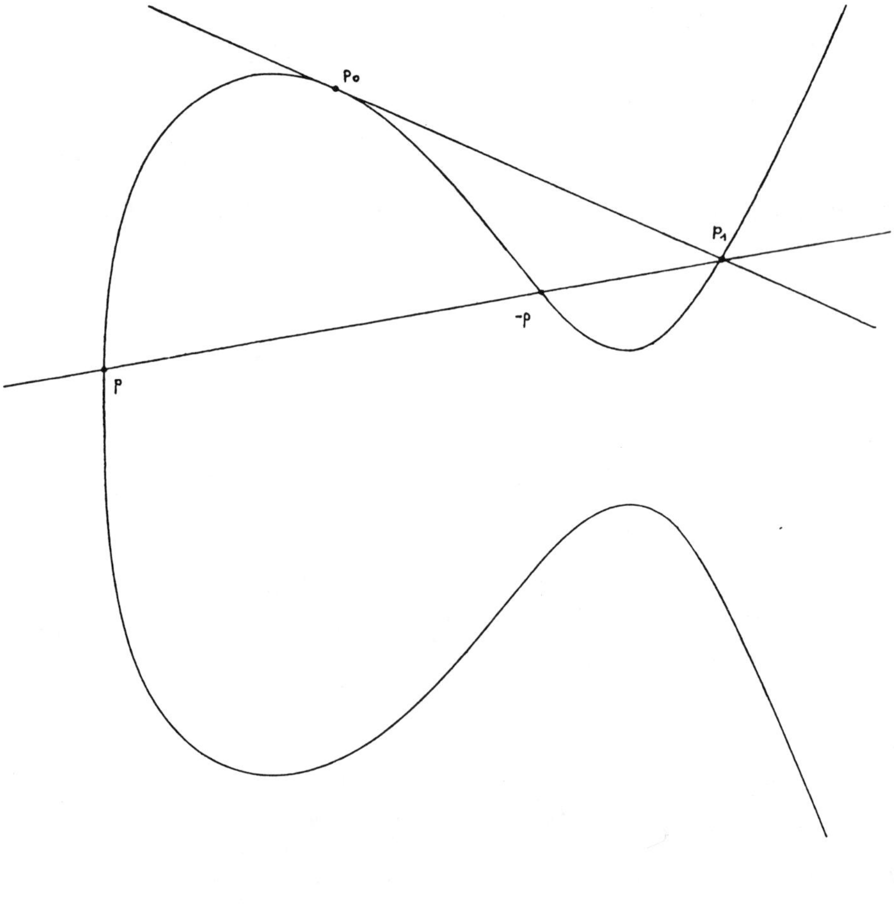

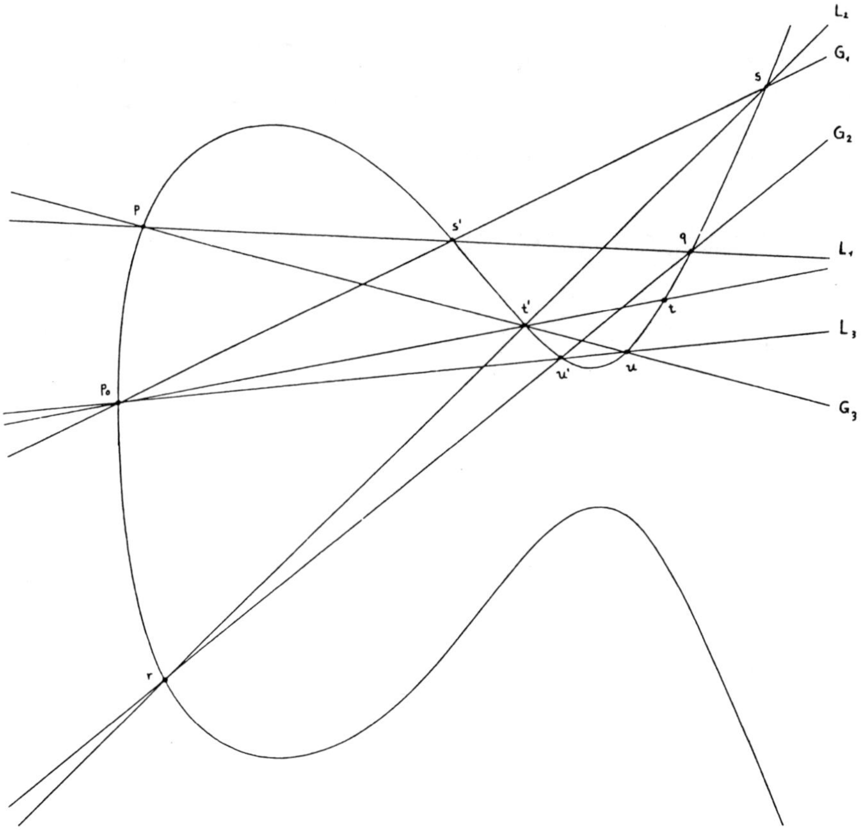

The second figure illustrates the following proof of the associative law $(p+q) + r = p + (q+r)$.

We set

$p + q = s$

$s + r = t$

$q + r = u.$

We have to show : $p + u = t$.

We denote the auxiliary points appearing in the construction of s, t, u by s', t', u'. In order to show $p + u = t$, we must prove that p, u and t' lie on a line. This happens as follows : let L_1 be the line through p, q, s', let L_2 be the line through s, r, t' and let L_3 be the line through p_0, u', u. Also, let G_1 be the line through p_0, s', s and let G_2 be the line through q, r, u'.

Now we apply Theorem 6.2.3 to the intersection of the cubics C and $C' = L_1 \cup L_2 \cup L_3$. This theorem says : if two curves C, C' of order n meet in exactly n^2 points, $n \cdot m$ of which lie on an irreducible curve C'' of order m, then the remainder lie on a curve C''' of order $n-m$.

We first assume that C and C' meet in exactly 9 distinct points. Thus we apply 6.2.3 to C and $C' = L_1 \cup L_2 \cup L_3$ as well as $C'' = G_1$, and it follows that the six points q, r, u', p, t', u lie on a quadric. But q, r, u' lie on the line G_2, and hence G_2 must be a component of this quadric. Let the other component be G_3. If p, t', u do not lie on G_2, then they must lie on G_3, which was to be proved.

Until now, the proof has assumed that C and C' meet in exactly 9 different points. But in our situation, some of the 9 intersection points can in fact coincide. Nevertheless, one can convince oneself that this is the case only for the (p,q,r) in a proper algebraic subset of $C \times C \times C$. Similarly for the last step of the proof, which we can only carry out when neither u, t' nor p lies on G_2. Again, the latter can happen only for the points (p,q,r) of a proper algebraic subset. Thus the two mappings

$$C \times C \times C \to C$$

defined by $(p,q,r) \mapsto (p+q) + r$ resp. $(p,q,r) \mapsto p + (q+r)$ coincide outside a proper algebraic subset. These mappings are regular mappings (see below), in particular, they are continuous in the usual sense. They agree on an open dense set (namely, the complement of a proper algebraic subset of $C \times C \times C$). Thus they agree everywhere. This completes the proof of the theorem.

We now have two kinds of structure on a nonsingular plane cubic C. On the one hand, C has the structure of a projective-algebraic manifold, on the other hand it has the structure of an abelian group.

Since the group structure relates in a natural way to the description of C as a plane curve of order 3, we hope of course that the two structures are compatible with each other.

This should mean the following : the group structure is characterised by two mappings,

$$C \times C \to C$$

resp. $C \to C$

which send (p,q) to $p+q$ resp. p to $-p$. On the other hand, one has a natural class of mappings for the projective-algebraic manifolds, namely the regular mappings. A mapping $f : X \to Y$ of projective-algebraic manifolds is called <u>regular</u> when X and Y can be covered with affine coordinate neighbourhoods U_i resp. V_j so that for each point in X there is a neighbourhood U_i, and a neighbourhood V_j of its image, such that $f : U_i \to V_j$ can be described in terms of the associated affine coordinates by means of rational functions with non-vanishing denominator. In this way one defines the concept of a regular mapping of algebraic manifolds. In particular, one can say when two such manifolds are <u>isomorphic</u>. This isomorphism concept concerns the structure of the manifolds as "abstract" algebraic manifolds, not their embedding in projective space. For example, a line and a non-singular plane quadric are isomorphic as abstract manifolds, but there is no collineation of the plane which carries one to the other.

Since $C \times C$, along with C, is also a projective algebraic manifold, it is meaningful to speak of regular mappings $C \times C \to C$ resp. $C \to C$. One now sees easily that addition resp. inversion can in fact be described in terms of affine coordinates by rational functions. If, say, one writes the equation of C in the Weierstrass normal form $y^2 = 4x^3 - g_2x - g_3$, and if (x_1,y_1) resp. (x_2,y_2) are the coordinates of p resp. q, then the line through p, q has equation

$$y = \frac{(y_1-y_2)}{(x_1-x_2)} (x-x_2) + y_2.$$

If one substitutes this in the equation to the curve then, dividing by $(x-x_1)(x-x_2)$, one obtains a linear equation in x with coefficients which are rational functions in x_1, y_1, x_2, y_2. The solution is the x-coordinate of the point r' at which the line through p, q meets the cubic C. One easily sees, by arguments of this type, that the group addition and inversion are defined by rational functions, and one can then convince oneself that they are regular.

Projective—algebraic manifolds with an abelian group structure for which addition and inversion are described by regular mappings are called abelian manifolds.

The preceding remarks have not completely proved, but have at least made understandable, the following :

Corollary to Theorem 1

The nonsingular plane cubics, with the group structure $(C,+)$ defined above, are one-dimensional abelian manifolds.

The definition of the group structure $(C,+)$ depends on the choice of the identity element p_0. However, it follows quite easily from the construction that the group structures resulting from different choices of the identity element are all conjugate by translation, so that the abelian manifold defined by a plane cubic is unique up to isomorphism.

In what follows we shall derive quite a different description of the abelian manifolds associated with nonsingular cubics. Namely, we shall identify them with the 1-dimensional complex tori. To do this we shall use function-theoretic arguments, for which we must refer to the function theory textbooks, say that of Cartan [C2].

We first explain what complex tori are. Let ω be a complex number with positive imaginary part, and let Γ_ω be the lattice in $\mathbb{C}$ consisting of all complex numbers $m + n\omega$, where m and n are integers. In general, a lattice $\Gamma \subset \mathbb{C}^n$ is a subgroup $\Gamma = \mathbb{Z}e_1 \oplus \ldots \oplus \mathbb{Z}e_{2n}$ where $e_1,\ldots,e_{2n} \in \mathbb{C}^n$ are real linearly independent.* We consider the quotient $\mathbb{C}/\Gamma_\omega$. Since Γ_ω is a subgroup of the additive group $\mathbb{C}$, $\mathbb{C}/\Gamma_\omega$ has the structure of an abelian group in a canonical way. In addition, $\mathbb{C}/\Gamma_\omega$ has the structure of a one-dimensional complex manifold in a canonical way : a function on an open set $U \subset \mathbb{C}/\Gamma_\omega$ is, by definition, holomorphic just in case its composition with the quotient mapping $\mathbb{C} \to \mathbb{C}/\Gamma_\omega$ from the preimage onto U is holomorphic. One sees immediately that addition and inversion in the group structure on $\mathbb{C}/\Gamma_\omega$ are described by holomorphic mappings.

---

* In the case $n = 1$ and $\Gamma \subset \mathbb{C}$ there is obviously a complex number a with $a\Gamma = \Gamma_\omega$ for suitable ω, so we can confine ourselves to lattices of this kind.

The abelian group and complex manifold structures defined on $\mathbb{C}/\Gamma_\omega$ are therefore compatible. As a topological manifold, $\mathbb{C}/\Gamma_\omega$ is obviously homeomorphic to $S^1 \times S^1$, a torus. Hence we call $\mathbb{C}/\Gamma_\omega$, with its structure as an abelian group and a complex manifold, a 1-dimensional <u>complex torus</u>. Correspondingly, one defines n-dimensional complex tori by $\mathbb{C}^n/\Gamma$, where Γ is a lattice in $\mathbb{C}^n$.

Naturally, one can also regard abelian manifolds as complex manifolds with an abelian group structure, and in this sense it is meaningful to ask whether the abelian manifolds are perhaps isomorphic to tori or not. In what follows we shall show that each 1-dimensional abelian manifold is isomorphic to a 1-dimensional complex torus, and conversely. In order to describe explicitly the isomorphism from complex tori to abelian manifolds, we have to use meromorphic functions on the complex tori, and hence it is necessary to concern ourselves with these meromorphic functions first.

When ϕ is a meromorphic function on the complex torus $\mathbb{C}/\Gamma_\omega$ then composition with the quotient mapping $\mathbb{C} \to \mathbb{C}/\Gamma_\omega$ gives a Γ-invariant function f meromorphic on the whole complex number plane $\mathbb{C}$. Γ-invariance means that

$$f(z+m+n\omega) = f(z),$$

which is equivalent to

$$f(z+1) = f(z),$$
$$f(z+\omega) = f(z).$$

Thus f is a <u>doubly periodic function</u>, i.e. one with two periods linearly independent over the reals, namely 1 and ω. Thus a meromorphic function on $\mathbb{C}/\Gamma_\omega$ defines a doubly periodic meromorphic function on $\mathbb{C}$, with periods 1 and ω, and conversely, such a function defines a meromorphic function on $\mathbb{C}/\Gamma_\omega$.

Now let ϕ be any meromorphic function on the complex torus $\mathbb{C}/\Gamma_\omega$ and let f be the associated meromorphic doubly periodic function on $\mathbb{C}$. The meromorphic function ϕ has of course only finitely many zeroes $p_1,\dots,p_r$ and finitely many poles $q_1,\dots,q_s$. Let $m_1,\dots,m_r$ be the orders of the zeroes and let $n_1,\dots,n_s$ be the orders of the poles. Then one calls the formal sum

$$(\phi) = m_1p_1 + \dots + m_rp_r - n_1q_1 - \dots - n_sq_s$$

the <u>principal divisor</u> of the function ϕ. Now in our case we are in quite a special situation, since the points p_i and q_j are elements of $\mathbb{C}/\Gamma_\omega$, and hence of an abelian group, so that we can not only construct the formal sum above, we can also form the real sum in $\mathbb{C}/\Gamma_\omega$. And for this we have the following theorem :

Theorem 2 (4. Liouville's Theorem)

A principal divisor on a 1-dimensional complex torus $\mathbb{C}/\Gamma_\omega$ satisfies

$$m_1p_1 + \dots + m_rp_r - n_1q_1 - \dots - n_sq_s = 0$$

in $\mathbb{C}/\Gamma_\omega$.

Proof : We reduce to the proof of the corresponding assertion about the doubly periodic function f on $\mathbb{C}$ which is associated with ϕ. The function f has infinitely many zeroes and poles : if $a_i \in \mathbb{C}$ resp. $b_j \in \mathbb{C}$ is a number with image p_i resp. q_j in $\mathbb{C}/\Gamma_\omega$, then the zeroes resp. poles of f are just $a_i + m + n\omega$ resp. $b_j + m + n\omega$. We now choose a z_0 so that no zeroes or poles lie on the edges of the parallelogram D with vertices z_0, $z_0 + 1$, $z_0 + \omega$, $z_0 + 1 + \omega$. Without loss of generality we can assume that $a_1,\dots,a_r$ and $b_1,\dots,b_s$ lie in this parallelogram.

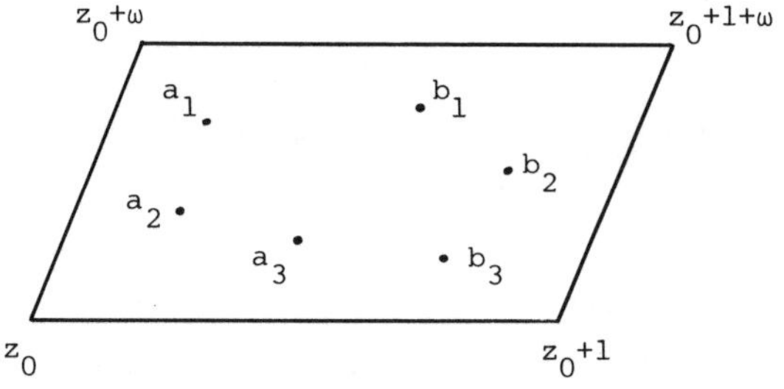

The proof of our theorem is now equivalent to the proof of the following assertion :

$$m_1a_1 + \dots + m_ra_r - n_1b_1 - \dots - n_sb_s \in \Gamma_\omega .$$

We shall prove this assertion by viewing the sum as the sum of residues of a suitable function, compute the latter by the residue theorem, and thereby establish that the integral in question is a number from Γ_ω. To do this we consider the function

$$z \cdot \frac{f'(z)}{f(z)} .$$

In our period parallelogram this function has poles of order 1 at

$a_1,\ldots,a_r$, $b_1,\ldots,b_s$, and no other poles. Its residue at a_i is $m_i a_i$ and its residue at b_j is $-n_j b_j$. Thus the sum of its residues in the period parallelogram is precisely

$$m_1a_1 + \ldots + m_ra_r - n_1b_1 - \ldots - n_sb_s.$$

By the residue theorem, this sum is equal to the integral

$$\frac{1}{2\pi i}\int_{\partial D} z\frac{f'(z)}{f(z)}dz$$

and hence equal to the sum of 4 integrals

$$\frac{1}{2\pi i}\left(\int_{z_0}^{z_0+1} z\frac{f'(z)}{f(z)}dz + \int_{z_0+1}^{z_0+1+\omega} z\frac{f'(z)}{f(z)}dz + \int_{z_0+1+\omega}^{z_0+\omega} z\frac{f'(z)}{f(z)}dz + \int_{z_0+\omega}^{z_0} z\frac{f'(z)}{f(z)}dz\right).$$

But f is doubly periodic with periods 1 and ω. For this reason, $\frac{f'(z)}{f(z)}$ has the same value at corresponding points on opposite sides, while the values of z differ by 1 resp. ω. Thus by corresponding substitutions for the variables of integration and combining integrals 1 and 3, as well as 2 and 4, one obtains the value of the integral as

$$\frac{1}{2\pi i}\left(-\omega\int_{z_0}^{z_0+1}\frac{f'(z)}{f(z)}dz + \int_{z_0}^{z_0+\omega}\frac{f'(z)}{f(z)}dz\right)$$

$$= \frac{1}{2\pi i}\left(-\omega\int_{z_0}^{z_0+1} d(\log f) + \int_{z_0}^{z_0+\omega} d(\log f)\right)$$

$$= \frac{1}{2\pi i}\left(-\omega(\log f(z_0+1) - \log f(z_0)) + (\log f(z_0+\omega) - \log f(z_0))\right).$$

Since ω and 1 are periods of f, one obtains the same value of f by continuation from the initial to the final point of integration, and hence for $\log f$ one obtains a value differing from the initial value by a multiple of $2\pi i$. Altogether, we obtain

$$m_1a_1 + \ldots + m_ra_r - n_1b_1 - \ldots - n_sb_s = \frac{1}{2\pi i}\int_{\partial D} z\frac{f'(z)}{f(z)}dz \in \Gamma_\omega ,$$

and this proves the theorem.

We shall now explicitly construct, for a given lattice Γ_ω in $\mathbb{C}$, two doubly periodic functions $\wp_\omega$ and $\wp'_\omega$ which will give us a mapping of the complex torus $\mathbb{C}/\Gamma_\omega$ onto a certain elliptic curve. In fact, we define the <u>Weierstrass</u> <u>$\wp$-function</u> by

$$\mathcal{P}_\omega(z) = \frac{1}{z^2} + \sum_{\gamma\in\Gamma_\omega}{}' \left(\frac{1}{(z-\gamma)^2} - \frac{1}{\gamma^2}\right).$$

The prime on the summation sign means that the sum is over all $\gamma \neq 0$ in Γ_ω. One can of course show that this infinite series of meromorphic functions converges. It is in fact normally convergent on each compact subregion of $\mathbb{C}$. The limit is therefore a meromorphic function, and one can differentiate term-by-term. (For this, and what follows, see Cartan [C2], Chap. V, sections 2.1 and 2.5.) Termwise differentiation yields the derivative

$$\mathcal{P}'_\omega(z) = -2 \sum_{\gamma\in\Gamma_\omega} \frac{1}{(z-\gamma)^3}.$$

It is obvious that $\mathcal{P}'_\omega$ is doubly periodic. It follows easily that $\mathcal{P}_\omega$ is doubly periodic. $\mathcal{P}_\omega$ has poles precisely at the points of Γ_ω, and in fact poles of order 2 ; $\mathcal{P}'_\omega$ likewise has its poles at the points of Γ_ω, and they are of order 3.

We shall now derive a relation between $\mathcal{P}_\omega$ and $\mathcal{P}'_\omega$. To do this we use the Laurent expansion of $\mathcal{P}_\omega$ in the neighbourhood of $z = 0$. Since $\mathcal{P}_\omega(z) = \mathcal{P}_\omega(-z)$ and $\mathcal{P}_\omega(z) - \frac{1}{z^2}$ is holomorphic in the neighbourhood of $z = 0$ and zero at $z = 0$, the Laurent expansion is of the form

$$\mathcal{P}_\omega(z) = \frac{1}{z^2} + a_2 z^2 + a_4 z^4 + \ldots \tag{1}$$

Differentiating and squaring gives:

$$(\mathcal{P}'_\omega(z))^2 = \frac{4}{z^6} - \frac{8a_2}{z^2} - 16a_4 + \ldots \tag{2}$$

Raising both sides of (1) to the power 3 gives

$$(\mathcal{P}_\omega(z))^3 = \frac{1}{z^6} + \frac{3a_2}{z^2} + 3a_4 + \ldots \tag{3}$$

It follows from (1), (2) and (3) that the function

$$\mathcal{P}'^2_\omega - 4\mathcal{P}^3_\omega + 20a_2\mathcal{P}_\omega + 28a_4$$

is holomorphic in the neighbourhood of the origin and zero at $z = 0$. On the other hand, this function is doubly periodic, with periods 1 and ω. It could have poles only in Γ_ω, but we have just seen that it has none there. Thus the function is holomorphic over the whole plane. Since it is also doubly-periodic, it is a holomorphic, bounded function over the whole plane, and hence constant by a theorem of Liouville, hence identically 0, since it vanishes at $z = 0$. We therefore have

$$\wp'^2_\omega - 4\wp^3_\omega + 20a_2\wp_\omega + 28a_4 = 0.$$

This is a first order differential equation which the $\wp$-function must satisfy. Of course, the coefficients a_2, a_4 of the differential equation – and with them the solution – depend on ω; a_2 and a_4 may be easily calculated by term-wise differentiation of $\wp_\omega(z) - \frac{1}{z^2}$ at $z = 0$. The results are

$$a_2 = 3 \sum_{\gamma \in \Gamma_\omega}{}' \frac{1}{\gamma^4}$$

$$a_4 = 5 \sum_{\gamma \in \Gamma_\omega}{}' \frac{1}{\gamma^6}.$$

Making the abbreviations

$$g_2(\omega) = 60 \sum_{\gamma \in \Gamma_\omega}{}' \frac{1}{\gamma^4}$$

$$g_3(\omega) = 140 \sum_{\gamma \in \Gamma_\omega}{}' \frac{1}{\gamma^6}$$

we have proved :

Theorem 3

Let ω be a complex number with nonvanishing imaginary part and let $\wp_\omega$ be the associated Weierstrass $\wp$-function. Also, let $g_2(\omega)$ and $g_3(\omega)$ be the functions of ω defined above. Then $\wp_\omega$ is a doubly periodic function of z with periods 1 and ω and it satisfies the differential equation

$$\wp'^2_\omega = 4\wp^3_\omega - g_2(\omega)\wp_\omega - g_3(\omega).$$

Remark : If we set $x = \wp_\omega(z)$, and $g_2 = g_2(\omega)$ as well as $g_3 = g_3(\omega)$, then the differential equation reads

$$\frac{dx}{dz} = \sqrt{4x^3 - g_2x - g_3}$$

or $$\frac{dz}{dx} = \frac{1}{\sqrt{4x^3 - g_2x - g_3}}$$

hence $$z = \int \frac{dx}{\sqrt{4x^3 - g_2x - g_3}}$$

The (many-valued) function z of x is therefore an (indefinite) elliptic integral (in Weierstrass normal form). The inverse functions of elliptic integrals are called elliptic functions. Thus the Weierstrass $\wp$-function and its derivative $\wp'$ are particular elliptic functions. In general the elliptic functions are just the doubly periodic functions meromorphic over the whole plane $\mathbb{C}$.

In view of the differential equation for the $\wp$-function it is clear how to set up a mapping from the torus $\mathbb{C}/\Gamma_\omega$ onto a cubic. Set

$$x(z) = \wp(z)$$
$$y(z) = \wp'(z).$$

Then

$$y^2 = 4x^3 - g_2x - g_3.$$

This is the equation of a cubic in Weierstrass normal form, and the existence of the differential equation just means that $z \mapsto (x(z),y(z))$ maps the plane $\mathbb{C}$ into the cubic. The mapping is also well defined at the poles : its image is the inflection point $(0,0,1)$ on the line at infinity. Since x and y are doubly periodic, the mapping of $\mathbb{C}$ into the cubic factorises over the torus $\mathbb{C}/\Gamma_\omega$. One can now prove the following assertions about this mapping :

Theorem 4

(i) Let ω be a complex number with positive imaginary part. Then the cubic C_ω with affine equation

$$y^2 = 4x^3 - g_2(\omega)x - g_3(\omega)$$

is nonsingular.

(ii) Each nonsingular cubic has, relative to suitable coordinates, an equation of the form

$$y^2 = 4x^3 - g_2(\omega)x - g_3(\omega).$$

(iii) The mapping

$$\mathbb{C}/\Gamma_\omega \to C_\omega$$

of the complex torus $\mathbb{C}/\Gamma_\omega$ onto the cubic C_ω induced by the Weierstrass $\wp$-function $x = \wp(z)$ and its derivative $y = \wp'(z)$ is biholomorphic (i.e. bijective and holomorphic, with holomorphic inverse).

Remark : As 1-dimensional complex manifolds, all nonsingular cubics are therefore biholomorphically equivalent to 1-dimensional complex tori, and conversely.

Since the function theory of these curves is just the theory of elliptic functions, one also calls them elliptic curves. For the proof of the above theorem we refer to the appropriate textbooks, e.g. Cartan [C2], V, §2, 5 and VI, §5, 3.

We shall now conclude by setting up a relation between the group structure of the complex tori $\mathbb{C}/\Gamma_\omega$ and the cubics C_ω, choosing the point $(0,0,1)$ as the identity element on C_ω in order to map the identity element of the torus to the identity element of the abelian manifold C_ω. Then we have :

Theorem 5

Setting $x = \wp_\omega(z)$, $y = \wp'_\omega(z)$ defines an isomorphism of the 1-dimensional complex torus $\mathbb{C}/\Gamma_\omega$ onto the 1-dimensional abelian manifold $(C_\omega,+)$.

Proof : Let $p,q,r \in C_\omega$ and $p + q = r$. Let $\Phi_1 = 0$ be the equation of the line through p, q, and let $\Phi_2 = 0$ be the equation of the line through p_0 and r. Then $\Phi = \Phi_1/\Phi_2$ defines a rational function on C_ω, and by composing with $\mathbb{C}/\Gamma_\omega \to C_\omega$ one obtains a rational function ϕ on $\mathbb{C}/\Gamma_\omega$. The two zeroes a_1, a_2 of ϕ are the preimages of p and q, the poles b_1 and b_2 are the preimages of r and p_0. We have $b_2 = 0$ by the choice of p_0. Hence it follows from Theorem 2 that :

$$a_1 + a_2 = b_1 ,$$

which was to be proved.

We now want to interpret our results on the inflection points of plane cubics from the previous paragraph in the light of our newly acquired knowledge of the connection between cubic curves and complex tori.

For this purpose we choose the distinguished point p_0 on the cubic C, i.e. the identity element of our abelian group, to be an inflection point. It is then clear from the construction of the inverse $-p$ to p, given earlier, that $-p$ is just the third intersection point with c of the line through p and p_0. But then it is clear that

Proposition 6

If one chooses the identity element on a plane nonsingular cubic C to be an inflection point, then one has

$$p + q + r = 0$$

in the associated group structure exactly when p, q, r lie on a line.

From this it follows that :

Corollary 7

If one chooses the identity element on a plane nonsingular cubic C to be an inflection point, then in the associated group structure we have : $p \in C$ is an inflection point just in case

$$3p = 0.$$

But now we know from Theorem 5 how to view the groups $(C,+)$: they are the complex tori $\mathbb{C}/\Gamma_\omega$. And it is a trivial problem to find the $p \in \mathbb{C}/\Gamma_\omega$ with $3p = 0$. They are the 9 points represented by the 9 complex numbers

$$\begin{array}{lll} 0 & \frac{1}{3} & \frac{2}{3} \\ \frac{\omega}{3} & \frac{1}{3} + \frac{\omega}{3} & \frac{2}{3} + \frac{\omega}{3} \\ \frac{2\omega}{3} & \frac{1}{3} + \frac{2\omega}{3} & \frac{2}{3} + \frac{2\omega}{3} . \end{array}$$

Thus we obtain anew the result that each nonsingular plane cubic has 9 inflection points (7.3.3). In addition, we immediately obtain the configuration of inflection points and lines through them described in (7.3.7) and (7.3.6) :

The 9 inflection points form a subgroup of the complex torus, isomorphic to the additive group $\mathbb{F}_3 \times \mathbb{F}_3$. By proposition 6, 3 of these inflection points, say p, q, r, lie on a line just in case $p + q + r = 0$. Thus two of them uniquely determine the third. This gives rise to $\binom{9}{2} \div 3 = 12$ lines, and one checks immediately that the configuration of these lines and the 9 inflection points is just the configuration $(9_4, 12_3)$ considered earlier. We see in this way that the inflection point configuration results quite simply from the description of the complex torus.

To conclude, we shall make a few remarks on higher-dimensional abelian varieties and complex tori. One can show that each higher-dimensional abelian manifold is isomorphic to a complex torus. But not every complex torus is a projective-algebraic manifold. One can prove : $\mathbb{C}^n/\Gamma$ is an abelian manifold just in case the following condition, which goes back essentially to Riemann, is satisfied. One chooses the coordinates in $\mathbb{C}^n$ so that Γ is generated by the vectors $(1,0,\ldots,0),\ldots,(0,\ldots,0,1)$, $(a_{11},\ldots,a_{1n}),\ldots,(a_{n1},\ldots,a_{nn})$. Let A be the matrix (a_{ij}). Then Riemann's condition reads : there are integers $d_1,\ldots,d_n \neq 0$ such that the matrix $Z = AD$, where $D = (\delta_{ij}d_i)$, satisfies

(1) Z is symmetric

(2) $\mathrm{Im}(Z)$ is positive definite.

The set of matrices Z which satisfy conditions (1) and (2) may be viewed as an open domain in $\mathbb{C}^{n(n+1)/2}$, more precisely as a certain bounded symmetric domain in the sense of E. Cartan.

For each D, the mapping $Z \mapsto ZD^{-1}$ maps this domain onto a locally analytic set of dimension $n(n+1)/2$ in the $\mathbb{C}^{n^2}$ of $(n \times n)$-matrices. If one constructs the union of these denumerably many subsets over all D, then one obtains the set of all matrices A for which the <u>Riemann period relations</u> (1), (2) are satisfied, and hence for which $\mathbb{C}^n/\Gamma$ is an abelian manifold. One sees : for $n > 1$, almost all complex tori are not abelian manifolds ! If, e.g.

$$A = i\begin{pmatrix} a & b \\ c & d \end{pmatrix}$$

where $a, b, c, d \in \mathbb{R}$ and a, b, c, d are linearly independent over $\mathbb{Q}$ with $ad - bc \notin \mathbb{Q}$, then A does not satisfy the Riemann period relations and the associated 2-dimensional torus is not algebraic. To give quite a concrete example : if $\Gamma \subset \mathbb{C}^2$ is the lattice generated by $(1,0)$, $(0,1)$, $(1,\sqrt{2})$, $(\sqrt{3},\sqrt{5})$, then $\mathbb{C}^2/\Gamma$ is not algebraic. (Cf. Chern [C3] §7, Mumford [M4] Chap. I, Siegel [S5] III, Chap. V, §11.)

It may seem that we have strayed far from our main theme, the theory of algebraic curves, with these remarks on higher-dimensional tori and abelian manifolds. However, this is not so. The theory of abelian manifolds is closely connected with the theory of plane curves. It is not possible to go further into this in the space of this course. But we would like now to give an idea of the connection between the two theories, even though we have to use some concepts we have not previously introduced.

Let X be a nonsingular algebraic curve of genus p. (The genus will be defined in §8.) One chooses a basis $\alpha_1,\ldots,\alpha_p$ for the vector space of holomorphic differential forms on X, and chooses 1-cycles $\gamma_1,\ldots,\gamma_{2p}$ on X which form a basis for the first homology group of X. Then one can define the period matrix (w_{ij}) by

$$w_{ij} = \int_{\gamma_j} \alpha_i \ .$$

Its columns generate a lattice $\Gamma \subset \mathbb{C}^p$. The complex torus $\mathbb{C}^p/\Gamma$ is an abelian manifold, the <u>Jacobian variety</u> of X. Now an important theorem of Torelli says that X, as an abstract algebraic curve, is determined up to isomorphism by its Jacobian variety. Using this theorem, D. Mumford 1961 showed that the set of isomorphism classes of algebraic curves of genus p carries, in a natural way, the structure of a complex space M_p with singularities, and that this <u>moduli space</u> M_p can be viewed as a Zariski-open subset of a projective algebraic variety. The dimension of M_p is 0 for $p = 0$, 1 for $p = 1$ and $(3p-3)$ for $p > 1$.

The number 3p-3 was already guessed by Riemann, who called it the <u>number of moduli</u>. For $p = 1$, i.e. for elliptic curves, the number of moduli is 1, and M_1 is the quotient of the upper half plane H under the action of $SL(2,\mathbb{Z})$ via linear fractional transformations, because $\mathbb{C}/\Gamma_\omega$ is isomorphic to $\mathbb{C}/\Gamma_{\omega'}$ just in case ω goes to ω' by an operation in $SL(2,\mathbb{Z})$. M_1 is isomorphic to $\mathbb{C}$, and the quotient mapping $H \to H/SL(2,\mathbb{Z}) \cong \mathbb{C}$ is given by none other than the J-invariant

$$J(\omega) = \frac{g_2^3(\omega)}{g_2^3(\omega)-27g_3^2(\omega)} \ .$$

With this glimpse, we conclude this section on special curves. The cubics are really very special curves. They have particularly nice properties, which one understands very well, and their study belongs to the most beautiful part of classical algebraic geometry and classical function theory. As for the general theory of algebraic curves of higher genus, we shall be able to take only the very first steps in the next chapter — more than this is not possible in the space of this course.

III. INVESTIGATION OF CURVES BY RESOLUTION OF SINGULARITIES

One of the most natural questions in the investigation of a particular class of mathematical objects is always the problem of classification of these objects. One tries to understand the structure of the objects under investigation as far as possible and, on the basis of this understanding, to divide the totality of objects into classes in a suitable way. The finer this classification is, the better. But in general, i.e. very often, the finest classification - up to isomorphism of the structure in question - is very difficult, and one often has to be content with a coarser classification.

Suppose that we consider the problem of classifying algebraic curves. The finest meaningful classification of algebraic curves in the complex projective plane would surely be up to equivalence under collineations of $P_2(\mathbb{C})$. In terms of this equivalence relation, we have classified the quadrics in 7.1 and the cubics in 7.3. In the case of irreducible cubics one has to distinguish between nonsingular and two types of singular curves - according to the type of singularity. Carrying out such a classification for curves of order greater than 3 becomes increasingly complicated as the order increases. One reason for this is that such curves can have ever more complicated singularities with increasing order,and the classification of these singularities alone is already an unsolved and in fact meaningless problem at this level of generality – even though recent work of Arnold [Al] and others has made much progress in this direction.

The conclusion to be drawn from this, regarding the classification problem, is that one must simplify the problem. One way of doing this is to pass from plane algebraic curves, by a suitable simplification process, to curves which no longer have singularities. Such a process is called resolution of singularities of the curve.

The curves which one obtains by this resolution process are, admittedly, no longer curves in the projective plane, but, depending on the kind of resolution process, they lie in various other algebraic manifolds or else are constructed only as abstract algebraic curves, without being embedded in a particular manifold. For this reason, the most natural way to view the classification problem is as the problem of classifying these nonsingular abstract algebraic curves up to isomorphism.

In this form the problem is still very difficult when one interprets it correctly. What is "classification up to isomorphism" to mean? Let us recall the classification of nonsingular plane cubics in 7.3. There we classified these curves up to equivalence under collineations by a number, the j-invariant, and one can show, in this case, that this is also the classification of the abstract algebraic curves up to isomorphism. The j-invariant can take each value in the complex number plane $\mathbb{C}$. In other words : the isomorphism classes of elliptic curves are parametrised by the points of $\mathbb{C}$. From the topological standpoint, the elliptic curves are all tori, and one can show that each nonsingular algebraic curve which is topologically a torus is an elliptic curve. Thus we see that the isomorphism classes of nonsingular curves which have genus 1 topologically are parametrised by the points of $\mathbb{C}$.

The classification of more complicated curves is the same in principle, but much more difficult. One first makes a very coarse topological classification : nonsingular curves are topologically 2-dimensional compact orientable manifolds, and the classification of such surfaces is known : they are classified by one number, the genus. The compact orientable surface of genus p is just the surface which results from the 2-sphere by attaching p handles.

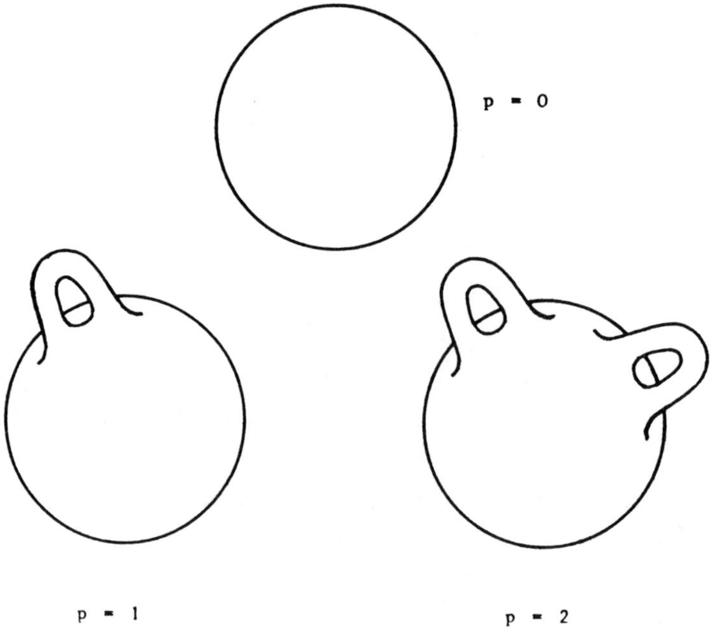

For elliptic curves, $p = 1$.

In analogy with the theory of elliptic curves, one again hopes that for each p there is a nice space M_p whose points classify the isomorphism classes of curves of genus p. One can in fact construct such a space, and Mumford has even shown that M_p can be given the structure of a quasi-projective algebraic variety of dimension $(3p-3)$ (cf. the remarks at the end of 7.4). However, the construction of M_p is very complicated and goes far beyond the scope of this course.

Here we shall be satisfied with much less : by a suitable analysis of singular points, we shall show how one resolves the singularities of each plane curve by a suitable modification process. We shall then define the genus of the plane curve to be the genus of this nonsingular model. Finally, we shall see how this invariant of the curve, the genus, can be computed from other invariants, namely the order of the curve and certain invariants of its singular points. In this way we shall arrive systematically at a comprehension of a phenomenon observed earlier in 3.4, that in a family of curves of the same order the singular curves have a topological type different from that of the non-singular ones. The systematic investigation of such phenomena by Lefschetz [L1], Griffiths [G2], Grothendieck and Deligne [D1] has led to very important developments in modern algebraic geometry (cf. also [D3]).

8. Local investigations

8.1 Localisation - local rings

In paragraph 8 we are going to carry out a local investigation of plane curves at their singular points. The results obtained will then be used in the next paragraph to make global statements about such curves.

In the present section 8.1 we want to give a heuristic discussion of what we mean by "local" in this context. As an example, consider the complex affine plane $\mathbb{C}^2$ and the curve C with affine equation

$$y^2 = x^2(x+1),$$

the plane cubic with an ordinary double point from 7.3. We first study this curve with the help of a parametrisation $t \mapsto (x(t),y(t))$. We easily obtain such a parametrisation by projecting C from the double point $(0,0)$ onto the line $x = 1$. The line with equation

$$y = tx$$

cuts the line $x = 1$ at $y = t$, and it cuts the cubic, outside of $(0,0)$, at $x = t^2 - 1$, $y = t^3 - t$.

Thus we obtain the parametrisation of C

$$x(t) = t^2 - 1$$
$$y(t) = t^3 - t.$$

This is a rational, everywhere defined mapping ϕ of the line $\mathbb{C}$ onto the curve C. It sends the two points $t = 1$ and $t = -1$ to the singular point $(0,0)$ of C. Apart from this, the mapping $\mathbb{C} - \{1,-1\} \to C - \{0\}$ defined by the parametrisation is bijective and the inverse mapping is rational and everywhere defined. Thus the singular point is replaced by two points in $\mathbb{C}$, and the resulting curve, namely $\mathbb{C}$, is nonsingular. Outside the singular point nothing has changed. Thus the mapping

$$\phi : \mathbb{C} \to C$$

gives us a particularly simple example of the passage from a singular to a nonsingular curve, illustrating a process we shall later call resolution of singularities.

Now why is it that we must replace the singular point by two points? Consider the intersection of our curve C with a neighbourhood U of the singular point, say with the polydisc

$$U = \{(x,y) \mid |x| < c, |y| < 2c\}$$

where $0 < c < 1$. One sees immediately that the condition $|y| < 2c$ is automatically satisfied for points of C with $|x| < c$. Now let $W = \phi^{-1}(U)$ be the preimage of U in $\mathbb{C}$. Obviously

$$W = \{t \in \mathbb{C} \mid |t^2-1| < c\}$$

consists of the two regions W^- and W^+, enclosed by Cassini ovals.

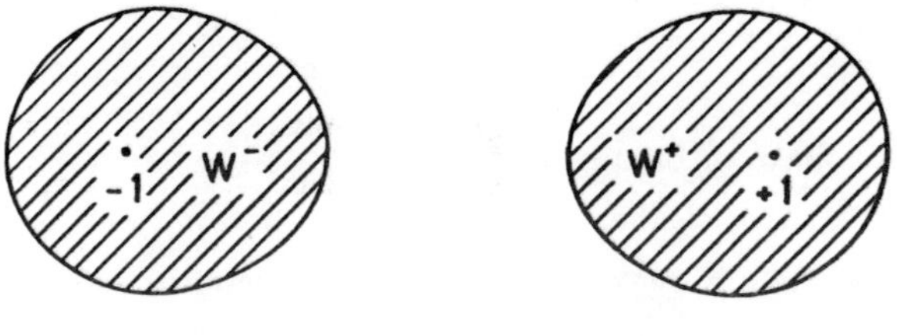

The mapping ϕ induces embeddings

$$\phi : W^{+} \to \mathbb{C}^{2}$$

$$\phi : W^{-} \to \mathbb{C}^{2},$$

whose images $V^{\pm} = \phi(W^{\pm})$, being biholomorphic images of $W^{\pm}$, look like $W^{\pm}$. V^{+} and V^{-} intersect only at the origin, transversely in fact, and their union is

$$V^{+} \cup V^{-} = C \cap U.$$

Thus the intersection of C with the neighbourhood U of the singular point falls into two components V^{+}, V^{-}, and since these "components" are nonsingular, resolution of singularities in this case consists simply of "separating" the two components.

An intuition for this situation is already given by the picture of the corresponding real curves :

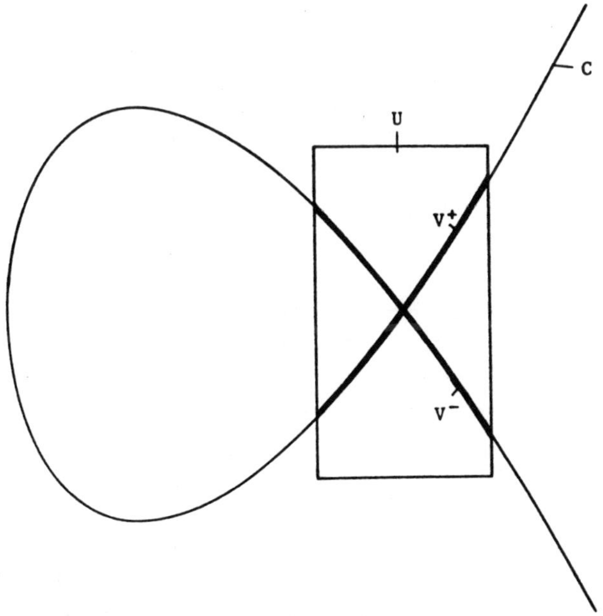

Naturally, one also wishes to interpret the decomposition of the curve C into two "components", or "branches" as they are also called, in the neighbourhood of the singular point in analytic terms, using the equation. In 5.1 we described the decomposition of curves in the projective plane into their irreducible components using the decomposition of polynomials, whose zero sets the curves were, into irreducible factors. In the present case our curve C is irreducible and the polynomial $f(x,y) = y^2 - x^2(x+1)$ is irreducible accordingly. The decomposition of the curve in a suitable neighbourhood of the singular point is therefore not reflected by the decomposition of the polynomial f into the product of two polynomials. Rightly so, because the curve C indeed does not decompose into two pieces over the whole plane, but only in a suitably chosen, sufficiently small, neighbourhood.

This raises the question of what a "sufficiently small neighbourhood" should mean. The most obvious neighbourhood concept in algebraic geometry is the Zariski-open neighbourhood. The Zariski-open neighbourhoods U of a point x_0 in the affine plane result, e.g., by removing from $\mathbb{C}^2$ some points and curves which do not meet x_0. As rational functions on U one admits all rational functions, i.e. quotients $\frac{p}{q}$ of polynomials, whose denominators are nonvanishing on U. The set of all these rational functions on all Zariski-open neighbourhoods of x_0 forms a ring : if $\mathfrak{M} \subset \mathbb{C}[x_1,x_2]$ is the maximal ideal of polynomials which vanish at x_0, then this ring is just the ring

$$\mathbb{C}[x_1,x_2]_{\mathfrak{M}} = \{\frac{p}{q} \mid p,q \in \mathbb{C}[x_1,x_2],\ q \notin \mathfrak{M}\}.$$

This ring is obviously a "local ring", i.e. it has exactly one maximal ideal, namely the ideal of the $\frac{p}{q}$ with $p \in \mathfrak{M}$, the functions vanishing at x_0. One also calls $\mathbb{C}[x_1,x_2]_{\mathfrak{M}}$ the localisation of $\mathbb{C}[x_1,x_2]$ at x_0 , or the algebraic local ring of $\mathbb{C}^2$ at x_0.

One could perhaps hope that the intersection of our irreducible curve C with a suitable Zariski-open neighbourhood U would decompose into two components. Algebraically, it would correspond to a factorisation $f = f_1 \cdot f_2$ of f in $\mathbb{C}[x_1,x_2]_{\mathfrak{M}}$, where f_1 and f_2 both lay in the maximal ideal of $\mathbb{C}[x_1,x_2]_{\mathfrak{M}}$. But this cannot be, otherwise we would have $f_i = \frac{p_i}{q_i}$ with $p_i \in \mathfrak{M}$, $q_i \notin \mathfrak{M}$, and hence $q_1 \cdot q_2 \cdot f = p_1 \cdot p_2$, and the irreducible factors in $\mathfrak{M}$ of p_1 and p_2 would both divide f, contradicting the irreducibility of f. One can also see geometrically that the intersection of C with a

Zariski-open neighbourhood cannot decompose into two pieces. Because $C \cap U$ results from C by removal of some points and intersections with curves, and hence just by the removal of finitely many points. Correspondingly, the preimage $V = \phi^{-1}(C \cap U)$ results from $\mathbb{C}$ by removal of finitely many points, and hence consists of just one component. Thus we see that the Zariski-neighbourhoods are too large, and that curves cannot decompose within them.

Thus we now consider neighbourhoods U of the singular point in $\mathbb{C}^2$ in the sense of the usual <u>classical topology</u>, e.g. our polydisc. Also, we admit more functions on these neighbourhoods U as possible factors of f, namely all functions which are <u>holomorphic</u>, i.e. locally expandable in convergent power series, in U.

For example, in the polydisc U considered above we can regard the function $\sqrt{x+1}$ (where we choose one of the two branches of the root function) as such a well defined holomorphic function, because $x+1$ is nonzero in U and U is simply connected. This immediately gives a factorisation of the polynomial $f = y^2 - x^2(x+1)$ into

$$f = f_1 \cdot f_2$$

with

$$f_1 = y + x\sqrt{x+1}$$

$$f_2 = y - x\sqrt{x+1}.$$

And this decomposition gives just what we want : if $\sqrt{x+1}$ is the branch with $\sqrt{1} = +1$, then the zero set of f_1 is just V^-, and the zero set of f_2 is just V^+. If we let $\mathcal{O}(U)$ denote the ring of all functions holomorphic in U, then in this ring we have the desired decomposition into factors

$$f = f_1 \cdot f_2 .$$

For many purposes this is an adequate description of the situation, especially when we are interested in the topological properties of a curve in the neighbourhood of a singular point. In this case we intersect the curve C with suitable neighbourhoods of x_0 , e.g. balls or polydiscs, and treat the intersection $V = C \cap U$ as the zero set of the holomorphic function $f \in \mathcal{O}(U)$.

For certain other purposes it can cause trouble to choose a particular neighbourhood U of x_0. Then it is more convenient, as we have already done in the case of the Zariski topology, to consider the system of all neighbourhoods U of x_0 , and the system of all holomorphic functions on all neighbourhoods of x_0. In this case one naturally identifies all holomorphic functions which agree on a suitable neighbourhood of x_0 , since one is interested only in the behaviour of the functions in arbitrarily small neighbourhoods of x_0. The resulting equivalence classes of holomorphic functions are called germs of holomorphic functions at x_0. Thus one has the

Definition :

A germ of holomorphic functions at a point x_0 is an equivalence class of holomorphic functions defined in neighbourhoods of x_0. Two such functions are equivalent, i.e. they define the same germ, when their restrictions to a suitable neighbourhood of x_0 coincide.

The set of all germs of holomorphic functions obviously constitutes a ring $\mathcal{O}_{\mathbb{C}^2,x_0}$, the ring of germs of holomorphic functions at x_0.

Now one can also describe this ring as follows : each function which is holomorphic in a neighbourhood of x_0 can be associated with its Taylor expansion at x_0. The latter is a convergent power series which converges to the function in a neighbourhood of x_0. Functions which define the same germ at x_0 naturally have the same Taylor expansion at x_0. Thus we obtain a mapping of $\mathcal{O}_{\mathbb{C}^2,x_0}$ into the ring $\mathbb{C}\{x,y\}$ of convergent power series. This mapping is of course a homomorphism because of the known rules for the power series expansion of sums and products of functions. It is injective, because the power series expansion of a function f in a neighbourhood of x_0 converges to f, and thereby determines the germ of f at x_0 uniquely. It is also surjective, because each convergent power series converges to a holomorphic function in some neighbourhood of the expansion point. Thus, by associating each germ at x_0 with its Taylor expansion at x_0 , we obtain an isomorphism

$$\mathcal{O}_{\mathbb{C}^2,x_0} \cong \mathbb{C}\{x,y\}.$$

Naturally, one can also define the ring $\mathcal{O}_{\mathbb{C}^n,x_0}$ of germs of holomorphic functions at a point $x_0 \in \mathbb{C}^n$ quite analogously, and one then has an isomorphism

$$\mathcal{O}_{\mathbb{C}^n,x_0} \cong \mathbb{C}\{x_1,\ldots,x_n\}$$

onto the ring of convergent power series in n variables, $\mathbb{C}\{x_1,\ldots,x_n\}$. The earlier considerations on the localisation of polynomial rings may be carried out in precisely the same way for n variables, and one has in particular the algebraic local ring $\mathbb{C}[x_1,\ldots,x_n]_{\mathfrak{M}}$, the rational functions defined in a Zariski neighbourhood of $x_0 \in \mathbb{C}^n$, where $\mathfrak{M}$ is the maximal ideal for x_0.

Now we return to our factorisation problem. The factorisation $f = f_1 \cdot f_2$ in $\mathcal{O}(U)$ obviously yields, by passing to germs, a factorisation $f = f_1 \cdot f_2$ in $\mathcal{O}_{\mathbb{C}^2,x_0}$, and hence in $\mathbb{C}\{x,y\}$. Conversely, when we have a decomposition of f into $f_1 \cdot f_2$ in $\mathcal{O}_{\mathbb{C}^2,x_0}$, we can represent the germs by holomorphic functions f, f_1, f_2 in a suitable neighbourhood U, so that we then have $f = f_1 \cdot f_2$ in $\mathcal{O}(U)$. Thus we see that the decomposition of the curve into several analytic components in a sufficiently small neighbourhood of x_0 just means that, for the equation $f(x,y) = 0$ of the curve, f is reducible in $\mathcal{O}_{\mathbb{C}^2,x_0}$.

The local ring $\mathcal{O}_{\mathbb{C}^2,x_0} \cong \mathbb{C}\{x,y\}$ therefore provides us with a ring which is independent of each choice of neighbourhood and contains enough functions to express the geometric properties of curves $f(x,y) = 0$ at a singular point x_0 by corresponding algebraic properties of $f \in \mathcal{O}_{\mathbb{C}^2,x_0}$.

Admittedly, algebraic geometers are not very glad to use the ring $\mathbb{C}\{x,y\}$. Among other reasons, this is because they want to consider zero sets of polynomials whose coefficients are not real or complex, but in any field k. In an arbitrary field it is no longer meaningful to speak of convergent power series. However, one can always consider the ring of formal power series $k[[x_1,\ldots,x_n]]$ in the indeterminates $x_1,\ldots,x_n$ with coefficients in k. For $k = \mathbb{C}$ one has of course $\mathbb{C}\{x_1,\ldots,x_n\} \subset \mathbb{C}[[x_1,\ldots,x_n]]$.

As far as the problem of decomposition into irreducible factors is concerned, it is of course clear that each decomposition $f = f_1 \cdot f_2$ in $\mathbb{C}\{x_1,\ldots,x_n\}$ leads to a corresponding decomposition in $\mathbb{C}[[x_1,\ldots,x_n]]$. Conversely, one can show that : if $f \in \mathbb{C}\{x_1,\ldots,x_n\}$ has a decomposition $f = f_1 \cdot f_2$ in $\mathbb{C}[[x_1,\ldots,x_n]]$, where f_1, f_2 are power series without constant terms, then it already has such a decomposition in $\mathbb{C}\{x_1,\ldots,x_n\}$. Thus the irreducibility of $f \in \mathbb{C}\{x_1,\ldots,x_n\}$ is equivalent to the irreducibility of f viewed as an element of $\mathbb{C}[[x_1,\ldots,x_n]]$. Thus the ring $\mathbb{C}[[x_1,\ldots,x_n]]$ is just as good as the ring $\mathbb{C}\{x_1,\ldots,x_n\}$, as far as the investigation of local decomposition into irreducible components is concerned.

It may appear to be a disadvantage that one cannot interpret formal power series as functions or germs to the same extent that one can interpret convergent power series. However, this is no problem for modern algebraic geometry. There one constructs a topological space X with a kind of Zariski topology for each commutative ring R, and for each open set U of X a class of regular functions on U, so that R itself is just the ring of functions regular on the whole of X. As a point set, X is just the set of all prime ideals of R. One calls this space the <u>spectrum of R</u> and writes $X = \mathrm{Spec}(R)$. By means of this construction one can interpret all algebraic statements about R as geometric statements about $\mathrm{Spec}\, R$.

This blending of algebra and geometry is the foundation on which algebraic geometry was rebuilt by Grothendieck around 20 years ago, and which led to a fantastic development of this field. However, these considerations are outside the scope of our course. Those who are interested will find an introduction in Chapter V of the book "Basic Algebraic Geometry" by Shafarevich, [S4]. The objects, such as Spec(R), which take the place of the classical varieties in this algebraic geometry, are also known as <u>schemes</u>.

We have seen that the algebraic local ring $\mathbb{C}[x_1,\ldots,x_n]_{\mathfrak{M}}$ contains too few functions to give an adequate analysis of the singularities of zero sets of polynomials, while the rings $\mathbb{C}\{x_1,\ldots,x_n\}$ and $\mathbb{C}[[x_1,\ldots,x_n]]$ do contain enough functions. We shall conclude by considering a further possibility for the algebraic geometer to provide himself with a ring containing "enough" functions. Let us return to our example $y^2 - x^2(x+1)$. The formula

$$y^2 - x^2(x+1) = (y+x\sqrt{x+1})(y-x\sqrt{x+1})$$

shows that we can carry out the desired factorisation as soon as we have the functions $y(x) = \pm\, x\sqrt{x+1}$ at our disposal, i.e. the two solutions of the equation

$$y^2 - x^2(x+1) = 0.$$

If we want to analyse other singularities in $\mathbb{C}^{n+1}$, then we are led to other such equations. In general, these equations are of the form

$$y^k + a_1(x)y^{k-1} + \dots + a_k(x) = 0$$

with $a_i(x) \in \mathbb{C}[x_1,\dots,x_n]_{\mathfrak{M}}$. This leads to consideration of the following ring $\mathbb{C}<x_1,\dots,x_n>$.

Definition :

$$\mathbb{C}<x_1,\dots,x_n> = \{y \in \mathbb{C}[[x_1,\dots,x_n]] \mid y \text{ is entirely algebraic over } \mathbb{C}[x_1,\dots,x_n]_{\mathfrak{M}}\}.$$

The condition that y be entirely algebraic over $\mathbb{C}[x_1,\dots,x_n]_{\mathfrak{M}}$ is just that y satisfy an equation

$$y^k + a_1 y^{k-1} + \dots + a_{k-1}y + a_k = 0$$

in $\mathbb{C}[[x_1,\dots,x_n]]$ with $a_i \in \mathbb{C}[x_1,\dots,x_n]_{\mathfrak{M}}$.

The local ring $\mathbb{C}<x_1,\dots,x_n>$ defined in this way contains just enough functions to enable all the necessary factorisations to be carried out. One can show (Nagata [N1] VII, 45.1, and exercise 4) :

$f \in \mathbb{C}<x_1,\dots,x_n>$ is irreducible if and only if it is irreducible in $\mathbb{C}[[x_1,\dots,x_n]]$.

One can deduce from this that $\mathbb{C}<x_1,\dots,x_n>$ also contains just enough functions to admit other important theorems for this ring, such as the implicit function theorem. We shall formulate this somewhat more precisely. In the following proposition we consider normalised polynomials with coefficients in the ring $A = \mathbb{C}<x_1,\dots,x_n>$, i.e. polynomials $f \in A[y]$ from the polynomial ring with leading coefficient 1, namely

$$y^k + a_1 y^{k-1} + \dots + a_k$$

with $a_i \in A$. For such a polynomial f we let $\bar{f}$ denote the polynomial in $\mathbb{C}[y]$ which results from f when we take the values of the coefficients a_i at $x = 0$, thus

$$\bar{f} = y^k + a_1(0)y^{k-1} + \dots + a_k(0).$$

We can view f as a function $f(x,y)$ in x and y and construct the partial derivative $\frac{\partial f}{\partial y}$ by formal differentiation. The value of this derivative at $x = 0$, $y = 0$ is a certain complex number $\frac{\partial f}{\partial y}(0,0)$. With these arrangements we have :

Proposition 1

Normalised polynomials $f \in A[y]$ satisfy :

(i) Let $y_1,\dots,y_r \in \mathbb{C}$ be the different zeroes of $\bar{f}(y)$, and suppose y_i has multiplicity k_i. Let $\bar{g}_i = (y-y_i)^{k_i}$, so that

$$\bar{f} = \bar{g}_1 \cdot \dots \cdot \bar{g}_r$$

holds in $\mathbb{C}[y]$. Then there are normalised polynomials $f_i \in A[y]$ with $\bar{f}_i = \bar{g}_i$, such that

$$f = f_1 \cdot \dots \cdot f_r$$

already holds in $A[y]$.

(ii) We view the normalised polynomial $f \in \mathbb{C}\langle x_1,\dots,x_n\rangle[y]$ as a function $f(x,y)$ in x and y. When $f(0,0) = 0$ and $\frac{\partial f}{\partial y}(0,0) \neq 0$ then there is a unique solution $y(x)$ of the implicit equation $f(x,y) = 0$ with $y(0) = 0$ in $\mathbb{C}\langle x_1,\dots,x_n\rangle$. I.e. there is an $a \in \mathbb{C}\langle x_1,\dots,x_n\rangle$ with $a(0) = 0$ such that $f(x,a) = 0$ in $\mathbb{C}\langle x_1,\dots,x_n\rangle$.

Remarks :

(i) Assertion (ii) says that the germs of functions in $\mathbb{C}\langle x_1,\dots,x_n\rangle$ satisfy the implicit function theorem. The proof is trivial : by the usual implicit function theorem for formal power series, $f(x,y) = 0$ has a unique solution $a \in \mathbb{C}[[x_1,\dots,x_n]]$, and by definition of $\mathbb{C}\langle x_1,\dots,x_n\rangle$ this zero a of $f \in \mathbb{C}[[x_1,\dots,x_n]][y]$ must already belong to $\mathbb{C}\langle x_1,\dots,x_n\rangle$, because a is entirely algebraic over $\mathbb{C}\langle x_1,\dots,x_n\rangle$, hence, since $\mathbb{C}\langle x_1,\dots,x_n\rangle$ is by definition entirely algebraic over $\mathbb{C}[x_1,\dots,x_n]_{\mathfrak{m}}$, a is also entirely algebraic over $\mathbb{C}[x_1,\dots,x_n]_{\mathfrak{m}}$.

(ii) One can show that assertion (ii) also implies assertion (i) of Proposition 1. We shall not give the proof. The geometric meaning of the assertion about the ring $A = \mathbb{C}\langle x_1,\dots,x_n\rangle$ is approximately the following : each finitely branched covering of Spec A decomposes into as many connected components as it has points over the point of Spec A corresponding to the maximal ideal $(x_1,\dots,x_n)$, i.e. over the point which corresponds to the origin of $\mathbb{C}^n$. (Cf. the book of Kurke on henselian rings, [K4].)

(iii) One can formulate assertions (i) and (ii) of Proposition 1 so that they are meaningful for each local ring A with maximal ideal $\mathfrak{M}$, by replacing in general the passage from $\mathbb{C}\langle x_1,\ldots,x_n\rangle$ to $\mathbb{C}$ by the passage from A to $A/\mathfrak{M}$. One then calls (i) Hensel's lemma for A, and (ii) the implicit function theorem. One can show (see, e.g., Kurke) that these assertions are equivalent. A local ring A for which these assertions hold is called a henselian local ring. Thus Proposition 1 just says that $\mathbb{C}\langle x_1,\ldots,x_n\rangle$ is henselian. $\mathbb{C}\{x_1,\ldots,x_n\}$ and $\mathbb{C}[[x_1,\ldots,x_n]]$ are also henselian, because they satisfy the implicit function theorem.

(iv) One can show (Nagata [N1], VII, §43, Kurke [K4]), that each local ring has a unique "smallest" henselian extension, its henselisation, a concept introduced by Nagata 1953. The henselisation of $\mathbb{C}[x_1,\ldots,x_n]_{\mathfrak{M}}$ is just $\mathbb{C}\langle x_1,\ldots,x_n\rangle$. In particular, since $\mathbb{C}\{x_1,\ldots,x_n\}$ is a henselian ring which contains $\mathbb{C}[x_1,\ldots,x_n]_{\mathfrak{M}}$, we have that

$$\mathbb{C}\langle x_1,\ldots,x_n\rangle \subset \mathbb{C}\{x_1,\ldots,x_n\}.$$

(v) In the case of the algebraic local ring $\mathbb{C}[x_1,\ldots,x_n]_{\mathfrak{M}}$ we have established what is the appropriate neighbourhood concept : it is that of Zariski-open neighbourhoods, which leads to the Zariski topology.

What is the neighbourhood concept for $\mathbb{C}\langle x_1,\ldots,x_n\rangle$? The function germs in $\mathbb{C}\langle x_1,\ldots,x_n\rangle$ can initially be regarded as germs of holomorphic functions. However, since they are defined as solutions of polynomial equations whose coefficients are defined on a whole Zariski-open set, these holomorphic functions can be analytically continued inside a whole Zariski-open set U. But they do not remain single-valued under this analytic continuation. One only obtains well-defined analytic functions by passing to a suitable finite covering of U. The "open sets" on which the functions of $\mathbb{C}\langle x_1,\ldots,x_n\rangle$ become properly defined are thus no longer subsets of $\mathbb{C}^n$, but finite coverings of Zariski-open subsets of $\mathbb{C}^n$. The associated "topology", the so-called étale topology (from the French étale-morphisme for unbranched covering) is thus no longer a topology in the usual sense, but in a more general sense. It is a "Grothendieck topology".

This étale topology is essentially more suitable than the Zariski topology for carrying over the methods of algebraic topology, developed for classical topology, to the investigation of algebraic manifolds over arbitrary fields.

What are the neighbourhoods for $\mathbb{C}\{x_1,\dots,x_n\}$?

There are no more finite neighbourhoods : we are interested only in the behaviour of analytic functions in arbitrarily small classical neighbourhoods of 0 in $\mathbb{C}^n$, i.e. the complex analytic neighbourhood germs of 0 in $\mathbb{C}^n$, an infinitesimal neighbourhood of 0 in $\mathbb{C}^n$, so to speak. The "formal neighbourhood associated with" $\mathbb{C}[[x_1,\dots,x_n]]$ of 0 in $\mathbb{C}^n$ is, finally, the inverse limit of the "finite infinitesimal neighbourhoods" U_i , where U_i is defined by saying that two power series $f, g \in \mathbb{C}[[x_1,\dots,x_n]]$ coincide on U_i when they coincide up to terms of order greater than i.

Summary :

In order to localise the investigation of algebraic geometric problems, one must define a suitable neighbourhood concept, and a suitable concept of admissible functions defined on these neighbourhoods. This has led us to the consideration of the following local rings, with the following containments :

$\mathbb{C}[x_1,\dots,x_n]_{\mathfrak{M}} \subset \mathbb{C}\langle x_1,\dots,x_n\rangle \subset \mathbb{C}\{x_1,\dots,x_n\} \subset \mathbb{C}[[x_1,\dots,x_n]]$.

$\mathbb{C}[x_1,\dots,x_n]_{\mathfrak{M}}$ is the localisation of $\mathbb{C}[x_1,\dots,x_n]$ by the maximal ideal $(x_1,\dots,x_n)$.

$\mathbb{C}\langle x_1,\dots,x_n\rangle$ is the ring of algebraic power series and is the henselisation of $\mathbb{C}[x_1,\dots,x_n]_{\mathfrak{M}}$.

$\mathbb{C}\{x_1,\dots,x_n\}$ is the ring of convergent power series.

$\mathbb{C}[[x_1,\dots,x_n]]$ is the ring of formal power series.

The solution of algebraic geometric problems is often done at several levels :

(1) One first attempts to solve a corresponding formal problem, i.e. a problem which ultimately leads to the consideration of formal power series.

(2) When one can solve the formal problem, one can attempt to show that there is also a convergent solution. An example of this is the proof of the implicit function theorem in the analytic case : one first shows the existence of a formal solution and then proves its convergence. An example which is much more important and difficult to prove is the Artin approximation theorem [A2] : if a

complex analytic implicit equation $f(x,y) = 0$ has a formal solution $\bar{y} = \bar{y}(x)$, then it also has convergent solutions y, and the formal solution $\bar{y}$ may be approximated arbitrarily well by convergent solutions y.

(3) When one can solve the convergent problem (or the formal one), one can investigate whether there are in fact algebraic solutions, i.e. solutions described by functions in the henselisations of algebraic local rings. This is possible for many problems in algebraic geometry on the basis of a theory developed by M. Artin ([A3]).

(4) Finally, when (3) admits a positive solution, one can attempt to show that the solutions can be described by rational functions, and that the spaces constructed are algebraic varieties in the usual sense. Of course, it usually turns out that this is definitely not the case. For this reason one is often forced, in treating such problems, to admit a larger class of "algebraic spaces" than the algebraic varieties. These were introduced by Artin and Knutson [K2].

In our course we shall operate in §8 mainly at the level of convergent power series. This has the advantage that passing to the consideration of zeroes of holomorphic functions in open sets in the usual topology is quite natural, and we can also bring to bear on these zero sets all the classical analytic and topological methods that we want to use.

8.2 Singularities as analytic set germs

In the previous section we have seen that the study of singularities of algebraic sets leads us to view the latter as analytic sets, i.e. as zero sets of analytic functions, and to investigate them by the methods of function theory. Therefore, we shall now develop the most important basic concepts from the theory of analytic sets and functions, in particular the properties of the ring $\mathbb{C}\{z_1,\ldots,z_n\}$ of convergent power series. These and further details can also be read in textbooks of complex analysis such as Gunning-Rossi [G6] or Grauert-Fritzsche [G4].

The following theorem, found by Weierstrass c. 1860, is basic in the function theory of several variables :

Theorem 1 : (Weierstrass preparation theorem)

Let $g(t,z) = g(t,z_1,\ldots,z_n)$ be a convergent power series from $\mathbb{C}\{t,z_1,\ldots,z_n\}$ and let g be t-general (or t-regular) of order k, i.e. suppose $g(t,0)$ is a power series with lowest power t^k.

Then there exist $u(t,z) \in \mathbb{C}\{t,z\}$ and $c_i(z) \in \mathbb{C}\{z\}$ such that

$$g(t,z) = (t^k + c_1(z)t^{k-1} + \ldots + c_k(z)) \cdot u(t,z)$$

with $c_i(0) = 0$ and $u(0,0) \neq 0$. This determines the c_i and u uniquely.

Remarks :

(i) The theorem is called the "preparation" theorem because the power series g is "prepared" thereby for the study of its zeroes.

Since u nowhere vanishes in a neighbourhood of $0 \in \mathbb{C}^{n+1}$, the zeroes of g there coincide with those of the polynomial

$$t^k + c_1(z)t^{k-1} + \ldots + c_k(z).$$

Such a normalised polynomial with coefficients in $\mathbb{C}\{z_1,\ldots,z_n\}$ is called a Weierstrass polynomial. The condition $u(0,0) \neq 0$ just means that u is a unit in the ring $\mathbb{C}\{t,z\}$. The representation of g as a product of a Weierstrass polynomial with a unit will also enable us to investigate divisibility questions in $\mathbb{C}\{t,z\}$ with the help of results we already know about polynomial rings.

(ii) Relative to suitable coordinates (t',z'), each $g \in \mathbb{C}\{t,z\}$ is t'-general : if $g = \sum_{j=k}^{\infty} p_j$ $(p_k \neq 0)$ is the expansion of g in homogeneous polynomials with

$$p_k(t,z) = \sum_{\nu_0+\ldots+\nu_n=k} a_{\nu_0\ldots\nu_n} t^{\nu_0} z_1^{\nu_1} \ldots z_n^{\nu_n}$$

and if we set $t' = t$, $z_i' = z_i - \varepsilon_i t$ (with $\varepsilon_i \in \mathbb{C}$), then the coefficient of t'^k in $p_k(t',z')$ is equal to

$$c = \sum_{\nu_0+\ldots+\nu_n=k} a_{\nu_0\ldots\nu_n} \varepsilon_1^{\nu_1} \ldots \varepsilon_n^{\nu_n}$$

and for almost every choice of the ε_i this expression does not vanish; c is also the coefficient of t'^k in $g(t',z')$ hence, by correct choice of the ε_i, g becomes t'-general of order k.

One can prove Theorem 1 by showing the existence of a formal solution by comparing coefficients and induction, then carrying out a convergence proof (cf. e.g. Brill [Bl2]). However, the attempt to find versions of the preparation theorem suitable for other theories (e.g. for differentiable functions) has led to more elegant methods, which we shall use for the proof here.

Theorem 1 follows from the somewhat more general Theorem 2, which represents a kind of division algorithm for convergent power series.

Theorem 2 : (Division theorem)

Let $f,g \in \mathbb{C}\{t,z\}$ and let g be t-general of order k. Then there exist $q \in \mathbb{C}\{t,z\}$ and a polynomial $r \in \mathbb{C}\{z\}[t]$ of degree $\leq k-1$ such that

$$r(t,z) = \sum_{i=1}^{k} a_i(z) \cdot t^{k-i} \qquad (a_i(z) \in \mathbb{C}\{z\})$$

with $f = q \cdot g + r$,

and q, r are uniquely determined thereby.

Theorem 2 is often called the Weierstrass formula ; however, it was first proved by Stickelberger 1887 and independently by Späth in the year 1929. The notation in the above formula suggests that q is the quotient after division of f by g, and r is the remainder.

Derivation of Theorem 1 from Theorem 2 :

We set $f(t,z) := t^k$. Then by Theorem 2,

$$t^k = q \cdot g + r$$

where $r = \sum_{i=1}^{k} a_i t^{k-i}$, or, written differently,

$$q(t,z)g(t,z) = t^k - \sum_{i=1}^{k} a_i(z) t^{k-i}.$$

If one sets $z = 0$ in this equation and compares the coefficients of t^k then the result – since $g(t,0) = ct^k$ + higher terms – is that $q(0,0) \neq 0$. Thus we can divide by the unit q and obtain, with $u := q^{-1}$, $c_i := -a_i$, that

$$g = (t^k + \sum_{i=1}^{k} c_i t^{k-i}) \cdot u.$$

Incidentally, it is also not too difficult to derive Theorem 2 from Theorem 1 (see, e.g., Grauert-Remmert [G5], p. 43).

The proof of Theorem 1 and Theorem 2 is obtained with the help of the following theorem, which is a special case of Theorem 2.

Theorem 3 (Special division theorem)

Let $p_k(t,y) \in \mathbb{C}\{y_1,\dots,y_k\}[t]$ be "the general polynomial of degree k", i.e.

$$p_k(t,y) = t^k + \sum_{i=1}^{k} y_i \cdot t^{k-i}.$$

Then, for each $f(t,z,y) \in \mathbb{C}\{t,z,y\}$ there exist $q \in \mathbb{C}\{t,z,y\}$ and a polynomial $r(t,z,y) = \sum_{i=1}^{k} A_i(z,y) \cdot t^{k-i}$ of degree $\leq k-1$ over $\mathbb{C}\{z,y\}$ such that

$$f = q \cdot p_k + r.$$

This is a special case of Theorem 2, because here we divide only by the "general polynomial", and not by an arbitrary t-general function from $\mathbb{C}\{t,z,y\}$.

Using the "Malgrange trick", one can reduce the preparation theorem and the division theorem to the special division theorem. This goes as follows :

Reduction of Theorem 1 to Theorem 3 :

Let $g \in \mathbb{C}\{t,z\}$ be t-general of order k, i.e. $g(t,0)$ is a power series of the form

$$g(t,0) = c \cdot t^k + \dots \text{ (higher terms in } t\text{) with } c \neq 0.$$

By Theorem 3,

$$g(t,z) = q(t,z,y)(t^k+y_1t^{k-1}+\dots+y_k) + r(t,z,y) \qquad (*)$$

with a polynomial $r(t,z,y) = A_1(z,y)t^{k-1} + \dots + A_k(z,y)$ and $q \in \mathbb{C}\{t,z,y\}$.

Our aim is to replace the "general coefficients" y_i of the polynomial p_k by suitable holomorphic functions $y_i(z)$ so that the remainder term r in (*) vanishes. With this in mind we first show :

Claim : $$\frac{\partial A_i}{\partial y_j}(0,0) = \begin{cases} 0 & \text{for } i > j \\ -c & \text{for } i = j. \end{cases}$$

Proof : If one sets $y = z = 0$ in equation (*) and compares the coefficients of $t^0,\dots,t^k$, then one obtains

$$A_i(0,0) = 0 \text{ and } q(0,0,0) = c.$$

If one differentiates both sides of (*) with respect to the variable

y_j, then the result for $y = z = 0$ is :

$$0 = \frac{\partial q}{\partial y_j}(t,0,0)\cdot t^k + q(t,0,0)\cdot t^{k-j} + \frac{\partial A_1}{\partial y_j}(0,0)\cdot t^{k-1} + \ldots + \frac{\partial A_k}{\partial y_j}(0,0).$$

If one compares the coefficients of $t^0, t^1, \ldots, t^{k-1}$, then it follows that

$$\frac{\partial A_k}{\partial y_j}(0,0) = 0,\ \frac{\partial A_{k-1}}{\partial y_j}(0,0) = 0, \ldots,\ \frac{\partial A_{j+1}}{\partial y_j}(0,0) = 0$$

and

$$\frac{\partial A_j}{\partial y_j}(0,0) = -q(0,0,0) = -c,$$

and the above claim is proved.

The matrix $\frac{\partial A_i}{\partial y_j}$ is therefore an upper triangular matrix with determinant $(-c)^k \neq 0$. The equations $A_i(z,y) = 0$ therefore satisfy the hypotheses of the implicit function theorem ; consequently there are $y_j(z) \in \mathbb{C}\{z\}$, with $y_j(0) = 0$, such that

$$A_i(z, y_1(z), \ldots, y_k(z)) = 0,\ i = 1,\ldots,k.$$

If we now substitute $y = y(z)$ in equation (*) and set $u(t,z) = q(t,z,y(z))$, then we obtain

$$g(t,z) = (t^k + y_1(z)\cdot t^{k-1} + \ldots + y_k(z))\cdot u(t,z)$$

and $u(0,0) = c \neq 0$. This shows that g is the product of a Weierstrass polynomial and a unit u.

Proof of uniqueness of the product decomposition :

$$\text{Let } g(t,z) = u\cdot(t^k + c_1 t^{k-1} + \ldots + c_k)$$
$$= \tilde{u}\cdot(t^k + \tilde{c}_1 t^{k-1} + \ldots + \tilde{c}_k)$$

and let $U = V \times W$ be a neighbourhood of $0 \in \mathbb{C} \times \mathbb{C}^n$ on which u and $\tilde{u}$ vanish nowhere. Since the roots of a polynomial depend continuously on the coefficients, all k zeroes (counting multiplicity) of the polynomials

$$t^k + c_1(z)t^{k-1} + \ldots + c_k(z),$$
$$t^k + \tilde{c}_1(z)t^{k-1} + \ldots + \tilde{c}_k(z)$$

lie in V for sufficiently small $z \in \mathbb{C}^n$. Since $u \neq 0$ these are just the zeroes of $g(t,z)$ in V, i.e. the two polynomials have the same zeroes and hence coincide. Thus, for all sufficiently small z, $c_i(z) = \tilde{c}_i(z)$, and it follows that $c_i = \tilde{c}_i$ and hence $u = \tilde{u}$ also.

Reduction of Theorem 2 to Theorem 3 :

Let g be t-regular of order k and let $f \in \mathbb{C}\{t,z\}$. By Theorem 3 we can write g and f in the form

$$g = \tilde{q}(t,z,y) \cdot p_k + \tilde{r}(t,z,y)$$

$$f = \tilde{\tilde{q}}(t,z,y) \cdot p_k + \tilde{\tilde{r}}(t,z,y),$$

where $\tilde{r}$ and $\tilde{\tilde{r}}$ are again polynomials of degree $k-1$ with coefficients in $\mathbb{C}\{z,y\}$. As in the above proof of Theorem 1, we substitute $y = y(z)$, so that

$$\tilde{r}(t,z,y(z)) \equiv 0.$$

Then we obtain

$$g(t,z) = \tilde{q}(t,z,y(z)) \cdot p_k \quad (\text{with } \tilde{q}(0,0,0) \neq 0)$$

$$\begin{aligned} f(t,z) &= \tilde{\tilde{q}}(t,z,y(z)) \cdot p_k + \tilde{\tilde{r}}(t,z,y(z)) \\ &= \tilde{\tilde{q}} \cdot \tilde{q}^{-1} \cdot g + \tilde{\tilde{r}}. \end{aligned}$$

If we set $q(t,z) := \tilde{\tilde{q}}(t,z,y(z)) \cdot \tilde{q}^{-1}(t,z,y(z))$ and $r(t,z) := \tilde{\tilde{r}}(t,z,y(z))$ then we have the desired representation

$$f = q \cdot g + r.$$

The uniqueness of this decomposition is shown in the same way as for Theorem 1 : if $f = q_1 \cdot g + r_1 = q_2 \cdot g + r_2$, then

$$r_1 - r_2 = (q_2 - q_1) \cdot g.$$

Theorem 1 shows that $g(t,z)$ has k zeroes, counting multiplicity, in a neighbourhood of $0 \in \mathbb{C}$ for sufficiently small z. Then the polynomial $r_1(t,z) - r_2(t,z)$ of degree $\leq k-1$ also has at least k zeroes, and hence vanishes identically. Thus $r_1 = r_2$ and hence $q_1 = q_2$.

We now come to the

Proof of Theorem 3 :

The idea is to divide the function f successively by the linear factors of the general polynomial p_k.

Step I : (Division by a general linear factor $t - x_i$).

For each $F \in \mathbb{C}\{t,z,x_1,\ldots,x_k\}$ there exist $Q \in \mathbb{C}\{t,z,x\}$ and $R \in \mathbb{C}\{z,x\}$ such that

$$F = Q(t-x_i) + R$$

<u>because</u> : if one sets $R(z,x) := F(x_i,z,x)$ then $(t-x_i)$ divides the series $F - R = F(t,z,x) - F(x_i,z,x)$. (This fact from function theory is easily seen by expanding $F - R$ in terms of the coordinates $(t-x_i)$, $(t+x_i)$, $z_1,\dots,z_n$, $x_1,\dots,\hat{x}_i,\dots,x_k$.)

<u>Step II</u> : (Division by a general product of linear factors)

For each $F \in \mathbb{C}\{t,z,x_1,\dots,x_k\}$ there exist $Q \in \mathbb{C}\{t,z,x\}$ and a polynomial $R \in \mathbb{C}\{z,x\}[t]$ of degree $< k$ such that

$$F = Q(t-x_1)(t-x_2)\dots(t-x_k) + R$$

This equation determines Q and R uniquely.

<u>Proof</u> : By Step I we have

$$F = Q_1(t-x_1) + R_1 \qquad (Q_1 \in \mathbb{C}\{t,z,x\},\ R_1 \in \mathbb{C}\{z,x\})$$

$$Q_1 = Q_2(t-x_2) + R_2 \qquad (Q_2 \in \mathbb{C}\{t,z,x\},\ R_2 \in \mathbb{C}\{z,x\})$$

$$\vdots$$

$$Q_{k-1} = Q_k(t-x_k) + R_k \qquad (Q_k \in \mathbb{C}\{t,z,x\},\ R_k \in \mathbb{C}\{z,x\}).$$

Successive substitutions give

$$F = Q_k(t-x_1)(t-x_2)\ \dots\ (t-x_k)$$
$$+ R_1 + (t-x_1)R_2 + \dots + (t-x_1)(t-x_2)\ \dots\ (t-x_{k-1})R_k.$$

Thus with $Q := Q_k$, $R := R_1 + (t-x_1)R_2 + \dots + (t-x_1)(t-x_2)\dots(t-x_{k-1})R_k$ we have

$$F = Q \cdot (t-x_1) \cdot \ \dots\ \cdot (t-x_k) + R.$$

One proves the uniqueness of Q and R exactly as we proved the uniqueness assertion in Theorem 2 above.

<u>Step III</u> : **<u>(The Lojaziewicz trick)</u>**

Let $\sigma_i(x)$ be the ith elementary symmetric function in $x_1,\dots,x_k$ (cf. van der Waerden [W1], §33). We substitute $y_i := \sigma_i(x)$. Then we obtain the "splitting" of the general polynomial p_k into its linear factors :

$$p_k(t,y) = t^k + y_1t^{k-1} + \dots + y_k$$
$$= (t-x_1)(t-x_2)\dots(t-x_k).$$

We want to divide the function $f(t,z,y)$ by the general polynomial, and so we set

$$F(t,z,x) := f(t,z,\sigma_1(x),\dots,\sigma_k(x))$$

and get, by Step II :

$$F(t,z,x) = Q(t,z,x)(t-x_1)\dots(t-x_k) + R(t,z,x).$$

Q and R are then symmetric, i.e. invariant under permutations of $x_1,\dots,x_k$, because such permutations leave F and the polynomial $(t-x_1)\cdot\ldots\cdot(t-x_k)$ fixed, and hence cannot change Q and R by the uniqueness assertion of Step II. By the symmetry of Q and R, and the fundamental theorem of symmetric functions, there is a holomorphic function $q(t,z,y) \in \mathbb{C}\{t,z,y\}$ and a polynomial $r(t,z,y)$ of degree $< k$ in t with coefficients in $\mathbb{C}\{z,y\}$ such that

$$q(t,z,\sigma_1(x),\dots,\sigma_k(x)) = Q(t,z,x) \quad \text{and}$$

$$r(t,z,\sigma_1(x),\dots,\sigma_k(x)) = R(t,z,x).$$

We therefore have

$$f(t,z,\sigma_1(x),\dots,\sigma_k(x)) =$$
$$q(t,z,\sigma_1(x),\dots,\sigma_k(x))\cdot(t^k+\sigma_1(x)t^{k-1}+\dots+\sigma_k(x))+r(t,z,\sigma_1(x),\dots,\sigma_k(x))$$

and since the mapping of $\mathbb{C}^k$ into $\mathbb{C}^k$ which sends $(x_1,\dots,x_k)$ to $(\sigma_1(x),\dots,\sigma_k(x))$ is surjective, Theorem 3 follows.

In the proof we have used the fundamental theorem of symmetric functions. This theorem says that a convergent power series $\Phi(x_1,\dots,x_k,z_1,\dots,z_m) \in \mathbb{C}\{x_1,\dots,x_k,z_1,\dots,z_m\}$ which is invariant under permutations of the x_i may be written as a holomorphic function in the elementary symmetric functions. This means that there is a convergent power series $\Psi(y_1,\dots,y_k,z_1,\dots,z_m) \in \mathbb{C}\{y_1,\dots,y_k,z_1,\dots,z_m\}$ such that

$$\Phi(x_1,\dots,x_k,z) = \Psi(\sigma_1(x),\dots,\sigma_k(x),z).$$

For polynomials, and $m = 0$, this theorem is proved in every algebra course (cf. van der Waerden [W1], §33), and from that one directly obtains the corresponding assertion for formal power series. A proof of the theorem for convergent power series may be found e.g. in Tougeron [T4], Chap. IX, 6.5.

Remark : The preparation theorem and division theorem are easy to prove for formal power series. The division theorem for real analytic functions follows from the "complex case" by passing to the real part :

from this one can proceed as above to the Weierstrass preparation theorem for real analytic functions. On the other hand, the analogue of the preparation theorem for germs of differentiable functions is a deep result of Malgrange 1964 [M5]. This "Malgrange preparation theorem" has in the meantime become one of the most important tools for the investigation of differentiable functions, above all in the work of Mather on finitely determined germs of differentiable functions [G1].

As an application of Theorem 1 and Theorem 2 we now prove some algebraic properties of the ring $\mathbb{C}\{z_1,\dots,z_n\}$ of convergent power series. The preparation and division theorems are used to reduce these results to the corresponding theorems about polynomial rings.

Proposition 4 : (Rückert basis theorem)

The ring $\mathbb{C}\{z_1,\dots,z_n\}$ of convergent power series in n variables is noetherian, i.e. each of its ideals is finitely generated (cf. van der Waerden [W1], §115).

Proof : (by induction on n)

The base step of the induction, $n = 0$, is trivial.

Let $I \in \mathbb{C}\{z_1,\dots,z_n\}$ be an ideal and let g be a non-zero element of the ideal. The remarks after Theorem 1 show that there are always coordinates $z_1',\dots,z_n'$ in $\mathbb{C}^n$ such that the germ g of holomorphic functions is z_n'-general. Since the assertion of the theorem is independent of the choice of coordinates, we can assume without loss of generality that g is z_n-general. $\mathbb{C}\{z_1,\dots,z_{n-1}\}$ is noetherian by the induction hypothesis. The Hilbert basis theorem (cf. van der Waerden [W1], §115) says that the polynomial ring over a noetherian ring is again noetherian ; thus the ideal $I \cap \mathbb{C}\{z_1,\dots,z_{n-1}\}[z_n]$ is generated by finitely many functions $g_1,\dots,g_k$. Now if $f \in I$, the division theorem allows f to be written as

$$f = q\cdot g+r \quad \text{with} \quad r \in \mathbb{C}\{z_1,\dots,z_{n-1}\}[z_n].$$

Since f and g lie in the ideal I , we also have $r = f-q\cdot g \in I$. Thus $r \in \mathbb{C}\{z_1,\dots,z_{n-1}\}[z_n] \cap I$ and r may be written as a linear combination of the g_i :

$$r = a_1 \cdot g_1 + \dots + a_k \cdot g_k.$$

This implies

$$f = q \cdot g + a_1 \cdot g_1 + \dots + a_k \cdot g_k.$$

Hence we have shown that $g, g_1,\dots,g_k$ generate the ideal I.

The geometric meaning of this theorem is that the zero sets of arbitrarily many analytic functions can always be described locally by finitely many equations.

The attempt to decompose algebraic sets locally into components has led us to investigate algebraic sets by analytic methods. In 5.1 we have seen that the unique decomposition of algebraic curves into irreducible components follows from the unique factorisation property of the ring $\mathbb{C}[x_1,x_2]$. Therefore, we now investigate the divisibility properties of $\mathbb{C}\{z_1,\dots,z_n\}$.

Lemma 5

A Weierstrass polynomial $h \in \mathbb{C}\{z_1,\dots,z_{n-1}\}[z_n]$ is reducible in $\mathbb{C}\{z_1,\dots,z_{n-1}\}[z_n]$ if and only if it is reducible in $\mathbb{C}\{z_1,\dots,z_n\}$. And when h is reducible, then all its factors are Weierstrass polynomials, up to units.

Proof :

First assume that h is reducible in $\mathbb{C}\{z_1,\dots,z_n\}$; say $h = g_1 \cdot g_2$ where $g_1,g_2 \in \mathbb{C}\{z_1,\dots,z_n\}$ are non-units. Since h is a Weierstrass polynomial, hence z_n-general, g_1 and g_2 are also z_n-general. By the Weierstrass preparation theorem, we then have $g_i = u_i \cdot h_i$, where u_i is a unit in $\mathbb{C}\{z_1,\dots,z_n\}$ and $h_i \in \mathbb{C}\{z_1,\dots,z_{n-1}\}[z_n]$ is a Weierstrass polynomial. Thus we have two decompositions of h as a product of a unit and a Weierstrass polynomial, namely

$$h = 1 \cdot h \quad\text{and}\quad h = (u_1u_2) \cdot (h_1h_2).$$

The uniqueness assertion of Theorem 1 then yields

$$h = h_1 \cdot h_2 ,$$

and h_1,h_2 are not units in $\mathbb{C}\{z_1,\dots,z_{n-1}\}[z_n]$, otherwise g_1,g_2 would also be units.

Conversely, if $h = g_1 \cdot g_2$ is a decomposition of h into the product of two non-units $g_1,g_2 \in \mathbb{C}\{z_1,\dots,z_{n-1}\}[z_n]$, then we still have to show that g_1 and g_2 are also not units in the larger ring $\mathbb{C}\{z_1,\dots,z_n\}$.

Let

$$g_1 = a \cdot z_n^r + \ldots \quad \text{(lower terms in } z_n)$$

$$g_2 = b \cdot z_n^s + \ldots \quad \text{(lower terms in } z_n)$$

with $a,b \in \mathbb{C}\{z\}$. Then comparison of coefficients in $h = g_1 \cdot g_2$ gives

$$ab = 1,$$

hence we can assume without loss of generality that $a = b = 1$. Now if g_1 were a unit in $\mathbb{C}\{z_1,\ldots,z_n\}$,

$$g_2 = g_1^{-1}h$$

would be a decomposition of the Weierstrass polynomial g_2 into the product of a unit with a Weierstrass polynomial. The uniqueness assertion of Theorem 1 again yields

$$g_1 = 1,\ h = g_2.$$

But this contradicts the assumption that g_1 is a non-unit in $\mathbb{C}\{z_1,\ldots,z_{n-1}\}[z_n]$.

In 4.1 we gave various characterisations of factorial rings. To prove the next theorem we shall use the result that a ring is factorial if and only if the decomposition of each element into irreducible factors is unique.

Proposition 6 :

$\mathbb{C}\{z_1,\ldots,z_n\}$ is factorial.

Proof : (by induction on n)

Let $f \in \mathbb{C}\{z_1,\ldots,z_n\}$. Again assume without loss of generality that f is z_n-general. By the Weierstrass preparation theorem, $f = u \cdot \tilde{f}$, where $\tilde{f} \in \mathbb{C}\{z_1,\ldots,z_{n-1}\}[z_n]$ is a Weierstrass polynomial and u is a unit. The ring $\mathbb{C}\{z_1,\ldots,z_{n-1}\}[z_n]$ is factorial by the induction hypothesis and the theorem of Gauss (4.2.2), hence in $\mathbb{C}\{z_1,\ldots,z_n\}[z_n]$ there is a unique decomposition, $\tilde{f} = f_1 \cdot f_2 \cdot \ldots \cdot f_k$ of $\tilde{f}$ into irreducible elements. Then by Lemma 5,

$$f = u \cdot f_1 \cdot \ldots \cdot f_k$$

is a decomposition of f in $\mathbb{C}\{z_1,\ldots,z_n\}$, unique up to units and the order of factors.

$\mathbb{C}\{z_1,\ldots,z_n\}$ is therefore a Noetherian, factorial, local ring.

Remarks :

(i) The Weierstrass preparation theorem immediately implies the implicit function theorem : if $f(t,z) \in \mathbb{C}\{t,z_1,\dots,z_n\}$ with $f(0,0) = 0$ and $\frac{\partial f}{\partial t}(0,0) \neq 0$, then this just says that f is t-general of order 1. Then by Theorem 1 there is a $t(z) \in \mathbb{C}\{z\}$ and a unit u such that

$$f(t,z) = (t-t(z)) \cdot u.$$

Since $u(0,0) \neq 0$, $t = t(z)$ is a unique solution of the implicit equation $f(t,z) = 0$ in a neighbourhood of zero. Conversely, one sees easily that the implicit function theorem is equivalent to the preparation theorem for t-general functions of order 1.

By remark (iii) of 8.1 it follows in particular that $\mathbb{C}\{z_1,\dots,z_n\}$ is a henselian local ring. One can also easily derive Hensel's lemma directly from the Weierstrass preparation theorem (Grauert-Fritzsche [G4], p. 88).

(ii) Let $a = (a_1,\dots,a_n) \in \mathbb{C}^n$ and let $\mathcal{O}_{\mathbb{C}^n,a}$ be the ring of germs of holomorphic functions at a (cf. 8.1). If one associates each such germ with its Taylor expansion at a, then one obtains a convergent power series in the variables $z_1 - a_1,\dots,z_n - a_n$. If we denote the ring of these convergent power series by $\mathbb{C}\{z_1-a_1,\dots,z_n-a_n\}$ then we obtain an isomorphism

$$\mathcal{O}_{\mathbb{C}^n,a} \cong \mathbb{C}\{z_1-a_1,\dots,z_n-a_n\}.$$

In particular, for $a = 0$ we have

$$\mathcal{O}_{\mathbb{C}^n,0} \cong \mathbb{C}\{z_1,\dots,z_n\}.$$

In what follows we shall frequently move back and forth between the notions of "germs of holomorphic functions" and "convergent power series" without always mentioning this identification explicitly.

After these investigations of germs of holomorphic functions we now return to our proper theme, the investigation of local properties of analytic sets.

Definition :

Let U be an open set in $\mathbb{C}^n$, $X \subset U$.

(i) If $x \in U$, then X is called analytic at x if there is a neighbourhood V of x in U and finitely many holomorphic functions $f_1,\ldots,f_r$ on V such that

$X \cap V = \{z \in V \mid f_1(z) = \ldots = f_r(z) = 0\}$

(ii) X is called an <u>analytic subset</u> of U when X is analytic at all $x \in U$.

Remark : If X is analytic in U, then X is closed in U.

Proof : Let $x \in U - X$ and V be as above. Then $X \cap V$ is closed in V, and hence there is a neighbourhood $W \subset V$ of x which does not meet X. Thus $U - X$ is open.

This proof uses the fact that X is also analytic at points in $U - X$. Sets which are analytic only at their own points are called <u>locally analytic</u>. For example, the set $\mathbb{C} - \{0\} \subset \mathbb{C}^2$ is locally analytic in $\mathbb{C}^2$, analytic in $\mathbb{C}^2 - \{0\}$, but not analytic in $\mathbb{C}^2$.

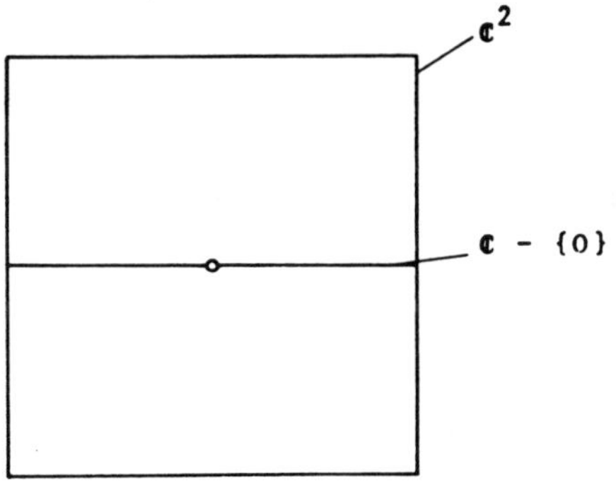

Just as we introduced germs of functions in 8.1 to investigate local properties of functions, we now introduce germs of analytic sets:

Definition : Let U and U' be open in $\mathbb{C}^n$, and let $X \subset U$, $X' \subset U'$ be analytic subsets. X and X' define the same <u>germ of analytic sets</u> at $x \in U \cap U'$ when there is an open neighbourhood $V \subset U \cap U'$ of x such that

$X \cap V = X' \cap V.$

We write (X,x) for the set germ of X at x ; X is called a <u>representative</u> of (X,x).

If one wants to be pedantically exact, one can define (X,x) to be the class of all pairs (X',U') equivalent to (X,U) under the equivalence relation defined above (U' an open neighbourhood of x, X' analytic in U').

The concept of set germs allows the economical formulation of assertions about X which depend only on the properties of the set X in arbitrarily small neighbourhoods of a point $x \in X$, namely, assertions about the singularity of X at x. Hence in what follows we shall frequently use the word "<u>singularity</u>" synonymously with "set germ".

We now want to investigate the relations between ideals in $\mathcal{O}_{\mathbb{C}^n,a}$ and germs of analytic sets. The process of describing a set by the ideal of functions which vanish on it comes from algebraic geometry, as developed in the '20's and '30's of this century. We have become partially acquainted with these algebraic-geometric methods already with the introduction of algebraic curves in Section 5.

If $I \subset \mathcal{O}_{\mathbb{C}^n,a}$ is an ideal, then by Proposition 4 it is generated by finitely many function germs $f_1,\ldots,f_r$. Let $\tilde{f}_1,\ldots,\tilde{f}_r$ be functions, on a neighbourhood U of a, whose germs at a are just $f_1,\ldots,f_r$. We consider the zero set of these functions :

$$\underline{X}(\tilde{f}_1,\ldots,\tilde{f}_r) := \{z \in U \mid \tilde{f}_1(x) = \ldots = \tilde{f}_r(z) = 0\}.$$

We shall show that the germ of this set at a does not depend on the choice of generating systems and representatives.

Let $(g_1,\ldots,g_s)$ be another generating system for I, and let $\tilde{g}_1,\ldots,\tilde{g}_s$ be representatives of $g_1,\ldots,g_s$ on a neighbourhood U' of a. Then there are germs $a_{ij} \in \mathcal{O}_{\mathbb{C}^n,a}$ with $f_i = \Sigma a_{ij} g_j$ and hence also representatives $\tilde{a}_{ij}$ of a_{ij} such that

$$\tilde{f}_i = \Sigma \tilde{a}_{ij} \cdot \tilde{g}_j$$

on a suitable neighbourhood W of a. Then obviously

$$\underline{X}(\tilde{g}_1,\ldots,\tilde{g}_s) \cap W \subset \underline{X}(\tilde{f}_1,\ldots,\tilde{f}_r) \cap W.$$

Analogously, there is a neighbourhood W' of a such that $\underline{X}(\tilde{f}) \cap W' \subset \underline{X}(\tilde{g}) \cap W'$. Thus in $W \cap W'$, $\underline{X}(\tilde{f})$ and $\underline{X}(\tilde{g})$ coincide. But that just means that the sets define the same germ at a.

Definition :

$\underline{X}(I) = (\underline{X}(\tilde{f}_1,\ldots,\tilde{f}_r),a)$ is called the set germ defined by the ideal I.

One can also state this definition somewhat roughly as follows : $\underline{X}(I)$ is the zero set of the ideal I. Conversely, if (X,a) is an analytic set germ, then one has the ideal of function germs which vanish on X.

Definition :

The ideal of the analytic set germ X is the set $\underline{J}(X)$ of all germs $f \in \mathcal{O}_{\mathbb{C}^n,a}$ which have a representative $\tilde{f}$, on a neighbourhood U of a, which vanishes on a representative $\tilde{X} \subset U$ of X.

We have already met a similar situation in 5.1, where we first defined plane curves as zero sets of arbitrary polynomials (this is analogous to the association $I \mapsto \underline{X}(I)$) and then constructed the equation of each curve (this is analogous to the association $X \mapsto \underline{J}(X)$).

A few trivialities :

(i) $I_1 \subset I_2 \Rightarrow \underline{X}(I_1) \supset \underline{X}(I_2)$

(ii) $X_1 \subset X_2 \Rightarrow \underline{J}(X_1) \supset \underline{J}(X_2)$

(iii) $\underline{X}(\underline{J}(X)) = X$

(iv) $\underline{J}(\underline{X}(I)) \supset I$.

It is not in general true that $\underline{J}(\underline{X}(I)) = I$. Example : if I is the ideal in $\mathcal{O}_{\mathbb{C},0} = \mathbb{C}\{z\}$ generated by z^2, then $\underline{X}(I) = \{0\}$ and $\underline{J}(\underline{X}(I)) = (z) \neq (z^2)$. Thus one must add to the ideal I all functions whose powers lie in I, i.e. all "roots" of functions from I.

Definition :

If R is a ring and $I \subset R$ is an ideal, then

$\mathrm{rad}(I) := \{f \in R \mid f^k \in I \text{ for some } k\}$

is called the radical of I.

Theorem 7 (Rückert's Nullstellensatz)

The ideal of an analytic set germ $\underline{X}(I)$ satisfies

$\underline{J}(\underline{X}(I)) = \mathrm{rad}(I)$.

Remark :

This is the analogue of Hilbert's Nullstellensatz, which expresses exactly the same relation between algebraic sets and ideals in the polynomial ring $\mathbb{C}[z_1,\ldots,z_n]$. We have not proved the Hilbert Nullstellensatz but have contented ourselves with a weaker form, Study's lemma (4.3, Theorem 2). Here, too, we shall prove the Rückert Nullstellensatz only for principal ideals ; the complete proof is more complicated, though based on the same idea (Gunning-Rossi [G6], p. 90-97).

Proof of Theorem 7 : (for the case in which $I = (f)$ is a principal ideal in $\mathcal{O}_{\mathbb{C}^n,0}$)

The inclusion $\mathrm{rad}(f) \subset \underline{J}(\underline{X}(f))$ is clear ; we have to show that $\underline{J}(\underline{X}(f)) \subset \mathrm{rad}(f)$, in other words that

$$g|_{\underline{X}(f)} = 0 \Rightarrow f \text{ divides } g^k \text{ for a suitable } k.$$

It suffices to show this for irreducible germs f, because if $f_1 \cdot \ldots \cdot f_r$ is the decomposition of f into irreducible factors, then g vanishes on each of the sets $\underline{X}(f_i)$. If the theorem is proved for irreducible functions, then f_i divides a power g^{k_i} of g. But then f divides $g^{k_1+\ldots+k_r}$.

So assume, without loss of generality, that f is irreducible.

Suppose that the assertion is false. Then f and g have no common divisor. As usual we choose coordinates $z_1,\ldots,z_n$ in $\mathbb{C}^n$ so that f and g are z_n-general. By the preparation theorem, f and g are each the product of a unit with a Weierstrass polynomial. Since we are interested only in the zero sets of f and g, and divisibility properties, assume without loss of generality that f and g are Weierstrass polynomials in $\mathcal{O}_{\mathbb{C}^{n-1},0}[z_n]$.

The functions f and g are relatively prime in $\mathcal{O}_{\mathbb{C}^n,0}$, and hence also in $\mathcal{O}_{\mathbb{C}^{n-1},0}[z_n]$, by Lemma 5. If K is the quotient field of $\mathcal{O}_{\mathbb{C}^{n-1},0}$, then it easily follows that f and g are also relatively prime in $K[z_n]$ (exercise!*). With the help of the euclidean algorithm, we have seen in 4.2 that there are then $\alpha, \beta \in K[z_n]$ such that

$$\alpha f + \beta g = 1.$$

---

*See [R1], Theorem 180.

If we write $\alpha = a/c$, $\beta = b/c$ with $a,b \in \mathcal{O}_{\mathbb{C}^{n-1},0}[z_n]$, $c \in \mathcal{O}_{\mathbb{C}^{n-1},0}$ and $c \neq 0$, then

$$a \cdot f + b \cdot g = c \in \mathcal{O}_{\mathbb{C}^{n-1},0}$$

i.e. this (nonvanishing) linear combination of f and g does not depend on the last variable z_n.

The function f is a Weierstrass polynomial, say

$$f = z_n^k + c_1(z_1,\ldots,z_{n-1})z_n^{k-1} + \ldots + c_k(z_1,\ldots,z_{n-1})$$

with $c_i(0,\ldots,0) = 0$. Since the roots of a polynomial depend continuously on the coefficients (Rouché's theorem, Behnke-Sommer [Bl] III, §4, Theorem 18), for any $\varepsilon > 0$ there is a neighbourhood

$$U = \{(z_1,\ldots,z_{n-1}) \mid \|(z_1,\ldots,z_{n-1})\| < \delta\}$$

of 0 in $\mathbb{C}^{n-1}$ such that for $(z_1,\ldots,z_{n-1}) \in U$ the polynomial $f(z_1,\ldots,z_{n-1},t)$ has k roots (counting multiplicity) with absolute value less than ε.

The zero set $\underline{X}(f)$ may be visualised in the real domain as follows :

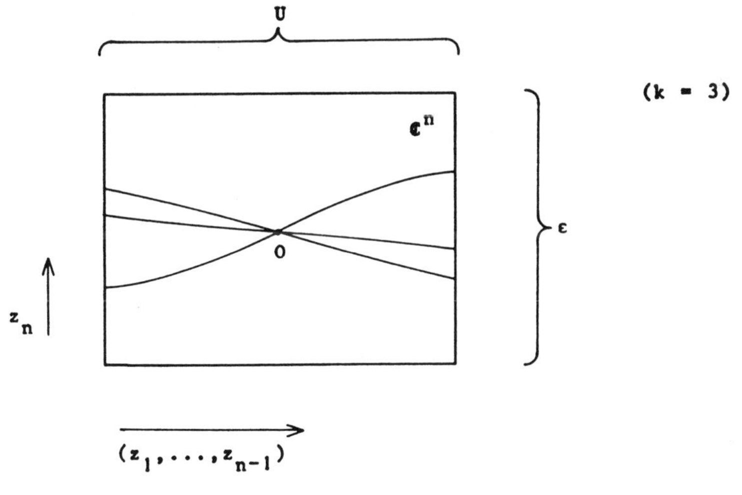

In particular, at least one zero of f lies over each $(z_1,\ldots,z_{n-1}) \in U$. Since g vanishes on $\underline{X}(f)$, it follows that for each $(z_1,\ldots,z_{n-1}) \in U$ there is a z_n such that

$c(z_1,\ldots,z_n) = a(z_1,\ldots,z_n)\cdot f(z_1,\ldots,z_n) + b(z_1,\ldots,z_n)\cdot g(z_1,\ldots,z_n) = 0.$ But since c does not depend on the variable z_n, $c \equiv 0$.

This is a contradiction. Hence the assumption that f divides no power of g is false, as was to be shown.

Remark : The description of the hypersurface $\underline{X}(f)$ as a "finitely branched covering" of the regular set $\mathbb{C}^{n-1}$, used in the above proof, is the real geometric content of the Weierstrass preparation theorem. This description may be generalised to all analytic set germs (Gunning-Rossi [G6], p. 98). Its analogue in algebraic geometry is the Noether normalisation lemma ([S4], I, §5.4).

Consequence of Theorem 7 : $X \mapsto \underline{J}(X)$ gives a bijection between analytic set germs at a and radical ideals (i.e. ideals I with $I = \text{rad}(I)$) in $\mathcal{O}_{\mathbb{C}^n,a}$.

In complex analysis one associates each analytic set germ (X,a) with the ring of germs at a of holomorphic functions on X. If $(X,a) \subset (\mathbb{C}^n,a)$ is such a set germ, then two holomorphic function germs f, g in $(\mathbb{C}^n,a)$ define the same function germ on X when $f - g$ vanishes on X, i.e. when $f - g \in \underline{J}(X)$. Thus the ring of all holomorphic function germs on (X,a) is just the ring $\mathcal{O}_{\mathbb{C}^n,a}/\underline{J}(X)$.

Definition :

(i) An analytic algebra is a $\mathbb{C}$-algebra of the form $\mathbb{C}\{z_1,\ldots,z_n\}/I$, where I is an ideal in $\mathbb{C}\{z_1,\ldots,z_n\}$.

(ii) An analytic algebra A is called reduced when it contains no non-zero nilpotent elements. An element $f \in A$ is called nilpotent when $f^k = 0$ for sufficiently large k.

Remarks :

(i) Obviously $\mathbb{C}\{z_1,\ldots,z_n\}/I$ is reduced just in case I is a radical ideal.

(ii) If X is an analytic set germ at $a \in \mathbb{C}^n$, then the analytic algebra $\mathcal{O}_{X,a} := \mathcal{O}_{\mathbb{C}^n,a}/\underline{J}(X)$ of germs at a of holomorphic functions on X is reduced.

(iii) It follows from the Rückert Nullstellensatz that $X \mapsto \mathcal{O}_{X,0} = \mathcal{O}_{\mathbb{C}^n,0}/\underline{J}(X)$ defines a bijection between the analytic set germs at 0 and the reduced analytic algebras $\mathbb{C}\{z_1,\ldots,z_n\}/I$.

(iv) Just as it was convenient in 6.2 to admit curves with multiple components, so it is sometimes necessary in complex analysis to consider non-reduced analytic algebras. However, we do not wish to go further into this.

We are mainly interested in the zero sets of single equations (curves in $\mathbb{C}^2$), hence in hypersurfaces.

Definition :

A set germ $\underline{X}(I)$ at $a \in \mathbb{C}^n$ is called a hypersurface when $I = (f)$ is a principal ideal in $\mathcal{O}_{\mathbb{C}^n,a}$.

Remark :

If $f = f_1^{k_1} \cdot \ldots \cdot f_r^{k_r}$ is the decomposition of f into different irreducible factors, then $\operatorname{rad}(f) = (f_1 \cdot \ldots \cdot f_r)$.

$f_1 \cdot \ldots \cdot f_r = 0$ is called "the equation" of the hypersurface ; it is uniquely determined up to units.

This definition is quite analogous to the definition of the equation of an algebraic curve which we gave earlier in 5.1 - except that we did not then formulate our statements ideal-theoretically.

Now that we have defined the concept of the germ of an analytic set, we shall also define morphisms between set germs.

Definition :

Let $(X,p) \subset (\mathbb{C}^m,p)$, $(Y,q) \subset (\mathbb{C}^n,q)$ be analytic set germs. Let U, V be neighbourhoods of p in $\mathbb{C}^m$ and let

$$f : U \to \mathbb{C}^n$$
$$g : V \to \mathbb{C}^n$$

be two analytic mappings with $f(p) = q$ and $g(p) = q$, which map representatives of X into representatives of Y. Then f and g define the same mapping germ $(X,p) \to (Y,q)$ when there is a neighbourhood $W \subset \mathbb{C}^m$ of p and a representative $\tilde{X}$ of X in W such that

$$f|_{\tilde{X}} = g|_{\tilde{X}} .$$

Naturally, the concept of a germ of holomorphic functions in remark (ii) above is a special case of the concept of a "mapping germ", namely that with $Y = \mathbb{C}^1$.

It is clear how one defines the composition of mapping germs, and one sees easily that the analytic set germs, together with the mapping germs, constitutes a category.

If $\phi : (X,p) \to (Y,q)$ is a mapping germ, then it induces a mapping $\phi^* : \mathcal{O}_{Y,q} \to \mathcal{O}_{X,p}$ of the associated local $\mathbb{C}$-algebras. The mapping ϕ^* is defined simply by $\phi^*(f) = f \circ \phi$. This relationship is functorial.

Conversely, if $\phi^* : \mathcal{O}_{Y,q} \to \mathcal{O}_{X,p}$ is a morphism of reduced analytic algebras, then ϕ^* comes from a mapping germ $\Psi : (X,p) \to (Y,q)$.

<u>**Proof**</u> : Let $(Y,q) \subset (\mathbb{C}^n,q)$ and $(X,p) \subset (\mathbb{C}^m,p)$ be the given set germs and let $z_1,\ldots,z_n$ be coordinate functions of $\mathbb{C}^n$ around q.

The point $\phi^*(z_i)$ lies in $\mathcal{O}_{X,p} = \mathcal{O}_{\mathbb{C}^m,p}/\underline{J}(X)$; let $z_i(x_1,\ldots,x_m)$ be functions in $\mathcal{O}_{\mathbb{C}^m,p}$ which represent the $\phi^*(z_i)$. We set :

$$\Psi : (\mathbb{C}^m,p) \to (\mathbb{C}^n,q)$$
$$(x_1,\ldots,x_m) \mapsto (z_1(x),\ldots,z_n(x)).$$

Then Ψ induces a mapping germ $\Psi : (X,p) \to (Y,q)$ (exercise!), and $\Psi^*(z_i) = \phi^*(z_i)$ by construction. Since each element of $\mathcal{O}_{Y,q}$ may be written as a power series in the z_i , the assertion follows with the help of the Artin-Rees lemma.

To sum up : the correspondence $(X,p) \mapsto \mathcal{O}_{X,p}$ defines an anti-equivalence between the category of analytic set germs and the category of reduced local analytic $\mathbb{C}$-algebras.

Thus the category of analytic algebras is an algebraic reflection of the category actually of interest to us, that of analytic set germs, and much work in complex analysis may also be expressed in the language of these algebras. However, we shall mainly stick to the geometric viewpoint.

As in any category, we naturally have the concept of isomorphism for analytic set germs :

<u>Definition</u> :

A mapping germ $f : (X,p) \to (Y,q)$ is called an <u>isomorphism</u> when there is a mapping germ $g : (Y,q) \to (X,p)$ such that

$g \circ f = \text{identity}, f \circ g = \text{identity}.$

From what was said above, f is an isomorphism if and only if $f^* : \mathcal{O}_{Y,q} \to \mathcal{O}_{X,p}$ is an isomorphism of local $\mathbb{C}$-algebras.

In particular, if $(X,0)$ and $(Y,0)$ are both set germs in $\mathbb{C}^n$, then each isomorphism $f : (X,0) \to (Y,0)$ is induced by a local isomorphism $F : (\mathbb{C}^n,0) \to (\mathbb{C}^n,0)$.

Proof :

Let $F : (\mathbb{C}^n,0) \to (\mathbb{C}^n,0)$ and $G : (\mathbb{C}^n,0) \to (\mathbb{C}^n,0)$ be mapping germs which induce morphisms $f : (X,0) \to (Y,0)$ and $g : (Y,0) \to (X,0)$, so that the corresponding homomorphisms $f^* : \mathcal{O}_{Y,0} \to \mathcal{O}_{X,0}$ and $g^* : \mathcal{O}_{X,0} \to \mathcal{O}_{Y,0}$ satisfy

$$f^* \circ g^* = \text{id} \quad \text{and} \quad g^* \circ f^* = \text{id}.$$

We would like to be able to show that F and G are isomorphisms, because then the above assertion will be proved.

In order to show this, we first make the following additional assumption (A) : for at least one of the set germs, say $(X,0)$, the embedding $(X,0) \subset (\mathbb{C}^n,0)$ is an embedding in an affine space with the smallest possible dimension n. That means that X does not lie in the hyperplane $z_i = 0$ for any local coordinate function in any coordinate system. Such local coordinate functions are of course just the elements of $\mathcal{M} - \mathcal{M}^2$, where $\mathcal{M}$ is the maximal ideal of $\mathbb{C}\{x_1,\ldots,x_n\}$. Thus assumption (A) is equivalent to

$$\underline{J}(X) \subset \mathcal{M}^2.$$

Under this assumption, we now prove : F and G are local isomorphisms. To do this we consider the composition $H = G \circ F$. In local coordinates, H is described by equations

$$x_i' = \sum_{j=1}^{n} c_{ij} x_j \qquad i = 1,\ldots,n$$

with $c_{ij} \in \mathbb{C}\{x_1,\ldots,x_n\}$. Since $f^* \circ g^* = \text{id}$, the relation $x_i' = x_i$ holds mod $\underline{J}(X)$, i.e.

$$\sum_j c_{ij} x_j \equiv x_i \mod \underline{J}(X).$$

Then since $\underline{J}(X) \subset \mathcal{M}^2$ we certainly have :

$$\sum_j c_{ij} x_j \equiv x_i \mod \mathcal{M}^2.$$

But this just means that the Jacobian matrix of H is the identity :

$$\left(\frac{\partial x_i'}{\partial x_j}(0)\right) = (c_{ij}(0)) = 1.$$

Hence, by the inverse function theorem, H is a local isomorphism!

The same then holds for F and G as well, as was to be shown.

To conclude, we still have to free ourselves of the additional assumption (A) about the minimal embedding dimension. If $(X,0) \subset (\mathbb{C}^n,0)$ and $(Y,0) \subset (\mathbb{C}^n,0)$ are embeddable in lower-dimensional smooth subspaces $(\tilde{\mathbb{C}}^{n-1},0)$ and $(\tilde{\tilde{\mathbb{C}}}^{n-1},0)$, then one can induce f^* by an isomorphism between $(\tilde{\mathbb{C}}^{n-1},0)$ and $(\tilde{\tilde{\mathbb{C}}}^{n-1},0)$. (By induction, with the above proof as base step.) This isomorphism is obviously extendable to an isomorphism $F : (\mathbb{C}^n,0) \to (\mathbb{C}^n,0)$, which induces f^*. The assertion is now completely proved.

We have now constructed the category of objects whose description, investigation and particularly classification will concern us in what follows : the category of germs of analytic sets - synonymously, singularities - or equivalently, the category of analytic algebras. Among the many possible viewpoints from which these objects can be investigated and classified, we shall naturally work out only a few. In particular, one must not expect that we will be able to find a meaningful classification of all germs of analytic sets up to isomorphism - isomorphism is much too fine an equivalence relation and there are far too many analytic sets for it. We can be well satisfied if we succeed merely in finding a suitable classification for a few classes of hypersurfaces.

In particular, we shall first deal with 1-dimensional hypersurfaces — the curves in the plane $\mathbb{C}^2$ — and hence germs of analytic sets X which are zero sets of holomorphic functions f of two variables. Previously we have studied affine algebraic curves in the plane $\mathbb{C}^2$, i.e. zero sets of polynomials. In considering the singularities of analytic sets in $\mathbb{C}^2$ we are, a priori, considering a larger class of objects than the singularities of algebraic curves. It would therefore seem that our problem has become more complicated. In truth it is quite the opposite. It turns out that the class has not in fact become any larger ; namely, N. Levinson proved the following theorem in 1960 ([L6], [L7]) :

<u>**Theorem 8**</u> :

Let $f \in \mathcal{O}_{\mathbb{C}^2,0}$ be any germ of holomorphic functions of two variables, without multiple factors. Then there are local coordinates z_1,z_2 in a neighbourhood of $0 \in \mathbb{C}^2$ for which $f(z_1,z_2)$ is a polynomial.

This means that, up to isomorphism of analytic set germs, all singularities which can occur with analytic curves in the plane already occur with algebraic curves, so that, in a certain sense, we have not enlarged our class of objects. On the other hand, we have enlarged the class of admissible mappings. Now, all mappings are defined by analytic functions, whereas for the singularities of algebraic curves the natural morphisms first appeared to be mappings given by rational functions. But we have already seen, through the discussion in 8.1, that we must go beyond this for an adequate treatment of the local properties of curves. The introduction of the category of analytic set germs is precisely what does this. For example, an irreducible cubic with an ordinary double point and a reducible quadric also with an ordinary double point have isomorphic analytic set germs corresponding to their singular points, but one can find no isomorphism described by rational functions. Quite generally, the problem of classifying singularities becomes simplified by the admission of the larger class of analytic isomorphisms. We shall see, though, that it remains interesting, and quite difficult enough.

We remark that N. Levinson's theorem has been considerably generalised in various directions. The condition that $f(z_1,z_2) \in \mathbb{C}\{z_1,z_2\}$ have no multiple components can also be expressed by saying that f has an isolated singularity at $0 \in \mathbb{C}^2$. In general, $f \in \mathbb{C}\{z_1,\ldots,z_n\}$ has an isolated singularity at $0 \in \mathbb{C}^n$ just in case the set germ defined by $\frac{\partial f}{\partial z_1} = 0,\ldots, \frac{\partial f}{\partial z_n} = 0$ is just the origin. By the Rückert Nullstellensatz, this is equivalent to

$$\mathfrak{M}^k \subset \left(\frac{\partial f}{\partial z_1},\ldots,\frac{\partial f}{\partial z_n}\right)$$

for a suitable natural number k, where $\mathfrak{M} = (z_1,\ldots,z_n)$ is the maximal ideal of $\mathbb{C}\{z_1,\ldots,z_n\}$. (Thus $\mathfrak{M}^k$ is the ideal of all power series of the form

$$\sum_{i_1+\ldots+i_n \geq k} a_{i_1,\ldots,i_n} z_1^{i_1} \cdot \ldots \cdot z_n^{i_n}.)$$

Theorem 9 (Mather)

Let $f \in \mathbb{C}\{z_1,\ldots,z_n\}$ be a power series with an isolated singularity at 0, so that

$$\mathfrak{M}^k \subset \left(\frac{\partial f}{\partial z_1},\ldots,\frac{\partial f}{\partial z_n}\right).$$

Then for each $g \in \mathbb{C}\{z_1,\dots,z_n\}$ with $g - f \in \mathfrak{M}^{k+2}$ there is a local isomorphism $\phi : (\mathbb{C}^n,0) \to (\mathbb{C}^n,0)$ with $g = f \circ \phi$.

If one applies this in particular to the case where g is the Taylor expansion of f up to order $k+1$, then the result is that f is transformed by a change of coordinates into g, and hence into a polynomial.

It follows in particular that hypersurfaces with isolated singularities are algebraic. Theorems of similar type had previously been proved by Hironaka and Samuel. For a proof of Theorem 9 see Wassermann : Stability of Unfoldings, [W6], Th. 2.6.

As far as zero sets are concerned, around 1968 an even more general result was proved ; the proof, in contrast to that of Theorem 9, goes essentially beyond the scope of this course (cf. M. Artin : The implicit function theorem in algebraic geometry, [A3]).

Theorem 10 : (M. Artin)

Each isolated singularity of an analytic set is algebraic, i.e. the associated analytic set germ is isomorphic to an analytic set germ of an affine algebraic set.

In the investigation of singularities we frequently want to go from the pure "point" description by set germs to a "local" description by representative analytic sets. For this it is important to grasp the relation between the ideal for the germ at a point x_0 of a given analytic set X and the ideal for a point $x \in X$ in a neighbourhood of x_0. This is done by means of the following theorem :

Theorem 11 :

Let X be an analytic set in a domain U of $\mathbb{C}^n$, and for each $x \in U$ let $J_x = \underline{J}(X,x) \subset \mathcal{O}_{\mathbb{C}^n,x}$ be the ideal of the set germ (X,x). If $f_1,\dots,f_r$ are generators of J_{x_0} and $\tilde{f}_1,\dots,\tilde{f}_r$ are representatives of them, then $\tilde{f}_1,\dots,\tilde{f}_r$ also represent generators of J_x for all x in a suitable neighbourhood of x_0.

This theorem is a corollary of a still stronger theorem which says somewhat more about the relation between the generators, and can be formulated as follows : the ideal sheaf $J(X) = \bigcup_{x \in U} J_x$ of an analytic set is coherent. For the difficult proof we refer to Gunning-Rossi [G6] IVD, Cor. 3, p. 141.

In what follows we shall rather briefly develop the most important concepts for the local investigation of analytic sets :

(1) Local decomposition into irreducible components. (2) Regular and singular points. (3) Dimension of analytic sets.

1. Local decomposition into irreducible components:

If $X = \underline{X}(f_1,\ldots,f_r)$ and $Y = \underline{X}(g_1,\ldots,g_s)$ are germs of analytic sets at $a \in \mathbb{C}^n$, then so too are

$$X \cap Y = \underline{X}(f_1,\ldots,f_r\ ,\ g_1,\ldots,g_s) \text{ and}$$

$$X \cup Y = \underline{X}(f_1g_1, f_1g_2,\ldots,f_1g_s, f_2g_1,\ldots,f_rg_s).$$

We can therefore attempt to represent an analytic set germ as the union of indecomposable germs.

Definition :

Let X be an analytic set germ at $a \in \mathbb{C}^n$. X is called <u>reducible</u> if there are germs $X_1 \subsetneqq X$, $X_2 \subsetneqq X$ such that $X = X_1 \cup X_2$. Otherwise X is called <u>irreducible</u>.

Proposition 12 :

An analytic set germ X is irreducible just in case its ideal $\underline{J}(X)$ is a prime ideal.

Proof : If $X = X_1 \cup X_2$ with $X_1 \subsetneqq X$, $X_2 \subsetneqq X$, then $\underline{J}(X_1) \supsetneqq \underline{J}(X)$ and $\underline{J}(X_2) \supsetneqq \underline{J}(X)$. We choose functions $f \in \underline{J}(X_1) - \underline{J}(X)$ and $g \in \underline{J}(X_2) - \underline{J}(X)$. Then $f \cdot g$ lies in $\underline{J}(X_1) \cap \underline{J}(X_2) = \underline{J}(X)$; thus $\underline{J}(X)$ is not a prime ideal.

Conversely, if $\underline{J}(X)$ is not a prime ideal, then there are $f, g \in \mathcal{O}_{\mathbb{C}^n,a} - \underline{J}(X)$ such that $f \cdot g \in \underline{J}(X)$. Since f and g both do not vanish on X, one obtains $X_1 := X \cap \underline{X}(f) \subsetneqq X$ and $X_2 := X \cap \underline{X}(g) \neq X$ with $X_1 \cup X_2 = X \cap \underline{X}(f\cdot g) = X$. Thus X is reducible.

Theorem 13 :

Each analytic set germ at $a \in \mathbb{C}^n$ has a unique decomposition

$$X = X_1 \cup \ldots \cup X_r$$

into irreducible germs (X_i,a) with $X_i \not\subset X_j$ for $i \neq j$. The X_i are called the <u>irreducible components</u> of X. (In the case of curves the X_i are also called <u>branches</u> of X.)

Proof : The existence of the decomposition follows from the fact that $\mathcal{O}_{\mathbb{C}^n,a}$ is noetherian. When X is not irreducible, one can decompose it : $X = X_1 \cup X_2$. If X_1 and X_2 are irreducible, then one is finished. If not, one decomposes further. This process must come to an end. If this were not so, one would have an infinite descending sequence of germs

$$Y_1 \supsetneqq Y_2 \supsetneqq \ldots$$

and hence an infinite strictly ascending sequence of ideals

$$\underline{J}(Y_1) \subsetneqq \underline{J}(Y_2) \subsetneqq \ldots$$

in $\mathcal{O}_{\mathbb{C}^n,a}$. This is impossible, because $\mathcal{O}_{\mathbb{C}^n,a}$ is noetherian (cf. van der Waerden [W1] §115).

If $X = X_1' \cup \ldots \cup X_s'$ is another decomposition of X into irreducible sets, then we now have to show that each component of one decomposition is contained in a component of the other. If, say, X_1' were not contained in any of the X_i , then

$$X_1' = (X_1' \cap X_1) \cup \ldots \cup (X_1' \cap X_r)$$

would be a proper decomposition of X_1' , contradicting the irreducibility of X_1'.

The geometric statement of Theorem 13 corresponds to the algebraic theorem of prime factorisation in the noetherian ring $\mathcal{O}_{\mathbb{C}^n,a}$ ([W1], §118). For arbitrary noetherian rings one can give the prime factorisation theorem an analogous geometric formulation with the help of the language of affine schemes.

Proposition 14 :

If $X = \underline{X}(f)$ is a hypersurface and $f = f_1^{k_1} \cdot \ldots \cdot f_r^{k_r}$ is the decomposition of f into distinct irreducible factors, then

$$\underline{X}(f) = \underline{X}(f_1) \cup \ldots \cup \underline{X}(f_r)$$

is the decomposition of X into irreducible components.

Proof : Since $\mathcal{O}_{\mathbb{C}^n,a}$ is factorial, the f_i are also prime elements (4.1.1), and hence the (f_i) are prime ideals. Therefore, the $\underline{X}(f_i)$ are irreducible.

Remark :

One can analogously define the concepts of "irreducible" and "decomposition into irreducible components" not only for germs of analytic sets, but also for the analytic subsets $X \subset G$ in a domain G of $\mathbb{C}^n$. In general, such a decomposition can then have infinitely many components. However, the system of irreducible components is always locally finite, and the connection with the decomposition of the germ (X,x) at a point x is the following :

For $\varepsilon > 0$ let U_ε be the ball or polydisc of radius ε around x. Then there is a number r such that for sufficiently small ε the intersection $U_\varepsilon \cap X$ decomposes into r irreducible components

$$U_\varepsilon \cap X = X_1 \cup \ldots \cup X_r$$

and the germs $(X_1,x),\ldots,(X_r,x)$ are the irreducible components of (X,x). The proof of this assertion about the connection between point and local decomposition is not simple (cf. Whitney, Complex Analytic Varieties, Theorem 85 [W9]).

2. Regular and singular points :

Definition : Let X be an analytic subset in a domain of $\mathbb{C}^n$ and let $x \in X$. Then x is a regular point of X when there is a neighbourhood U of x in $\mathbb{C}^n$ such that $X \cap U$ is a complex analytic submanifold of U. Otherwise x is called a singular point of X. We denote the set of singular points of X by $S(X)$.

In what follows we shall derive criteria for regularity. The most important criterion characterises singular points as those at which the Jacobian matrix of partial derivatives degenerates, i.e. has a lower rank. We shall now make this precise :

Let X be an analytic set in $\mathbb{C}^n$ and let $x \in X$. Let J_x be the ideal of the germ (X,x) and let $f_1,\ldots,f_r$ be generators of J_x. Then we set

$$\rho_{X,x} := \operatorname{rank} \left(\frac{\partial f_i}{\partial z_j}(x)\right).$$

One easily checks that $\rho_{X,x}$ does not depend on the choice of generators.

If X is regular at x and if $X \cap U$ is a ρ-codimensional complex submanifold in a neighbourhood U of x, then of course $\rho_{X,y} = \rho$ for all $y \in X \cap U$.

In general one can say only the following : if $x \in X$, and if

$f_1,\dots,f_r$ generate the ideal J_x, then of course $\operatorname{rank}(\frac{\partial f_i}{\partial z_j}(y)) \geq \operatorname{rank}(\frac{\partial f_i}{\partial z_j}(x))$ for all y in a neighbourhood $U(x)$ of x. Since $f_1,\dots,f_r$ are part of a generator system of J_y $(y \in U(x))$, one also has

$$\rho_{X,y} \geq \rho_{X,x} \quad \text{for } y \in U(x).$$

The example of singularities of algebraic curves (cf. 5.3) suggests the conjecture that constancy of rank is characteristic of regular points. This is in fact the case :

Theorem 15 :

Let X be an analytic subset in a domain of $\mathbb{C}^n$, and let $x \in X$. Then the following statements are equivalent :

(i) (X,x) is regular

(ii) (X,x) is isomorphic to $(\mathbb{C}^{n-\rho},0) \subset (\mathbb{C}^n,0)$

(iii) $\mathcal{O}_{X,x}$ is isomorphic to $\mathbb{C}\{z_1,\dots,z_{n-\rho}\}$

(iv) $\rho_{X,y} = \rho$ for all y in a neighbourhood of x.

Proof :The implications (i) $\Rightarrow$ (ii) $\Rightarrow$ (iii) $\Rightarrow$ (iv) are clear. We have to show (iv) $\Rightarrow$ (i).

Let J_x be the ideal of X at x and let $f_1,\dots,f_r$ be representatives of a generator system for J_x, numbered, without loss of generality, so that

$$\operatorname{rank}(\frac{\partial f_i}{\partial z_j})_{i,j=1,\dots,\rho} = \rho.$$

By the implicit function theorem one is then free to assume that

$$f_i(z) = z_i \qquad i = 1,\dots,\rho$$

in a neighbourhood U of $x = 0$ in $\mathbb{C}^n$. Let $\tilde{X} = \{z \in U \mid z_1 = \dots = z_\rho = 0\}$. Trivially, we have $U \cap X \subset \tilde{X}$. To prove the theorem it suffices to show that $U \cap X$ coincides with the regular set $\tilde{X}$, i.e. to show that

$$f_i|_{\tilde{X}} = 0 \quad \text{for } i = \rho+1,\dots,r.$$

It follows from the constancy $\rho_{X,y} = \rho$ for all $y \in V \cap X$ (where $V \subset U$ is a suitable neighbourhood of x) that

$$\frac{\partial f_i}{\partial z_j}\Big|_{X \cap V} = 0 \quad \text{for } i,j = \rho+1,\dots,r,$$

hence these $\frac{\partial f_i}{\partial z_j}$ are linear combinations of the f_k $(k = 1,\ldots,r)$:

$$\frac{\partial f_i}{\partial z_j} = \sum_{k=1}^{r} a_{ijk} f_k \quad \text{for } i,j = \rho+1,\ldots,r.$$

By repeated differentiation and induction it follows that all partial derivatives of f_i $(i \geq \rho+1)$ with respect to variables z_j $(j \geq \rho+1)$ vanish when restricted to $X \cap V$. In particular, the Taylor series of f_i at the point 0 lies in the ideal

$$(z_1,\ldots,z_\rho) \cdot \mathbb{C}\{z_1,\ldots,z_n\}.$$

Thus $f_i|_{\tilde{X}\cap V} = 0$, as was to be shown.

Proposition 16 :

Let $f \neq 0$ be a holomorphic function in a domain U of $\mathbb{C}^n$ and let X be the hypersurface $\{z \in U \mid f(z) = 0\}$. The germs of f at the points of X are assumed, in addition, to have no multiple factors. Then the set of singular points is equal to

$$S(X) = \{z \in X \mid \frac{\partial f}{\partial z_1}(z) = \ldots = \frac{\partial f}{\partial z_n}(z) = 0\}.$$

Proof : If a partial derivative $\frac{\partial f}{\partial z_i}(z) \neq 0$, then z is a regular point of X by the implicit function theorem. Conversely, if all partial derivatives $\frac{\partial f}{\partial z_i}(x) = 0$ at a point $x \in X$, then by Theorem 15(iv) we have to show that there is no neighbourhood U of x such that $\frac{\partial f}{\partial z_i}(y) = 0$ for all $y \in U \cap X$ and $i = 1,\ldots,n$. Suppose there were such a U. Then the germs of the partial derivatives would satisfy:

$$\frac{\partial f}{\partial z_i} \in \underline{J}(X,x) \quad \text{for } i = 1,\ldots,n.$$

Thus there would be germs a_i such that

$$\frac{\partial f}{\partial z_i} = a_i f \quad \text{for } i = 1,\ldots,n.$$

Differentiation gives :

$$\frac{\partial^2 f}{\partial z_i \partial z_j} = \frac{\partial a_i}{\partial z_j} \cdot f + a_i \cdot \frac{\partial f}{\partial z_j} \in \underline{J}(X,x).$$

It follows by repeated differentiation and induction that all partial derivatives of f lie in $\underline{J}(X,x)$, and hence all terms of the Taylor series of f vanish at x. But this implies $f \equiv 0$, contrary to hypothesis.

Remarks :

(i) For algebraic curves this shows that the present definition coincides with the concept of singular given in 5.3.

(ii) Proposition 16 shows that for hypersurfaces X the singularity set $S(X)$ is a proper analytic subset. This holds quite generally :

Theorem 17 :

The singularity set $S(X)$ of an analytic set X is a proper analytic subset of X.

The proof makes essential use of the coherence of the ideal sheaf (Theorem 11) and a regularity criterion different from those in Theorem 15. See Gunning-Rossi [G6], Chap. IV, D, Cor. 4, p. 141.

Remarks : (i) One can show, in addition, that $X - S(X)$ is open and dense in X. If X is irreducible, then $X - S(X)$ is even connected. Thus the singularity set $S(X)$ is a "thin subset" of X, while almost all points of X belong to the manifold of regular points. Regularity is the normal case, singularity the exception.

(ii) If we set

$$X^1 := X - S(X)$$
$$X^2 := S(X) - S(S(X))$$
$$X^3 := S(S(X)) - S(S(S(X)))$$
$$\cdots\cdots\cdots$$

then we obtain a decomposition $X = \cup X^i$ into non-singular manifolds X^i with

$$X^i \cap X^j = \emptyset \text{ for } i \neq j \text{ and}$$
$$X^{i+1} \subset \overline{X^i}.$$

Decompositions of this and similar kinds are called <u>stratifications</u> ; the X^i are the <u>strata</u>. Stratification is a method of reducing the investigation of singularities to the investigation of non-singular objects. The other two most important methods are <u>resolution</u> and <u>deformation</u>. Later in this chapter (§8.4) we shall develop the method of resolution in detail.

3. Dimension of analytic sets :

Definition : Let X be an analytic subset of an open set in $\mathbb{C}^n$.

(i) If $x \in X$ is a regular point, then there is a neighbourhood U of x such that $U \cap X$ is a connected complex submanifold. Then x is called a <u>regular point of dimension</u> d when $X \cap U$ has (complex) dimension d.

(ii) If $x \in X$ is arbitrary, then each neighbourhood of x contains regular points of X. The <u>dimension of the set germ</u> (X,x) is the greatest number d such that each neighbourhood of x contains regular points of X of dimension d. We write $d = \dim_x X$.

(iii) The <u>dimension of X</u> is

$$\dim X := \max_{x \in X} \dim_x X.$$

X is called <u>pure-dimensional</u> if $\dim X = \dim_x X$ for all $x \in X$.

(iv) The germ (X,x) is called <u>pure-dimensional</u> when it has a pure-dimensional representative.

Remark : If X is irreducible, then the set of regular points is connected, hence X is pure-dimensional. Thus an analytic set Y is pure-dimensional just in case all irreducible components have the same dimension. The same holds for germs of analytic sets.

For example, if $X_1 = \{z \in \mathbb{C}^3 \mid z_3 = 0\}$, $X_2 = \{z \in \mathbb{C}^3 \mid z_1 = z_2 = 0\}$ and $X = X_1 \cup X_2$, then $\dim X = 2$, $\dim_z X = 1$ if $z \in X_2$ and $z \neq 0$, and $\dim_z X = 2$ if $z \in X_1$. In particular, $\dim_0 X = 2$.

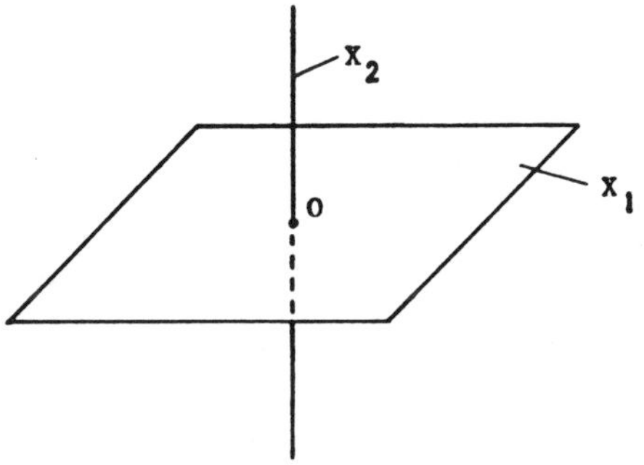

In books on the function theory of several variables, the dimension of analytic set germs is often introduced algebraically, i.e. with the help of the associated analytic algebra. These definitions of dimension are perhaps not quite so intuitive, but they have the advantage that one can more easily compute with them. Here are a few such definitions :

<u>**The Chevalley dimension**</u> : (cf. Grauert-Remmert [G5], p. 109)

If A is an analytic $\mathbb{C}$-algebra, then the Chevalley dimension of A is the smallest number d for which there are functions $f_1,\ldots,f_d$ with

$$\mathfrak{M}^k \subset (f_1,\ldots,f_d) \text{ for a suitable } k.$$

Here $\mathfrak{M}$ is the maximal ideal in A.

If $A = \mathcal{O}_{X,0}$ is the algebra of the set germ $(X,0)$ then $\mathfrak{M}^k \subset (f_1,\ldots,f_d)$ says, by the Rückert Nullstellensatz, that the intersection of X with the d hyperplanes $\{f_1 = 0\},\ldots,\{f_d = 0\}$ is a single point. The geometric idea is the following : in the d-dimensional space X the zero set of an equation is in general a $(d-1)$-dimensional subspace ; one therefore needs d equations in order to describe a zero-dimensional subspace.

<u>**The Weierstrass dimension**</u> : (cf. Gunning-Rossi [G6], p. 110)

In the remark after Theorem 7 we mentioned that each analytic set germ X may be described as a finitely branched covering of a regular set. The dimension of this regular set is well-defined and is called the Weierstrass dimension of X.

$\pi : X \to \mathbb{C}^d$ is such a branched covering just in case $\mathcal{O}_{X,0}$ is a finite $\mathbb{C}\{z_1,\ldots,z_d\}$-module by virtue of π^*. This makes it possible to view the concept of Weierstrass dimension algebraically.

<u>**The Krull dimension**</u> : (Grauert-Remmert [G5], §6).

Let A be an analytic algebra. A <u>prime ideal chain</u> of length ℓ in A is an ascending sequence

$$\mathcal{P}_0 \subsetneq \mathcal{P}_1 \subsetneq \cdots \subsetneq \mathcal{P}_\ell \subsetneq A$$

of prime ideals. The Krull dimension of A is the maximal length of a prime ideal chain in A.

A prime ideal chain in $\mathcal{O}_{X,0}$ corresponds to an ascending sequence of irreducible subspaces of X

$$X_\ell \subsetneq X_{\ell-1} \subsetneq \cdots \subsetneq X_0 \subset X.$$

The concept of Krull dimension is based on the idea that through the point $X_d = \{0\}$ one can put a one-dimensional curve X_{d-1}, through this one can put a two-dimensional surface X_{d-2}, etc. In this way one obtains an ascending chain of irreducible subspaces of dimensions $0, 1, 2, \ldots, d$.

In commutative algebra one defines the Krull dimension of an arbitrary ring R quite generally to be the maximal length of a prime ideal chain. The Krull dimension concept is therefore the most general of those presented here.

Theorem 18 :

If (X,x) is an analytic set germ and $A = \mathcal{O}_{X,x}$ is the algebra of germs of holomorphic functions on X, then all these concepts of dimension coincide.

The coincidence of the Chevalley, Krull and Weierstrass dimensions is shown in Grauert-Remmert [G5] §4 ; the coincidence of our concept of dimension with the Weierstrass dimension is established in Gunning-Rossi [G6] (III.C.3, p. 111).

An important tool for the investigation of dimension is the following result of Krull :

Theorem 19 : **(Krull principal ideal theorem)**

Let A be an analytic algebra of dimension d and let $f \in A$. Then $\dim A/f\cdot A \geq d-1$, and if $f \neq 0$ is not a zero divisor in A, then $\dim A/f\cdot A = d-1$.

A proof may be found in Grauert-Remmert [G5], p. 129. The Krull principal ideal theorem means that the intersection of a hypersurface H with a germ X in general has codimension 1 in X. But when H contains a component of X, this need not be so.

For example, if $X \subset \mathbb{C}^2$ is the union of the two coordinate axes $X = \{z \in \mathbb{C}^2 \mid z_1 \cdot z_2 = 0\}$, and $f(z) = z_2$, then $A = \mathcal{O}_{X,0} = \mathbb{C}\{z_1,z_2\}/(z_1\cdot z_2)$, i.e. $\dim A = 1$, but also $\dim A/f\cdot A = \dim \mathbb{C}\{z_1\} = 1$.

Corollary : If A is an analytic algebra of dimension d, and if $f_1,\ldots,f_k \in A$, then $\dim A/(f_1,\ldots,f_k) \geq d-k$.

Remark : In order to describe an analytic set germ $(X,0) \subset (\mathbb{C}^n,0)$ of codimension k, one therefore needs at least k functions. Set germs for which k functions really suffice are called complete intersections. Many methods of investigating hypersurface singularities may be generalised to complete intersections. (See Grauert-Remmert [G5], p. 114). Complete intersections are, in a certain sense, particularly simple singularities, since they may be investigated particularly easily. However, many interesting singularities are not complete intersections, and it is easy to give examples. Set germs of pure codimension 1 are always complete intersections.

Proposition 20 :

An analytic set germ $(X,x) \subset (\mathbb{C}^n,x)$ is pure 1-codimensional just in case (X,x) is a hypersurface.

Proof : Without loss of generality, assume X is irreducible. If X is a hypersurface, then $X - S(X)$ is obviously an $(n-1)$-dimensional submanifold, and hence X is 1-codimensional.

Conversely, if X is a 1-codimensional irreducible germ, then we choose a function $f \neq 0$ from the vanishing ideal $\underline{J}(X)$ of X. Let $f = f_1^{k_1} \cdot \ldots \cdot f_r^{k_r}$ be its decomposition into irreducible factors. Since $\underline{J}(X)$ is a prime ideal, one of the factors, say f_1, lies in $\underline{J}(X)$.

We want to show that f_1 generates the ideal $\underline{J}(X)$. Suppose this were not so. Then there would be a $g \in \underline{J}(X) - (f_1)$. This g would define a non-zero divisor in the analytic algebra $A = \mathcal{O}_{\mathbb{C}^n,x}/(f_1)$, since A is an integral domain. The Krull principal ideal theorem would then give : $\dim A/g\cdot A = n-2$. Then, since $f_1, g \in \underline{J}(X)$ it would follow that $\dim \mathcal{O}_{X,x} \leq n-2$, contrary to the assumption that $\dim \mathcal{O}_{X,x} = n-1$. Hence $\underline{J}(X) = (f_1)$.

8.3 Newton polygons and Puiseux expansions

In the previous section we have developed a few general concepts with which we can investigate the zero sets of analytic functions of several variables. In the present we shall in particular make a local investigation of plane curves, i.e. the analytic sets in a domain of $\mathbb{C}^2$ which are the local zero sets of functions $f(x,y)$ of two variables. Since we want to carry out only purely local investigations, i.e. only the germ of the curve at a point, say the origin of $\mathbb{C}^2$, is of interest to us, we shall be interested only in the power series expansion of the function $f(x,y)$ at this point. Thus we begin with a convergent power series $f \in \mathbb{C}\{x,y\}$ with $f(0,0) = 0$.

We now want to describe the zero set of f, i.e. the set of solutions of the equation $f(x,y) = 0$, in a suitable neighbourhood of the origin. We shall see that in a certain sense one can solve this equation quite explicitly. To do this one uses a method which goes back to Newton and was developed by him in correspondence with Leibniz and Oldenburg (letters from Newton to Oldenburg, 13 June 1676 and 24 October 1676, see [N4]). The second letter gives a very detailed account of methods of definition and handling of infinite series, in particular power series. We reproduce here the relevant passages of both letters.

NEWTON TO OLDENBURG

13 JUNE 1676

From the original in the University Library, Cambridge.(1)

Cambridge
June 13 1676

Translation

Most worthy Sir,

Though the modesty of Mr Leibniz, in the extracts(2) from his letter which you have lately sent me, pays great tribute to our countrymen for a certain theory of infinite series, about which there now begins to be some talk, yet I have no doubt that he has discovered not only a method for reducing any quantities whatever to such series, as he asserts, but also various shortened forms, perhaps like our own, if not even better. Since, however, he very much wants to know what has been discovered in this subject by the English, and since I myself fell upon this theory some years ago, I have sent you some of those things which occurred to me in order to satisfy his wishes, at any rate in part.

...

The extractions of affected roots, of equations with several literal terms,(7) resemble in form their extractions in numbers, but the method of Vieta and our fellow-countryman Oughtred is less suitable for this purpose. Therefore I have been led to devise another, an example of which the following diagrams(8) display, where the right-hand column exhibits the results of substituting in the middle column the values of y, p, q, r, etc. shown in the left-hand column. The first diagram displays the solution of this numerical equation, $y^3-2y-5=0$; and here at the top of the column the negative part of the root, subtracted from the positive part, gives the actual root 2·09455148; and the second diagram displays the solution of this literal equation,

$$y^3+axy+a^2y-x^3-2a^3=0.$$

		$\left(a-\frac{x}{4}+\frac{xx}{64a}+\frac{131x^3}{512aa}+\frac{509x^4}{16384a^3}\right.$ &c
$a+p=y$	y^3	$a^3+3aap+3app+p^3$
	$+axy$	$+aax+axp$
	$+aay$	$+a^3+aap$
	$-x^3$	$-x^3$
	$-2a^3$	$-2a^3$
$-\frac{1}{4}x+q=p$	p^3	$-\frac{1}{64}x^3+\frac{3}{16}xxq$ &c
	$+3app$	$+\frac{3}{16}axx-\frac{3}{2}axq+3aqq$
	$+axp$	$-\frac{1}{4}axx+axq$
	$+4aap$	$-aax+4aaq$
	$+aax$	$+aax$
	$-x^3$	$-x^3$
$+\frac{xx}{64a}+r=q$	$3aqq$	$+\frac{3x^4}{4096a}$ &c
	$+\frac{3}{16}xxq$	$+\frac{3x^4}{1024a}$ &c
	$-\frac{1}{2}axq$	$-\frac{1}{128}x^3-\frac{1}{2}axr$
	$+4aaq$	$+\frac{1}{16}axx+4aar$
	$-\frac{65}{64}x^3$	$-\frac{65}{64}x^3$
	$-\frac{1}{16}axx$	$-\frac{1}{16}axx$

$$\left.+4aa-\tfrac{1}{2}ax\right)+\frac{131}{128}x^3-\frac{15x^4}{4096a}\left(+\frac{131x^3}{512aa}+\frac{509x^4}{16384a^3}\right..$$

		$\left(\begin{array}{l}+2{,}10000000\\ -0{,}00544852\end{array}\right.$ 2,09455148
$2+p=y$	y^3 $-2y$ -5	$+8+12p+6pp+p^3$ $-4-2p$ -5
	summa	$-1+10p+6pp+p^3$
$+0{,}1+q=p$	$+p^3$ $+6pp$ $+10p$ -1	$+0{,}001+0{,}03q+0{,}3qq+q^3$ $+0{,}06\quad +1{,}2\quad +6$ $+1\qquad +10,$ -1
	summa	$0{,}061\quad +11{,}23q+6{,}3qq+q^3$
$-0{,}0054+r=q$	$+q^3$ $+6{,}3qq$ $+11{,}23q$ $+0{,}061$	$-0{,}0000001+0{,}000r$ &c $+0{,}0001837-0{,}068$ $-0{,}060642\quad +11{,}23$ $+0{,}061$
	summa	$+0{,}0005416+11{,}162r$
$-0{,}00004852+s=r$		

In the first diagram the first term of the value of p, q, r in the first column is found by dividing the first term of the sum given in the line next above by the coefficient of the second term of the same sum, [as -1 by 10, or 0·061 by 11·23, and by changing the sign of the quotient];[10] and the same term is found in almost the same way in the second diagram. Here to be sure the chief difficulty is in finding the first term of the root; a general method[11] by which this is effected I pass over here for the sake of brevity, as also some other things which tidy up the operation. And as there is not time here to explain the ways of abbreviating the process I shall merely say generally that the root of any equation once extracted can be kept as a rule for solving similar equations; and that from several such rules it is usually possible to form a more general rule; and that all roots, whether they be simple or affected, can be extracted in limitless ways, and on that account the simpler of the ways must always be considered.

...

Translation

Cambridge October 24 1676

Most worthy Sir,

I can hardly tell with what pleasure I have read the letters of those very distinguished men Leibniz[2] and Tschirnhaus.[3] Leibniz's method for obtaining convergent series is certainly very elegant, and it would have sufficiently revealed the genius of its author, even if he had written nothing else. But what he has scattered elsewhere throughout his letter is most worthy of his reputation—it leads us also to hope for very great things from him. The variety of ways by which the same goal is approached has given me the greater pleasure, because three methods of arriving at series of that kind had already become known to me, so[4] that I could scarcely expect a new one to be communicated to us. One of mine I have described before; I now add another, namely, that by which I first chanced on these series—for I chanced on them before I knew the divisions[5] and extractions of roots which I now use. And an explanation of this will serve to lay bare, what Leibniz desires from me, the basis of the theorem set forth near the beginning of the former letter.[6] ...

What the celebrated Leibniz wants me to explain I have partly described above. But as to finding the terms p, q, r, in the extraction of an affected root, first I get p thus.[66] Having described the right angle BAC, I divide its sides BA, AC into equal parts, and then draw normals dividing the angular space into equal parallelograms or squares, which I suppose to be designated by the dimensions of two indefinite kinds, say x and y, ascending in order from the end A, as is seen inscribed in fig. 1; where y denotes the root to be extracted and x the other indefinite quantity, from powers of which a series is to be

B

x^4	x^4y	x^4y^2	x^4y^3	x^4y^4
x^3	x^3y	x^3y^2	x^3y^3	x^3y^4
x^2	x^2y	x^2y^2	x^2y^3	x^2y^4
x	xy	xy^2	xy^3	xy^4
0	y	y^2	y^3	y^4

A C

Fig. 1

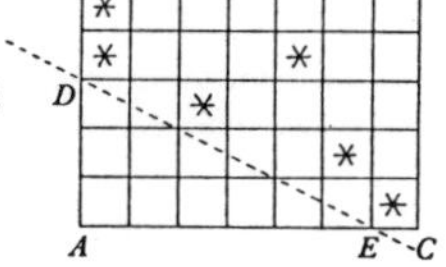

Fig. 2

constructed. Then, when some equation is proposed, I distinguish the parallelograms corresponding to each of its terms with some mark, and apply a ruler to two or perhaps more of the marked parallelograms, of which one is the lowest in the left-hand column next AB, and others situated to the right of the ruler, while all the rest not touching the ruler lie above it. I pick out the terms of the equation distinguished by the parallelograms in contact with the ruler, and thence get the quantity to be added to the quotient.

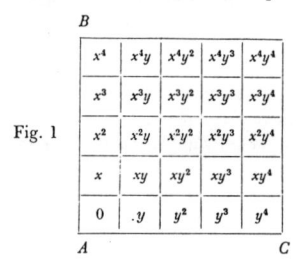

Thus to extract the root y from

$$y^6 - 5xy^5 + (x^3/a)\,y^4 - 7a^2x^2y^2 + 6a^3x^3 + b^2x^4 = 0;$$

the parallelograms answering to the terms of this equation I denote by some mark $*$ as seen in fig. 2. Then I apply the ruler DE to the lower of the places marked in the left-hand column, and rotate it from the lower to the higher to the right till it begins to reach likewise another or perhaps several of the remaining marked places. And I see that the places x^3, x^2y^2 and y^6 are thus reached. And so from the terms $y^6 - 7a^2x^2y^2 + 6a^3x^3$ as though equal to zero (and further reduced if desired to $v^6 - 7v^2 + 6 = 0$ by putting $y = v\sqrt{(ax)}$), I seek the value of y, and find four, namely

$$+\sqrt{(ax)}, \quad -\sqrt{(ax)}, \quad +\sqrt{(2ax)}, \quad -\sqrt{(2ax)},^{(68)}$$

of which any one may be taken as the first term of the quotient, according as it has been decided to extract one or other of the roots. Thus the equation

$$y^3 + axy + a^2y - x^3 - 2a^3 = 0,$$

which I solved in my former letter, gives $-2a^3 + a^2y + y^3 = 0$, and hence $y = a$ very nearly. And so since a is the first term of the value of y, I put p for all the rest to infinity, and substitute $a + p$ for y. Here some difficulties will sometimes arise, but Leibniz I think will need no help to extricate himself from them. But the ensuing terms q, r, s, are obtained, from the second and third equations and the rest, in the same way as the first term p from the first equation, only with less trouble, because the remaining terms of the value of y commonly result from dividing the term involving the lowest power of the variable x by the coefficient of the root p, q, r or s.

Very roughly speaking, Newton begins by viewing $f(x,y) = 0$ as an implicit equation for y as a function of x and computes the solution y by an approximation process which yields an expansion of the solution y in powers of x. When the conditions of the implicit function theorem are satisfied, one knows that the solution y can in fact be represented by a convergent power series in x. But in general one cannot expect this. This is already seen in the following quite trivial example : $f(x,y) = x^p - y^q$.

$$x^p - y^q = 0$$

has the solution $y = x^{p/q}$.

One sees from this that fractional powers of x have to be admitted in order to represent the solution y, at any rate. Somewhat more generally, we can find a solution of the form

$$y = tx^{\mu},$$

where t is a complex number and $\mu = p/q$ is a positive rational number, whenever $f(x,y)$ is a polynomial of the following form :

$$f(x,y) = \sum_{\alpha+\mu\beta=\nu} a_{\alpha\beta} x^{\alpha} y^{\beta}.$$

Such an f is a quasi-homogeneous polynomial, as defined earlier in 4.4. In this case one can find a solution $y = tx^{\mu}$ by substituting $y = tx^{\mu}$ in $f(x,y)$ and obtaining

$$\begin{aligned} f(x,tx^{\mu}) &= \Sigma a_{\alpha\beta} x^{\alpha+\mu\beta} t^{\beta} \\ &= x^{\nu} \Sigma a_{\alpha\beta} t^{\beta} \\ &= x^{\nu} g(t). \end{aligned}$$

If t_0 is a zero of the polynomial $g(t)$, then $y = t_0 x^{\mu}$ is a solution of $f(x,y) = 0$. We can choose $t_0 \neq 0$ just in case $g(t) \neq c \cdot t^m$, i.e. when $f(x,y)$ contains at least two distinct monomials.

The condition that $f(x,y)$ contain only monomials $x^{\alpha}y^{\beta}$ for which $\alpha + \mu\beta = \nu$ can also be described – and this is an essential part of Newton's idea – in the following intuitive geometric way :

Each monomial $x^{\alpha}y^{\beta}$ corresponds to the pair (α,β) of natural numbers, and hence to a point of the lattice $\mathbb{N}^2$ of points (α,β) with integer coordinates in the plane $\mathbb{R}^2$. Now when $f(x,y) = \Sigma a_{\alpha\beta} x^{\alpha} y^{\beta}$ is an arbitrary power series, we consider the set of lattice points (α,β) whose monomials $x^\alpha y^\beta$ really appear in the power series, i.e. those for which the coefficients $a_{\alpha\beta} \neq 0$. We call this set of lattice points the carrier $\Delta(f)$ of f, thus

$$\Delta(f) = \{(\alpha,\beta) \in \mathbb{N}^2 \mid a_{\alpha\beta} \neq 0\}.$$

We first consider two examples.

Example 1 : $f(x,y) = y^4 - 2x^3y^2 + x^6$

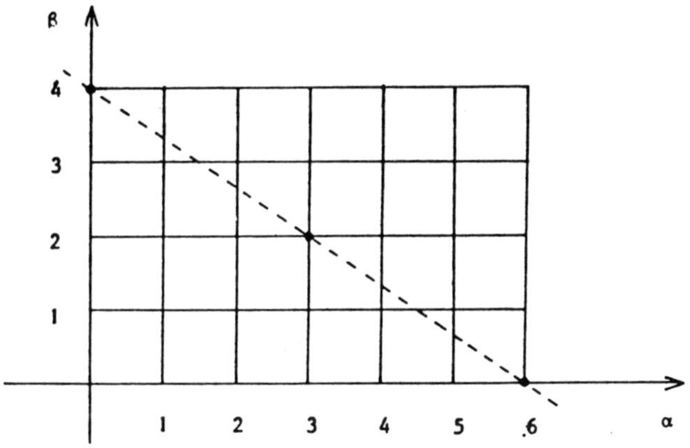

Example 2 : $f(x,y) = y^4 - 2x^3y^2 - 4x^5y + x^6 - x^7$.

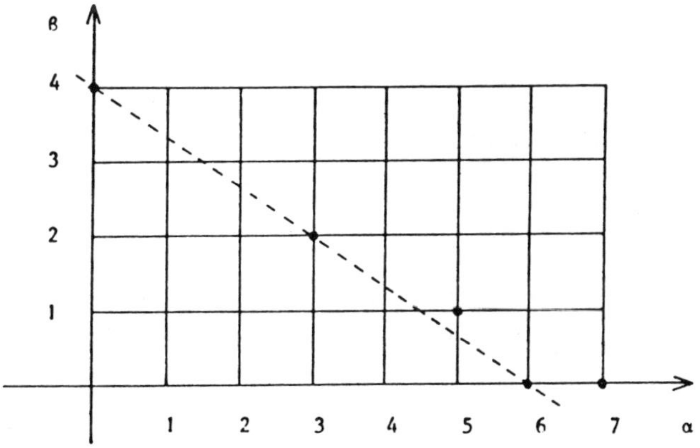

The condition that there be rational numbers μ and ν such that $\alpha + \mu\beta = \nu$ for all $(\alpha,\beta) \in \Delta(f)$ obviously means that all points of $\Delta(f)$ lie on a line. And in fact this line then has slope $-\frac{1}{\mu}$ and it meets the α-axis at $\alpha = \nu$.

<u>**In Example 1**</u> this condition is satisfied with $\mu = \frac{3}{2}$, $\nu = 6$. We can therefore find a solution of the form $y = tx^{3/2}$. Substitution in f yields

$$f(x,tx^{3/2}) = x^6(t^4 - 2t^2 + 1).$$

The zeroes of $g(t) = t^4 - 2t^2 + 1 = (t^2-1)^2$ are $t = \pm 1$. Thus we obtain two solutions of our equation, $y = x^{3/2}$ and $y = -x^{3/2}$.

<u>**In Example 2**</u> the points of $\Delta(f)$ do not all lie on a line and we therefore cannot find a solution as simply as in Example 1. However, we can view the function f in Example 2 as a sum

$$f = \tilde{f} + h,$$

where $\tilde{f}$ is the function $f(x,y) = y^4 - 2x^3y^2 + x^6$ of Example 1 and $h(x,y) = -4x^5y - x^7$ is a higher order perturbation term. Here, the order of a monomial $x^\alpha y^\beta$ is not $\alpha + \beta$, as usual, but $\alpha + \mu\beta$. With respect to this definition of order, all monomials of $\tilde{f}$ in our example have order 6, and the terms in h have higher order, namely x^5y has order $6\frac{1}{2}$ and x^7 has order 7.

Now when we simply substitute the solution $y = x^{3/2}$ of $\tilde{f}(x,y) = 0$ in $f(x,y)$ then f does not vanish, but at least the terms of lower order do. This means that we can regard the solution $y = x^{3/2}$ of $\tilde{f}(x,y) = 0$ as an approximate solution to $f(x,y) = 0$, and hope that the true solution differs from the approximate solution only by terms of higher order. We express the true solution as

$$\begin{aligned} y &= x^{3/2} + \text{terms of higher order} \\ &= x^{3/2}(1+y_1). \end{aligned}$$

In order to avoid calculating with fractional powers of x from now on, we also make the substitution $x^{\frac{1}{2}} = x_1$.

Altogether, we substitute in $f(x,y)$

$$y = x_1^3(1+y_1)$$
$$x = x_1^2.$$

Then the result is

$$f(x_1^2, x_1^3(1+y_1)) = x_1^{12} \cdot f_1(x_1,y_1), \text{ where}$$

$$f_1(x_1,y_1) = y_1^4 + 4y_1^3 + 4y_1^2 - 4x_1y_1 - 4x_1 - x_1^2.$$

Again we consider the carrier, $\Delta(f_1)$:

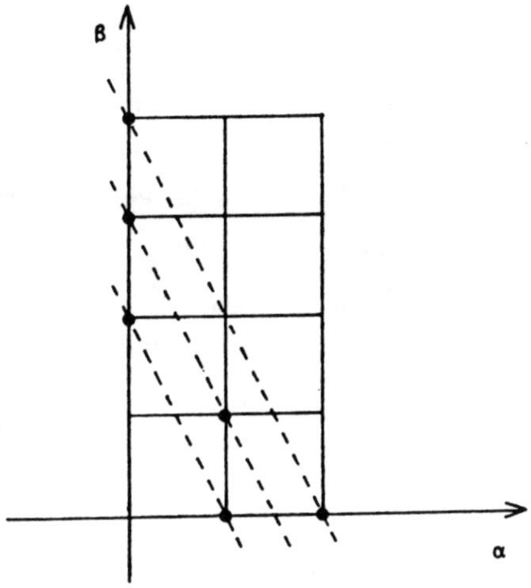

We see that $\Delta(f_1)$ does not lie on one line, however it lies on a system of parallel lines of slope -2. For this reason, we seek a solution of $f_1(x_1,y_1) = 0$ of the form

$$y_1 = t_1 x_1^{\frac{1}{2}} .$$

Substitution in the terms of lower order yields

$$4y_1^2 - 4x_1 = x_1(4t_1^2-4) = 0,$$

hence $t_1 = \pm 1$. If we choose $t_1 = 1$ and substitute $y_1 = x_1^{\frac{1}{2}}$ in $f_1(x_1,y_1)$, then we obtain

$$f_1(x_1,x_1^{\frac{1}{2}}) = x_1^2 + 4x_1^{3/2} + 4x_1 - 4x_1^{3/2} - 4x_1 - x_1^2 = 0.$$

Thus $y_1 = x_1^{\frac{1}{2}}$ is a solution of $f_1(x_1,y_1) = 0$. Because of this,

$$y = x^{3/2}(1+y_1) = x^{3/2}(1+x_1^{\frac{1}{2}}) = x^{3/2}(1+x^{\frac{1}{4}})$$

is a solution of $f(x,y) = 0$. Thus with

$$y = x^{3/2} + x^{7/4}$$

we have found a solution of the original equation $f(x,y) = 0$!

After these examples, we now describe the general Newton process :

Let $f(x,y) = \Sigma a_{\alpha\beta} x^{\alpha} y^{\beta}$ be a convergent power series and, without loss of generality, let f be y-general (say of order $m > 0$, i.e. $a_{0m} \neq 0$ and $a_{0i} = 0$ for $i < m$). As above,

$$\Delta(f) := \{(\alpha,\beta) \in \mathbb{N}^2 \mid a_{\alpha\beta} \neq 0\}$$

is the carrier of f.

We now want to find the line at which we begin the process. It is the "steepest possible line" through the lowest point of the carrier on the β - axis. In order to make this precise, we introduce a few concepts, which we shall also need later.

For each point p of the carrier of f we consider the positive quadrant $p + (\mathbb{R}^+)^2$ moved up to p. From the union of all these

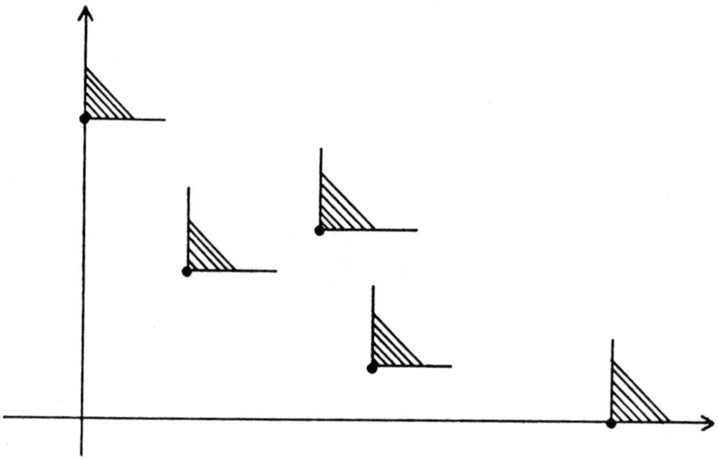

displaced quadrants we construct the convex hull

$$\operatorname{conv}\Big(\bigcup_{p \in \Delta(f)} (p+(\mathbb{R}^+)^2)\Big).$$

The boundary consists of a compact polygonal path (where all the segments have negative slope) and two half lines. This compact polygonal path is called the <u>Newton polygon</u> of f.

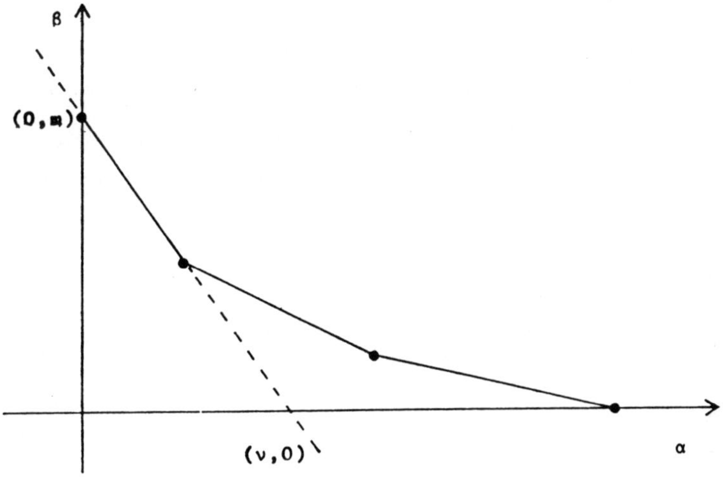

If the Newton polygon is a single point

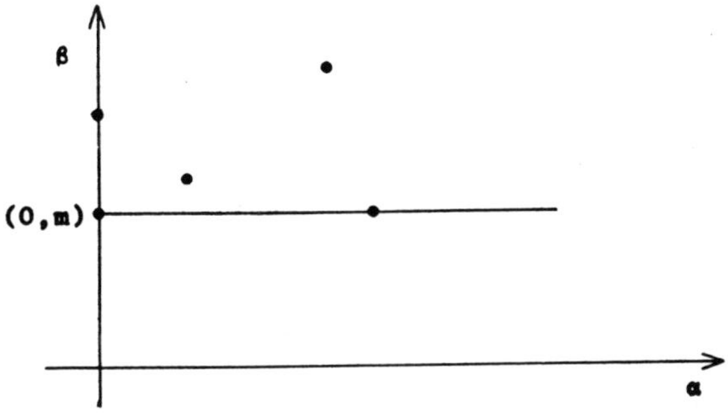

then $f = y^m \cdot \tilde{g}(x,y)$ with $\tilde{g}(0,0) \neq 0$. In this case $y \equiv 0$ is a solution of $f(x,y) = 0$.

Otherwise, the steepest segment of the Newton polygon is the starting line which we want. We therefore let $-\frac{1}{\mu_0}$ be the slope of the steepest segment of the Newton polygon. Then we can partition f into separate parts according to the weight given by μ_0 :

$$f = \sum_{\alpha+\mu_0\beta=\nu} a_{\alpha\beta}x^{\alpha}y^{\beta} + \sum_{\alpha+\mu_0\beta>\nu} a_{\alpha\beta}x^{\alpha}y^{\beta} \tag{1}$$

Here ν is the intercept on the α-axis of the line through $(0,m)$ with slope $-\frac{1}{\mu_0}$ (hence $\nu = m\mu_0$), and by construction there are at least two number pairs $(\alpha,\beta) \in \Delta(f)$ with $\alpha + \mu_0\beta = \nu$.

The first approximate solution of $f(x,y) = 0$ that we construct is a solution of the quasi-homogeneous equation

$$\tilde{f}(x,y) = \sum_{\alpha+\mu_0\beta=\nu} a_{\alpha\beta}x^{\alpha}y^{\beta} = 0.$$

We substitute $y = tx^{\mu_0}$ and obtain :

$$\tilde{f}(x,y) = x^{\nu}\left(\sum_{\alpha+\mu_0\beta=\mu_0 m} a_{\alpha\beta}t^{\beta}\right) = x^{\nu}g(t).$$

Here g is a polynomial of degree m. Since at least two $a_{\alpha\beta}$ are nonzero, the polynomial

$$g(t) = \sum_{\alpha+\mu_0\beta=\mu_0 m} a_{\alpha\beta}t^{\beta}$$

has a nonzero root t_0. Then

$$y_0 = t_0x^{\mu_0}$$

is the first approximate solution of our implicit equation $f(x,y) = 0$. The exponent μ_0 is rational, say

$$\mu_0 = \frac{p_0}{q_0}$$

where p_0, q_0 are relatively prime natural numbers.

We substitute $x_1 := x^{1/q_0}$, so that the first approximate solution is $y_0 = t_0x_1^{p_0}$. In order to improve this approximate solution, we make a fresh start with

$$y = x_1^{p_0}(t_0+y_1)$$

and substitute this throughout the equation $f(x,y) = 0$. This gives a new power series in x_1 and y_1 :

$$f(x_1^{q_0}, x_1^{p_0}(t_0+y_1)).$$

By (1), $x_1^{\nu q_0}$ divides this power series, i.e.

$$f(x_1^{q_0}, x_1^{p_0}(t_0+y_1)) = x_1^{\nu p_0} \cdot f_1(x_1,y_1)$$

and f_1 is again y_1-general of an order $m_1 \leq m$. Now we iterate the process just described :

We construct the Newton polygon of f_1 , let $-\frac{1}{\mu_1}$ be its steepest negative slope $(\mu_1 = \frac{p_1}{q_1})$, obtain an approximate solution $y_1 = t_1 x_1^{\mu_1}$, substitute $x_2 := x_1^{1/q_1}$, again make an improved start $y_1 = x_2^{p_1}(t_1+y_2)$ which is substituted throughout $f_1(x_1,y_1) = 0$ and pull out all possible powers of x_2 :

$$f_1(x_2^{q_1}, x_2^{p_1}(t_1+y_2)) = x_2^{\nu_1 q_1} f_2(x_2,y_2)$$

with a y_2-general f_2 of order $m_2 \leq m_1$, etc.

Altogether, we obtain a sequence of convergent power series $f_i(x_i,y_i)$ (with $x_{i+1} = x_i^{1/q_i}$), each y_i-general of order m_i where

$$m = m_0 \geq m_1 \geq m_2 \geq \dots ,$$

and a sequence of approximate solutions

$$y = x^{\mu_0}(t_0+y_1)$$

$$y_1 = x_1^{\mu_1}(t_1+y_2)$$

$$y_2 = x_2^{\mu_2}(t_2+y_3)$$

$$\vdots$$

Thus the result is that

$$\begin{aligned} y &= x^{\mu_0}(t_0+x_1^{\mu_1}(t_1+x_2^{\mu_2}(t_2+\dots))) \\ &= t_0 x^{\mu_0} + t_1 x^{\mu_0} x_1^{\mu_1} + t_2 x^{\mu_0} x_1^{\mu_1} x_2^{\mu_2} + \dots \qquad (2) \\ &= t_0 x^{\mu_0} + t_1 x^{\mu_0+\mu_1/q_0} + t_2 x^{\mu_0+\mu_1/q_0+\mu_2/q_0 q_1} + \dots \end{aligned}$$

- an expansion of y as a series in ascending fractional powers of x.

We now want to show that this series is meaningful in the sense that the denominators in the exponents do not increase indefinitely. Apart from this, the convergence of the series still remains to be proved later.

In certain cases the process breaks off with $y_i = 0$ for some i (e.g. when the Newton polygon of f_i is a single point), and then the series (2) is a polynomial in $x^{1/n}$, for some n, which satisfies the equation $f(x,y) = 0$.

For the general case we show the following :

Assertion :

There is an index i_0 such that μ_i is always an integer for $i \geq i_0$.

Then $q_i = 1$ and $x_{i+1} = x_i$ for $i \geq i_0$, hence $n := q_0 q_1 \ldots q_{i_0}$ is a common denominator for all exponents in the series (2). In other words, y may be represented as a power series in $x^{1/n}$.

Proof :

The series f_i are y_i-general of order m_i, and the m_i form a descending sequence of natural numbers :

$$m_0 \geq m_1 \geq m_2 \geq \ldots$$

We shall show in a moment that μ_i fails to be an integer only when $m_i > m_{i+1}$. This happens only finitely often, hence from a certain i_0 onwards all the μ_i are integers.

Now it remains to show that :

If $m_i = m_{i+1}$, then $\mu_i \in \mathbb{N}$.

Proof : (without loss of generality, for the case $i = 0$)

Making the substitution $x = x_1^{q_0}$, $y = x_1^{p_0}(t_0+y_1)$ in the equation (1), one obtains

$$x_1^{\nu q_0} f_1(x_1,y_1) = f(x_1^{q_0}, x_1^{p_0}(t_0+y_1))$$

$$= x_1^{\nu q_0}\Big(\sum_{\alpha+\mu_0\beta=\mu_0 m} a_{\alpha\beta}(t_0+y_1)^{\beta} + x_1(\ldots)\Big)$$

Hence it follows that

$$f_1(0,y_1) = \sum_{\alpha+\mu_0\beta=\mu_0 m} a_{\alpha\beta}(t_0+y_1)^{\beta} = g(t_0+y_1).$$

Here t_0 is a nonzero solution of the equation $g(t) = 0$, and g is a polynomial of degree $m = m_0$. The number m_1 is just the order of the zero $y_1 = 0$ in $f_1(0,y_1)$, and hence the order of the zero t_0 of g.

If – as we have assumed – $m_1 = m_0 = m$, then g has the form

$$g(t) = c(t-t_0)^m \qquad (c \neq 0).$$

In particular, the coefficient of t^{m-1} in the polynomial

$$g(t) = \sum_{\alpha+\mu_0\beta=\mu_0 m} a_{\alpha\beta} t^{\beta}$$

does not vanish, i.e. $a_{\alpha,m-1} \neq 0$ for some $\alpha \in \mathbb{N}$ with $\alpha + \mu_0(m-1) = \mu_0 m$. It follows that $\mu_0 = \alpha \in \mathbb{N}$, as was to be shown.

Thus we have now expanded y as a formal power series in $x^{1/n}$.

$$y(x) = \Sigma a_i x^{i/n}.$$

By construction, $f(x,y(x))$ vanishes to arbitrarily high order. Thus $f(x,y(x)) \equiv 0$ and $y(x)$ is a formal solution of the equation $f(x,y) = 0$.

Definition :

The series $y(x) = \Sigma a_i x^{i/n}$ is a Puiseux expansion for the curve with equation $f(x,y) = 0$.

The series of course depends on the choice of zeroes t_i. Nevertheless, we shall soon see that at irreducible singularities it is legitimate to speak of "the" Puiseux expansion. The naming of the series after Puiseux rather than Newton is based upon the fact that Puiseux investigated this series expansion more thoroughly in his work, as we shall see later.

If one does not want to work with fractional exponents, i.e. with many-valued functions, then one can set $z = x^{1/n}$. This substitution converts the Puiseux expansion into a power series $y(z) = \Sigma a_i z^i$ which solves the implicit equation $f(z^n,y(z)) = 0$. In this way one can, as we often shall in what follows, work as usual with formal and convergent power series.

Newton said nothing about the convergence of the series in the letters we have cited. A convergence proof by direct estimation of the coefficients obtained by the process would probably be awkward; we therefore take another approach. We first prove the existence of a convergent solution. One can immediately derive the existence of a convergent solution from the existence of a formal solution just proved by the Artin approximation theorem (cf. 8.1). However, it is not necessary to use this deep theorem ; for this simple problem our knowledge of analytic sets suffices. When $\varepsilon, \delta > 0$ we let $U_{\varepsilon,\delta}$ denote the polydisc

$$U_{\varepsilon,\delta} := \{(x,y) \in \mathbb{C}^2 \mid |y| < \varepsilon, |x| < \delta\}.$$

Theorem 1 :

Let $f(x,y) \in \mathbb{C}\{x,y\}$ be y-general of order $m > 0$ and irreducible. Then there is an $\varepsilon_0 > 0$ such that for each $0 < \varepsilon < \varepsilon_0$ there is a $\delta > 0$ with the following properties :

If

$$X := \{(x,y) \in U_{\varepsilon,\delta} \mid f(x,y) = 0\}$$

is the zero set of f, then there is a convergent power series $y(z) \in \mathbb{C}\{z\}$ for which the mapping $\pi : B \to \mathbb{C}^2$ from the disc $B := \{z \in \mathbb{C} \mid |z| < \delta^{1/m}\}$ to $\mathbb{C}^2$ with $\pi(z) = (z^m, y(z))$ is holomorphic and onto X, i.e.

$$\pi : B \twoheadrightarrow X.$$

The restriction $\pi : B - \{0\} \to X - \{0\}$ is biholomorphic and $\pi^{-1}(0) = \{0\}$.

Proof : Without loss of generality let f be a Weierstrass polynomial

$$f(x,y) = y^m + c_1(x)y^{m-1} + \ldots + c_m(x).$$

We have seen earlier (say, in the proof of 8.2.7) that for sufficiently small ε there is a δ such that for $x_0 \in D_\delta = \{x \in \mathbb{C} \mid |x| < \delta\}$ the polynomial $f(x_0,y)$ has exactly m zeroes (counting multiplicity) in the polydisc $U_{\varepsilon,\delta}$.

Since f is irreducible, f and $\frac{\partial f}{\partial y}$ are relatively prime as Weierstrass polynomials and by Lemma 8.2.5 they are also relatively prime in $\mathbb{C}\{x,y\}$. Thus $\frac{\partial f}{\partial y}$ is not a zero divisor in $\mathcal{O}_{X,0}$ and, by the Krull principal ideal theorem 8.2.19, the set of common zeroes of f and $\frac{\partial f}{\partial y}$ is zero-dimensional, and hence consists of isolated points.

Thus if δ is sufficiently small and if $x_0 \in D_\delta - \{0\}$, then $\frac{\partial f}{\partial y}$ is nonzero on the zero set of $f(x_0,y)$; i.e. the equation $f(x_0,y) = 0$ has m simple zeroes. If y_0 is such a zero then the implicit function theorem gives a holomorphic function $\bar{y}(x)$ and a neighbourhood in which it locally parametrises X. X is therefore locally a manifold around (x_0,y_0) and the projection

$$X \to D_\delta$$
$$(x,y) \mapsto x$$

is biholomorphic locally around (x_0,y_0).

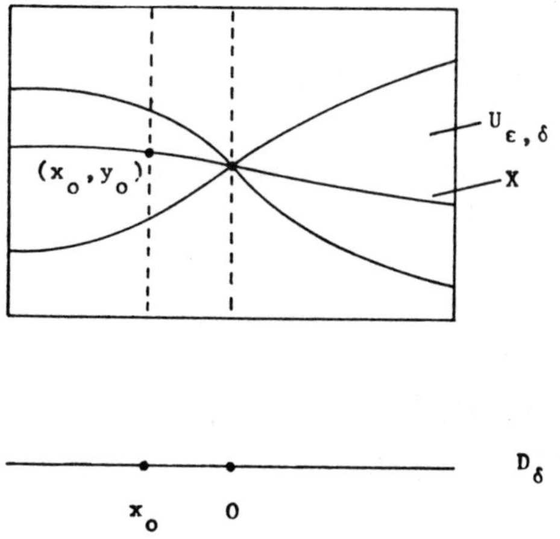

Thus the projection $X - \{0\} \to D_\delta - \{0\}$ is a covering in the topological sense. The only possible singular point of X is 0 ; by remark (i) following 8.2.17, $X - \{0\}$ is connected. But one knows from topology* that such a covering of $D_\delta - \{0\}$ looks like the covering of the punctured disc

$$B - \{0\} \to D_\delta - \{0\}$$

given by $z \mapsto z^m$. More precisely : if one chooses $z_0 \in B - \{0\}$ and $y_0 \in \mathbb{C}$ with $f(z_0^m, y_0) = 0$, then there is a unique homeomorphism

$$\pi : B - \{0\} \to X - \{0\}$$

which sends z_0 to (z_0^m, y_0) and makes the following diagram commute :

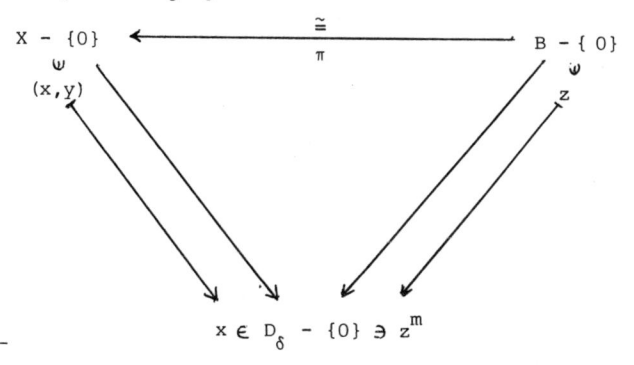

$$\begin{array}{ccc} X - \{0\} \ni (x,y) & \xleftarrow[\pi]{\cong} & B - \{0\} \ni z \\ & \searrow \quad \swarrow & \\ & x \in D_\delta - \{0\} \ni z^m & \end{array}$$

---

*cf., e.g., [S3], §55.

π has the form $\pi(z) = (z^m, y(z))$, and $y(z)$ is a well defined continuous function on $B - \{0\}$. Thus $y(z)$ is holomorphic, because it coincides locally with the composite of the mapping $z \mapsto z^m$ and the solution $\bar{y}$ constructed above by the implicit function theorem. The mapping π is biholomorphic from $B - \{0\}$ to $X - \{0\}$ because the inverse mapping is given locally by $\pi^{-1}(x,y) = \sqrt[m]{x}$, with a suitable branch of the mth root.

Since the zeroes of a polynomial depend continuously on the coefficients, $y(z)$ tends to zero with z (since $y(z)$ is the solution of the equation $f(z^m,y) = 0$). The function $y(z)$ with $y(0) = 0$ is therefore holomorphic in $B - \{0\}$ and continuous at 0, and hence a holomorphic function on B.

$\pi(z) = (z^m, y(z))$ is the desired parametrisation of X.

We can now show the convergence of the series obtained by the Newton process (the Puiseux series) :

Let our $f(x,y) \in \mathbb{C}\{x,y\}$ be, without loss of generality, a Weierstrass polynomial of degree m. If f is irreducible, then Theorem 1 yields a convergent series $y(x^{1/m})$ which satisfies the equation $f(x,y) = 0$. But then the "functions"

$$y(x^{1/m}),\ y(\varepsilon x^{1/m}),\dots,y(\varepsilon^{m-1}\, x^{1/m}) \qquad (\varepsilon = e^{2\pi i/m})$$

also satisfy this equation. Thus we have already found m distinct roots of the Weierstrass polynomial $f(x,y)$ of degree m, and they are all convergent series in $\mathbb{C}\{x^{1/m}\}$. But the series obtained by the Newton process also satisfies the equation $f(x,y) = 0$. As a polynomial of degree m has at most m roots, it coincides with one of the series above. In particular, it is convergent.

In general, f has a decomposition into irreducible factors, $f_1^{r_1},\dots,f_n^{r_n}$. The Puiseux series then satisfies one of the equations $f_i(x,y) = 0$, and hence, by what was said above, it is also convergent in this case. We have seen that each irreducible curve with an equation $f(x,y) = 0$ admits a solution $y = y(x)$ describable by a series expansion in powers of $x^{1/m}$, determined uniquely up to the choice of branch of $x^{1/m}$, where m is the degree of the Weierstrass polynomial f. Equivalent to this description of the solution is the parametrisation $z \mapsto (z^m, y(z))$ of a disc of the z-line in the (x,y)-plane. Conversely, each such parametrisation $z \mapsto (z^m, y(z))$ yields a curve.

Theorem 2 :

Let m be a natural number, and $y(z) \in \mathbb{C}\{z\}$ a convergent power series. The image X of the mapping $z \mapsto (z^m, y(z))$ is then the zero set of the analytic function

$$f(x,y) = \prod_{\varepsilon^m=1} (y - y(\varepsilon x^{1/m}))$$

in a neighbourhood of $0 \in \mathbb{C}^2$.

Proof : On the set $\{(x,y) \in \mathbb{C}^2 \mid x \neq 0\}$, f is obviously, in the neighbourhood of the origin, a well-defined, bounded holomorphic function which may be continuously extended on the y-axis. Thus f is holomorphic on $\mathbb{C}^2$ and X is obviously just the zero set of f.

In Theorem 1, the curve X was parametrised by the mapping $\pi : \mathbb{C} \to X$ of the regular analytic set $\mathbb{C}$ onto the possibly singular set X. We have thereby met the second important method (after stratification in 8.2) for the reduction of singular objects to regular objects - resolution of singularities.

Definition

Let X be an analytic set and let $S(X)$ be the set of its singular points. A resolution of singularities of X is a proper, surjective, holomorphic mapping $\pi : \tilde{X} \to X$ of a complex manifold $\tilde{X}$ onto X such that

$$\pi : \tilde{X} - \pi^{-1}(S(X)) \to X - S(X)$$

is biholomorphic and $\pi^{-1}(S(X))$ is a nowhere dense analytic subset in $\tilde{X}$.

Remarks :

(i) It is important to understand this definition precisely. In this regard, it must first be stressed that a resolution of singularities of X is given not simply by a nonsingular $\tilde{X}$, but only by the mapping $\pi : \tilde{X} \to X$ with the properties listed in the above definition. For example, when one resolves the local irreducible singularities of curves, then for suitable representatives X, like those previously chosen, the manifold $\tilde{X}$ is the same for all curves, namely a disc. The difference between the curve singularities first appears in the mapping π, and hence in the Puiseux expansion.

The idea of resolution of singularities by purely algebraic or algebraic-geometric processes is about 100 years old and goes back to Kronecker and, independently, Noether. (More on the history is in [B2], p. 362-363.) Both use certain birational transformations, including quadratic transformations, which we shall say more about in section 8.4. These quadratic transformations can already be found in the work of Newton, Maclaurin and Braikenridge on the organic generation of curves, and later (1750) in Cramer's investigation of singularities.

"However, what these authors are concerned with is not in fact a theory of reduction of singularities but merely the simplification of curve constructions by simplification of curve equations, so as to recognise their form in the neighbourhood of singular points, be they finite or at infinity." (cf. L. Berzolari: "Algebraische Transformationen und Korrespondenzen". Encyclopädie der mathematischen Wissenschaften, Band IIIC 11, especially section 66.)

For irreducible plane algebraic curve germs, Theorem 1 gives a resolution of singularities by means of the Puiseux expansion. And one can resolve reducible curve germs by resolving the reducible components individually and then taking the "disjoint union" of all these resolutions. From there it is not far to a global resolution of singularities for an algebraic curve. In the next section we shall study the resolution of singularities of plane curves in detail, namely the resolution by quadratic transformations. Resolving the singularities of surfaces is much more difficult; a general process for the resolution of two-dimensional singularities was found by Hirzebruch [H5] 1952. For algebraic varieties of arbitrary dimension, the existence of a resolution of singularities was proved by Hironaka [H3] in the year 1964 ; and after some years he was able to extend it to arbitrary complex spaces. Hironaka's proof is among the most difficult in all mathematics, but the basic idea is again the application of generalisations of quadratic transformations.

In the case of resolution of irreducible singularities of curves the mapping $\pi : B \to X$, $z \mapsto (z^m, y(z))$ is even a homeomorphism. The associated mapping $\pi^* : \mathcal{O}_{X,0} = \mathbb{C}\{x,y\}/(f) \to \mathbb{C}\{z\}$, which associates each function $\phi \in \mathcal{O}_{X,0}$ with the function $\phi \circ \pi$, is therefore injective. However, it is not in general surjective.

<u>Example</u> :

If $f(x,y) = y^2 - x^3$, then we have the parametrisation $x = z^2$,

$y = z^3$ (the parametrisation is $y(x) = x^{3/2}$). For a power series $\gamma(x,y) = \Sigma a_{ij} x^i y^j \in \mathbb{C}\{x,y\}/(f)$ we therefore have $\pi^*(\gamma) = \Sigma a_{ij} z^{2i+3j}$; i.e. the image of π^* is the set

$$\pi^*(\mathcal{O}_{X,0}) = \{\Sigma c_i z^i \in \mathbb{C}\{z\} \mid c_1 = 0\}.$$

We shall now describe the function theoretic difference between B and X more precisely. By the Riemann extension theorem, a function which is continuous on the disc B and holomorphic on B - {0} is already holomorphic on B. However, no analogous assertion holds for the above semicubical parabola X with equation $y^2 - x^3 = 0$. E.g. the function $\phi(x,y) = \frac{y}{x}$ is continuous on $X = \{(x,y) \mid y^2 = x^3\}$ and holomorphic on X - {0}, but not holomorphic on the whole of X. We note, incidentally, that in the above notation $z = \frac{y}{x}$ is precisely the power missing in $\pi^*(\mathcal{O}_{X,0})$, which one must add in order to obtain resolution in the local ring $\mathbb{C}\{z\}$.

One calls an irreducible analytic set germ (X,y) <u>normal</u> when each continuous function on X which is holomorphic on the set X - S(X) of regular points is already holomorphic on X.

If X is an analytic space, then a proper surjective holomorphic mapping $\pi : \tilde{X} \to X$ of a normal analytic space $\tilde{X}$ (i.e. $\tilde{X}$ is normal at each of its points) onto X is called a <u>normalisation</u> of X when π induces a biholomorphic mapping $\pi : \tilde{X} - \pi^{-1}(S(X)) \to X - S(X)$ and $\pi^{-1}(S(X))$ is nowhere dense in $\tilde{X}$. One can show that there is always a unique smallest normalisation of X. This is called <u>the normalisation</u> of X.

In particular, each resolution of singularities is a normalisation: for curves these concepts coincide completely. In the above example of the cuspidal cubic we have indeed seen how the local ring $\mathbb{C}\{z\}$ results from resolution by adjunction of those function germs which are continuous on X and holomorphic on X - {0}. For higher-dimensional spaces it is much simpler to find a normalisation rather than a resolution (Narasimhan [N3], Ch. VI). For example, a hypersurface $X \subset \mathbb{C}^n$ is itself normal when the singularity set S(X) has codimension at least 2 in X.

In Theorem 1 we have resolved the singularity $X = \{(x,y) \mid f(x,y) = 0\}$ by studying the branched covering $X \to D_\delta =: D$, $(x,y) \mapsto x$. The resolution $\pi(z) = (z^m, y(z))$ depends not only on the abstract topological covering $X - \{0\} \to D - \{0\}$, but also quite essentially on the way this covering is "embedded" in $\mathbb{C}^2$.

$$\begin{array}{ccc} & \mathbb{C}^2 & \\ & \cup & \\ (x,y) \in X & \xleftarrow{\pi} & \tilde{X} \ni z \\ & & \\ & \downarrow & \\ & x \in D \ni z^m & \end{array}$$

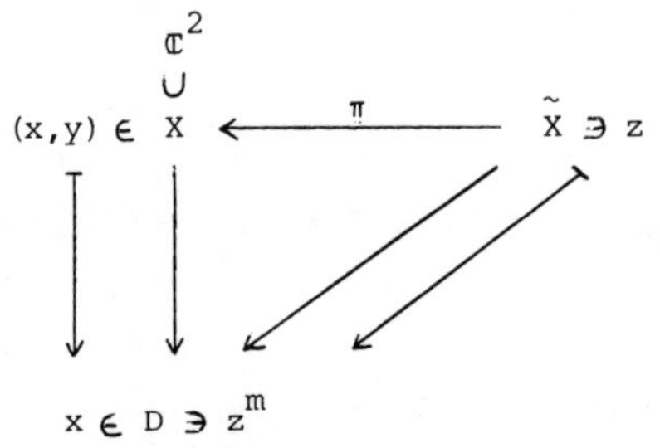

We shall now analyse in more detail how the Puiseux expansion $y(z)$ is related to the covering $X \to D$.

Over each point $x_0 \in D - \{0\}$ lie m different solutions $y_i(x_0) = y(\varepsilon^i x_0^{1/m})$ of the equation $f(x_0,y) = 0$ (where $\varepsilon = e^{2\pi i/m}$). The covering $X - \{0\} \to D - \{0\}$ is unbranched, hence each path $x(t)$

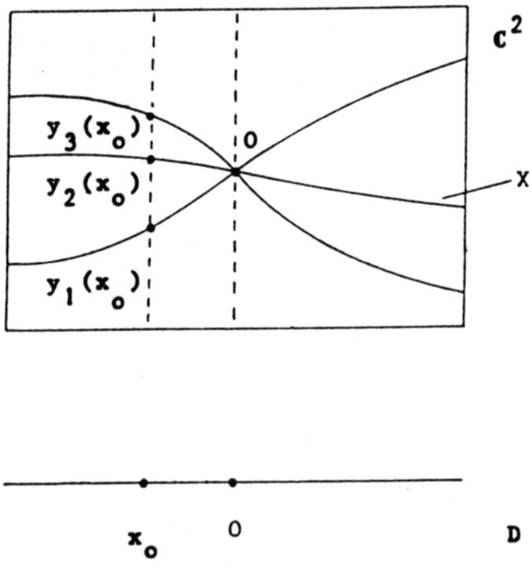

in $D - \{0\}$ $(0 \leq t \leq 1)$ may be lifted to m distinct paths.

$(x(t),y_1(t)), \ldots, (x(t),y_m(t))$

in X. The y_i are then well-defined complex-valued continuous functions, on the interval $[0,1]$, such that $y_i(t) \neq y_j(t)$ for $i \neq j$.

Example : We consider the familiar semicubical parabola (cf. 5.3).

$$X = \{(x,y) \in \mathbb{C}^2 \mid y^2 = x^3\}.$$

The resolution of the singularity is $\pi(z) = (z^2,z^3)$, thus $m = 2$, $y(z) = z^3$. Let the path $x(t)$ be one circuit of the unit circle

$$x(t) = e^{2\pi i t}.$$

The solutions of $f(x(t),y)$ are then

$$y_1(t) = e^{2\pi i(3/2)t}$$

$$y_2(t) = -e^{2\pi i(3/2)t}.$$

The graphs of these two mappings lie on the cylinder

$$[0,1] \times S^1 \subset [0,1] \times \mathbb{C}.$$

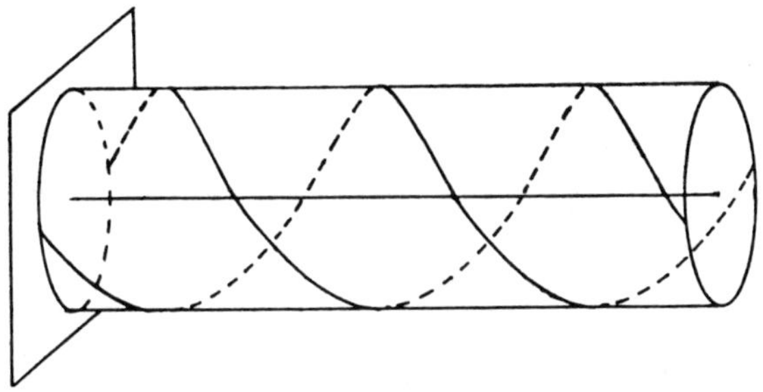

In future we shall represent such graphs by suitable plane projections:

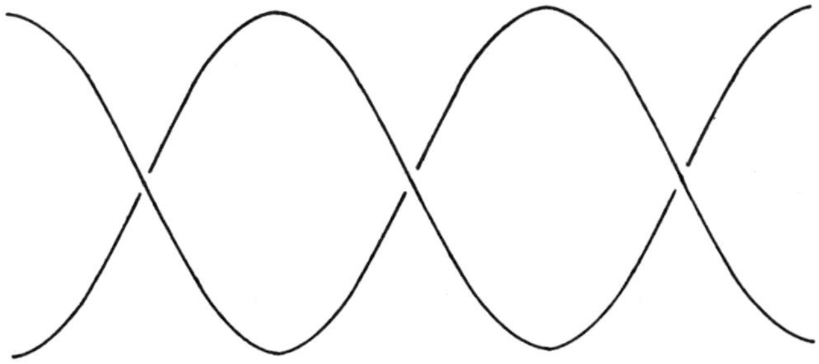

In general, for a closed curve $x(t)$ in $D - \{0\}$ we have m continuous functions $y_i : [0,1] \to \mathbb{C}$ with $y_i(t) \neq y_j(t)$ for $i \neq j$ and $\{y_1(0),\dots,y_m(0)\} = \{y_1(1),\dots,y_m(1)\}$. The graph then looks somewhat as follows :

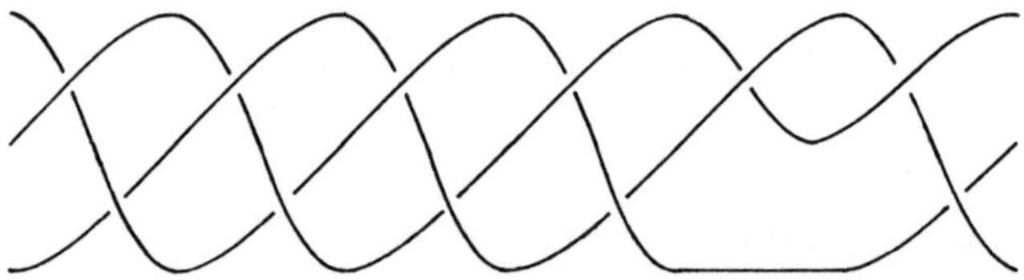

— a braid!

Intuitively, it is perfectly clear what one means by a braid. A braid results when one associates each point in $[0,1]$ with m different complex numbers, in a continuous way, and so that the same set of complex numbers is associated with both 0 and 1. "Pushing the threads" with fixed initial and final point does not amount to anything.

One can make this precise as follows (cf. Birman [B10]) : let the symmetric group S_m act on $\mathbb{C}^m$ by exchanging coordinates. Then

$$Y_m := \{(z_1,\dots,z_m) \in \mathbb{C}^m \mid z_i \neq z_j \text{ for } i \neq j\}/S_m$$

is the set of all unordered m-tuples of distinct complex numbers. (The set Y_m can be identified with the space of all complex polynomials of mth degree with distinct roots, by associating each $\bar{z} = \overline{(z_1,\dots,z_m)} \in Y_m$ with the polynomial $x^m + \sigma_1(z)x^{m-1} + \dots + \sigma_m(z)$, which is the polynomial with roots $z_1,\dots,z_m$. Thus Y_m is homeomorphic, even biholomorphically equivalent, to the complement of the discriminant surface in $\mathbb{C}^m$, which we have considered in 4.2).

Definition :

A braid on m strings (and initial point $\bar{y}$) is a homotopy class of a closed path in Y_m with initial and final point $\bar{y}$, i.e. a braid is an element of $\pi_1(Y_m,\bar{y})$.

$\pi_1(Y_m,\bar{y})$ is a group, and multiplication of braids is simply "joining together".

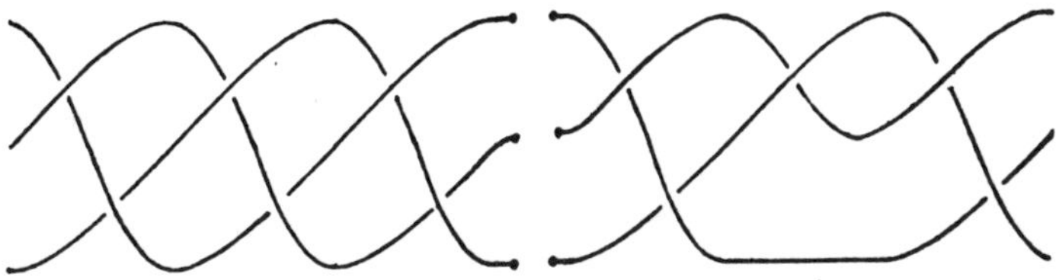

One also calls this group the <u>Artin braid group</u> B(m).

The process for solving the equation $f(x,y) = 0$ by a power series $y = y(x)$ goes back to Newton, as we have said, was further developed by Cramer (1750) and perfected by Puiseux [P3] 1850. In Puiseux one also finds the idea of investigating singularities with the help of braids. Admittedly, topology was not sufficiently developed in the time of Puiseux to enable the concept of braid group to be made precise as above, or in any similar way. (The fundamental group was first developed by Poincaré, nearly fifty years later.) Nevertheless, it seems to me that Puiseux's investigation must be implicitly based on an idea similar to braids. The concept of "braid" was first made explicit by E. Artin [A8] 1925. (Previously it had also been implicit in Hurwitz.)

Braids and braidlike weaving of course appear very early in human culture, at least since the Middle Stone Age. On the following pages we reproduce, from [S11], a description of one very old special weaving technique – tablet weaving – for which the resulting weaves are really braids in the mathematical sense.

Weaving

ON A LOOM

Weaving is the interlacing of two sets of yarns at right angles to each other to form a plane or fabric. One set, the *warp,* is stretched out flat and held taut so that each warp yarn, or *end,* is held with the same degree of tension. A *loom* is a device designed for stretching all the warp ends next to each other—a device that also holds them taut and allows for adjusting their tension.

The other set of yarns, really a single strand, is called the *weft,* or filler. In the simplest weaving, a plain weave, the weft yarn is passed over and under successive warp ends across the width of the warp, and then passed back under and over alternate ends. The sides of the weaving, where the weft turns and passes back across the warp, are called the *selvages.*

Speed and ease in weaving are accomplished by various shedding mechanisms. The alternate warp ends, the ones the filler is to go under, are lifted together, forming a space between them and the warp ends that the weft is to go over. The triangular space formed by separating the warp this way is called a *shed.* Ordinarily, on looms, shedding is achieved with heddles and harnesses; in tablet weaving, the shedding mechanism is the tablets.

When a shed is formed, the weft can be passed through the shed, across the warp, with the aid of a flat tool, called a *shuttle,* that has the filler wrapped around it. This weft "shot," as it is called, is then forced, or *beaten down,* against the body of the weaving.

As the weaving progresses, the shed is changed so that different warp ends are on top of a new shed. Then, with the shuttle, another shot of weft is passed back through the new shed and beaten down.

WITH TABLETS

The tablets used in tablet weaving are made of any rigid material, cut to a regular geometrical shape (most often square), with a hole punched near each corner. The corners of the tablets are rounded to facilitate turning.

A warp end is threaded through each hole of a tablet in one direction, either all up from back to front, or all down from front to back. When an entire warp is threaded this way with an appropriate number of identical tablets, and the tablets are held together in a deck, a shed is formed between the ends threaded through the top holes of the tablets and those threaded through the bottom holes.

Notice that each tablet can be turned around freely. This turning twists the warp ends of one tablet together and changes the shed by raising and lowering the ends in rotation.

Ordinarily, tablet weaving is limited in width by the number of tablets that can be held and turned in your hands, and tablet weaving is most suitable for making long narrow strips of fabric.

Loom-Weaving Terms
a. warp — d. heddle
b. weft — e. shed
c. plain weave — f. shuttle

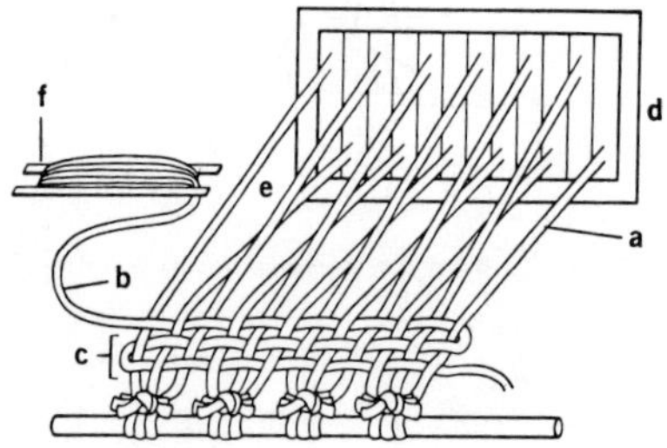

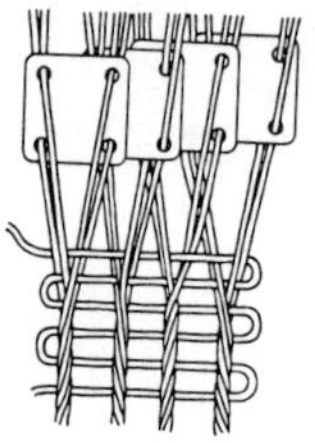

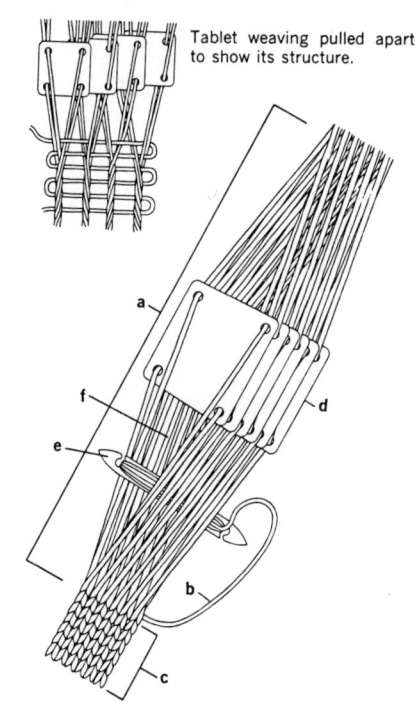

Tablet weaving pulled apart to show its structure.

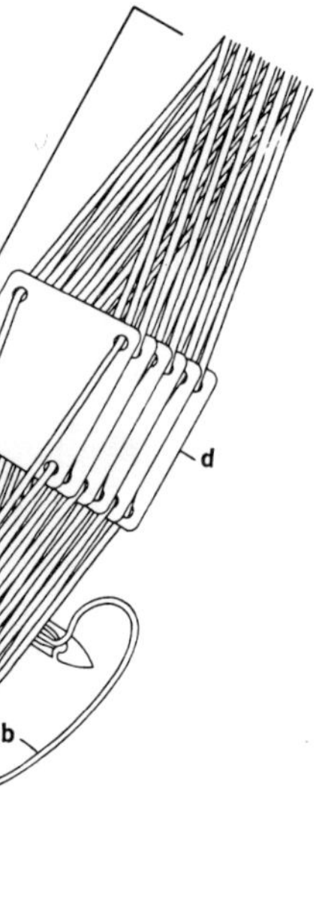

Tablet-Weaving Terms
a. warp — d. tablets
b. weft — e. shed
c. plain tablet weave — f. shuttle

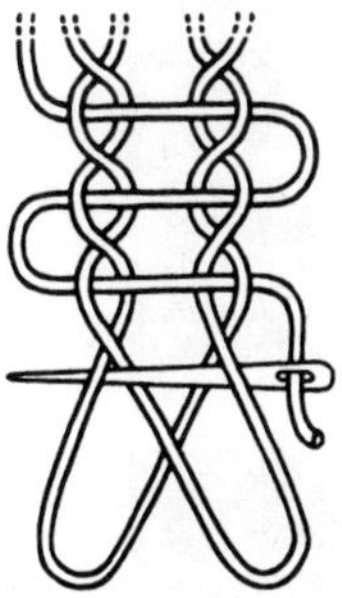

This diagram, after Margrethe Hald, shows how the structure of 2-hole tablet weaving could be reproduced without tablets.

The fabric is woven by passing the filler, or weft, through the shed formed, beating down, and then turning the tablets to form a new shed. The warp ends of each tablet twist around the weft. Hence, tablet-woven bands are *warp-faced* (that is, the weft only shows at the selvages of the band) and *warp-twisted* (that is, the warp ends of one tablet are twisted around each other). This twisted warp makes for an exceptionally strong fabric. One can also weave a fabric that is not warp-twisted, a plain weave, by turning the tablets in a slightly different manner.

A *ply* is the number of yarns twisted together. With 4-holed tablets, a 4-ply fabric is obtained: 1 ply (or yarn) for each threaded hole. This means the tablet-woven band is many plied and very thick.

Pattern in tablet weaving is determined by threading different colored threads through the holes in the tablets, and by the sequence and direction of turning them.

The structure of tablet-woven fabric raises the question of how the method was probably developed. It is possible to produce identical weaves without tablets, twisting two or more warp ends by hand before inserting a weft. The use of tablets with holes to twist the warp while forming a shed appears an ingenious solution to the difficulty of performing both operations by hand. The value of the twist is clear: the twisting together of warp yarns makes for a thicker, stronger fabric, and is related to the process of twisting yarns to make rope.

Types of weaves in the tablet weaving of Quechua Indians from Titicachi, Munecas Province, Bolivia. The patterns from this rich cultural tradition show, among other things, condor, llama, horse, deer and highly stylised tomatoes.

Each closed path $x(t)$ in $D - \{0\}$ yields a braid in the way described above. Since this braid results from the covering $X - \{0\} \to D - \{0\}$, homotopic paths yield the same braid.

Now we must pay some attention to the initial and final points of the braids. If $x(t)$ and $x'(t)$ are closed paths in $D - \{0\}$ with different initial points, then in general the associated braids γ and γ' have different initial points. Now if x and x' are free homotopic (i.e. if there is a homotopy h_t between x and x' such that the paths $h_t(0)$ and $h_t(1)$ coincide), then the associated braids γ and γ' are also free homotopic as paths in Y_m. Intuitively, this means that the threads of γ may be deformed into those of γ', with the initial and final point of each thread being moved in the same way.

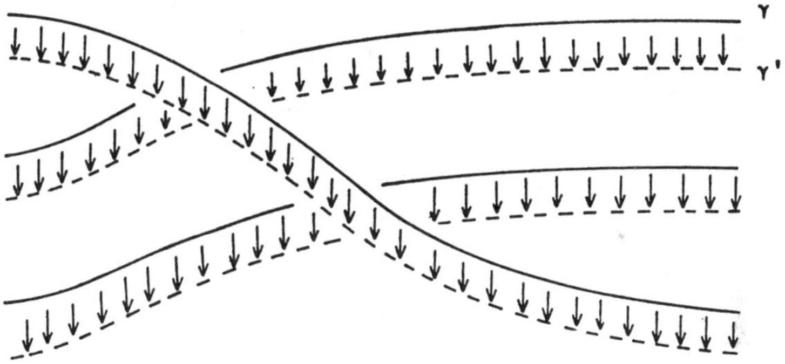

We call two braids (topologically) <u>equivalent</u> when they are free homotopic as paths in Y_m.

Thus we have proved : free homotopic paths in $D - \{0\}$ yield equivalent braids.

Because of this fact, and because the fundamental group of $D - \{0\}$ is generated by a circuit around 0, it suffices to confine ourselves to the standard path

$$x(t) = \delta e^{2\pi i t} \quad (\delta \text{ sufficiently small})$$

when investigating braids. We call this braid "the braid of the singularity". We shall now study how the braid of a singularity is related to the Puiseux expansion $y(x)$. First, a couple of examples.

Example 1 :

Suppose the Puiseux expansion of f is $y = x^{3/2} + x^{7/4}$. When one chooses δ sufficiently small, the term $x^{7/4}$ is small in comparison with $x^{3/2}$, i.e., to a first approximation the points lying over $x = \delta$ are the two points $y = \pm \delta^{3/2}$. And as x rotates around the circle $|x| = \delta$ they turn one and a half times around the origin in $\mathbb{C}$.

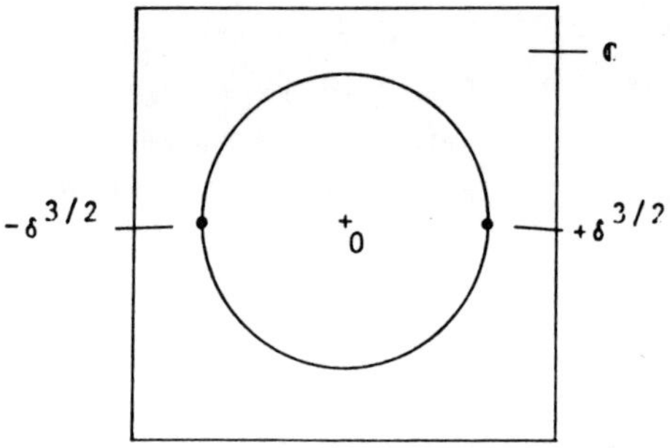

The braid is one we have already seen previously :

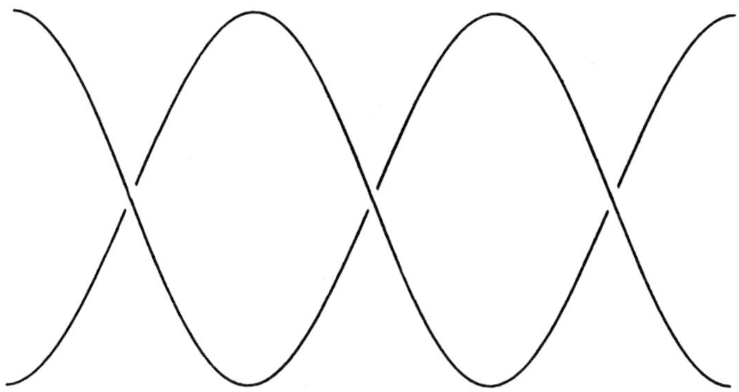

Now we consider the perturbation term $x^{7/4}$. Over the point $x = \delta$ there are now four points $y_1,\ldots,y_4$, grouped in pairs around the points $y = \delta^{3/2}$ and $y = -\delta^{3/2}$ previously considered :

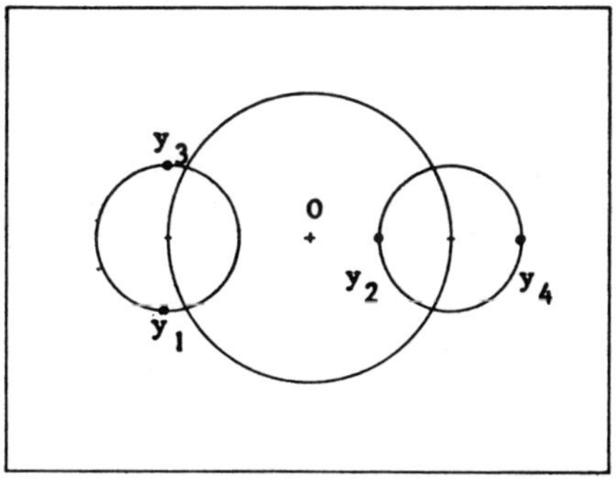

During a circuit of x around the circle $|x| = \delta$, the points $\delta^{3/2}$ and $-\delta^{3/2}$, as before, move one and a half times around the big circle; and the points actually of interest to us, y_1, y_2, y_3, y_4 each rotate one and three quarter times around the strings of the first braid. The resulting braid is as follows.

$x = \delta e^{2\pi i t}$

$y = \eta + i\zeta$

$y = x^{3/2} + x^{7/4}$ Projection of the graph on the (t,η)-plane

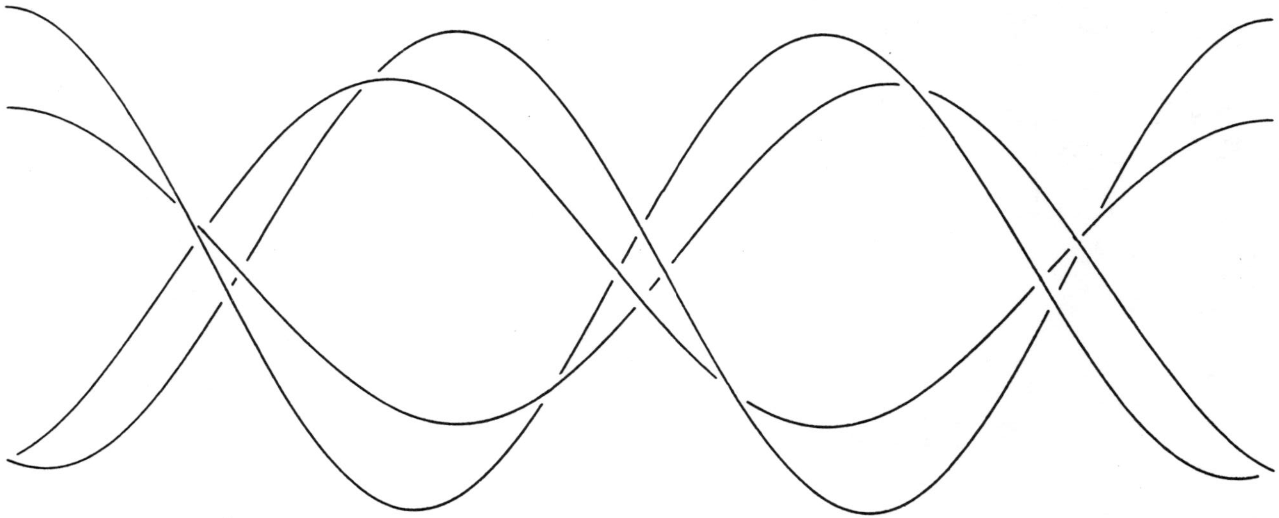

Example 2 :

Suppose the Puiseux expansion of f is $y = x^{3/2} + x^{37/2}$. To a first approximation, we again have $y = x^{3/2}$:

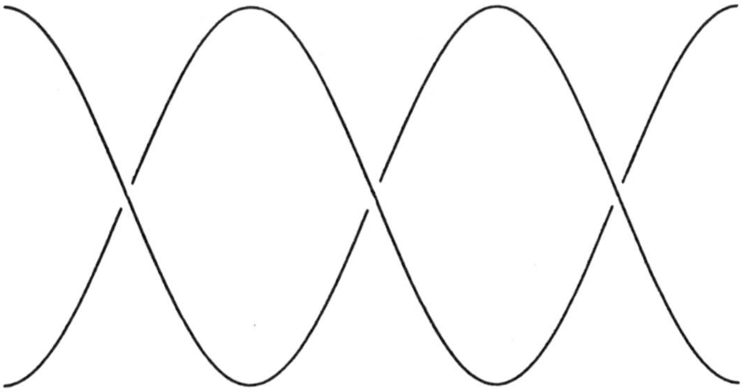

But this time the perturbation term $x^{37/2}$ does not alter the number of strings of the braid. The braid for $x^{3/2} + x^{37/2}$ just oscillates around the first approximation :

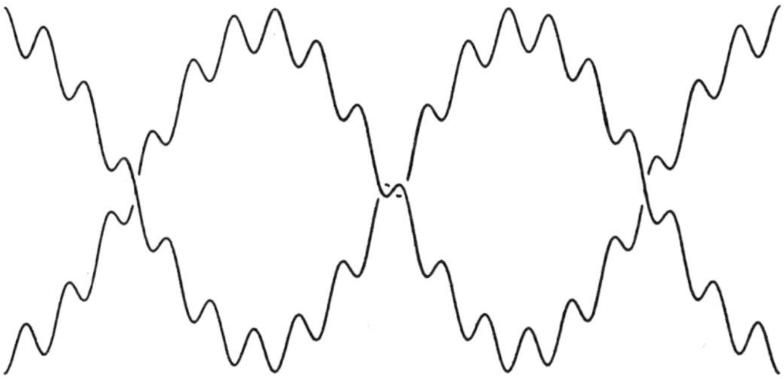

This braid may be smoothed out so as to give the original braid, thus the braids for $x^{3/2}$ and $x^{3/2} + x^{37/2}$ are equivalent.

This example suggests the conjecture that for a function f with the Puiseux expansion $y(x) = \Sigma a_\kappa x^\kappa$ $(a_\kappa \neq 0)$ the braid of f depends only on the terms $a_\kappa x^\kappa$ for which the denominators of the exponents increase. We shall formulate this more precisely shortly.

In order to be able to associate a singularity with a braid which is uniquely determined, up to equivalence, we must choose the coordinates in the plane suitably, because the latter enter essentially into the definition of the projection mapping $(x,y) \mapsto x$ and hence into the definition of the braid. For example, with the singularity $y^2 - x^3 = 0$ we have associated the Puiseux series $y = x^{3/2}$ and the following braid :

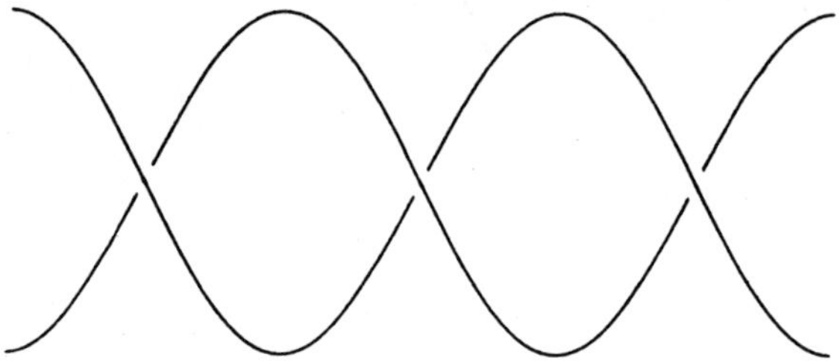

But if we exchange the rôles of the coordinates x and y in the construction, then we obtain the singularity $y^3 - x^2 = 0$ with the Puiseux series $y = x^{2/3}$ and the three-stringed braid

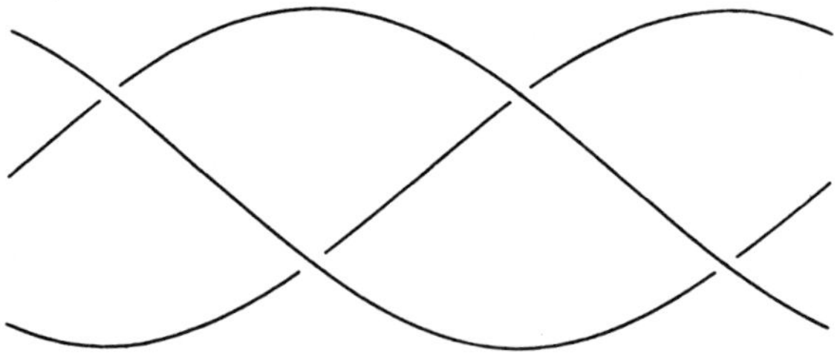

This is because we have obtained the braid by study of the covering $X - \{0\} \to D - \{0\}$, $(x,y) \mapsto x$, and this covering is essentially different in the two cases, as the real picture already shows :

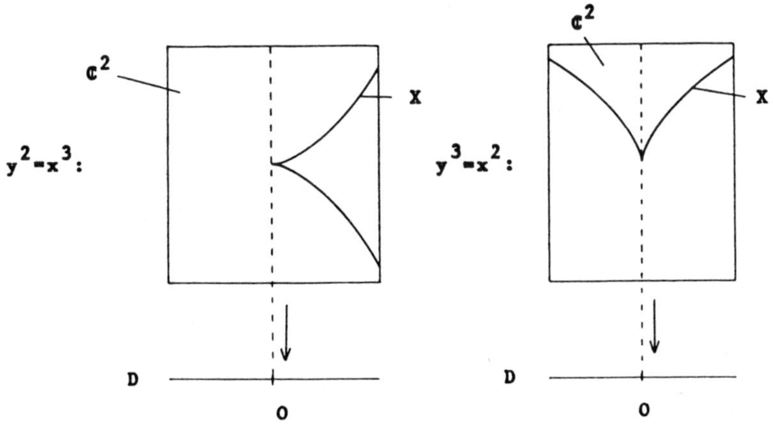

The covering is given by projection along the y-axis; and in the first case the y-axis cuts the singularity with multiplicity 2, in the second case with multiplicity 3. As we have seen in 5.3, this second case is exceptional. We therefore introduce the following convention for the choice of coordinates x, y, and we shall make it the basis for all our future investigation of Puiseux expansions and the associated braids, unless the context demands otherwise.

Let $f(x,y)$ be irreducible. If m is the multiplicity of f at the origin (cf. 5.3), let f be y-general of order m. I.e., suppose the y-axis is not a tangent at the singular point of the curve.

This is equivalent to assuming that the Puiseux expansion of f has the form

$$y = cx^{\alpha} + \text{higher terms} \quad (c \neq 0)$$

with $\alpha \geq 1$.

We now define the Puiseux pairs of f, a sequence of pairs of integers which describes exactly those positions in the Puiseux series where the denominator of the exponent increases, and what the exponents are at those positions.

We write the Puiseux expansion of f in the form

$$y = \Sigma a_{\kappa} x^{\kappa}$$

with $a_\kappa \neq 0$, $\kappa \in \mathbb{Q}$, $\kappa \geq 1$. Then f is regular just in case all κ are integers. In this case no Puiseux pairs are defined.

Otherwise, there is a smallest κ_1 which is not an integer.

$$\kappa_1 = \frac{n_1}{m_1} \quad (n_1 > m_1)$$

with relatively prime m_1, n_1. The number pair (m_1, n_1) is the first Puiseux pair of f. Some of the exponents which follow may be of the form $\frac{q}{m_1}$ $(q > n_1)$, but if not all of them are, we shall come to a κ_2 which is not so representable. We then write κ_2 in the form

$$\kappa_2 = \frac{n_2}{m_1 \cdot m_2}, \quad \gcd(n_2, m_2) = 1,\ m_2 > 1.$$

(If necessary we must multiply the fraction for κ_2 on top and bottom by a divisor of m_1.) The numbers n_2 and m_2 are uniquely determined, and the pair (m_2, n_2) is the second Puiseux pair of f.

In general, if the Puiseux pairs $(m_1,n_1),\ldots,(m_j,n_j)$ are already defined, let κ_{j+1} be the smallest exponent for which the preceding exponents are all expressible in the form

$$\kappa = \frac{q}{m_1 \cdot \ldots \cdot m_j},$$

while κ_{j+1} itself is not. Then let

$$\kappa_{j+1} = \frac{n_{j+1}}{m_1 \cdot \ldots \cdot m_{j+1}} \quad \text{with } \gcd(n_{j+1}, m_{j+1}) = 1.$$

Then (m_{j+1}, n_{j+1}) is the next Puiseux pair. Eventually this process terminates, i.e. there is a g such that $m_1 \ldots m_g = m$ is a common denominator of all exponents in the Puiseux series. Hence we obtain in this way a finite sequence $(m_1,n_1),\ldots,(m_g,n_g)$ of pairs of integers.

Definition :

The pairs $(m_1,n_1),\ldots,(m_g,n_g)$ defined in this way are called the Puiseux pairs of f.

Examples : If $y = x^{3/2} + x^{7/4}$, then the Puiseux pairs are

$$(m_1,n_1) = (2,3), \quad (m_2,n_2) = (2,7).$$

If $y = x^{3/2} + x^{5/3} + x^{37/2} = x^{3/2} + x^{10/2\cdot 3} + x^{111/2\cdot 3}$ then $(m_1,n_1) = (2,3)$, $(m_2,n_2) = (3,10)$.

Since the exponents κ_j are monotonically increasing and greater than 1, the Puiseux pairs satisfy the following conditions :

$$m_1 < n_1$$
$$n_{j-1}m_j < n_j \quad \text{for} \quad j \geq 2 \qquad (1)$$
$$\gcd(n_j, m_j) = 1 \quad \text{for} \quad j = 1,\ldots,g.$$

Conversely, any given sequence of pairs of natural numbers, $(m_1,n_1),\ldots,(m_g,n_g)$, which satisfies the conditions (1) is the sequence of Puiseux pairs of a certain Puiseux expansion, say the "standard expansion"

$$y(x) = x^{n_1/m_1} + x^{n_2/m_1 \cdot m_2} + \ldots + x^{n_g/m_1 \cdots m_g}$$
$$= x^{\kappa_1} + x^{\kappa_2} + \ldots + x^{\kappa_g}.$$

What does the braid of such a standard expansion look like? In this general case we can describe the braid in a way quite similar to that used already for the special example 1. The braid results from letting x run once around a small circle

$$x(t) = \delta e^{2\pi i t}$$

centred on the origin. We arrive at the braid via finitely many approximations. For the first approximation we consider only the term x^{κ_1}. Over the initial point $x = \delta$ of our path we then have the points

$$y_k = \delta^{\kappa_1} e^{2\pi i \kappa_1 . k} \qquad k = 1,\ldots,m_1$$

on a circle with centre 0 and radius δ^{κ_1}.

The second term x^{κ_2} improves the approximation ; now one has sets of m_2 points grouped of circles of radius δ^{κ_2} around the points $y_1,\ldots,y_{m_1}$ found first, etc. The points on the last and smallest circles (radius δ^{κ_g}) are those whose paths we want to investigate, because when x describes its circle once, these paths give us the desired braid.

If δ is chosen sufficiently small, none of these circles interfere with each other. If we let $x(t)$ describe the circle $|x| = \delta$, then all these circles begin to rotate around each other with different velocities. Thus each $y_\kappa(t)$ describes an iterated epicyclic curve, like those we have already met (for $g = 2$) in our report on the epicycles of Hipparchus and Ptolemy (cf. 1.7). Here the angular velocity of the ith circle relative to the complex plane equals κ_i. The points on the smallest circles then describe a braid of $m_1 \cdot \ldots \cdot m_g$

threads, like a cable, resulting from the twisting of m_1 thick strands of $m_2 \cdot \ldots \cdot m_g$ threads, where each of the latter strands results from twisting of m_2 finer strands of $m_3 \cdot \ldots \cdot m_g$ threads, which in turn results from twisting m_3 still finer strands. All of these strands are twisted in the same sense, whereas most cables used in practice involve alternating directions of twist. (Particulars may be found e.g. in The Ashley Book of Knots, from which we have reproduced a few pages here and at the end of this section.)

ON KNOTS

103. *Yarn:* Is a number of fibers twisted together, "right-handed." *Thread:* In ropemaking is the same as yarn.

104. *Sewing thread:* May be two, three, or more small yarns twisted together. *Sailmaker's sewing thread:* Consists of a number of cotton or linen yarns loose-twisted and is often called *sewing twine.*

105. *Strand:* Is two or more yarns or threads twisted together, generally *left*-handed.

106. *Rope:* Is three or more left-handed strands twisted together, right-handed, called plain-laid rope.

107. *Hawser:* Large plain-laid rope generally over 5″ in circumference is called hawser-laid.

108. *Cable* or *cable-laid rope:* Three plain- or hawser-laid ropes laid up together, left-handed; also called water-laid because it was presumed to be less pervious to moisture than plain-laid rope. Four-strand cable has been used for stays.

109. *Four-strand rope:* Right-handed, is used for lanyards, bucket bails, manropes, and sometimes for the running rigging of yachts.

110. *Shroud-laid rope:* Right-handed, four strands with a center core or *heart* (formerly termed a *goke*) was used for standing rigging before the days of wire rope. The heart is of plain-laid rope about half the size of one of the strands.

111. *Six-strand rope:* Right-handed with a heart, very hard-laid, was formerly used for tiller rope. The best was made of hide.

Six-strand "limber rope" was formerly laid along a keel and used to clear the limbers when they became clogged. It was made of horsehair, which resists moisture and decay better than vegetable fiber. Nowadays six-strand rope with wire cores in each strand is made for mooring cable and buoy ropes for small craft.

112. *Backhanded* or *reverse-laid rope:* In this material the yarns and the strands are *both* right-handed. It may be either three- or four-strand and is more pliant than plain-laid rope and less liable to kink when new, but it does not wear so well, is difficult to splice, and takes up moisture readily. Formerly it was used in the Navy for gun tackle and braces. Nowadays (in cotton) it is sometimes used for yacht running rigging. Lang-laid wire rope is somewhat similar in structure.

113. *Left-handed* or *left-laid rope:* The yarns are left-handed, the strands are right-handed, and the rope left-handed, the direct opposite of right-handed rope. Coupled with a right-handed or plain-laid rope of equal size, this is now used in roping seines and nets. The opposite twists compensate, so that wet seines have no tendency to twist and roll up at the edges.

In ropemaking, strictly speaking, yarns are "spun," strands are "formed," ropes are "laid," and cables are "closed," but these terms are often used indiscriminately.

Formerly plain-laid and hawser-laid meant the same thing. Now the term *hawser-laid* refers only to large plain-laid ropes suitable for towing, warping, and mooring.

It is a common mistake of recent years to use the terms *hawser-laid* and *cable-laid* interchangeably. This leaves two totally different products without distinguishing names, and it is no longer certain when either name is applied just what thing is referred to.

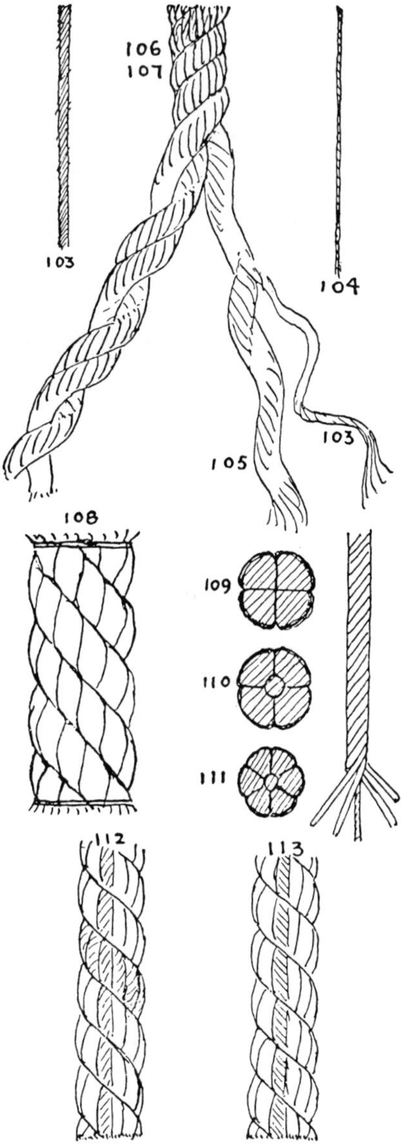

Example : $y = x^{7/6} + x^{37/24} + x^{20/9}$.

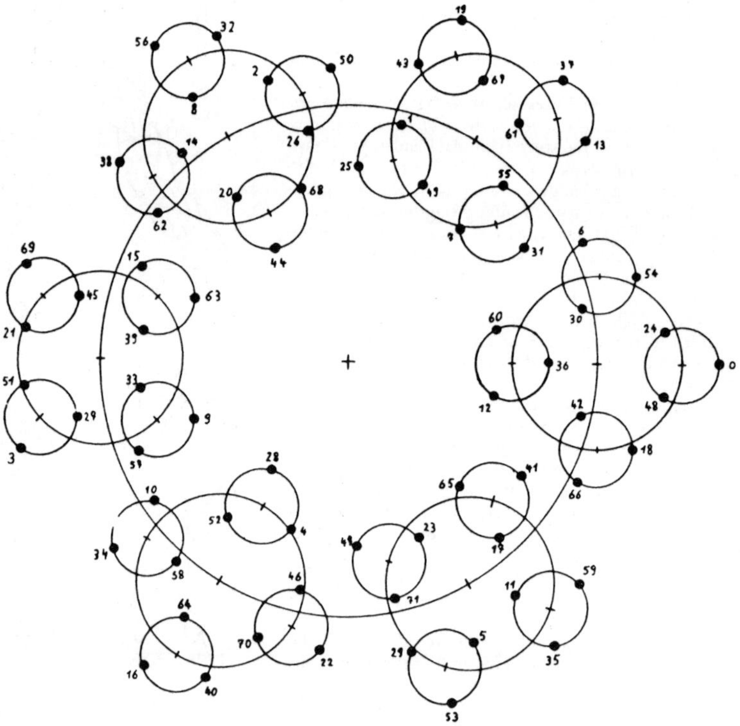

In this example the point with number k goes to the point with number $k+1$ (mod 72) after one rotation.

Thus for standard expansions we can now construct the associated braid. The next theorem shows that the braid of an arbitrary irreducible singularity looks exactly like that of the standard expansion with the same Puiseux pairs.

<u>**Theorem 3**</u> :

Puiseux expansions with the same Puiseux pairs yield equivalent braids.

<u>**Proof**</u> : A complete proof would be lengthy and technical (cf. Pham [Pl]); we want only to give the geometric idea here.

Let $y(x) = \Sigma a_\kappa x^\kappa$ be the Puiseux expansion of f, and let $y_1(x) = \sum_{i=1}^{g} x^{\kappa_i}$ be the standard expansion corresponding to the Puiseux pairs of f.

We split y into

$$y = \tilde{y} + \tilde{g}$$

where $\tilde{y} = \sum_{i=1}^{g} a_{\kappa_i} x^{\kappa_i}$ contains just the terms which correspond to the Puiseux pairs of f, and $\tilde{g}$ is the remainder.

The terms of the Puiseux expansion corresponding to the Puiseux pairs are also called <u>essential</u> (or <u>characteristic</u>) terms of the Puiseux series.

The terms of $\tilde{g}$ lead only to "useless oscillations" of the threads, which can be smoothed out again, in the construction of the braid by the process described above (cf. example 2!). Thus y and $\tilde{y}$ yield equivalent braids.

In the construction of the braids for $\tilde{y}$ and y_1, similar configurations of circles appear, except that the radii are different and the points on the circles are shifted.

However, if δ is sufficiently small, these radii no longer interfere with each other, and the braids for

$$\sum_{j=1}^{g} a_{\kappa_j} x^{\kappa_j} \quad \text{and} \quad \sum_{j=1}^{g} \frac{a_{\kappa_j}}{|a_{\kappa_j}|} x^{\kappa_j}$$

are equivalent (a deformation yields, e.g., the family of braids corresponding to $\sum \frac{a_{\kappa_j}}{1-t+t|a_{\kappa_j}|} x^{\kappa_j}$).

Thus we can suppose, without loss of generality, that all a_{κ_j} have the value 1. Now the radii of the circles are all equal, and only the initial points are somewhat rotated relative to each other. By means of a suitable family of iterated rotations one can then construct a free homotopy between the braids in question, and thus one obtains the complete equivalence between the braids of y and its associated standard expansion y_1.

With Theorem 3 we have qualitatively captured the many-valued behaviour of the many-valued "functions" $y = y(x)$ which we have defined as solutions of the analytic equations $f(x,y) = 0$ and described quantitatively as expansions $y = \Sigma a_\kappa x^\kappa$ in fractional powers x^κ by means of the Puiseux expansion, which goes essentially back to Newton. Theorem 3 says that the numbers $(m_1,n_1),\ldots,(m_g,n_g)$ suffice to qualitatively describe this many-valued behaviour. Thus we have first described the solution "functions" $y(x)$ of $f(x,y) = 0$ quantitatively, then qualitatively, and finally – though more coarsely than the original quantitative description – quantitatively again, namely by the invariants (m_j,n_j).

We shall now find a similar description, at first qualitative and then quantitative, for the zero set X of $f(x,y) = 0$ in the neighbourhood of the origin. Since the plane curve X is indeed parametrised in a neighbourhood of the origin by the many-valued "mapping" $x \mapsto (x,y(x))$, where $y(x)$ is the many-valued "solution function", this qualitative description of the zero set is essentially immediate from that of the solution function, and the principal result again is that the zero set is already uniquely characterised, qualitatively, by the invariants $(m_1,n_1),\ldots,(m_g,n_g)$.

The qualitative description of the zero set X is best carried out by topological means. Indeed, topology was intentionally developed by Poincaré as a kind of qualitative substitute for geometry, because purely quantitative-analytic description became too complicated in many situations in modern mathematics, and the equipment of classical geometry was inadequate for the necessary qualitative-geometric description.

The simplest definition we could make to enable the local qualitative description of analytic sets would perhaps be the following :

Definition :

An analytic set X is homeomorphic at $x \in X$ to an analytic set Y at $y \in Y$ when there is a homeomorphism $(U,x) \approx (V,y)$ for suitable neighbourhoods U of x in X and V of y in Y. Since this property obviously depends only on the germs (X,x) and (Y,y) – in fact only on their isomorphism classes – we then say the germ (X,x)

is <u>homeomorphic</u> to (Y,y).

What does this definition achieve? It certainly allows us to make a coarse qualitative distinction between analytic set germs. E.g. two pure-dimensional set germs of different dimensions could not, of course, be homeomorphic, because of Brouwer's theorem on the invariance of dimension under homeomorphisms. But it also permits set germs of the same dimension to be distinguished. As an example, we consider the n-dimensional hypersurface germ X_a in $\mathbb{C}^{n+1}$ which is given by the equation

$$x_1^{a_1} + \ldots + x_{n+1}^{a_{n+1}} = 0,$$

where $a = (a_1,\ldots,a_{n+1})$ is an n-tuple of natural numbers $a_i > 1$. The hypersurface has an isolated singular point at $0 \in \mathbb{C}^{n+1}$, and we can ask whether our qualitative concept of homeomorphism of set germs can distinguish between this singular point and a nonsingular point, in other words, we ask : is $(X_a,0)$ always non-homeomorphic to $(\mathbb{C}^n,0)$? It turns out that the answer depends on the dimension.

Theorem 4 :

For $n > 2$: $(X_a,0)$ is homeomorphic to $(\mathbb{C}^n,0)$ just in case one of the two following conditions is satisfied :

(i) There are at least two indices i, j such that $(a_i,a_k) = 1$ for $k \neq i$ and $(a_j,a_k) = 1$ for $k \neq j$.

(ii) There is an index i with $(a_i,a_k) = 1$ for $k \neq i$, where a_i is odd, and for even a_j, a_k one has $(a_j,a_k) = 2$ for $j \neq k$, and the number of these even exponents is odd.

For $n = 2$: $(X_a,0)$ is never homeomorphic to $(\mathbb{C}^2,0)$.

For $n = 1$: $(X_a,0)$ is always homeomorphic to $(\mathbb{C},0)$ when $(X_a,0)$ is irreducible, i.e. when $(a_1,a_2) = 1$.

Examples :

(i) For $a = (2,2,2,2,3)$, X_a is not homeomorphic to $\mathbb{C}^4$, but for $a = (2,2,2,3)$, X_a is homeomorphic to $\mathbb{C}^3$.

(ii) For $a = (2,2,2,3,5)$ and $a = (2,2,3,5)$, X_a is homeomorphic to $\mathbb{C}^4$ resp. $\mathbb{C}^3$, but for $a = (2,3,5)$, X_a is not homeomorphic to $\mathbb{C}^2$, but to a cone over the spherical dodecahedral space (cf. [B6]).

The example $x_1^{a_1} + \ldots + x_{n+1}^{a_{n+1}}$ has played an important rôle stimulating progress in the topological investigation of singularities

([B4], [W7]). Mumford [M6] showed in 1961 that a two-dimensional normal analytic set germ which has a proper singularity is never homeomorphic to $(\mathbb{C}^2,0)$. This result led to the conjecture that, in higher dimensions too, the set germ of a singularity could not be homeomorphic to $(\mathbb{C}^n,0)$. The theorem above, and the realisation that one could construct "exotic spheres" with the help of the sets X_a, has greatly stimulated interest in the topological investigation of singularities.

Theorem 4 completely answers the qualitative question we have posed for this example. Its proof for the cases $n > 2$ ([B4]) and $n = 2$ ([M6]) goes beyond the framework of this course. Nevertheless, we have cited the theorem because it shows the scope and limits of our first attempt at the local qualitative description of analytic sets : for sets of dimension $n > 1$ this attempt is very interesting, but for the case $n = 1$, the one we are concerned with in this course, it is uninteresting.

In fact, 1-dimensional analytic set germs satisfy

Proposition 5 :

All irreducible 1-dimensional analytic set germs are homeomorphic to $(\mathbb{C},0)$. Two reducible 1-dimensional set germs are homeomorphic just in case they have the same number of irreducible components.

Proof : We prove the theorem only for irreducible analytic set germs in $\mathbb{C}^2$. In this case it follows from what we have said earlier. In particular, for an irreducible 1-dimensional germ (X,x), the resolution $(\tilde{X},\tilde{x}) \to (X,x)$ which we have described in terms of the Puiseux expansion, yields a homeomorphism between $(\tilde{X},\tilde{x})$ and (X,x). But $(\tilde{X},\tilde{x})$ is simply equal to $(\mathbb{C},0)$.

Thus if we want to capture the local qualitative properties of plane curves, we cannot consider them simply as abstract curves, but as curves embedded in $\mathbb{C}^2$. This leads to the following :

Definition :

Let X and Y be analytic subsets in domains of $\mathbb{C}^n$ and suppose $x \in X$, $y \in Y$. Then the germ (X,x) is <u>topologically equivalent</u> to (Y,y) when there are neighbourhoods U and V of x resp. y in $\mathbb{C}^n$ and a homeomorphism $\phi : U \to V$ such that $\phi(x) = y$ and $\phi(X \cap U) = Y \cap V$.

Naturally, (X,x) and (Y,y) are homeomorphic if they are

topologically equivalent, but the converse does not hold. We have already seen this in our first discussion on the topology of singularities of plane curves in 5.3. Among other things, we found there that for a curve $C \subset \mathbb{C}^2$ with equation $x^p - y^q = 0$ the pair $(\mathbb{C}^2, C)$ was homeomorphic to the pair $(\mathbb{R}^4, \tilde{C})$, where $\tilde{C}$ is the cone with vertex 0 over a torus knot of type (p,q) in $S^3 \subset \mathbb{R}^4$, and it follows in particular that $(C,0)$ is not a topological submanifold of $\mathbb{C}^2$, i.e. it is not topologically equivalent to $(\mathbb{C},0)$. We shall now continue the discussion of the topology of singularities of plane curves begun in 5.3, and extend it to a complete classification of plane curve germs up to topological equivalence.

This classification was carried out essentially in the '20's and '30's of the present century, in particular by the mathematicians Brauner, Kähler, Burau and Zariski. For a more detailed historical survey we refer to the work of Reeve [R2].

In the definition of topological equivalence of (X,x) and (Y,y) we have left completely open the nature of the neighbourhoods U of x and V of y for which $X \cap U$ is homeomorphic to $Y \cap V$. However, when one wants to prove topological equivalence of analytic sets X and Y given concretely by equations, or when one wants to describe the local topological type of X at x, it is useful and necessary to choose the neighbourhoods U and V in a suitable and sufficiently simple way. One tries to choose the class of neighbourhoods so that, for any two sufficiently small neighbourhoods U' and U'' of $x \in X$ which are in the class we have : $X \cap U'$ is homeomorphic to $X \cap U''$. If one wants something like this, then the neighbourhoods admitted previously cannot be completely arbitrary, with holes or ugly boundaries, but one must choose nice simple neighbourhoods.

It turns out that for most such purposes balls or polydiscs are the most suitable. The open resp. closed ball B_ε resp. $\bar{B}_\varepsilon$ of radius ε around the origin in $\mathbb{C}^n$ is the set

$$B_\varepsilon = \{z \in \mathbb{C}^n \mid \|z\| < \varepsilon\}$$

resp.

$$\bar{B}_\varepsilon = \{z \in \mathbb{C}^n \mid \|z\| \leq \varepsilon\}.$$

The open resp. closed polydisc with multiradius $\delta = (\delta_1, \ldots, \delta_n)$ is given by

$$D_\delta = \{z \in \mathbb{C}^n \mid |z_i| < \delta_i\} \text{ resp.}$$

$$\bar{D}_\delta = \{z \in \mathbb{C}^n \mid |z_i| \leq \delta_i\}.$$

The advantage of working with balls is of course that the closed balls are smooth manifolds with boundary – the boundary is a $(2n-1)$-sphere

$$S_\varepsilon = \partial \bar{B}_\varepsilon.$$

The polydiscs are indeed also topologically manifolds with boundary, but the boundary is not smooth – it has edges. For example, a two-dimensional polydisc $\bar{D} \subset \mathbb{C}^2 = \mathbb{C} \times \mathbb{C}$, namely

$$\bar{D} = \{(x,y) \in \mathbb{C}^2 \mid |x| \leq \delta,\ |y| \leq \eta\},$$

is the product of the two circular discs,

$$\bar{D} = \{x \in \mathbb{C} \mid |x| \leq \delta\} \times \{y \in \mathbb{C} \mid |y| \leq \eta\},$$

and this has the boundary

$$\partial\bar{D} = \{x \mid |x| = \delta\} \times \{y \mid |y| \leq \eta\} \cup \{x \mid |x| \leq \delta\} \times \{y \mid |y| = \eta\}.$$

Thus $\partial\bar{D}$ is the union of the two solid tori

$$T^+ = \{x \mid |x| = \delta\} \times \{y \mid |y| \leq \eta\} \approx S^1 \times D^2$$

$$T^- = \{x \mid |x| \leq \delta\} \times \{y \mid |y| = \eta\} \approx D^2 \times S^1.$$

These two solid tori meet along the 2-dimensional torus

$$T^+ \cap T^- = \{(x,y) \in \mathbb{C}^2 \mid |x| = \delta,\ |y| = \eta\} \approx S^1 \times S^1,$$

and along this torus the boundary $\partial\bar{D}$ has an edge. In spite of the edge, it is convenient in many cases to work with polydiscs, especially when certain variables are distinguished, as they are with Puiseux expansions. The product structure of the polydisc is particularly suited to such a product decomposition of $\mathbb{C}^n$.

Of course, from a topological standpoint – in contrast to the function theoretic standpoint by the way – balls and polydiscs are not essentially different. They are homeomorphic to each other, and thus the boundary of a polydisc is homeomorphic to a sphere : for the 3-sphere S_ε in $\mathbb{C}^2$, $\varepsilon^2 = \delta^2 + \eta^2$ gives a natural decomposition into two solid tori

$$S_\varepsilon = T_+ \cup T_-$$

$$T_+ = \{(x,y) \in S_\varepsilon \mid |y| \leq \eta\}$$

$$T_- = \{(x,y) \in S_\varepsilon \mid |x| \leq \delta\}$$

which we have already considered in 5.3. One obtains homeomorphisms

$$T_+ \to T^+$$
$$T_- \to T^-$$

by $(x,y) \mapsto (\delta x/|x|, y)$

resp. $(x,y) \mapsto (x, \eta y/|y|)$

and these define a homeomorphism

$$\psi : T_+ \cup T_- \to T^+ \cup T^- .$$

The boundary of the polydisc, $\partial\bar{D}$, is therefore a sphere with edges and for that reason we also denote it by

$$\Sigma := \partial\bar{D}$$

in what follows, and we have just defined a homeomorphism

$$\psi : S_\varepsilon = T_+ \cup T_- \to T^+ \cup T^- = \Sigma .$$

The following picture illustrates this situation :

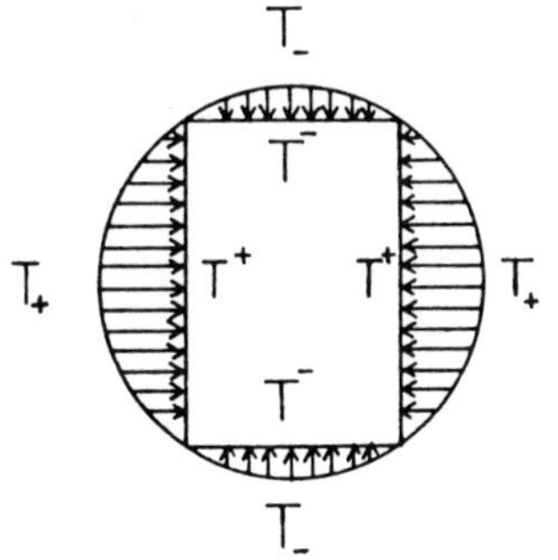

Correspondingly, one can construct homeomorphisms of the solid ball B_ε onto the polydisc D. This is a trivial exercise. However, one can prove an even better result in this direction. Roughly speaking, it says that the intersection of a curve with a small ball around a singular point looks topologically exactly like the intersection with a small polydisc. We now want to make this precise.

Let $f(x,y) \in \mathbb{C}\{x,y\}$ be a convergent power series without multiple factors, suppose $f(0,0) = 0$, and let X be the zero set of $f(x,y) = 0$ in the domain of convergence of f. Let the coordinates x, y be chosen in such a way that the y-axis is not a tangent to X at the origin, i.e. y is m-general and m is the multiplicity of f at the origin. We consider the intersection of X with a small polydisc

$$\bar{D} = \{(x,y) \in \mathbb{C}^2 \mid |x| \le \delta,\ |y| \le \eta\}$$

resp. a small ball $\bar{B}_\varepsilon$.

"Small" here means the following :
Let $t \mapsto (t^{m_k}, y_k(t))$ be the resolution of the singularity of a branch of X by means of the Puiseux expansion. Then $\bar{D}$ shall first be chosen so small that $\bar{D} \cap X$ is the union of images of the discs of radius δ^{1/m_k} under this resolution, and so that the origin is the only singular point of $\bar{D} \cap X$. Further, δ shall be chosen so small that $|y_k(t)| < \eta$, hence so that $X \cap \Sigma \subset \overset{\circ}{T}{}^+$. Finally, δ shall also be chosen so small that

$$\mathrm{Re}(m_k t \bar{t}^{m_k} + \overline{y_k(t)} \cdot y_k'(t) \cdot t) \neq 0 \qquad (*)$$

for $0 < |t| \le \delta^{1/m_k}$, and ε shall be chosen so small that $\bar{B} = \bar{B}_\varepsilon$ lies in $\bar{D}$. The condition (*) then guarantees that X will cut the sphere $S_\varepsilon = \partial\bar{B}_\varepsilon$ transversely. When all these conditions are satisfied, we say that $\bar{D}$ is a small polydisc and $\bar{B}$ is a small ball around the singular point 0 of X.

Then one can prove the following theorem : (cf. Pham [Pl])

Theorem 6

Let X be a curve as above through $0 \in \mathbb{C}^2$. Then :

(i) For any two small balls $\bar{B}'$, $\bar{B}''$ around the point $0 \in X$ there is a homeomorphism $\phi : \bar{B}' \to \bar{B}''$ with $\phi(0) = 0$ and $\phi(X \cap \bar{B}') = X \cap \bar{B}''$.

An analogous assertion holds for any two small polydiscs $\bar{D}'$, $\bar{D}''$.

(ii) If $\bar{B}$ is a small ball and $\bar{D}$ is a small polydisc around $0 \in X$, then there is a homeomorphism $\phi : \bar{B} \to \bar{D}$ with $\phi(0) = 0$ and $\phi(X \cap \bar{B}) = X \cap \bar{D}$.

We shall not carry out the proof here, but merely suggest the idea of it by a picture. The meaning of the picture is that one obtains the desired mapping by allowing the points of $\bar{B}$ to move along the integral curves of the suitably chosen vector field. In order to map $X \cap \bar{B}$ into $X \cap \bar{D}$ resp. $X \cap \bar{B}'$ into $X \cap \bar{B}''$, one must choose the vector field so that it is tangential to X at the points of $X - \{0\}$. We

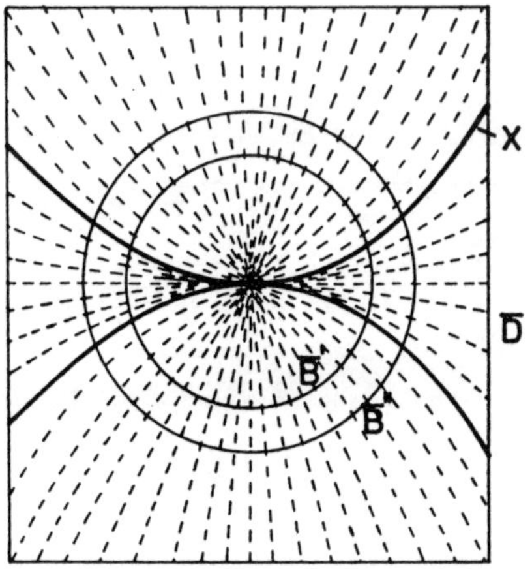

refer to Pham [Pl] and also Milnor [Ml] for the details.

Corollary 7

For all small balls $\bar{B}$ or polydiscs $\bar{D}$ with boundary $\partial\bar{B} = S$ resp. $\partial\bar{D} = \Sigma$, the pairs

$$(S, S \cap X) \quad \text{and} \quad (\Sigma, \Sigma \cap X)$$

resulting from intersection with the curve are homeomorphic to each other. $S \cap X$ is the disjoint union of k circles in the 3-sphere S, where k is the number of irreducible components of X at the origin. Thus we see that the intersection of the curve with the 3-sphere which one obtains as the boundary of a good small neighbourhood of the singular point, consists of finitely many circles in such an S^3, possibly linked with each other. This means :

$$K = X \cap S$$

is a link. When X is irreducible at the singular point, K is a knot in S^3.

Definition :

A knot is a subset $K \subset S^3$ which is homeomorphic to S^1 and provided with an orientation. Two knots K' and K'' in S^3 are equivalent when there is an orientation preserving homeomorphism $\phi : S^3 \to S^3$ which maps K' onto K'' with preservation of their orientations. Links and their equivalence are defined correspondingly.

<u>**Remarks**</u> :

(i) One could place additional conditions on the knot K, such as its being a differentiable, or at least topological, submanifold of S^3. But in our investigations these are the only kinds of knot which appear anyway.

(ii) Our knot $K = X \cap S$ is the image of a circle $S^1 = \{t \in \mathbb{C} \mid |t| = c\}$ under the resolution of the singularity. We orient S^1 in the positive sense and K correspondingly.

The simplest knot is the trivial knot :

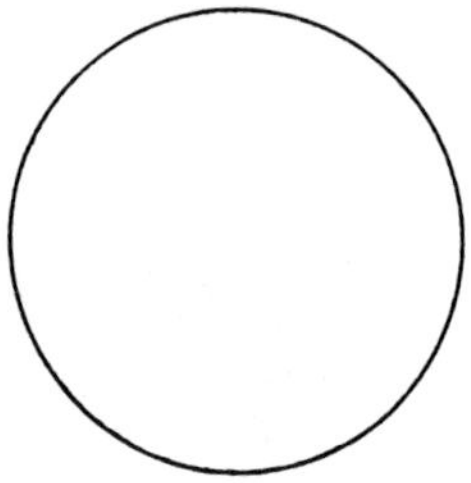

The simplest non-trivial knot is the trefoil knot.

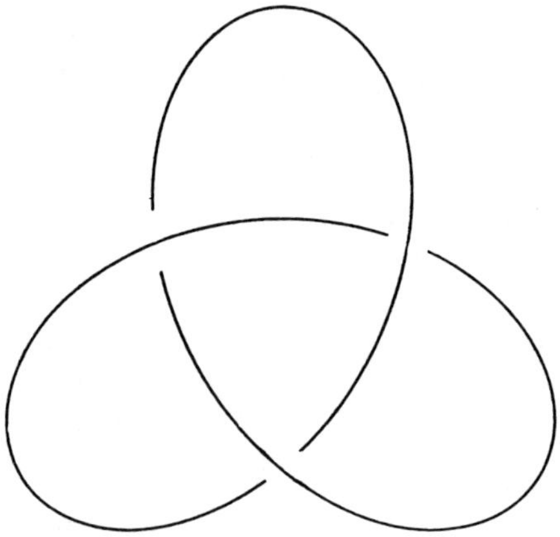

It results from joining the ends of the simplest knot in the colloquial sense,  , so that the figure cannot be undone. The trefoil knot is a favourite ornament in Romanesque churches. One finds it, e.g., on some columns at the crossing of Bonn cathedral :

The link of the trefoil knot with a circle shown below (Dreischenkel or Triquetra) is probably a symbol of the Holy Trinity.

Knots and links have indeed always been very popular themes for ornament. In Celtic and Scandinavian ornament in particular one finds very artistic knot patterns. On the pages which follow we reproduce some figures from the book of A. Speltz [S12].

Tafel 59.

Fig. 1. **Manuskriptmalerei aus dem 10. Jahrhundert** (Dolmetsch).
„ 2. **Initiale aus einem Psalter von Ricemarchus.** Jetzt im Trinity College, Dublin; aus dem 11. Jahrhundert (Owen Jones).
„ 3 u. 4. **Manuskriptmalereien aus dem 10. Jahrhundert** (Owen Jones).
„ 5. **Das Kreuz von Aberlemno** (Owen Jones).
„ 6. **Initiale aus dem 7. Jahrhundert** (Dolmetsch).
„ 7—11. **Manuskriptmalereien keltisch-angelsächsischen Ursprungs** (Owen Jones).
„ 12. **Ornament vom Sockel eines Kreuzes in der Kirche von Eassie, Angusshire** (Owen Jones).
„ 13. **Ornament vom Sockel eines Kreuzes an der Kirche von St. Vigean, Angusshire** (Owen Jones).
„ 14. **Ornament vom Sockel eines Kreuzes an der Kirche von Meigle, Angusshire** (Owen Jones).

Fig. 1. **10th century manuscript painting** (Dolmetsch).

Fig. 2. **Initial from a psalter of Ricemarchus**, 11th century. Now in Trinity College, Dublin (Owen Jones).

Figs. 3 and 4. **10th century manuscript painting** (Owen Jones).

Fig. 5. **The cross of Aberlemno** (Owen Jones).

Fig. 6. **Initial, 7th century** (Dolmetsch).

Figs. 7-11. **Manuscript painting of Celtic-Anglosaxon origin** (Owen Jones).

Fig. 12. **Ornament from the base of a cross in the church of Eassie, Angusshire** (Owen Jones).

Fig. 13. **Ornament from the base of a cross in the church of St. Vigean, Angusshire** (Owen Jones).

Fig. 14. **Ornament from the base of a cross in the church of Meigle, Angusshire** (Owen Jones).

Das keltische Ornament.

Tafel 60.

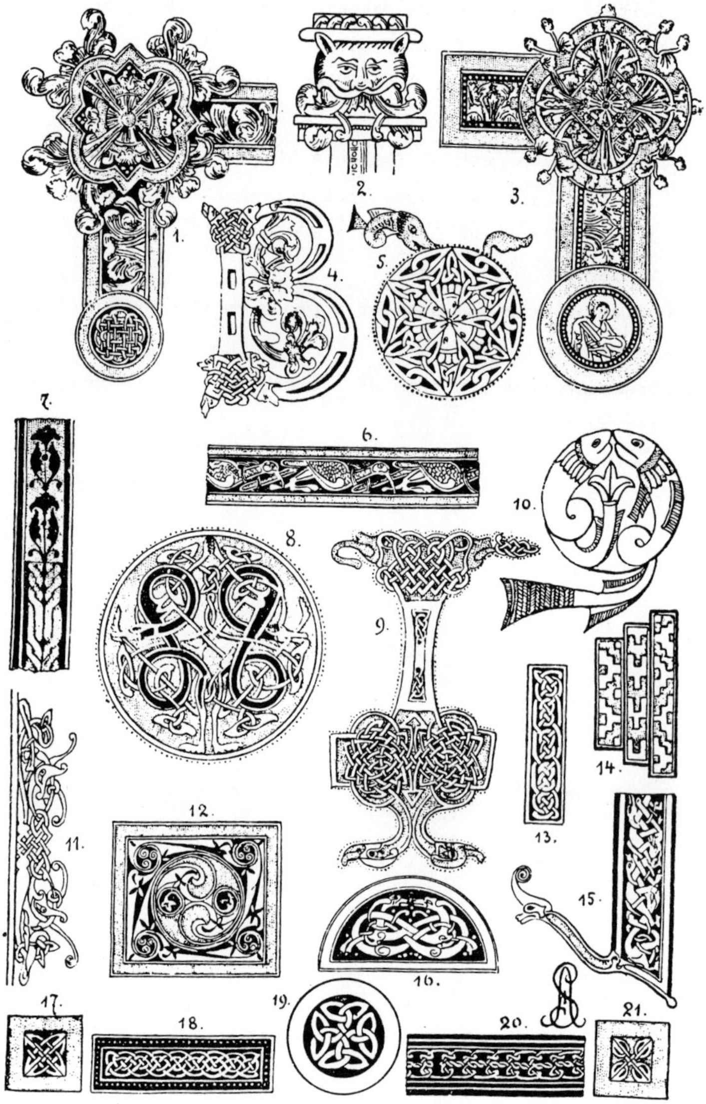

8*

Fig. 1. 3. 4 u. 8. Von Manuskriptmalereien aus dem 10. Jahrhundert (Dolmetsch u Owen Jones).
„ 2. Desgleichen aus dem 11. Jahrhundert (Dolmetsch).
„ 5. 6 u. 10. Desgleichen aus dem 8. Jahrhundert (Dolmetsch).
„ 7. Desgleichen aus dem 9. Jahrhundert (Dolmetsch).
„ 9. Initiale aus der franko-sächsischen Bibel von St. Denis aus dem 9. Jahrhundert (Owen Jones).
„ 11—21. Aus Manuskriptmalereien keltisch-angelsächsischen Ursprungs (Owen Jones).

Figs. 1, 3, 4 and 8. **From 10th century manuscript painting** (Dolmetsch and Owen Jones).

Fig. 2. **The same, from 11th century** (Dolmetsch).

Figs. 5. 6 and 10. **The same, from 8th century** (Dolmetsch).

Fig. 7. **The same, from 9th century** (Dolmetsch).

Fig. 9. **Initial from the 9th century Franco-Saxon bible of St. Denis** (Owen Jones).

Figs. 11-21. **From manuscript painting of Celtic-Anglosaxon origin** (Owen Jones).

At least as old as these ornaments, and much more important in practice are knots which were developed for connecting, fastening and holding things in place, especially by sailors but also in many other crafts. Most of these knots are "open knots" whose ends are not connected together (as was done above to produce the trefoil knot from an ordinary knot). Friction prevents these knots from coming undone – an aspect which is completely ignored by mathematical knot theory. Some "closed knots" (hence knots in the mathematical sense) also occur in practice, particularly as ornamental forms such as the so-called Turk's head. Braids also have practical significance, even today. Many cords, ropes and cables are today plaited mechanically as braids, e.g. heavy anchor ropes. The reproductions from "The Ashley Book of Knots" [A9] at the end of this chapter give an impression of the practical significance of knots.

Corollary 7 tells us that each singular point of an irreducible curve $f(x,y) = 0$ in $\mathbb{C}^2$ is associated with a knot which is unique up to equivalence. One can show, incidentally, that the equivalence class of this knot also does not depend on the choice of coordinates in $\mathbb{C}^2$. In what follows we shall precisely determine the nature of this knot. But first we want to establish that this knot already determines the topology of the singularity uniquely.

Theorem 8 :

Let $\bar{B}$ be a small ball around the singular point x of the curve X with boundary $\partial\bar{B} = S$ and let $K = S \cap X$. Also, let $C(K) \subset \bar{B}$ be the cone with base K and vertex x. Then :

$(\bar{B}, \bar{B} \cap X)$ and $(\bar{B}, C(K))$ are homeomorphic.

The proof uses the same basic idea as Theorem 6. For the details we refer to Pham [P1] or Milnor [M1] 2.10.

Corollary :

Theorem 8 shows that curve germs with equivalent links are topologically equivalent. One can prove that the converse also holds : thus curve germs are topologically equivalent if and only if they have equivalent links.

Theorem 8 and its corollary show that if we are to classify irreducible curve germs up to topological equivalence, then we need only classify their knots $K \subset S$ up to equivalence. Corollary 7 shows us that we may just as well consider knots $K \subset \Sigma$ in the sphere Σ with edges, and the homeomorphism $\Psi : S \to \Sigma$ allows us to view such knots as knots in the ordinary sphere S again.

Thus we consider knots $K \subset \Sigma$. In principle we can easily obtain a grasp of these knots with the help of the braids which were used in the qualitative description of the many-valued solutions $y_i(x)$ of the equation $f(x,y) = 0$ to the curve X with $X \cap \Sigma = K$. The associated braid was the mapping of the unit interval into the symmetric product Y_m which associated each $t \in [0,1]$ with the m complex numbers $y_i(t)$, $i = 1,\dots,m$ obtained as solutions of the equation $f(x,y) = 0$ for $x = \delta e^{2\pi it}$.

This can also be expressed as follows :

The projection $(x,y) \mapsto x$ defines a mapping $K \to S^1$ onto the circle $S^1 = \{x \mid |x| = \delta\}$, and the points of K over the point $x = \delta e^{2\pi it}$ are just the points $(\delta e^{2\pi it}, y_i(t))$, $i = 1,\dots,m$.

Thus the knot K is uniquely determined by the functions $y_i(t)$ which describe the braid. We can also describe this construction of the knot from the braid intuitively as follows :

Because of the choice of our small polydisc $D = \{(x,y) \in \mathbb{C}^2 \mid |x| \leq \delta, |y| \leq \eta\}$, the function values $y_i(t)$ lie entirely inside the disc $D^2 = \{y \in \mathbb{C} \mid |y| < \eta\}$. This means that the knot K lies entirely inside one of the solid tori T^+, T^- into which the edged sphere Σ has been divided, namely

$$K \subset S^1 \times D^2 = T^+$$

We can construct this solid torus by identifying the discs at the end of the solid cylinder $[0,1] \times D^2$ by the equivalence relation $(0,y) \sim (1,y)$. This means that we consider the mapping

$$[0,1] \times D^2 \longrightarrow S^1 \times D^2$$

$$(t,y) \longmapsto (e^{2\pi it}, y).$$

Under this mapping, the points $(t, y_i(t))$ making up the graph of the many-valued "function" $t \mapsto y_1(t),\dots,y_m(t)$ go precisely to the knot K.

Intuitively speaking : the knot K results from the braid when we identify corresponding initial and final points. The link resulting from identification of initial and final points of a braid is also called a closed braid. The following picture illustrates the description of the trefoil knot as a closed braid :

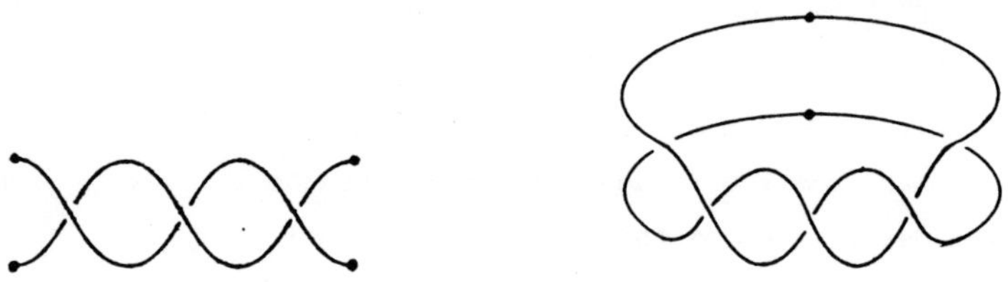

This is indeed the trefoil knot :

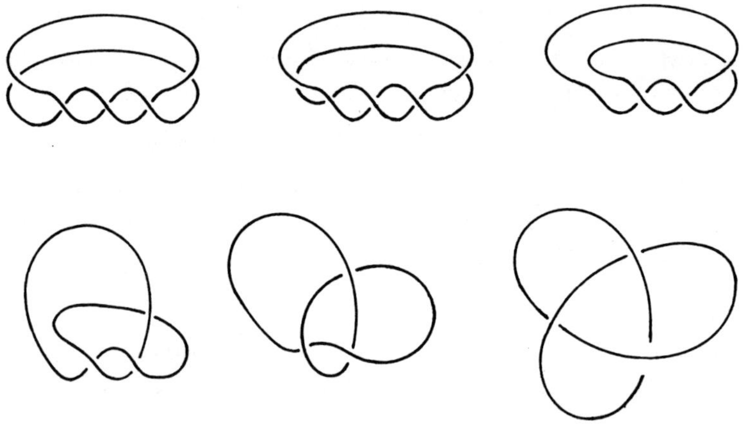

Here is an example of a braid which yields a link rather than a knot.

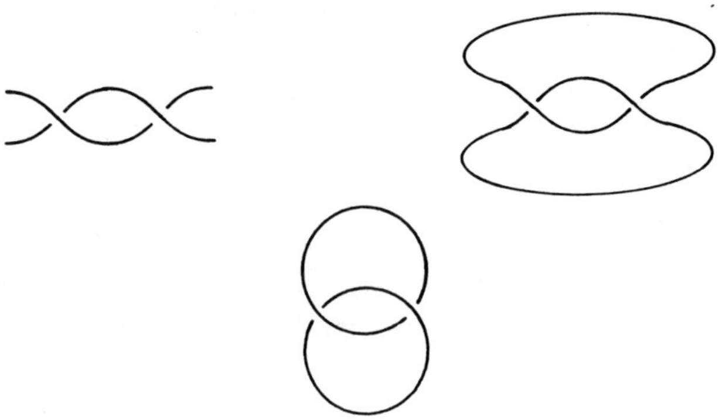

One can usually obtain a knot in many different ways as a closed braid. Here is another generation of the trefoil knot, this time from a closed braid with 3 strings :

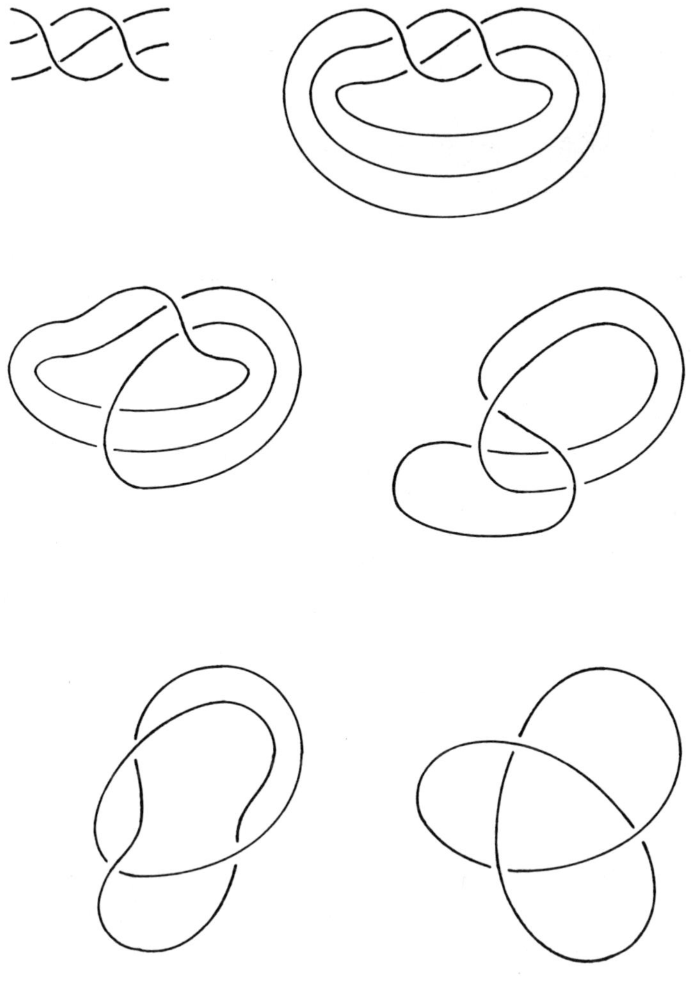

2241. A *covering for a ring or grommet.* Make a loose *three-strand grommet* by winding a cord in a widely and evenly spaced helix three times around the circuit of the ring that is to be covered. Tie the two ends together, leaving them long enough for doubling. Count the turns and take another cord of another color or size material and wind it the *same number of turns* in the opposite direction three times around the ring. This is to act merely as a clue. Take a longer cord of the first material and with it follow parallel with the *second* cord but tucking alternately over and under at the crossings. Next remove the clue and double both strands throughout or triple and quadruple them if necessary to cover the ring. Use a wire needle (#99L) to tie the knot and work it taut with a pricker (#99A).

2241

Aus: The Ashley Book of Knots

It can be proved (Alexander 1923) that each knot and link is obtainable as a closed braid. This enables the theory of braid groups to be applied to the (unsolved*) problem of classifying knots.

Intuitively, it is clear that the closure of a given braid is determined, up to knot equivalence, by the group element representing the braid, and indeed by just the conjugacy class of the element. Because when one precedes the given braid by any other, and succeeds it by the inverse of the latter, identification of initial and final points leads to cancellation of the initial braid by the final one.

Proposition 9 :

Closed braids which correspond to topologically equivalent (open) braids are topologically equivalent as links.

By a theorem of Markov (1935), one can even say precisely when two braids from different braid groups B_m and B_n define equivalent links. The condition is of the type saying that there is a sequence of certain algebraic operations converting one braid to the other. However, one does not have an algorithm which establishes, for two given braids, whether such a sequence of operations exists. One part of the

---

*This problem has been solved, in a certain sense, by G. Hemion : Acta Math. 142 (1979), 123-155 (Translator's note.)

problem is the conjugacy problem for braid groups, i.e. the problem of deciding algorithmically whether two braids in the braid group B_n are conjugate. This problem was solved by Garside in 1969*, but the general problem of classifying knots and links is still unsolved. (For literature, cf. Birman [B10].)

In what follows we shall see that the knots we have to deal with in connection with singularities of plane curves are of a very special kind, which admit a very nice description and for which the classification problem is completely solved.

First we look at the knot which corresponds to an irreducible curve with only one Puiseux pair (m,n), say the generalised parabola $x^n - y^m = 0$. The Puiseux expansion is $y = x^{n/m}$ and the associated braid is given by the functions

$$y_k(t) = \eta e^{2\pi i \frac{k+t}{m} n}, \quad k = 1,\dots,m \quad \text{with} \quad \eta = \delta^{n/m},$$

hence it is a braid with m strings which twist at an angle $e^{2\pi i . n/m}$ for $0 \leq t \leq 1$. The following picture shows the braid for $(m,n) = (3,5)$.

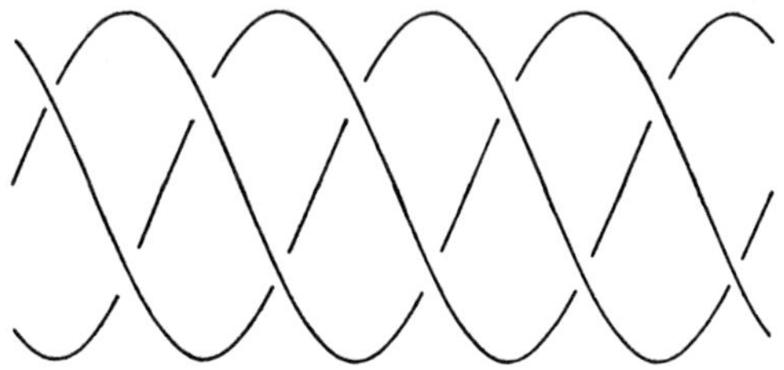

The corresponding knot is the homeomorphic image of the circle $S^1 = \{t \in \mathbb{C} \mid |t| = 1\}$ under the mapping

$$S^1 \to T_{\pm}$$
$$t \mapsto (\delta t^m, \eta t^n).$$

Thus it is a circle which lies on the torus

---

*The same result was announced in 1968 by G.S. Makanin (Makanin, G.S.: The conjugacy problem in the braid group. Soviet Math. Dokl. 9 (1968), 1156-1157).

$$S^1 \times S^1 = \{(x,y) \mid |x| = \delta, |y| = \eta\} \subset T_{\pm}$$

and winds m times in one direction and n times in the other. Here is a picture for the example above, $(m,n) = (3,5)$:

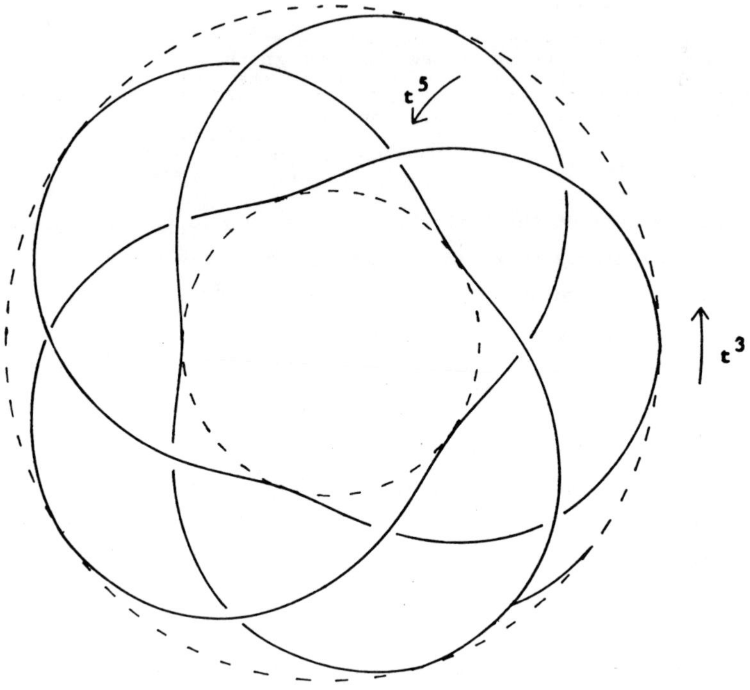

Such knots are called <u>torus knots</u>, and we have already seen in 5.3 that such knots appear in the topological description of singularities. The "winding around" can be made precise as follows. We consider the two projections

$$\pi_i : S^1 \times S^1 \to S^1 , \quad i = 1,2$$

of the factors of $S^1 \times S^1$. Then the restriction to $K \subset S^1 \times S^1$

$$\pi_i : K \to S^1$$

yields coverings of the circle S^1 by the circle K of degree m for $i = 1$ resp. n for $i = 2$, and in this way the winding numbers are characterised. In general, for any circle $K \subset S^1 \times S^1$, one can define two mappings $\pi_i : K \to S^1$ by restriction of the projections, and with them two numbers, m and n, which are the degrees of these mappings. One can prove that for such a knot K there is a homeomorphism of $S_1 \times S_1$, homotopic to the identity, which carries K into the standard knot with parametrisation $t \mapsto (t^m, t^n)$ described above. This leads us to the following :

__Definition__ :

Let F be a torus surface and let $h : F \to S^1 \times S^1$ be a homeomorphism. Suppose $K \subset F$ is homeomorphic to S^1 and oriented. Then K is called a torus knot of type (m,n) on the torus surface F with trivialisation h when the projections $\pi_i \circ h : T \to S^1$, for $i = 1,2$, are such that $\pi_i \circ h : F \to S^1$ has mapping degree m for $i = 1$ and n for $i = 2$.

__Remark__ :

It is important to understand that the type (m,n) of the torus knot $K \subset F$ is defined only when the trivialisation $h : F \to S^1 \times S^1$ is given. One easily sees this, for example as follows. Suppose we have the standard knot $K \subset S^1 \times S^1$ of type (m,n). Now let $h : S^1 \times S^1 \to S^1 \times S^1$ be the mapping $(t_1,t_2) \mapsto (t_1, t_1^k \cdot t_2)$, where k is any integer. (Intuitive description of h : cut $S^1 \times S^1$ along $\{1\} \times S^1$, twist through $2\pi k$, and paste together again.) This h yields a new trivialisation of $S^1 \times S^1$, and relative to this new trivialisation K has type $(m,km+n)$. In general, if $h : S^1 \times S^1 \to S^1 \times S^1$ is the mapping $(t_1,t_2) \mapsto (t_1^a t_2^b, t_1^c t_2^d)$, where $a,b,c,d \in \mathbb{Z}$ and $\det\begin{pmatrix} a & b \\ c & d \end{pmatrix} = \pm 1$, then a knot K of type (m,n) goes to a knot of type $(am+bn, cm+dn)$. One easily sees that with the help of such automorphisms of $S^1 \times S^1$ the knot K is convertible to torus knots of all other types (r,s) $(\gcd(r,s) = 1)$. Thus one sees that the numbers describing the type of the knot really do depend on the trivialisation of F, i.e. on the choice of latitude circle $S^1 \times \{1\}$ and longitude circle $\{1\} \times S^1$ on F.

When F is any abstract surface, for which one only knows that it is homeomorphic to a torus, then there is no obvious distinguished trivialisation of F. However, the torus surfaces that we have to consider are always boundaries of solid tori in the 3-sphere S resp. Σ, moreover they are also embedded in a particular way in the solid tori $T^{\pm}$ resp. $T_{\pm}$, as we shall describe shortly. In this situation we have two distinct methods for defining a trivialisation of F, leading in general to genuinely distinct trivialisations.

Method I :

Let $K_0 \subset S^3$ be a differentiable oriented submanifold homeomorphic to S^1, and let T be a tubular neighbourhood of K_0, homeomorphic to a closed solid torus. Let F be the torus surface $F = \partial T$. Then we consider the trivialisation $F \approx S^1 \times S^1$ of F such that :

(i) A longitude circle $\{t_1\} \times S^1$ results from cutting F by a plane transverse to K_0.

(ii) A latitude circle $S^1 \times \{t_2\}$ has linking number* 0 with K_0, and the projection $F \to K_0$ induces an orientation-preserving covering $S^1 \times \{t_2\} \to K_0$.

Example 1 :

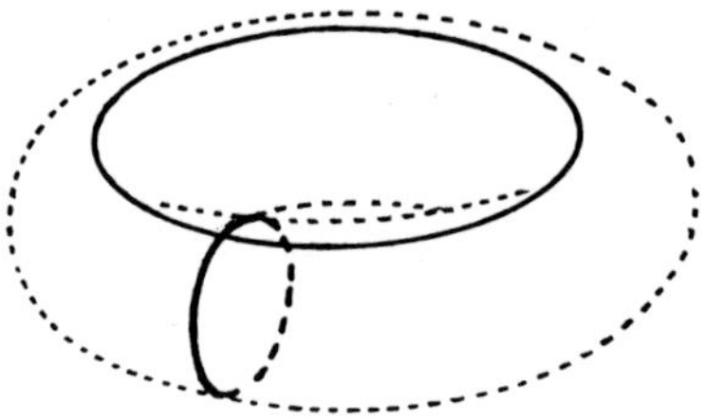

---

*The linking number of two knots K and K' in S^3 is obtained as follows : one chooses a compact orientable surface, $F \subset S^3$ whose boundary ∂F is just K. The knot K' can be deformed so that it cuts the surface $\mathring{F}$ transversely. The linking number of K and K' is then the intersection number of K' and $\mathring{F}$ (cf. 6.3). For further details see Seifert-Threlfall [S3] §77 (see also Chapter 8.5, Theorem 1).

Example 2 :

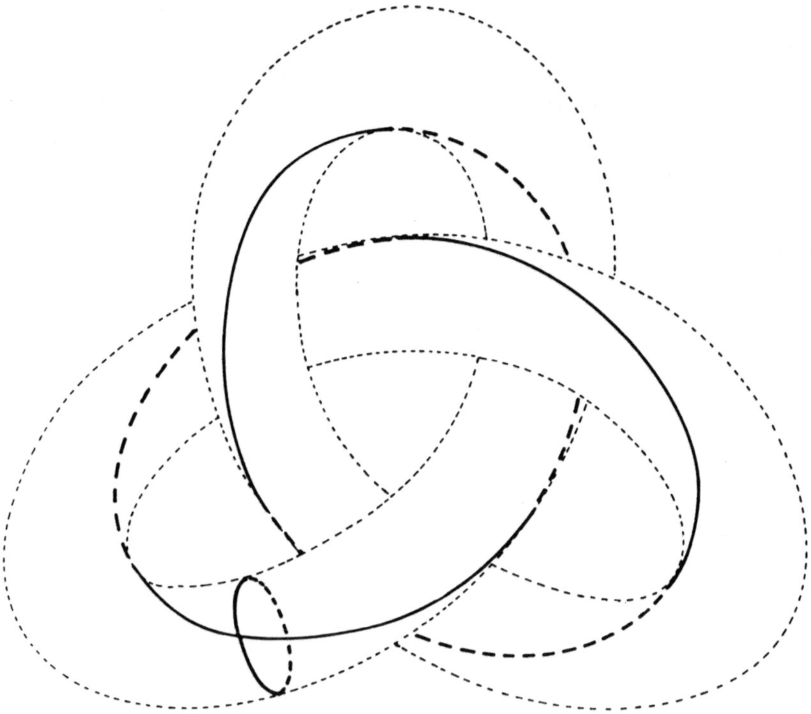

When one represents the 3-sphere as $\mathbb{R}^3 \cup \{\infty\}$, one can interpret method I as a method of introducing coordinates on the surface F which give position relative to the knot K_0, but not relative to a fixed space $\mathbb{R}^3$. The second method, on the other hand, addresses itself to the coordinates of the space $\mathbb{R}^3$. To make this work, we have to make special assumptions about the position of the spine K_0 of the solid torus T we begin with.

Method II :

Let $S^3 = T_+ \cup T_-$ be the decomposition of the sphere into two solid tori.

$$T_+ = S^1 \times D^2$$
$$T_- = D^2 \times S^1$$

previously considered. We call a differentiable knot K_0 <u>regularly embedded</u> - ad hoc definition - when $K_0 \subset S^1 \times \mathring{D}^2$ and the projection $K_0 \to S^1$ is a differentiable orientation-preserving covering. Then we choose a tubular neighbourhood T of K_0 such that for each $t \in S^1$ the intersection $T \cap (\{t\} \times D^2)$ consists of disjoint discs $D_i(t)$ around the points $p_i(t)$ of $K_0 \cap (\{t\} \times D^2)$. We choose the boundary of such a disc as longitude circle on $F = \partial T$. As latitude circle on F we choose a curve whose points $q_i(t) \in D_i(t)$ are such that $q_i(t) - p_i(t)$ has constant direction in D^2.

Example :

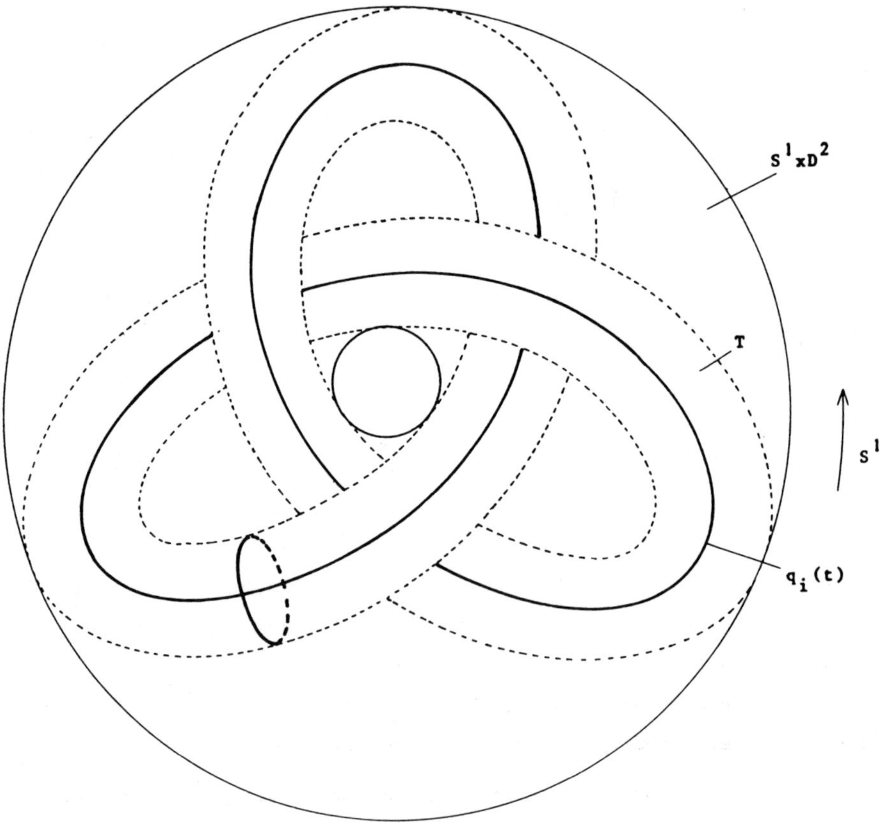

In what follows we shall always use method II for the choice of trivialisation. This means that when K_1 is a regularly embedded torus knot of type (m,n) on F relative to this trivialisation, and when the projection onto the spine of T_+, $K_0 \to S^1$, is a k-fold covering, then for K_1 the projection $K_1 \to S^1$ onto the spine of T_+ is an $(m \cdot k)$-fold covering, and when one traverses K_0 once the m points of K_1 over the moving point of K_0 turn in D^2 through $2\pi n/m$ relative to this point.

If now the knot K_0 is an ordinary torus knot in the sense defined above, then the torus knot K_1 on a tubular neighbourhood of K_0 is in a certain sense a torus knot "of higher order". We can make this idea for the construction of higher order torus knots precise as follows.

Definition :

The trivial knot $S^1 \times \{0\} \subset S^1 \times D^2 \subset S^3$ is a torus knot of $\underline{0^{th}}$ order.

A torus knot of type (m_1,n_1) on the boundary of a tubular neighbourhood of the trivial knot (with trivialisation by method II) is a torus knot of 1st order and type (m_1,n_1).

If $K_i \subset S^1 \times D^2 \subset S^3$ is a regularly embedded torus knot of ith order and type $(m_1,n_1),\ldots,(m_i,n_i)$, and if K_{i+1} is a torus knot of type (m_{i+1},n_{i+1}) on a tubular neighbourhood of K_i (with trivialisation by method II), then K_{i+1} is called a torus knot of $(i+1)^{th}$ order and type $(m_1,n_1),\ldots,(m_{i+1},n_{i+1})$.

Torus knots of higher order are also called iterated torus knots. In the English literature such knots are often called "cable knots" because the braids from which they are obtained resemble the cables.

Remark :

Many authors define iterated torus knots using trivialisation by method I, which results in quite different numbers for the type. Certain authors complete the confusion by defining the (m_i,n_i) by one method and erroneously computing them by the other.

Now we can formulate our results on the topology of singularities of plane curves.

Proposition 10 :

The knot corresponding to the Puiseux expansion

$$y = x^{n_1/m_1} + x^{n_2/m_1m_2} + \ldots + x^{n_g/m_1\cdots m_g}$$

with Puiseux pairs $(m_1,n_1),\ldots,(m_g,n_g)$ is an iterated torus knot of order g and type $(m_1,n_1),\ldots,(m_g,n_g)$.

Proof :

This follows immediately from the earlier description of the braid corresponding to the Puiseux expansion and the definition of iterated torus knots, or also directly, by inductively considering the knots which result from breaking off the Puiseux expansion at the ith term.

Remark :

If one uses method I for the trivialisation of the tori, then one finds the type numbers of the iterated torus knots to be

$$(m_1,\lambda_1),\ldots,(m_g,\lambda_g)$$

where the λ_i are determined recursively as follows :

$$\lambda_1 = n_1$$
$$\lambda_i = n_i - n_{i-1}m_i + \lambda_{i-1}m_{i-1}m_i.$$

One can obtain this computational formula by geometric considerations (cf. Lê Dũng Trang [L5]) ; in 8.5 we shall at least indicate another proof of it.

Thus we see that the Puiseux pairs play a decisive rôle, not only in the qualitative description of the solutions $y(x)$ of $f(x,y) = 0$, but also in the qualitative description of the zero set. In fact, they characterise the situation completely. Using methods of knot theory, which we do not want to give here, the iterated torus knots can be completely classified, and this leads to the following result :

Proposition 11 :

Two iterated torus knots corresponding to Puiseux expansions are equivalent just in case they have the same type, i.e. when the Puiseux pairs of the two expansions are equal.

Now we summarise the essentials of our previous results :

Theorem 12 :

Let (X,x) be an irreducible plane curve germ, parametrised by a

Puiseux expansion with the Puiseux pairs $(m_1,n_1),\ldots,(m_g,n_g)$. Then the intersection K of X with the boundary S (or Σ) of a small ball (or polydisc) around x is an iterated torus knot of type $(m_1,n_1),\ldots,(m_g,n_g)$. Two such germs are topologically equivalent just in case their Puiseux pairs coincide.

Proof : Theorem 3, Propositions 9, 10, 11, Theorem 8 and Corollary.

With this, the irreducible curve germs are completely classified from a topological standpoint. We now want to describe briefly how one can extend this result to the case of reducible curve germs. More details may be found, e.g., in the work of Reeve [R2].

If X is a reducible curve germ, then the intersection of X with a small sphere S is the union of linked knots. Each knot belongs to an irreducible component of X. Certain information about the kind of linking is obtained from a knowledge of the linking number of any two knots.

As in 6.1, one can define intersection multiplicity for curve germs, and then one has :

Proposition 13 :

Let X_1 and X_2 be irreducible curve germs at a point x of $\mathbb{C}^2$, and let $K_1 = X_1 \cap S$ resp. $K_2 = X_2 \cap S$ be the knots which result from intersecting them with a small sphere S around x. Then the linking number of K_1 and K_2 is just the intersection multiplicity of X_1 and X_2.

E.g. if X is the union of two transversely intersecting lines, then $X \cap S$ consists of two circles with linking number 1 :

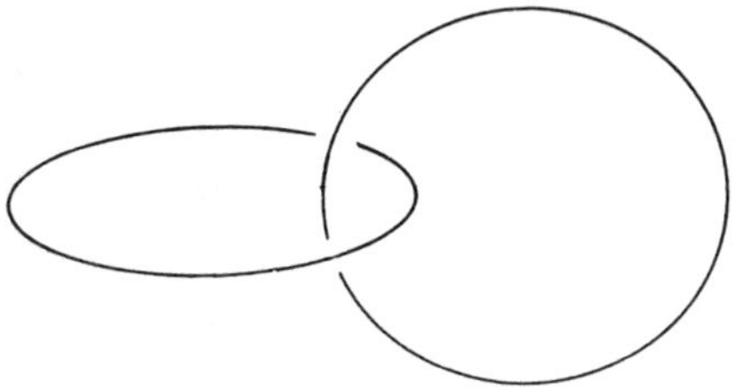

If X is the union of three different lines, then one obtains a link of the following form :

Since the intersection multiplicity of two curve germs at a point is always positive, a link such as the following cannot occur as the link of a plane singularity.

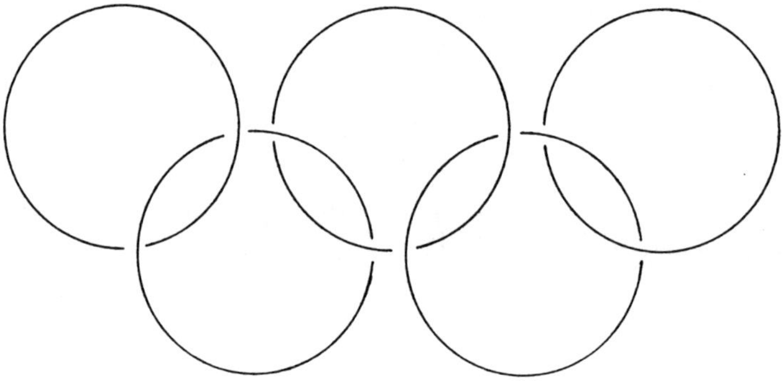

Since we shall often be calculating with intersection multiplicities, we give a few more characterisations of the intersection multiplicity of two curve germs (cf. e.g. Fulton [Fl]).

Theorem 14 :

Suppose (X,x) and (X',x) are two different irreducible curve germs in $\mathbb{C}^2$ and let ν be the intersection multiplicity of X and X'. Then :

(i) ν equals the linking number of the knots of (X,x) and (X',x).

(ii) If f resp. g are the function germs defining X resp. X', then the vector space $\mathcal{O}_{\mathbb{C}^2,x}/(f,g)$ is finite-dimensional by the Rückert Nullstellensatz (8.2.7). The intersection multiplicity ν is just the dimension $\dim_{\mathbb{C}}\mathcal{O}_{\mathbb{C}^2,x}/(f,g)$ of this vector space.

(iii) If $\pi : (\mathbb{C},0) \to (X,x)$ is the resolution of the singularity of X by the Puiseux expansion (Theorem 1), and g is again the function germ defining X', then ν equals the order of the mapping

$$g \circ \pi : (\mathbb{C},0) \to (\mathbb{C},0).$$

The method of computing the intersection multiplicity with the help of the resultant (6.1), while it allows a very simple proof of Bézout's theorem, leads to very laborious calculations even in simple examples. Assertions (ii) and (iii) of Theorem 14 now give us a simpler method of computation.

Example :

Let $f(x,y) = x^n - y^m$ and $g(x,y) = x^p - y^q$, with $\gcd(m,n) = 1$, $\gcd(p,q) = 1$, $(n,m) \neq (p,q)$. The resolution of the singularity of X is $\pi(z) = (z^m,z^n)$, hence

$$g \circ \pi(z) = z^{mp} - z^{nq}$$

and the intersection multiplicity ν of X and X' is

$$\nu = \min\{mp,nq\}.$$

Proposition 13 alone does not yield a complete description of the link of a reducible singularity. E.g. three knots can be linked together in various ways even though the individual linking numbers coincide. This is already shown by the following example, the Borromean rings :

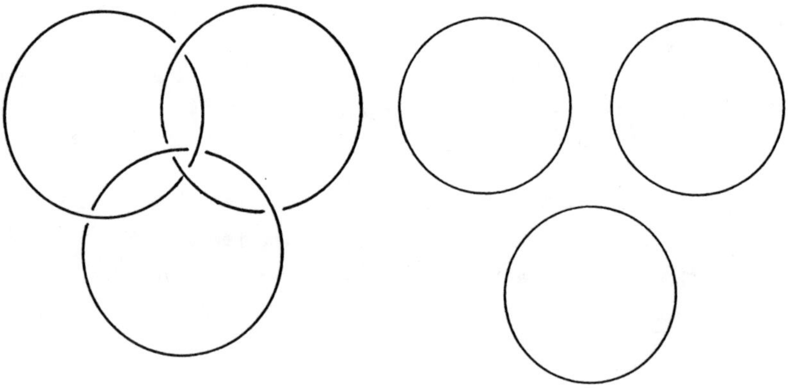

Nevertheless, the knots together with their linking numbers do suffice to characterise the singularity.

<u>Theorem 15</u> :

Two curve germs (X,x) and (X',x) are topologically equivalent just in case there is a bijection between the irreducible components of X and X' such that

(i) the Puiseux pairs of corresponding components are the same,

(ii) the intersection numbers of corresponding components coincide.

With this, we have reached a complete topological classification of the plane singularities. We now want to say a little about the connection with the analytic classification.

For the topological type of a plane irreducible singularity, only the terms corresponding to the Puiseux pairs are essential, and the topological type is also independent of the coefficients of these terms (as long as they are nonzero). Now how much can one alter the "inessential terms" of the Puiseux series without altering the analytic type of the singularity?

We know, at any rate, by Mather's theorem (8.2.9) that only finitely many terms of the equation $f(x,y) = 0$ are relevant for the analytic type of its singularity. Hence only finitely many terms of the

Puiseux expansion can influence the analytic type. More precisely, we have (cf. [H2] for what follows) :

Theorem 16 :

Let (X,x) be a plane irreducible singularity with multiplicity m and Puiseux expansion $y = \Sigma a_i x^{i/m}$. Let $\pi : (\mathbb{C},0) \to (X,x)$ be the resolution of the singularity from Theorem 1 ; π induces an embedding $\pi^* : \mathcal{O}_{X,x} \hookrightarrow \mathcal{O}_{\mathbb{C},0}$. Let N be the number

$$N := 2 \dim \mathcal{O}_{\mathbb{C},0}/\pi^*(\mathcal{O}_{X,x}).$$

Then a singularity (X',x') with Puiseux expansion $y = \Sigma b_i x^{i/m}$ is analytically equivalent to (X,x) when $a_i = b_i$ for $i \leq N$. N depends only on the topological type of the singularity.

Remark :

N may be computed from just the Puiseux pairs of X, or other topological invariants of the singularity (cf. 8.5).

The extent to which the coefficients a_i $(i \leq N)$ can still vary is difficult to say in general, as the following analysis of two examples shows.

Example 1 :

An irreducible singularity with (3,7) as its single Puiseux pair is analytically equivalent to the singularity

$$y = x^{7/3}$$

if the coefficient of $x^{8/3}$ vanishes in the Puiseux expansion. Otherwise the singularity is analytically equivalent to

$$y = x^{7/3} + x^{8/3}.$$

The singularities $y = x^{7/3}$ and $y = x^{7/3} + x^{8/3}$ are not analytically equivalent.

Example 2 :

The singularities with the Puiseux series

$$y = x^{9/4} + x^{10/4} + tx^{11/4} \quad (t \in \mathbb{C})$$

are all topologically equivalent (the single Puiseux pair is (4,9)), but analytically inequivalent.

To prove these assertions we apply the following general fact :

<u>**Lemma 17**</u> :

Suppose (X_1,x_1) resp. (X_2,x_2) are irreducible analytic curve germs and let $\pi_1 : (\tilde{X}_1,\tilde{x}_1) \to (X_1,x_1)$ resp. $\pi_2 : (\tilde{X}_2,\tilde{x}_2) \to (X_2,x_2)$ be resolutions of their singularities. Then (X_1,x_1) is analytically equivalent to (X_2,x_2) just in case there is an isomorphism $\tilde{\phi} : (\tilde{X}_1,\tilde{x}_1) \to (\tilde{X}_2,\tilde{x}_2)$ such that the associated homomorphism of local rings, $\tilde{\phi}^* : \mathcal{O}_{\tilde{X}_2,\tilde{x}_2} \to \mathcal{O}_{\tilde{X}_1,\tilde{x}_1}$ maps the subring $\pi_2^*(\mathcal{O}_{X_2,x_2})$ bijectively onto $\pi_1^*(\mathcal{O}_{X_1,x_1})$.

<u>**Proof of the lemma**</u> :

If $\phi : (X_1,x_1) \to (X_2,x_2)$ is an isomorphism of analytic set germs, then ϕ may be lifted to a homeomorphism $\tilde{\phi} : \tilde{X}_1 \to \tilde{X}_2$ so that the following diagram commutes :

$$\begin{array}{ccc} (\tilde{X}_1,\tilde{x}_1) & \xrightarrow{\tilde{\phi}} & (\tilde{X}_2,\tilde{x}_2) \\ \downarrow \pi_1 & & \downarrow \pi_2 \\ (X_1,x_1) & \xrightarrow{\phi} & (X_2,x_2) \end{array}$$

Since $\tilde{\phi}$ is analytic away from $\tilde{x}_1$, $\tilde{\phi}$ is an analytic isomorphism by the Riemann extension theorem. Obviously $\tilde{\phi}$ is the desired mapping.

Conversely, such a $\tilde{\phi}$ yields the isomorphism

$$\pi_1^{*-1} \circ \tilde{\phi}^* \circ \pi_2^* : \mathcal{O}_{X_2,x_2} \to \mathcal{O}_{X_1,x_1}$$

of analytic rings. (This is well defined because π_1^* is injective.) As we have seen in 8.2, the two set germs are then analytically equivalent.

<u>**Analysis of example 1**</u> :

Since the number N in Theorem 16 is a topological invariant, we can compute it with the help of the standard expansion with given Puiseux pairs. In the case of the singularity of example 1 the standard expansion is $y = x^{7/3}$, hence by Theorem 1 the resolution of the singularity is of the form

$$\pi_0 : \mathbb{C} \to \mathbb{C}^2 , \; z \mapsto (z^3,z^7)$$

and $\pi_0^*(\mathcal{O}_{X,0}) = \{\Sigma a_k z^k \mid a_1=a_2=a_4=a_5=a_8=a_{11} = 0\}$. Consequently $N = 12$, and by Theorem 16 we may break off the Puiseux expansion of the singularity in example 1 after the term $x^{12/3}$. Moreover, we make the

Assertion :

The singularity with the Puiseux series

$$y = x^{7/3} + a_8x^{8/3} + a_9x^{9/3} + a_{10}x^{10/3} + a_{11}x^{11/3} + \dots$$

is equivalent to the singularity

$$y = x^{7/3} + a_8x^{8/3}.$$

Proof :

We have to show that the subring R' of $\mathbb{C}\{z\}$ consisting of the power series in

$$z^3 \quad \text{and} \quad z^7 + a_8z^8 + a_9z^9 + a_{10}z^{10} + a_{11}z^{11} + \dots$$

can be carried into the ring generated by

$$z^3 \quad \text{and} \quad z^7 + a_8z^8$$

by coordinate transformation. By a theorem of Gorenstein, the ideal generated by z^{12} lies in both rings, hence one can work modulo (z^{12}) (cf. Hironaka [H2], p. 155). Thus let $R = R'/(z^{12})$. Then R is the ring in $\mathbb{C}\{z\}/(z^{12})$ generated by

$$z^3 \quad \text{and} \quad z^7 + a_8z^8 + a_9z^9 + a_{10}z^{10} + a_{11}z^{11}.$$

Since $z^9 \in R$, R is also generated by

$$z^3 \quad \text{and} \quad z^7 + a_8z^8 + a_{10}z^{10} + a_{11}z^{11}.$$

Since $z^3(z^7+a_8z^8+a_{10}z^{10}+a_{11}z^{11}) = z^{10} + a_8z^{11}$ in R, the ring R is also generated by

$$u = z^3 \text{ and } v = (z^7+a_8z^8+a_{10}z^{10}+a_{11}z^{11}) - a_{10}(z^{10}+a_8z^{11})$$
$$= z^7 + a_8z^8 + cz^{11}.$$

Making the first coordinate transformation, $z \mapsto z + \lambda z^5$, with $\lambda = -c/7$, u and v go respectively to

$$u' = (z+\lambda z^5)^3 = z^3 + 3\lambda z^7 + 3\lambda^2 z^{11}$$

$$v' = z^7 + a_8z^8 + 7\lambda z^{11} + cz^{11} = z^7 + a_8z^8.$$

The ring generated by u' and v' is also generated by

$$u' - 3\lambda v' = z^3 - 3\lambda a_8z^8 + 3\lambda^2 z^{11} \quad \text{and} \quad v' = z^7 + a_8z^8.$$

The coordinate transformation $z \mapsto z + \lambda a_8z^6$ converts this ring into the one generated by

$$z^3 + 3\lambda^2 z^{11} \quad \text{and} \quad z^7 + a_8z^8.$$

With the coordinate transformation $z \mapsto z - \lambda^2 z^9$ we finally come to the ring generated by

$$z^3 \text{ and } z^7 + a_8 z^8,$$

and this proves the assertion.

Now let $(X_\lambda, 0)$ be the singularity with the Puiseux expansion $y = x^{7/3} + \lambda x^{8/3}$, so that $x = z^3$ and $y = z^7 + \lambda z^8$. By Lemma 17, the coordinate transformation $z \mapsto \lambda^{-1} z$ immediately converts the singularity $(X_\lambda, 0)$ into $(X_1, 0)$ for $\lambda \neq 0$. This proves that an irreducible singularity with only one Puiseux pair is analytically equivalent to either $(X_0, 0)$ or $(X_1, 0)$. It remains to show only that $(X_0, 0)$ and $(X_1, 0)$ are not isomorphic.

Suppose now that the singularities $(X_0, 0)$ and $(X_1, 0)$ were isomorphic. Then by Lemma 17 there would be an automorphism ϕ of the analytic algebra $\mathbb{C}\{z\}$ carrying the ring $R_0 := \pi_0^*(\mathcal{O}_{X_0,0})$ of all series in z^3 and z^7 into the ring $R_1 := \pi_1^*(\mathcal{O}_{X_1,0})$ of all series in z^3 and $z^7 + z^8$. The automorphism ϕ carries the maximal ideal (z) of $\mathbb{C}\{z\}$ into itself, and hence also induces an isomorphism of the ring $\mathbb{C}\{z\}/(z^r)$ for all $r \geq 0$. Obviously, ϕ is already determined by the image of z,

$$\phi(z) = \sum_{k \geq 1} a_k z^k \quad \text{(without loss of generality, } a_1 = 1\text{)}.$$

The series z^3 lies in R_1, hence there is an $x \in R_0$ with $\phi(x) = z^3$. Computing modulo (z^5), we obtain

$$x \equiv c \cdot z^3 \mod (z^5) \quad (\text{with } c \in \mathbb{C})$$

and it follows that

$$z^3 = \phi(x) \equiv c \cdot \phi(z^3) \equiv c(z^3 + 3a_2 z^4) \mod (z^5),$$

hence $a_2 = 0$.

On the other hand, $z^7 + z^8$ is also in the image of R_0 under ϕ, say $z^7 + z^8 = \phi(y)$ with $y \in R_0$. If we compute modulo (z^9), we obtain

$$y \equiv c' \cdot z^7 \mod (z^9),$$

hence

$$z^7 + z^8 \equiv c' \cdot \phi(z^7) \equiv c' \cdot (z^7 + 7a_2 z^8) \mod (z^9).$$

But this is impossible because $a_2 = 0$.

Analysis of example 2 :

We show that the singularities X_t with Puiseux expansions

$$y = x^{9/4} + x^{10/4} + tx^{11/4}$$

are all equivalent topologically, but distinct analytically.

Proof :

The topological equivalence is clear, because all these singularities have just the single Puiseux pair (4,9). To prove the analytic distinctness we again use Lemma 17. Thus let $\pi_t : (\mathbb{C},0) \to (X_t,0)$ be the resolution of the singularity by the Puiseux expansion

$$\pi_t(z) = (z^4, z^9+z^{10}+tz^{11}),$$

and let $R_t := \pi_t^*(\mathcal{O}_{X_t,0}) \subset \mathbb{C}\{z\}$ be the subalgebra of power series in z^4 and $z^9 + z^{10} + tz^{11}$. We now assume that there is an automorphism ϕ of $\mathbb{C}\{z\}$ which carries R_s into R_t. We shall show that this implies $s = t$, and then we will be finished.

Let $x,y \in R_s$ be elements such that

$$\phi(x) = z^4 \tag{1}$$

$$\phi(y) = z^9 + z^{10} + tz^{11}. \tag{2}$$

Such elements must exist by the hypothesis on ϕ. The automorphism ϕ is determined by $\phi(z)$. Suppose

$$\phi(z) = k_0 z + k_1 z^2 + k_2 z^3 + \ldots \quad \text{with } k_0 \neq 0. \tag{3}$$

Since x lies in the ring R_s, which is generated by z^4 and $z^9 + z^{10} + sz^{11}$, x has no terms of order 5, 6 or 7 as a power series in z. Thus if we compute modulo (z^7) we obtain :

$$x \equiv cz^4 \bmod (z^7), \quad 0 \neq c \in \mathbb{C}. \tag{4}$$

It follows from (3) and (4) that

$$\begin{aligned}\phi(x) &\equiv c\cdot\phi(z^4)\\ &\equiv c(k_0^4 z^4 + 4k_0^3 k_1 z^5 + (4k_0^3 k_2 + 6k_0^2 k_1^2) z^6) \bmod (z^7).\end{aligned} \tag{5}$$

Comparing coefficients in (1) and (5) yields

$$k_1 = k_2 = 0. \tag{6}$$

We argue similarly for the element y. Since $y \in R_s$,

$$y \equiv d(z^9+z^{10}+sz^{11}) \quad \bmod (z^{12}) \tag{7}$$

for a nonvanishing constant d. It follows from (3), (6) and (7) that:

$$\phi(y) = d\cdot\phi(z^9+z^{10}+sz^{11})$$

$$\equiv dk_0^9\cdot(z^9+k_0z^{10}+sk_0^2z^{11}) \mod(z^{12}). \qquad (8)$$

Comparing coefficients in (2) and (8) first yields $dk_0^9 = 1$, then $k_0 = 1$ and finally the assertion claimed : $s = t$.

Thus examples 1 and 2 together show that a given topological type may consist of infinitely many analytically distinct singularities, which form a complex analytically parametrised family of curves, or of only finitely many analytically distinct singularities, depending on the topological type. The Puiseux pair (4,9) is an example for the first case, and (3,7) is an example for the second.

CHAPTER 17: THE TURK'S-HEAD

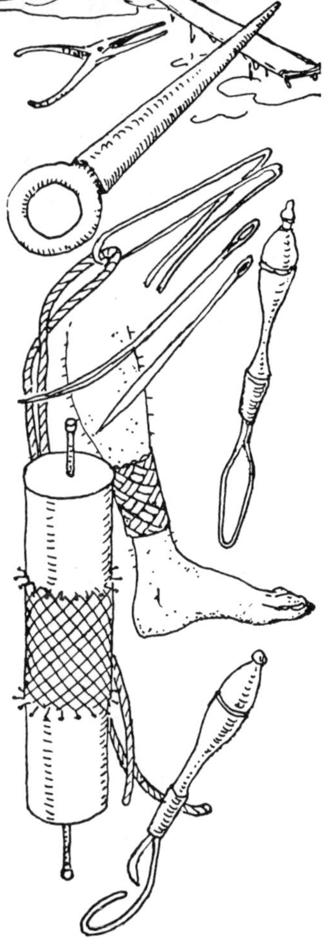

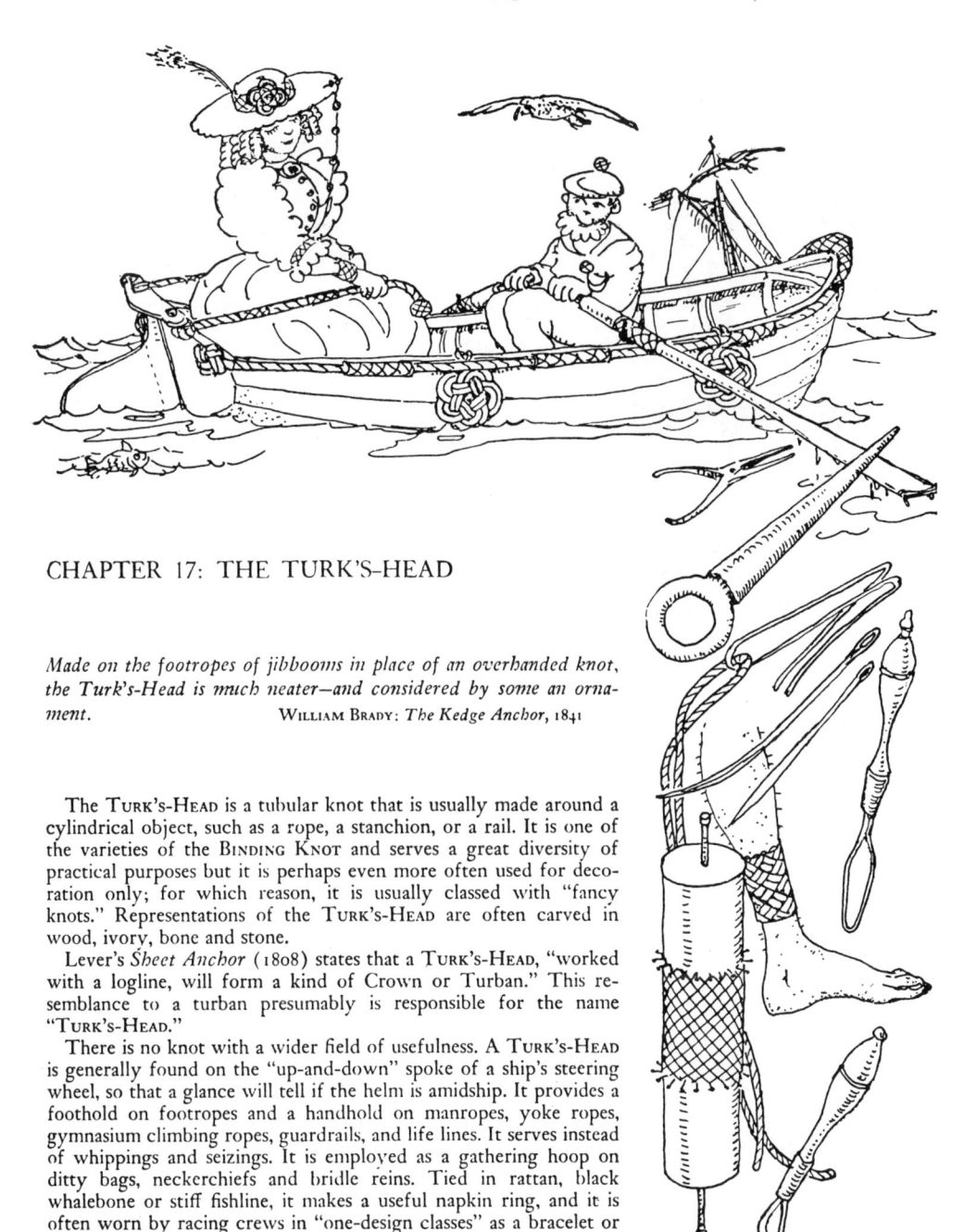

Made on the footropes of jibbooms in place of an overhanded knot, the Turk's-Head is much neater—and considered by some an ornament. William Brady: *The Kedge Anchor*, 1841

The Turk's-Head is a tubular knot that is usually made around a cylindrical object, such as a rope, a stanchion, or a rail. It is one of the varieties of the Binding Knot and serves a great diversity of practical purposes but it is perhaps even more often used for decoration only; for which reason, it is usually classed with "fancy knots." Representations of the Turk's-Head are often carved in wood, ivory, bone and stone.

Lever's *Sheet Anchor* (1808) states that a Turk's-Head, "worked with a logline, will form a kind of Crown or Turban." This resemblance to a turban presumably is responsible for the name "Turk's-Head."

There is no knot with a wider field of usefulness. A Turk's-Head is generally found on the "up-and-down" spoke of a ship's steering wheel, so that a glance will tell if the helm is amidship. It provides a foothold on footropes and a handhold on manropes, yoke ropes, gymnasium climbing ropes, guardrails, and life lines. It serves instead of whippings and seizings. It is employed as a gathering hoop on ditty bags, neckerchiefs and bridle reins. Tied in rattan, black whalebone or stiff fishline, it makes a useful napkin ring, and it is often worn by racing crews in "one-design classes" as a bracelet or anklet. It will cover loose ends in sinnets and splices. It furnishes a handgrip on fishing rods, archery bows and vaulting poles. It will stiffen sprung vaulting poles, fishing rods, spars, oars and paddles. On a pole or rope it will raise a bole big enough to prevent a hitch

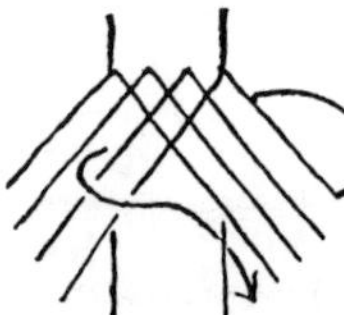

1300

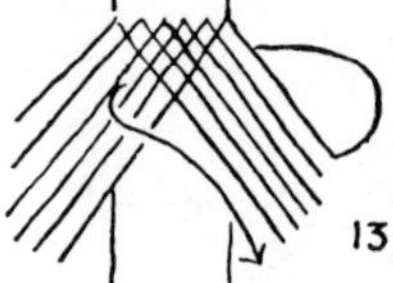

1301

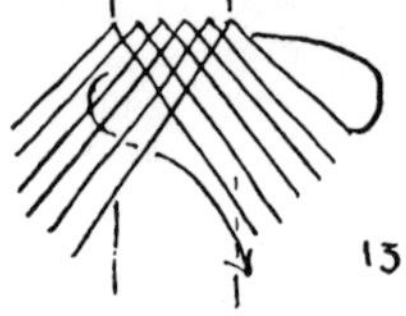

1302

1303

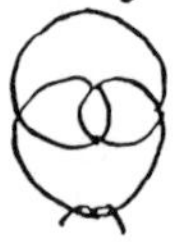

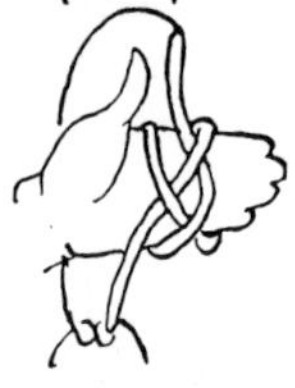

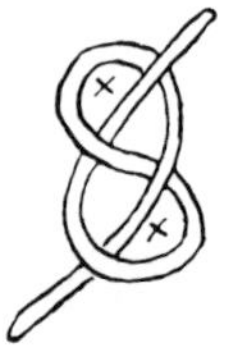

1304

1305

1300. Coachwhipping, based on Square Sinnet, makes a herringbone weave. The directions for Square Sinnet are given on page 493. This may be made with eight strands around a rope or rail, and gives four lengthwise rows of "herringboning." The legs may be left long enough for sticking back at both ends, which is done in the manner shown as #1290.

1301. Square Sinnet of twelve and sixteen strands can be employed in the same way, using three or four strands to each unit, as the case may be.

1302. Six rows of herringboning will result if the strands are led as shown here. Care must be exercised in these last two to arrange the seizings so that the rims will be symmetrical. The ends should be stuck back with a needle before removing the seizings. Some of the ends are stuck once and trimmed, others are led back two and three tucks in order to scatter them. Coachwhipping ordinarily is not doubled; it is completed in one operation. But if the surface has not been completely covered, double the knot, using a needle.

The common Turk's-Head is made of a single continuous line and is an older knot than the multi-strand one. Sometimes it is called the Running Turk's-Head, a term which may have been applied in contradistinction to Standing Turk's-Head, or it may be descriptive of the sailor's use of the knot as a gathering hoop or puckering ring to slide up and down on bag lanyards, neckerchiefs, etc. It should be understood that whenever the name "Turk's-Head" is applied by sailors without qualification, the single-line knot is always the one that is referred to.

The name "Turk's-Head" first appears in Darcy Lever's *The Sheet Anchor* (1808), but the knot is much older. I have a powder horn dated 1676 which has several Turk's-Heads carved around it, and Leonardo da Vinci (1452–1519) shows a number in disk form, in a drawing that is reproduced by Öhrvall in *Om Knutar* (1916).

In discussing the Single-Strand Turk's-Head the use of the word *strand* will be avoided as it is ambiguous. *Cord* or *line* will designate the material of the knot and the word *lead* will designate a single circuit of the cord around the cylinder or barrel. The size of a knot is designated by the number of its leads and bights. Bights are the scallops or coves formed by the cord where it changes direction at the rims. The total number of leads denotes the width of a knot along the cylinder, and the total number of bights denotes the length of a knot *around* the barrel or cylinder.

Each reappearance of the *cord* or *lead* on the surface will be termed a *part*. Only one part, the upper one, is in evidence at each crossing in the finished knot. To *follow* a cord or lead is to parallel it with identical over-and-under sequence, which alternates in the common Turk's-Head. When a lead has been followed throughout a whole knot, the knot is said to have been *doubled*.

The sailor interprets the word *double* in his own way. When a finished knot consists of two parallel cords the sailor describes it as having been *doubled twice*; when it exhibits three parallel cords throughout, it has been *doubled three times*.

A knot that is doubled three times is said by sailors to have three lays. It is also called a Three-Ply Knot.

Tucking over a cord is the same as *passing* or *crossing over*. A sailor may *tuck* either under and over, or over and under.

1303, 1305. Ordinarily the sailor ties a Turk's-Head directly around his fingers. When it has been formed it is placed around the object that is to be its permanent support.

THE TURK'S-HEAD

There are two sizes that the sailor commonly ties in this direct manual way: #1303, which has three leads and two bights; and #1305, which has three leads and four bights.

1304. An unusual but simple method of tying the THREE-LEAD, TWO-BIGHT TURK'S-HEAD is to first make the FIGURE-EIGHT KNOT, then insert thumb and finger into two compartments as shown, and pinch them together. When the two ends meet the knot is complete.

1306. The sailor also ties the THREE-LEAD BY FIVE-BIGHT KNOT, either directly or more often by lengthening #1305, a process that is later described as #1316.

1307. Occasionally he ties directly the FIVE-LEAD BY THREE-BIGHT KNOT as shown here. After reaching the position of the left diagram, the left turn of the two center leads is shifted to the right over the next one to assume the position of the right diagram. To complete the knot, follow the line indicated by the arrow. Any of the TURK'S-HEADS may be doubled or tripled by paralleling one end with the other.

1308, 1309, 1310, 1311. There are several manual methods of tying the FOUR-LEAD by THREE-BIGHT KNOT. No particular technique is required. After reaching the position shown in any final diagram the knot is placed around its permanent support and "faired," but not drawn up. The lay is then paralleled as many times as wished by "following the lead" that has been established. To do this tuck in one end beside its opposing end, and continue to tuck contrariwise and parallel with the other end, following the lead with identical over-and-under sequence. The second lead must be kept always on the same side of the first lead, either right or left according to how it was started. When the knot has as many plies as desired it is worked snug with a pricker. This is done by progressing from one end of the cord to the other through the whole knot, back and forth, gradually pricking up and hauling out the slack. The knot must not at any time be distorted by pulling too strongly on any one part. When completed it should be so snug around its support that it will not slip. To tie #1311: Start as if you were making KNIFE LANYARD KNOT #781.

I have known several sailors who could tie directly in hand 4L × 5B and 5L × 4B TURK'S-HEADS but in each case their methods were individual and often too cumbersome to be generally practical. They were also perhaps unnecessary, as it is easier to tie large knots by *raising* smaller ones to larger dimensions. For this purpose there are several different methods to follow.

There is but one actual limitation to the size and proportions of SINGLE-LINE TURK'S-HEADS: *A knot of one line is impossible in which the number of leads and the number of bights have a common divisor.* All others are possible if the knot tier has sufficient time and cord at his disposal.

This "Law of the Common Divisor" was discovered at the same time by George H. Taber and the author.

The operation of the Law of the Common Divisor is quite simple. For example, within the limits of twenty-four leads and twenty-four bights there are 576 combinations. Of these combinations, 240 have a common divisor and cannot be tied as a TURK'S-HEAD, and 336 have no common divisor and can be tied. If a knot is attempted in one cord with dimensions that possess a common divisor, the working end and the standing end will meet before the desired knot is complete. Such a knot, being composed of more than one line, can be tied only as a MULTI-STRAND KNOT.

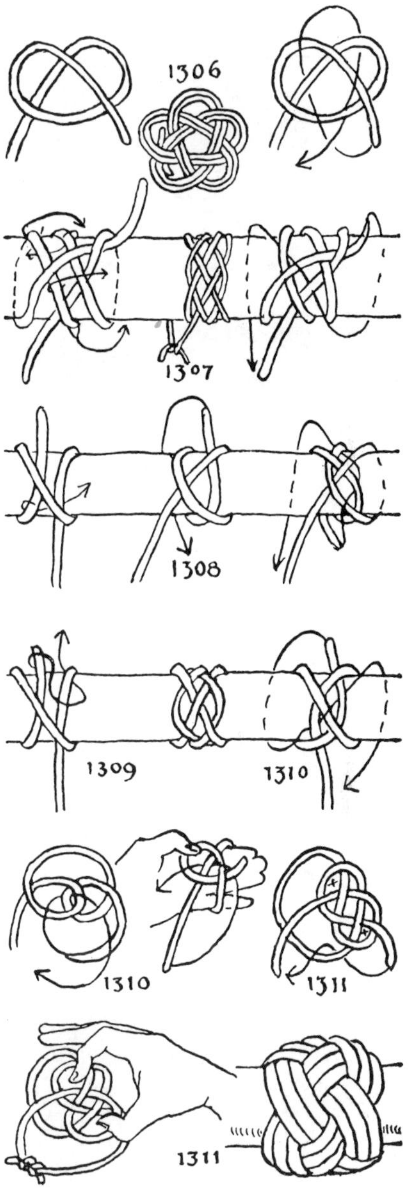

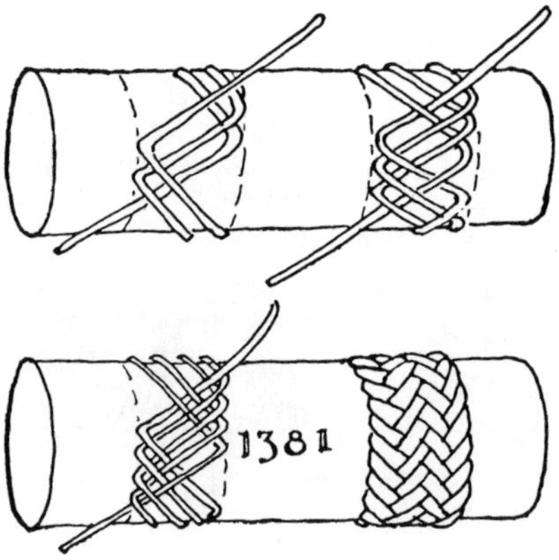

1381

HERRINGBONE WEAVE

1381. The four Square Turk's-Heads of page 236 lend themselves readily to different weaves, of which over-two-and-under-two and over-three-and-under-three are the simplest. The knots are tied on the barrel in much the same manner that has been described already for Knots #1325–28. With an over-two-under-two lead, a knot is completed each time four bights are added to each rim, and with over-three-under-three, a knot is completed each time six bights are added to each rim.

To tie an "Over-Two-Under-Two" Knot: Start as in first diagram in Knot #1325. Take all crossings *over* until three *parallel* leads are encountered. Tuck *under* the first one in each group of three, until a group of four parallel leads is met. Thereafter tuck over two and under two. A knot is completed at any time when the lead runs over-two-under-two throughout. Then tie the two ends together.

To tie an "Over-Three-Under-Three" Knot: When the knot has progressed as far as the first diagram, make one more circuit *over-all.* When four parallel leads are encountered, tuck under the first one of them; when five parallel leads are encountered, tuck under the first two of them; and when six parallel leads are encountered, tuck under the first three of them and over the second three. When completing the knot, butt the ends together as in #1329 and #1330, and withdraw them into the middle of the knot without "doubling."

1382

A ROUND SINNET TURK'S-HEAD

1382. Turk's-Heads of French Sinnet, Chain Sinnet and Flat Sinnet have been shown, and only Solid Sinnets have been left unconsidered. The first attempted was the Round Sinnet of six strands, which makes a Turk's-Head of two cords. A working drawing was made in circular form, with strands widely separated so that all crossings were clearly depicted. This proved feasible, but a more practical method suggested itself. A very loose grommet (#2864) was made. Into this the ends of three shoestrings were tucked over and under, exactly as in short splicing, until they encircled the grommet, after making the same number of turns but in the opposite direction as the banding. Opposite ends were then knotted together, taking care that two ends of the same string were not bent together. The shoestrings serve merely as a clue. Next untie one of the three knots and replace the shoestrings with a long cord of the same material as the grommet. Double the knot that has been made, using a wire needle. Half knot, and "bury" opposing ends as in Long Splicing #2697.

1382

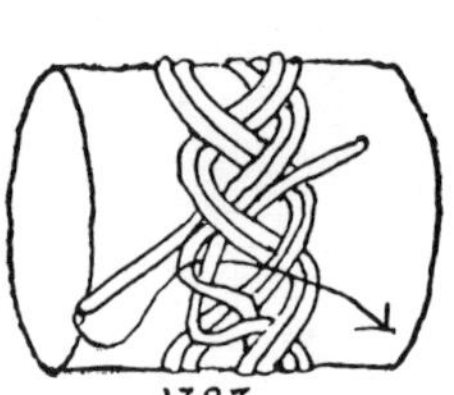

1383

A core consisting of an ordinary grommet (#2864) is advisable, if the knot is to be doubled or tripled.

1383

1383. To make a Turk's-Head employing a *continuous length* of Three-Strand Flat Sinnet for the basic material: First form an ordinary Three-Lead Turk's-Head (#1306) and double it, leaving one long end. With this end and the two parallel leads already established proceed to plat a Three-Strand Flat Sinnet in the ordinary way (#1315 and #1316) but following the line of the Turk's-Head that has been formed.

THE ASHLEY BOOK OF KNOTS

2425

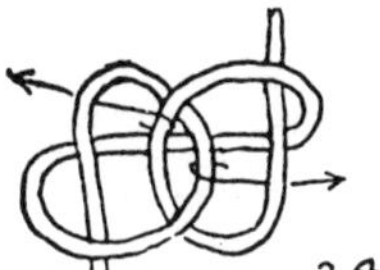

2428

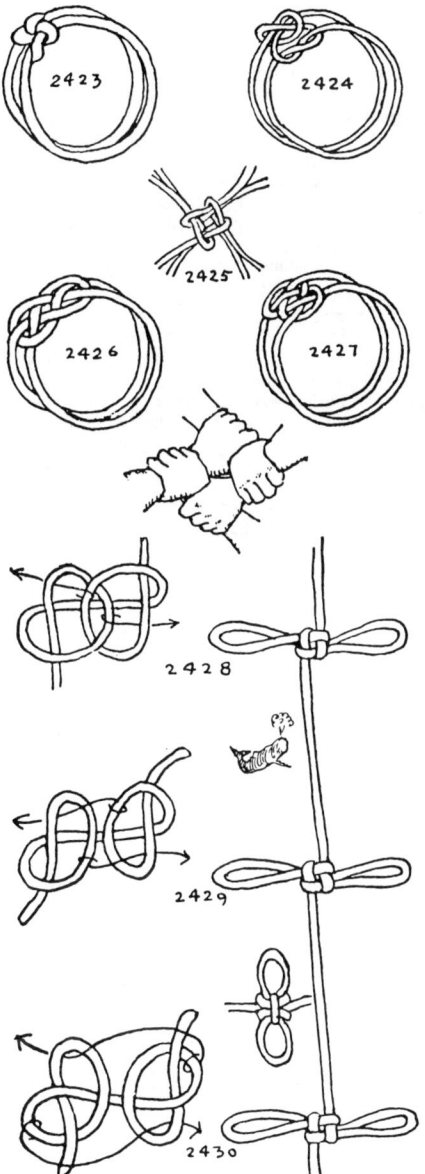

2429

2430

2423. This is Knot #2418 tied with gold wires. Formerly this was a very characteristic gift from a sailor to his sweetheart. The two rings could move independently but could not be separated, which undoubtedly carried a meaning to the young couple that made the long separation entailed by a sea voyage more tolerable.

2424. Jewelers not conversant with the symbolism of the previous knot were apt to make the True-Lover's Knot in this form, which, while it superficially seemed about the same, consisted of but one wire and held the two rings rigidly together.

2425. A True-Lover's Knot tied in the bights of four short strands symbolizes the four clasped hands of two lovers arranged in the manner called a "hand chair." The knot was shown to me by Mrs. Osborn W. Bright.

2426. A True-Lover's Knot in ring form, based on the Carrick Bend. This, like #2424, is made of a single wire, and I cannot but feel that it was less popular than the one to follow, which does not submerge all individuality. One reason that I believe that this knot is held in lower esteem by the sailor is that the only two specimens I have ever seen were both of silver. Certainly the baser metal must bear some significance. Moreover one of them was exposed for sale in a pawnshop window.

2427. A Carrick Bend in two separate rings which interlock harmoniously, perhaps the handsomest of the four rings shown.

There are probably other knots that are called by the name True-Lover's, but these are all I have found that fill the specifications that were somewhat arbitrarily adopted. These are outlined on page 383.

2428. The following knots, through page 389, are Single-Cord Lanyard Knots with a loop at either side and a four-part crown at the center. Two-Strand Knot #2451 is the well-known Chinese Crown Knot. I set out to find a knot of similar appearance for a *single-cord lanyard* and this series resulted before I was through. The first five are of similar aspect. The simplest is #2432, which is the common Sheepshank Knot with the parts pulled together. This

2976

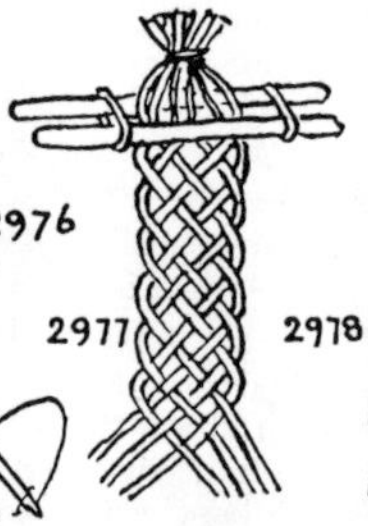

2977

2978

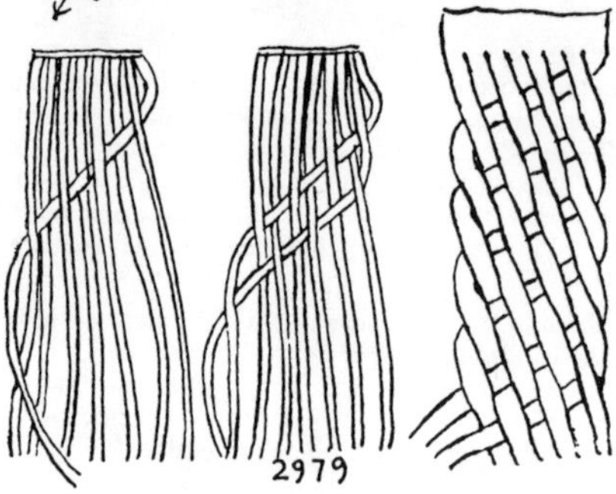

2979

2980

2981

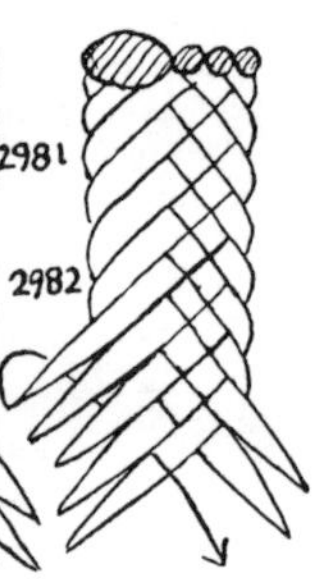

2982

2983

2984

2985

2976. **French Sinnet** is a flat plat in which the strands have a regular over-one-and-under-one weave. The French call it "Tresse Anglaise." As this has occasionally been translated literally in English books, it has caused some confusion. French Sinnet is generally tied with an odd number of strands but may be tied quite as satisfactorily with an even number. In working it the strands are customarily divided so that if there is an odd one it is placed with the group that is held in the left hand.

French Sinnet (1). The example given here is of seven strands. With four strands in the left hand, take the upper left strand and lead it diagonally down to the center, crossing its three sister strands alternately under, over and under. It has now become the lowest member of the right-hand group. Next take the upper member of the right-hand group (which now has four strands) and lead it to the left in the same order as before, under one, over one, under one. Repeat these two operations until sufficient sinnet is made.

2977. Six-Strand French Sinnet with three strands in the left hand and three in the right. Lead the top left strand to the center over one and under one. Follow with the top right strand under one, over one and under one, and repeat from the start.

2978. Double French Sinnet. This drawing illustrates a French Sinnet of five leads platted with doubled strands, ten in all.

2979. French Sinnet (2). In the braid and trimming trades ashore a different technique from #2976 is used, which gives a different character to the product, although it is structurally the same. All the strands are held in the left hand and only the top right strand is worked. This strand is led almost horizontally across *all the other* strands, in alternate over-and-under sequence. Then the next top right strand is treated likewise, and the process continued. The much shorter diagonal of the right strands is responsible for the changed appearance.

2980. The texture of sinnets based on Flat and French Sinnets has a range of possibilities analogous to weaving and the range widens rapidly as the number of the strands is increased. But the comparative narrowness of the sinnet limits the size of the patterns employed.

With eight strands (five in the left hand) lead the top left strand under three, and over one to the center. Then lead the top right strand over three to the center. Repeat the two movements.

2981. Six strands (three in the left hand). Lead the left strand over one and under one. Lead the right strand under one and over two. Repeat.

2982. Seven strands (five in the left hand). Move the left under three, over one. Move the right under one, over one. Repeat.

2983. Seven strands (four in left hand). Move left outer strand under two, over one, move right strand under two, over one. Repeat.

8.4 Resolution of singularities by quadratic transformations

In the previous section we have already given an algorithm for the resolution of a plane singularity Y (8.3.1). With this resolution $\pi : B \to Y$, B is an abstract manifold, while the singularity Y is embedded in a two-dimensional space. In this section we want to construct a resolution $\pi' : \tilde{Y} \to Y$ for which $\tilde{Y}$ is likewise embedded in a surface, and indeed in such a way as to reflect the embedding of Y in the plane.

If a plane projective-algebraic curve is given, then one can alter its form and position by suitable transformations of the projective plane $P_2(\mathbb{C})$. We already know that linear transformations make no essential change in the singularities of the curve. The next most complicated transformations are quadratic transformations. These were already used by Newton and Maclaurin (cf. the remarks in 8.3).

Consider for example the quadratic transformation T of $P_2(\mathbb{C})$ given by*

$$T : (x_0, x_1, x_2) \mapsto (x_1x_2, x_0x_2, x_0x_1).$$

We apply the transformation to the quartic with the equation

$$x_1^2x_2^2 + x_0^2x_2^2 + x_0^2x_1^2 = 0.$$

This quartic has three ordinary double points, at the "fundamental points" $(1,0,0)$, $(0,1,0)$ and $(0,0,1)$ of $P_2(\mathbb{C})$. Application of the quadratic transformation T yields the equation

$$(x_0x_2x_0x_1)^2 + (x_1x_2x_0x_1)^2 + (x_1x_2x_0x_2)^2 = 0$$

or $$(x_0x_1x_2)^2(x_0^2+x_1^2+x_2^2) = 0.$$

I.e., the given curve goes over to the union of three lines and a conic. The three lines are in a certain sense insignificant ; essentially, we have transformed our quartic into a nonsingular quadric.

Conversely, the quadric $x_0^2 + x_1^2 + x_2^2 = 0$ goes over to the original quartic under T. Outside a certain point set, T is an involution : $T \circ T = 1$.

Now what is the geometric meaning of these transformations?

---

*We shall not at first worry about the fact that this mapping is not well defined on the whole of $P_2(\mathbb{C})$ – one can in any case compute algebraically with these substitutions.

Recall how we interpreted the homogeneous coordinates of $P_2(\mathbb{R})$ in 3.2, following Plücker, as triangle coordinates in the plane. A triangle is given in the plane, and the coordinates x_0, x_1, x_2 measure the distances from the three sides.

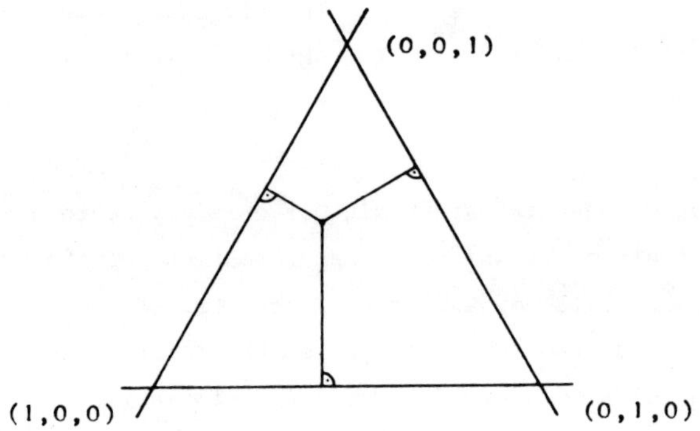

Lines through the fundamental points $p_0 = (1,0,0)$, $p_1 = (0,1,0)$ resp. $p_2 = (0,0,1)$ are determined by the ratios $x_1 : x_2$, $x_0 : x_2$ resp. $x_0 : x_1$. These ratios go to their inverses under the transformation T. Thus the transformation T sends a line through a vertex to its reflection in the corresponding angle bisector. This enables us to construct the image of each point (except the vertices) with ruler and compasses :

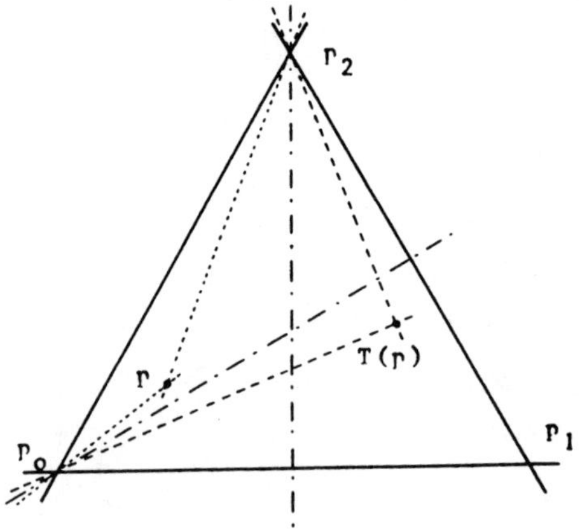

In this way one sees again, very clearly, that T is an involution. Now we construct the image of a circle under the transformation T. The form of the curve obtained depends on the position of the circle relative to the fundamental triangle. The following picture shows this curve for the case where the triangle is equilateral and the circle lies symmetrically with the triangle inside it. In the picture, some points on the circle are numbered in the positive direction, and their images on the curve are numbered correspondingly.

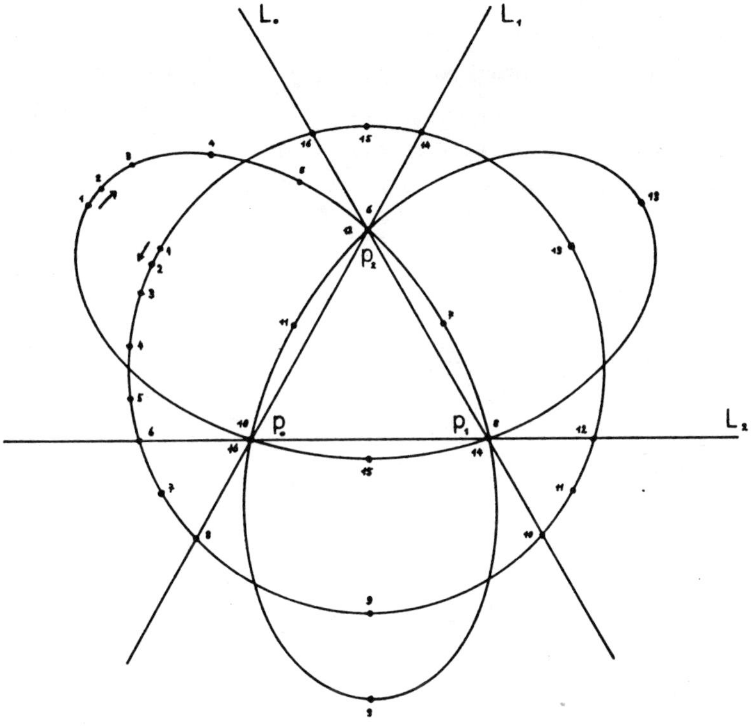

Quadratic transformation of a quadric into a quartic and vice versa : resolution of singularities by quadratic transformation.

The construction yields a quartic with three double points, lying precisely at the three fundamental points p_0, p_1, p_2. Conversely, this quartic goes back to the circle under T.

One observes the following : outside the fundamental triangle, T is bijective, and the sides $L_i := \{x \in P_2(\mathbb{C}) \mid x_i = 0\}$ of the fundamental triangle are mapped to the opposite vertices. Our circle cuts each side of the triangle twice, resulting in a double point of the image at a vertex of the triangle.

Now what should one consider the images of the fundamental points p_0, p_1, p_2 to be ? The images should be sensible enough to lie on the opposite sides. E.g., corresponding to the point p_1 of the quartic there are two different points of the circle, and indeed these two image points are just the limits of points of the quartic near the singular point. The image of the point p_1 of the quartic therefore depends on the direction in which the point is approached. For the point p_1 of the plane there is an image point for each direction – i.e. for each line g through p_1 – on the opposite side $L_1 : x_1 = 0$, namely its intersection with the reflection g' of g in the corresponding angle bisector (this can also be the point at infinity on L_1).

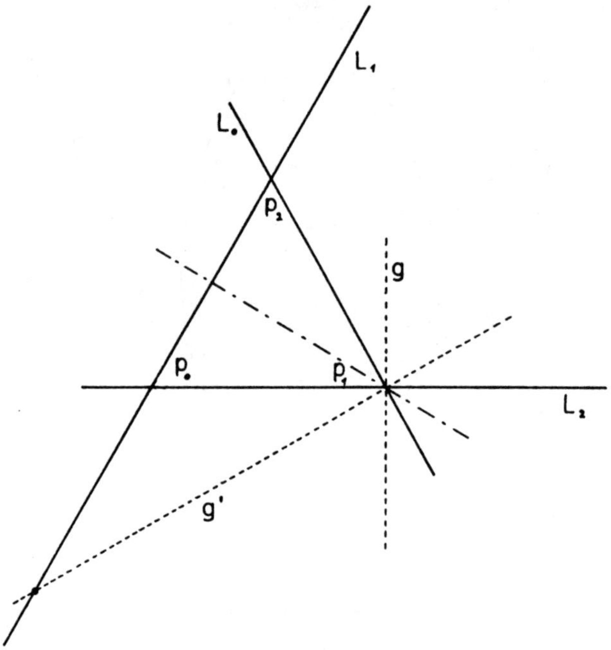

This is how the two branches of the quartic at the singular point are separated.

The transformation T causes singularities to vanish for certain curves, such as the quartic considered above. On the other hand, it can also produce singularities – as with the circle – in fact it always does when a curve cuts a triangle side in more than one point. We therefore want to abstract the process which has led to the vanishing of singularities, so as to find a similar process of a purely local nature. We shall then use the latter process for the resolution of singularities. Under the transformation T, each line through a fundamental point p_i yields an image point of p_i. Thus the essence of the process consists in replacing the point p_i by the set of all lines through p_i. This may be formulated with set-theoretic precision as follows :

Let $X = \mathbb{C}^2$, $p = 0$.

The set of all lines through p is a one-dimensional projective space (cf. 3.2).

$$\{g \mid g \text{ is a line through } p\} = P_1(\mathbb{C}).$$

We construct the set

$$X' := \{(x,g) \in \mathbb{C}^2 \times P_1(\mathbb{C}) \mid x \in g\}.$$

If $x \neq p$, then there is exactly one line through x and p, if $x = p$, then all lines of $P_1(\mathbb{C})$ go through x.

The projection

$$\pi : X' \to X$$
$$(x,g) \mapsto x$$

is therefore a homeomorphism over $X - \{p\}$, while the fibre $\pi^{-1}(p)$ of p is the projective line $P_1(\mathbb{C})$. Also, π is a proper mapping.

We want to investigate this projection $\pi : X' \to X$ and the space X' in more detail.

X' is a submanifold of $\mathbb{C}^2 \times P_1(\mathbb{C})$, because it is described by a regular equation :

$$X' = \{(x_0,x_1 ; z_0,z_1) \in \mathbb{C}^2 \times P_1(\mathbb{C}) \mid x_0 z_1 = x_1 z_0\}.$$

The usual coordinate covering of $P_1(\mathbb{C})$ yields one for $X \times P_1(\mathbb{C})$ and also for X' :

$$U_0 = \{(x;z) \in X' \mid z_0 \neq 0\}$$
$$U_1 = \{(x;z) \in X' \mid z_1 \neq 0\}$$

with the coordinates :

$$v_0 = x_0 \qquad u_0 = {}^{z_1}/_{z_0} \quad \text{resp.}$$
$$v_1 = x_1 \qquad u_1 = {}^{z_0}/_{z_1} .$$

The two coordinate neighbourhoods U_0 and U_1 are biholomorphically equivalent to $\mathbb{C}^2$ and the coordinate exchange is

$$u_1 = u_0^{-1}$$
$$v_1 = u_0 v_0 .$$

In the first component the transition function is just the coordinate exchange for the Riemann number sphere S^2. Hence $(u_0,v_0) \mapsto u_0$ resp. $(u_1,v_1) \mapsto u_1$ defines a holomorphic mapping $X' \to P_1(\mathbb{C})$, namely the mapping induced by the projection $\mathbb{C}^2 \times P_1(\mathbb{C}) \to P_1(\mathbb{C})$. Over each point of S^2 lies a complex line with v_0 resp. v_1 as coordinate. Thus X' is the total space of a complex line bundle* over S^2. Over the northern resp. southern hemisphere, D^+ resp. D^-, this bundle is trivial (i.e. isomorphic to $D^+ \times \mathbb{C}$ resp. $D^- \times \mathbb{C}$).

One can also think of S^2 as put together from a northern and southern hemisphere ; then the "coordinate exchange" along the equator is again $u_1 = u_0^{-1}$.

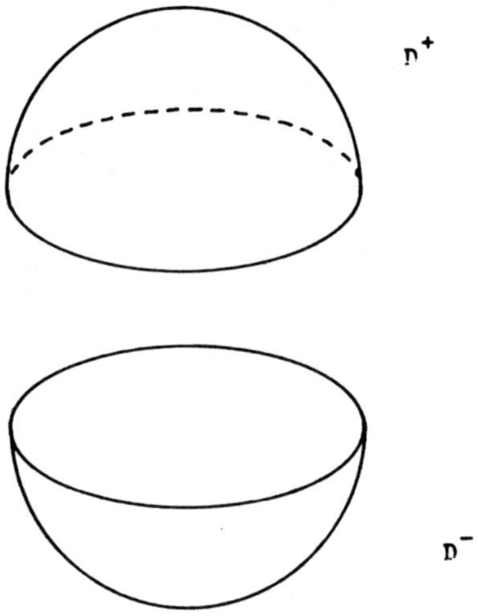

The two trivial bundles over D^+ resp. D^- are then pasted together along the equator by the "pasting function" u_0 which describes the coordinate exchange in the second component. The bundle described

---

*The basic facts about vector bundles may be found, say, in Bröcker-Jänich [B8] §3.

here is known as the Hopf bundle*.

The fact that X' is a line bundle over S^2 can of course also be read off directly from the definition. The projection

$$X' \to P_1(\mathbb{C})$$
$$(x,g) \mapsto g$$

has, as fibre over each line g, all the points of the line g, and we know that $P_1(\mathbb{C})$ is homeomorphic to S^2.

Most important in practice is the description of the projection $\pi : X' = U_0 \cup U_1 \to X$ in local coordinates

$$\pi : (u_0,v_0) \mapsto (v_0,u_0v_0) \quad \text{on } U_0$$
$$\pi : (u_1,v_1) \mapsto (u_1v_1,v_1) \quad \text{on } U_1.$$

This description of the mapping by a quadratic transformation should be remembered without fail.

We now try to give a real picture of the projection $\pi : X' \to X$.

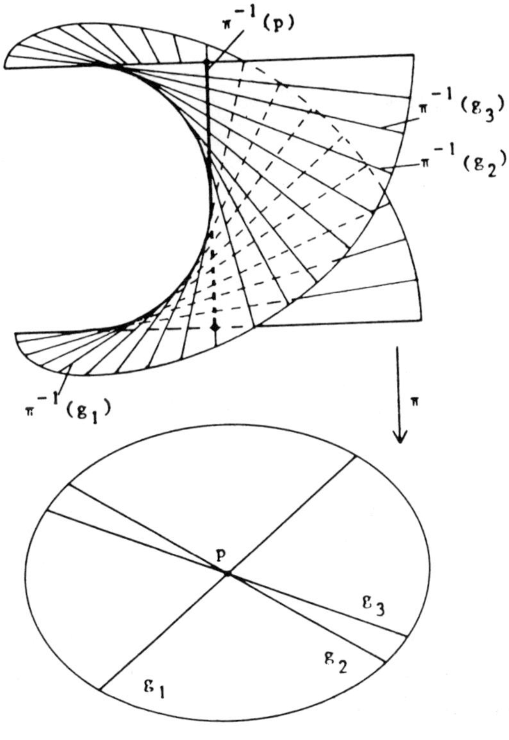

---

*The sphere bundle of the Hopf bundle, i.e. the set of all points $\{(x,g) \in X' \mid |x| = 1\}$ with the induced mapping on $P_1(\mathbb{C})$ is the Hopf fibration described in 3.4, incidentally.

The picture perhaps does not give quite the right impression, inasmuch as there is not a line over the point p, but a one-dimensional real projective space, and hence a circle S^1. Thus the "spiral staircase" surface shown above should really be a Möbius band.

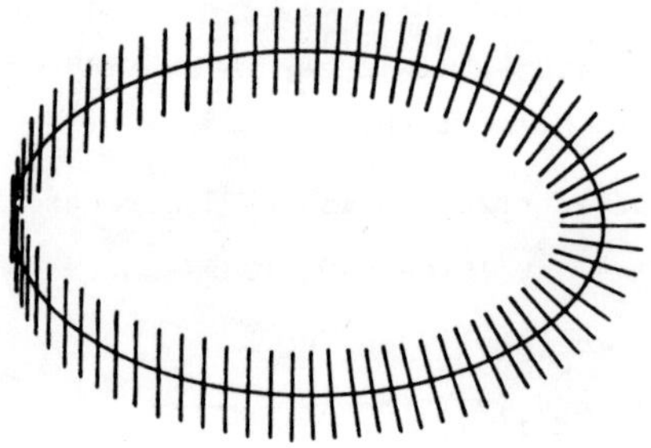

(The Möbius band is also the real analogue of the Hopf bundle described above.)

Thus in the reals the passage from X to X' looks as follows :

One cuts a disc with midpoint $p = 0$ out of $X = \mathbb{R}^2$ and replaces it by a (compact) Möbius band. We have already described this process in more detail in 3.3.

Here is yet another description of X'. X' is the closure in $\mathbb{C}^2 \times P_1(\mathbb{C})$ of the graph of the mapping

$$\mathbb{C}^2 - \{0\} \to P_1(\mathbb{C}) \qquad (x_1,x_2) \mapsto (x_1,x_2)$$

which sends each point of $\mathbb{C}^2 - \{0\}$ to the line through this point and 0. One proves this easily from the definition of X'.

Definition :

The process described above for passing from X to $\pi : X' \to X$ is called a σ-process with centre $p \in X$, or a quadratic transformation at p or blowing up the point p.

The words "blowing up" are appropriate inasmuch as the point p is replaced by a sphere $S^2 = P_1(\mathbb{C})$. A second interpretation of the words as "enlargement" would also be reasonable, because the process can separate curve branches which previously met, or reduce their contact (intersection number). The separation of things not previously separable to the eye is typical of enlargement. The term "quadratic transformation" comes about because π is a quadratic mapping in local coordinates. The newly introduced line $\pi^{-1}(0)$ is also called the exceptional line. In general, a curve C in a surface X' is called exceptional when it admits blowing down, i.e. when contraction of all the points of C to a single point results in a complex surface X. To distinguish the line introduced by the σ-process from these general exceptional curves, one also calls it an exceptional curve of the first kind.

The σ-process is a purely local process. For that reason we can quite easily generalise the blowing up of the 0-point in $\mathbb{C}^2$ as follows, and define the blowing up of an arbitrary point p in a 2-dimensional complex manifold M. Let $\phi : U \to V$ be a chart mapping of a neighbourhood U of p in M onto a neighbourhood V of 0 in $\mathbb{C}^2$, and let $\pi : X' \to \mathbb{C}^2$ be the σ-process at the origin of $\mathbb{C}^2$. Let V' be the complex manifold $V' = \pi^{-1}(V)$. The blowing up of p in M now takes place, intuitively speaking, by replacing the open set U biholomorphically equivalent to V by V'. More precisely : we define a complex manifold M' by pasting $M - \{p\}$ and V' together along $U - \{p\}$ resp. $V' - \pi^{-1}(0)$ by means of the biholomorphic mapping $\pi^{-1} \circ \phi : U - \{p\} \to V' - \pi^{-1}(0)$. One has a natural holomorphic mapping $\pi' : M' \to M$ given by the identity on $M' - V'$ and by $\phi^{-1} \circ \pi$ on V', and

$$\pi' : M' \to M$$

is the desired σ-process on M at p.

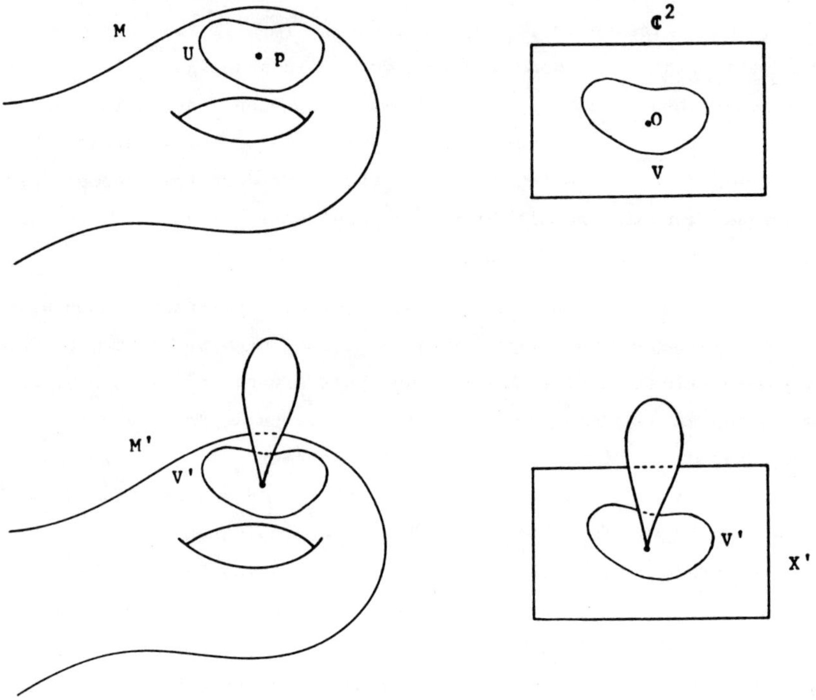

If ϕ' is another chart of M around p and if one carries out the blowing up operation as above with the chart ϕ', then one obtains a "blown up" manifold $\pi'' : M'' \to M$. One easily sees that then there is a biholomorphic equivalence $M' \overset{\cong}{\to} M''$ such that the diagram

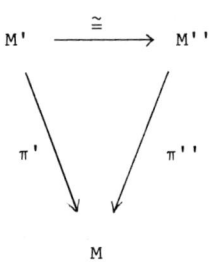

$$\begin{array}{ccc} M' & \xrightarrow{\cong} & M'' \\ & {}_{\pi'}\searrow \quad \swarrow_{\pi''} & \\ & M & \end{array}$$

commutes. The "blown up manifold" M' is therefore essentially unique.

<u>Remark</u> :

One can define σ-processes in just the same way in n dimensions. If one wants to blow up the origin in $\mathbb{C}^n$, then one sets

$X' := \{(x_1,\ldots,x_n ; z_1,\ldots,z_n) \in \mathbb{C}^n \times P_{n-1}(\mathbb{C}) \mid \exists \lambda \in \mathbb{C} \text{ with } x_i = \lambda z_i\}.$

Again one has n coordinate neighbourhoods isomorphic to $\mathbb{C}^n$:

$$U_i = \{(x_1,\ldots,x_n ; z_1,\ldots,z_n) \in X' \mid z_i \neq 0\}$$

with coordinates

$$u_1 = \frac{z_1}{z_i}, \ldots, u_{i-1} = \frac{z_{i-1}}{z_i}, u_i = x_i, u_{i+1} = \frac{z_{i+1}}{z_i}, \ldots, u_n = \frac{z_n}{z_i}.$$

The projection $\pi : U_i \to \mathbb{C}^n$ then has the form

$(u_1,\ldots,u_n) \mapsto (u_1u_i,\ldots,u_{i-1}u_i, u_i, u_{i+1}u_i ,\ldots, u_nu_i)$.

One can even define the concept of blowing up for essentially more general objects; and one actually needs these "monoidal transformations" for the resolution of arbitrary singularities of higher dimensional analytic sets. However, we do not wish to go into this here.

We shall now investigate the connection between the quadratic transformation T of $P_2(\mathbb{C})$ studied at the beginning, and the process of blowing up.

T was the transformation

$$T : P_2(\mathbb{C}) \to P_2(\mathbb{C})$$
$$(x_0,x_1,x_2) \mapsto (x_1x_2,x_0x_2,x_0x_1).$$

As above, let $p_0 = (1,0,0)$, $p_1 = (0,1,0)$, $p_2 = (0,0,1)$ be the fundamental points and let $L_i := \{x_i = 0\}$ be the sides of the fundamental triangle of $P_2(\mathbb{C})$.

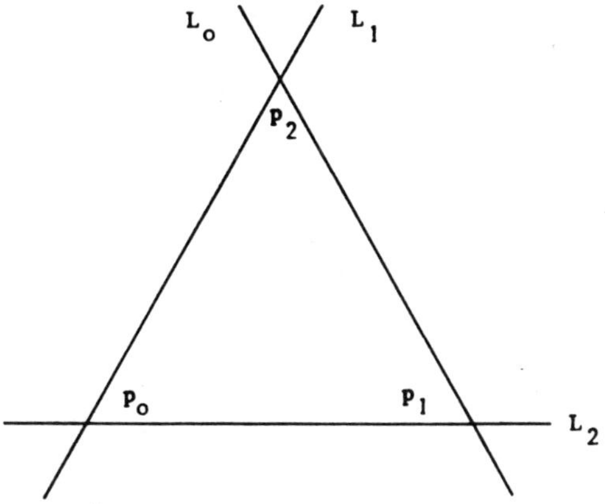

In order to save writing, let us agree that in this context the indices i, j, k appearing in a formula will always be pairwise distinct.

Thus :

$$T : x \mapsto z \quad \text{with} \quad z_i = x_j x_k .$$

T is not defined when two of the coordinates x_0, x_1, x_2 are simultaneously zero. This happens just at the fundamental points

$$p_0 = (1,0,0),\ p_1 = (0,1,0),\ p_2 = (0,0,1).$$

$$T : P_2(\mathbb{C}) - \{p_0, p_1, p_2\} \to P_2(\mathbb{C})$$

is therefore a well-defined mapping. Its graph is a certain subset of $P_2(\mathbb{C}) \times P_2(\mathbb{C})$ and its closure is the submanifold

$$\tilde{X} = \{(x_0, x_1, x_2; z_0, z_1, z_2) \in P_2(\mathbb{C}) \times P_2(\mathbb{C}) \mid z_i x_i = z_j x_j\}.$$

One has the two projections π_1 and π_2 of $\tilde{X} \to P_2(\mathbb{C})$. T carries a point $x \in P_2(\mathbb{C})$ over to $\pi_2 \circ \pi_1^{-1}(x)$:

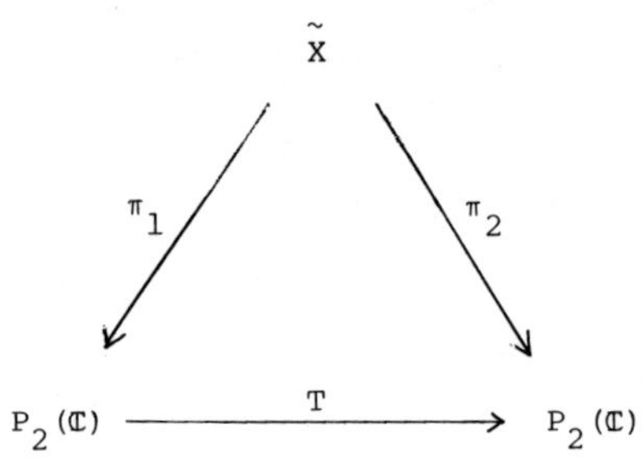

To study π_1 and π_2 we choose a covering of $\tilde{X}$ by the coordinate neighbourhoods

$$U_{ij} := \{(x;z) \in \tilde{X} \mid x_i \neq 0,\ z_j \neq 0\}$$

with the coordinates

$$u_{ij} = \frac{z_k}{z_j},\ v_{ij} = \frac{x_k}{x_i} .$$

On U_{ij}, the projection π_1 has the form

$$x_i = 1,\ x_j = u_{ij} v_{ij},\ x_k = v_{ij} \qquad (*)$$

π_2 has the form

$$z_i = u_{ij} v_{ij},\ z_j = 1,\ z_k = u_{ij}, \qquad (*)$$

and these are – relative to the usual coordinate neighbourhoods of $P_2(\mathbb{C})$ – just the formulae we have found for blowing up the point p_i.

More precisely, one sees that π_1 and π_2 are biholomorphic over $P_2(\mathbb{C}) - \{p_1, p_2, p_0\}$. Under π_1, the point p_i is blown up to the line

$$L_i' := \{(x,z) \in \tilde{X} \mid x_j = x_k = 0\}$$

and P_j is blown up analogously by π_2 to

$$L_j'' := \{(x,z) \in \tilde{X} \mid z_i = z_k = 0\}$$

where $\pi_1^{-1}(P_i) = L_i'$, $\pi_1^{-1}(L_i) = L_i''$
$\pi_2^{-1}(P_i) = L_i''$, $\pi_2^{-1}(L_i) = L_i'$

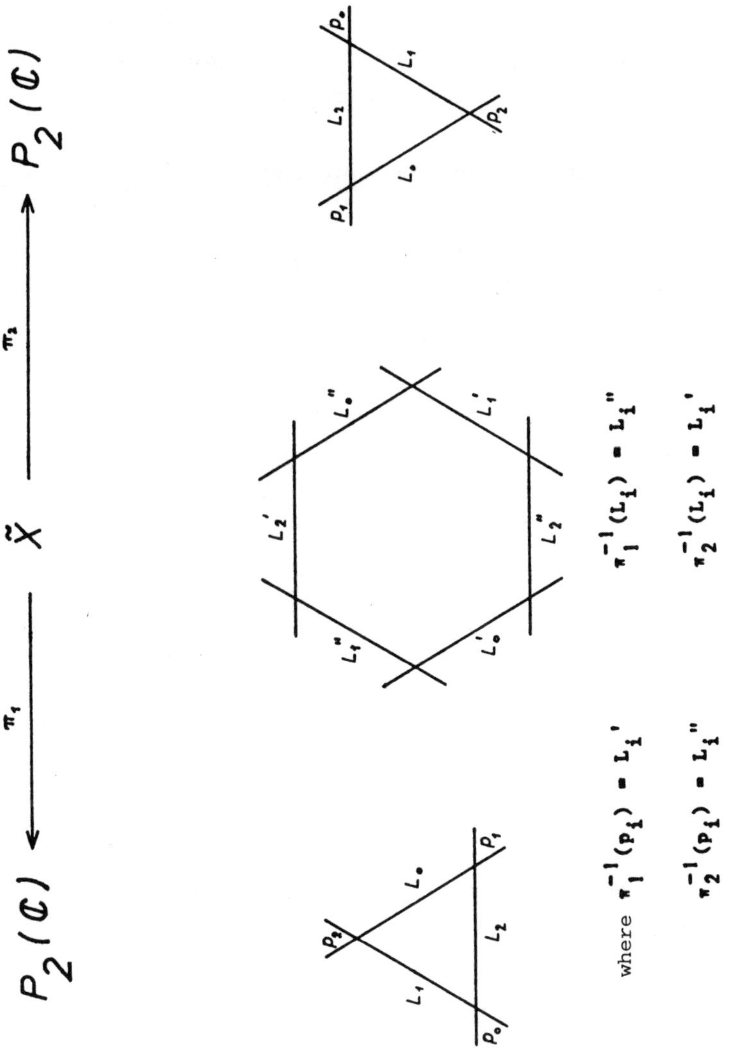

Summary :

Under the transformation T the fundamental points P_1, P_2, P_0 are blown up, and then the three lines L_1, L_2, L_0 are blown down.

From the standpoint of σ-processes, the transformation is already quite complicated. More generally, one can consider transformations of $P_2(\mathbb{C})$ which result from repeated blowing up and blowing down (as long as the manifold one ends with is again isomorphic to $P_2(\mathbb{C})$). This gives the group of <u>birational transformations</u> or <u>Cremona transformations</u> which was first investigated in full generality by L. Cremona around 1863-65.

In algebraic geometry, birational isomorphisms are usually defined somewhat differently : a <u>rational mapping</u> from $P_2(\mathbb{C})$ to $P_2(\mathbb{C})$ is a transformation $(x_0,x_1,x_2) \mapsto (F_0(x_0,x_1,x_2),F_1(x_0,x_1,x_2),F_2(x_0,x_1,x_2))$, where the F_i are homogeneous polynomials of the same degree in the x_j. If a rational mapping has a rational inverse, then it is called a birational isomorphism. It is a theorem that these two definitions coincide (Shafarevich [S4] IV, §3.4).

More precisely : one can obtain each birational isomorphism of the plane by first blowing up a certain number of points in succession, then blowing down the same number of exceptional curves of the first kind.

The birational transformations of the projective plane onto itself form a group, the Cremona group. This is in a certain sense an infinite-dimensional group, but not as much so as an infinite-dimensional Lie group. The infinitely many dimensions arise because polynomials of arbitrarily high degree can occur in the description of transformations by homogeneous polynomials. (Or, what comes to the same thing, arbitrarily many points can be blown up.) The structure of the Cremona group is not easily understood, and has given rise to many investigations which today are, unfortunately, almost forgotten. (Cf. Berzolari [B14], especially sections 46-66.)

The relation between the birational transformations of the plane onto itself and the quadratic transformations is the following. First, some remarks on quadratic transformations : at the beginning of this section we defined the special quadratic transformation T by

$$T(x_0,x_1,x_2) = (x_1x_2,x_0x_2,x_0x_1).$$

More generally, one can consider arbitrary birational quadratic transformations of the plane, i.e. those given by homogeneous polynomials of degree 2. Each such transformation results from blowing up three points and subsequently blowing down three exceptional curves of the first kind. In general, the blown up points are distinct. In this general case the quadratic transformation results from the transformation T considered above by composing with projective linear transformations before and after. In the singular case either two or all three points coincide, i.e. one first blows up one point of the plane, and then further points on the exceptional lines. The standard equations for this are

$$(x_0,x_1,x_2) \mapsto (x_0^2,x_0x_1,x_1x_2) \quad \text{resp.}$$

$$(x_0,x_1,x_2) \mapsto (x_0x_1+cx_2^2,x_1^2,x_1x_2) \quad \text{with} \quad c \neq 0.$$

So much for quadratic birational transformations. The connection with arbitrary birational transformations is now given by theorem from around 1870 found by M. Noether (and independently by Clifford and Posanes) :

<u>**Theorem**</u> :

Each birational transformation of the plane is a product of quadratic transformations.

Now that we have obtained the concept of quadratic transformation, we next want to investigate how a hypersurface Y in $\mathbb{C}^{n+1}$ behaves under a quadratic transformation.

Suppose $\pi : X' \to X := \mathbb{C}^{n+1}$ is a σ-process at the origin. Since $\pi : X' - \pi^{-1}(0) \to X - \{0\}$ is biholomorphic, we need only consider hypersurfaces which go through the origin. Assume then that Y is defined in a neighbourhood of the origin by the affine equation $f(x_0,\ldots,x_n) = 0$, where

$$f(x_0,\ldots,x_n) = \Sigma a_{p_0,\ldots,p_n} x_0^{p_0} \cdot \ldots \cdot x_n^{p_n} \quad \text{and} \quad f(0) = 0.$$

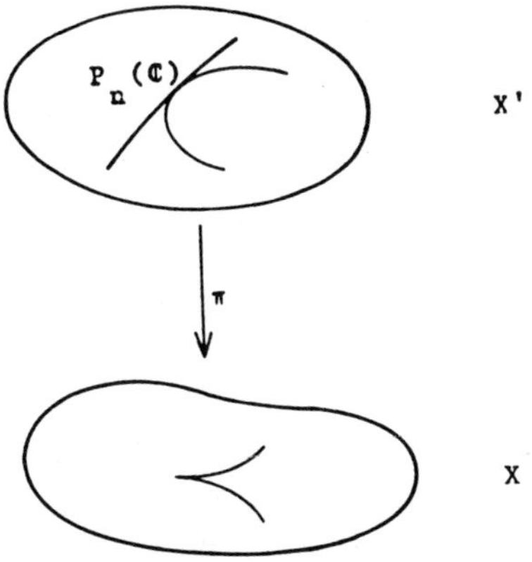

X' has a covering by coordinate neighbourhoods $X' = \bigcup_{i=0}^{n} U_i$ with $U_i \cong \mathbb{C}^{n+1}$. We choose a fixed U_i with coordinates $u_0, \ldots, u_n$. Relative to these coordinates, π may be described as follows :

$$\pi : U_i \to X = \mathbb{C}^{n+1}$$

$$(u_0, \ldots, u_n) \mapsto (u_0 u_i, \ldots, u_{i-1} u_i, u_i, u_{i+1} u_i, \ldots, u_n u_i).$$

Let $\nu := \nu_0(f)$ be the order of f at 0. Then $\pi^{-1}(Y) \cap U_i$ is given by

$$0 = \Sigma a_{p_0 \ldots p_n} u_i^{p_0+\ldots+p_n} u_0^{p_0} \cdot \ldots \cdot u_{i-1}^{p_{i-1}} u_{i+1}^{p_{i+1}} \cdot \ldots \cdot u_n^{p_n}$$

$$= u_i^{\nu} (\Sigma a_{p_0 \ldots p_n} u_i^{(p_0+\ldots+p_n)-\nu} u_0^{p_0} \cdot \ldots \cdot u_{i-1}^{p_{i-1}} \cdot u_{i+1}^{p_{i+1}} \cdot \ldots \cdot u_n^{p_n})$$

$$= u_i^{\nu} f_i'(u_0, \ldots, u_n).$$

Thus $\pi^{-1}(Y) \cap U_i$ consists of $P_n(\mathbb{C}) \cap U_i$, and a hypersurface Y_i' with equation $f_i'(u_0, \ldots, u_n) = 0$, where $f_i(u_0, \ldots, u_n)$ does not vanish identically on $\{u_i = 0\}$.

<u>Definition</u> :

The <u>strict</u> (or <u>proper</u>) <u>preimage</u> Y' of a hypersurface Y in X' under the σ-process $\pi : X' \to X$ is the union of the hypersurfaces $Y_i' = \{f_i' = 0\}$, namely $Y' = \bigcup_{i=0}^{n} Y_i'$. $\pi^{-1}(0)$ is called the <u>exceptional hypersurface</u>, or for $n = 1$ the <u>exceptional curve</u>.

We now consider a few examples (for $n = 1$).

<u>Example 1</u> :

$$f(x_0, x_1) = x_0 \cdot x_1,$$

Y therefore consists of just the two coordinate axes :

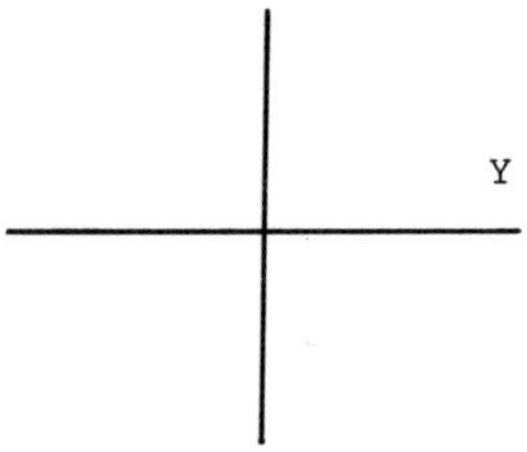

$$\pi : U_0 \to \mathbb{C}^2 \qquad \pi : U_1 \to \mathbb{C}^2$$

$$(v_0, u_0) \mapsto (v_0, u_0 v_0) \qquad (u_1, v_1) \mapsto (u_1 v_1, v_1)$$

Thus $\pi^{-1}(Y) \cap U_0$ and $\pi^{-1}(Y) \cap U_1$ are given by the equations $u_0^2 v_0 = v_0(u_0^2) = 0$ resp. $u_1 v_1^2 = v_1^2(u_1) = 0$.

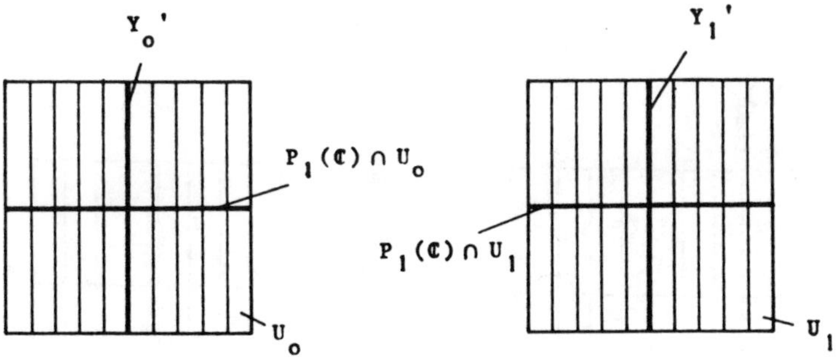

(Shaded parts are identified by $u_1 = v_0^{-1}$, $v_1 = u_0v_0$.)

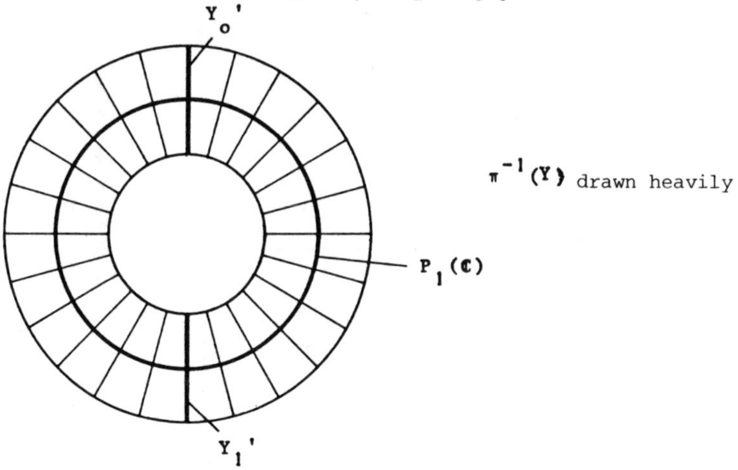

The strict preimage $Y' = Y_0' \cup Y_1'$ of Y therefore consists of two lines, which no longer intersect. This was to be expected, because the σ-process is set up precisely to separate all lines through the origin. (The picture above must of course be understood as symbolic only – in truth, blowing up yields a Möbius band in the real case, and a Hopf bundle in the complex.)

Example 2 :

$$f(x_0,x_1) = x_0^p + x_1^q \, , \; p \leq q.$$

Equation of $\pi^{-1}(Y) \cap U_0$: $v_0^p + (u_0v_0)^q = v_0^p(1+u_0^q v_0^{q-p}) = 0.$

Equation of $\pi^{-1}(Y) \cap U_1$: $(u_1v_1)^p + v_1^q = v_1^p(u_1^p+v_1^{q-p}) = 0.$

We consider the situation in the two coordinate neighbourhoods U_0 and U_1.

First the case $q = p$: in this case Y_0' and Y_1' each consist of p parallel lines :

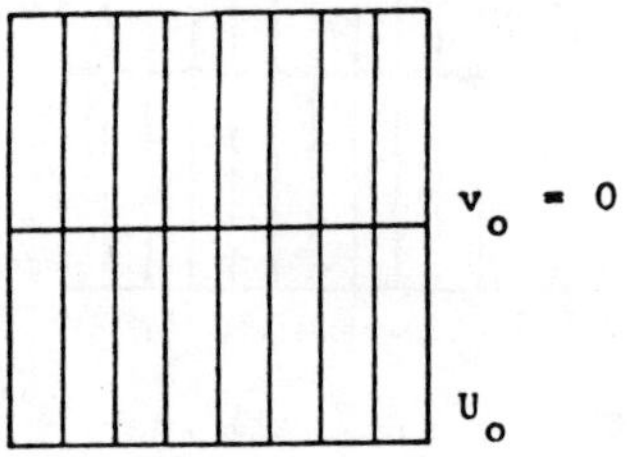

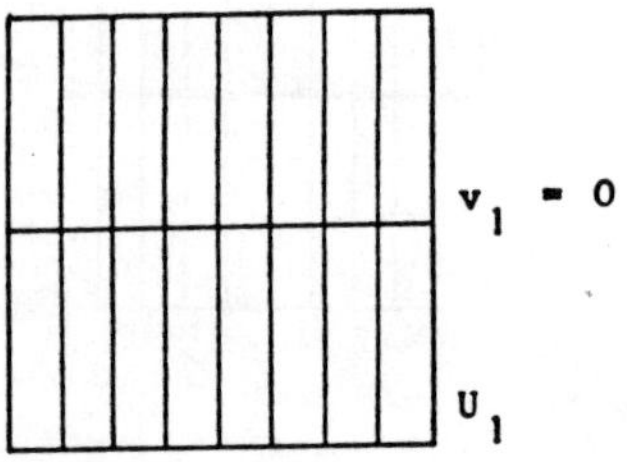

This is because Y in this case is the union of p lines through the origin.

Now let $q > p$:

Then Y_0' does not meet the line $\{v_0 = 0\}$. Since π is biholomorphic away from the origin, Y_0' in this case is nonsingular.

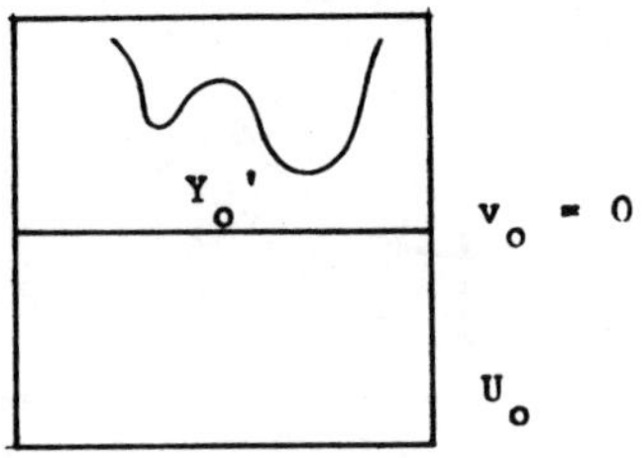

For our analysis in U_1 we distinguish three cases : If $q - p < p$, Y_1' has the line $v_1 = 0$ as tangent, for example like this :

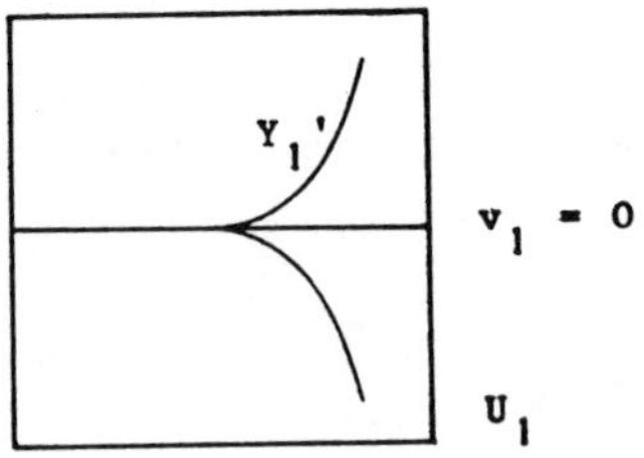

or (for $p = 2$, $q = 3$, hence in the case of Neil's parabola) :

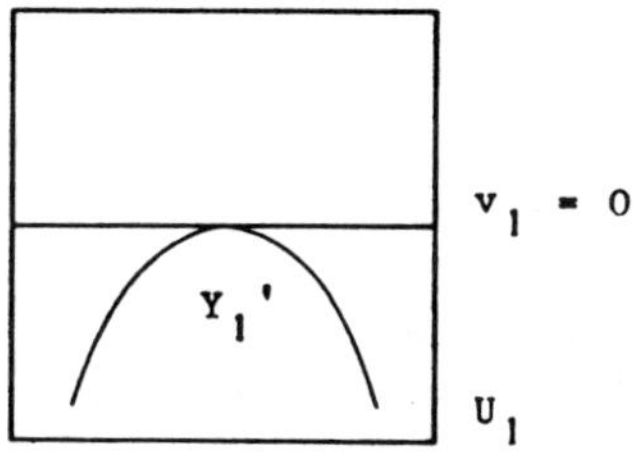

Thus the multiplicity of Y_1' at the origin is smaller than that of Y. In the case of Neil's parabola the singularity is even resolved.

If $q - p = p$, Y_1' consists of p lines which meet at $0 \in U_1$:

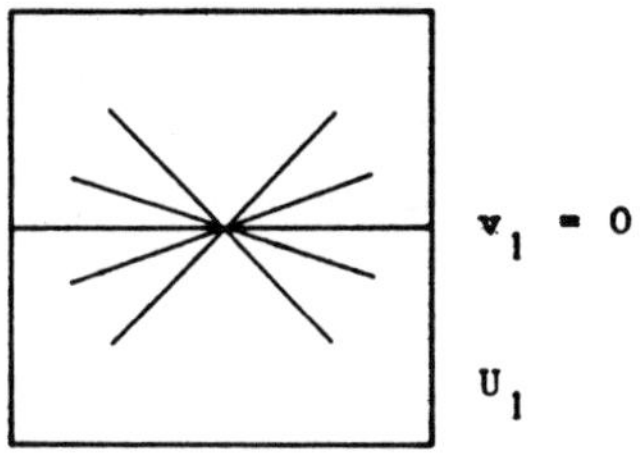

As we have seen above, this singularity of Y_1' may be resolved by a further σ-process.

If $q - p > p$, then Y_1' has a tangent at $0 \in U_1$ which is perpendicular to $v_1 = 0$, e.g. in the case $p = 2$, $q = 5$:

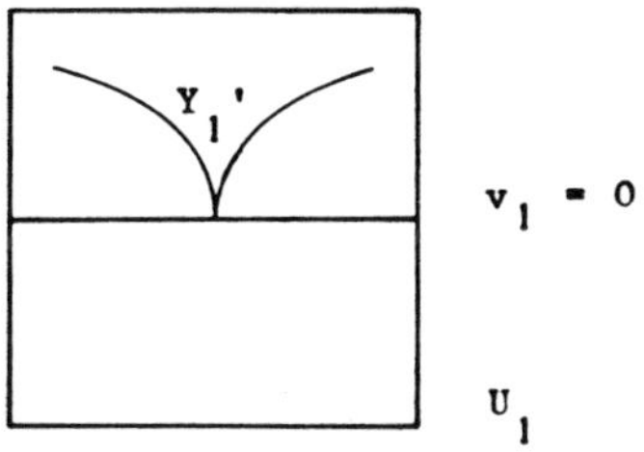

This singularity has also been improved, in a certain way, by blowing up, because $Y_i' = \{u_1^2 + v_1^3 = 0\}$ is Neil's parabola, whose singularity we already know to be removable by a σ-process.

These observations suggest the conjecture that any singularity is improved in a certain way by blowing up. The concepts necessary for measuring this improvement will be developed in what follows. First, however, we convince ourselves that a singularity of a plane curve is at least not worsened by blowing up, in the sense that its multiplicity is not increased.

Lemma 1 :

Let X be a complex surface, $Y \subset X$ a curve in X, and $p \in Y$ a point of Y. Also let $\pi : X' \to X$ be a σ-process at p, and let Y' be the strict preimage of Y under this process. Further, let $p_1,\dots,p_s$ be the points of Y' on the exceptional curve $E = \pi^{-1}(p)$. Then :

(i) $\sum_{i=1}^{s} \nu_{p_i}(Y',E) = \nu_p(Y)$

(ii) $\sum_{i=1}^{s} \nu_{p_i}(Y') \leq \nu_p(Y)$, in particular $\nu_{p_i}(Y') \leq \nu_p(Y)$ for all i.

(iii) If $\nu_{p_i}(Y') = \nu_p(Y)$ for some p_i, then $s = 1$, i.e. p_i is the single point of $Y' \cap E$.

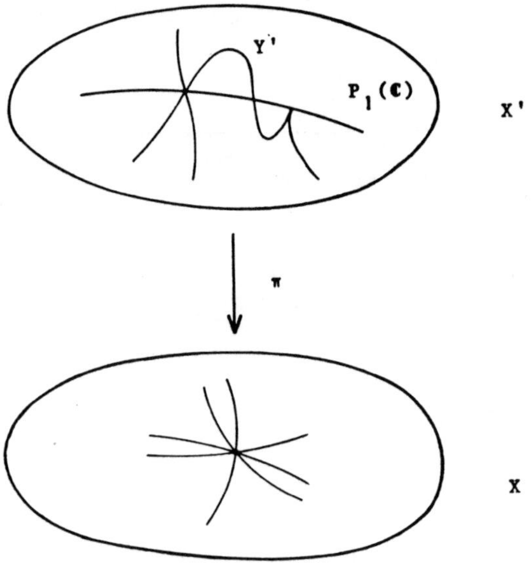

Proof :

We can assume without loss of generality that $p = 0$ and X is a neighbourhood of 0 in $\mathbb{C}^2$. Coordinates x, y in $\mathbb{C}^2$ are chosen so that neither of the coordinate axes is a tangent to Y at p.

Let $f = f^{(\nu)} + f^{(\nu+1)} + \dots$ be the power series expansion of f around 0, where $f^{(\mu)}$ consists of all terms of μth order and $\nu := \nu_p(f)$.

If we substitute the formula for the quadratic transformation in f, then we obtain :

$$f(v_0,u_0v_0) = f^{(\nu)}(v_0,u_0v_0) + f^{(\nu+1)}(v_0,u_0v_0) + \dots$$

$$= v_0^{\nu}f^{(\nu)}(1,u_0) + v_0^{\nu+1}f^{(\nu+1)}(1,u_0) + \dots$$

$$= v_0^{\nu}(f^{(\nu)}(1,u_0) + v_0f^{(\nu+1)}(1,u_0) + \dots)$$

$$f(u_1v_1,v_1) = v_1^{\nu}(f^{(\nu)}(u_1,1) + v_1f^{(\nu+1)}(u_1,1) + \dots).$$

Thus $Y' \cap E = \{(0,u_0) \mid f^{(\nu)}(1,u_0) = 0\} \cup \{(u_1,0) \mid f^{(\nu)}(u_1,1) = 0\}$. But since $u_1 = 0$ is not a zero of $f^{(\nu)}(u_1,1)$ by our choice of coordinates, all zeroes of $f^{(\nu)}(u_1,1)$ already lie in U_0, i.e. $Y' \cap \pi^{-1}(p)$ is the zero set of $f^{(\nu)}(1,u_0)$ on the u_0-axis.

Since $f^{(\nu)}(1,u_0)$ is a polynomial of degree ν, the sum of the multiplicities at the zeroes $p_1,\dots,p_s$ $(s \leq \nu)$ of $f^{(\nu)}(1,u_0)$ is just ν. By definition of intersection multiplicity, the order of such a zero equals $\nu_{p_i}(Y',\pi^{-1}(p)) \geq \nu_{p_i}(Y') \cdot \nu_{p_i}(E) = \nu_{p_i}(Y')$. This implies, first, assertion (i), and then also (ii) and (iii), trivially.

Our goal is to show that each singularity of a plane curve Y may be resolved by a finite sequence of σ-processes. To do this, it suffices to show that repeated blowing up will at some stage reduce the multiplicity of the strict preimage of Y. Thus if the multiplicity is not at first reduced by application of a quadratic transformation, we must guarantee that the singularity has nevertheless been improved in a certain way. To measure the improvement we shall use the Newton polygon of the singularity at the point in question. For this purpose we have the following preparations.

<u>Definition</u> :

Let $Y \subset X$ be a curve in the surface X and suppose $p \in Y$. For each finite sequence of σ-processes $\psi_i : X_i \xrightarrow{\phi_i} X_{i-1} \to \dots \to X_1 \xrightarrow{\phi_1} X$, where ϕ_1 is a σ-process at p, let Y_i be the strict preimage of Y in X_i under ψ_i, and let $E_i := \psi_i^{-1}(p)$. For $i \geq 2$, ϕ_i can also consist of several σ-processes at different points.

We now assume that for all j, $2 \leq j \leq i$, X_j results from X_{j-1} by blowing up the intersection points of E_{j-1} with Y_{j-1}. Then the points of $E_i \cap Y_i$ are called <u>infinitely near points of</u> p <u>in</u> X_i relative to ψ_i.

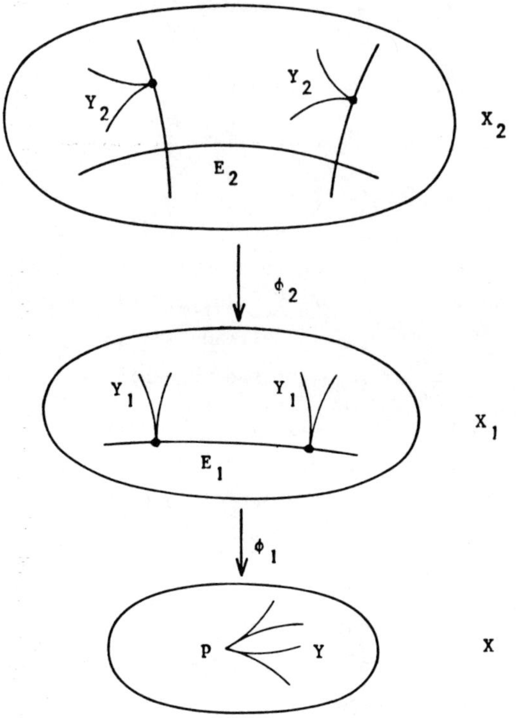

(The infinitely near points of p are drawn heavily.)

In particular, the points $x \in E_i \cap Y_i$ with $\nu_x(Y_i) = \nu_p(Y) =: \nu$ are called <u>ν-tuple infinitely near points of</u> p <u>in</u> X_i.

By Lemma 1, each X_j contains at most one ν-tuple infinitely near point x_j of p. Thus the ν-tuple infinitely near points of p constitute a well-defined sequence. In order to show that the multiplicity of the strict preimage becomes smaller than ν after finitely many σ-processes, we must prove that this sequence is finite.

To do this we shall construct a smooth curve W which hugs Y so well at the singular point p that its strict preimages at the ν-tuple infinitely near points p_i always remain tangential to Y_i. In other words : W accompanies Y under the σ-processes, up to multiplicity at the infinitely near points of p one level smaller. After each σ-process we shall choose the strict preimage of W and the new exceptional line as coordinate axes, and consider the Newton polygon of the singularity relative to these coordinates. The slopes of the segments of the Newton polygon will measure how much the singularity has been improved. For this purpose we now need a few more general results about the Newton polygon.

Supplementary results on the Newton polygon :

Lemma 2 :

For each irreducible curve germ $(Y,0)$ in $(\mathbb{C}^2,0)$, different from the coordinate axes, the Newton polygon consists of a single segment.

Proof :

For each segment of the Newton polygon we construct a monomial mapping $\mathbb{C}^2 \to \mathbb{C}^2$ and consider the preimage of Y under this mapping.

Let ℓ_i be the ith segment of the Newton polygon. Let the end-points of ℓ_i be (p_i,q_i) and (p_{i+1},q_{i+1}), and let (p_i+a_i,q_i-b_i) be the next lattice point on ℓ_i after (p_i,q_i).

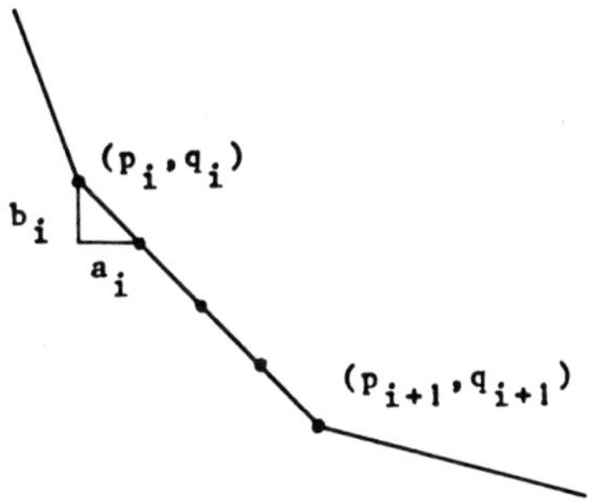

Then a_i and b_i are relatively prime natural numbers and the line through ℓ_i has the equation $b_ip + a_iq = e_i$, where $e_i := b_ip_i+a_iq_i$. The lattice points on ℓ_i are the points $(p_i+\kappa a_i,\ q_i-\kappa b_i)$, $\kappa = 0,1,\ldots,k_i$.

Since $(a_i,b_i) = 1$, we can of course find natural numbers c_i,d_i with

$$\frac{d_i}{c_i} > \frac{b_i}{a_i} \quad \text{and} \quad a_id_i - b_ic_i = 1. \qquad (*)$$

These relations continue to hold when we add positive multiples of (a_i,b_i) to (c_i,d_i). Thus we can choose c_i,d_i to have the additional property that the line $d_ip + c_iq = f_i$ through (p_i,q_i) with

$f_i := d_ip_i + c_iq_i$ lies below the Newton polygon. (For $i = 1$ this is clear in any case and for $i > 1$ we can always attain $\frac{b_{i-1}}{a_{i-1}} > \frac{d_i}{c_i} > \frac{b_i}{a_i}$.)

Now let $U_i := \mathbb{C} \times \mathbb{C}$ with coordinates u_i, v_i and let $\pi_i : U_i \to \mathbb{C}^2$ be the mapping

$$(u_i,v_i) \mapsto (u_i^{d_i}v_i^{b_i}, u_i^{c_i}v_i^{a_i}).$$

Let $U_i^* := \mathbb{C}^* \times \mathbb{C}^*$. Then $\pi_i\big|_{U_i^*} : U_i^* \to \mathbb{C}^* \times \mathbb{C}^*$ is biholomorphic by (*). If $f(x,y) = 0$ is the local equation for Y, where $f(x,y) = \Sigma a_{pq}x^py^q$, then

$$\begin{aligned}(f \circ \pi_i)(u_i,v_i) &= \Sigma a_{pq}u_i^{d_ip+c_iq}v_i^{b_ip+a_iq}\\ &= v_i^{e_i}u_i^{f_i}\Big(\sum_{p,q\in \ell_i} a_{pq}u_i^{d_i(p-p_i)+c_i(q-q_i)} + v_ih(u_i,v_i)\Big)\\ &= v_i^{e_i}u_i^{f_i}\Big(\sum_{\kappa=1}^{k_i} a_{p_i+\kappa a_i,q_i-\kappa b_i}u_i^{\kappa}+v_ih(u_i,v_i)\Big)\\ &= v_i^{e_i}u_i^{f_i}(g_i(u_i) + v_ih(u_i,v_i)).\end{aligned}$$

Here, the polynomial g_i has just the coefficients of f which correspond to the lattice points of ℓ_i. Thus the preimage of Y in U_i^* is the curve Y_i with the equation

$$g_i(u_i) + v_ih(u_i,v_i) = 0.$$

Y_i cuts the u_i-axis at the zeroes of $g_i(u_i) = 0$. Since $a_{p_iq_i} \neq 0$, $u_i = 0$ is not a zero. Let r_i be the number of distinct zeroes of g_i, $0 \leq r_i \leq k_i$.

Assertion 1 :

Y has at least r_i branches, whose preimages in U_i cut the u_i-axis at the r_i zeroes of g_i.

Proof :

We consider the intersection of Y with a small polydisc $P_\varepsilon = \{|x| < \varepsilon, |y| < \varepsilon\}$. The preimage of P_ε is contained in a neighbourhood $W_\delta = \{|u_i| < \delta$ or $|v_i| < \delta\}$ of the axes.

Then for sufficiently small δ, $Y_i \cap W_\delta$ has a connected component Y_{ij} for each zero x_j of g_i, meeting the u_i-axis at this zero $(j = 1,\ldots,r_j)$. If one constructs the intersection with U_i^*, then Y_{ij} possibly decomposes into several connected components, namely, when x_j is reducible and δ is sufficiently small.

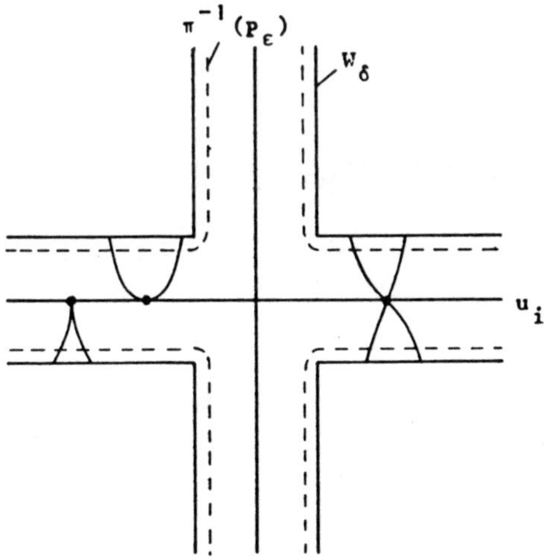

Since $\pi_i : U_i^* \to \mathbb{C}^* \times \mathbb{C}^*$ is a homeomorphism, one obtains, for sufficiently small ε, at least r_i connected components in $(Y \cap P_\varepsilon) - \{0\}$, corresponding to at least r_i distinct branches of the curve Y.

Assertion 2 :

The branches of Y which correspond to different segments of the Newton polygon are all distinct.

Proof :

Consider the following diagram (without loss of generality let $i > k$).

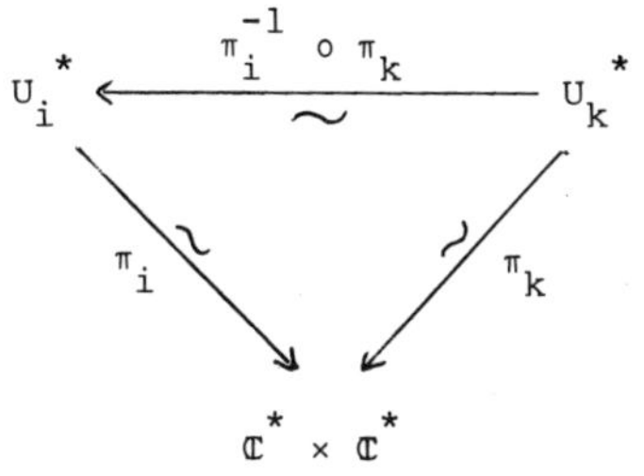

All the mappings present in it are biholomorphic. $\pi_i^{-1} \circ \pi_k : U_k^* \to U_i^*$ is a monomial mapping.

$$u_i = u_k^\alpha v_k^\beta$$
$$v_i = u_k^\gamma v_k^\delta$$

where $\alpha,\beta,\gamma,\delta$ are defined by $\begin{pmatrix}\alpha & \beta\\ \gamma & \delta\end{pmatrix} = \begin{pmatrix}a_i & -b_i\\ -c_i & d_i\end{pmatrix}\begin{pmatrix}d_k & b_k\\ c_k & a_k\end{pmatrix}$.

Since $\frac{d_k}{c_k} > \frac{b_k}{a_k} > \frac{d_i}{c_i} > \frac{b_i}{a_i}$, we have $\alpha,\beta > 0$; $\gamma,\delta < 0$. Therefore $u_i \to 0$ and $v_i \to \infty$ as $v_k \to 0$ and $u_k \to$ const. Hence a connected component of $(Y \cap P_\varepsilon) - \{0\}$ cannot simultaneously be the image of a Y_{ij} from U_i and a $Y_{k,s}$ from U_k. This proves the lemma.

Remark :

Assertions 1 and 2 together give a better result:

If r_i is the number of zeroes of the polynomial g_i, then Y has at least Σr_i branches. Since $r_i \geq 1$ one has in particular :

Corollary :

Y has at least as many branches as the Newton polygon has segments.

Lemma 3 :

Let X be a curve germ in $(\mathbb{C}^2,0)$, with irreducible components $X_1,\ldots,X_r$ (all different from the coordinate axes). Let ℓ_i be the Newton polygon of X_i, which we know to be a segment by Lemma 2. Let the components be numbered in such a way that for $i > k$ the slope of ℓ_i is greater than the slope of ℓ_k.

Then the Newton polygon of X results from joining together the different segments ℓ_i, suitably displaced. More precisely : if ℓ_i has initial point $(0,p_i)$ and final point $(q_i,0)$ $(i = 1,\ldots,r)$, then the Newton polygon of X consists of the segments $\ell_1',\ldots,\ell_r'$, where ℓ_i' is the segment with initial point $(\sum_{k<i} q_k , \sum_{j\geq i} p_j)$ and final point $(\sum_{k\leq i} q_k , \sum_{j>i} p_j)$.

Example :

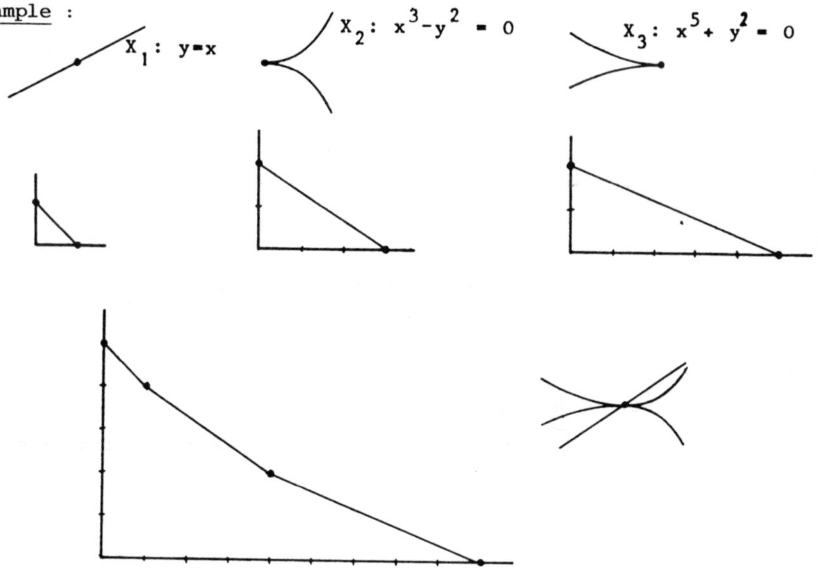

Proof :

One can prove that for two curves X', X'' with equations $f = 0$ resp. $g = 0$ the Newton polygon of $X = X' \cup X''$ results from joining together the segments of the Newton polygons for X' and X'' (the assertion of the lemma then follows immediately by induction). To prove this more general assertion one shows :

(i) The monomial $x^p y^q$ corresponding to a vertex (p,q) of the composite polygon can be written in exactly one way as the product of a monomial in f and a monomial in g.

(ii) No point of $\Delta(f \cdot g)$ lies below this polygon, because $\Delta(f \cdot g) \subset \Delta(f) + \Delta(g)$ and $\Delta(f) + \Delta(g)$ contains no points which lie below the composite polygon.

The assertion then follows immediately from the definition of Newton polygon.

Curves with maximal contact

Suppose X is a complex surface, $Y \subset X$ a curve on X and $p \in Y$ a point on the curve. On the basis of the program we have already drawn up for the resolution of singularities, we would like to have at our disposal, in this situation, a smooth curve W through p on X which hugs the curve Y as well as possible at p. In other words : W is to have the "best possible" contact with Y. For this purpose we must first define "best possible contact" precisely.

Several definitions of "best possible contact" are conceivable, not necessarily equivalent to each other. We choose the concept of "maximal contact" as the most suitable for our purpose. In order to motivate the technical form of the definition, we first give a short heuristic discussion.

When a curve Y is irreducible at p and hence has a unique tangent there, one would want a smooth curve W with maximal contact to have at least the same tangent at p as Y. One would want the same when all irreducible branches of Y have the same tangent at p. On the other hand, when there are branches with different tangents, then the smooth curve W cuts at least one of these branches transversely and hence does not have especially good contact with it. Since this situation occurs for any W with at least one branch, in this sense all W are equally bad – or, if one prefers, equally good. Thus in the case of branches with different tangents one can say that all smooth curves W have the maximal possible contact.

Now we return to the case where all branches of Y have the same tangent, which is the case, in particular, when Y only has one branch. Then W must have the same tangent. But it is still conceivable that different W hug Y differently. One already sees this in a trivial example : when Y is the x-axis in $\mathbb{C}^2$ with coordinates x, y, and W_k is the parabola $y = x^k$, $k > 1$, then W_k obviously hugs Y better than W_m when $k > m$. The question is, how is one to make "better" hugging precise? There are several possibilities. We describe two different definitions, which one can prove to be equivalent with the help of theorems proved later.

The first definition of maximal contact between Y and W uses σ-processes. The σ-process, applied repeatedly, has the facility to separate curves which previously were in contact, and hence to reduce their contact. Thus the curves Y and W have maximal contact when they stay together as long as possible under repeated σ-processes. "As long as possible" may be defined as follows. W has maximal contact with Y when the strict preimage W' of W cuts the strict preimage Y' of Y tangentially as long as possible under σ-processes, i.e. as long as all branches of Y' still go through the same point, have the same tangent there and as long as the latter is transverse to the exceptional curve last introduced. One can also describe this condition, perhaps more concretely, with the help of the multiplicities of the infinitely near points of the singular points :

Let m be the multiplicity of Y at p. Then W has maximal contact with Y when the strict preimage W' goes through all m-tuple infinitely near points of p on Y' under iterated σ-processes, and also has the same tangent as the last m-tuple point when all branches of Y' have the same tangent. (Then W' and Y' automatically have the same tangent for the earlier m-tuple points.)

We see that this first geometric definition of maximal contact – per definitionem – does what we want : curves with maximal contact stay together as long as possible under iterated σ-processes. But in order to prove later the existence of such curves of maximal contact, we need a criterion for maximal contact, if possible a numerical one. I.e. we need a quantitative measure of the contact between W and Y, and we must be able to compute this measure, this invariant, with the help of algebraic data, i.e. the equations of W and Y.

The intersection number $\nu_p(W,Y)$ of course offers itself to measure contact between W and Y. And in fact one can show for a curve Y irreducible at p that W and Y have maximal contact (in the sense of the first definition above) when $\nu_p(W,Y)$ has the maximum possible value, i.e. when $\nu_p(W,Y) = \sup \nu_p(\tilde{W},y)$, where the supremum is taken over all smooth $\tilde{W}$ through p which are different from Y.

When Y has several branches Y_i at p, the situation is not quite so simple. First of all, one sees in simple examples that in this case the condition that $\nu_p(W,Y)$ be maximal is too sharp. This condition always implies maximal contact, but not conversely. For example, if Y consists of two ordinary cusps with different tangents, then each smooth W has maximal contact according to our first definition. But $\nu_p(W,Y)$ has the maximal value 5 only when the tangent of W equals the tangent of one of the two cusps, while for all other W we have only $\nu_p(W,Y) = 4$. Moreover, it is clear that simple comparison of the intersection numbers $\nu_p(W,Y_i)$ does not permit a comparison of the contact of W with the different branches, because by Proposition 6.1.3 we have $\nu_p(W,Y_i) \geq \nu_p(Y_i)$, with equality just in case of transversality. Thus it can very well happen that
$\nu_p(W,Y_i) > \nu_p(W,Y_j)$ because $\nu_p(Y_i) > \nu_p(Y_j)$ even though W meets the

branch Y_i transversely, and hence with bad contact, while it has a tangent in common with Y_j. This fault can be remedied by considering the quotient $\delta_p(W,Y_i) = \nu_p(W,Y_i)/\nu_p(Y_i)$ instead of $\nu_p(W,Y_i)$. This "contact exponent" of W and Y_i yields a better measure of the contact of W with the different branches of Y.

Now how does one obtain a measure of contact for the whole curve Y from these measures of contact with the individual branches? If one took the maximum of the $\delta_p(W,Y_i)$ then for $\Delta_p(W,Y) := \max \delta_p(W,Y_i)$ the assertion $\Delta_p(W,Y) = k$ would mean that W has contact with at least one branch Y_i such that $\delta_p(W,Y_i) = k$, and that its contact with no other branch is better. Conversely, for $\delta_p(W,Y) := \min \delta_p(W,Y_i)$ the assertion $\delta_p(W,Y) = k$ would mean that W has contact with all branches Y_j such that $\delta_p(W,Y_j) \geq k$, and that for one branch only $\delta_p(W,Y_j) = k$ holds. In both cases we could now try to define maximal contact between W and Y by requiring $\Delta_p(W,Y)$ resp. $\delta_p(W,Y)$ to be maximal. In the first case, however, this would be too strong a requirement. One sees this again from the above example of the two cusps transverse to each other when W is tangential to one of the two cusps, $\Delta_p(W,Y) = 3/2$, and otherwise $\Delta_p(W,Y) = 1$. Thus $\Delta_p(W,Y)$ would be maximal only for W with this special tangent, whereas with our first definition all W should have maximal contact. On the other hand, $\delta_p(W,Y)$ gives the desired result in this example : $\delta_p(W,Y) = 1$ for all W, and hence $\delta_p(W,Y)$ is maximal.

All these considerations motivate the following sequence of defitions :

<u>Definition</u> :

Suppose X is a complex surface, $Y \subset X$ a curve on X and $p \in Y$ a point on Y. Also, let Y_i be the irreducible components of Y at p.

(i) If $W \subset X$ is a smooth curve through p, then the <u>contact exponent of</u> W <u>and</u> Y <u>at</u> p is the rational number

$$\delta_p(W,Y) := \min_i \frac{\nu_p(W,Y_i)}{\nu_p(Y_i)} .$$

(ii) The <u>first characteristic exponent of</u> Y <u>at</u> p is

$$\delta_p(Y) := \sup_W \delta_p(W,Y),$$

where the supremum is taken over all smooth $W \neq Y_i$ through p.

(iii) A smooth curve W through p has <u>maximal contact</u> with Y at p just in case :

$$\delta_p(W,Y) = \delta_p(Y).$$

Examples :

(1) Let $Y \subset \mathbb{C}^2$ be the curve with the equation $y^2 - x^3 = 0$ and let p be the origin. Then $\delta_p(Y) = 3/2$, and each W with the x-axis as tangent has maximal contact.

(2) Let $Y \subset \mathbb{C}^2$ be the curve with equation $y^2 - x^5 = 0$ and let p be the origin. Then $\delta_p(Y) = 5/2$. The x-axis has maximal contact with Y. The curve W with equation $y = x^2$ has non-maximal contact with Y, because $\delta_p(W,Y) = 4/2 = 2$.

(3) Let $Y \subset \mathbb{C}^2$ be the curve with equation $(y^2-x^3)(x^2-y^5) = 0$ and let p be the origin. Then $\delta_p(Y) = 1$. Each smooth W through p has maximal contact.

(4) Let $Y \subset \mathbb{C}^2$ be the curve with the equation $(y^2-x^3)(x^5-y^2) = 0$ and let p be the origin. Then $\delta_p(Y) = 3/2$ and each smooth W through p with the x-axis as tangent has maximal contact with Y.

(5) Let $Y \subset \mathbb{C}^2$ be the x-axis and let p be the origin. Then $\delta_p(Y) = \infty$, and there is no W with maximal contact.

In what follows we shall see that for each singular Y the first characteristic exponent $\delta_p(Y)$ is finite, and hence a rational number. We shall investigate the behaviour of $\delta_p(Y)$ under quadratic transformations and compute the integer part $[\delta_p(Y)]$ from the number of infinitely near m-tuple points, $m = \nu_p(Y)$. In order to be able to do this, we must first develop a method for computing the contact exponent $\delta_p(W)$. We shall see that we can derive this from the Newton polygon by suitable choice of coordinates.

First we describe the connection between contact exponents and Newton polygons. We consider a curve germ $(Y,0)$ in $\mathbb{C}^2$, described by the equation $f(x,y) = \sum_{\alpha,\beta} a_{\alpha\beta}x^\alpha y^\beta = 0$. Here f is y-general of order $m = \nu_0(Y)$, i.e. the y-axis is not a tangent to Y. In addition, coordinates are chosen so that the smooth curve W we want to compare with Y is just the x-axis, i.e. W has the equation $y = 0$.

Lemma 4 :

The slope of the steepest segment of the Newton polygon is just $\frac{-1}{\delta_0(W,Y)}$.

Proof :

Let $f = f_1 \cdot \ldots \cdot f_r$ be the decomposition of f into irreducible factors, $Y = \bigcup_{i=1}^{r} Y_i$ the associated decomposition of Y into irreducible components. The Newton polygon of f results from "joining together" the Newton polygons for the f_i , as we have described in Lemma 3. But by Lemma 2 the Newton polygons for the f_i are simply segments ; suppose their slopes are $-\frac{1}{\delta_i}$. The slope of the steepest segment of the Newton polygon for f is then equal to $-\frac{1}{\delta}$ with $\delta = \min_i \delta_i$. On the other hand, $\delta_0(W,Y) = \min_i \delta_0(W,Y_i)$. Thus we need only show that $\delta_i = \delta_0(W,Y_i)$ for all i.

Since the y-axis is obviously not a tangent of Y_i , f_i is again y-general (cf. 8.2.5), and in fact of order $m_i := \nu_0(Y_i)$. Thus the Newton polygon for f_i has the form

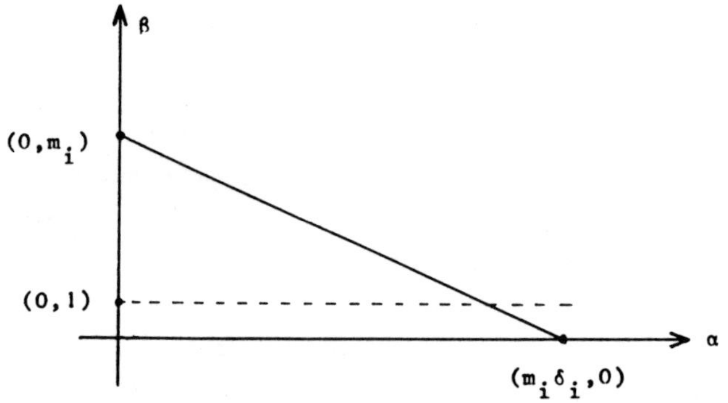

We want to compute $\nu_0(W,Y_i) = \dim_{\mathbb{C}} \mathbb{C}\{x,y\}/(y,f_i)$ (cf. 8.3.14) : $f_i(x,y) - f_i(x,0)$ is a power series divisible by y, so in any case it lies in the ideal (y,f_i). Hence y and $f_i(x,0)$ also generate this ideal. Now $f_i(x,0)$ is of the form

$$f_i(x,0) = x^{m_i\delta_i}h(x)$$

with $h(0) \neq 0$. Thus h is invertible in $\mathbb{C}\{x,y\}$, and so

$$(y,f_i) = (y,f_i(x,0)) = (y,x^{m_i\delta_i}).$$

It follows that $\nu_0(W,Y_i) = \dim_{\mathbb{C}}\mathbb{C}\{x,y\}/(y,f_i) = m_i\delta_i$. This gives

$$\delta_0(W,Y_i) = \frac{\nu_0(W,Y_i)}{\nu_0(Y_i)} = \frac{m_i\delta_i}{m_i} = \delta_i \text{ , which was to be shown.}$$

With this formula we can give a criterion for W to have maximal contact with Y. Let ℓ_1 again be the steepest segment in the Newton polygon. As in 8.3 we decompose f as follows :

$$f(x,y) = F(x,y) + \sum_{\alpha+\delta\beta>m\delta} a_{\alpha\beta}x^\alpha y^\beta$$

where $F = \sum_{\alpha+\delta\beta=\delta m} a_{\alpha\beta}x^\alpha y^\beta$ is the quasihomogeneous part of smallest order relative to the weight given by ℓ_1 :

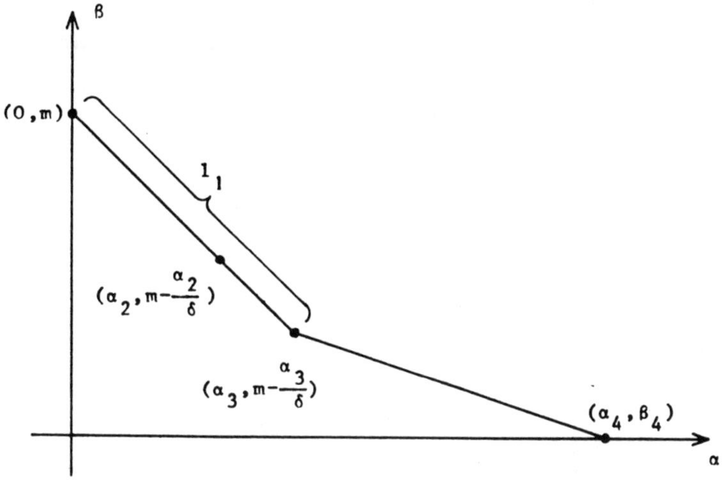

<u>Lemma 5</u> :

W has <u>non</u>-maximal contact with Y just in case $F(x,y)$ has the form

$$F(x,y) = c(y-\lambda x^\delta)^m \quad (m \in \mathbb{N},\ c \neq 0).$$

<u>Proof</u> :

(i) If F has the binomial form $F(x,y) = c(y-\lambda x^\delta)^m$, let V be the curve given by $y - \lambda x^\delta = 0$. Then V is at any rate a smooth curve through the origin. We make the coordinate change

$$y' = y - \lambda x^\delta$$
$$x' = x.$$

In the new coordinates V has the equation $y' = 0$, and the monomial $a_{\alpha\beta}x^\alpha y^\beta$ of $f = \Sigma a_{\alpha\beta}x^\alpha y^\beta$ becomes

$$a_{\alpha\beta}x'^\alpha(y'+\lambda x'^\delta)^\beta.$$

The transformed power series therefore has the form

$$f' = a_{0,m}y'^m + \sum_{\alpha+\delta\beta>\delta m} a_{\alpha\beta}x'^{\alpha}(y'+\lambda x'^{\delta})^{\beta},$$

in particular, f is again y'-general of order m.

Now what does the new Newton polygon look like?

The points of the carrier $\Delta(f')$ of f', which correspond to the terms $a_{\alpha\beta}x'^{\alpha}(y'+\lambda x'^{\delta})^{\beta}$ (with $\alpha + \delta\beta > \delta m$), all lie strictly above the line ℓ through $(0,m)$ with slope $-1/\delta$. Namely, these terms yield only points of the form $(\alpha+k\delta,\beta-k)$ for the carrier of f', hence points on the line through (α,β) with slope $-1/\delta$. But (α,β) already lies above the line ℓ, hence so do all the new points.

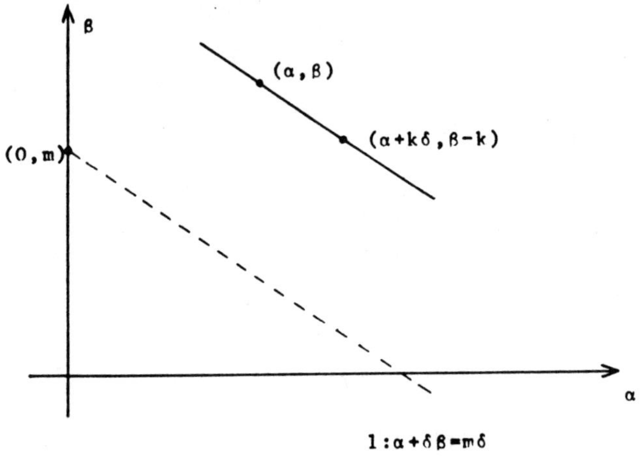

Thus the line ℓ meets the convex hull of $\Delta(f')$ only at $(0,m)$, i.e. the slope of the steepest segment of the Newton polygon is more gentle than $|1/\delta|$. Then by Lemma 4

$$\delta_0(V,Y) > \delta = \delta_0(W,Y),$$

and W does not have maximal contact.

(ii) Conversely, suppose that W does not have maximal contact with Y. Thus there is a smooth curve V with $\delta_0(V,Y) > \delta_0(W,Y)$. V is not tangential to the y-axis, otherwise we would have $\delta_0(V,Y) = 1$, since the y-axis is not tangent to Y. Hence it follows from the implicit function theorem that V is described by an equation

$$y - \sum_{i=1}^{\infty} b_i x^i = 0.$$

One infers immediately from the Newton polygon of V that $b_i = 0$ for

$i < \delta_0(W,V) := d$ and $b_{\delta_0(W,V)} \neq 0$. We first prove indirectly that $d = \delta_0(W,V) = \delta_0(W,Y)$.

Assumption :

$$d = \delta_0(W,V) < \delta_0(W,Y).$$

In order to apply Lemma 4, we make the coordinate change

$$\begin{array}{l} x' = x \\ \qquad\qquad f'(x',y') := f(x',y'+\Sigma b_i x'^i) \qquad (*) \\ y' = y - \Sigma b_i x^i \end{array}$$

Then V again has the required form $V = \{(x',y') \mid y' = 0\}$. We consider the effect of the coordinate change on the Newton polygon. A monomial $x^\alpha y^\beta$ is transformed into

$$x'^\alpha (y'+\Sigma b_i x'^i)^\beta.$$

Since $b_i = 0$ for $i < d$, all points of $\Delta(f')$ which originate from the monomial $a_{\alpha\beta}x^\alpha y^\beta$ lie either on the line through (α,β) with slope $-1/d$, or to the right of it. For this reason, the monomial $x'^{md}y'^0$ appears only in the expression which results from transformation of y^m, but there it has the coefficient $b_d^m \neq 0$.

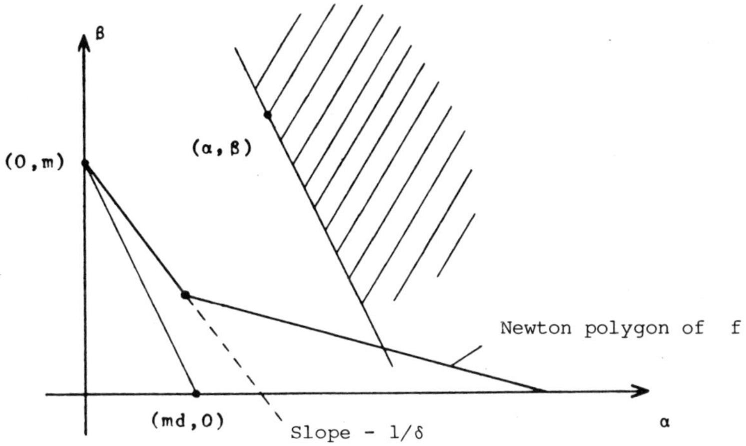

Thus the Newton polygon of f' is the segment connecting $(0,m)$ and $(md,0)$, and this segment has slope $-1/d$. Then by Lemma 4, $\delta_0(V,Y) = d < \delta_0(W,Y)$, contrary to assumption.

Assumption :

$$d > \delta_0(W,Y).$$

We again make the coordinate transformation (*)

$$x' = x$$
$$y' = y - \Sigma b_i x^i.$$

Again, the points of $\Delta(f')$ which originate from the monomial $a_{\alpha\beta}x^\alpha y^\beta$ of f lie on or above the line through (α,β) with slope $-1/d$. But this time these lines are flatter than the first segment of the Newton polygon of f.

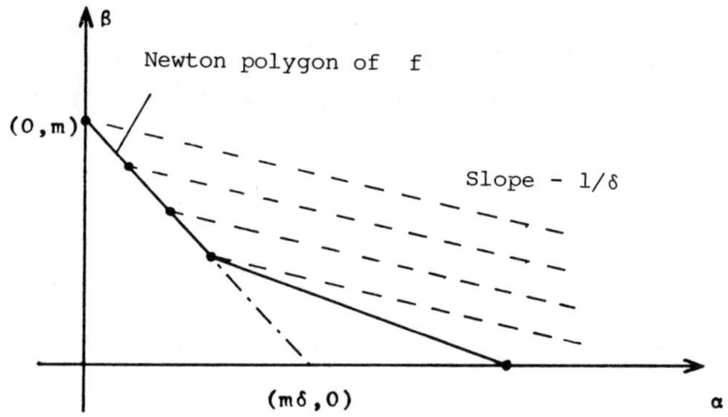

The points of the carrier of f on the steepest segment of the Newton polygon for f also lie in $\Delta(f')$, hence the slope of the first segment of the Newton polygon for f' is again equal to $-1/\delta$. By Lemma 4, this yields $\delta_0(V,Y) = \delta_0(W,Y)$, contrary to assumption.

Thus $\delta_0(W,V) = \delta_0(W,Y) = \delta$. Since $\nu_0(V) = 1$, δ is a whole number. Let us see what this implies for the coordinate transformation (*) ! Under the coordinate transformation (*), the monomial $a_{\alpha\beta}x^\alpha y^\beta$ this time yields points in the carrier of f' which lie on or above the line through (α,β) with slope $-1/d = -1/\delta$.

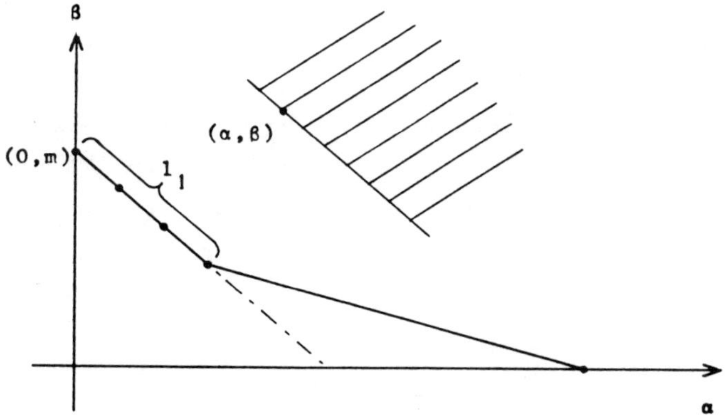

Since $\delta_0(V,Y) > \delta_0(W,Y)$, the steepest segment of the Newton polygon of f' is flatter than ℓ_1. Thus the coordinate transformation (*) causes all points of $\Delta(f)$ on ℓ_1 (except $(0,m)$) to vanish.

This means :

$F(x,y) = cy'^m$ + terms of higher order, relative to the weight given by ℓ_1.

Now

$F(x,y) = F(x',y'+b_\delta x'^\delta)$ + terms of higher order

and $F(x',y'+b_\delta x'^\delta)$ is quasihomogeneous of order m relative to our weight.

Thus

$F(x',y'+b_\delta x'^\delta) = cy'^m$

or $F(x,y) = c(y-b_\delta x^\delta)^m$

and Lemma 5 is proved.

We now come to the behaviour of the contact exponent under blowing up.

Lemma 6 : **(Stability of maximal contact)**

Let Y be singular at $0 \in \mathbb{C}^2$, and suppose W has maximal contact with Y at 0. Let $\pi : X_1 \to \mathbb{C}^2$ result from blowing up $0 \in \mathbb{C}^2$, let W_1 be the strict preimage of W, and let Y_1 be the strict preimage of Y. Then :

There is a point $p_1 \in Y_1$ with $\pi(p_1) = 0$ and $\nu_{p_1}(Y_1) = \nu_0(Y) = m$ just in case $\delta_0(W,Y) \geq 2$. In this case $p_1 \in W_1$ and W_1 has maximal contact with Y_1 at p_1. In addition, $\delta_{p_1}(W_1,Y_1) = \delta_0(W,Y) - 1$.

<u>**Proof**</u> :

Without loss of generality, let W again be the x-axis and let the equation of Y be y-general of order $m = \nu_0(Y)$. Suppose the σ-process is described by $(u,v) \mapsto (v,uv) = (x,y)$ in the neighbourhood of an infinitely near point p_1 of 0, where $p_1 = (0,0)$. Then Y_1 is described by

$$f_1(v,u) = \Sigma a_{\alpha\beta} v^{\alpha+\beta-m} u^{\beta}$$

where $f(x,y) = \Sigma a_{\alpha\beta} x^{\alpha} y^{\beta}$ is the power series which describes Y. The degree of the monomial $v^{\alpha+\beta-m} u^{\beta}$ is $\alpha + 2\beta - m$, hence f_1 has multiplicity m at p_1 just in case $a_{\alpha\beta} = 0$ for $\alpha + 2\beta < 2m$, and this is so precisely when the steepest line of the Newton polygon for f has slope less than $1/2$.

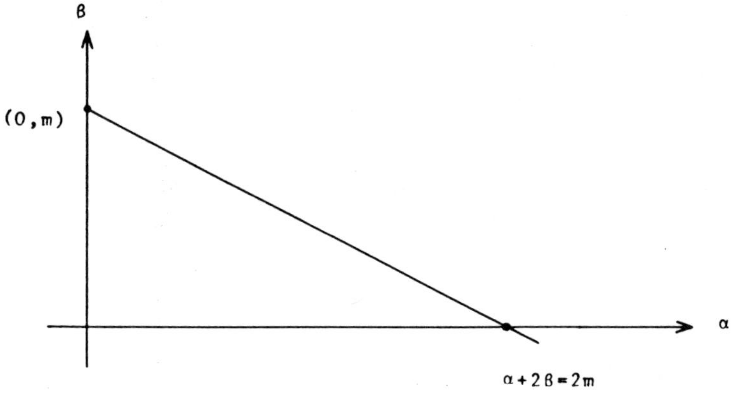

The first assertion of the lemma then follows.

We now look at the transformation of the Newton polygon in this case. A segment through the points $(0,m)$ and (α,β) in the carrier of f goes over to the segment between $(0,m)$ and $(\alpha+\beta-m,\beta)$.

In this diagram the exponent of v is carried by the horizontal axis, and the exponent of u by the vertical.

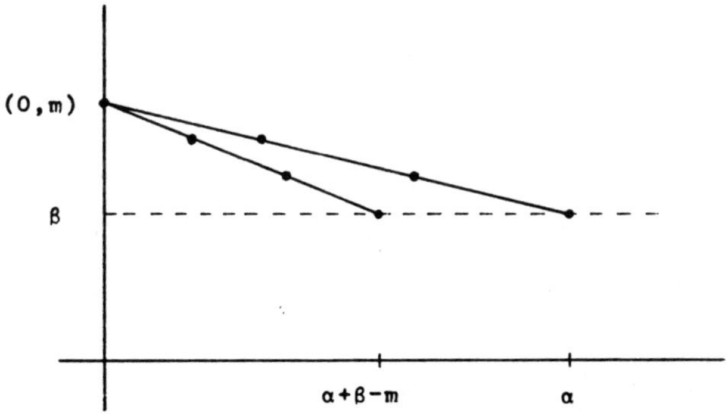

The original segment had slope $\frac{-1}{\alpha/(m-\beta)}$, while the new one has slope $\frac{-1}{(\alpha+\beta-m)/(m-\beta)} = \frac{-1}{\alpha/(m-\beta)-1}$. If $-1/\delta$ is the slope of the steepest segment of the Newton polygon for f, then the slope of the steepest segment of the Newton polygon for f_1 equals $-1/(\delta-1)$. By Lemma 4 this is just $\delta(W_1,Y_1) = \delta(W,Y) - 1$. Thus it remains to show only that W_1 has maximal contact with Y_1.

It follows from the considerations above that the "quasihomogeneous part" is computed from the steepest segment of the Newton polygon for f_1 as follows :

$$F_1(v,u) = \frac{F(v,u)}{v^m} .$$

If W_1 has non-maximal contact with Y_1 , then F_1 is binomial (Lemma 5) :

$$F_1(v,u) = (u-\lambda v^{\delta})^m.$$

But then

$$F(x,y) = x^m(y/x-\lambda x^{\delta})^m = (y-\lambda x^{\delta+1})^m$$

is also binomial, and W has non-maximal contact with Y. This contradicts the hypothesis, and hence Lemma 6 is proved.

<u>Lemma 7</u> :

If Y has a proper singularity at 0, then $\delta_0(Y)$ is finite.

<u>Proof</u> :

Suppose $\delta_0(Y) = \infty$. We have to show that Y is regular at 0.

Let W_0 be any smooth curve through 0. We choose coordinates x, y so that $W_0 = \{y=0\}$ and $f(x,y)$ is a Weierstrass polynomial in y of degree m. The carrier of f therefore lies – with the exception of the point $(0,m)$ – entirely beneath the line $\beta = m - 1$.

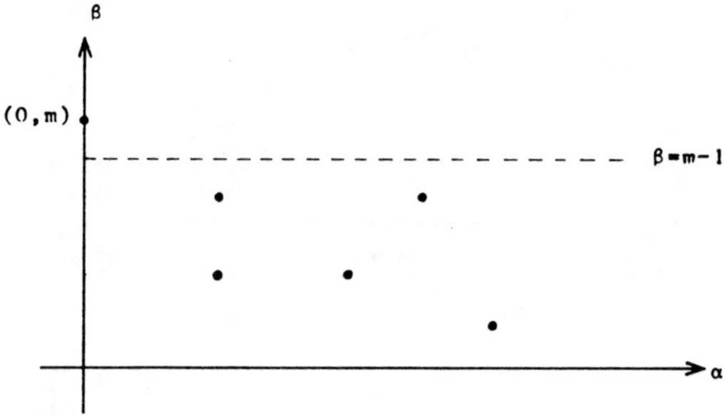

If $W_0 = Y$, then we are finished. Otherwise W_0 has non-maximal contact with Y, hence by Lemma 5 f has the form

$$f(x,y) = a_{0,m}(y-b_0x^{\delta_0})^m + \sum_{\alpha+\delta_0\beta>\delta_0 m} a_{\alpha\beta}x^\alpha y^\beta \qquad (\delta_0 \in \mathbb{N}).$$

In the proof of Lemma 5 (part (i)) we have seen that $W_1 := \{(x,y) \mid y - b_0x^{\delta_0} = 0\}$ has better contact with Y than W_0 has. We again make the coordinate transformation

$$x_1 = x$$
$$y_1 = y - b_0x^{\delta_0}$$

(thus $W_1 = \{y_1=0\}$). It follows from these formulae that the carrier of the transformed equation again has the above form, i.e. with the exception of the point $(0,m)$ it lies beneath the line $\beta = m - 1$.

If $W_1 = Y$ then we are finished ; otherwise f again has, in the new coordinates, the form

$$f(x,y) = (y_1-b_1x_1^{\delta_1})^m + \sum_{\alpha+\delta_1\beta>\delta_1 m} a^{(1)}_{\alpha\beta}x_1^\alpha y_1^\beta \qquad (\delta_1 \in \mathbb{N})$$
$$= (y-b_0x^{\delta_0}-b_1x^{\delta_1})^m + \sum_{\alpha+\delta_1\beta>\delta_1 m} a^{(1)}_{\alpha\beta}x^\alpha y_1^\beta$$

with $\delta_1 > \delta_0$ by Lemma 4, and $\delta_1 \in \mathbb{N}$.

Again, $W_2 = \{y_1-b_1x_1^{\delta_1} = 0\}$ has better contact, we make the corresponding coordinate transformation, etc. ... .

If this process ever terminates, i.e. if $W_i = Y$ for some i, then there is no more to show. If not we have, for each r, the representation

$$f(x,y) = a_{0,m}(y_r - b_r x^{\delta_r})^m + \sum_{\alpha+\delta_r\beta>\delta_r m} a^{(r)}_{\alpha\beta} x^\alpha y_r^\beta \quad (\delta_r \in \mathbb{N},\ \delta_r > \delta_{r-1})$$

$$= a_{0,m}(y - b_0 x^{\delta_0} - b_1 x^{\delta_1} - \ldots - b_r x^{\delta_r})^m + \sum_{\alpha+\delta_r\beta>\delta_r m} a^{(r)}_{\alpha\beta} x^\alpha y_r^\beta .$$

The points of the carrier of the remainder term $\sum_{\alpha+\delta_r\beta>\delta_r m} a^{(r)}_{\alpha\beta} x^\alpha y_r^\beta$ lie above the line through $(0,m)$ with slope $-1/\delta_r$, and below the line $\beta = m - 1$.

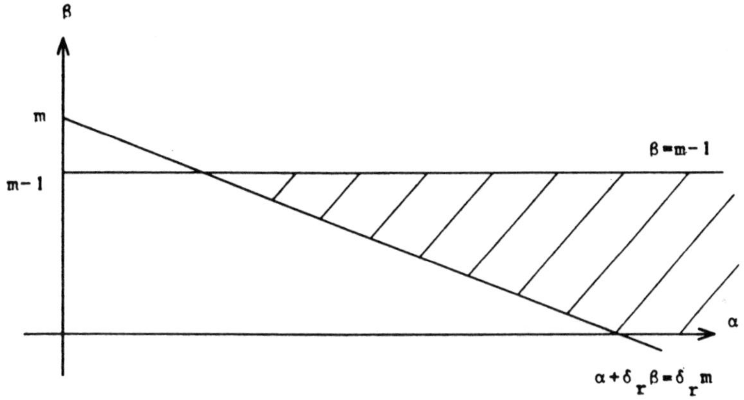

The δ_i are monotonically increasing, hence the remainder term becomes divisible by an arbitrarily high power of x when r is sufficiently large. Hence the sequence of remainder terms tends to zero and we have

$$f(x,y) = a_{0,m}(y - \sum_{i=0}^{\infty} b_i x^{\delta_i})^m$$

and the curve Y is regular at 0.

(Remark : This proof is essentially an explicit implementation of the Newton process, and $y = \Sigma b_i x^{\delta_i}$ is the Puiseux series for f.)

We can now prove the theorem already announced :

Theorem 8 :

Let Y be a curve on a complex surface and let $p \in Y$ be a point with multiplicity $\nu_p(Y) = m$. Then the number of infinitely near m-tuple points equals the integer part, $[\delta_p(Y)]$, of the first characteristic exponent of Y at p.

Proof :

If $(Y,0)$ is regular (i.e. $m = 1$ and $\delta_0(Y) = \infty$), then iterated blowing up always yields regular curve germs, and the assertion in this case is trivial. If $(Y,0)$ is singular, then by the preceding lemma there is a smooth curve W which has maximal contact with Y. Lemma 6 shows that one comes to a singularity (Y',p') of multiplicity m, whose contact exponent is less than two, in precisely $[\delta_0(Y)] - 1$ blowing up operations. It also shows that singularities which result from blowing up (Y',p') have multiplicity less than m. Thus by Lemma 1(iii) the number of m-tuple infinitely near points exactly equals $[\delta_0(Y)]$.

We now have the tools to prove the central result of this section.

Theorem 9 :

The singularities of each plane algebraic curve, and more generally, the singularities of any curve on a compact complex surface, may be resolved by a finite sequence of σ-processes.

Proof :

The curve has only finitely many singular points ; it suffices to show that each individual singularity can be resolved by finitely many blowing up operations. By Theorem 8, one obtains singularities of lower multiplicity by finitely many σ-processes (more precisely : by as many σ-processes as the integer part of the contact exponent). Thus the assertion follows by induction on the multiplicity of the singularity.

Now that we have secured the existence of a resolution of singularities by σ-processes, we want to investigate these resolutions more closely, and to describe the different kinds.

First we consider an example! We look at Neil's parabola, given by the equation $x^2 + y^3 = 0$. If we blow up the origin once, then the singularity is resolved immediately.

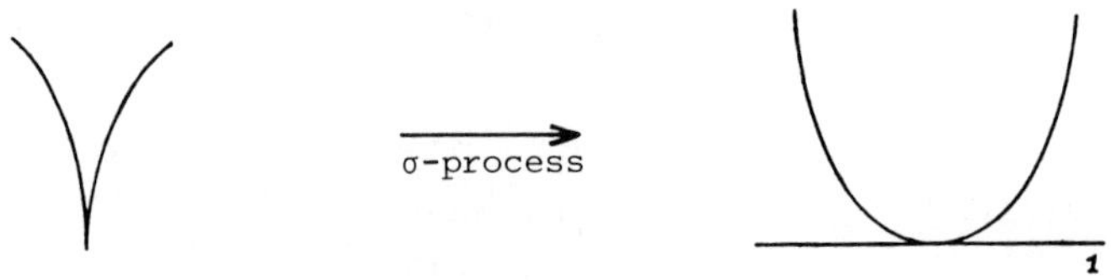

However, one is still not satisfied with this situation : the exceptional curve is not cut transversely. By blowing up again we obtain a better situation, though it is still not optimal : the proper preimage of the curve (dotted) indeed cuts each of the exceptional curves transversely, but at the point at which these two curves meet.

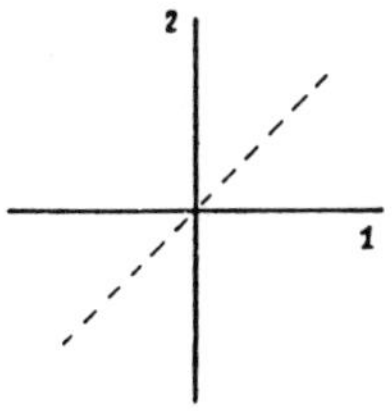

However, when we apply one more σ-process, we finally reach a situation in which the proper preimage of the curve meets only one of the exceptional curves, and that transversely :

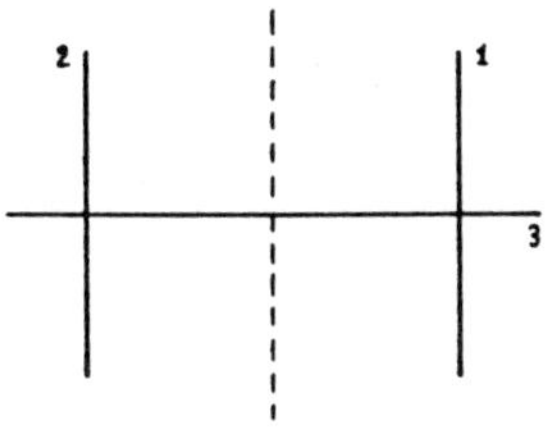

This example is typical, inasmuch as such a situation may be reached quite generally, for any singularity.

Corollary 10 :

Each singularity of a plane curve may be resolved by finitely many σ-processes so that the proper preimage of the curve meets the preimage of the singular point – i.e. the system of exceptional curves – transversely at a regular point.

Proof :

First we resolve the singularity. If we then have non-transversality, but higher contact, at a point, we blow up at this point. Each blowing up lowers the contact by Lemma 6. If more than two curves meet at a point, then the curves may be separated by blowing up at this point.

An analogue of Corollary 10 also holds in higher dimensions. This analogue says that each singularity may always be resolved to the point where the components of the preimages are manifolds which meet transversely in pairs and have only normal crossings, i.e. they meet locally like coordinate hyperplanes in $\mathbb{C}^n$. This theorem is a very frequently used result, because it permits the investigation of singularities to be reduced to the investigation of very simple singular standard situations — the normal crossings.

In higher dimensions a singularity can have several essentially different such resolutions. For plane curves, too, there are several such resolutions, because when one has a resolution with normal crossings, one can blow up again and obtain another resolution with normal crossings. But in the case of curves there is obviously a unique "smallest" resolution with normal crossings, namely, the one obtained when we blow up as few times as possible.

Definition :

For a singularity of a plane curve, the smallest resolution with normal crossings is called the standard resolution of the given singularity.

Theorem 9 makes an assertion about what one can attain with a finite sequence of σ-processes. Similarly, one can ask what is attainable by a finite sequence of quadratic transformations. (Here we mean the quadratic transformations considered in the introduction to this chapter, namely birational transformations

$$T : P_2(\mathbb{C}) \to P_2(\mathbb{C})$$
$$x = (x_0,x_1,x_2) \mapsto (f_0(x),f_1(x),f_2(x))$$

where the f_i are homogeneous polynomials of degree 2 in the x_i). The answer is given by the following :

Corollary 11 (M. Noether) :

By means of a finite sequence of quadratic transformations, each curve in the projective plane may be transformed into another curve of the projective plane with only ordinary singularities.

Proof :

Suppose a curve in the plane is given. We place our fundamental triangle so that only one vertex lies at a singularity and the three lines are general lines, i.e. no further singularities lie on the lines and the lines meet the curve transversely everywhere.

In particular, the lines must be different from the tangents to the branches of the curve at the chosen point.

As we have seen in 8.4, the quadratic transformation with this fundamental triangle corresponds to blowing up the three vertices and blowing down the lines. Thus the quadratic transformation chosen blows up the singularity at the vertex of the triangle, and thereby simplifies it. On the other hand, blowing down the lines results in multiple points, and in fact ordinary singularities.

No other singularities result from blowing down the three lines. In particular, the singularity which results from blowing up the chosen vertex is not altered by blowing down the lines, since, by the assumption of general position for these lines, the strict preimages of the lines after blowing up the vertex do not meet the resulting singular points at all. Hence, for each singular point p of the given curve Y in the plane, one can choose a sequence of quadratic transformations of the plane which have the same effect on this point as a given sequence of σ-processes. Since one can resolve each singularity by a sequence of σ-processes, by Theorem 9, one can choose a sequence of quadratic transformations converting a plane curve Y and a singular point p into a plane curve Y' for which the points corresponding to p are nonsingular. The remaining singular points remain unaltered under the birational transformations, and only ordinary singularities appear as new singularities on Y'. If one applies similar transformations successively to the finitely many (non-ordinary) singular points of Y, one finally obtains a curve with only ordinary singularities. This proves Corollary 11.

The above sequence of quadratic transformations yields a birational transformation of $P_2(\mathbb{C})$ onto itself. Corollary 11 therefore says that each curve in $P_2(\mathbb{C})$ is birationally equivalent to a curve in $P_2(\mathbb{C})$ with only ordinary singularities.

When one wants to retain the embedding in $P_2(\mathbb{C})$, then this result cannot be improved, since the ordinary singularities cannot be removed. Certainly, we can do this in a resolution by σ-processes, but then we must take into account that our curve lies in a really complicated surface, which itself is embedded in a high-dimensional projective space.

To conclude this section we shall therefore discuss whether this curve can be embedded as a non-singular curve in a space of lower dimension. We shall present only the ideas, and omit the details.

We know that when a point of a plane curve is blown up, the new curve lies in $P_1(\mathbb{C}) \times P_2(\mathbb{C})$.

This transforms the plane X into the surface X' in $P_1(\mathbb{C}) \times P_2(\mathbb{C})$. Now one can embed each product of projective spaces, $P_N(\mathbb{C}) \times P_M(\mathbb{C})$, as a submanifold in a projective space $P_L(\mathbb{C})$ with $L = (N+1)(M+1) - 1$, e.g. by the <u>Segre embedding</u> $((x_i),(y_j)) \mapsto (x_i \cdot y_j)$, where x_i resp. y_j are the homogeneous coordinates in $P_N(\mathbb{C})$ resp. $P_M(\mathbb{C})$ (cf. Mumford [M8]). This makes X' also a surface in $P_5(\mathbb{C})$. Thus blowing up a point of the plane X results in a projective algebraic surface X'. More generally : if $X \subset P_K(\mathbb{C})$ is a projective algebraic surface, then the surface X' which results from blowing up a point p of X is again projective algebraic. Proof : X' lies in the k-dimensional manifold which results from blowing up p in $P_K(\mathbb{C})$. The latter lies in $P_K(\mathbb{C}) \times P_{K-1}(\mathbb{C})$, which is embedded in $P_L(\mathbb{C})$ with $L = K(K+1) - 1$ by the Segre embedding. Thus X' is also embedded in $P_L(\mathbb{C})$. It follows by induction that repeated σ-processes on projective algebraic surfaces always yield projective algebraic surfaces.

Thus the curve C resulting from a plane curve by resolution of singularities can be embedded in a high-dimensional $P_N(\mathbb{C})$, where N is larger, the more σ-processes are performed. Now we want to project this curve onto a space of lower dimension. In order to clarify what can happen, we first consider the case of projection of a curve C in space onto a plane E from a centre of projection p.

The curve C' in the plane E which we obtain as the image of C can acquire singularities in various ways. For example, it can happen that a ray of projection, i.e. a line through the centre of projection p and a point q of E meets the curve C more than once. When that happens – and only then – C' has several branches at q, one for each point in which the ray meets C. As another example, it can happen that the ray of projection meets the curve non-transversely at some point, and hence is a tangent to the curve. When that happens — and only then — the corresponding branch of C' has a singular point at q. We illustrate both these possibilities for the generation of singularities with a picture of the corresponding real situation.

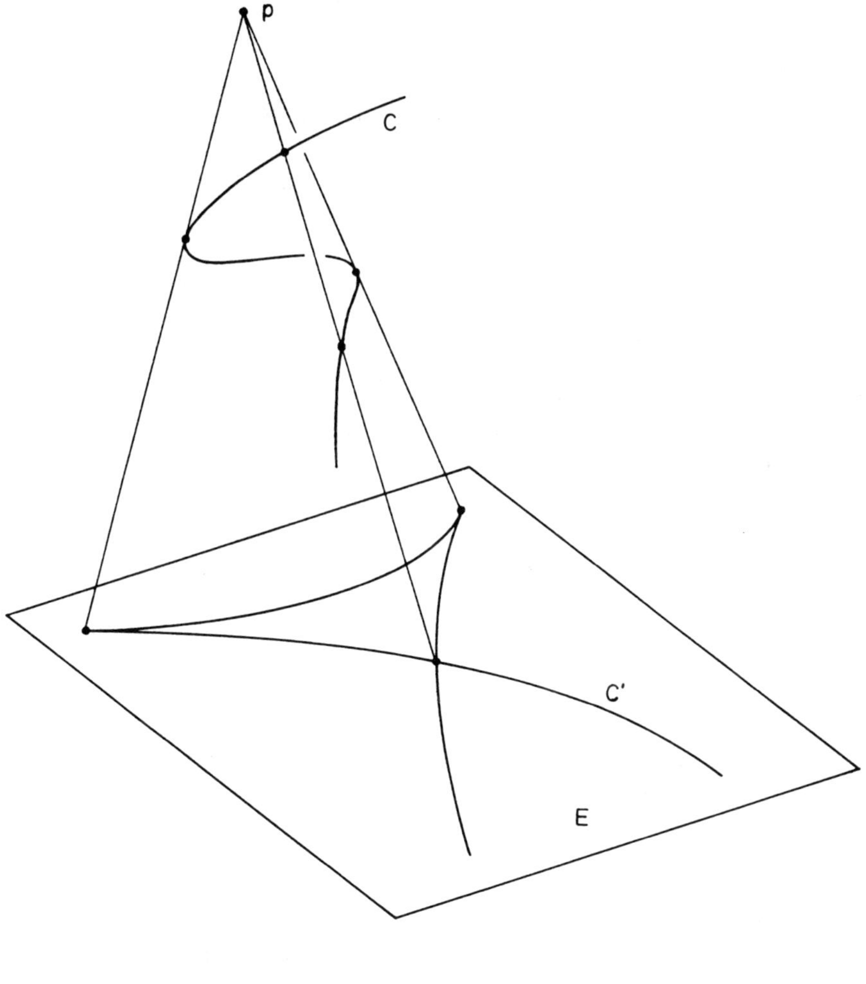

At bottom, we have seen all this before, namely with the consideration of singularities of outlines in sections 1.8 and 4.2. (Admittedly only the curve C' appeared in the pictures there, as the projected outline of a surface F. The curve C itself is not shown in the pictures in 1.8. It is the "apparent outline" on the surface F itself which appears with projection from p – the line which separates "visible" and "invisible" parts of F. In addition, the centre of projection in the pictures there has been shifted to infinity.)

The situation is analogous for the projection of a curve C in a higher-dimensional projective space $P_N(\mathbb{C})$ onto a hyperplane $P_{N-1}(\mathbb{C})$: the image has no singularities just in case there are no rays of projection which are tangents or secants of the curve.

Now the set of all points which lie on the tangents to an algebraic curve is an algebraic subset of dimension ≤ 2, and the set of all points which lie on secants is an algebraic subset of dimension ≤ 3. Hence in $P_N(\mathbb{C})$, $N > 3$, we can always find a point which does not lie on a tangent or a secant. If we project from this point onto a $P_{N-1}(\mathbb{C})$, then no ray of projection coincides with a tangent or a secant, and hence we generate no new singularities. If we now begin with a nonsingular curve C which we have embedded in a $P_N(\mathbb{C})$, and project successively onto subspaces whose dimensions decrease by one each time, then we finally obtain the result : C can be embedded as a nonsingular curve in $P_3(\mathbb{C})$. In a space of still lower dimension, i.e. in the projective plane, we cannot in general project C onto a nonsingular image, since we can no longer avoid the secant set in $P_3(\mathbb{C})$. However, we can avoid the set of points on triple secants, i.e. on the secants which meet C in more than two points, because this set is at most two-dimensional. In addition, we can, for the same reason, avoid the points on secants which meet C at two points where the tangents are coplanar with the secant. And finally we can, as before, avoid the set of points on tangents of C. When we choose our centre of projection $p \in P_3(\mathbb{C})$ in this way generally, we obtain as image of C a curve C' which has only ordinary double points. Thus we have :

Each algebraic curve is birationally equivalent to a plane algebraic curve which has at most ordinary double points as singularities.

This theorem was first proved by Clebsch.

<u>Description of the resolution of singularities of plane curves</u> :

We now want to consider resolution in more detail. There are various methods for describing resolution. Here we shall present two of them, namely

a) the multiplicity sequence and

b) resolution graphs.

We already know that Puiseux pairs describe singularities topologically. The goal of the present section is to explain how these different sets of data are related to each other, and how one may be computed from the other.

We begin with the definition of the multiplicity sequence. This is not difficult for irreducible curves. Suppose we are given an irreducible curve germ with an isolated singularity. Let ν_0 be the multiplicity of this curve germ at this point. If we blow up once, then we again find at most one singularity. Let ν_1 be the multiplicity of the curve germ blown up once, ν_2 the multiplicity of the curve germ blown up twice, and continue to the standard resolution. The sequence ends with a sequence of ones. The sequence of these multiplicities, $(\nu_0,\nu_1,\ldots,\nu_{N-1})$, where the last one is not counted, is then the <u>multiplicity sequence</u>.

With reducible curves we must proceed somewhat more carefully since we must describe, among other things, how long the preimages of the individual components stay together, and when they separate. Now we come to the definition of the multiplicity sequence in the general case.

Let X be a surface and let Y be a curve in X with an isolated singularity $p \in Y$. For the curve germ $(Y,p) \subset (X,p)$ we consider the sequence of mappings

$$X = X_0 \xleftarrow{\phi_1} X_1 \xleftarrow{\phi_2} X_2 \xleftarrow{\phi_3} \cdots \xleftarrow{\phi_i} X_i \longleftarrow \cdots \xleftarrow{\phi_N} X_N ,$$

defined recursively as follows. Suppose ϕ_1 to ϕ_i have already been defined. Let $\psi_i : X_i \to X_0$ be the product of the mappings ϕ_j, $j = 1,\ldots,i$, thus $\psi_i = \phi_1 \circ \phi_2 \circ \cdots \circ \phi_i$ and $\psi_0 : X_0 \to X_0$ is the identity. Let $E_i = \psi_i^{-1}(p)$ be the exceptional curve in X_i, and let Y_i be the proper preimage of Y under ψ_i in X_i. Then let $\phi_{i+1} : X_{i+1} \to X_i$ be the blowing up of precisely the points of $Y_i \cap E_i$ which are still singular or else non-transverse intersections of Y_i with the exceptional curve E_i at regular points of E_i.

By Corollary 11 to the main theorem on the resolution of singularities, we know that after finitely many such blowing up processes ϕ_i the proper preimage of the curve becomes nonsingular and it meets the exceptional curve transversely at regular points. Let N be the minimal number of such blowing up processes ϕ_i. Then $\psi_N : X_N \to X_0$ is the standard resolution of the singularity $(Y,p) \subset (X,p)$, and $\phi_1,\dots,\phi_N$ is the sequence of mappings we wished to define.

The points in $Y_i \cap E_i$ are the <u>infinitely near points of the singular point</u> $p \in Y$ <u>in</u> X_i. Let (Y^r,p) be the irreducible components of (Y,p), $r = 1,\dots,s$. Since we need only consider a neighbourhood of p in X sufficiently small that Y decomposes into the irreducible components $Y^1,\dots,Y^s$, we can assume without loss of generality that Y already decomposes in X. Let $Y_i^r \subset X_i$ be the proper preimage of Y^r under ψ_i. When we resolve an irreducible curve, we have exactly one infinitely near point of the singular point on each level. Thus exactly one of the infinitely near points of $p \in Y$ in X_i lies on Y_i^r. Let $Y_i^r \cap E_i = \{p_i^r\}$.

<u>Definition</u> :

Let

$$\nu_i(Y^r) := \nu_{p_i^r}(Y_i^r)$$

be the multiplicity of Y_i^r at p_i^r. The sequence of numbers $(\nu_0(Y^r),\nu_1(Y^r),\dots,\nu_{N-1}(Y^r))$ is the <u>multiplicity sequence of the branch</u> (Y^r,p) <u>of</u> (Y,p) with respect to the standard resolution $\psi_N : X_N \to X_0$ of (Y,p) when $N > 0$. If p is regular, we associate p with the empty sequence.

<u>Remark</u> :

We exclude the multiplicity $\nu_N(Y^r)$ in order to be able to formulate our later Theorem 12 more simply.

We obtain s sequences of numbers in this way. Certain numbers in them correspond to the same points. Thus we also obtain, for each i with $0 \leq i \leq N-1$, a partition of the index set $\{1,\dots,s\}$ into subsets such that : r and r' are in the same subset just in case $p_i^r = p_i^{r'}$, i.e., when Y_i^r and $Y_i^{r'}$ go through the same infinitely near point of p in X_i.

<u>Definition</u> :

The multiplicity sequences

$$(\nu_0(Y^r),\nu_1(Y^r),\ldots,\nu_{N-1}(Y^r)) \qquad r = 1,\ldots,s$$

of the branches (Y^r,p) of (Y,p), together with the above partitions of the index set which describe the coincidences of infinitely near points on the branches, is called the <u>system of multiplicity sequences of</u> (Y,p) with respect to the standard resolution.

We now want to give an intuitive graphical description of these systems. Since infinitely near points of p which belong to two different components of Y and do not coincide in X_i also do not coincide in X_{i+1}, the partitions become increasingly refined. Hence when we number the irreducible components suitably, we have a partition of the index set $\{1,\ldots,s\}$ into intervals on each level. We shall assume that our components have been numbered in this way. We now associate the numbers in the multiplicity sequence of a branch Y^r with points which we draw one above the other and connect by lines. We label the points from top to bottom with the multiplicities $\nu_{N-1}(Y^r),\ldots,\nu_0(Y^r)$. The schemata for the different branches are drawn side by side, beginning on the left with Y^1 in the numbering. When we finally indicate the multi-element subsets in the partition by encircling their elements, we obtain a graphical schema such as

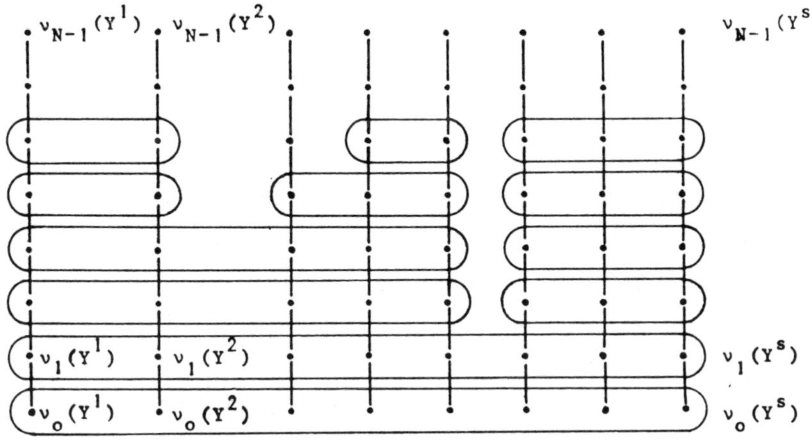

This schema contains the same information as the system of multiplicity sequences. We can also encode the same information by graphical schemata like the following :

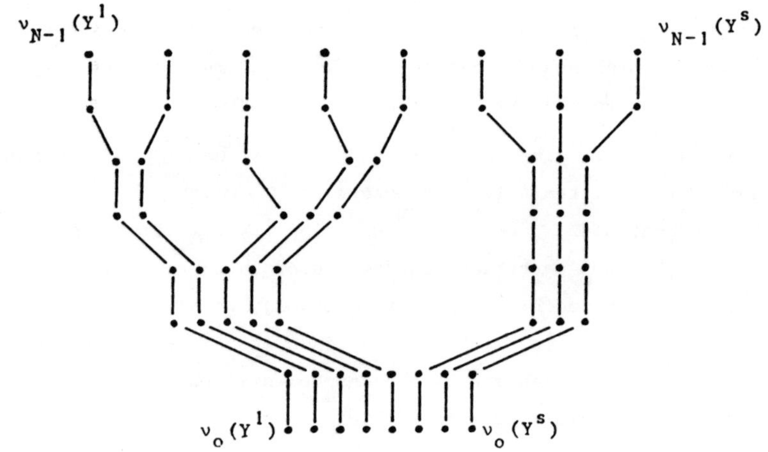

Here, the incidences are indicated by close proximity of the corresponding points. We cannot simplify the schema still further. If we fuse incident points into a single point, then we lose information. Because at such a point we must then write different multiplicities, and we no longer know which multiplicity belongs to which sequence. Thus a schema such as the following gives too little information about the singularity :

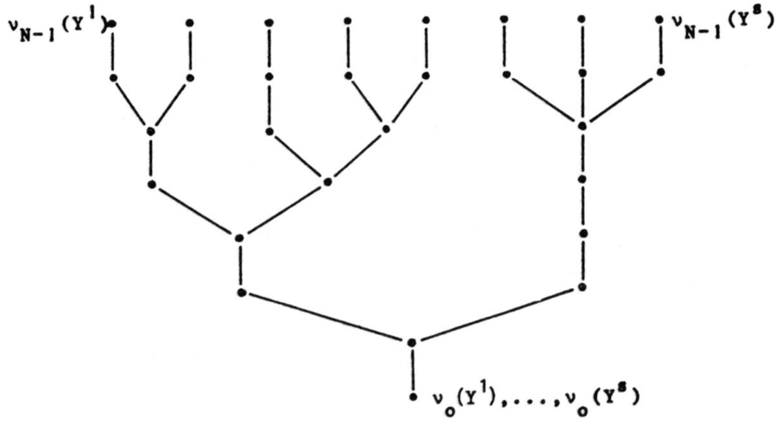

Notation :

For the sake of convenience, we have defined the multiplicity sequence of a curve so that the sequences of all branches have the same length N. As a rule, however, the proper preimage of a branch Y^r of the curve will already meet the exceptional curve transversely, and will no longer meet the proper preimages of the other branches, after $L_r < N$ steps. The smallest number L_r with this property lies between the length N_r of the standard resolution of the single branch Y^r and the length N of the standard resolution of the whole curve Y. We call L_r the proper length of the multiplicity sequence of Y^r in the system of multiplicity sequences of Y.

We now address ourselves to the second way of describing the resolution. As above, let $\psi_N : X_N \to X_0$ be the standard resolution of the curve germ $(Y,p) \subset (X,p)$, the product of the mappings $X_N \to X_{N-1} \to \dots \to X_1 \to X_0 = X$. Let $E_N = \psi_N^{-1}(p)$ be the exceptional curve. The irreducible components of E_N are nonsingular and biholomorphically equivalent to $P_1(\mathbb{C})$, because blowing up introduces a $P_1(\mathbb{C})$ and outside the blown up point nothing changes. Any two components meet transversely at at most one point, and at most two components of E_N go through any point of X_N. One sees this inductively. We trivially obtain such a situation with the first blowing up. Now suppose that any two components of $E_i = \psi_i^{-1}(p)$ meet transversely in X_i at at most one point, and that at most two components of E_i go through any point of X_i. Then only two cases are possible for a point of E_i: either it lies on only one component or it is a crossing point of two components. The following sketches indicate what a σ-process does to such a point.

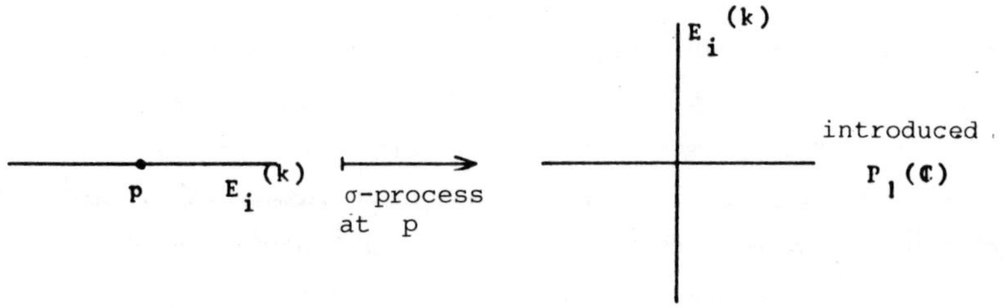

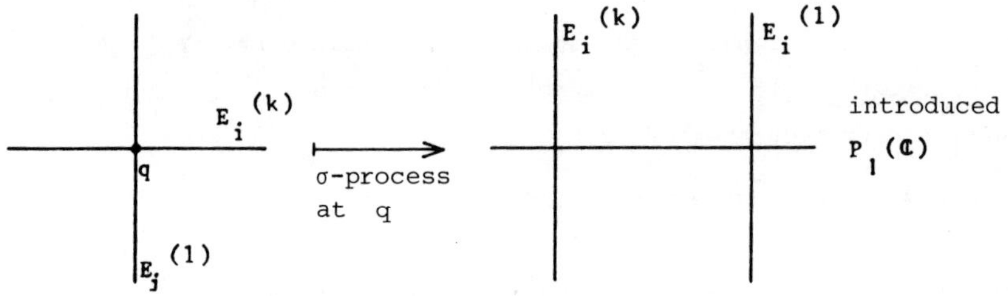

In both cases the $P_1(\mathbb{C})$ introduced meets each component of E_i transversely at at most one point, and different components at different points. But the situation remains unaltered away from the blown up point, hence our assertion holds for E_{i+1}.

Let Y_N be the proper preimage of Y in X_N. Again let $X_0 = X$ be so small that Y decomposes into s connected components in X_0, where s is the number of irreducible components of Y. Then Y_N decomposes into s smooth connected components $Y_N^1, \ldots, Y_N^s$, each of which meets E_N transversely at one point – this is precisely how the standard resolution was chosen. In summary, we have a situation which we can describe schematically by pictures such as the following :

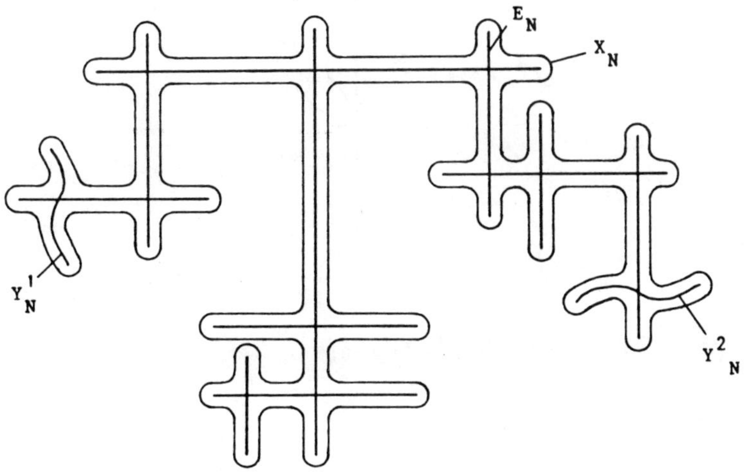

We remark that no cycles appear in these schemata, and hence we can always represent the projective line $P_1(\mathbb{C})$ by straight line segments. Now we come to the definition of resolution graphs (also known as dual graphs).

<u>Definition</u> :

A <u>resolution graph</u> is a graph marked with values in the following way :

Each component of E_n is associated with a point •, each component of Y_N with a point *. Two points are connected by a segment just in case the corresponding components intersect. The point • corresponding to the component D of E_N is given the value $i(D)$, where $1 \leq i(D) \leq N$ is the smallest index i for which the image of D in X_i under the product of the mappings $X_N \to \ldots \to X_i$ is still not a point.

The assignment of values is necessary, since one must know which blowing up processes $\phi_i : X_i \to X_{i-1}$ the curves arise from. We note that different points can have the same indices, since, in the case of reducible curves, several points may be blown up at once by $\phi_i : X_i \to X_{i-1}$.

We now want to illustrate our definitions by an example :

Example :

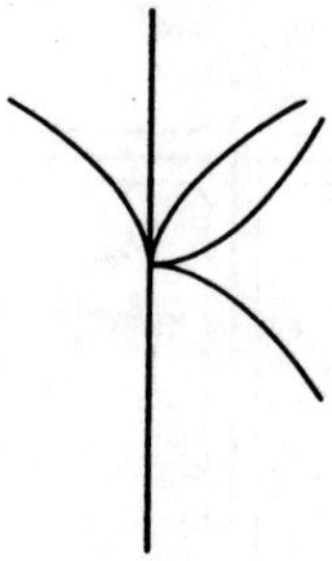

$Y = \{(x,y) \in \mathbb{C}^2 \mid x(x^2-y^3)(x^3-y^2) = 0\}$, $p = (0,0)$.

Standard resolution :

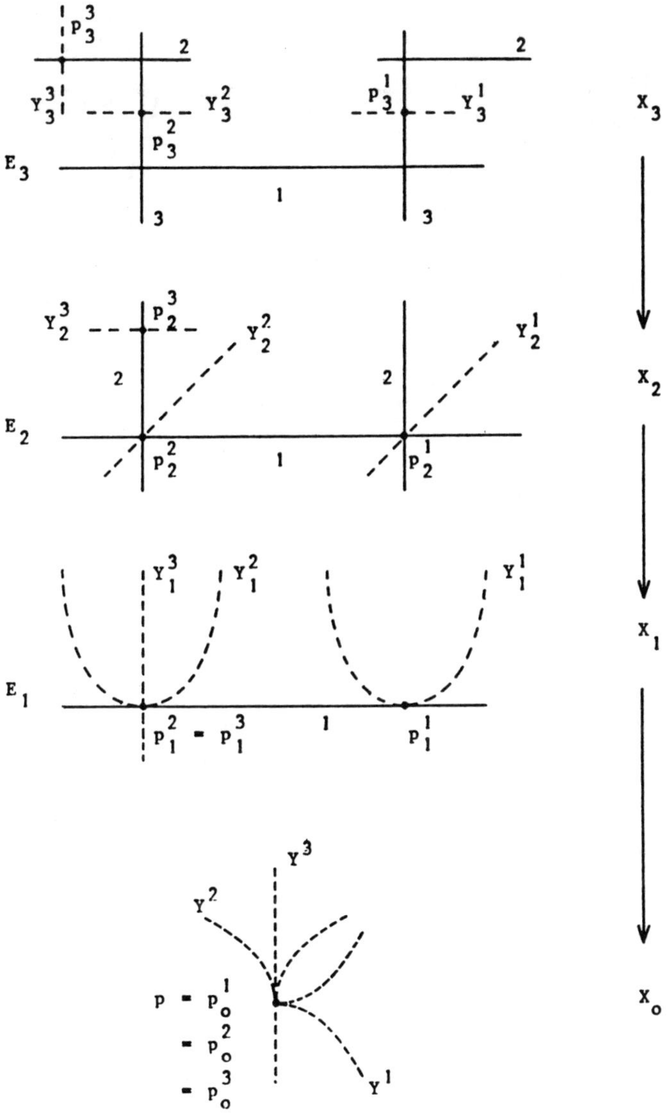

(The numbers on the components D of E_i correspond to the indices i(D).)

Multiplicity sequence :

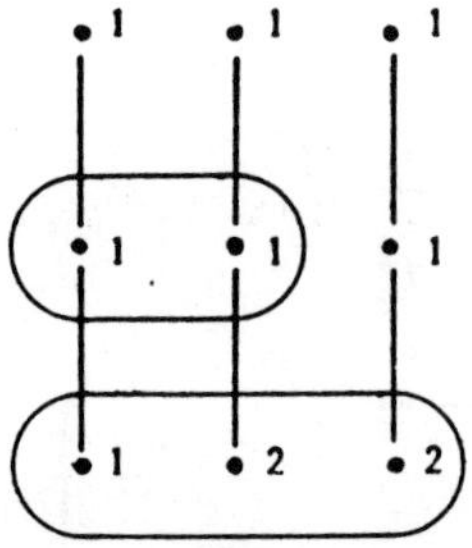

or

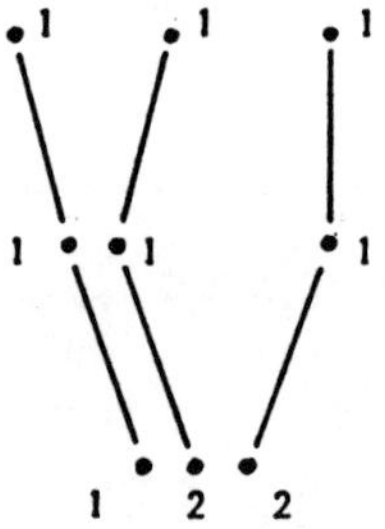

Resolution graph :

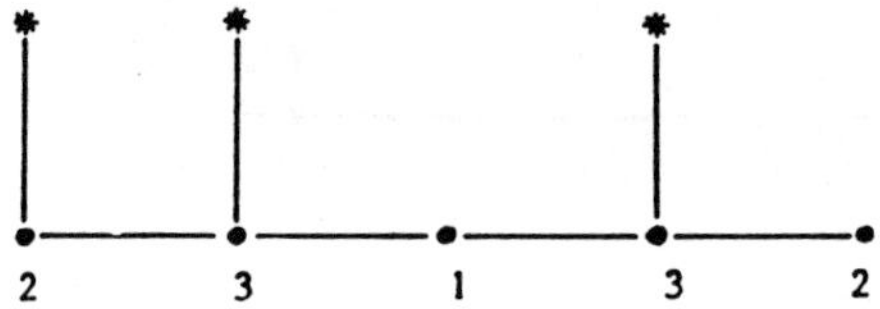

We have now introduced different sets of data which describe singularities. We know the following invariants : Puiseux pairs and intersection numbers, the multiplicity sequence, and finally the resolution graph. Now we are interested in the connection between all these sets of data. We first show how the multiplicity sequence may be computed from the Puiseux pairs and intersection numbers.

To do this we begin with the case of an irreducible curve Y with an isolated singularity at 0. Let $\nu_0 = \nu_0(Y)$ be the multiplicity at this point. Let

$$x = t^m$$
$$y = a_1 t^{k_1} + a_2 t^{k_2} + \ldots, \quad a_i \neq 0, \ k_i \in \mathbb{N} \cup \{0\}$$

be the Puiseux expansion. The inessential terms are admitted and it is not assumed that $k_1 > m$. We want to express the multiplicity ν_0 in terms of m and the k_i. We recall that the multiplicity is the intersection multiplicity with a general line. But we obtain the intersection multiplicity with a general line from Theorem 8.3.14, by substituting the Puiseux expansion of the curve in the general line equation. To do this for the curve Y, we must distinguish 4 cases :

(i) $m \le k_1 \Rightarrow \nu_0 = m$

(ii) $m > k_1 > 0 \Rightarrow \nu_0 = k_1$.

Now, the case $k_1 = 0$ can also occur. This means that the curve branch described by the Puiseux expansion does not go through the origin. We must therefore make the following coordinate change :

$$x' = x$$
$$y' = y - a_1.$$

After this, k_2 plays the rôle of k_1, and we have the further cases :

(iii) $k_1 = 0,\ m \le k_2 \Rightarrow \nu_0 = m$

(iv) $k_1 = 0,\ m > k_2 \Rightarrow \nu_0 = k_2$.

Now we perform a σ-process at $(x(0),y(0))$. Let Y_1 be the proper preimage of Y under this process. Let ν_1 be the multiplicity of Y_1 at the intersection with the exceptional curve. We now investigate what the Puiseux expansion looks like in the individual cases for Y_1.

Case (i) :

In this case we look at the Puiseux expansion in the chart given by the transformation

$$x = v$$
$$y = uv,$$

namely :

$$u = a_1 t^{k_1 - m} + a_2 t^{k_2 - m} + \dots$$
$$v = t^m.$$

Case (ii)

Transformation : $x = uv$

$$y = v$$

$$u = b_1 t^{m-k_1} + b_2 t^{k_2+m-2k_1} + \dots$$
$$v = a_1 t^{k_1} + a_2 t^{k_2} + \dots$$

Case (iii)

Transformation : $x = v$

$$y - a_1 = uv$$

$$u = a_2 t^{k_2 - m} + a_3 t^{k_3 - m} + \dots$$
$$v = t^m.$$

Case (iv)

Transformation : $x = uv$

$$y - a_1 = v$$
$$u = b_1 t^{m-k_2} + \dots$$
$$v = a_2 t^{k_2} + a_3 t^{k_3} + \dots .$$

In each case, the Puiseux expansion obtained for Y_1 can again be assigned to cases (i) - (iv), and the next multiplicity ν_1 in the multiplicity sequence can be computed accordingly. By iteration one obtains an algorithm for computing the multiplicity sequence of an irreducible curve.

We want to analyse this algorithm in still more detail.

Suppose we take case (i), i.e. $m \leq k_1$. It then follows that $\nu_0 = m$. If we take case (i) again for Y_1, then $m \leq k_1 - m$, i.e. $2m \leq k_1$. Again it follows that $\nu_1 = m$. If we also take case (i) for the next transform Y_2, so that $3m \leq k_1$, then it also follows that $\nu_2 = m$. In general, it follows from $\mu m \leq k_1$ that the first μ multiplicities in the multiplicity sequence satisfy :
$\nu_0 = \dots = \nu_{\mu-1} = m$.

Thus, let

$$k_1 = \mu m + r, \ 0 \leq r < m$$

(If $r > 0$, $\mu = [\delta(Y)]$ by Theorem 8.) Then we obtain

$$\nu_0 = \dots = \nu_{\mu-1} = m.$$

Now if $r > 0$, the next multiplicity $\nu_\mu = r$. At this stage m plays the rôle that k_1 played previously, and r plays the rôle previously played by m. Hence if we again write

$$m = \mu' r + r', \ 0 \leq r' < r,$$

then the next μ' multiplicities in the multiplicity sequence satisfy

$$\nu_\mu = \dots = \nu_{\mu+\mu'-1} = r.$$

What is happening here is obviously a euclidean algorithm for the numbers m, k_1.

$k_1 = \mu m + r \Rightarrow \mu$ times the multiplicity m

$m = \mu' r + r' \Rightarrow \mu'$ times the multiplicity r

$r = \mu'' r' + r'' \Rightarrow \mu''$ times the multiplicity r'

$r' = \mu''' r'' + r''' \Rightarrow \mu'''$ times the multiplicity r''

$$\cdots\cdots\cdots$$

$$r^{(k-1)} = \mu^{(k+1)} r^{(k)} .$$

We have obtained this algorithm under the assumption that case (i) holds at the beginning, so that $k_1 \geq m$. However, case (ii) also leads to such an algorithm, except that $\mu = 0$ and $r = k_1$.

As long as the algorithm runs on m and k_1, case (i) or (ii) always appears. But now what happens when the euclidean algorithm comes to an end for the numbers m, k_1 ?

One easily convinces oneself that the case now is (iii) or (iv), and that the two power series of the Puiseux expansion begin as follows:

$$t^{r^{(k)}} + \text{higher terms}$$
$$c \cdot t^{k_2 - k_1} + \text{higher terms.}$$

(In order to see this, one sets $m = k_1$ in case (i) resp. (ii).) Thus a new euclidean algorithm now begins for the numbers $k_2 - k_1$ and $r^{(k)}$, and so it continues.

Thus one sees that altogether the multiplicity sequence is obtained from the Puiseux expansion with the help of a chain of euclidean algorithms, where each algorithm begins with the difference $k_i - k_{i-1}$ between two successive exponents and the last multiplicity computed from the preceding algorithm.

We now claim that one need only pay attention to the characteristic terms in the power series

$$y = a_1 t^{k_1} + \ldots + a_m t^{k_m} + \ldots .$$

Suppose, e.g., that the expansion does not begin with a characteristic term. Then k_1 is a multiple of m, $k_1 = \mu m$. Thus m appears at least μ times at the beginning of the multiplicity sequence. But we would also have obtained this by omitting $a_1 t^{k_1}$, because $k_2 > k_1$ and hence $k_2 > \mu m$. Carrying out the algorithm first for k_1 and m and then for $k_2 - k_1$ and m, or just for k_2 and m, amounts to the same thing. The same argument holds for k_2 etc. to the first characteristic term, with which we therefore can begin our algorithm equally well. One similarly convinces oneself that the other non-characteristic terms are irrelevant.

Thus we have essentially proved the following result :

Theorem 12 (Enriques-Chisini)

(i) For an irreducible curve with Puiseux expansion

$$x = t^m$$
$$y = a_1 t^{k_1} + a_2 t^{k_2} + \dots + a_g t^{k_g},$$

in which only essential (characteristic) terms appear, the multiplicity sequence is determined by the following chain of g euclidean algorithms :

$$\begin{aligned} \kappa_i &= \mu_{i,1} r_{i,1} + r_{i,2} \\ r_{i,1} &= \mu_{i,2} r_{i,2} + r_{i,3} \\ &\vdots \qquad\qquad\qquad\qquad i = 1,\dots,g \\ r_{i,w(i)-1} &= \mu_{i,w(i)} r_{i,w(i)} \end{aligned}$$

with $0 \leq r_{i,j+1} < r_{i,j}$

$$\kappa_i = k_i - k_{i-1} \ \text{(where } k_0 = 0\text{)}, \ i = 1,\dots,g$$
$$r_{i,1} = r_{i-1,w(i-1)} \quad \text{for } i > 1, \ r_{1,1} = m.$$

In the multiplicity sequence, the multiplicity r_{ij} then appears μ_{ij} times, where $i = 1,\dots,g$; $j = 1,\dots,w(i)$. (If a certain multiplicity arises from several successive algorithms, then it is also counted multiply.)

(ii) For an arbitrary irreducible curve one obtains the multiplicity sequence by omitting all non-characteristic terms from the Puiseux expansion and then applying the algorithm above.

(iii) Conversely, one can reconstruct the exponents of the characteristic terms of the Puiseux expansion of an irreducible curve, i.e. the Puiseux pairs of the curve, from the multiplicity sequence, by the chain of euclidean algorithms.

The upshot of this theorem is that, for an irreducible curve, knowledge of the Puiseux pairs is equivalent to knowledge of the multiplicity sequence.

Examples :

a) Let Y^1 be given by the Puiseux expansion

$x_1 = t^{100}$

$y_1 = t^{250} + t^{375} + t^{410} + t^{417}$

$250 = 2.100 + 50$

$100 = 2.50$

$125 = 2.50 + 25$

$50 = 2.25$

$35 = 1.25 + 10$

$25 = 2.10 + 5$

$10 = 2.5$

$7 = 1.5 + 2$

$5 = 2.2 + 1$

$2 = 2.1$

Multiplicity sequence :

(100,100, 50,50,50,50, 25,25,25, 10,10, 5,5,5, 2,2, 1,1)

b) Let Y^2 be given by

$x_2 = t^{100}$

$y_2 = t^{250} + t^{375} + t^{390} + t^{391}$

$250 = 2.100 + 50$

$100 = 2.50$

$125 = 2.50 + 25$

$50 = 2.25$

$15 = 0.25 + 15$

$25 = 1.15 + 10$

$15 = 1.10 + 5$

$10 = 2.5$

$1 = 0.5 + 1$

$5 = 5.1$

Multiplicity sequence :

(100,100, 50,50,50,50, 25,25, 15, 10, 5,5, 1,1,1,1,1).

c) Let Y^3 be given by

$$x_3 = t^{100}$$

$$y_3 = t^{250} + t^{375} + t^{390} - t^{393}.$$

The first three algorithms are the same as in b). What follows then is

$3 = 0.5 + 3$

$5 = 1.3 + 2$

$3 = 1.2 + 1$

$2 = 2.1$

Multiplicity sequence :

(100,100, 50,50,50,50, 25,25, 15, 10, 5,5, 3, 2, 1,1)

In the case of a reducible curve we must know, in addition to the multiplicity sequences of the irreducible components, the incidence schemata of the infinitely near points. In the description by Puiseux pairs we must know, in the addition to the Puiseux pairs, the intersection numbers of the individual components. The following theorem gives information on the relation between these additional pieces of data.

Theorem 13 (Max Noether)

Let $(Y,0)$ and $(Y',0)$ be distinct irreducible curve germs in $(\mathbb{C}^2,0)$. Let the multiplicity sequences of their union be

$$(\nu_0,\nu_1,\dots,\nu_{N-1})$$

$$(\nu'_0,\nu'_1,\dots,\nu'_{N-1})$$

where, for $0 \leq j \leq \rho$, the multiplicities ν_j, ν'_j belong to the same infinitely near point in the standard resolution sequence of $Y \cup Y'$, and for $j > \rho$ they do not.

Then the intersection multiplicities satisfy

$$\nu_0(Y,Y') = \sum_{j=0}^{\rho} \nu_j \nu'_j .$$

Proof :

Let $f(x,y) = 0$ be the equation of Y, and let ν_0 be the multiplicity of Y. Coordinates x, y may be chosen so that

$$f(x,y) = y^{\nu_0} + \text{terms of higher order.}$$

(Since the curve germ is irreducible, we have only one tangent at 0. We can therefore choose the coordinates so that the line $y = 0$ is the tangent. But the homogeneous part of f of degree ν_0 is just the equation of the tangent. This gives the form above.)

Suppose Y' has the Puiseux expansion

$$x(t) = t^{\nu'_0}$$
$$y(t) = a_1 t^{k_1} + a_2 t^{k_2} + \dots .$$

Then, by Theorem 8.3.14 :

$$\nu_0(Y,Y') = \text{ord } f(x(t),y(t)). \qquad (*)$$

We now perform a σ-process at $0 \in \mathbb{C}^2$ and consider the proper preimages Y_1 resp. Y'_1 of Y resp. Y' in the chart obtained by the coordinate transformation

$$x = v$$
$$y = uv.$$

We have

$$f(x,y) = f(v,uv) = v^{\nu_0}\tilde{f}(u,v)$$

and $\tilde{f}(u,v) = 0$ is the equation for Y_1.

$$u(t) = a_1 t^{k_1-\nu'_0} + \dots$$
$$v(t) = t^{\nu'_0}$$

is the Puiseux expansion for Y'_1 and, when Y_1 and Y'_1 meet at p_1 on the exceptional curve :

$$\nu_{p_1}(Y_1,Y'_1) = \text{ord } \tilde{f}(u(t),v(t)) \qquad (**)$$

Moreover,

$$f(x(t),y(t) = v(t)^{\nu_0}\tilde{f}(u(t),v(t))$$
$$= t^{\nu_0\nu'_0}\tilde{f}(u(t),v(t)).$$

Combined with (*) and (**), this gives

$$\nu_0(Y,Y') = \nu_{p_1}(Y_1,Y_1') + \nu_0\nu_0' ,$$

and the formula asserted in the theorem then follows by induction.

Corollary 14 :

The system of multiplicity sequences of a plane curve is determined by the multiplicity sequences of its branches and the intersection multiplicities of all pairs of branches. Conversely, this data is determined by the system of multiplicity sequences. (The proof gives the method of determination.)

Proof :

Suppose we are given the multiplicity sequences $(\nu_0(Y^r),\nu_1(Y^r),\ldots,\nu_{N_r-1}(Y^r))$ of the branches Y^r relative to the standard resolutions of the individual branches, and the intersection multiplicities $\nu_0(Y^q,Y^r)$. First we formally extend these sequences to infinite sequences, by setting all subsequent terms equal to 1. We do this because it is possible that in the standard resolution of the whole curve the individual branches have to be blown up beyond the stages in their own standard resolutions. Thus for any two branches Y^q, Y^r of the curve we have two sequences :

$$\nu_0(Y^q),\nu_1(Y^q),\nu_2(Y^q),\ldots \quad\text{and}\quad \nu_0(Y^r),\nu_1(Y^r),\nu_2(Y^r),\ldots$$

of which all but finitely many terms are ones. Then let $\rho = \rho(Y^q,Y^r)$ be the unique number such that

$$\nu_0(Y^q,Y^r) = \sum_{j=0}^{\rho} \nu_j(Y^q)\,\nu_j(Y^r) .$$

Such a ρ exists by Max Noether's theorem, and it is obviously unique, because for $\rho' < \rho$ resp. $\rho' > \rho$ the right side of the equation becomes smaller resp. greater than the left. This $\rho(Y^q,Y^r)$ is just the number ρ up to which the infinitely near points of the two branches coincide. Let N be the maximum of all $\rho(Y^q,Y^r)$ for all pairs of branches and all lengths N_r of multiplicity sequences of the branches for their own standard resolutions. Then N is obviously the length of the multiplicity sequences in the system of multiplicity sequences of the curve $Y = \bigcup_{r=1}^{s} Y^r$, and one obtains these sequences by terminating the sequences for the branches after the Nth place. These sequences, together with the numbers $\rho(Y^q, Y^r)$ give the system of multiplicity sequences.

The proof of the converse is trivial. Because the multiplicity sequences of the branches, relative to their own standard resolutions, result from the multiplicity sequences of the system, possibly with the omission of a few ones. And it is clear, on the basis of the description of the multiplicity sequence of an irreducible curve by a chain of euclidean algorithms, how many ones remain after these omissions: their number equals the immediately preceding multiplicity in the sequence, and when the sequence already begins with ones, their number equals 1.

Example :

We continue the above example and determine the system of multiplicity sequences of $Y = Y^1 \cup Y^2 \cup Y^3$. If one sets up the equations $f_i(x,y) = 0$, $i = 1,2,3$, for the curves Y^i by the rule given earlier (obviously without actually multiplying out the product), and replaces x, y by $x_j(t)$, $y_j(t)$, then one obtains the intersection multiplicities $\nu_0(Y^i, Y^j)$ from the orders of these power series by Theorem 8.3.14 :

$$\nu_0(Y^1, Y^2) = 31\ 625$$
$$\nu_0(Y^1, Y^3) = 31\ 625$$
$$\nu_0(Y^2, Y^3) = 31\ 630.$$

Then one obtains the system of multiplicity sequences of $Y = Y^1 \cup Y^2 \cup Y^3$ by the process in the proof of Corollary 14 :

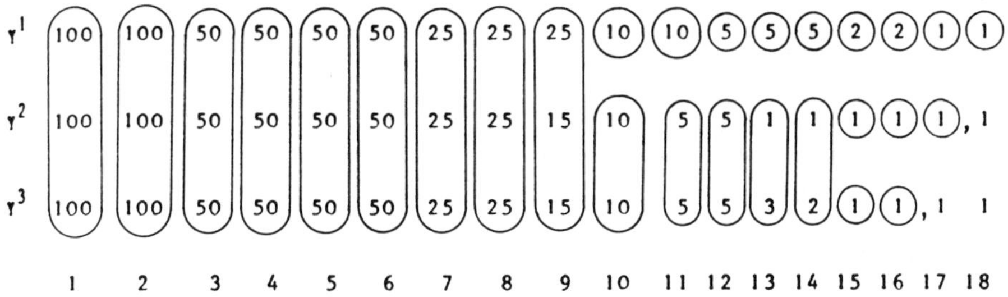

Construction of resolution graphs in the irreducible case

Let Y be an irreducible curve, given by its Puiseux expansion, and let

$$\kappa_i = \mu_{i1} r_{i1} + r_{i2}$$
$$r_{i,1} = \mu_{i2} r_{i2} + r_{i3}$$
$$\cdots\cdots\cdots \qquad i = 1,\dots,g$$
$$r_{i,w(i)-1} = \mu_{i,w(i)} r_{i,w(i)}$$

be the chain of euclidean algorithms, obtained from the g essential terms of the Puiseux expansion, which we have used to describe the multiplicity sequence in Theorem 12.

We shall now use these algorithms to describe the resolution graph of Y.

The resolution graph is put together from g Puiseux chains P_i, $i = 1,\dots,g$, and each Puiseux chain P_i is in turn constructed from $w(i)$ elementary chains E_{ij}, $j = 1,\dots,w(i)$. The latter elementary chain is the subgraph of the resolution graph consisting of the μ_{ij} curves which result from blowing up the μ_{ij} points of multiplicity r_{ij} arising from the ith algorithm. This subgraph looks like :

μ_{ij} points

Here the values of the points are successive, i.e. if C is the curve in this chain with smallest value $i_1 = i(C)$, then the valuation is

$i_1 \quad i_1+1 \qquad\qquad\qquad i_1+\mu_{ij} - 1.$

Thus it suffices to know the initial and final point of the chain, and the value of the initial point. Since we shall be able to derive the value from the length and order of preceding chains when we later describe the construction of the graph, it suffices to mark the initial and final point of the chain.

Also, since we shall be able to find the positions of the final points by marking the initial points, it suffices just to mark the initial point, say by O.

E_{ij}

We now construct the Puiseux chains P_i from these elementary chains as follows. The points of P_i are the points of the E_{ij}, $1 \leq j \leq w(i)$, with the appropriate connections. Points from the same elementary chain E_{ij} are connected as in E_{ij}. The final point of each elementary chain E_{ij} is connected to the initial point of the next-but-one elementary chain $E_{i,j+2}$, as long as there is one, namely for $j \leq w(i) - 2$.

The final point of $E_{i,w(i)-1}$ is connected to the final point of $E_{i,w(i)}$. We call this final point of $E_{i,w(i)}$ the contact point of the Puiseux chain P_i and mark it by $\odot$. The two endpoints of the chain constructed by this rule are : first, the initial point of $E_{i,1}$, if there is one, i.e. if $\mu_{i1} \neq 0$, otherwise the initial point of $E_{i,3}$, if there is one, i.e. if $w(i) \geq 3$, resp. the final point of $E_{i,2}$ if $w(i) = 2$. We mark this point by **O** and call it the initial point of P_i. Second, the other endpoint of P_i is the initial point of $E_{i,2}$.

The Puiseux chain P_i therefore looks like this :

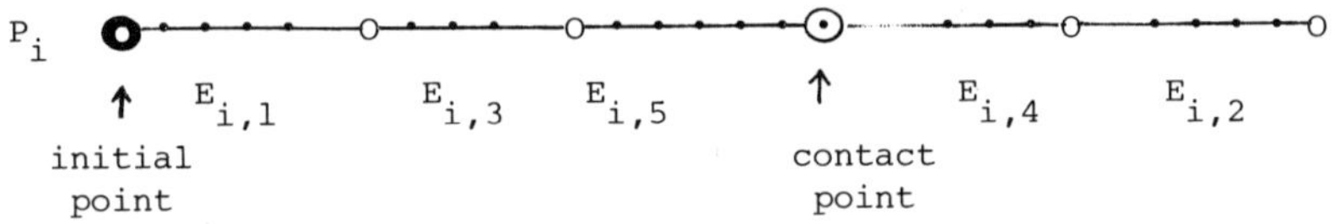

We now construct the resolution graph from the Puiseux chains P_i, $i = 1,\ldots,g$, and the point * which represents the proper preimage of Y in the standard resolution, by the following rule :

(1) For $i < g$ the contact point of P_i is connected to the initial point of P_{i+1}.

(2) The contact point of P_g is connected to *.

The resolution graph therefore looks like a mobile suspended from *.

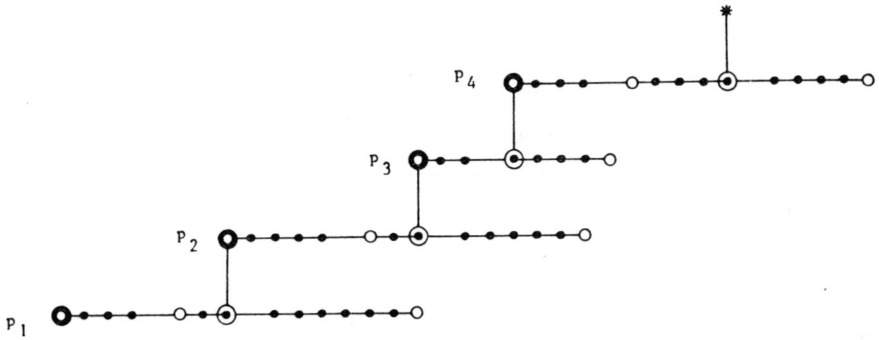

Theorem 15 :

The resolution graph of an irreducible plane curve is uniquely determined by the characteristic exponents of the Puiseux expansion, or equivalently, by the multiplicity sequence. It is the valuated graph described by the rule above. Conversely, the resolution graph uniquely determines the characteristic exponents of the Puiseux expansion.

Proof :

We start with the curve Y given by a Puiseux expansion $x = x(t)$, $y = y(t)$ and resolve the singularity by successive quadratic transformations $x = u$, $y = uv$ etc.. As in the proof of Theorem 12, we obtain Puiseux expansions for the proper preimages of Y. We have derived the multiplicity sequence from these Puiseux expansions in the proof of Theorem 12, and formulated the result with the help of the chain of euclidean algorithms described there.

But one can also use these Puiseux expansions to derive the tangents at the singular points of the proper preimages. Various possibilities arise : one or two previously introduced curves may pass through the singular point in question. The tangent may now coincide with the tangent of one of these curves, or it may not. After blowing up the point the tangent may continue to coincide with the tangent of this curve – this happens as long as the multiplicity within the multiplicity sequence of the same euclidean algorithm remains the same. As long as this happens, the result is an elementary chain of resolution graphs. But the tangent can be a tangent of the newly blown up curve. Then a new elementary chain of the same euclidean algorithm begins. Finally, the tangent can differ from the tangents of both exceptional curves. Then a new Puiseux chain begins.

In this way one obtains the rule for the construction of resolution graphs. One can easily persuade oneself of the details (cf. also the Diplomarbeit of Mr. Schulze-Röbbecke [S9]). Conversely, if the resolution graph is given, one easily reconstructs the chain of euclidean algorithms. Because one knows the multiplicities of the last elementary chain $E_{g,w(g)}$: they are 1, and one knows the lengths μ_{ij} of all E_{ij} , because these are immediately determined from the valuation of the graph. From this one can reconstruct the gth algorithm from P_g , then from P_{g-1} the $(g-1)^{th}$ etc. From the algorithms one immediately obtains the characteristic exponents of the Puiseux expansion.

One can put together the resolution graph of a reducible curve recursively from those of its irreducible components when one knows the system of multiplicity sequences.

Construction of resolution graphs in the reducible case

Let the curve germ (Y,p) have the irreducible components $Y^1,\dots,Y^s$, and suppose that the resolution graph G' of $Y' = Y^1 \cup \dots \cup Y^{s-1}$ has already been constructed. Let G'' be the resolution graph of the irreducible curve germ $Y'' = Y^s$. One obtains the system of multiplicity sequences of Y' resp. Y'' as a part of the system of multiplicity sequences of Y. These sequences give the proper lengths $L'_1,\dots,L'_{s-1}$ of the branches $Y^1,\dots,Y^{s-1}$ of Y' and the length L_s for Y'' – the smallest numbers of quadratic transformations after which the proper preimage of the branch is regular, transverse to a single exceptional curve, and disjoint from other branches of Y' resp. Y''. Now when the curves Y' and Y'' are combined the proper length of a branch Y^r $(1 \leq r \leq s-1)$ or of $Y'' = Y^s$ can grow, namely, when the number of common infinitely near points of Y^r and Y'' is greater than L'_r (resp. when the contact of Y'' and Y^r, for an $r \leq s - 1$, is greater than the length of the multiplicity sequence of Y''). If q_r is the difference between the new and old proper lengths of the branch Y^r $(r = 1,\dots,s)$, then we extend the graph G' resp. G'' by introducing a chain of q_r points between the point * , which corresponds to Y^r, and the point C^r of G' resp. G'' connected to *. The values on the graph are extended accordingly.

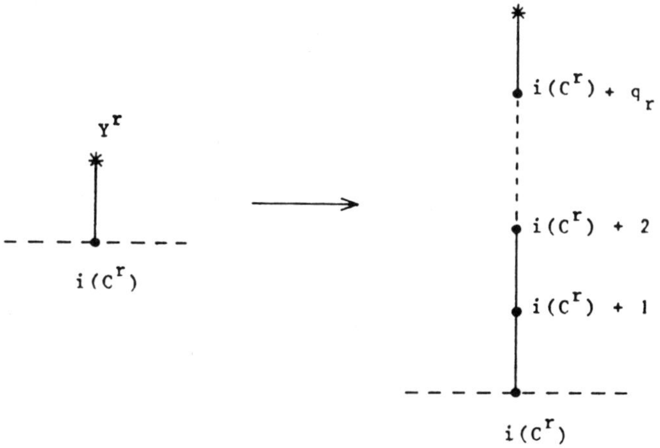

We call the resulting graphs $\tilde{G}'$ resp. $\tilde{G}''$. We index the points C'' of $\tilde{G}''$ apart from * by their values* $i(C'')$, and denote them correspondingly by C_i'', $1 \leq i \leq k$. Now let κ be the greatest index for which blowing down C_κ'' results in the infinitely near points of Y'' also being infinitely near points of Y'. Then each C_i'', $i \leq \kappa$, must be identified with a uniquely determined point C_i' of $\tilde{G}'$ (namely, so that the infinitely near points of Y' and Y'' resulting from blowing down C_i' and C_i'' coincide). This identification may be derived immediately from the system of multiplicity sequences of Y.

The resolution graph G of $Y = Y' \cup Y''$ is then constructed from the modified graphs $\tilde{G}'$, $\tilde{G}''$ of Y' and Y'' by the following rule :

(1) The points of G are the points of $\tilde{G}'$ and the points C_i'' with $i > \kappa$, together with * from $\tilde{G}''$.

(2) If a point C_i'', $i > \kappa$, resp. the point * of $\tilde{G}''$ is connected in $\tilde{G}''$ to a C_j'', then it remains connected to C_j'' in G for $j > \kappa$, and for $j \leq \kappa$ it is connected in G to the corresponding C_j'.

(3) Points of $\tilde{G}'$ also remain connected in G when C_j', for $j < \kappa$, is connected to C_κ' in $\tilde{G}'$, but C_j'' is not connected to C_κ'' in $\tilde{G}'$.

Proposition 16 :

The resolution graph of a reducible singularity of a plane curve is uniquely determined by the system of multiplicity sequences. The resolution graphs of the individual branches are determined by Theorem 15, and the resolution graph of the curve itself is determined from these graphs and the system of multiplicity sequences by the above rule.

Proof :

Let $X_k \to X_{k-1} \to \dots \to X_0$ be the sequence of iterated σ-processes which gives the standard resolution of the curve Y in the surface X_0. Let the number κ be defined as above. We consider the situation in X_κ. In X_κ there is exactly one exceptional curve C_κ'', obtained by the blowing up $X_\kappa \to X_{\kappa-1}$, and cut by the proper preimages of Y' and Y'' at distinct points. On the curves C_j'' with $j \leq \kappa$ in X_j, containing the infinitely near points of Y'', there are also infinitely near points of Y', and hence the C_j'', $j \leq \kappa$, on $\tilde{G}''$ must be

---

*Since Y'' is irreducible, each value appears only once.

identified with the corresponding C'_j on $\tilde{G}'$. On the other hand, the C''_j from $\tilde{G}''$ with $j > \kappa$ are different from curves of $\tilde{G}'$. This gives rule (1) for the description of the points of G. Rule (2) again comes from the identification of C'_j and C''_j for $j \leq \kappa$. One sees rule (3) as follows : when a point C'_j is connected in $\tilde{G}'$ to C'_κ, but C''_j is not connected to C''_κ in $\tilde{G}''$, this just means that the curve in X_κ which corresponds to C'_j and C''_j goes through the infinitely near point p of Y'' on C''_κ. Blowing up separates these points of C''_κ in $X_{\kappa+1}$, and hence C''_κ must not be connected to C'_κ in G. Other connections between points of $\tilde{G}'$ are obviously not disturbed by blowing up p.

As an example, we want to construct the resolution graph of the curve $Y = Y^1 \cup Y^2 \cup Y^3$, which we have already considered after Theorem 12 and Corollary 14. Its system of multiplicity sequences is :

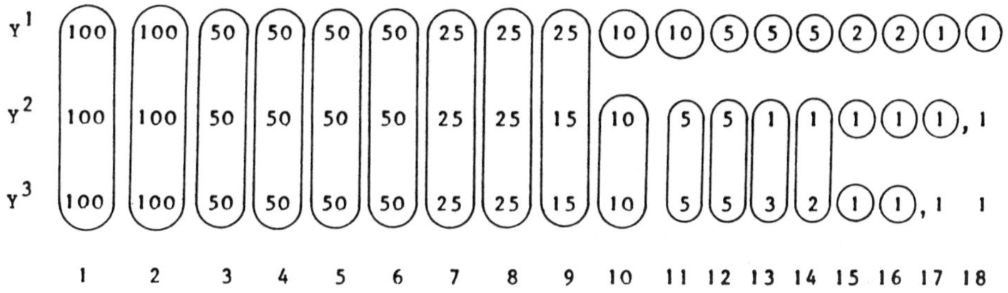

By Theorem 15, the resolution graphs of the individual irreducible branches have the following form :

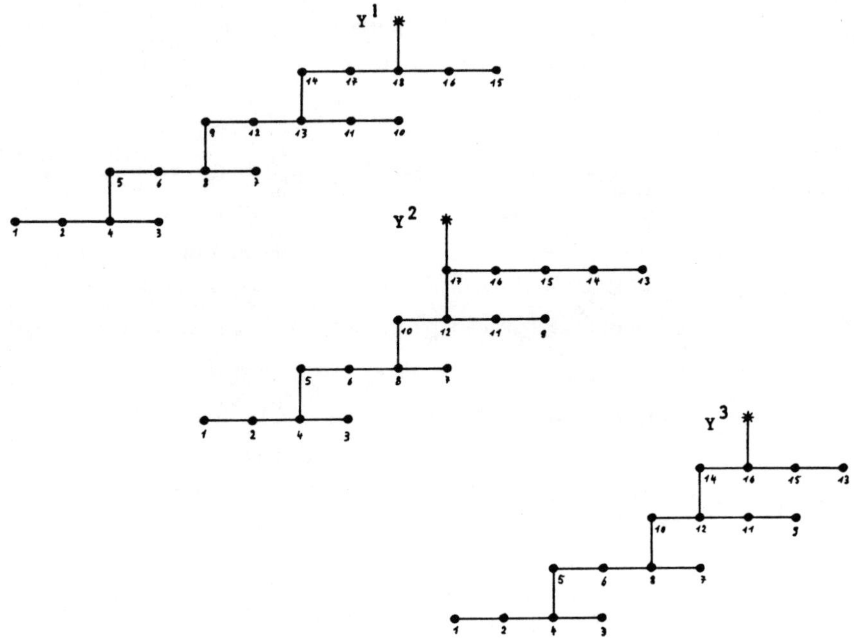

From them, we construct the resolution graph of Y with the help of Proposition 16.

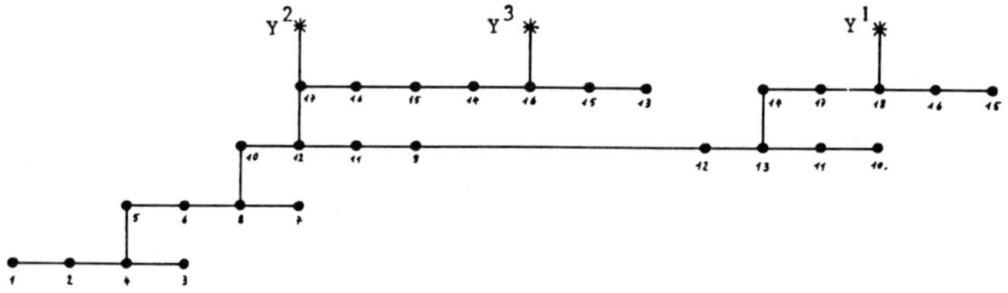

Now we want to proceed conversely, to obtain the system of multiplicity sequences from the resolution graph of a possibly reducible curve Y. In view of the results already obtained, it is clear that it suffices to use the resolution graph G of Y to reconstruct the (extended) resolution graphs $\tilde{G}_i$ of the branches Y^i of Y, in such a way that it is clear which points of G correspond to the points of the $\tilde{G}_i$. To do this we subject the graph G repeatedly to a contraction operation which corresponds to blowing down of curves. We want to call a point in a valuated graph contractible when it is connected only to points of lower value (in particular, it is not connected to the valuated points *). Contraction of this point consists in removing the point and connecting in pairs all the points connected to it. Then we obtain the following

Rule for reconstruction of the subgraphs $\tilde{G}_i$:

One removes from G all marked points * except that which corresponds to Y^i. In the resulting graph one repeatedly applies all possible contractions as long as possible. The non-contractible valuated graph finally obtained is $\tilde{G}_i$.

It is easy to see that we have the following :

Corollary to Proposition 16

The system of multiplicity sequences may be determined from the resolution graph by the above rules.

Construction of the total preimages of the standard resolution

Let $(Y,0)$ be a plane curve with the square-free equation $f(x,y) = 0$ in $(\mathbb{C}^2,0)$ and let $\pi : X \to \mathbb{C}^2$ be the standard resolution. Let $\tilde{f} = f \circ \pi$ and let $\tilde{Y}$ be the zero set of $\tilde{f}$. As a subset of X, $\tilde{Y}$ consists of the proper preimage Y' of Y and the exceptional curve $E = \pi^{-1}(0)$. Then $\tilde{f}$ vanishes along each irreducible component C_i of E with a certain multiplicity m_i. We speak of the total preimage of Y to mean the divisor of $\tilde{f}$, i.e. the curve with multiple components

$$Y' + \Sigma m_i C_i .$$

The passage from the singular curve Y to this divisor with normal crossings is often described as resolution of the singularity, especially when one views the curve Y as the fibre over 0 of the mapping $f : \mathbb{C}^2 \to \mathbb{C}$. This passage to a divisor, whose only singularities are normal crossings represents, as we remarked earlier, a considerable simplification of the situation. This reduction permits many geometric and algebraic problems about singularities to be solved in a simple way, and it is altogether the most important method for the investigation of singularities.

It is therefore of interest to find the multiplicities m_i of the components of the exceptional curve. In principle, these multiplicities can be determined recursively as follows. Each curve C_i in the resolution graph results from blowing up an infinitely near point p_i of Y, and this point has a certain multiplicity ν_i. This ν_i is the sum of the multiplicities of the proper preimages of branches of Y which go through p_i, and can be derived from the multiplicity sequence. Now there are two possibilities :

<u>Case I</u> :

Only one (previously appearing) exceptional curve $C_{j(i)}$ goes through p_i.

<u>Case II</u> :

Two (previously appearing) exceptional curves go through p_i. Let $C_{j(i)}$ be the first, and $C_{k(i)}$ the second, to appear.

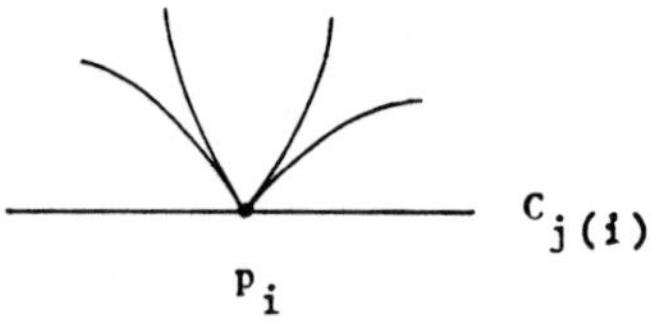

Case I

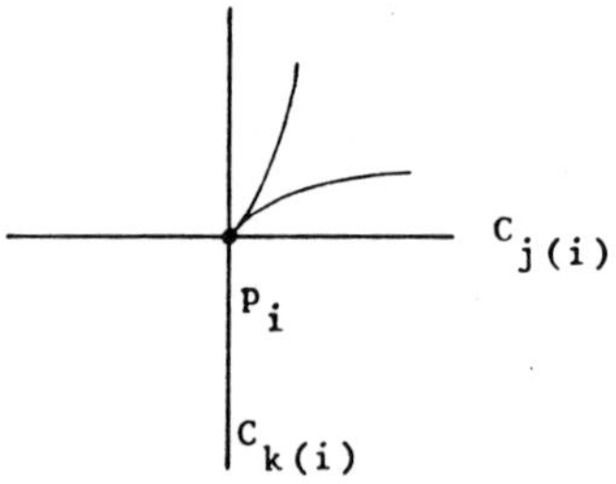

Case II

The indices $j(i)$ and $k(i)$ can be determined from the resolution graph : one removes all points marked * and contracts all points with greater values than C_i has. After that, C_i is connected to $C_{j(i)}$ in case I, and to $C_{j(i)}$ and $C_{k(i)}$ in case II.

<u>Proposition 17</u>

The multiplicities m_i of the exceptional curves C_i in the standard resolution of the curve Y may be determined recursively as follows :

(i) For the curve C_0 first introduced,

$m_0 = \nu_0(Y)$

(ii) $$m_i = \begin{cases} \nu_i + m_{j(i)} + m_{k(i)} & \text{in case II} \\ \nu_i + m_{j(i)} & \text{in case I.} \end{cases}$$

Proof : Trivial.

The topology of resolution

It is important to have a clear view of what the topology of resolution looks like. We have already seen, in connection with the definition of σ-process, what a neighbourhood of a curve created by a σ-process looks like : like a neighbourhood of the 0-section in the Hopf line bundle. We shall now describe what the neighbourhoods look like after further points on them have been blown up. To do this we consider the complex manifold L^k which results from identification of two copies of $\mathbb{C}^2$, with coordinates (u_1,v_1) resp. (u_2,v_2), along $\mathbb{C}^* \times \mathbb{C}$ by means of the following coordinate transformation :

$$u_2 = u_1^{-1}$$
$$v_2 = u_1^k v_1.$$

We view $P_1(\mathbb{C})$ as the complex manifold which results from identification of two copies of $\mathbb{C}$, with coordinates u_1 resp. u_2 , along $\mathbb{C}^*$ by means of the coordinate transformation $u_2 = u_1^{-1}$. Then $(u_i,v_i) \mapsto u_i$ defines a holomorphic mapping

$$L^k \to P_1(\mathbb{C}).$$

The fibre over each point of $P_1(\mathbb{C})$ is a complex line $\mathbb{C}$. Thus the manifold L^k is the total space of a complex line bundle over $P_1(\mathbb{C})$, and in fact it is the kth power of the Hopf line bundle $L = L^1$ (i.e. the k-fold tensor product). The bundle has a unique zero section $v_i = 0$, and the restriction of $L^k \to P_1(\mathbb{C})$ to this curve yields a biholomorphic mapping of the zero section onto $P_1(\mathbb{C})$. One can show, incidentally, that the self-intersection number of the zero section in L^k is precisely $-k$. We now assert

Proposition 18 :

Let C be the exceptional curve of a σ-process $\pi_1 : X_1 \to X_0$ and let $\pi_i : X_i \to X_{i-1}$, $i = 2,\ldots,k$, be further σ-processes, in which the points blown up are always points on C or on the proper preimage of C. Let $\tilde{C}$ be the proper preimage of C in X_k. Then there is a neighbourhood U of $\tilde{C}$ in X_k and a biholomorphic mapping of U onto a neighbourhood of the zero section in L^k, and $\tilde{C}$ is mapped onto the zero section.

Proof :

The proof is by induction on k. The basis of the induction, $k = 1$, has already been proved when σ-processes were introduced. Thus it suffices to show the following : if one blows up a point on the zero section of L^k, then a neighbourhood of the proper preimage of the zero section is biholomorphic to a neighbourhood of the zero section in L^{k+1}. But this follows by application of the coordinate transformation formulae for L^k, L^{k+1}, and quadratic transformations, because

$$u_2 = u_1^{-1}$$
$$v_2 = u_1^k v_1$$

and

$$u_1 = \tilde{u}_1$$
$$v_1 = \tilde{u}_1 \tilde{v}_1$$

imply

$$u_2 = \tilde{u}_1^{-1}$$
$$v_2 = \tilde{u}_1^{k+1} \tilde{v}_1 .$$

With this, we have proved

Proposition 19 :

If $\phi : \tilde{X} \to X$ is an iterated σ-process with exceptional curve $E = \phi^{-1}(x)$, and if C_i , $i = 1,\ldots,s$, are the irreducible components of E and $-k_i$ is the self-intersection number of C_i , then a suitable neighbourhood of E in $\tilde{X}$ results from pasting together suitable neighbourhoods of the zero sections in L^{k_i}, $i = 1,\ldots,s$.

This brings us to the following topological construction of a neighbourhood U of the exceptional curve $E = \bigcup C_i$. For each C_i we consider a neighbourhood U_i of the zero section in L^{k_i} with smooth boundary, so that U_i meets every line of the line bundle L^{k_i} in a disc. Thus U_i is a locally trivial fibre bundle over C_i with fibre

$$D^2 = \{v \in \mathbb{C} \mid \|v\| \leq \rho\}.$$

The U_i are now pasted together as follows : when C_i and C_j meet at $p \in C_i \cap C_j$, one chooses a small disc D_i resp. D_j around p

in $C_i \cap C_j$. Then the part of U_i resp. U_j over D_i resp. D_j is fibre-faithfully homeomorphic to $D_i \times D$ resp. $D_j \times D$. We choose homeomorphisms of D_i and D_j onto D and use them to define a homeomorphism

$$h_{ij} : D_i \times D \approx D_j \times D$$

which exchanges the factors. Then we identify U_i and U_j along $D_i \times D$ resp. $D_j \times D$ by h_{ij}. Doing this for each pair (i,j), we finally obtain a space

$$U = \bigcup_{i,j} U_{ij}$$

homeomorphic to a neighbourhood of E in $\tilde{X}$. This process of pasting together disc bundles over manifolds, sometimes known as "plumbing", permits many topological problems in the resolution of singularities to be reduced to simpler problems. One can gain a good intuitive grasp of the process by considering its real analogue :

The C_i , being 1-dimensional real projective spaces, are homeomorphic to S^1. U_i is homeomorphic to a k_i-tuply twisted interval bundle, hence homeomorphic to $S^1 \times I$ for even k_i and to the Möbius band $S^1 \tilde{\times} I$ for odd k_i. The bands are pasted together along squares.

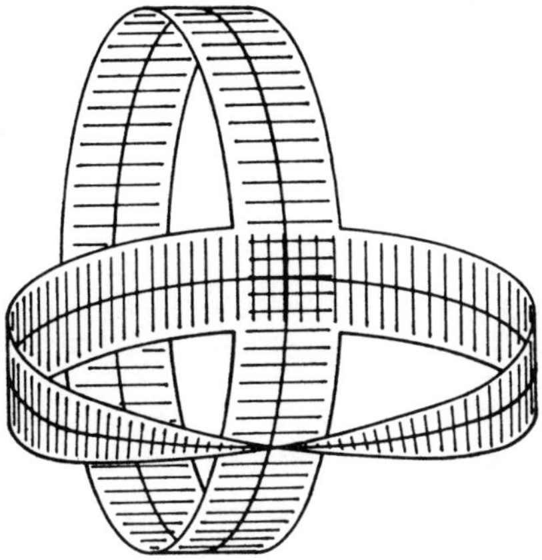

One can also describe this construction of a neighbourhood of the exceptional curve E schematically by pictures such as the following :

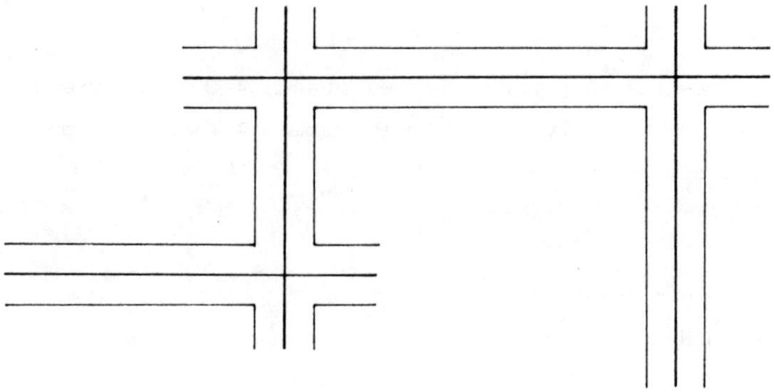

The proper preimages Y_i' of the branches of the resolved curve singularity cut the exceptional curve E transversely in the standard resolution. Hence one chooses a neighbourhood U of E, pasted together from disc bundles, so that the intersection of each Y_i' with U is just a disc.

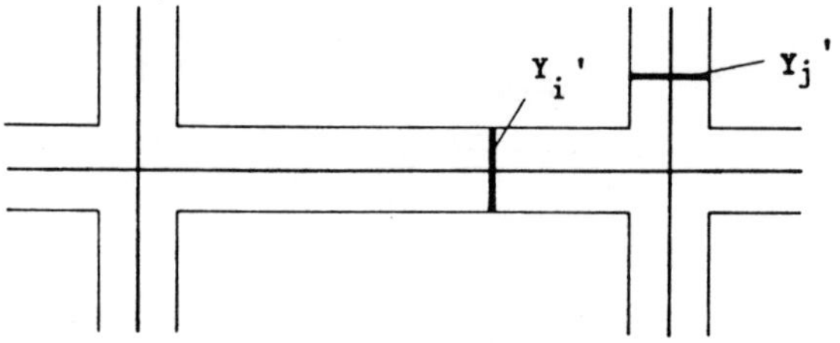

It is clear, and not difficult to prove, that from the topological standpoint it is irrelevant just which disc is Y_i'. Essentially it is a matter of which component of E meets Y_i', and this is precisely what the resolution graph describes. If one now contracts the set $\bigcup_i C_i$ in our neighbourhood U constructed by plumbing disc bundles to a point p, then U becomes a neighbourhood of the singular point p in $\mathbb{C}^2$, homeomorphic to a ball, and the Y_i'-discs become the branches of the curve. Thus we see

Proposition 20 :

Singularities with the same resolution graph are topologically equivalent.

Supplement to Proposition 20 :

The converse also holds.

This corollary will not be proved here, because it would require additional considerations from knot theory or something similar (cf. also our remarks on this in 8.3).

We now collect our previous results into a theorem.

Theorem 21 :

The following pieces of data concerning the singularity of a plane curve are equivalent :

(1) The topological type – characterised by the iterated torus knots corresponding to the branches and their linking numbers.
(2) The Puiseux pairs of the branches and the intersection numbers of the branches with each other.
(3) The system of multiplicity sequences.
(4) The resolution graph.

This list of data does not exhaust the possibilities for characterising the topological type. However, we want to leave it at this and instead go on to a more refined investigation of the topology of singularities in the next section.

8.5 Topology of singularities

In this course we have frequently been concerned with the topology of singularities. In 3.4 we saw that the appearance of a singularity in a family of curves leads to a change in genus, and we announced that we would prove formulae – the formulae of Clebsch and Noether – which describe this phenomenon precisely. As preparation for this, we shall derive a corresponding local formula in this paragraph.

Outside of 3.4, we have also investigated the topology of singularities in 5.3, 8.3 and 8.4. Among other things, we have seen that intersecting a locally irreducible plane curve with a small sphere around a singular point yields an iterated torus knot and, in the case of a reducible singularity, several linked iterated torus knots, one for each branch.

In the previous section we have just seen that this topological situation is determined by the Puiseux pairs and intersection numbers, or by the system of multiplicity sequences, or by the resolution graph.

We shall now treat an important property of the knots and links involved in singularities, which will lead us to the desired local Plücker formula. To do this we commence with a theorem of Seifert (1934) [S17].

Theorem 1 :

Each knot may be spanned by an orientable surface.

Proof : (after R.H. Fox, A quick trip through knot theory in : Topology of 3-manifolds [F2]) :

We project the knot on the plane so that the projection has only normal crossings. E.g. here is a projection of a trefoil knot :

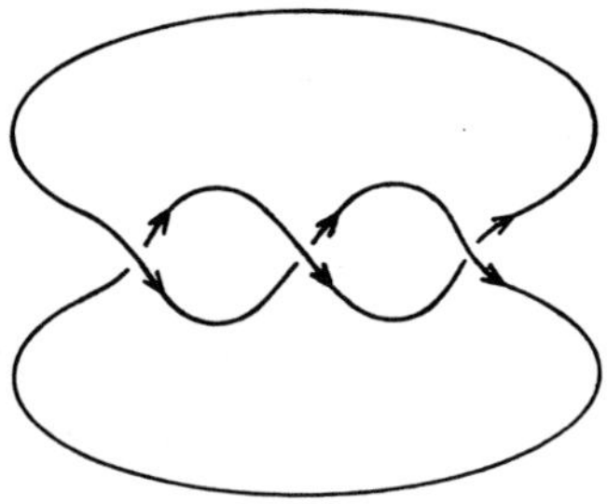

We span each crossing by a twisted rectangle, like this :

and not like this :

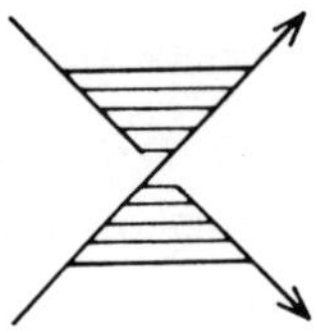

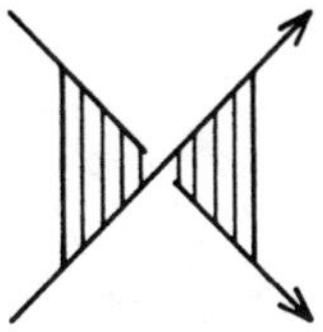

When we remove the interiors of these rectangles and the parts of their boundaries which lie on the projection of the knot, what remains is the projection of a disjoint system of circles (Seifert circles).

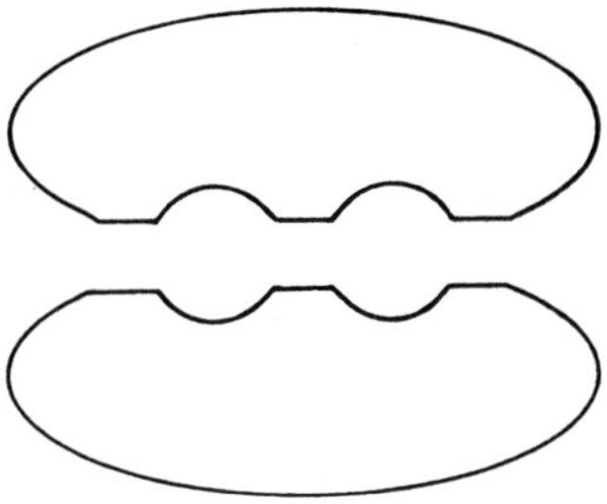

These circles are unknotted in three-dimensional space and not linked with each other. We can therefore span each of these circles by a disc in space, so that the discs are disjoint.

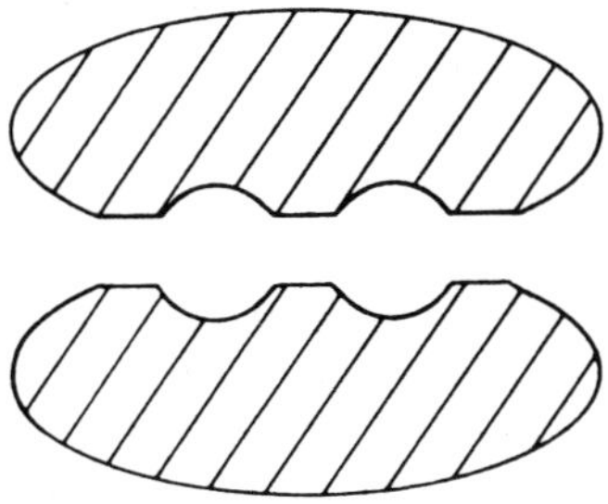

The union of these discs with the spanning rectangles is the desired orientable surface.

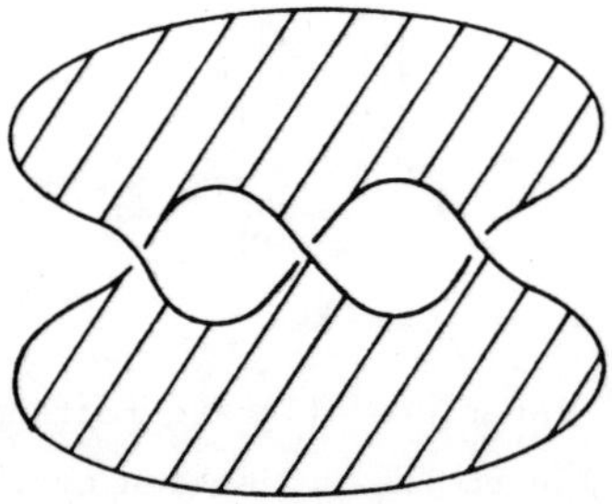

Here is another example, with the usual projection of the trefoil knot, where one must think of the "outer" surface as closed.

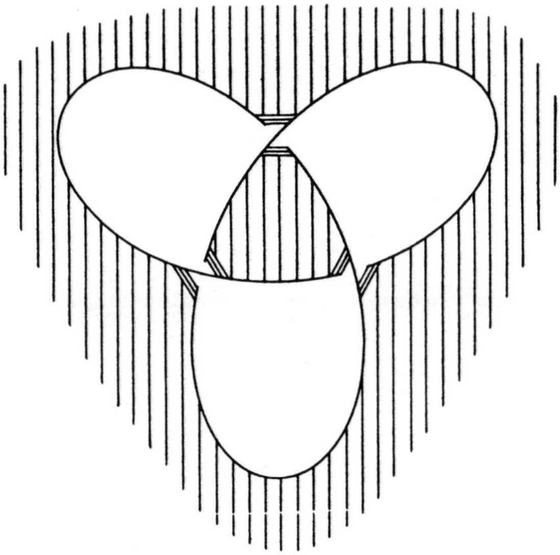

The following two pictures show different views of the figure which results from spanning the trefoil knot by two different surfaces (plain and dotted).

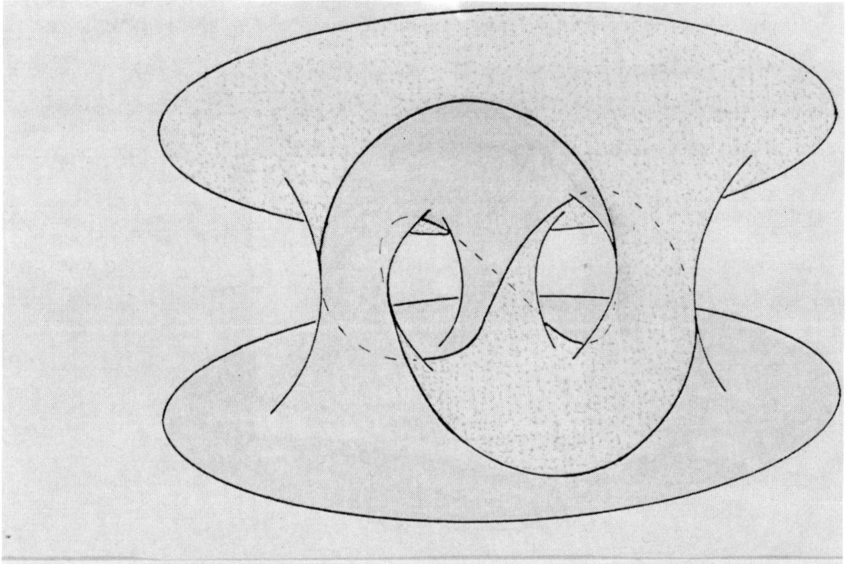

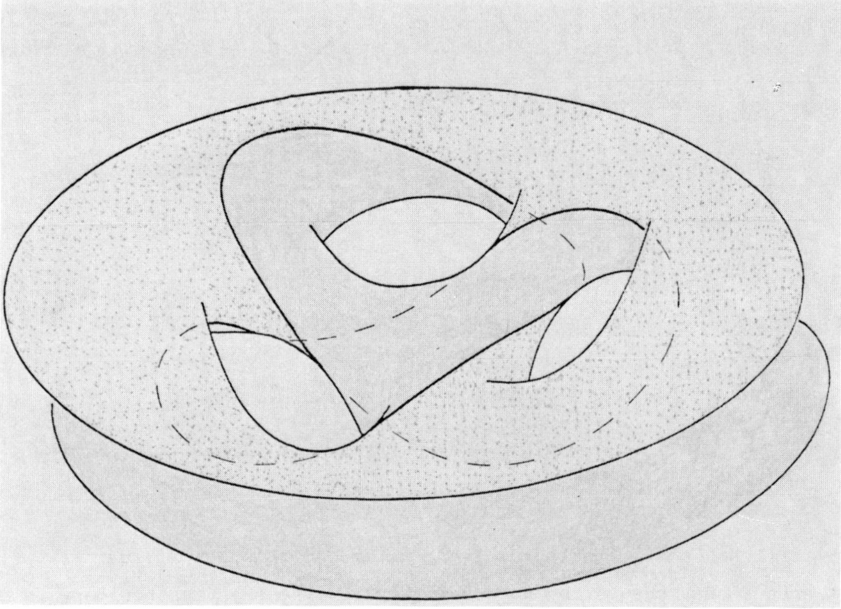

It is not difficult to calculate the genus g of the spanning surface. (By the genus of a connected orientable surface with boundary we mean the genus of the connected orientable surface without boundary which results from replacing each boundary component by a disc.)

If d is the number of crossings in the projection and f is the number of Seifert circles, then the Euler characteristic χ of the spanning surface obviously satisfies $\chi = d + f - 2d = f - d$, and hence

$$g = \frac{d-f+1}{2} .$$

In the example of the trefoil knot above, $g = \frac{3-2+1}{2} = 1$, hence the spanning surface is a once-perforated torus

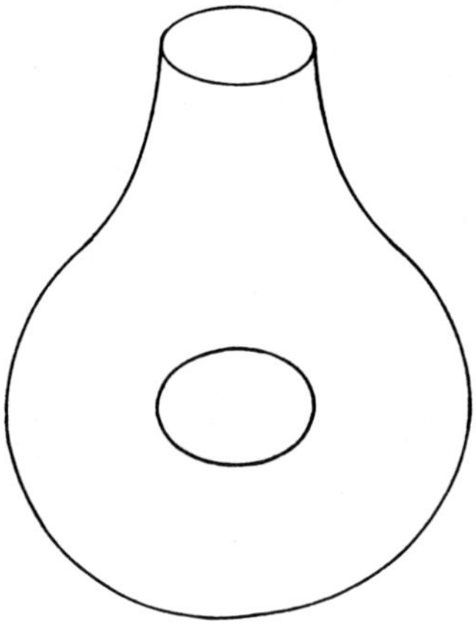

However, this perforated torus is not embedded in space in the trivial way just shown, but in the way shown earlier, so that its boundary is a trefoil knot.

Of course, a knot may be spanned by several possible surfaces — since one can always add superfluous handles.

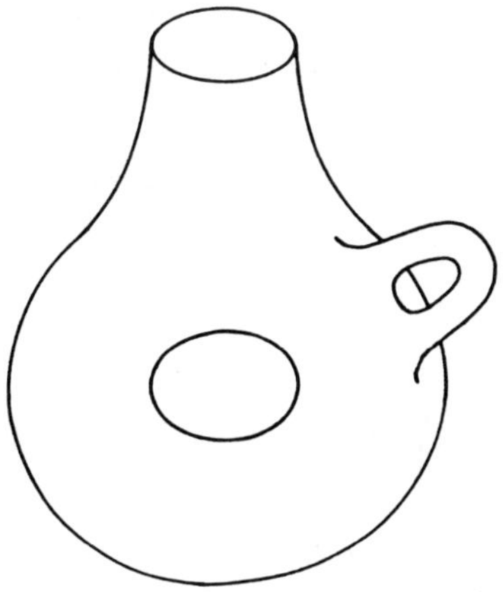

But we can try to find a "smallest possible" spanning surface. At any rate, such a surface should have the smallest possible genus. This smallest possible genus of a spanning surface is called the genus of the knot, and it is obviously an important invariant. In the above example of the trefoil knot the genus is 1, because when a knot has genus 0 it is obviously unknotted. Thus the spanning surface we have constructed has minimal genus. In general, a knot can have several essentially different spanning surfaces of the same genus.

Example : (H.F. Trotter [T5]) :

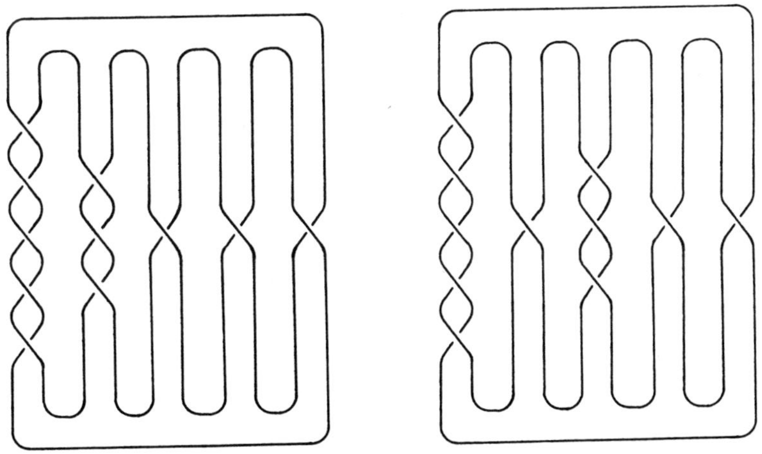

It is easy to see that these two knots are equivalent (see Figure 2 of Trotter). But one can show that there is no homeomorphism of S^3 which carries one spanning surface to the other.

Now there is an interesting class of knots, the <u>fibred knots</u>, for which such problems do not appear. Roughly speaking, a fibred knot is a knot $K \subset S^3$ for which $S^3 - K$ is fibred over S^1 by connected surfaces.

Definition :

Let $K \subset S^3$ be a knot or, more generally, a link in S^3. Let $\phi : S^3 - K \to S^1$ be a differentiable mapping with the following properties :

(i) $\phi : S^3 - K \to S^1$ is a locally trivial differentiable fibre bundle[*]

(ii) The fibre $F_t = \phi^{-1}(t)$ and its closure $\bar{F}_t$ in S^3 are such that F_t is the interior of the compact orientable differentiable surface $\bar{F}_t$ with boundary $\partial \bar{F}_t = K$.

---

[*] If E, B are differentiable manifolds, then a differentiable mapping $\phi : E \to B$ is called <u>locally trivial differentiable fibre bundle</u> with typical fibre F if each point $p \in B$ has a neighbourhood U for which $\phi : \phi^{-1}(U) \to U$ looks like the projection $\pi_1 : U \times F \to U$ of $U \times F$ onto the first factor, i.e. there is a diffeomorphism $h_U : U \times F \to \phi^{-1}(U)$ such that the following diagram commutes :

$$\begin{array}{ccc} U \times F & \xrightarrow{h_U} & \phi^{-1}(U) \\ & \searrow \pi_1 \quad \phi \swarrow & \\ & U & \end{array}$$

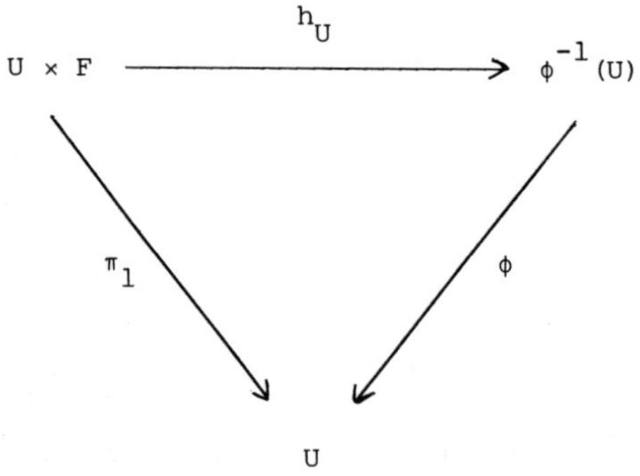

(iii) A differentiable tubular neighbourhood U of K in S^3 may be identified with a solid torus $S^1 \times D^2$ in such a way that

$$F_t \cap S^1 \times D^2 = \{(s,z) \in S^1 \times D^2 - \{0\} \mid z/\|z\| = t\}.$$

Thus the fibres F_t meet along K as shown in the following picture ("open book" structure) :

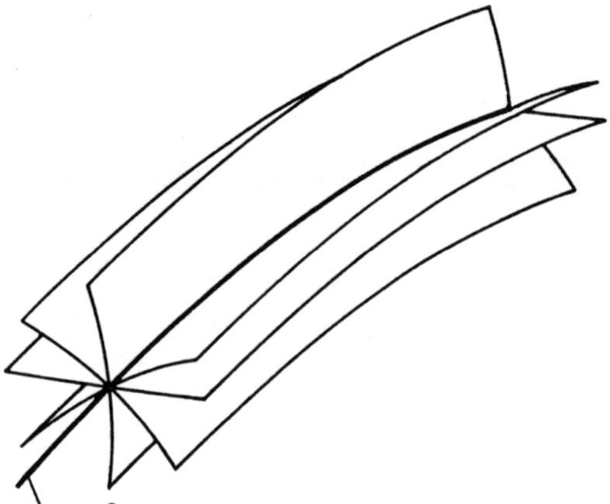

A link $K \subset S^3$ with such a fibration $S^3 - K \to S^1$ is called a <u>fibred knot</u>.

Stallings [S13] showed in 1962 that a knot $K \subset S^3$ admits such a fibration just in case the knot group $\pi_1(S^3-K)$ has a finitely generated commutator subgroup. Knots with this property were investigated by Neuwirth in particular. For that reason they are also known as <u>Neuwirth knots</u> or <u>Neuwirth-Stallings knots</u>.

These fibred knots are of interest to us because the knots associated with singularities of plane curves possess such fibrations in quite a natural way.

Theorem (Milnor) :

Let $(X,0) \subset (\mathbb{C}^2,0)$ be a plane curve with equation $f(z) = 0$, let $S^3 \subset \mathbb{C}^2$ be a sphere around the origin with sufficiently small radius, and let $K \subset S^3$ be the link $K = S^3 \cap X$. (By Theorem 8.3.12, K consists of linked iterated torus knots.) Let

$$\phi : S^3 - K \to S^1$$

be the mapping defined by $\phi(z) = \frac{f(z)}{\|f(z)\|}$. Then K is a fibred knot with this mapping.

<u>Proof</u> :

See Milnor [M1], Singular points of complex hypersurfaces, Theorems 4.8 and 6.1.

<u>Remark</u> :

The fibration defined above is often called the Milnor fibration. Of course it was known long before Milnor that, e.g., the complement $S^3 - K$ of the trefoil knot could be fibred.

For a fibred knot K the spanning surface of minimal genus is uniquely determined and isotopic to the closed fibre (Burde and Zieschang [B13], 1967). In fact Giffen proved that the fibration $S^3 - K \to S^1$ is uniquely determined by K up to an isotopy of S^3. Thus if we can find any fibration at all for the complement of our knot, then we already know automatically that it essentially coincides with the Milnor fibration. In what follows we set ourselves the problem of determining this fibration, or at least computing the genus of the fibre.

We can start from the equation of the curve, the Puiseux pairs, the multiplicity sequence or the resolution graph, since all these pieces of data determine the topology of the singularity and hence of the fibration also. Each of these approaches has its own advantages, and we shall pursue each of them, at least initially, without necessarily carrying them to a conclusion.

First we shall develop a description of the Milnor fibration for an irreducible curve, proceeding from the fact that the knot associated with the curve is an iterated torus knot characterised by the Puiseux pairs. This description of the Milnor fibration was found by A'Campo [A4] in 1973.

Just as the iterated torus knots result from iteration of a construction, the Milnor fibration results from iteration of a corresponding construction. We first consider the simplest conceivable case, from which we shall obtain the basic building block for the construction of the Milnor fibration.

This simplest case is that of a curve $x^p - y^q = 0$. We consider the Milnor fibration of this curve. We have defined the Milnor fibration as a fibration of the complement $S^3 - K \to S^1$, where S^3 is a small sphere, i.e. the boundary of a ball, around the singular point and K is the intersection of S^3 with the curve. Now, however, we want to use a polydisc Δ instead of a sphere, to simplify the description. Let K' be the intersection of $\Sigma = \partial\Delta$ with the curve. We have already established earlier, in 8.3.6, that the pairs (S^3, K) and (Σ, K') are homeomorphic. Thus, in view of the uniqueness of the fibration for Neuwirth-Stallings knots, we can describe the Milnor fibration just as well by describing a fibration of $\Sigma - K'$ over S^1.

Let Δ be the polydisc

$$\Delta = \{(x,y) \in \mathbb{C}^2 \mid |x| \leq 1, |y| \leq 1\}.$$

As we have seen earlier,

$$\Sigma = T^+ \cup T^-$$

where T^+ and T^- are both solid tori :

$$T^+ = \{(x,y) \in \mathbb{C}^2 \mid |x| = 1, |y| \leq 1\}$$
$$T^- = \{(x,y) \in \mathbb{C}^2 \mid |x| \leq 1, |y| = 1\}.$$

Now let

$$f(x,y) = x^p - y^q$$

and

$$K' = \{(x,y) \in \Sigma \mid f(x,y) = 0\}.$$

Provided p and q are relatively prime, K' is a torus knot of type (p,q), lying on the torus

$$T = T^+ \cap T^-.$$

We fibre $\Sigma - K'$ over S^1 by the mapping

$$\phi : \Sigma - K' \to S^1$$

with

$$\phi(x,y) = \frac{f(x,y)}{\|f(x,y)\|} .$$

Let

$$F_t = \phi^{-1}(t) \quad \text{for} \quad t \in S^1$$

be the fibre over t and let

$$F_t^+ = F_t \cap T^+$$
$$F_t^- = F_t \cap T^-.$$

The surfaces F_t^+ resp. F_t^- in the two solid tori are explicitly described by a very simple equation, and we can use this equation to analyse them completely. We do this for F_t^+. We view T^+ as $S^1 \times D^2$, where $D^2 = \{y \in \mathbb{C} \mid |y| \leq 1\}$. For $x \in S^1$ we consider the intersection of $D = \{x\} \times D^2$ with all the fibres F_t^+. One obtains the intersection as certain curves in D, in fact just the preimages of all lines through the point x^p under the mapping $D \to D$ with $y \mapsto y^q$. For example, when $q = 3$ the partition of D into curves looks like this:

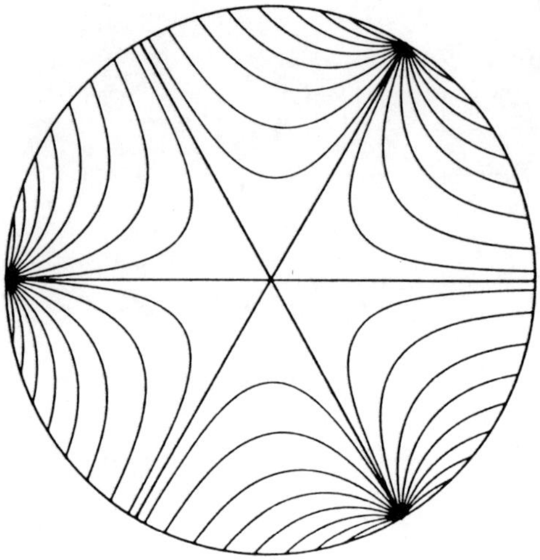

In general, D is divided into $2q$ sectors in a way which one can easily visualise by considering the picture above.

If the value of x^p now varies, the partition of $\{x\} \times D^2$ into curves changes. The following pictures show the partition of $\{x\} \times D^2$ for $q = 3$ and various values of x^p.

Sections of $\{x\} \times D^2$ with the fibres of F_t^+ for $x^p = e^{\pi i/3}$, $e^{\pi i/4}$, $e^{\pi i/12}$, 1, $e^{-\pi i/12}$ and $e^{-\pi i/4}$. The intersection with the fibre F_1^+ is drawn heavily in each case.

On the basis of this explicit description of the intersection of the fibre F_t^+ with the discs one easily becomes convinced that the fibre F_t^+ looks as follows :

F_t^+ consists of p connected components. Each connected component is a differentiable surface with piecewise differentiable boundary and diffeomorphic to a $2q$-gon. The boundary lies on the boundary of the solid torus, and every second vertex of the boundary lies on the knot. The next picture shows how F_t^+ lies in the torus.

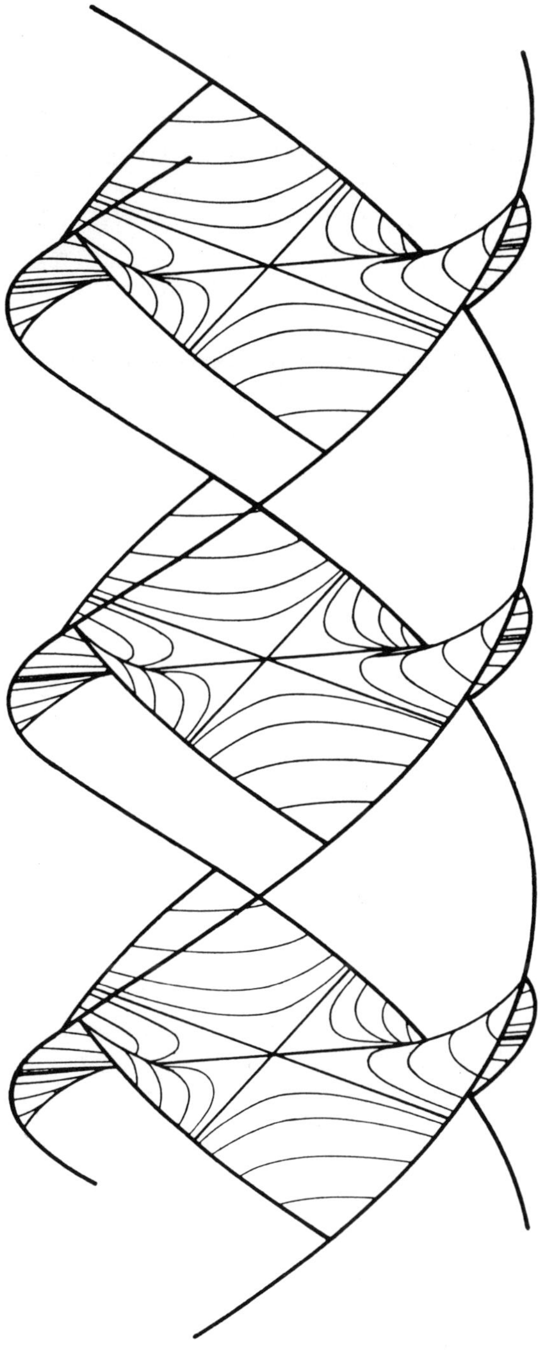

(The torus is stretched in the form of a cylinder. The picture shows the case $q = 3$. One sees that the components of F_t^+ look something like "monkey saddles".)

One shows analogously that F_t^- consists of q 2p-gons in T^-.

$$F_t = F_t^+ \cup F_t^-$$

results from the p 2q-gons and the q 2p-gons by identifying each edge of a 2q-gon which does not lie on K with an edge of a 2p-gon which does not lie on K. Each 2p-gon has exactly one edge in common with each 2q-gon. One can decide from the above description which edges are to be identified with each other, and with which orientation. We do not want to go further into this for the moment. The above description immediately gives the Euler characteristic of the fibre :

$$\chi = p + q - pq.$$

Hence the first Betti number of the fibre is

$$\mu = (p-1)(q-1).$$

We now want to present a few other models of the fibre. To do this we view K as a knot in $\mathbb{R}^3$ and describe a spanning surface for K of minimal genus. Because of the uniqueness of the Milnor fibration up to isotopy, and the fact that the spanning surface of minimal genus is uniquely determined up to isotopy and isotopic to a surface in the Milnor fibration, the surface we describe is isotopic to the previously described surfaces of the Milnor fibration. It is quite easy now to see what must be done, by intuitive topology.

Here are the two models for the spanning surface. The first proceeds from the nature of K as a torus knot of type (p,q). The torus is embedded in the usual trivial way in $\mathbb{R}^3$, and projected onto the plane through its spine. One obtains a projection of the knot as in the following picture for $p = 5$, $q = 3$.

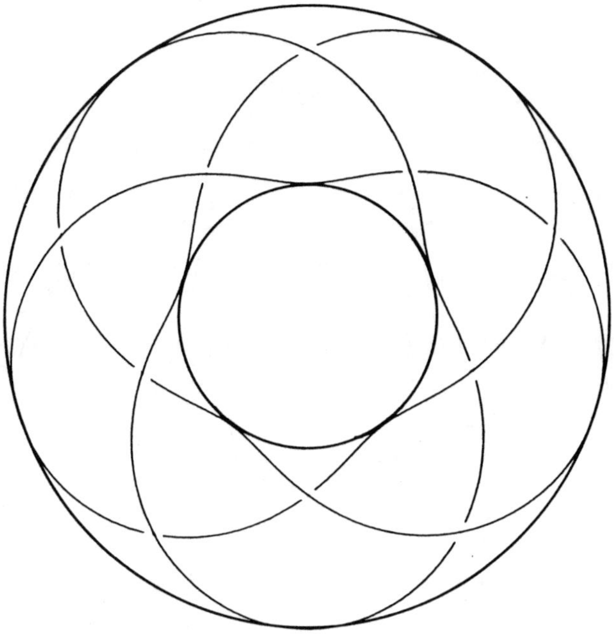

In general K is a closed braid whose projection results from cyclically joining projections of p braids of q strings, each of which has the appearance shown in the following picture for q = 7 :

q = 7

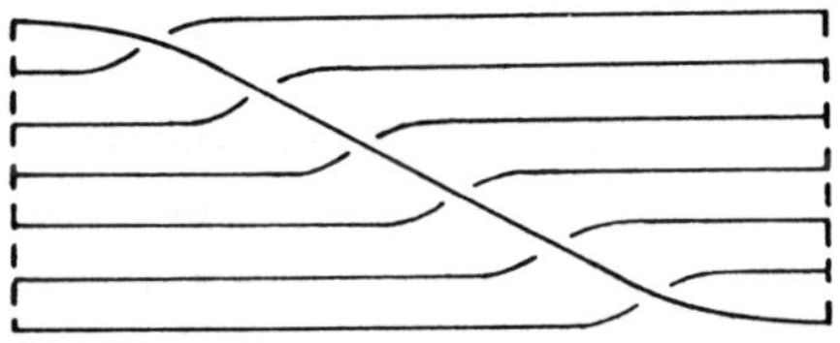

To construct the spanning surface we use the process we have learned in the proof of Seifert's Theorem 1. Thus we span each crossing by a small quadrangular strip. The resulting Seifert circles then form a system of q (topologically) concentric circles. We insert q discs in these q circles and obtain the desired surface F. In the following picture with p = 5 and q = 3 only two of the three discs have been shown for the sake of clarity, and even then not the full discs, but only the annuli which result from omitting small discs.

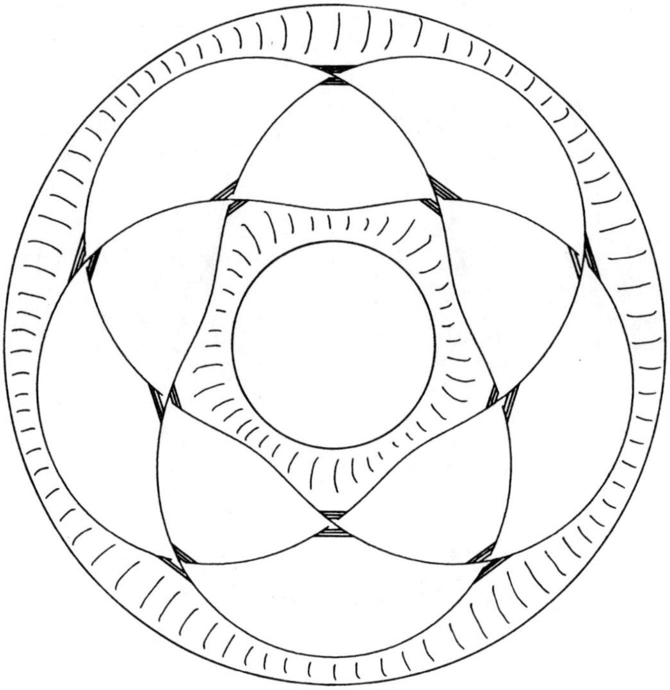

The q attached discs correspond essentially, though not exactly, to the q 2p-gons in the outer torus appearing in the previous construction. One sees immediately that F has first Betti number (p-1)(q-1), hence it is in fact a surface of minimal genus with boundary K.

We obtain our second model immediately from the one just constructed : obviously F is, up to isotopy, obtainable from q parallel discs (cross-sections of a cylinder), when neighbouring discs are connected by p bands, each twisted once and in the same sense, whose spines are parallel to the axis of the cylinder. The boundary of each disc is connected alternately to the preceding and succeeding disc by these bands. The following picture shows the resulting surface for q = 3, p = 5.

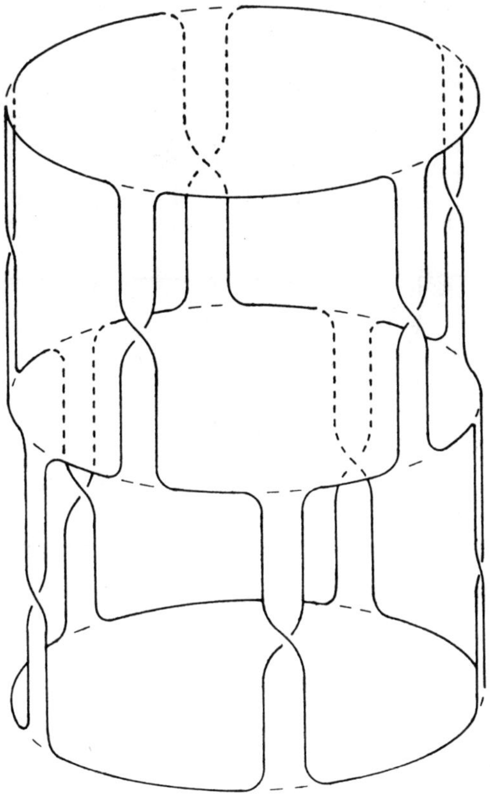

One obtains a plane projection which is possibly even simpler by deforming the discs to strips isotopically and pulling the connecting lines isotopically to one side of the strips.

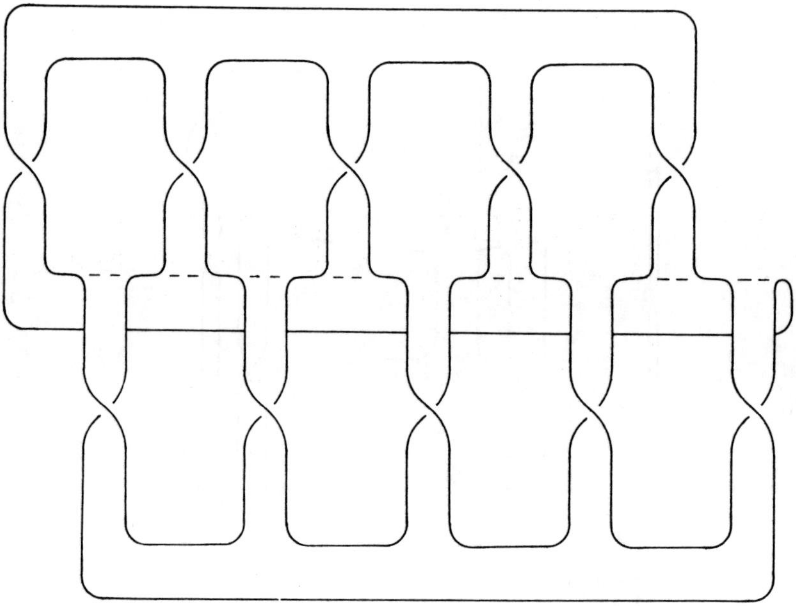

Of course, one can imagine still more models of the Milnor fibre, but we want to leave it at this.

However, we still want to secure a property of the fibration $\phi : \Sigma - K \to S^1$. In T^+ resp. T^- we consider the somewhat smaller tori

$$T_\rho^+ := \{(x,y) \in \mathbb{C}^2 \mid |x| = 1,\ |y| \le \rho\}$$

$$T_\rho^- := \{(x,y) \in \mathbb{C}^2 \mid |x| \le \rho,\ |y| = 1\}$$

with $\rho < 1$. Then the fibres F_t^+ (resp. F_t^-) intersect these solid tori T_ρ^+ resp. T_ρ^- in p (resp. q) disjoint discs which are isotopic to the discs $\{x\} \times D^2$ (resp. $D^2 \times \{y\}$) in F_t^+ resp. F_t^-. If one restricts the mapping $\phi : \Sigma - K \to S^1$ to the tori ∂T^+ resp. ∂T^-, then these fibrations of S^1 are isotopic to the projections $S^1 \times \partial D^2 \to S^1$ resp. $\partial D^2 \times S^1 \to S^1$. Now we consider the complement space of $\mathring{T}_\rho^-$ in Σ. This is a solid torus T with the same spine as T^+, and containing T^+ in its interior. We identify the pair (T,T^+) with a pair of solid tori $(S^1 \times D_\delta,\ S^1 \times D_1)$, where

$D_\delta = \{z \in \mathbb{C} \mid |z| \leq \delta\}$ and $\delta > 1$. On T^+ in T lies the torus knot K, with parametrisation $x = e^{2\pi i\phi/q}$, $y = e^{2\pi i\phi/p}$ relative to the chosen trivialisation of $\partial T^+ = S^1 \times S^1 = \{(x,y) \in \mathbb{C}^2 \mid |x| = 1, |y| = 1\}$ (in particular, the linking number of K with the spine equals p).

We consider the restriction of ϕ to $T - K$

$$\phi_{p,q} : T - K \to S^1.$$

From what we have said above it is clear that the mapping on the boundary $S^1 \times \partial D_\delta$ induced by $\phi_{p,q}$ is isotopic to the projection on the second factor ∂D_δ. Hence we can replace the mapping $\phi_{p,q}$ by an isotopic mapping

$$\psi_{p,q} : T - K \to S^1$$

with the following properties :

(i) The fibres of the mapping meet the boundary in q circles parallel to the spine (i.e. circles $S^1 \times \{y_i\}$).

(ii) $\psi_{p,q}$ is equivalent to the restriction of the Milnor fibration $\Sigma - K \to S^1$ to a solid torus somewhat larger than T^+, namely the complement space of $\mathring{T}^-_\rho$.

These mappings $\psi_{p,q}$ are the blocks from which we shall build the Milnor fibration for a singularity with several Puiseux pairs.

One obtains a good intuitive grasp of the way the fibres of $\psi_{p,q}$ lie in the solid torus T when one looks again at the model of the Milnor fibre on p. 552, which we obtained by projection of the torus knot. In this picture we have in fact shown the part of the Milnor fibre which lies in the somewhat larger torus T. One obtains the whole fibre by attaching to each of the q "latitude circles" a disc in the complementary solid torus.

Now we give an iterative description of the construction of the Milnor fibre for an iterated torus knot. The construction follows the iterative generation of the knot itself. As we have seen in 8.3, this construction takes a "small" solid torus T_i^+ around a previously constructed torus knot K_{i-1} of $(i-1)^{th}$ order, and draws on it a torus knot K_i (of first order on ∂T_i^+). The resulting torus knot is then of ith order. The type of K_i is given by two numbers. However, these numbers are only determined when a trivialisation of the boundary of T_i^+ is chosen. We have seen that there are two possibilities for carrying out such a trivialisation consistently. The first works with parallels (= latitude circles) and meridians, the latter of which bound discs in T_i^+ transverse to the spine K_{i-1}. The second method uses the fact that each K_i is embedded in a solid torus in a special way. We have used this second method because K_i is then simply of type (m_i,n_i), where (m_i,n_i) is the ith Puiseux pair. However, in describing the Milnor fibration it is more convenient to go over to the other description.

We consider a Puiseux expansion with only essential terms :

$$y = x^{n_1/m_1} + x^{n_2/m_1 \cdot m_2} + \ldots + x^{n_g/m_1 \ldots m_g}.$$

We define the numbers p_i and q_i recursively by

$$p_i = n_i - n_{i-1}m_i + p_{i-1}m_i m_{i-1}\ ,\ p_1 = n_1$$
$$q_i = m_i .$$

Then, by the remark in 8.3.10, K_i is (relative to the first method of trivialisation of tori) a torus knot on ∂T_i^+ of type (q_i,p_i), where p_i is the linking number of K_i with the spine K_{i-1} of T_i^+. (Remark : this linking number is the intersection number of the curves which result from breaking off the Puiseux expansion at the $(i-1)^{th}$ resp. i^{th} places, and can be computed by the methods of the previous section.)

Now we suppose that we have already constructed the Milnor fibration $\phi_{i-1} : S^3 - K_{i-1} \to S^1$. We consider a small tubular neighbourhood T_i of K_{i-1}. T_i is a solid torus, and the fibres $F_{i-1,t}$ of ϕ_{i-1} meet T_i in curves which obviously have linking number 0 with $K_{i-1} = \partial \bar{F}_{i-1,t}$. Hence one can – by the first method – trivialise the boundary ∂T_i so that each fibre $F_{i-1,t}$ cuts the torus ∂T_i transversely in a parallel circle. Now we define our fibration :

$$\phi_i : S^3 - K_i \to S^1$$

as follows :

$$\phi_i \big|_{T_i - K_i} = \psi_{p_i,q_i}$$

$$\phi_i \big|_{S^3 - \mathring{T}_i} = \phi_{i-1}^{q_i}$$

(Here we have identified T_i with the solid torus T on which we previously constructed the fibration $\psi_{p_i,q_i} : T - K \to S^1$, T_i^+ with the smaller torus T^+ and K_i with K.)

These mappings fit together, because they induce the same fibration on the boundary ∂T_i.

What does a fibre $F_{i,t}$ of ϕ_i look like ? Let $F^*_{p_i,q_i,t}$ be the fibre of ψ_{p_i,q_i}. Then $F^*_{p_i,q_i,t}$ results from the Milnor fibre of $x^{p_i} - y^{q_i}$ by removal of q_i discs. On the resulting q_i boundary components of $\tilde{F}^*_{p_i,q_i,t}$ we attach the q_i connected components of the fibre of ϕ_{i-1} over t along their boundaries. The latter are obtained as follows : let $\tau_1,\ldots,\tau_{q_i}$ be the q_i^{th} roots of t, and let F_{i-1,τ_i} be the fibres of $\phi_{i-1}: S^3 - K_{i-1} \to S^1$ over τ_i. Let $\tilde{F}_{i-1,\tau_i} = F_{i-1,\tau_i} - \mathring{T}_i$. This surface is of course isotopic to $\bar{F}_{i-1,\tau_i}$ in an obvious sense, because only a collar has been removed from F_{i-1,τ_i}. The $\tilde{F}_{i-1,\tau_i}$ are the components of $\phi^{q_i}_{i-1} : S^3 - \mathring{T}_i \to S^1$, and hence we obtain the full description of the Milnor fibre $F_{i,t}$:

$$F_{i,t} = F^*_{p_i,q_i,t} \cup \tilde{F}_{i-1,\tau_1} \cup \ldots \cup \tilde{F}_{i-1,\tau_{q_i}}.$$

<u>**Example**</u> :

$$y = x^{3/2} + x^{7/4}$$

$$p_1 = 3 \quad p_2 = 13$$

$$q_1 = 2 \quad q_2 = 2.$$

The Milnor fibre of $x^2 + y^{13}$ has genus 6, hence the Milnor fibre $F_{2,t}$ looks like this as an abstract surface (i.e., disregarding embedding in S^3) :

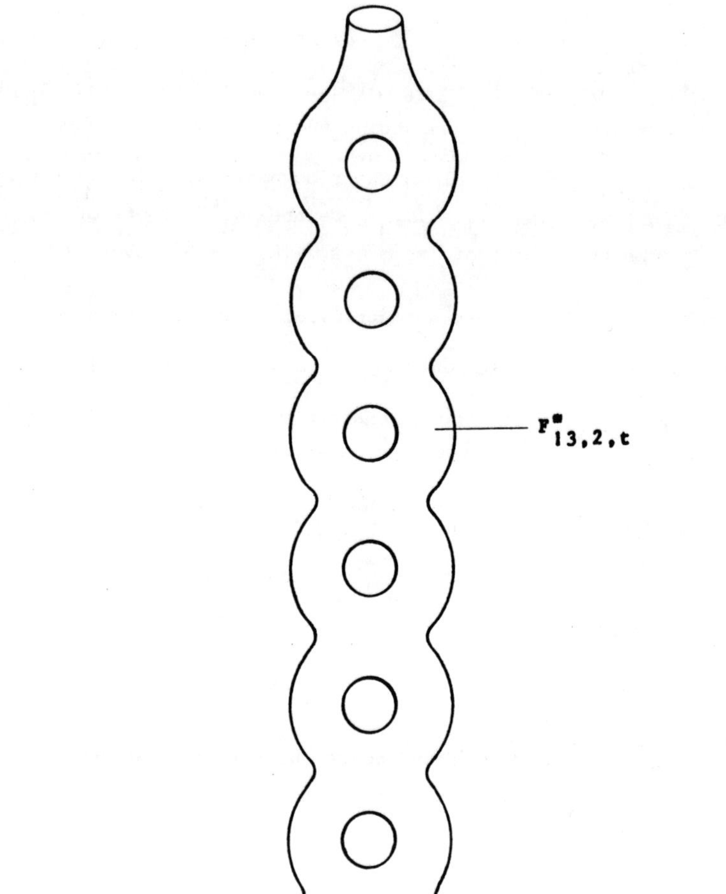
$F^{*}_{13,2,t}$
$\tilde{F}_{1,\tau_2}$
$\tilde{F}_{1,\tau_1}$

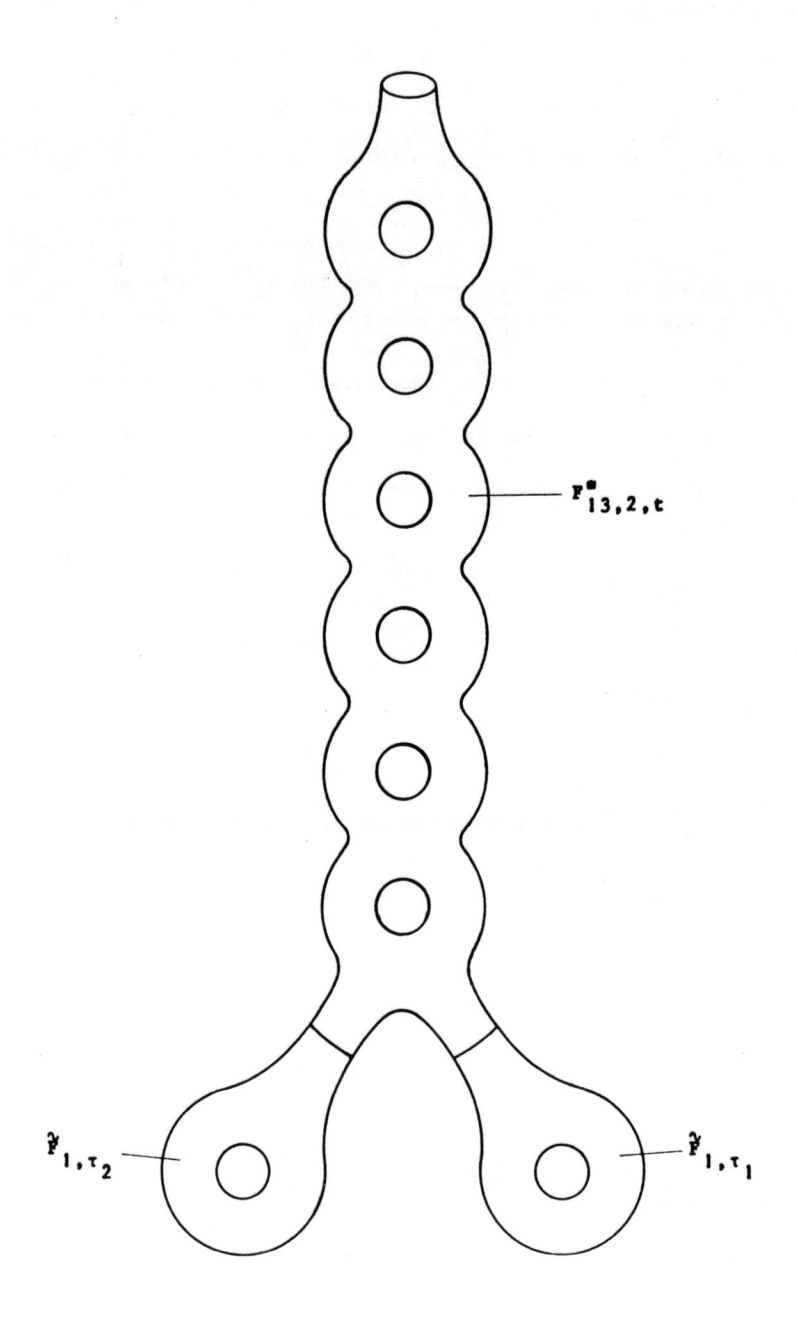

We have just described the Milnor fibre abstractly, as well as with its embedding, but we shall not attempt to draw an embedding for this example here, though with a suitable projection of the knot it would certainly be possible.

So much for the description of the Milnor fibration using Puiseux pairs.

To describe the Milnor fibration by means of resolution or the multiplicity sequence it is convenient to consider, not the Milnor fibration itself, but an equivalent fibration which is in any case more natural from the standpoint of analytic geometry. We now want to describe this fibration. As always, we investigate a plane singularity $(Y,0) \subset (\mathbb{C}^2,0)$ with equation $f(z) = 0$. Let B be a ball around 0 with sufficiently small radius.

<u>Lemma 1</u> :

For sufficiently small $\delta > 0$ the mapping

$$f : B \cap \{z \in \mathbb{C}^2 \mid |f(z)| = \delta\} \to S_\delta$$

onto the circle $S_\delta = \{\tau \in \mathbb{C} \mid |\tau| = \delta\}$ is a locally trivial differentiable fibre bundle, and this fibration is equivalent to the Milnor fibration of Y.

<u>Proof</u> :

A complete proof may be found in the book of Milnor [Ml], p. 52 ff; here we give only a rough outline of the idea.

The Milnor fibre F_t $(t \in S^1)$ is the intersection of the preimage of the ray through 0 and t with the boundary $S = \partial B$ of the ball B :

$$F_t = f^{-1}(\{\lambda t \mid \lambda > 0\}) \cap \partial B.$$

One constructs a vector field on $B - f^{-1}(0)$ which is always tangential to the surfaces $f^{-1}(\{\lambda t \mid \lambda > 0\})$ and directed "outwards".

Here is a schematic picture :

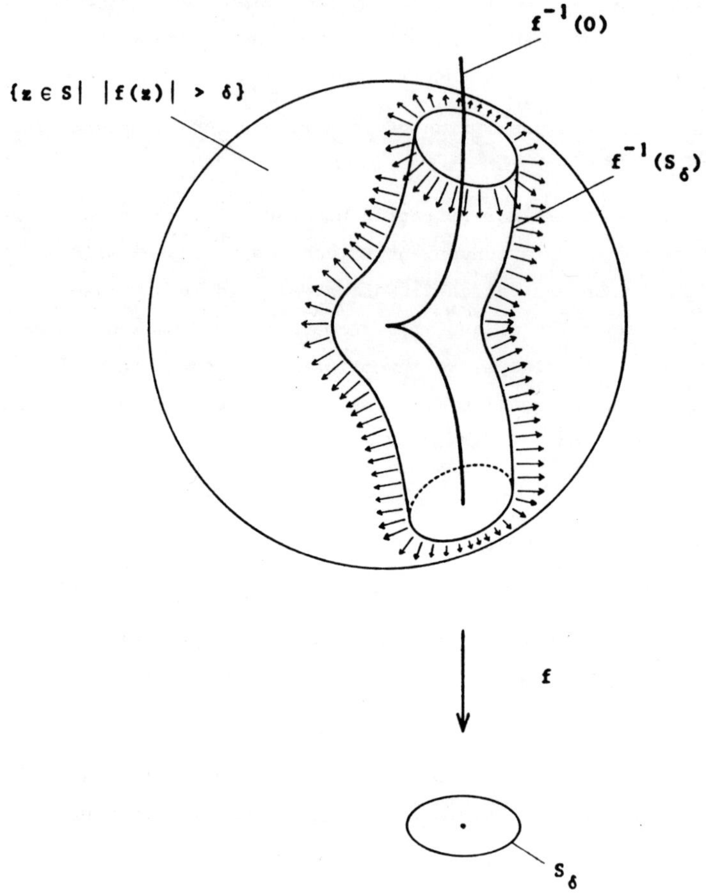

By integrating this vector field one transports $f^{-1}(S_\delta) \cap B$ into the boundary $S = \partial B$ of B and thus obtains a diffeomorphism Φ between $f^{-1}(S_\delta) \cap B$ and the complement $S - \{z \in S \mid |f(z)| < \delta\}$ of a tubular neighbourhood $\{z \in S \mid |f(z)| < \delta\}$ of the knot $K = Y \cap S$ such that the following diagram commutes.

$$\begin{array}{ccc}
f^{-1}(S_\delta) \cap B & \xrightarrow{\quad\Phi\quad} & S - \{z \in S \mid |f(z)| < \delta\} \\
f \Big\downarrow & & \Big\downarrow \phi = \dfrac{f}{|f|} \\
S_\delta & \longrightarrow & S^1 \\
\tau & \longmapsto & \tau/\delta
\end{array}$$

If δ is sufficiently small, then the fibration δ on $\{z \in S \mid |f(z)| < \delta\}$ has an "open book" structure as we have described in the definition of fibred knots, and $\phi : S - \{z \in S \mid |f(z)| \leq \delta\} \to S^1$ is equivalent to the fibration $\phi : S - K \to S^1$. The equivalence of fibrations asserted in the lemma then follows.

The description of the Milnor fibration $S - K \to S^1$ originally given, by $z \mapsto f(z)/|f(z)|$, is a typical topological, or more precisely, differential topological, construction in the category of differentiable manifolds. The second description of this fibration, which is equivalent to the first by Lemma 1, takes the actual situation better into account, since we are analysing a complex analytic situation here. With this construction the fibres

$$X_t = \{z \in B \mid f(z) = t\}$$

are not only differentiable manifolds diffeomorphic to the Milnor fibres, but they are even complex-analytic manifolds. This yields an important relation between the local complex analytic situation studied here and a frequently studied global algebraic-geometric situation : when $f(z)$ is a polynomial in two variables without multiple factors, one can also consider the family of plane affine-algebraic curves

$$C_t = \{z \in \mathbb{C}^2 \mid f(z) = t\}.$$

Suppose that C_0 has a singularity at the origin. If we choose B to be a ball around 0 in $\mathbb{C}^2$ of sufficiently small radius ε, then a sufficiently small δ and $t \neq 0$ such that $|t| < \delta$, then the intersection

$$C_t \cap B = X_t$$

is just the Milnor fibre. Thus the curve C_t , $0 < |t| < \delta$ resulting from C_0 by deformation differs from C_0 in B the way the Milnor fibre X_t differs from the singular analytic set X_0. This connection between the local and global situation is the basis for the later application of our results to the proof of the Noether formulae.

Next we describe the Milnor fibration with the help of the standard resolution of the singularity. Let $\pi : \tilde{X} \to \mathbb{C}^2$ be the standard resolution of the singularity $(Y,0) \subset (\mathbb{C}^2,0)$.

We let $D_1,\ldots,D_r$ denote the lines of the exceptional divisor $\pi^{-1}(0)$ and let $\tilde{Y}_1,\ldots,\tilde{Y}_k$ denote the components of the strict preimage of Y. Let m_i be the order of vanishing of the function $f \circ \pi$ on D_i (i.e.

m_i is the multiplicity of D_i in the divisor $\pi^{-1}(0)$). In the previous chapter we have described how all this information may be obtained from the resolution graph of the singularity.

Under resolution, the ball B becomes a neighbourhood $\tilde{B}$ of the exceptional divisor $\bigcup_{i=1}^{r} D_i$ and the Milnor fibration

$$f : \{z \in B \mid |f(z)| = \delta\} \to S_\delta$$

is equivalent to the fibration

$$f \circ \pi : \{z \in \tilde{B} \mid |f \circ \pi(z)| = \delta\} \to S_\delta .$$

In particular, the Milnor fibre $F = f^{-1}(\delta) \cap B$ is homeomorphic to

$$\tilde{F} := \{z \in \tilde{B} \mid f \circ \pi(z) = \delta\}.$$

The following picture illustrates this in a schematic way :

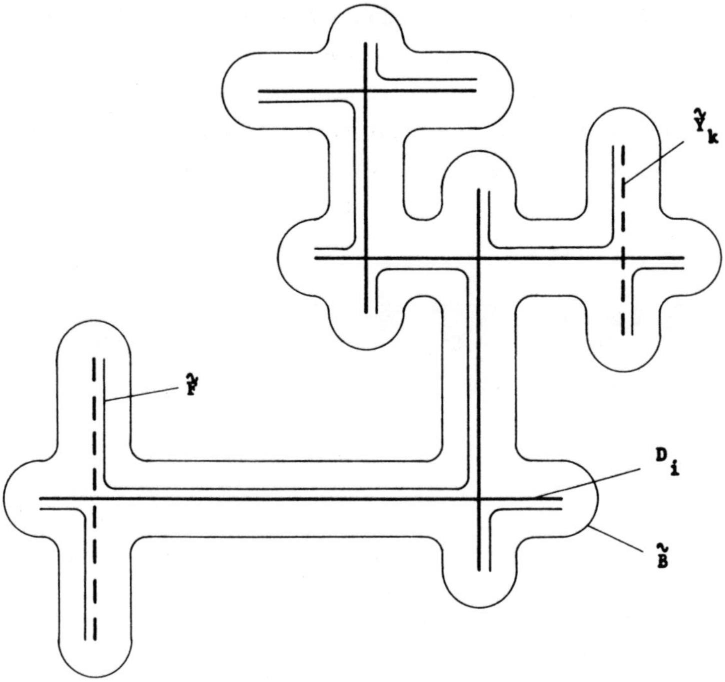

This is indeed a very schematic picture : $\tilde{F}$ is a real 2-dimensional connected manifold which hugs closely to the real 2-dimensional total preimage $\pi^{-1}(Y)$ in the real 4-dimensional space $\tilde{X}$.

In a neighbourhood of a simple point x_0 of the total preimage of Y, $f \circ \pi$ looks, in suitable coordinates, like

$$(z_1,z_2) \mapsto z_1^m$$

where m is the order of vanishing of $f \circ \pi$ at x_0. The fibre $\tilde{F}$ is locally homeomorphic to the set

$$\{(z_1,z_2) \mid z_1^m = \delta\},$$

hence, around x_0, $\tilde{F}$ is locally an m-fold covering of the component D_i resp. $\tilde{Y}_k$ through x_0 of the total preimage of Y.

At a crossing point x_1 of two components D and E of the total preimage, $f \circ \pi$ looks, in suitable coordinates, like

$$(z_1,z_2) \mapsto z_1^m z_2^n ,$$

where m resp. n is the multiplicity of D resp. E in the divisor $\pi^{-1}(Y)$. Around x_1 we construct the polydisc

$$\Delta := \{(z_1,z_2) \mid |z_1| \leq 1, |z_2| \leq 1\}$$

whose boundary consists of the two solid tori

$$T^+ = \{(z_1,z_2) \in \Delta \mid |z_1| = 1\}$$

and $$T^- = \{(z_1,z_2) \in \Delta \mid |z_2| = 1\}.$$

We assume without loss of generality that $\delta < 1$.

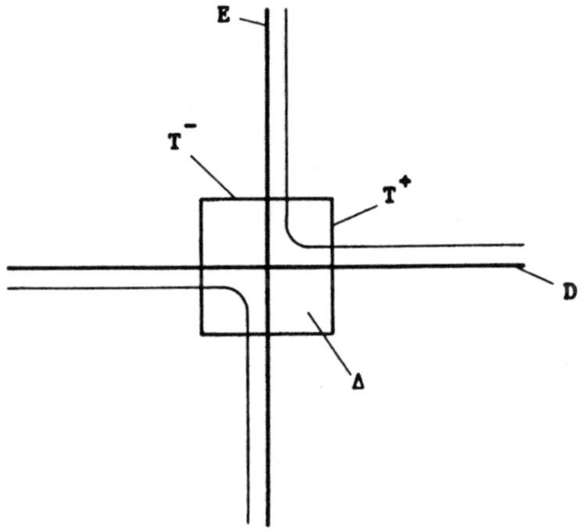

$\tilde{F}$ then meets T^+ and T^- each in $\gcd(m,n)$ circles, because

$$\tilde{F} \cap T^+ = \{(z_1,z_2) \mid |z_1| = 1 \text{ and } z_1^m z_2^n = \delta\}$$

is obviously the union of $\gcd(m,n)$ torus knots of type $\left(\frac{m}{\gcd(m,n)}, \frac{n}{\gcd(m,n)}\right)$, and analogously for $\tilde{F} \cap T^-$. Also, one easily sees the following :

$\tilde{F} \cap \Delta$ consists of $\gcd(m,n)$ cylinders spanning the circles of $\tilde{F} \cap T^+$ and $\tilde{F} \cap T^-$. In particular, the Euler characteristic of $\tilde{F} \cap \Delta$ equals zero.

We now want to put together $\tilde{F}$ from the pieces just described. Around each of the double points of

$$\pi^{-1}(Y) = \bigcup_{i=1}^{r} D_i \cup \bigcup_{k=1}^{s} \tilde{Y}_k$$

we choose a small polydisc as above. Let K be the union of all these polydiscs. Thus if the exceptional curve D_i meets exactly r_i other curves of the total preimage, then $D_i - K$ is an r_i-tuply perforated 2-sphere.

$\tilde{F}$ is made up of pieces of $\tilde{F} - \mathring{K} \cap \tilde{F}$ which cover the perforated spheres $D_i - K$ — we shall call these pieces M_i — together with the intersection of $\tilde{F}$ with K (which is a union of cylinders) and the parts of $\tilde{F} - \mathring{K} \cap \tilde{F}$ which lie over the components $\tilde{Y}_k - K$. The latter are simple coverings of the perforated disc $\tilde{Y}_k - K$, and hence also cylinders.

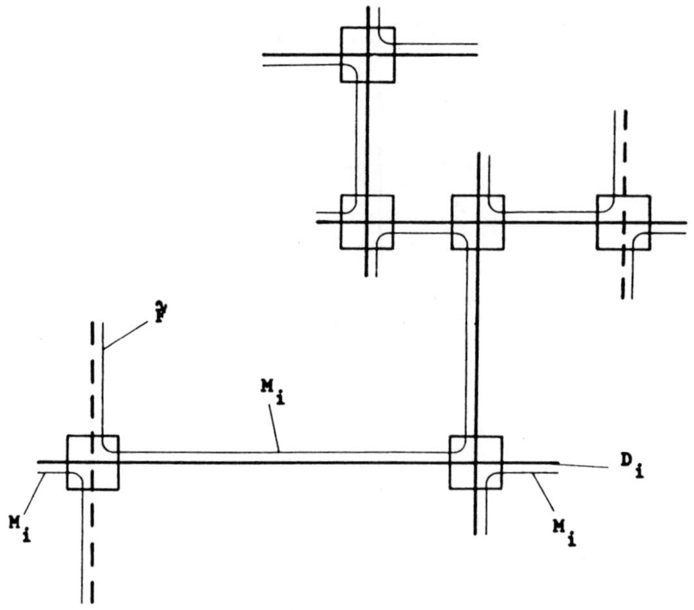

Each M_i is an m_i-fold covering of $D_i - K$, and hence a Riemann surface with holes. Cylinders of $\tilde{F} \cap K$ are attached to these holes ; as we have seen above, $\gcd(m_i, m_j)$ cylinders correspond to an intersection point of D_i with D_j. M_j is attached to the other end of the cylinder. One proceeds analogously for the intersection of D_i with a component $\tilde{Y}_k$.

<u>**Example**</u> :

We consider the singularity $(x^2+y^3)(x^3+y^2) = 0$. The standard resolution looks like this :

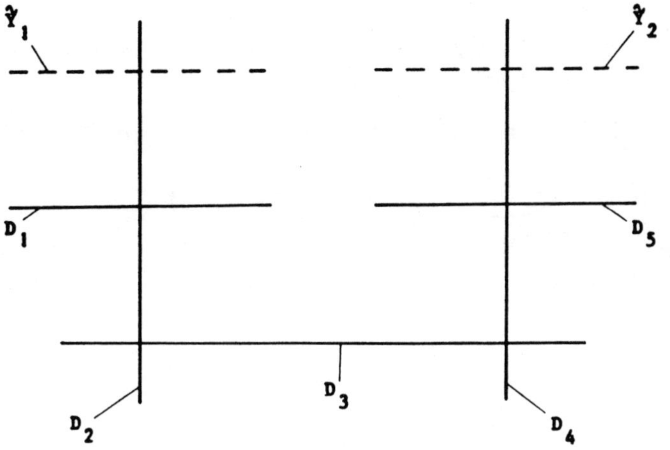

and $m_1 = 5$, $m_2 = 10$, $m_3 = 4$, $m_4 = 10$, $m_5 = 5$.

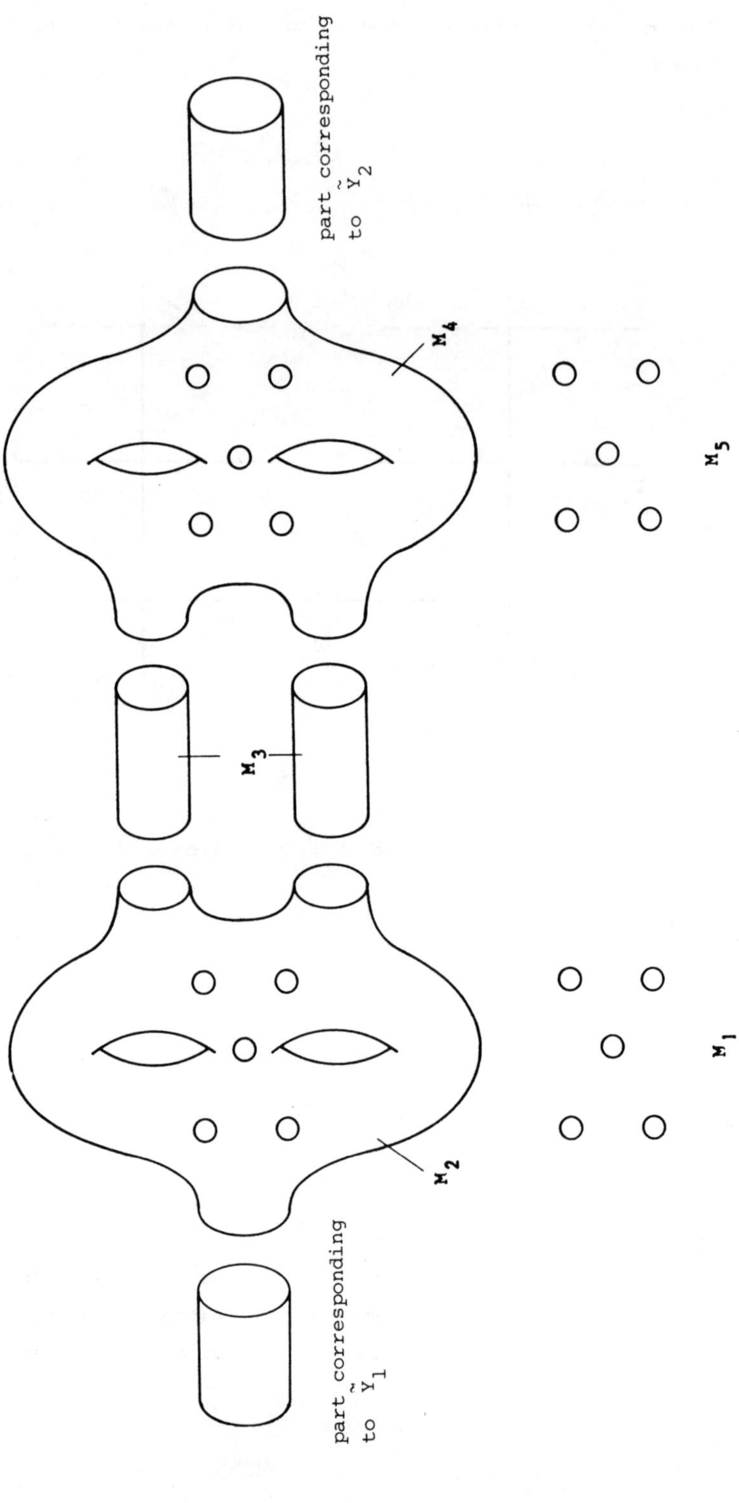

part corresponding to $\tilde{Y}_2$
M_4
M_5
M_3
M_1
M_2
part corresponding to $\tilde{Y}_1$

Putting this together :

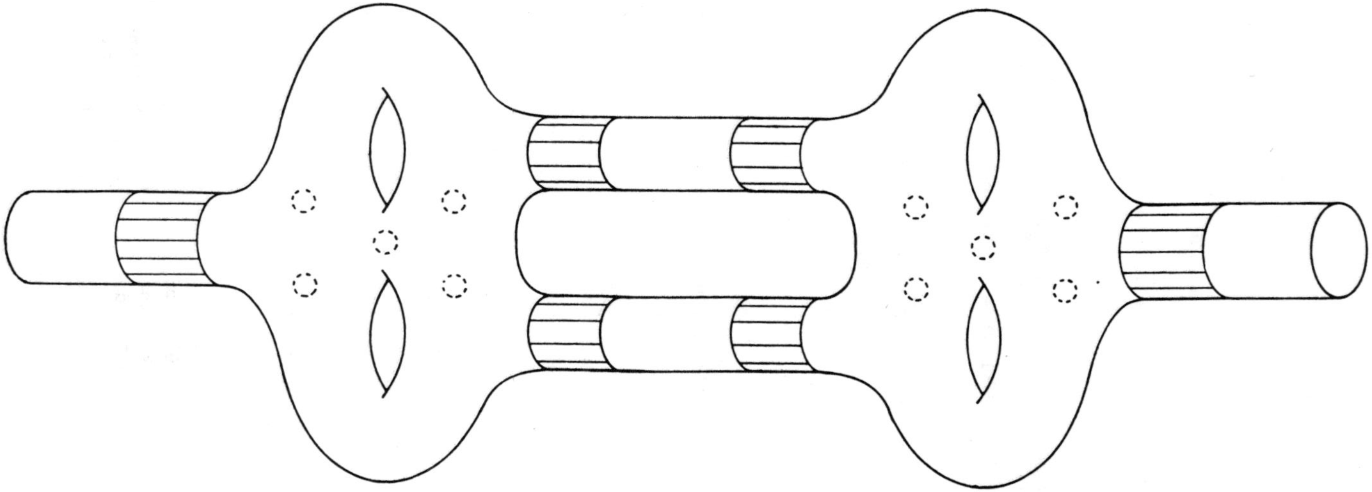

(the cylinders resulting from $\tilde{F} \cap K$ are drawn shaded)

To compute the Euler characteristic of $\tilde{F}$ we apply two elementary properties of Euler characteristic :

Lemma 2 :

(i) If A and B are subcomplexes of the simplicial complex $A \cup B$, then

$$\chi(A \cup B) = \chi(A) + \chi(B) - \chi(A \cap B).$$

(ii) If $B \to A$ is an m-fold unbranched covering, then

$$\chi(B) = m \cdot \chi(A).$$

Proof :

(i) The numbers of n-simplexes satisfy

$$\begin{aligned} &\#(n\text{-simplexes of } A \cup B) \\ = &\#(n\text{-simplexes of } A) + \#(n\text{-simplexes of } B) \\ &\qquad - \#(n\text{-simplexes of } A \cap B). \end{aligned}$$

(ii) A sufficiently fine triangulation of A may be lifted to a triangulation of B. For each n, the number of n-simplexes is multiplied by exactly m.

In our case, $\tilde{F}$ is the union of the M_i, the components of $\tilde{F} \cap K$, and the parts of F lying over the $\tilde{Y}_k$. Apart from the M_i, all of these are parts of the cylinder $S^1 \times [0,1]$, and hence of Euler characteristic 0. The intersections of the individual parts are circles, and hence also of Euler characteristic 0. Hence by Lemma 2(i) :

$$\chi(\tilde{F}) = \sum_{i=1}^{r} \chi(M_i).$$

Each M_i is an m_i-tuple covering of the r_i-tuply perforated sphere $D_i - K$, hence by Lemma 2(ii) :

$$\chi(M_i) = m_i(2-r_i).$$

Altogether, the Euler characteristic of the Milnor fibre F is given by :

Lemma 3 :

$$\chi(F) = \sum_{i=1}^{r} m_i(2-r_i).$$

As already mentioned, the multiplicities m_i of the lines D_i of the exceptional divisor, and the numbers r_i of double points of $\pi^{-1}(Y)$ on D_i, may be easily determined from the resolution graph of the singularity, and for that reason Lemma 3 allows the genus of the Milnor fibre to be calculated without difficulty from the resolution graph. We omit the derivation of an explicit formula.

Example :

For the singularity $(x^2+y^3)(x^3+y^2) = 0$ just considered

$\chi(F) = 5.1 + 10.(-1) + 4.0 + 10.(-1) + 5.1 = -10.$

F is therefore a twice-perforated surface of genus 5, in agreement with the geometric description of F already found.

So much for the description of the Milnor fibre with the help of the resolution graph for the standard resolution.

In conclusion, we now give a formula for the computation of the Euler characteristic of the Milnor fibre from the multiplicity sequence. We shall apply this formula in the next paragraph to prove the formulae of Clebsch and Noether.

Theorem 4 :

Let $(Y,0) \subset (\mathbb{C}^2,0)$ be a singularity with r irreducible components. Then the Euler characteristic of the Milnor fibre F of Y is

$\chi(F) = r - \Sigma\nu_i(\nu_i-1),$

where ν_i runs through the multiplicities of the strict preimages of Y at all infinitely near points of $0 \in Y$.

Examples :

(1) When Y is an ordinary double point, $r = 2$, $\nu_1 = 2$ and there are no further ν_i. Therefore

$\chi(F) = 2 - 2.1 = 0,$

in agreement with the fact, long known to us, that in this case the Milnor fibre is a cylinder $S^1 \times [0,1]$.

(2) When Y is an ordinary cusp, $r = 1$, $\nu_1 = 2$ and the other ν_i equal 1. Therefore

$\chi(F) = 1 - 2 = -1$

in agreement with the fact – which we have already found – that the Milnor fibre is a perforated torus in this case.

(3) For the singularity $(y^2+x^3)(y^3+x^2)$, the tree of infinitely near points (weighted by multiplicities) looks like this :

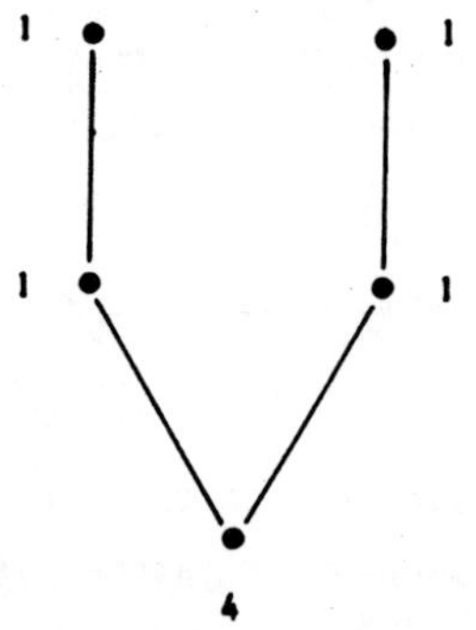

Thus $\chi(F) = 2 - 4.3 - 1.0 - 1.0 - 1.0 - 1.0 = -10$, in agreement with our earlier calculation.

<u>**Proof of Theorem 4**</u> :

If Y is regular at 0, then the theorem is certainly correct. We carry out the proof by induction on the number of σ-processes which are necessary to resolve the singularity. As in Lemma 1 we choose a small ball B around 0 and represent F by a regular fibre $F = B \cap f^{-1}(t)$. Now we blow up the point $0 \in \mathbb{C}^2$ once by

$$\pi : \tilde{X} \to \mathbb{C}^2.$$

Suppose the strict preimage $\tilde{Y}$ of Y consists of s components $Y_1, \ldots, Y_s$, and let E be the exceptional line $\pi^{-1}(0)$.

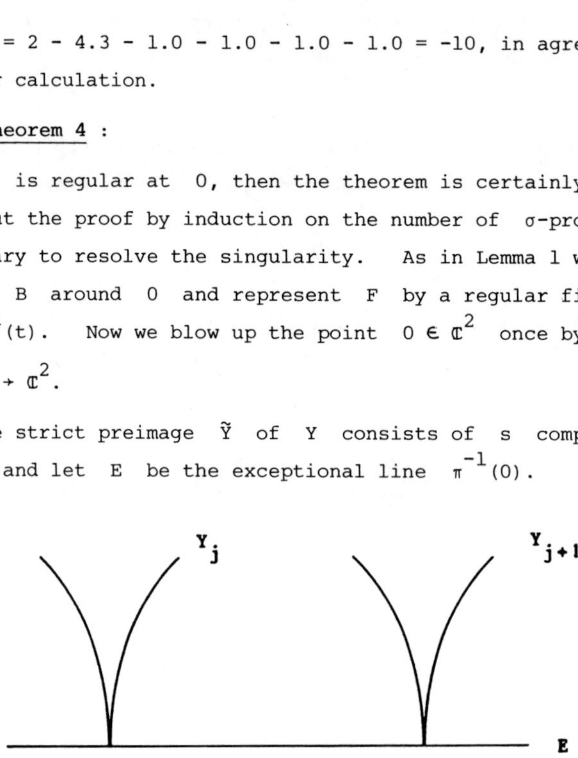

By Lemma 8.4.1, $\tilde{Y}$ meets the exceptional line E with multiplicity

$$\nu := \nu_0(Y)$$

i.e. $$\sum_{j=1}^{s} \nu(Y_j, E) = \nu. \qquad (1)$$

By slightly pushing the Y_j one can obtain analytic sets Y_j' which meet E transversely in $\nu(Y_j,E)$ points and give analytic curves $\bar{Y}_j$ in the ball B when blown down.

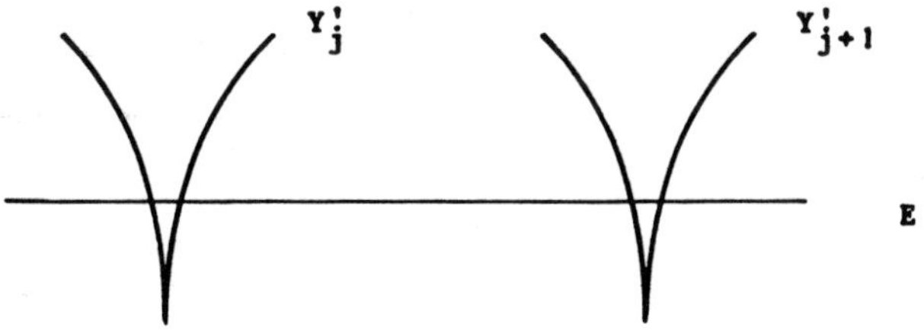

Let f' be the equation of $\bar{Y} = \bigcup \bar{Y}_j$ (it is not trivial to show that there is such an equation f' which describes the blown down curve $\bar{Y}$ in the whole of B!). $\bar{F} := f'^{-1}(t) \cap B$ is homeomorphic to the original Milnor fibre F if all the deformations are chosen to be sufficiently small. The strict preimage F' of $\bar{F}$ is homeomorphic to $\bar{F}$ (and hence also to F) and hugs the total preimage $E \cup \bigcup_{j=1}^{s} Y'_j$ of $\bar{Y}$ in a similar way, as we have seen in the proof of Lemma 3.

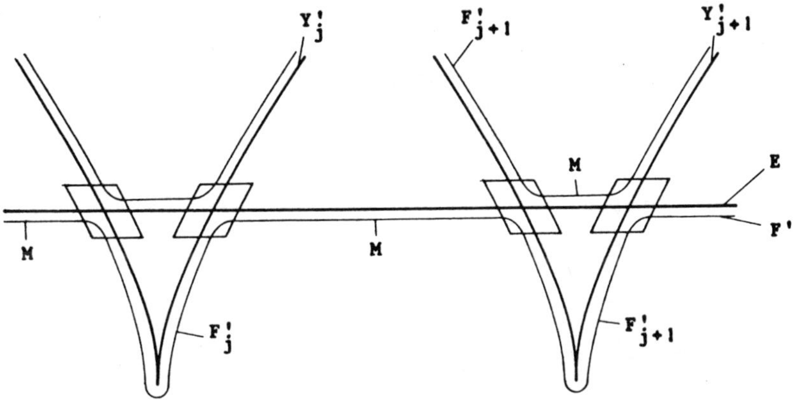

Around each of the intersection points of Y'_j and E we choose a small polydisc. F' is now composed of a part M which covers the perforated line E as above, the intersections with the polydiscs just chosen, and the parts of F' which lie over the Y'_j.

M is a ν-tuple covering of the $(\sum_{j=1}^{s} \nu(Y_j,E) = \nu)$-tuply perforated projective line E. Hence it follows by Lemma 2 that

$$\chi(M) = \nu(2-\nu). \tag{2}$$

The intersections of F' with the polydiscs are again cylinders, and the intersections with the individual pieces are again circles.

Thus it remains to compute the Euler characteristics of the parts F'_j of F' which lie over the Y'_j.

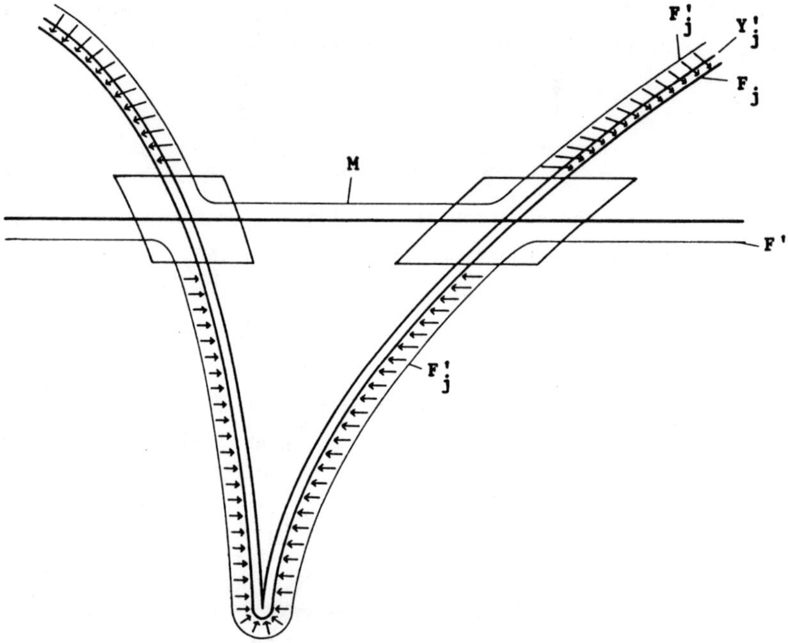

Outside the polydiscs, F'_j is deformable into the Milnor fibre F_j of Y'_j, as the arrows in the above picture indicate. Thus F'_j is homeomorphic to the $\nu(Y_j,E)$-tuply perforated Milnor fibre F_j of Y'_j (and, since Y'_j results just from pushing Y_j, F_j is also the Milnor fibre of Y_j). Consequently

$$\chi(F'_j) = \chi(F_j) - \nu(Y_j,E). \tag{3}$$

Altogether, it follows by Lemma 2 from (1), (2), (3) that

$$\begin{aligned}\chi(F') &= \chi(M) + \sum_{j=1}^{s} \chi(F'_j)\\ &= -\nu^2 + 2\nu + \sum_{j=1}^{s} \chi(F_j) - \sum_{j=1}^{s} \nu(Y_j,E)\\ &= \sum_{j=1}^{s} \chi(F_j) - \nu(\nu-1).\end{aligned}$$

Since $F' \approx F$, we obtain

$$\chi(F) = \sum_{j=1}^{s} \chi(F_j) - \nu(\nu-1)$$

where F_j is the Milnor fibre of the component Y_j of the strict pre-image of Y. The formula of Theorem 4 then follows by induction.

Remark :

Besides Lemma 3 and Theorem 4 there are further useful formulae for computing the homology of the Milnor fibre F :

(i) Let (Y,p) be an analytic curve germ and let $\pi : \tilde{Y} \to Y$ be an (abstract) resolution of singularities. If $\pi^{-1}(0) = \{p_1,\ldots,p_r\}$, then π induces an injective mapping :

$$\pi^* : \mathcal{O}_{Y,p} \to \bigoplus_{i=1}^{r} \mathcal{O}_{\tilde{Y},p_i}$$

via $f \mapsto f \circ \pi$. In [Ml], §10, Milnor proves the following formula for the Euler characteristic of the Milnor fibre F :

$$\chi(F) = r - 2\delta_p \text{ , where}$$

$$\delta_p = \dim_{\mathbb{C}}\left(\bigoplus_{i=1}^{r} \mathcal{O}_{\tilde{Y},p_i}/\pi^*(\mathcal{O}_{Y,p})\right).$$

Incidentally, this number $2\delta_p$ has already surfaced in Theorem 8.3.16 in connection with the deformation theory of the singularity, as an upper bound for the number of essential parameters for the deformations with the same topological type as a given singularity. More precisely, one can show the following : the number of essential parameters for the perturbation or, as one says, unfolding of a holomorphic function f with an isolated singularity (without preservation of topological type) equals the first Betti number of the Milnor fibre associated with the singularity (cf., e.g., [Al]).

The first Betti number of the Milnor fibre F is also called the Milnor number μ. Of course, $\chi = 1-\mu$ because F, being a connected bounded manifold, has second Betti number 0.

(ii) If f is the equation of the singularity, then

$$\mu = \dim_{\mathbb{C}}\mathbb{C}\{x,y\}/\left(\frac{\partial f}{\partial x}, \frac{\partial f}{\partial y}\right).$$

For a proof, see [Bl1], Appendix B.

In all, we have obtained the following formulae for μ :

$$\mu = \dim \mathbb{C}\{x,y\}/\left(\frac{\partial f}{\partial x}, \frac{\partial f}{\partial y}\right)$$

$$\mu = 2 \dim \bigoplus_{i=1}^{r} \mathcal{O}_{\tilde{Y},p_i} / \pi^* \mathcal{O}_{Y,p} - r + 1$$

$$\mu = \Sigma \nu_i (\nu_i - 1) - r + 1.$$

(iii) A beautiful description of the Milnor fibre from the Newton polygon is given by F. Ehlers in [El].

In our investigation of the Milnor fibration $\phi = \frac{f}{|f|} : S - K \to S^1$ we have mainly described the structure of the fibre $F_t = \phi^{-1}(t)$. But still more information is lodged in the Milnor fibration : if one runs once around the circle $[0,1] \to S^1$ with the parametrisation $\tau \mapsto e^{2\pi i \tau}$, then this path may be lifted to a homotopy (h_τ)

$$h_\tau : F_1 \to S - K.$$

Each h_τ is a diffeomorphism of F_1 onto $F_{e^{2\pi i \tau}}$.

The diffeomorphism

$$h := h_1 : F_1 \to F_1$$

is called the (geometric) monodromy of the singularity. The Milnor fibration is determined, essentially uniquely, by the fibre $F = F_1$ and the monodromy h, because it results from the trivial fibration $F \times [0,1] \to [0,1]$ by pasting $F \times \{0\}$ to $F \times \{1\}$ by h.

The monodromy of singularities was already investigated by Picard-Simart [P4] and then especially by Lefschetz [L1] around 1924. Investigation of the connection between monodromy and Feynman integrals in physics by Pham [P2] and the related work [B4] has greatly strengthened interest in the study of singularities. (See also : Hwa-Teplitz, Homology and Feynman Integrals [H7].)

In a similar direction, there are important works of Grothendieck, Deligne, P.A. Griffiths, Thom, Mather, Arnold.

In the meantime, numerous methods for the investigation of the monodromy of singularities have been developed, and they could in themselves be the subject of a course. Some of the processes introduced above to describe the Milnor fibre may be applied, with a little more effort, to obtain statements about the monodromy. Time prevents us from going further into this, unfortunately, and we refer instead to the works [A4], [A5], [A6], [A7] of A'Campo in particular, where the monodromy of curves is investigated by three quite different and beautiful geometric methods. The first method uses the iterative construction of the Milnor fibration, which we have developed at the beginning of this section following A'Campo. The second uses a wonderful combination of resolution and deformation of singularities, real and complex algebraic geometry, and Picard-Lefschetz theory. The third uses the description of the Milnor fibre by resolution which we have developed in this section and which goes back essentially to Clemens.

With this we leave the local study of singularities of plane curves, which has been our theme for this whole long chapter, and go to global investigations in the last chapter. There, we shall find the local results very useful.

9. Global Investigations

9.1 The Plücker formulae

Now that we have investigated the local properties of plane curves and their singular points from various viewpoints, in the previous paragraph, we next want to derive global assertions about plane complex projective algebraic curves. Above all, we shall calculate global invariants of such curves from the local invariants of their singular points which we have investigated earlier. The global invariants in question are the order, class and genus of curves. Formulae for the order and class were first presented by J. Plücker in 1834, and later generalised by M. Noether 1875 and 1883. We want to deal with these formulae in the present section. Then in the next section we shall derive the formula for the genus, which in special cases goes back to Riemann 1857 and Clebsch 1864, and in its general form to M. Noether 1874 and Weierstrass.

We have already defined the order of a plane curve C without multiple components, at the beginning of our analytic-algebraic investigations of curves in 2.3 and 5.1, as the degree of the polynomial describing the curve C. The order is the simplest conceivable invariant, and its significance was already seen by Newton. It is an invariant, not of an abstract curve, but of the curve C in the projective plane. The local analogue of this invariant is the multiplicity of a curve germ, which we have defined in 5.3.

As soon as the idea of duality emerged in the development of projective geometry, one could associate further invariants with a curve C, by considering the known invariants, but for the dual curve C' of C. In 6.2 we have defined the dual curve C' as the set of tangents of C, where these tangents are regarded as points in the dual projective plane. The order of this dual curve C' is then a new invariant of C, the <u>class</u> of C. It is important that this invariant likewise depends on the embedding in the projective plane, and that to define it one needs the geometry of the plane, since one must be able to view tangents as lines in the plane and points in the dual plane.

It is clear how one must define a corresponding local invariant for the branch of a plane algebraic curve : the <u>class</u> of such an irreducible branch is the multiplicity of the corresponding point on the dual curve. For a reducible curve germ the class is the sum of the classes of the branches.

<u>**Example**</u> :

Let C be an irreducible cubic with a cusp. In suitable coordinates (x_0,x_1,x_2), C has the equation :

$$x_0x_1^2 + x_2^3 = 0.$$

The point (x_0',x_1',x_2') of the dual curve which corresponds to the point (x_0,x_1,x_2) of C satisfies

$$(x_0',x_1',x_2') = (x_1^2, 2x_0x_1, 3x_2^2).$$

By eliminating x_0, x_1, x_2 from this and the equation of C one obtains the equation

$$27x_0'x_1'^2 - 4x_2'^3 = 0$$

for C'. Thus the dual curve C' is again a cubic with a cusp. The cusp $(1,0,0)$ of C corresponds to the inflection point $(0,1,0)$ of C', and the inflection point $(0,1,0)$ of C corresponds to the cusp $(1,0,0)$ of C'.

It is very pleasant to see the relation between the form of a curve and its dual, for suitably chosen examples, illustrated by pictures of the corresponding real curves.

In order to construct the dual of a given curve C without a lot of calculation, one can describe the duality which associates points of the plane E with lines of the dual plane E', and lines of E with points of E', in terms of the pole-polar relationship for a circle : each point p is associated with its polar in the circle, each line with its pole.

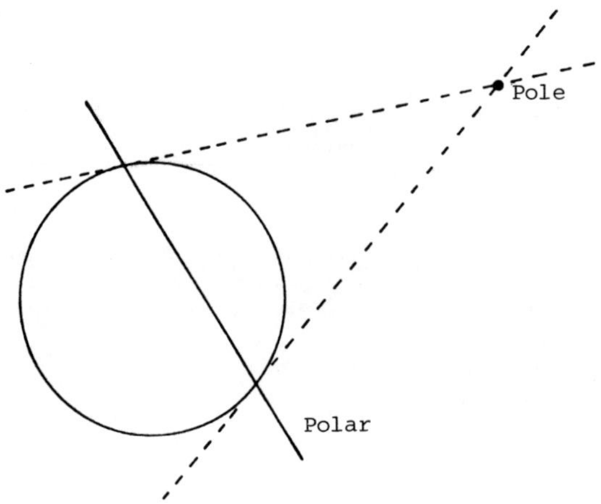

The <u>polar</u> of p with respect to a circle is defined by elementary geometry, as the line connecting the two contact points of the tangents from p to the circle. When the circle has the equation

$$x^2 + y^2 = 1$$

in cartesian coordinates, then the polar of the point $p = (a,b)$ has the equation

$$ax + by = 1.$$

Thus the pole-polar relationship associates the line with affine equation $ax + by = 1$ with the point whose affine coordinates are (a,b), and this is precisely the analytic description of duality.

On the basis of this analytic description it is also clear how one must extend the elementary geometric definition of polar when the pole p lies inside the circle.

One constructs the chord through p perpendicular to the radius through p, and then constructs the tangents at the points where this chord meets the circle. The parallel to the chord through the intersection of the tangents is the desired polar of the pole p.

The following pictures show some pairs of dual curves, constructed with the help of the pole-polar relationship in a circle.

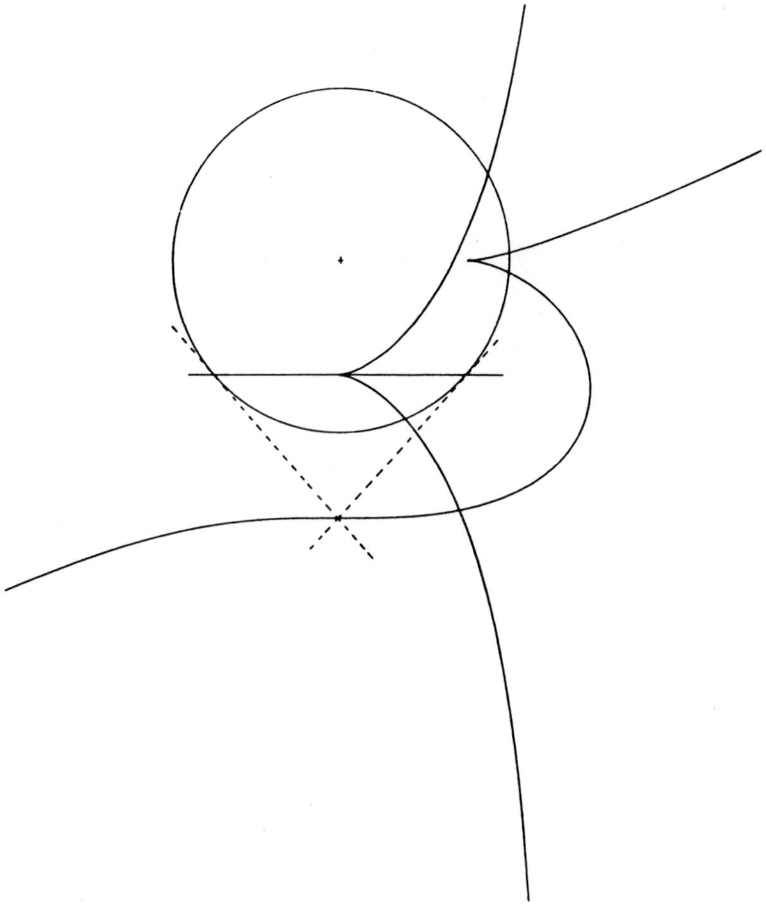

Construction of the dual of the cissoid of Diocles using the pole-polar relation in a circle. The cusp of the cissoid corresponds to the inflection point of the dual curve, and the cusp of the dual curve corresponds to the inflection point (at infinity) of the cissoid.

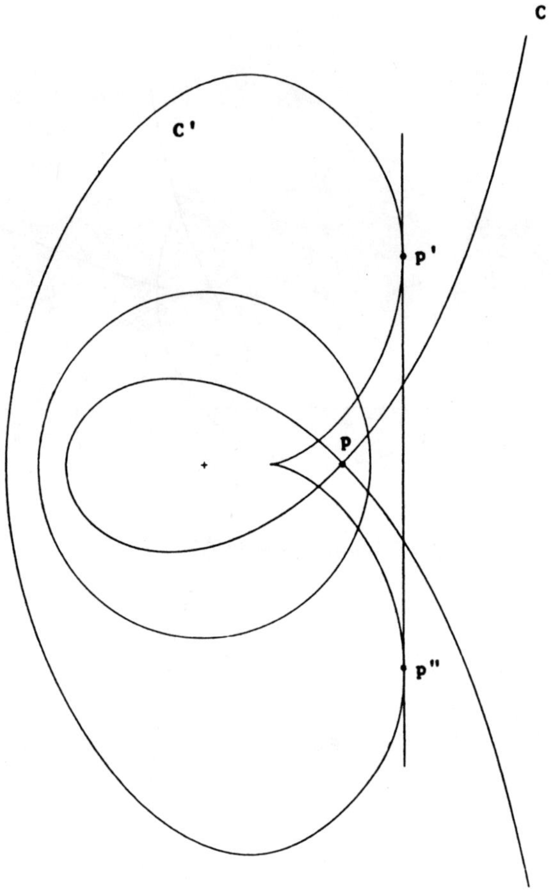

Construction of the dual C' of the cubic C with an ordinary double point p. The two tangents through p are the polars to the points p' and p'' on C' relative to the circle. The line through p' and p'' is the polar of p. This line is a double tangent. The cusp of C' corresponds to the inflection point, at infinity, of C.

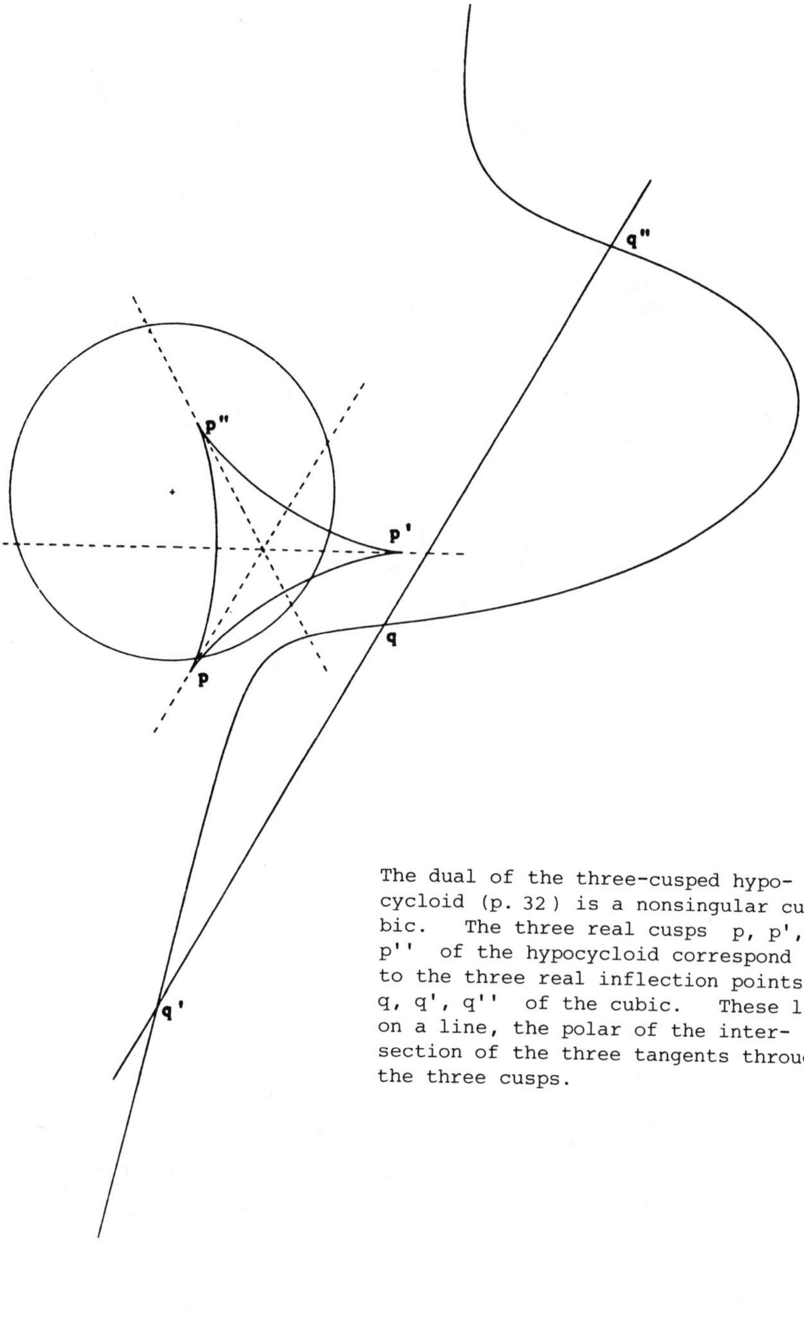

The dual of the three-cusped hypocycloid (p. 32) is a nonsingular cubic. The three real cusps p, p', p'' of the hypocycloid correspond to the three real inflection points q, q', q'' of the cubic. These lie on a line, the polar of the intersection of the three tangents through the three cusps.

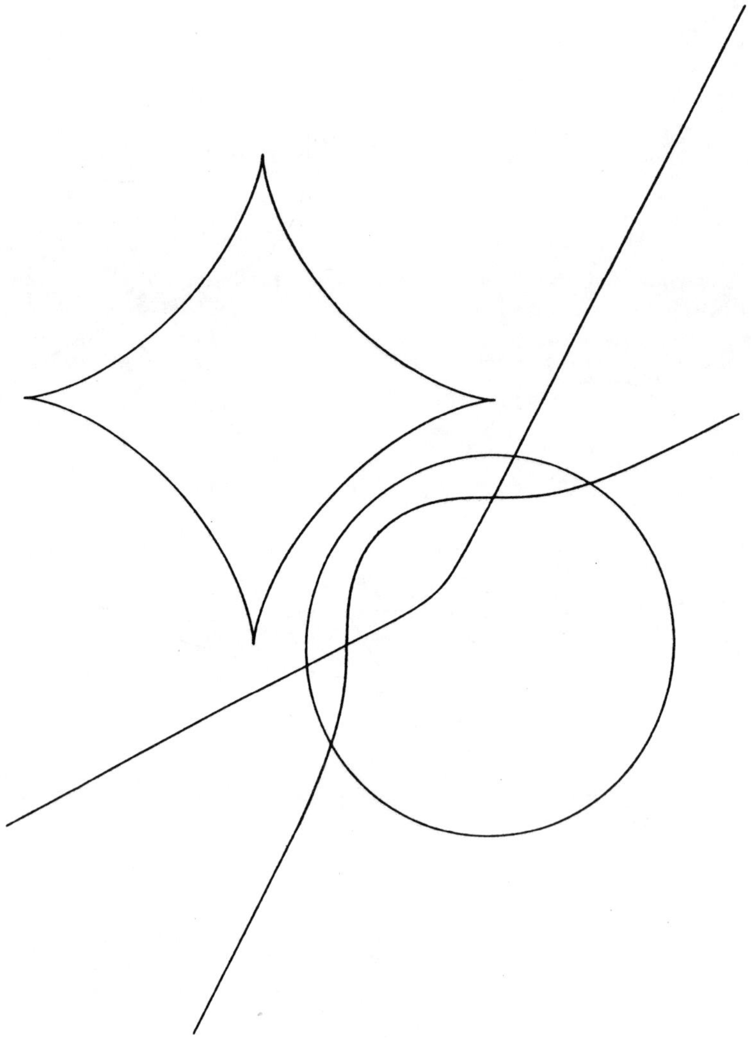

Dual of the astroid. The duality is expressed by the pole-polar relationship in the circle. The four cusps of the astroid correspond to the four inflection points of the dual curve. Since the tangents through opposite cusps of the astroid coincide, the dual curve has two double points, and for each branch of the dual curve through such a point the double point is an inflection point.

From the analysis of these examples we can already learn something about the relation between a curve C and its dual C'. The simple cusps of C correspond to the simple inflection points of C', and dually, the simple inflection points of C correspond to the simple cusps of C'. The ordinary double points of C correspond to the ordinary double tangents of C', and the ordinary double tangents to double points of C'. The ordinary tangents of C (i.e. those which have a contact point of multiplicity 2 with C and are otherwise transverse) correspond to the regular points of C', and the regular points of C correspond to the ordinary tangents of C'. We omit the general proof of all this – the proof is not difficult.

The class of a curve, as the examples show, is in general quite different from its order. We have already obtained some preliminary results on the relation between class and order. In 6.2.7 we proved, with the help of Bézout's theorem, that a nonsingular curve of order $m > 1$ has class $m(m-1)$. For singular curves the class is smaller than $m(m-1)$, and each singular point contributes to the lowering of the class. In what follows we shall calculate precisely how much a singular point lowers the class.

Previously we have also obtained results on the number of inflection points of a curve C of order m (7.4.1 and 7.4.2). We have shown that the regular inflection points of C are cut by the Hessian curve H_C, that their multiplicities equal the intersection multiplicities of C and H_C at the points in question, and that for this reason the number of inflection points of a nonsingular curve is $3m(m-2)$ by Bézout's theorem. In what follows we also want to compute this number for singular curves.

In an important special case, all the questions just raised are answered by the Plücker formulae.

Theorem 1 (Plücker formulae)

Let C be a plane curve without multiple components and without lines as components, and let C' be the dual curve of C. Assume that the curves C and C' have at most ordinary double points and simple cusps as singularities. This is equivalent to C's having only ordinary double points and cusps as singularities, and only ordinary tangents, ordinary double tangents and simple inflection tangents. Suppose that

n is the order of C

n' is the order of C'

d is the number of ordinary double points of C

d' is the number of ordinary double points of C'

s is the number of cusps of C

s' is the number of cusps of C'.

So n' is the class of C

d' is the number of double tangents of C

s' is the number of inflection tangents of C.

These projective invariants are subject to the following relations:

(i) $n' = n(n-1) - 2d - 3s$

(ii) $s' = 3n(n-2) - 6d - 8s$

(iii) $n = n'(n'-1) - 2d' - 3s'$

(iv) $s = 3n'(n'-2) - 6d' - 8s'$.

Remarks :

(1) One immediately deduces from the formulae that they are not independent : from any three of them one can derive the fourth by a simple calculation.

(2) The formulae easily yield the following assertion : any three of the six numbers n, d, s, n', d', s' are determined by the remaining three. For example, n, d, s determine the numbers n' and s' by the formulae (i) and (ii), and d' is obtained from (i), (ii), (iii) by the formula :

$$d' = \tfrac{1}{2}n(n-2)(n^2-9) - (2d+3s)(n^2-n-6) + 2d(d-1) + \frac{9}{2}s(s-1) + 6ds.$$

(3) The formulae (iii) and (iv) in Theorem 1 are dual to the formulae (i) and (ii). Hence to prove Theorem 1 it suffices to show (i) and (ii). But these formulae are simple special cases of the corresponding formulae (i) and (ii) in Theorem 2 which follows !

Examples :

(1) As our first example we consider the irreducible cubics, which we have already classified up to projective equivalence in 7.4, and divided into three types. Since these curves can have at most one ordinary double point or one simple cusp, one knows the numbers n, d, s. The Plücker formulae then yield n', d', s'. The results are collected in the following table and are in agreement with

propositions 7.3.3, 7.3.12 and 7.3.13 and the constructions of dual curves carried out previously.

curve	n	d	s	n'	d'	s'	dual curve
nonsingular cubic	3	0	0	6	0	9	sextic with 9 cusps
cubic with ordinary double point	3	1	0	4	0	3	quartic with 3 cusps, 1 double tangent
cubic with cusp	3	0	1	3	0	1	irreducible cubic with cusp

(2) As our second example we consider the astroid, which we defined in 1.7 and have already investigated in 2.4. The equation was

$$(x^2+y^2-1)^3 + 27x^2y^2 = 0.$$

Thus $n = 6$. For the affine curve we have found 8 singular points : the 4 cusps $(0,\pm 1)$, $(\pm 1,0)$ and the 4 points $(\pm i,\pm i)$. The latter 4 points are ordinary double points, because at these points the equation has a nondegenerate quadratic term. We find 2 more singular points at infinity. These are likewise cusps. One sees this without calculation, on symmetry grounds, since introduction of suitable homogeneous coordinates gives the homogeneous equation

$$(x^2 + y^2 + z^2)^3 + 27x^2y^2z^2 = 0.$$

The Plücker formulae then give the following invariants for the astroid :

$n = 6 \quad n' = 4$

$d = 4 \quad d' = 3$

$s = 6 \quad s' = 0.$

The dual curve is a quartic with 3 ordinary double points, of which two are real, and with 6 branches having inflection tangents, of which 4 are real. Thus we have applied Plücker formulae under somewhat more general assumptions than in Theorem 1 : after all, the tangents through opposite cusps of the astroid do coincide.

(3) As our third example we compute the number d' of double tangents of a nonsingular curve C of order n. It follows from the remark (2) following the Plücker formulae that

$$d' = \tfrac{1}{2}n(n-2)(n^2-9).$$

For $n = 1,2,3$ we have $d' = 0$, which is clear in any case by Bézout's theorem, because a double tangent must cut the curve in at least 4 points (counting multiplicity). Thus the smallest order for which double tangents can appear is $n = 4$. Here the result is

$$d' = 28.$$

These 28 double tangents of a plane quartic form an interesting configuration, which has been extensively investigated, first by Steiner and Plücker, then by Hesse, Aronhold and Noether, among others. It is related to other interesting configurations, e.g. the Kummer configuration and the configuration of 27 lines on a nonsingular cubic surface in $P_3(\mathbb{C})$. A survey of the literature on this may be found in Pascal's Repertorium der höheren Mathematik [P7], Band II.1, Chap. XVIII, §5.

Now we want to give the generalisation of the Plücker formulae found by Weierstrass and M. Noether.

Theorem 2

Let C be a plane algebraic curve without multiple components and without lines as components. Let n be the order of C and let n' be the class of C. Also, let s' be the number of inflection points of C. Here, all inflection points of nonsingular branches of C must be counted, and with the correct multiplicity in the sense of 7.3. Let p_i be the singular points of C and let ν_{ij} be the multiplicities of the infinitely near singular points of (C,p_i). Finally, let ν_k resp. ν_k' be the orders resp. classes of the singular branches of C. Then the following formulae hold for the class n' and the number s' of inflection points :

(i) $n' = n(n-1) - \Sigma\nu_{ij}(\nu_{ij}-1) - \Sigma(\nu_k-1)$

(ii) $s' = 3n(n-2) - 3 \cdot \Sigma\nu_{ij}(\nu_{ij}-1) - \Sigma(2\nu_k+\nu_k'-3)$.

The basic idea for proving the formulae in Theorem 2 is quite similar for both formulae. For the first formula we want to determine the number n' of intersections of the dual curve C' with a line L' in the dual projective plane, and for the second formula we want the number of regular inflection points of C. We already know from 7.3.1 and 7.3.2 that the inflection points on C are its intersections with the Hessian curve H_C. Thus we only need to determine the number of intersections of C with H_C, and it equals $n\cdot 3(n-2)$ by Bézout's theorem. However, since we want only the number s' of smooth inflection points, we must subtract the sum of the intersection numbers of C and H_C at the singular points of C. Thus the proof of formula (ii) reduces to computation of the intersection numbers at the singular points, from the multiplicities of their infinitely near points.

In a similar way, we also reduce the proof of formula (i) to the computation of the intersection numbers of C with a suitable curve, namely a polar. To do this we first go from the problem of determining intersections of C' and L' to the dual problem : the points of the line L' in the dual plane correspond to the lines of the pencil in the original plane through the point p dual to L'. An intersection point of L' and C' corresponds to a tangent to C which goes through p. We describe the situation analytically. Let x_0,x_1,x_2 be homogeneous coordinates, let $p = (a_0,a_1,a_2)$, and let $F(x_0,x_1,x_2)$ be the equation of C. By 5.3.4, the equation of the tangent to C at a regular point (x_0,x_1,x_2) reads :

$$\xi_0 \cdot \frac{\partial F}{\partial x_0}(x_0,x_1,x_2)+\xi_1 \cdot \frac{\partial F}{\partial x_1}(x_0,x_1,x_2)+\xi_2 \cdot \frac{\partial F}{\partial x_2}(x_0,x_1,x_2) = 0.$$

Here x_0,x_1,x_2 are constant – for the moment – and ξ_0,ξ_1,ξ_2 are variable. If p is to be a point on this tangent, then it must satisfy this linear equation. Thus one obtains the linear equation

$$a_0 \cdot \frac{\partial F}{\partial x_0}(x_0,x_1,x_2)+a_1 \frac{\partial F}{\partial x_1}(x_0,x_1,x_2)+a_2 \cdot \frac{\partial F}{\partial x_2}(x_0,x_1,x_2) = 0. \qquad (*)$$

If we regard (a_0,a_1,a_2) as constant in this equation, and (x_0,x_1,x_2) as variable, i.e. if we consider (*) to be the equation of a plane curve $\tilde{C}$ of order $n-1$, then the intersections of this curve with C are just the points of C whose tangents go through p. More precisely : they are the points q_i of C for which the line through q_i and p meets the curve at p with multiplicity greater than one. Among these points are the points q_j for which the line through q_j and p is a tangent, and the number of these points, counted with the correct multiplicity, is just the number n' to be determined. However, the singular points q_k of C also appear. When we have chosen the point p in sufficient generality, the tangents to C through the singular points do not go through p, and hence the singular points need not be counted in the computation of n'. Thus we find the class n' to be the intersection number of C and $\tilde{C}$, which by Bézout is $n(n-1)$, minus the sum of the intersection multiplicities of C and $\tilde{C}$ at the singular points of C. In this way, the proof of formula (i) of Theorem 2 is also reduced to the determination of the local intersection multiplicities of C with a suitable curve $\tilde{C}$. This curve $\tilde{C}$ is called the first polar of C relative to p, and in what follows we must study such polars in somewhat more detail.

Definition :

Let C be a plane curve of order n with homogeneous equation $F(x_0,x_1,x_2) = 0$. Let $1 \le r \le n$ and let $y = (y_0,y_1,y_2) \in P_2(\mathbb{C})$ be a point for which the homogeneous polynomial

$$D_y^r F(x_0,x_1,x_2) := \sum_{i_1,i_2,\ldots,i_r=0}^{2} y_{i_1}\cdots y_{i_r} \frac{\partial^r F}{\partial x_{i_1}\cdots \partial x_{i_r}}(x_0,x_1,x_2)$$

does not vanish identically. Then the curve of order $n-r$ with equation $D_y^r F(x_0,x_1,x_2) = 0$ is called the <u>r^{th} polar</u> of C relative to the point y. In particular, the 1st polar, the curve of order $n-1$ with equation

$$\sum_{i=0}^{2} y_i \frac{\partial F}{\partial x_i}(x_0,x_1,x_2) = 0$$

is simply called the <u>polar of</u> C <u>relative to the point</u> y.

A few remarks are called for in explanation of this definition :

It follows, by comparing coefficients of $\lambda^{n-r}\mu^r$ in the Taylor expansions

$$F(\lambda x+\mu y) = \sum_{r=0}^{n} \frac{1}{r!} \lambda^{n-r}\mu^r D_y^r F(x) = \sum_{r=0}^{n} \frac{1}{(n-r)!} \lambda^{n-r}\mu^r D_x^{n-r} F(y)$$

that

$$\frac{1}{r!} D_y^r F(x) = \frac{1}{(n-r)!} D_x^{n-r} F(y).$$

Hence $D_y^r F(x) \equiv 0$ just in case all $(n-r)$-tuple partial derivatives of F vanish at y, i.e. when y is at least an $(n-r+1)$-tuple point of C (cf. p.212). Thus it is precisely in this case that the r-tuple polar relative to y is not defined. In particular, the 1st polar relative to y is undefined only when y is an n-tuple point of C.

One computes immediately that the above definition does not depend on the choice of homogeneous coordinates, so the polars of a curve relative to a point are projectively invariantly defined.

The above definition is an algebraic definition. We want to investigate whether a more geometric presentation can be found, at least for the first polar.

To do this we simplify by choosing the homogeneous coordinates (x_0, x_1, x_2) so that the point p, relative to which we shall describe the first polar of C, has the coordinates $(0,0,1)$. Suppose C has the equation $F(x_0, x_1, x_2) = 0$. Then the first polar of C relative to p has the equation

$$\frac{\partial F}{\partial x_2}(x_0, x_1, x_2) = 0.$$

If we introduce affine coordinates x, y with

$$x = \frac{x_1}{x_0},$$

$$y = \frac{x_2}{x_0}$$

then the pencil of lines through p (with the exception of the line at infinity) becomes the family of parallel lines

$$x = c, \quad c \in \mathbb{C}$$

and the curve C has the affine equation

$$f(x,y) = 0$$

where $f(x,y) = F(1,x,y)$. The affine equation of the polar then reads

$$\frac{\partial f}{\partial y}(x,y) = 0.$$

Thus we obtain the following <u>description of the polar</u> :

The affine equation of the polar, $\frac{\partial f}{\partial y}(x,y) = 0$, describes the affine curve of those points in the affine plane at which the curve C_t of the family $f(x,y) = t$, $t \in \mathbb{C}$, has a vertical tangent $x = c$ or a singular point. In particular, the polar of C relative to p cuts the curve C at precisely the regular points of C whose tangents go through p, and at the singular points.

The following picture illustrates this situation for the folium of Descartes with the affine equation $x^3 + y^3 - 3xy = 0$. The polar is the parabola $x = y^2$.

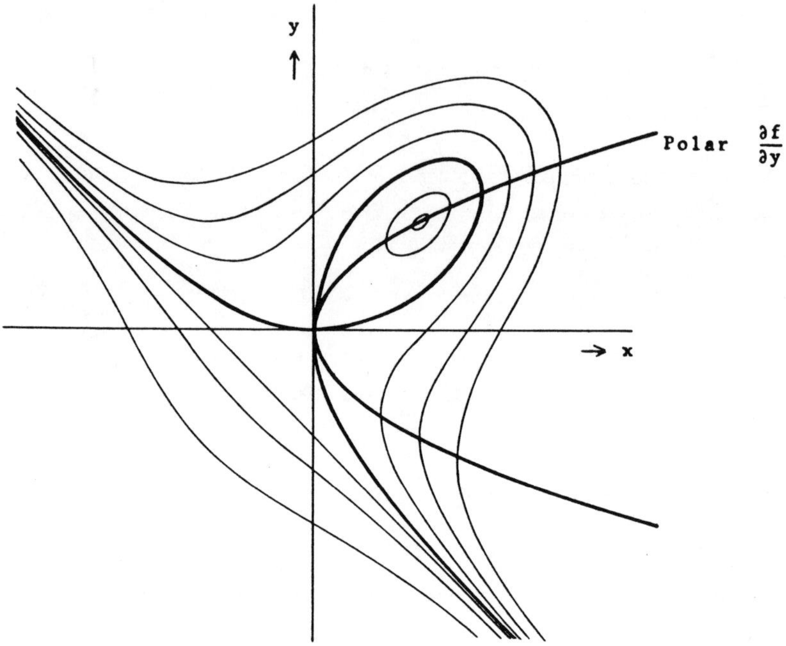

A polar of the folium of Descartes
The pole lies at infinity on the y-axis.

With the above choice of coordinates we have excluded the affine description of the polar in the neighbourhood of the pole. For general coordinates (x_0, x_1, x_2) the polar equation

$$a_0 \cdot \frac{\partial F}{\partial x_0} + a_1 \cdot \frac{\partial F}{\partial x_1} + a_2 \cdot \frac{\partial F}{\partial x_2} = 0$$

goes over to the affine polar equation

$$a_0(nf - x\frac{\partial f}{\partial x} - y\frac{\partial f}{\partial y}) + a_1 \frac{\partial f}{\partial x} + a_2 \frac{\partial f}{\partial y} = 0$$

for the affine coordinates $x = \frac{x_1}{x_0}$, $y = \frac{x_2}{x_0}$. Of course it is clear that for a quadric C this description of the polar coincides with the elementary geometric one. Because when C is a quadric, $n = 2$, and hence the polar is of order 1, i.e. a line, and this line must go through the contact points of both tangents from the pole to C.

One needs to be aware that this geometric description of the polar contains an arbitrary element, namely the choice of the line at infinity, $x_0 = 0$. The curves C_t belong, in the projective description, to the linear system

$$\lambda_0 F(x_0, x_1, x_2) + \lambda_1 x_0^n = 0,$$

where n is the order of C. Another choice of homogeneous coordinate system would give another line at infinity and thereby lead to another linear system of curves. However, the polar would still be the same, since it is defined in a projectively invariant manner.

For those who do not want to come to terms with this contradiction between the invariant algebraic definition and the non-invariant geometric description of polars, I shall give yet another description of the first polar of C relative to $p \notin C$. Again we consider the pencil of lines through p. Let L be a line in this pencil. We want to describe the n-1 intersections of L with the polar of C relative to p (n is the order of C). When we have done that for each L, then we have described the polar. The line L cuts C in n points. If we introduce an affine coordinate y on the line L so that $p = \infty$, then we obtain n numbers $y_1, \ldots, y_n$ as the coordinates of these n intersections with C. Now we construct the polynomial with these zeroes

$$\phi(y) = (y-y_1)(y-y_2)\ldots(y-y_n).$$

Then the n-1 intersections of L with the polar we seek are the n-1 critical points of ϕ, the solutions of the equation $\phi'(y) = 0$, i.e. the solutions of

$$\sum_{i=1}^{n} (y-y_1)\ldots\widehat{(y-y_i)}\ldots(y-y_n) = 0.$$

Of course, it is trivial to convert this description of the polar into the one above : if $f(x,y) = 0$ is the affine equation of C, and if $x = c$ is the equation of L, then $\phi(y) = \alpha f(c,y)$ with a constant factor α, and the critical points of ϕ' are the zeroes of $\frac{\partial f}{\partial y}(c,y)$, i.e. the intersections of L with the polar $\frac{\partial f}{\partial y}(x,y) = 0$. In 1857 de Jonquières named these n-1 points defined by $y_1, \ldots, y_n$ and $p = \infty$ the "harmonic mean points" of $y_1, \ldots, y_n$ relative to p. For $n = 2$, of course, one has the ordinary midpoint $\frac{1}{2}(y_1+y_2)$.

The independence of this "new" description of the polar from the choice of affine coordinate y on L is obvious, and even though it too is not very geometrical, it perhaps contributes to the elucidation of the concept.

The description of the polar as the locus of points with vertical tangents in a family of curves may be generalised to families of analytic curves. This is what we want to look at next.

Let V be a complex surface and let $f : V \to \mathbb{C}$ be a holomorphic function whose fibres $C_t = f^{-1}(t)$, $t \in \mathbb{C}$, are curves in V without multiple components. In addition to this family, let a smooth family of curves L_c on V be given, i.e. a holomorphic mapping $x : V \to \mathbb{C}$ which everywhere has rank 1. Its fibres $L_c = x^{-1}(c)$ are then non-singular curves in V.

<u>Definition</u> :

<u>The polar</u> of the family of curves $\{C_t\}$ relative to the family of smooth curves $\{L_c\}$ is the locus of those points q of V at which the curve C_t through q is singular or tangential to the curve L_c through q.

It is clear how to describe these polars analytically. If q is any point in V then, because of the rank condition on x, one can use the function x as one coordinate function and find a further holomorphic function y so that x and y are complex coordinates in this neighbourhood. By definition, the point q lies on the polar when the curve C_t through this point is singular there, i.e. when $\frac{\partial f}{\partial x}(q) = \frac{\partial f}{\partial y}(q) = 0$, or when the curves C_t and L_c through q are not transverse, i.e. when the Jacobian of f and x vanishes, and hence

$$\det \begin{pmatrix} \frac{\partial f}{\partial x} & \frac{\partial f}{\partial y} \\ 1 & 0 \end{pmatrix} = 0.$$

In this way we obtain the local equation of the polar to be

$$\frac{\partial f}{\partial y} = 0.$$

The definition obviously generalises the above affine description of the polar of a plane projective algebraic curve C : the family of smooth curves L_c is the family of parallel lines through the pole at infinity, and is given canonically thereby. The family $\{C_t\}$ indeed depends on a choice, but the polar is uniquely determined by $C = C_0$.

In contrast to this, in the general situation, which is the basis for the definition above, the polar is certainly not determined by C_0 alone, but depends on the choice of families $\{C_t\}$ and $\{L_c\}$.

Now that we have defined polars in general, we want to compute their intersection numbers with the curves C_t, because it is to this we have reduced the proof of formula (i) of Theorem 2, in the remarks after its statement. First we prove the following simple result :

Lemma 3 :

Let $\tilde{C}$ be the polar of the family of curves $\{C_t\}$ relative to the family of smooth curves $\{L_c\}$ on the surface V, and let q be any point of V. Let C_{t_0} resp. L_{c_0} be the curves of the two families through q. Then the difference of the intersection numbers,

$$\nu_q(C_{t_0},\tilde{C}) - \nu_q(C_{t_0},L_{c_0})$$

depends only on the singularity of the curve C_{t_0} at q and not on the choice of the families $\{C_t\}$ and $\{L_c\}$.

Proof :

We compute the intersection numbers using 8.3.14(iii). Let C_i be the branch of C_{t_0} through q and let $\pi_i : (\mathbb{C},0) \to (V,q)$ be the resolution of singularities. The intersection multiplicity of a curve with equation $\phi = 0$ with the branch C_i at q is then the order $0(\phi \circ \pi_i)$ at the origin.

First we show the – trivial – independence of the intersection numbers from the choice of the family $\{C_t\}$. Suppose the family $\{C_t\}$ is described locally by the equations $f(x,y) = t$, where local coordinates are chosen as above ; then $\tilde{C}$ has equation $\frac{\partial f}{\partial y} = 0$. If we now go to another family $\{C'_s\}$ with local equation $f'(x,y) = s$, then $f' = uf$, for some nonvanishing function u, holds in a neighbourhood of q. Thus the new polar $\tilde{C}'$ has equation $\frac{\partial f'}{\partial y} = u \cdot \frac{\partial f}{\partial y} + f \cdot \frac{\partial u}{\partial y}$. Therefore $\frac{\partial f'}{\partial y} \circ \pi_i = u \cdot \frac{\partial f}{\partial y} \circ \pi_i$, and it follows that $\nu_q(C_i,\tilde{C}') = \nu_q(C_i,\tilde{C})$, which shows the independence of the intersection numbers in question from the choice of $\{C_t\}$.

Now we prove the independence of the difference of intersection numbers, $\nu_q(C_{t_0},\tilde{C}) - \nu_q(C_{t_0},L_{c_0})$, from the family $\{L_c\}$. Here the individual intersection numbers may very well depend on the choice of $\{L_c\}$, but their difference is invariant. Let $\{L'_b\}$ be a second family of smooth curves, and let $\tilde{C}'$ be the associated polar. Since one can always compare each family with a third locally defined family whose curves cut both the others transversely, we can assume without loss of generality that the curves L_{c_0} and L'_{b_0} through q already meet transversely at q. Then we can choose local coordinates x, y in a neighbourhood of q so that L_c is given by $x = c$ and L'_b is given by $y = b$. If $f(x,y) = t$ is the equation of C_t, then the two polars $\tilde{C}$ and $\tilde{C}'$ have the equations $\frac{\partial f}{\partial y} = 0$ resp. $\frac{\partial f}{\partial x} = 0$. Suppose the resolution π_i is described by two power series $x_i(t)$ and $y_i(t)$. Of course,

$$f(x_i(t), y_i(t)) \equiv 0.$$

Differentiation with respect to t then yields :

$$\frac{\partial f}{\partial x}(x_i(t), y_i(t)) \cdot x_i'(t) + \frac{\partial f}{\partial y}(x_i(t), y_i(t)) \cdot y_i'(t) \equiv 0,$$

whence it follows trivially that

$$O\left(\frac{\partial f}{\partial x} \circ \pi_i\right) + O((x-c_0) \circ \pi_i) = O\left(\frac{\partial f}{\partial y} \circ \pi_i\right) + O((y-b_0) \circ \pi_i).$$

If we interpret these orders by 8.3.14, as above, as intersection numbers, then it follows that

$$\nu_q(C_i, \tilde{C}') + \nu_q(C_i, L_{c_0}) = \nu_q(C_i, \tilde{C}) + \nu_q(C_i, L'_{b_0})$$

and the invariance of the difference of intersection numbers is proved.

In the following lemma we compute the number $\nu_q(C_{t_0}, \tilde{C}) - \nu_q(C_{t_0}, L_{c_0})$.

<u>Lemma 4</u> :

Under the same general assumptions as in Lemma 3, the intersection number of C_{t_0} with the polar $\tilde{C}$ relative to the family of smooth curves $\{L_c\}$ satisfies

$$\nu_q(C_{t_0}, \tilde{C}) - \nu_q(C_{t_0}, L_{c_0}) = \sum_{j \geq 0} \nu_j(\nu_j - 1) - \rho,$$

where ρ is the number of branches of C_{t_0} at q and the sum is of the multiplicities ν_j of all infinitely near singularities of (C_{t_0}, q).

<u>Proof</u> :

By Lemma 3 we can assume, without loss of generality, that $\tilde{C}$ is a "general" polar, i.e. that the tangent of L_c does not coincide with the tangent of a branch of (C_{t_0}, q). Then $\nu_q(C_{t_0}, L_c) = \nu_0$, where

ν_0 is the multiplicity of C_{t_0} at q, and $\nu_q(\tilde{C}) = \nu_0 - 1$. We introduce a σ-process at q and investigate how the intersection multiplicities of the curves in question behave under this process, so as to carry out a proof by induction.

Let x, y be local coordinates around q, such that C_{t_0} has equation $f(x,y) = 0$, L_{c_0} has equation $x = 0$ and $\tilde{C}$ has equation $\frac{\partial f}{\partial y}(x,y) = 0$. Let C' resp. $\tilde{C}'$ be the proper preimages of C_{t_0} resp. $\tilde{C}$ under the σ-process. The infinitely near points of (C_{t_0}, q) on C' lie in a coordinate neighbourhood on the blown-up surface with coordinates u, v, where

$$x = v$$
$$y = uv.$$

The exceptional curve E has the equation $v = 0$.

If we expand f in a power series :

$$f(x,y) = \sum_{i+j\geq\nu_0} a_{ij}x^i y^j,$$

then C' has equation $\tilde{f}(u,v) = 0$ with

$$\tilde{f}(u,v) = \sum_{i+j\geq\nu_0} a_{ij}v^{i+j-\nu_0}u^j.$$

Since $\nu_q(\tilde{C}) = \nu_0 - 1$, $\tilde{C}'$ has equation $\phi(u,v) = 0$ where $\phi(u,v) = v^{-(\nu_0-1)} \cdot \frac{\partial f}{\partial y}(v,uv)$, hence

$$\phi(u,v) = \sum_{i+j\geq\nu_0} ja_{ij}v^{i+j-\nu_0}u^{j-1}.$$

Thus we see that $\phi = \frac{\partial \tilde{f}}{\partial u}$. This means that $\tilde{C}'$ is the polar of the system of curves $\tilde{f}(u,v) = s$ relative to the system of smooth curves L'_b defined by $v = b$. And the curve L'_0 through the singular points of C' is just the exceptional curve E. In computing the intersection numbers of C' and $\tilde{C}'$ at the singular points of C' we can therefore assume the formula of Lemma 4 by induction. If we sum over the intersection numbers thus computed for all singular points of C' on E, then we obtain

$$\nu(C',\tilde{C}') = \sum_{j>0} \nu_j(\nu_j-1) - \rho + \nu(C',E)$$
$$= \sum_{j>0} \nu_j(\nu_j-1) - \rho + \nu_0. \qquad (1)$$

On the other hand, the Noether method for computing the intersection numbers gives (cf. proof of 8.4.13)

$$\nu_p(C_{t_0},\tilde{C}) = \nu(C',\tilde{C}') + \nu_0(\nu_0-1). \qquad (2)$$

Equations (1) and (2) immediately yield the formula claimed :

$$\nu_p(C_{t_0},\tilde{C}) - \nu_0 = \sum_{j\geq 0} \nu_j(\nu_j-1) - \rho.$$

With the help of the formula just proved we can now easily prove the general formula for the reduction of the class due to singular points.

Proof of Theorem 2 :

Proof of (i) :

In the remarks following the statement of Theorem 2 we have already shown the following : if C is a curve without multiple components and without lines as components, of order n and class n', and if $\tilde{C}$ is the polar of C relative to a sufficiently general point, then

$$n' = n(n-1) - \Sigma\nu_{p_i}(C,\tilde{C}), \qquad (1)$$

where the summation is over all singular points of C. But we have just computed the numbers $\nu_{p_i}(C,\tilde{C})$ in Lemma 4 :

$$\nu_{p_i}(C,\tilde{C}) = \Sigma\nu_{ij}(\nu_{ij}-1) + \Sigma(\nu_k-1) \qquad (2)$$

where the first summation involves the multiplicities ν_{ij} of all infinitely near singular points of (C,p_i) and the second involves the multiplicities ν_k of the branches of C going through p_i. The formula (i) to be proved follows immediately from (1) and (2).

Proof of (ii) :

We could derive (ii) in a similar way to (i) by computation of the intersection multiplicities of the curve C with its Hessian, as we have already indicated earlier. But, in order to save work, we shall instead derive (ii) from (i) and the genus formula which will be proved in the next section.

The genus formula says that the genus p of C satisfies :

$$2p-2 = n(n-3) - \Sigma\nu_{ij}(\nu_{ij}-1). \qquad (1)$$

The generalised Plücker formula (i) yields the equation

$$0 = n(n-1) - n' - \Sigma\nu_{ij}(\nu_{ij}-1) - \Sigma(\nu_k-1). \qquad (2)$$

If one subtracts (2) from (1) then the result is

$$2p - 2 = n' - 2n + \Sigma(\nu_k - 1). \qquad (3)$$

Correspondingly, the genus p' of the dual curve C' obviously satisfies

$$2p' - 2 = n - 2n' + \Sigma(\nu'_\kappa - 1) \qquad (4)$$

where the summation is over the multiplicities ν'_κ of all singular branches of C', i.e. over the classes ν'_κ of all branches of C with $\nu'_\kappa > 1$.

Now genus is a birational invariant. This is evident from our definition in the next section. If one follows Max Noether in taking (1) as the definition of genus, then one can also prove the birational invariance purely algebraic-geometrically. Moreover, the curves C and C' are obviously birationally equivalent : one obtains a birational equivalence when one associates the points of C with their tangents, considered as points of C'. Thus we certainly have

$$p = p'. \qquad (5)$$

If one uses (5) to eliminate n' from (3) and (4), then one obtains

$$3(2p-2) = -3n + 2 \cdot \Sigma(\nu_k - 1) + \Sigma(\nu'_\kappa - 1). \qquad (6)$$

If one substitutes the expression for $2p-2$ given by (1), then the result is

$$3n(n-2) - 3 \cdot \Sigma\nu_{ij}(\nu_{ij} - 1) - \Sigma(2\nu_k - 2) = \Sigma(\nu'_\kappa - 1). \qquad (7)$$

In (7), we split the sum $\Sigma(\nu'_\kappa - 1)$ in two summands. One contains all the terms $\nu'_\kappa - 1$ which correspond to regular branches of C with $\nu'_\kappa > 1$. This sum equals the desired number, s', of inflection points. The second summand contains all terms $\nu'_\kappa - 1$ which correspond to singular branches of C. We take this sum to the other side of the equation. Then the result is just the formula to be proved :

$$s' = 3n(n-2) - 3 \cdot \Sigma\nu_{ij}(\nu_{ij} - 1) - \Sigma(2\nu_k + \nu'_\kappa - 3).$$

This proves the generalised Plücker formulae, and here we want to leave this theme for the time being. We shall return to it once again in the last chapter, in connection with the term "Plücker equivalent".

We conclude this section with a few supplementary remarks on the concept of polar and the related concept of adjoint curves. Polars are, as we shall see, special cases of adjoint curves. The concept of adjoint curve goes back to Riemann's fundamental work on abelian functions

[R3], and in the general form given here, probably to Brill and Noether (cf. e.g. [N5], IV).

In order to express the definition of adjoint curve precisely, we must first generalise our considerations of 6.2 somewhat. There we obtained conditions on linear systems of curves, under which curves of a system through certain points p_i of the plane have multiplicities at these points greater than or equal to ν_i. We shall now generalise these conditions by requiring the curves to have certain multiplicities at infinitely near points of the p_i. In what follows we always understand "curves" to include curves with multiple components.

<u>Definition</u>:

Let $X_0 \xleftarrow{\phi_1} X_1 \xleftarrow{\phi_2} X_2 \leftarrow \cdots \xleftarrow{\phi_N} X_N$ be a sequence of holomorphic mappings of complex surfaces, where $\phi_{i+1} : X_{i+1} \to X_i$ is the blowing up of finitely many distinct points $p_{ik} \in X_i$, and where the points blown up by ϕ_{i+2} lie on the exceptional lines $E_{ik} = \phi_{i+1}^{-1}(p_{ik})$ introduced by ϕ_{i+1}. For each point p_{ik} let a natural number ν_{ik} be given. We make the following definition by recursion on the length N of the sequence of blowing up processes :

A curve C_0 in X_0 has <u>multiplicity at least</u> ν_{ik} <u>at the infinitely near points</u> p_{ik} when

(i) C_0 has multiplicity at least ν_{0k} at the points $p_{0k} \in X_0$.

(ii) Let C_1 in X_1 be the curve $C_1 = \phi_1^{-1}(C_0) - \Sigma\nu_{0k}E_{0k}$, where $\phi_1^{-1}(C_0)$ denotes the total preimage of C_0. Then C_1 has multiplicity at least ν_{ik} at the infinitely near points p_{ik} with $i \geq 1$.

One easily convinces oneself that here, as in 6.2, we are dealing with linear conditions and consequently, as a generalisation of 6.2.1, one has put at most

$$\tfrac{1}{2}\Sigma\nu_{ik}(\nu_{ik}+1)$$

linearly independent conditions on the curve C_0.

<u>Definition</u> :

Let C be a plane curve without multiple components. A plane curve $\tilde{C}$ is called <u>adjoint</u> to C when $\tilde{C}$ has, at all infinitely near singular points p_{ij} of C with multiplicity ν_{ij}, multiplicity at least $\nu_{ij} - 1$. $\tilde{C}$ may have multiple components.

For each curve there are adjoint curves, because :

<u>Proposition 5</u> :

Each polar of C is an adjoint curve of C.

<u>Proof</u> :

If $f(x,y) = 0$ is an affine equation of C then, by an earlier remark, the equation of the polar has the form

$$a \cdot \frac{\partial f}{\partial x} + b \cdot \frac{\partial f}{\partial y} + c(nf - x \cdot \frac{\partial f}{\partial x} - y \cdot \frac{\partial f}{\partial y}) = 0,$$

with arbitrary constants a, b, c. Thus it suffices to prove the following more general local statement : let C_0 be an analytic curve in a neighbourhood of the origin in $\mathbb{C}^2$ with equation $f_0(x,y) = 0$, and let $\tilde{C}_0$ be a curve with equation $g_0(x,y) = 0$, where g_0 is of the form

$$g_0 = a_0 \cdot \frac{\partial f_0}{\partial x} + b_0 \cdot \frac{\partial f_0}{\partial y} + c_0 f_0$$

where a_0, b_0, c_0 are any analytic functions. Then $\tilde{C}_0$ has, at each infinitely near singular point of $(C_0, 0)$ with multiplicity ν, multiplicity at least $\nu - 1$. Proof of this assertion : for the point $0 \in \mathbb{C}^2$ it is clear that $\nu_0(\tilde{C}_0) \geq \nu_0(C_0) - 1$ holds there. For the remaining infinitely near points the assertion follows by induction on the length of the resolution process for the singularity :

If we blow up the origin, and hence set

$$x = uv$$
$$y = v$$
$$f_1(u,v) = v^{-n} f_0(uv,v)$$
$$g_1(u,v) = v^{-n+1} g_0(uv,v)$$

where $n = \nu_0(C_0)$, then it follows by an easy calculation that

$$g_1(u,v) = (a_0 - b_0 u) \cdot \frac{\partial f_1}{\partial u} + (b_0 v) \cdot \frac{\partial f_1}{\partial v} + (c_0 v + n b_0) \cdot f_1 .$$

The new function g_1 is therefore again of the form

$$g_1 = a_1 \cdot \frac{\partial f_1}{\partial u} + b_1 \cdot \frac{\partial f_1}{\partial v} + c_1 f_1$$

with suitable functions a_1, b_1, c_1 of u, v. The assertion about multiplicities then follows from the induction hypothesis, because $f_1(u,v) = 0$ is the equation of the proper preimage C_1 of C_0 and $g_1(u,v) = 0$ is the equation of the curve $\tilde{C}_1$, which is the total preimage of $\tilde{C}_0$ minus the n-tuple exceptional curve.

The polars of a curve C of order n therefore form a linear system of adjoint curves of order $n-1$. This linear system of polars has dimension ≤ 2, whereas the dimension of the linear system of all adjoint curves of order $n-1$ is greater, in general. Because it follows from the above-mentioned generalisation of 6.2.1 that the curves of order m adjoint to C form a linear system of dimension at least

$$\frac{m(m+3)}{2} - \tfrac{1}{2} \cdot \Sigma\nu_i(\nu_i-1),$$

where the ν_i are the multiplicities of the infinitely near points of C. It then follows, using the generalised Plücker formulae to make estimates when $m = n-1$, that the dimension is greater than $n-2$. The smaller the order m is, the smaller the dimension of the linear system of adjoint curves. This raises the question : from which m upwards does each curve of order n have at least one adjoint curve? If we also admit reducible curves, then the answer is : $m = n-1$. This is shown by the example of a curve C consisting of n lines in general position : on each line there are $n-1$ double points, so that an adjoint curve of order $m < n-1$ must contain all these n lines, contrary to $m < n-1$. In order to exclude such trivialities, we consider only irreducible curves from now on. In this case the answer to our question is : $m = n-3$, provided we exclude the rational curves, which have no adjoint curves of order $n-3$.

We first show, by means of an example, that there is an irreducible nonrational curve of order $n = 6$ for which no adjoint curve of order $m = n - 4 = 2$ exists. Our example is the dual curve C of a nonsingular cubic.

C is birationally equivalent to the cubic, and hence itself irreducible and non-rational. We know that C is a sextic with 9 cusps, which correspond to the nine inflection tangents of the cubic. The coordinates of these 9 cusps (for a cubic in normal form) are well-known from our earlier investigations in 7.3. One sees immediately that these 9 cusps cannot lie on a quadric. (Proof : they lie in threes on each of the three lines $x_0 = 0$, $x_1 = 0$, $x_2 = 0$. If they also lay on a quadric, then the latter would have to contain each of the three lines.) But an adjoint curve of order 2 would be a quadric through the cusps, and hence it does not exist.

Thus we consider the adjoint curves of order $n-3$. In [N5], section 30, Noether showed that for an irreducible curve C of order n the multiplicity conditions for adjoint curves of order $m \geq n-3$ are linearly independent. In particular, one obtains the dimension of the vector space of homogeneous polynomials of degree n-3 which describe adjoint curves as

$$p = \frac{n(n-3)}{2} + 1 - \tfrac{1}{2}\,\Sigma\nu_i(\nu_i-1),$$

where the summation extends over the multiplicities of all infinitely near singular points of C. In the next section we shall prove that this number is the genus of C, so $p > 0$ in our particular case and $p = 0$ only for rational curves. If we agree, in the case of a non-singular cubic, where $n = 3$ and $p = 1$, to admit an adjoint curve which is the zero set of a nonvanishing homogeneous polynomial of degree $n - 3 = 0$, and hence empty, then we have the following result.

Proposition 6 :

For each irreducible curve C of order n, the maximum number p of linearly independent homogeneous polynomials $\phi_1,\ldots,\phi_p$ of degree n-3 which describe adjoint curves of C equals

$$p = \frac{n(n-3)}{2} + 1 - \tfrac{1}{2}\,\Sigma\nu_i(\nu_i-1),$$

where the summation is over the multiplicities of all infinitely near singular points of C. (This number is the genus of C.)

In [N5], Noether showed how one could explicitly compute the homogeneous polynomials $\phi_1,\ldots,\phi_p$ for a given curve equation, and indeed by rational operations, which ultimately depend only on the solution of linear systems of equations. In what follows we shall call such polynomials "adjoint polynomials" for short.

The significance of the adjoint polynomials $\phi_1,\ldots,\phi_p$ of degree n-3 for a curve C of degree n and genus p lies in the fact that one can use them to describe explicitly the holomorphic differential forms on the Riemann surface C' which results from resolution of singularities. This was already seen by Riemann, and was the reason he introduced adjoint polynomials in [R3], section 9.

We shall be concerned with this description of holomorphic differential forms in the section after next.

9.2 The formulae of Clebsch and Noether

In 3.4 we have already shown, by an elementary analysis of a simple example, that the topological type within a family of curves changes in the passage from regular to singular curves. We shall now get a quantitative hold on this phenomenon, by means of formulae which go back to Clebsch and M. Noether.

Let $f : X \to S$ be a family of curves; more precisely, let f be a proper holomorphic mapping of the complex manifold X onto the disc $S := \{t \in \mathbb{C} \mid |t| < 1\}$. For $t \neq 0$ let the fibres $X_t = f^{-1}(t)$ be smooth irreducible curves, and suppose $X_0 = f^{-1}(0)$ has only finitely many singular points $p_1, \ldots, p_r$. For each of these points p_i we choose local complex coordinates and a small ball B_i around p_i, where the ball is taken with respect to the local coordinates and is small in the sense of 8.3.6. We want to compare the singular fibre X_0 with the neighbouring regular fibres X_t, $t \neq 0$. To do this we choose $|t|$ sufficiently small that X_t cuts the balls B in the way described in 8.5.1.

Our basic idea for investigating the passage from the general, regular, fibre X_t to the special, singular, fibre X_0 is quite simple. We decompose X_t and X_0 into their parts X_t' resp. X_0' outside the balls B_i and their parts Y_t resp. Y_0 inside the balls B_i. By 8.5.1, the connected components of Y_t are diffeomorphic to a Milnor fibre F_i of the singularity (X_0, p_i).

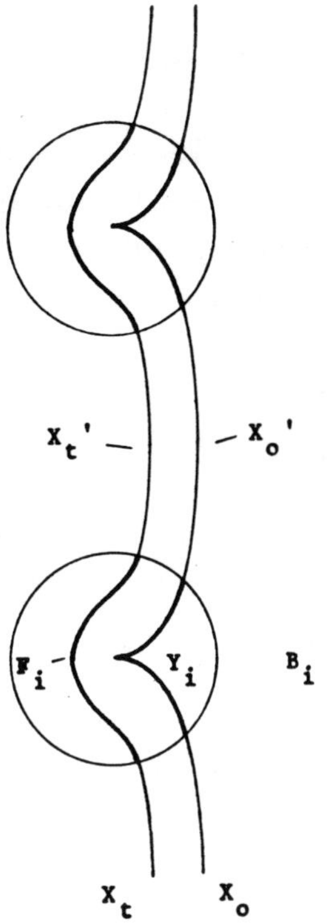

If we put this information together, then we obtain the following result on the relation between the Euler characteristics of the singular and nonsingular fibres :

<u>Lemma 1</u> :

Under the above hypotheses :

$$\chi(X_0) = \chi(X_t) - \sum_{i=1}^{r} (\chi(F_i)-1).$$

<u>Proof</u> : We set

$$X_t' := X_t - \bigcup_{i=1}^{r} \mathring{B}_i$$

$$X_0' := X_0 - \bigcup_{i=1}^{r} \mathring{B}_i.$$

Also let

$Y_i := B_i \cap X_0$ and

$F_i := B_i \cap X_t$.

By Theorem 8.3.8, Y_i is diffeomorphic to the cone over $Y_i \cap \partial B_i$, hence contractible, while F_i – as we have already said – is diffeomorphic to a Milnor fibre of the singularity of X_0 at p_i. The intersections $F_i \cap \partial B$ resp. $Y_i \cap \partial B$ are disjoint unions of circles, and hence have Euler characteristic 0. It therefore follows from the additivity of the Euler characteristic (8.5.2) that

$$\chi(X_t) = \chi(X_t') + \sum_{i=1}^{r} \chi(F_i)$$

$$\chi(X_0) = \chi(X_0') + \sum_{i=1}^{r} \chi(Y_i).$$

Now X_t' may be deformed into X_0', and hence $\chi(X_t') = \chi(X_0')$. This gives us the desired relation between $\chi(X_0)$ and $\chi(X_t)$:

$$\chi(X_0) = \chi(X_t) - \sum_{i=1}^{r} (\chi(F_i) - \chi(Y_i)) = \chi(X_t) - \sum_{i=1}^{r} (\chi(F_i) - 1).$$

The formula of Lemma 1 is good for computing $\chi(X_0)$, because in 8.5 we have given a series of ways for computing the Euler characteristics $\chi(F_i)$.

Later we shall show that one can view each plane algebraic curve C as the fibre of a family of curves in the way described above, and indeed in such a way that each of these curves may be embedded as a plane curve in $P_2(\mathbb{C})$ with the same order as C.

With the help of Lemma 1 we can then obtain information on the topological structure of C. In this sense (but not in the historical sense), Lemma 1 is the "prototype" of the genus formulae of Clebsch and Noether.

In this course we shall confine ourselves to determining the topological structure of plane algebraic curves C. Thus in future we shall say no more about the embedding of C as a topological space in the projective plane, even though there are many interesting problems here, some still unsolved. For example, in past years many mathematicians were interested in the question of when the fundamental group $\pi_1(P_2(\mathbb{C})-C)$ is abelian, and we want to say at least a few words about this problem here.

In 1933, E. van Kampen developed the following process for obtaining a presentation of the fundamental group by generators and relations (see [Z2], VIII.1).

Let L be a line in $P_2(\mathbb{C})$ which is transverse to C, i.e. for which $L \cap C$ consists of n distinct points $p_1,\ldots,p_n$, where n is the order of C. One chooses a base point $q \in L - C$ and paths γ_i in $L - C$ which go from q, without self-crossings, near to the point p_i, run once around p_i on a small circle, and then return to q on the same path. One can choose these paths so that their product $\gamma_1 \ldots \gamma_n$ is null homotopic on the n-tuply perforated 2-sphere ; then $\pi_1(L-C,q)$ is freely generated by the homotopy classes β_i of the γ_i $(i = 1,\ldots,n-1)$. One can now show that the inclusion $L - C \subset P_2(\mathbb{C}) - C$ induces a surjection of the fundamental groups $\pi_1(L-C,q) \to \pi_1(P_2(\mathbb{C})-C,q)$. Thus the images α_i of the β_i are generators of $\pi_1(P_2(\mathbb{C})-C,q)$.

One obtains relations between the α_i by letting the line L vary. Let $\mathcal{L}$ be the set of all lines through q which meet C in exactly n points. If L_t $(t \in [0,1])$ is a closed path in $\mathcal{L}$ with $L_0 = L_1 = L$, then there are continuous mappings $p_i : [0,1] \to C$ such that $p_i(0) = p_i$ and $L_t \cap C = \{p_1(t),\ldots,p_n(t)\}$ for all $t \in [0,1]$. Just as in 8.3 the curves $p_i(t)$ define a braid z, but this time not in the complex plane $\mathbb{C}$ but in the one-dimensional complex projective space L. One can now construct a family h_t of diffeomorphisms $h_t : L \to L_t$ such that $h_t(p_i) = p_i(t)$, $h_t(q) = q$ and $h_0 : L \to L$ is the identity ([B8], 9.5). Obviously the path $h_1 \circ \gamma_i$ in $P_2(\mathbb{C}) - C$ is then homotopic to γ_i. The homotopy class of the path $h_1 \circ \gamma_i$, which runs wholly in $L - C$, may be expressed as a word $w_i(\beta_1,\ldots,\beta_{n-1})$ in the generators β_i of $\pi_1(L-C,q)$. Then in $\pi_1(P_2(\mathbb{C})-C,q)$ we obviously have the relations $w_i(\alpha_1,\ldots,\alpha_{n-1}) = \alpha_i$. The word w_i is determined from the braid z algorithmically*, and in 8.3 we have described in principle how one can "compute" the braid z.

The construction just described yields a relation between the generators α_i for each closed path in $\mathcal{L}$, and van Kampen's theorem says that these relations already suffice to determine $\pi_1(P_2(\mathbb{C})-C,q)$.

If one applies van Kampen's method to a curve C which consists of a union of n lines, no three of which meet at a point, then one can

---

*See, e.g. : Looijenga, E. : The complement of the bifurcation variety of a simple singularity. Inv. Math. 23 (1973), 105-116, §3.

show that in this case $\pi_1(P_2(\mathbb{C})-C,q)$ is a free abelian group of rank n-1. Generalising this result, O. Zariski formulated the following theorem in 1929 : if C is a plane algebraic curve with only ordinary double points as singularities, then $\pi_1(P_2(\mathbb{C})-C,q)$ is abelian. His proof later turned out to have a gap (see [Z2], VIII.2, and the footnote there), and this theorem was first given a complete proof recently (Deligne, P. : Le group fondamentale du complément d'une courbe plane n'ayant que des points doubles ordinaires est abélien. Séminaire Bourbaki 1979/80, exposé 543).

After this short digression on the subtle questions concerning the embedding of a plane curve C in $P_2(\mathbb{C})$, we now confine ourselves entirely to the investigation of the topology of C itself.

Our examples in 3.4, and the reflections in connection with local resolution of singularities in 8.3, have already shown us how one has to analyse the topology of a curve C. The different irreducible components of C meet in finitely many points, and hence it is sufficient here to investigate only irreducible curves C. We shall therefore confine ourselves to this case in what follows.

It then becomes useful to understand a little more clearly what irreducibility of curves means. Accordingly, we prove the following two lemmas.

Lemma 2 :

A complex projective algebraic curve in $P_2(\mathbb{C})$ is irreducible as an algebraic curve just in case it is irreducible as an analytic set in $P_2(\mathbb{C})$.

Proof :

The following would be a very simple proof : it is trivial that an analytic irreducible curve is also algebraically irreducible. Conversely, if C is an analytically reducible curve, and hence the union of two proper analytic subsets C' and C'', then C' and C'' are projective algebraic by Chow's theorem, and hence C is also reducible as an algebraic curve.

However, we have not proved Chow's theorem in this course, and for that reason we want to prove directly that algebraic reducibility follows from analytic reducibility.

By 4.4.6 and 5.1.2 it suffices to show that analytically reducible complex affine algebraic curves are also algebraically reducible. Let C be a curve of order k which is analytically reducible. For suitable coordinates x, y in $\mathbb{C}^2$, C is the zero set of a polynomial

$$f(x,y) = y^k + c_1(x)y^{k-1} + \dots + c_k(x).$$

Let $x_1,\dots,x_r \in C$ be the points at which this polynomial has multiple zeroes, let $\tilde{\mathbb{C}} = \mathbb{C} - \{x_1,\dots,x_r\}$, and let $\tilde{C}$ be the preimage of $\tilde{\mathbb{C}}$ under the mapping $C \to \mathbb{C}$ defined by $(x,y) \mapsto x$.

$$\tilde{C} \to \tilde{\mathbb{C}}$$

is an unbranched covering of degree k. For each $x_0 \in \tilde{\mathbb{C}}$ there is a neighbourhood U such that exactly k disjoint sheets of $\tilde{C}$, whose y-coordinates are holomorphic functions y_i , $i = 1,\dots,k$, lie over U. For $x \in U$ the values $y_i(x)$ are the zeroes of the polynomial, and the functions y_i are uniquely determined up to order. By analytic continuation along paths in $\tilde{\mathbb{C}}$, the y_i define many-valued algebraic functions, and when we continue the y_i along a closed path with initial and final point x_0 , the y_i are permuted among themselves. The coefficients c_j of the polynomial are the elementary symmetric functions of the y_i. So much for the geometric and analytic description of the covering given by the mapping $C \to \mathbb{C}$ of our curve.

Now we suppose that C splits analytically into two complex analytic curves C' and C'' : $C = C' \cup C''$. We want to show, first, that this leads to a corresponding decomposition of the set of roots $\{y_1,\dots,y_k\}$, and then to a decomposition $f = g \cdot h$ into two polynomials g and h. It is clear that for each neighbourhood U of an $x_0 \in \tilde{\mathbb{C}}$ the set of roots $y_1,\dots,y_k$ divides into two subsets : the functions which describe sheets over C', and those which describe sheets over C''. Now this decomposition is invariant under analytic continuation along paths in $\tilde{\mathbb{C}}$. This is because $\tilde{C}$ decomposes into the disjoint open sets $\tilde{C}' = \tilde{C} \cap C'$ and $\tilde{C}'' = \tilde{C} \cap C''$, so that lifting a path from $\tilde{\mathbb{C}}$ to $\tilde{C}$ results in a path which lies wholly in $\tilde{C}'$ or wholly in $\tilde{C}''$. In this way we obtain a decomposition of the set $\{y_1,\dots,y_k\}$ of many-valued algebraic functions on $\tilde{\mathbb{C}}$ into two subsets, $\{y_1,\dots,y_m\}$ and $\{y_{m+1},\dots,y_k\}$, which is invariant under analytic continuation. Because of this invariance, the elementary symmetric functions $a_1,\dots,a_m$ of $y_1,\dots,y_k$ and $b_1,\dots,b_{k-m}$ of $y_{m+1},\dots,y_k$ are single-valued functions on C. But more than this is true : they are even polynomials.

Because they are entire functions whose absolute value increases with x at most as fast as a certain power of $|x|$, and one knows from function theory that functions with this rate of growth are polynomials. (The growth property follows from the fact that the y_i are algebraic functions. Because the root y of any polynomial $y^k + c_1y^{k-1}+\ldots+c_k = 0$ satisfies $|y| < \max_j \sqrt[j]{k|c_j|}$, since the y with greatest absolute value satisfies

$$|c_1 \cdot \frac{1}{y} + \ldots + c_k \cdot \frac{1}{y^k}| < \frac{1}{k} + \ldots + \frac{1}{k} = 1$$

and hence $|c_1y^{k-1} + \ldots + c_k| < |y^k|$. It follows from this bound that when the $c_j(x)$ are polynomials, and hence growing at most like powers of x, then the same is true for the algebraic functions $y_i(x)$.)

With the help of the polynomials $a_1,\ldots,a_m$ and $b_1,\ldots,b_{k-m}$ just obtained, we now define polynomials g and h in the variables x and y by

$$g(x,y) = y^m + a_1(x)y^{m-1} + \ldots + a_m(x)$$

$$h(x,y) = y^{k-m} + b_1(x)y^{k-m-1} + \ldots + b_{k-m}(x).$$

By construction, these polynomials satisfy

$$f = g\cdot h,$$

since the zeroes of g and h together are just the zeroes of f, hence we have found the desired product decomposition of f, and Lemma 2 is proved.

Now let $\pi : \tilde{C} \to C$ be a resolution of singularities of C. We know, by 8.3 and 8.4.9, that such a resolution of singularities always exists, and it follows immediately from the Riemann extension theorem that it is essentially unique. We also know, from our local description of the resolution in 8.3 and 8.4, to what extent C and $\tilde{C}$ differ topologically : C results from $\tilde{C}$ when the preimage points of each singular point p in C, which correspond to the branches of $\tilde{C}$ at p, are identified. $\tilde{C}$ is a 2-dimensional compact real manifold, hence the positions of the points to be identified do not matter from the topological standpoint, provided only that they lie in the right connected component. As far as the connectivity is concerned, we now have the following :

<u>Lemma 3</u> :

Let C be a plane algebraic curve and let $\pi : \tilde{C} \to C$ be the resolution of singularities. Then the connected components of $\tilde{C}$ correspond bijectively to the irreducible components of C. In particular, $\tilde{C}$ is connected just in case C is irreducible.

<u>Proof</u> :

The image of each connected component $\tilde{C}_i$ of $\tilde{C}$ under π is a compact, and hence closed, 1-dimensional analytic subset C_i of $P_2(\mathbb{C})$. Different connected components give different image curves. The C_i are analytically irreducible, since $\tilde{C}_i \to C_i$ is a resolution and $\tilde{C}_i$ is connected. By Lemma 2, the irreducible analytic components of C are also algebraically irreducible components, and so everything is proved.

The resolution $\tilde{C}$ of an irreducible curve C is therefore a compact, connected, orientable surface, and the topology of C is known as soon as one knows the topology of $\tilde{C}$ and the number of branches at the singular points of C.

But it is well known how one classifies the compact, orientable, connected surfaces (cf., e.g., [S3] or [B6]). Each such surface is homeomorphic to a surface which one obtains by attaching a certain number of handles to a 2-sphere, and the number p of these handles is an invariant which classifies the compact, orientable, connected surfaces up to homeomorphism.

p=0 **p=1** **p=2**

The number p is called the <u>genus</u> of the surface. It equals half the middle Betti number b_1. Since, of course, $b_0 = b_2 = 1$, it follows that the Euler characteristic

$$\chi = 2 - 2p.$$

After all these preparations it is clear that the genus of the resolution must be an important invariant for an irreducible plane algebraic curve C, since it contains essential qualitative information about C.

Definition :

If $\tilde{C} \to C$ is a resolution of singularities of an irreducible projective algebraic curve C, then the genus p(C) is defined to be the topological genus of $\tilde{C}$.

This definition of genus goes back essentially to Riemann 1857. Riemann defined the "connectivity" of a surface with the help of the number of "loop cuts" which one needs to render the surface simply connected. For a surface of genus p one needs 2p cuts, and Riemann's "connectivity" is 2p+1. This definition is admittedly not identical with the one we have given, but it is likewise a topological definition, and its equivalence with our definition is easily proved.

Riemann recognised that the number p of a Riemann surface could also be described in an analytic way, namely as the number of linearly independent integrals of the first kind. In somewhat more modern terminology : the genus p equals the number of linearly independent differential forms of degree 1. This form of the definition of genus can be generalised to any n-dimensional nonsingular projective algebraic manifold V : let h^q be the dimension of the complex vector space of holomorphic q-forms on V. Then one defines the arithmetic genus χ_a of V by

$$\chi_a(V) := \sum_{q=0}^{n} (-1)^q h^q.$$

Thus for a curve C this definition gives $\chi_a(C) = 1-p$.

Riemann also realised the fundamental significance of genus for geometry on the curve and for function theory on the associated Riemann surface. In particular, he saw that genus is a birational invariant : when there is a birational transformation T of a curve C into a curve C', then C and C' have the same genus. (For plane curves, such a birational transformation need not come from a birational transformation of the plane.) From our standpoint, the birational invariance of genus is easy to see : a birational transformation $T : C \to C'$ lifts to a birational transformation $\tilde{T} : \tilde{C} \to \tilde{C}'$. But, for the complex 1-dimensional nonsingular algebraic manifolds $\tilde{C}$, $\tilde{C}'$, such a birational transformation $\tilde{T}$ must be biregular. In particular, $\tilde{C}$ and $\tilde{C}'$ are homeomorphic, and hence they have the same genus.

After Riemann it was Clebsch, above all, who established the significance of genus for geometry on the curve and function theory on the associated surface. The word "genus" ("Geschlecht") is also due to him (1864). An important example of this is the theorem which Brill and Noether named the Riemann-Roch theorem. A special case of this theorem essentially expresses the equivalence of the two definitions of genus given above : the topological one and the one involving holomorphic differential forms. This theorem was later generalised in different stages and in different ways by Hirzebruch, Grothendieck, Atiyah and Singer. A special case of the Riemann-Roch-Hirzebruch theorem says e.g. that for any projective-algebraic manifold V the analytically defined genus $\chi_a(V)$ equals the purely topological one, the Todd genus $T(V)$, defined as a certain linear combination of the Chern numbers. For projective-algebraic manifolds one can also define $\chi_a(V)$ purely algebraically, when one chooses a somewhat different kind of definition. For the latter definition, Grothendieck has given a purely algebraic-geometric proof of a theorem of the Riemann-Roch type. On the other hand, $\chi_a(V)$ may also be interpreted as the index of a certain elliptic differential operator on V, and Atiyah and Singer have generalised this interpretation to a general index theorem for elliptic differential operators on manifolds, which expresses the analytically defined index by a topological invariant.

These historical remarks should certainly suffice to show what an important invariant genus is.

In what follows, our goal is to compute the genus of a curve from its degree and the invariants of its singularities alone. For a smooth curve this is quite simple.

Lemma 4 :

Let C be a smooth plane projective-algebraic curve of order d. Then the genus of C is

$$p(C) = \frac{(d-1)(d-2)}{2} .$$

Proof :

In $P_2(\mathbb{C})$ we choose a point $p \notin C$ and a line L which does not meet p. Let π be the projection of C on L from centre p.

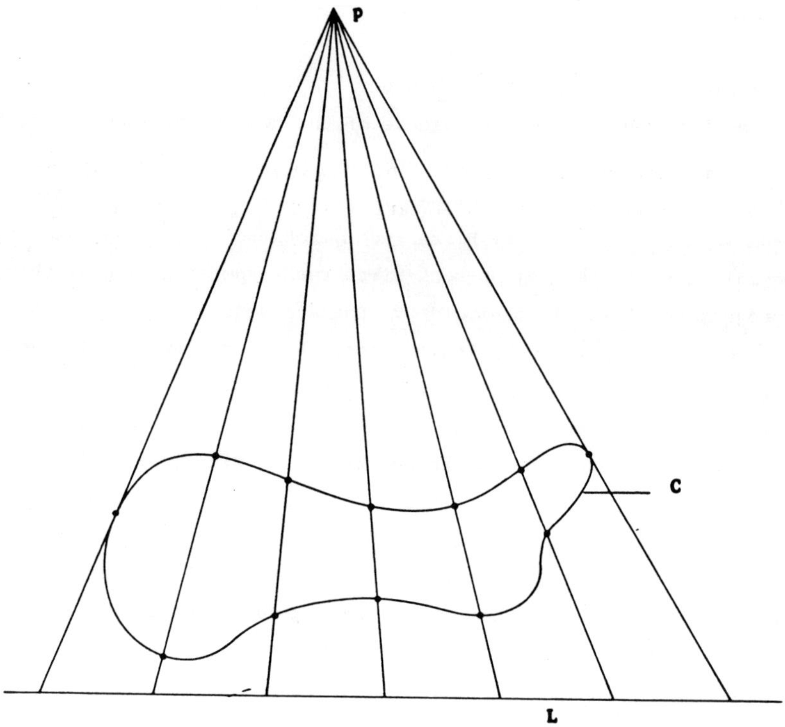

Then the projection is branched precisely where the ray of projection is tangential to the curve, and indeed the order of branching is just one less than the intersection multiplicity of the ray of projection with the curve C.

The number of branch points, counted with multiplicity in the sense of branching order, is therefore the number of intersections of the dual curve C^* (the curve in the dual projective plane consisting of all the tangents, see 6.2.5 and 9.1) with the line dual to p. But this is just the order of C^*, and hence the same as the class of C. In 6.2.7 we have computed the class of a smooth curve, it equals $d(d-1)$.

Thus we have represented C as a branched covering of the projective line L. The degree of the covering is d (a generic line of projection meets C in d points), and the number of branch points is d(d-1). Now we use the Hurwitz formula.

The <u>Hurwitz formula</u> says the following : if A is a branched covering of Riemann surface B, with covering degree n and r branch points (counting multiplicities), then

$$\chi(A) = n \cdot \chi(B) - r.$$

(Proof : one triangulates B in such a way that the images of the branch points are vertices of the triangulation. The triangulation may be lifted to a triangulation of A. In the process, the numbers of vertices, edges and faces are all multiplied by n, except that at a branch point $q \in B$, of order ν, $\nu+1$ vertices of the triangulation of A coincide. This explains the correction term in the Hurwitz formula.)

In our situation we obtain :

$$\chi(C) = d \cdot \chi(L) - d(d-1).$$

L is a projective line, hence diffeomorphic to the 2-sphere, so that $\chi(L) = 2$. It follows that

$$\chi(C) = 2d - d^2 + d = -d^2 + 3d.$$

Since $p(C) = \frac{2-\chi(C)}{2}$, it follows that :

$$p(C) = \frac{(d-1)\cdot(d-2)}{2}$$

and Lemma 4 is proved.

This enables us to compute the genus of a smooth curve. If $C \subset P_2(\mathbb{C})$ is now an arbitrary, not necessarily smooth, irreducible curve of degree d, then one calls the number

$$\pi(C) := \frac{(d-1)(d-2)}{2} = \frac{d(d-3)}{2} + 1$$

the <u>virtual genus</u> of C.

If C is regular, then genus and virtual genus coincide. If C has singularities, then a correction term appears, depending only on the local topological properties of C at the singular points. This is the content of the following formula of Max Noether (1875), which at the same time shows how the correction term depends on the resolution of singularities.

If C is globally reducible, one can also give an analogous formula. However, since the topological genus of C cannot be meaningfully defined in this case*, we use the Euler characteristic instead as global topological invariant of C.

<u>Theorem 5 (M. Noether formula**)</u> :

Let $C \subset P_2(\mathbb{C})$ be a plane complex projective-algebraic curve, without multiple components, of degree d. Then the Euler characteristic is

$$\chi(C) = d(3-d) - \sum_{j=1}^{s} (r_j - 1) + \sum_i \nu_i(\nu_i - 1).$$

Here r_j denotes the number of branches of the germ (C, p_j), at the singular points $p_1, \ldots, p_s$ of C. The second sum extends over the multiplicities of the proper preimages at all infinitely near points of all singular points $p_1, \ldots, p_s$ of C.

In particular, an irreducible curve C has genus

$$p(C) = \pi(C) - \tfrac{1}{2} \sum_i \nu_i(\nu_i - 1) = \frac{d(d-3)}{2} + 1 - \tfrac{1}{2} \sum_i \nu_i(\nu_i - 1).$$

<u>Proof</u> :

The proof takes place in two steps. The first step is to construct a family C_t of curves in $P_2(\mathbb{C})$ such that C_0 is isomorphic to C and C_t is nonsingular of degree d for $t \neq 0$. The second step applies previously proved theorems to this situation.

1) Suppose the curve C is given as the zero set of a homogeneous polynomial ϕ_0 of degree d without multiple factors. Thus C has the equation

$$\phi_0(x_0, x_1, x_2) = 0.$$

One obtains embeddings of C in 1-parameter families of curves as follows: let C_1 be any other plane projective-algebraic curve of degree d, given by the equation ϕ_1, which may have multiple factors.

$$\phi_1(x_0, x_1, x_2) = 0.$$

We consider the family $C_{(t_0, t_1)}$ of curves given by the following

---

*If C is reducible, then the resolved curve $\tilde{C}$ is not connected. However one wants to define genus for non-connected manifolds, there will always be a connected manifold of the same genus. But then genus would lose its characteristic property, namely, that of classifying manifolds.

**In the literature, this formula is frequently listed among the Plücker formulae; however it does not appear in Plücker's own work (cf. also [D3] VI.17).

equations :

$$t_0 \cdot \phi_0(x_0,x_1,x_2) + t_1 \cdot \phi_1(x_0,x_1,x_2) = 0 \qquad (t_0,t_1) \in P_1(\mathbb{C}).$$

Such a family is called a linear system or linear family ; the points of the intersection $C_0 \cap C_1$ are called the base points of the linear system (cf. 6.2).

In order to be able to apply Lemma 1, we want to go from a suitable linear system and construct a mapping like that in Lemma 1, whose fibres are the curves $C_{(t_0,t_1)}$. When one simply tries to define a mapping $P_2(\mathbb{C}) \to P_1(\mathbb{C})$ associating each point $t \in C_{(t_0,t_1)}$ with the point $(t_0,t_1) \in P_1(\mathbb{C})$, then one obtains no well-defined mapping, since all curves $C_{(t_0,t_1)}$ of the linear system go through a base point z, but only a rational "mapping" $\Phi : P_2(\mathbb{C}) \to P_1(\mathbb{C})$. To obtain a well-defined mapping, the different curves must first be separated by blowing up the base point. (When we choose C_1 so that C_0 and C_1 have no common component, there are at any rate only finitely many base points.)

To be able to remove these base points conveniently by blowing up, we choose C_1 so that C_1 meets the curve C_0 in d^2 different points. (For example, we can choose C_1 to be the union of d lines in general position.) Since the entire intersection multiplicity is d^2, C_0 and C_1 meet transversely (at regular points).

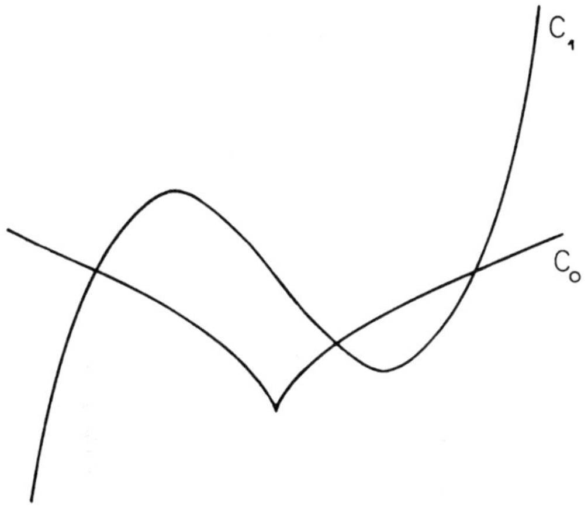

We blow up each of the d^2 intersection points once and thus obtain a mapping

$$\pi : Y \to P_2(\mathbb{C})$$

where Y is a compact algebraic manifold. We define

$$f : Y \to P_1(\mathbb{C})$$

by $f = \Phi \circ \pi$.

As a composition of rational "mappings", f is a priori only a rational "mapping". But the definition has been set up so that f is in fact an everywhere well-defined holomorphic mapping. This is seen immediately from the description of f in local coordinates :

Let (z_0, z_1, z_2) be homogeneous coordinates in $P_2(\mathbb{C})$ and let $p \in P_2(\mathbb{C})$ be a base point of the linear system. One of the homogeneous coordinates of p is non-zero ; without loss of generality suppose it is z_0. Then $y_1 = \frac{z_1}{z_0}$, $y_2 = \frac{z_2}{z_0}$ are local complex-analytic coordinates in a neighbourhood of p. But the functions

$$\phi_0(y_1, y_2) = \Phi_0(1, y_1, y_2)$$

$$\phi_1(y_1, y_2) = \Phi_1(1, y_1, y_2)$$

are also local complex-analytic coordinates in a neighbourhood of p, and indeed coordinates which vanish at p. For $\phi_0(y_1, y_2) = 0$ and $\phi_1(y_1, y_2) = 0$ are just affine equations for C and C_1 . and these curves are nonsingular and transverse at p. Thus we can use the latter coordinates to describe the blowing up of $P_2(\mathbb{C})$ at p in a suitable neighbourhood U of p. The preimage of U is covered by two coordinate neighbourhoods U_0 and U_1 , and $\pi : U_i \to U$ is described by the well-known formulae :

$$\begin{aligned} \phi_0 &= v_0 \\ \phi_1 &= v_0 u_0 \end{aligned} \quad \text{resp.} \quad \begin{aligned} \phi_0 &= u_1 v_1 \\ \phi_1 &= v_1 . \end{aligned}$$

Thus $f = \Phi \circ \pi$ is described in U_0 resp. U_1 by

$$f(u_0, v_0) = (1, u_0) \quad \text{resp.}$$

$$f(u_1, v_1) = (u_1, 1) \ ;$$

and these are of course well-defined holomorphic mappings. To those who have correctly understood σ-processes, all this is clear in any case.

The mapping $f : Y \to P_1(\mathbb{C})$ is therefore well defined and holomorphic, and the fibres $f^{-1}(t_0,t_1)$ are isomorphic to $C_{(t_0,t_1)}$, because C and C_1 are regular and transverse at the base points. Consequently, the curves $C_{(t_0,t_1)}$ are also regular at each base point, and for that reason π induces an isomorphism of the fibre $f^{-1}(t_0,t_1)$ onto the curve $C_{(t_0,t_1)}$. In particular, $f^{-1}(1,0)$ is isomorphic to C.

It then follows from the two Bertini theorems (cf., e.g., van der Waerden [W2] §47), that with the exception of finitely many $(t_0,t_1) \in P_1(\mathbb{C})$, the fibre $f^{-1}(t_0,t_1)$ is nonsingular and irreducible, and hence connected. Our family is therefore constructed!

In this course we have not proved the Bertini theorems, and hence we shall briefly indicate how one can prove the assertion directly. Let $E \subset Y$ be the degeneration set of the mapping $f : Y \to P_1(\mathbb{C})$, i.e. the set of points at which the mapping has nonmaximal rank, and hence rank 0. This means that if one describes f in local coordinates, then both partial derivatives of f vanish on E. The analytic set E can consist of finitely many points and finitely many one-dimensional irreducible components. The restriction of f to these one-dimensional components is constant, because, on the regular part of such a component, f is a function with identically vanishing derivative. This means that each of the finitely many components of the degeneration set E lies entirely within a fibre of f. It already follows from this that only finitely many fibres of f are singular. All other fibres are regular. But when they are regular, they are also irreducible, because a reducible curve always has singularities at the intersections of the components. Thus all but finitely many fibres of f are nonsingular and irreducible, as was to be proved.

2) Let S be a small disc around $0 = (1,0) \in P_1(\mathbb{C})$, so that $C_t := f^{-1}(1,t)$ is smooth and connected for $t \in S - \{0\}$. Let $X := f^{-1}(S)$ and let

$$f : X \to S$$

be the restriction of $f : Y \to P_1(\mathbb{C})$. Then by Lemma 1 :

$$\chi(C) = \chi(C_0) = \chi(C_t) - \sum_{j=1}^{s} (\chi(F_j)-1).$$

Here F_j is the Milnor fibre of (C_0,p_j). It then follows from the local formula 8.5.4 that

$$\chi(C) = \chi(C_t) - \sum_{j=1}^{s} (r_j - 1) + \sum_i \nu_i(\nu_i - 1).$$

Since C_t is isomorphic to a smooth plane projective curve of degree d, the formula for the Euler characteristic in the theorem follows from Lemma 4.

If C is irreducible, then we consider the resolution $\tilde{C}$ of C. We have

$$\chi(\tilde{C}) = \chi(C) + \sum_{j=1}^{s} (r_j - 1)$$

because C results from $\tilde{C}$ by identifying the r_i preimage points of a singular point p_i of C, corresponding to the r_i branches of C at p_i, into a single point, as the following schematic picture indicates :

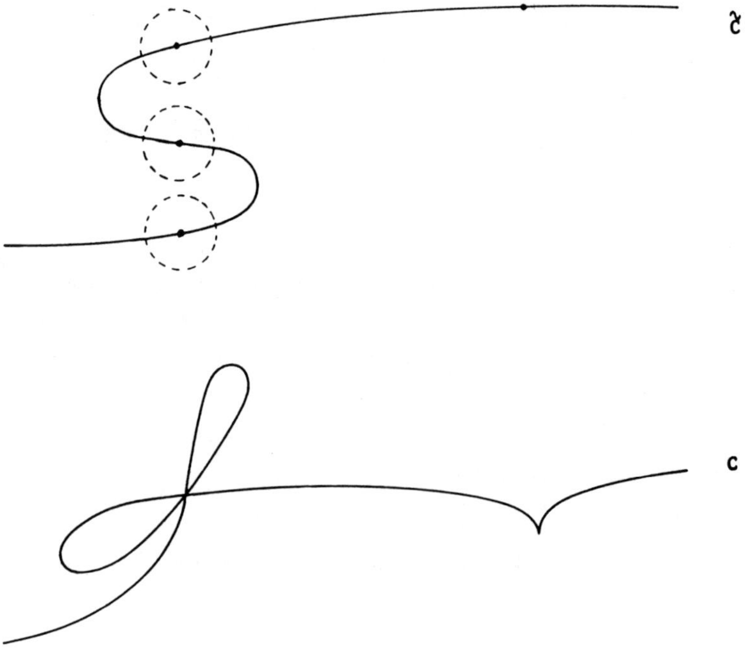

The assertion of the theorem then follows from

$p(C) = p(\tilde{C}) = 1 - \frac{1}{2}\cdot\chi(\tilde{C})$ and $p(C_t) = 1 - \frac{1}{2}\cdot\chi(C_t)$.

As a corollary, we obtain the following formula proved by Clebsch 1864 :

Corollary 6 (Clebsch's formula) :

Let $C \subset P_2(\mathbb{C})$ be an irreducible curve of degree d, which has only double points and cusps as singularities. Let r be the number of double points and let s be the number of cusps. Then

$$p(C) = \frac{d(d-3)}{2} + 1 - r - s.$$

Proof :

At a double point (resp. cusp) C has the local equation $xy = 0$ (resp. $x^2 - y^3 = 0$), and hence multiplicity 2 in either case. The curve becomes smooth after once blowing up such a point. Thus each double point resp. cusp makes a contribution -1 in the formula of Theorem 5.

Examples and remarks :

1) For a projective line L we have $d = 1$, $r = s = 0$, hence $p(L) = 0$. Of course, we have already established in 3.4 that L is homeomorphic to the 2-sphere S^2.

2) An irreducible quadric Q is nonsingular ; one obtains $p(Q) = \frac{2\cdot(-1)}{2} + 1 = 0$ for the genus. This agrees with our results in 7.1, where we constructed an isomorphism between Q and $P_1(\mathbb{C})$.

3) For the three types of irreducible cubic we obtain the following table :

curve	d	r	s	genus
nonsingular cubic	3	0	0	1
cubic with ordinary double point	3	1	0	0
cubic with cusp	3	0	1	0

This should also be compared with the topological investigations of 3.4.

4) The astroid $(x^2+y^2-1)^3 + 27x^2y^2 = 0$ (cf. p. 33, 81, 86) has degree 6. In the preceding section 9.1 we established that the astroid has 4 double points and 6 cusps. Thus one finds the genus of the astroid to be :

$$p = 9 + 1 - 6 - 4 = 0.$$

5) We consider the curve C given by the equation

$$y^2 = h(x) = x^{2p+2} + a_1 x^{2p+1} + \dots + a_{2p+2} ,$$

where $h(x)$ is a polynomial of degree $2p+2$ without multiple zeroes. Curves of this type are also called <u>hyperelliptic curves</u>. The projective equation of C is :

$$y^2 z^{2p} = x^{2p+2} + a_1 x^{2p+1} z + \dots + a_{2p+2} z^{2p+2} .$$

C has only one singularity, at the point $(0,1,0)$, and it is described in the coordinates x, z by

$$z^{2p} = x^{2p+2} + a_1 x^{2p+1} z + \dots + a_{2p+2} z^{2p+2} .$$

One easily computes that the sequence of multiplicities of proper preimages at the infinitely near points is $2p, \underbrace{2,\dots,2}_{p}$, and hence one obtains from Theorem 3 that :

$$p(C) = \frac{(2p+2)(2p-1)}{2} + 1 - \tfrac{1}{2} \cdot [2p(2p-1)+p\cdot 2\cdot (2-1)] = p.$$

This example shows that there is a plane irreducible curve of genus p for each natural number p.

6) The proof of Noether's formula showed that one can go from a singular curve C to a nonsingular curve not only by resolution, but also by "deformation", and that this formula also may be interpreted as follows : it describes the difference between two different kinds of nonsingular models of C, those of the resolved curve $\tilde{C}$ and the deformed curve C_t. Here the difference is given depending on the resolution.

7) If one uses the form of the local formula which we have given in remark (i) at the end of 8.5, then one obtains

$$\chi(C) = d(3-d) - \sum_{j=1}^{s} (r_j - 1) + 2 \sum_{j=1}^{s} \delta_{p_j}$$

resp.

$$p(C) = \pi(C) - \sum_{j=1}^{s} \delta_{p_j}$$

for irreducible curves in Theorem 5. The correction term in this formula may be interpreted in the sense of deformation theory as follows : if $(C,p) \subset (\mathbb{C}^2, p)$ is a reduced analytic curve germ, then C may be deformed, in a small neighbourhood U of p, by an arbitrarily small deformation into a curve C' which has only double points as singularities. The number of these double points is at most δ_p.

8) It follows from the formulae of Clebsch and Noether that the genus $p(C)$ of a nonsingular plane projective algebraic curve is of the form $p(C) = \frac{d(d-3)}{2} + 1$ for a natural number d (the degree of the curve). In particular, $P_2(\mathbb{C})$ contains no nonsingular algebraic curves of genus 2, 4, 5, 7, 8, 9, 11, 12, 13, 14, 16, ...

The above condition for the genus is therefore a necessary condition for an irreducible curve to be birationally equivalent to a smooth plane curve. However, this condition is not sufficient : for example, the hyperelliptic curves of example 5 have no nonsingular plane model for $p > 2$ ([S4] III §5 ex. 15).

To conclude this section it will be shown how the genus formula generalises when curves are considered in arbitrary smooth compact complex surfaces V instead of in $P_2(\mathbb{C})$. Since we can no longer speak of the degree of a curve, we must first develop the concepts which make a generalisation possible. We observe that, in the formula $p = \frac{1}{2} \cdot (d^2-3d) + 1$, the number d^2 is the self-intersection number $C \cdot C$ of the curve C. (For this and what follows, cf. §6.3). We shall see that $-3d$ can also be interpreted as the intersection number of C with a divisor in V - the "canonical divisor" of V. This canonical divisor will now be defined.

Let V be a connected compact complex manifold. A <u>divisor</u> on V is a finite linear combination

$$\sum n_i D_i \ , \ n_i \in \mathbb{Z}$$

where the D_i are closed irreducible 1-codimensional analytic subsets of V.

Divisors come locally from meromorphic functions : let $\{f_i, U_i\}$ be a system of locally meromorphic functions such that

(i) $\{U_i\}$ is an open covering of V,

(ii) f_i is a nowhere identically vanishing meromorphic function on U_i,

(iii) f_i/f_j has neither zeroes nor poles in $U_i \cap U_j$.

Then the zeroes, resp. poles, of $\{f_i, U_i\}$ form well-defined 1-codimensional analytic subsets of V. Let D_j be their irreducible components, with multiplicities m_j (negative for a pole, positive for a zero). Then

$\sum_j m_j D_j$

is the divisor determined by $\{f_i, U_i\}$. Each divisor arises in this way, and two systems $\{f_i, U_i\}$ and $\{f'_j, U'_j\}$ of locally meromorphic functions determine the same divisor (or are "equivalent") just in case f_i/f'_j has neither zeroes nor poles in $U_i \cap U'_j$.*

Divisors constitute an abelian group (written additively). If D and D' are divisors, induced by $\{f_i, U_i\}$ and $\{f'_j, U'_j\}$ respectively, then $D + D'$ is induced by $\{f_i \cdot f'_j, U_i \cap U'_j\}$. A divisor (induced by $\{f_i, U_i\}$) determines a line bundle (unique up to isomorphism) with the transition function f_i/f_j. Two divisors $\{f_i, U_i\}$ and $\{f'_j, U'_j\}$ induce isomorphic line bundles when there is a global meromorphic function f such that $f_i = f \cdot f'_j$ in $U_i \cap U_j$.

One can show (cf. [M9] §9) that if V is algebraic (i.e. if V is isomorphic to a closed submanifold in a $P_N(\mathbb{C})$), then each holomorphic line bundle over V comes from a divisor. This does not hold for arbitrary complex manifolds. Namely, there are smooth compact complex surfaces which contain no complex curves at all.

We therefore define the canonical line bundle of V generally first, and later restrict to the case in which it is induced by a divisor. The <u>canonical line bundle</u> ω_V on the n-dimensional complex manifold V is the bundle of holomorphic n-forms on V :

$$\omega_V = \Lambda^n T^*(V).$$

The transition functions are given by the determinants of the coordinate transformations.

It is clear that ω_V is induced by a divisor just in case there is a global, nonconstant, meromorphic n-form on V. The divisor of each such n-form induces the same bundle ω_V. The divisor of a meromorphic n-form is called the <u>canonical divisor</u> and denoted by K_V. The way in which such meromorphic forms can be explicitly described for a curve V will be investigated in detail in 9.3.

If E is a continuous complex m-dimensional vector bundle over V,

---

*More generally, a finite formal linear combination of analytic subsets on an analytic space is called a <u>Weil divisor</u>, a system of locally meromorphic functions with properties (i) to (iii) is called a <u>Cartier divisor</u>. On complex manifolds the two definitions coincide.

then one can associate a <u>Chern class</u> $c_i(E) \in H^{2i}(V, \mathbb{Z})$ with E for each i, $0 \leq i \leq m$. The element $c(E) = \sum_{i \geq 0} c_i(E)$ in the cohomology ring $H^*(V, \mathbb{Z})$ is called the total Chern class. The characteristic properties of Chern classes are :

(I) $c_0(E)$ is the identity element in $H^*(V, \mathbb{Z})$

(II) If $f : V \to W$ is continuous, then

$$c(f^*(E)) = f^*c(E)$$

(f^*E = induced bundle, while f^* on the right-hand side denotes the induced mapping in the cohomology).

(III) $c(E \oplus E') = c(E) \cdot c(E')$

(IV) Let $(z_0, \ldots, z_n)$ be homogeneous coordinates in $P_n(\mathbb{C})$ and let U_i be the coordinate neighbourhoods $z_i \neq 0$. Let $P_{n-1}(\mathbb{C}) \subset P_n(\mathbb{C})$ be the divisor defined by $z_0 = 0$. Then $P_{n-1}(\mathbb{C})$ defines a generator of $H_{2n-2}(P_n(\mathbb{C}), \mathbb{Z})$ (cf. 6.3.5) and, by Poincaré duality, a generator h_n of $H^2(P_n(\mathbb{C}), \mathbb{Z})$. Let η_n be the line bundle over $P_n(\mathbb{C})$ defined by $P_{n-1}(\mathbb{C})$ (transition function $g_{ij} = z_i \cdot z_j^{-1}$). Then

$$c(\eta_n) = 1 + h_n.$$

For the definition, and proof of these properties, cf. Hirzebruch [H6] §I.4 !

Now let V be a compact connected complex manifold of dimension 2. If E is a complex line bundle over V, induced by a divisor D, then

$$c_1(E) = h[D],$$

where $[D] \in H_{2n-2}(V, \mathbb{Z})$ is the class of D (cf. §6.3) and $h : H_{2n-2}(V, \mathbb{Z}) \to H^2(V, \mathbb{Z})$ denotes the Poincaré isomorphism. If D and D' are two divisors, then their intersection number is defined by

$$D \cdot D' := h(D)([D'])$$

i.e. the value of the cohomology class associated with D on the homology class of D'. Or, what comes to the same thing :

$$D \,.\, D' = \langle [D], [D'] \rangle [V] = (h(D) \cup h(D'))[V],$$

where $\langle\, ,\, \rangle$ denotes the product in the intersection ring and $\cup$ denotes the cup product in the cohomology ring (cf. §6.3). In §6.3 we have mentioned that, in the case of an algebraic manifold, the algebraically defined intersection multiplicity coincides with the topological intersection number defined here.

For the sake of simplicity, we assume in what follows that the canonical bundle of our complex surface V is induced by a canonical divisor K_V. Let $C \subset V$ be an irreducible complex curve, i.e. an irreducible 1-dimensional closed analytic subset of V. The generalised genus formula now runs as follows :

Theorem 7 :

$$p(C) = \tfrac{1}{2}(C\cdot C+C\cdot K_V) + 1 - \tfrac{1}{2}\sum_i \nu_i(\nu_i-1).$$

The sum again is over all the multiplicities of proper preimages at all infinitely near points of all singular points of C.

Before we prove the theorem, we show that our old formula is contained in Theorem 7 as a special case :

Example :

Let $V = P_2(\mathbb{C})$, covered by the coordinate neighbourhoods $U_i = \{z_i \neq 0\}$. The determinant of the transition functions is $\delta_{ij} = (z_j/z_i)^3 = (g_{ij})^{-3}$ (cf. IV). The canonical bundle is therefore

$$\omega_V = L^3$$

where L is the Hopf line bundle over $P_2(\mathbb{C})$ (cf. p.461). In general, $\omega_{P_n(\mathbb{C})} = L^{n+1}$.

A canonical divisor K_V is given, e.g., by $\{(z_0/z_i)^{-3}, U_i\}$. It follows that

$$K_V = -3[P_{n-1}(\mathbb{C})].$$

If $C \subset V = P_2(\mathbb{C})$ is a curve of degree d we therefore have

$$C\cdot C = d^2$$

$$C\cdot K_V = -3[P_{n-1}(\mathbb{C})] \cdot C = -3d.$$

Proof of Theorem 7 :

As with the topological proof of Bézout's theorem, we cannot really speak of a proof, since we use many facts not proved in the course.

We first prove Theorem 7 for smooth curves, in order to show the general case by induction on the number of σ-processes necessary to resolve the singularities of the curve.

(1) Let $C \subset V$ be smooth, and let $N(C)$ denote the normal bundle of C in V. We have

$$T(V)\big|_C = T(C) \oplus N(C).$$

It follows from properties I, II and III of the Chern classes that

$$i^*c_1(T(V)) = c_1(T(C)) + c_1(N(C)),$$

where $i : C \to V$ denotes the inclusion. Since the restriction to C of the bundle induced by the divisor C is just $N(C)$, it follows that $c_1(N(C)) = i^*h[C]$. In addition, one can show that $c_1(T(C))$ is just the Euler class of C (cf. [H6] §I.4.11). Hence if we apply the above equation to the fundamental cycle [C], it follows that

$$c_1(T(V))\cdot C = \chi(C) + C\cdot C .$$

Since $\Lambda^2 T(V)$ is dual to the canonical bundle, it follows (cf. [H6] §I.4.4.3) that

$$c_1(T(V)) = -c_1(T^*(V)) = -c_1(\Lambda^2 T^*(V)) = -c_1(\omega_V)$$

and we obtain

$$-K_V \cdot C = 2 - 2p(C) + C \cdot C,$$

as was to be shown.

(2) Now let $C \subset V$ be arbitrary, let $p \in C$ be a singular point, and let ν be the multiplicity of (C,p). In order to be able to carry out an induction it is enough, by part (1), to show the following :

If $\pi : \tilde{V} \to V$ is a quadratic transformation at $p \in V$ and if $\tilde{C}$ is the proper preimage of C, then

$$\tfrac{1}{2}(\tilde{C}\cdot\tilde{C}+K_{\tilde{V}}\cdot\tilde{C}) = \tfrac{1}{2}(C\cdot C+K_V\cdot C) - \tfrac{1}{2}\nu(\nu-1).$$

The equation results from the following identities (E is the exceptional divisor of π)

$$\begin{aligned}
\tfrac{1}{2}(C\cdot C+K_V\cdot C) &= \tfrac{1}{2}(\pi^{-1}(C)\cdot\pi^{-1}(C) + \pi^{-1}(K_V)\cdot\pi^{-1}(C)) \\
&= \tfrac{1}{2}((\tilde{C}+\nu E)(\tilde{C}+\nu E) + (K_{\tilde{V}}-E)(\tilde{C}+\nu E)) \\
&= \tfrac{1}{2}(\tilde{C}\tilde{C}+2\nu\tilde{C}E+\nu^2 EE+K_{\tilde{V}}\tilde{C}-E\tilde{C}+\nu K_{\tilde{V}}E-\nu EE) \\
&= \tfrac{1}{2}(\tilde{C}\tilde{C}+K_{\tilde{V}}\tilde{C}) + \tfrac{1}{2}(\nu^2-\nu).
\end{aligned}$$

Here we have used :

1) Invariance of the intersection numbers under passage to total preimage :

$D \cdot D' = \pi^{-1}(D) \cdot \pi^{-1}(D')$.

This follows from Noether's theorem (8.4.13) and the fact that the intersection number equals the sum of the local intersection multiplicities.

2) The generalised Hurwitz formula :

$K_{\tilde{V}} = E + \pi^{-1}(K_V)$.

Proof :

Suppose K_V is described by the meromorphic 2-form ω on V. Then $\omega \circ \pi$ is a meromorphic 2-form on $\tilde{V}$ which describes $K_{\tilde{V}}$. We consider coordinates (x,y) in $U \cap V$ and $(\tilde{x},\tilde{y})$ in $\tilde{U} \cap \tilde{V}$ with $\pi(\tilde{U}) \subset U$. If $\omega = f \cdot dx \wedge dy$ in U, then $\omega \circ \pi = \det\left(\frac{\partial^2 \pi}{\partial\tilde{x}\partial\tilde{y}}\right) f \circ \pi d\tilde{x} \wedge d\tilde{y}$ holds in $\tilde{U}$, i.e. $K_{\tilde{V}}$ is described in $\tilde{U}$ by $\det\left(\frac{\partial^2 \pi}{\partial\tilde{x}\partial\tilde{y}}\right) f \circ \pi$. But since E is described by $\det\left(\frac{\partial^2 \pi}{\partial\tilde{x}\partial\tilde{y}}\right)$ and $\pi^{-1}(K_V)$ is described by $f \circ \pi$, the assertion follows. (The proof shows that the assertion holds for arbitrary mappings $\pi : \tilde{V} \to V$ between manifolds of the same dimension when the functional determinant of π does not vanish identically and when E denotes the divisor associated with this functional determinant.)

3) $E \cdot E = -1$

Proof : (see also 8.4.18)

Let z_1, z_2 be local coordinates in a neighbourhood W around p in V. As usual, $\pi^{-1}(W)$ is covered by the two coordinate neighbourhoods U_1 (coordinates u_1, v_1) and U_2 (coordinates u_2, v_2), so that π is described relative to these coordinates by the equations

$$\begin{matrix} z_1 = v_1 \\ z_2 = u_1 v_1 \end{matrix} \quad \text{in } U_1 \qquad \begin{matrix} z_1 = u_2 v_2 \\ z_2 = v_2 \end{matrix} \quad \text{in } U_2$$

The exceptional divisor E is described in U_1 by $\{v_1=0\}$ and in U_2 by $\{v_2=0\}$.

Let Z be the divisor given by the holomorphic function $z_1 \circ \pi$ on $\pi^{-1}(W)$. Z is described by $v_1 = 0$ in U_1 and by $u_2 \cdot v_2 = 0$ in U_2, hence $Z = E + U$, where U is the divisor given by $u_2 = 0$.

Since Z comes from a holomorphic function defined globally in $\pi^{-1}(W)$, the bundle ζ over $\pi^{-1}(W)$ associated with Z is trivial and hence $c_1(\zeta) = 0$. It follows that :

$$0 = Z\cdot E = (E+U)\cdot E = E\cdot E + U\cdot E = E\cdot E + 1,$$

and hence $E\cdot E = -1$.

4) $K_{\tilde{V}}\cdot E = -1$.

<u>Proof</u> :

We apply the assertion about nonsingular curves, already proved in part (1), to the exceptional curve E and obtain :

$$0 = p(E) = \tfrac{1}{2}(E\cdot E+K_{\tilde{V}}\cdot E) + 1 = \tfrac{1}{2}(-1+K_{\tilde{V}}\cdot E) + 1,$$

hence $K_{\tilde{V}}\cdot E = -1$.

5) $\tilde{C}\cdot E = \nu$ by 8.4.1.

This completes the proof of Theorem 7.

9.3 Differential forms on Riemann surfaces and their periods

The theory of differential forms on Riemann surfaces, and their integrals, has its historical origin in investigations of elliptic integrals, e.g. integrals which describe arc length on ellipses or lemniscates. These investigations began with the discovery of Fagnano 1714 on the doubling of the lemniscate arc, were carried forward by Euler, Gauss and above all Abel, then brilliantly raised to a new level by Riemann in his work of 1857 [R3]. We shall not go into the historical aspect of this development here – for that we refer to the books of Dieudonné [D3], Chap. V, or Siegel [S5] – instead we shall begin immediately with the description of differential forms on a Riemann surface.

Suppose then that C' is an abstract closed Riemann surface, i.e. a complex one-dimensional connected compact complex manifold. (This formulation of the concept is due to H. Weyl.) We have spoken about the connection between abstract Riemann surfaces, concrete Riemann surfaces and algebraic curves earlier in 3.4 (cf. especially p.156). The closed Riemann surfaces are – as we claimed there without proof – biholomorphically equivalent as complex manifolds to the manifolds which result from blowing up singularities of irreducible plane projective algebraic curves (cf. also 9.2). We shall use this concrete description in the present section, in order to describe the differential forms on such a Riemann surface C'.

Since C' is complex one-dimensional, only the 1-forms are

interesting from an analytic standpoint, and when we speak of differential forms on C' we always mean 1-forms. We want to investigate the meromorphic differential forms on C'. Locally, such a differential form always has a representation

$$\omega = g(z)dz$$

where z is a local complex coordinate on the 1-dimensional complex manifold C', and $g(z)$ is a meromorphic function. We assume that the reader is already familiar with the calculus of differential forms (cf., say, the book of Narasimhan, [N6]).

Differential forms are there to be integrated. If we integrate a meromorphic 1-form ω over a closed curve, a "cycle" γ on C', which avoids the poles of ω, then we obtain a complex number

$$\int_\gamma d\omega .$$

These numbers are called <u>periods</u> – in 7.4, p. 322 , we have already spoken briefly about the great significance they have in relation to the moduli problem. The periods are the definite integrals of ω. Now we consider the indefinite integral ! If one integrates ω over a variable path with fixed initial point z_0 and variable endpoint z, then one obtains a many-valued function of z

$$y(z) = \int_{z_0}^{z} \omega$$

These functions are called <u>abelian integrals</u>. Special cases are the elliptic integrals, which we have studied in a little more detail in 7.4. Abelian integrals are in general holomorphic on C'. However, they have singularities at the poles of ω.

In this regard we can distinguish three cases :

I. The differential form ω has no poles, and hence is a holomorphic differential form. Then $\int_{z_0}^{z} \omega$ is a many-valued function without singularities. The many-valuedness is obviously described precisely by the periods, i.e. the difference between two branches of $\int_{z_0}^{z} \omega$ is a period $\int_\gamma \omega$. Such integrals are called <u>integrals of 1st kind</u>, and the holomorphic differential forms are called <u>differential forms of 1st kind</u> (older terminology : differentials of 1st kind).

II. The differential form ω has poles, but the residues of ω at the poles are all zero (i.e. in local coordinates ω has the form $\left(\frac{a_{-n}}{z^n} + \ldots + \frac{a_{-2}}{z^2} + f(z)\right)dz$ where $f(z)$ is a holomorphic function and $n \geq 2$). In this case $\int_{z_0}^{z} \omega$ is a many-valued meromorphic function on C', with poles at the poles of ω. The many-valuedness does not come from the singular points : in describing a small circle around a singular point one comes back to the same branch – this follows immediately from the residue theorem. Differential forms of this kind are called differential forms of the 2nd kind, the corresponding indefinite integrals are called integrals of the 2nd kind.

III. The differential form ω has poles, at which the residue of ω does not vanish. In this case the abelian integral $\int_{z_0}^{z} \omega$ has logarithmic singularities at these poles. Additional many-valuedness therefore appears as a result of the logarithmic singularity. With integrals of the 1st and 2nd kind the periods $\int_{\gamma} \omega$ are non-zero only for non-null homologous cycles γ in C', whereas in the 3rd case one must also consider periods $\int_{\gamma} \omega$ where ω is a small circle around such a pole, because this integral is just the residue of ω at this point – up to a factor $2\pi i$ – by the residue theorem (such periods are also called residual periods). Differential forms with such poles are called differential forms of the 3rd kind, and the associated integrals are called integrals of the 3rd kind.

This important division of the meromorphic differential forms on a Riemann surface is due to Riemann 1857.

We now want to describe the differentials of 1st kind quite explicitly. To do this we assume that $\pi : C' \to C$ is the resolution of singularities of a plane projective algebraic curve C with affine equation $f(x,y) = 0$. The affine coordinates x, y are rational functions on C, and hence give rational functions on C', which we again denote by x, y, when composed with π. Likewise, each rational function $R(x,y)$ of x and y gives a meromorphic function on C'. Finally, instead of the meromorphic functions x and y, we can consider their differentials dx and dy – the latter are certain meromorphic differential forms on C'. More generally, one obtains well defined meromorphic differential forms $R(x,y)dx + S(x,y)dy$ in this way. This is the form in which we shall first describe the differential forms of 1st kind.

The description we give already appears in Abel — though set up in a complicated way — and with clear proofs by Riemann in the cited work of 1857.

Theorem 1 :

Let C be an irreducible plane algebraic curve of order n with affine equation $f(x,y) = 0$, and let the coordinates be chosen so that $\frac{\partial f}{\partial y}$ does not vanish identically. Let C' be the Riemann surface which results from C by resolution of singularities. Then the nonvanishing differentials of 1st kind on C' are just the differential forms

$$\frac{\phi(x,y)dx}{\frac{\partial f}{\partial y}(x,y)}$$

where $\phi(x,y) = 0$ is the equation of a curve of order n-3 adjoint to C. Thus the vector space of differentials of 1st kind on C' has dimension p, where p is the genus of C'.

Proof : We first show that the differential forms ω of the form

$$\omega = \frac{\phi(x,y)dx}{\frac{\partial f}{\partial y}(x,y)},$$

where $\phi(x,y)$ is the equation of a curve $\tilde{C}$ of order n-3 adjoint to C, are in fact differentials of 1st kind on C', i.e. they have no poles. In order to avoid a separate discussion of possible poles at infinity, we use the substitution $x = \frac{x_1}{x_0}$, $y = \frac{x_2}{x_0}$ to write ω in the homogeneous form :

$$\omega = \frac{\Phi(x_0,x_1,x_2)(x_0dx_1-x_1dx_0)}{\frac{\partial F}{\partial x_2}(x_0,x_1,x_2)},$$

where $F(x_0,x_1,x_2) = 0$ is the homogeneous equation of C and $\Phi(x_0,x_1,x_2) = 0$ is the homogeneous equation of $\tilde{C}$.

Thus we see that poles result from the vanishing of $\frac{\partial F}{\partial x_2}$. The differential form ω therefore has poles, if at all, at the points of C' lying over the intersections of C with the polar $\tilde{\tilde{C}}$ which is described by the equation $\frac{\partial F}{\partial x_2}(x_0,x_1,x_2) = 0$. The intersections of the polar $\tilde{\tilde{C}}$ with C are, on the one hand, the singular points of C, on the other hand the smooth points at which the tangent to C goes through the pole. We consider this second case first. At these points ω has no pole. Proof : the function $f(x,y)$, regarded as a function on C', vanishes identically there. Hence its differential satisfies :

$$\frac{\partial f}{\partial x}dx + \frac{\partial f}{\partial y}dy = 0.$$

Therefore

$$\frac{dx}{\partial f/\partial y} = - \frac{dy}{\partial f/\partial x} .$$

Thus if we write ω in the form

$$\omega = \frac{\phi \, dx}{\partial f/\partial y}$$

we can write it equally well in the form

$$\omega = \frac{-\phi \, dy}{\partial f/\partial x} .$$

At a smooth point of C where $\frac{\partial f}{\partial y}$ vanishes, $\frac{\partial f}{\partial x}$ does not vanish, and for that reason ω has no pole there.

Now we consider a singular point p of C. We describe the situation in affine coordinates x, y. Let t be a local coordinate on C' in a neighbourhood of a point q which is mapped onto p by the resolution $\pi : C' \to C$. Let the resolution be described in the neighbourhood of q by two functions

$$x = x(t)$$
$$y = y(t).$$

In terms of the local coordinate t, the differential form ω is

$$\omega = \frac{\phi(x(t),y(t))x'(t)dt}{\frac{\partial f}{\partial y}(x(t),y(t))}$$

hence ω has no pole at q iff the orders of these functions of t at p satisfy :

$$O(\phi \circ \pi) + (O(x \circ \pi)-1) - O(\frac{\partial f}{\partial y} \circ \pi) \geq 0. \qquad (*)$$

By 8.3.14, we can interpret these numbers as intersection numbers : if C_q is the branch of C at p corresponding to q and if L is the perpendicular line $x = c$ through p, then we obtain the condition for ω to have no pole at q as :

$$\nu_p(C_q,\tilde{C}) + (\nu_p(C_q,L)-1) - \nu_p(C_q,\tilde{\tilde{C}}) \geq 0. \qquad (**)$$

For the case in which C is irreducible at p, so that $(C,p) = (C_q,p)$, we can use the fact that we have calculated the intersection multiplicity of C with the polar $\tilde{\tilde{C}}$ in 9.1.4 : in this case

$$\nu_p(C_q,\tilde{\tilde{C}}) = \sum \nu_j(\nu_j-1) - 1 + \nu_p(C_q,L)$$

and then the condition (**) equivalent to (*) reads

$$\nu_p(C,\tilde{C}) \geq \sum \nu_j(\nu_j-1),$$

where the summation is over all the multiplicities ν_j of infinitely near points of (C,p). But this inequality follows from an easy generalisation of Theorem 8.4.13 using the multiplicity conditions introduced in 9.1 with the definition of adjoint curves, because, at each infinitely near point of C with multiplicity ν_j, $\tilde{C}$ has multiplicity at least ν_j-1, and hence Max Noether's formula for calculating the intersection number yields a number $\geq \sum \nu_j(\nu_j-1)$. Thus one sees that the adjoint condition is formulated just so that the forms ω have no poles.

In the general case, where C is reducible at p, we cannot reach the conclusion quite so simply, but then the inequality (*) or (**) follows by a simple induction as in the proof of 9.1.4. We omit doing this here.

The proof of (**) shows that the differential forms ω of the form

$$\omega = \frac{\phi\, dx}{\partial f/\partial y},$$

where ϕ is the equation of an adjoint curve of order $n-3$, are holomorphic. Since, by 9.1.6, there are just p such equations which are linearly independent of each other, where p is the genus of C, we have thereby constructed p linearly independent differentials of the 1st kind. In 9.1.6 we obtained the number p in the following form :

$$p = \frac{n(n-3)}{2} + 1 - \tfrac{1}{2} \sum \nu_j(\nu_j-1).$$

But in 9.2 we proved that this number equals the genus of the Riemann surface C', i.e. the first Betti number b_1 of C' is

$$b_1 = 2p.$$

Using this fact, we now prove that there are no holomorphic 1-forms on C' other than those described above.

Let V be the vector space of holomorphic 1-forms on C' and let $m = \dim_{\mathbb{C}} V$. Then V is a real vector space of dimension $2m$. Now let $\gamma_1,\ldots,\gamma_{2p}$ be cycles on C' which form a basis for the first homology group of C'. Each γ_k represents a complex linear form ℓ_k on V, defined by

$$\ell_k(\omega) = \int_{\gamma_k} \omega \ .$$

The real part h_k of ℓ_k is a real linear form on V :

$$h_k(\omega) = \mathrm{Re} \int_{\gamma_k} \omega \ .$$

For $m > p$ to hold, the real subspace of V defined by the linear equations $h_1(\omega) = \ldots = h_{2p}(\omega) = 0$ would have to have positive dimension. Hence there would be a nonvanishing holomorphic 1-form ω with

$$\mathrm{Re} \int_{\gamma_k} \omega = 0 \ , \quad k = 1,\ldots,2p.$$

Now we consider the indefinite integral

$$y(z) = \int_{z_0}^{z} \omega \ .$$

Since ω has no pole, $y(z)$ would be a many-valued holomorphic function on C'. Because of the vanishing of the real parts of the periods, the real part of y would even be a single-valued function on C', and as the real part of a holomorphic function, $\mathrm{Re}(y(z))$ would be a harmonic function. Since C' is compact, this function would have to attain a maximum somewhere on C'. But, by the maximum principle for harmonic functions, a nonconstant function which is harmonic in some domain does not attain its maximum in the interior of the domain (cf. e.g. Behnke-Sommer [Bl], Chap. II, Appendix, Theorem 52). Since $\mathrm{Re}(y)$ has to attain a maximum, it must therefore be constant. Then, by the Cauchy-Riemann equations, the imaginary part would also have to be constant. Hence the function y would be constant, and therefore its differential

$$\omega = dy = 0,$$

contradicting the assumption $\omega \neq 0$. Thus the assumption $m > p$ has led to a contradiction and Theorem 1 is proved.

In Theorem 1 we have written down the holomorphic 1-forms in a simplest possible form. The numerator is the product of dx with an affine adjoint of order $n-3$, and the denominator is $\frac{\partial f}{\partial y}$. The equation $\frac{\partial f}{\partial y} = 0$ is the equation of a special polar. In proving the theorem we have already seen that the essential point is the comparison of intersection multiplicities of adjoint and polar. The special choice of polar is inessential.

It is therefore closer to the true state of affairs when we use, for the denominator, a polynomial which describes an arbitrary polar. In addition, it is more natural to use the equation of this general polar in homogeneous form

$$a_0 \frac{\partial F}{\partial x_0} + a_1 \frac{\partial F}{\partial x_1} + a_2 \frac{\partial F}{\partial x_2} = 0,$$

because only in this form is it symmetric. In this way one comes to the following description of differentials of the 1st kind in homogeneous form, which goes back to Aronhold (1861).

<u>Corollary</u> :

Let C be an irreducible plane curve of order n with homogeneous equation $F(x_0,x_1,x_2) = 0$, and let C' be the abstract Riemann surface which results from C by resolution of singularities. Then the holomorphic 1-forms on C' may be described in the following homogeneous form

$$\frac{\phi(x_0,x_1,x_2) \cdot \Sigma(-1)^{\operatorname{sign}(i,j,k)} a_i x_j dx_k}{\Sigma a_i \frac{\partial F}{\partial x_i}},$$

where $\sum a_i \frac{\partial F}{\partial x_i} = 0$ is the equation of an arbitrary, but fixed, polar, and $\phi(x_0,x_1,x_2) = 0$ is the homogeneous equation of an $(n-3)^{\text{th}}$ order curve adjoint to C. The vector space of these polynomials ϕ has dimension p, where p is the genus of the Riemann surface.

We could derive this corollary from Theorem 1, or we could confirm it directly in the same way as Theorem 1. Instead of this, we shall systematically explain, by a sequence of theorems which follows, how one can arrive at this form of differential forms. The roots of this go back to Poincaré, Picard and Lefschetz. These roots have been very essentially extended and deepened in recent years in a series of works by P.A. Griffiths, which we follow in our presentation (cf. Griffiths [G7]).

In what follows we shall assume, for the sake of simplicity, that C is a nonsingular curve in $P_2(\mathbb{C})$. Then C is a 1-dimensional complex manifold in particular. To describe the holomorphic differential forms on C we can in principle use local complex coordinates, and relative to such a coordinate z, each 1-form has the form $g(z)dz$ when restricted to the coordinate neighbourhood, where g is a holomorphic function. Passing to another coordinate t with $z = z(t)$, one has the transformation $g(z)dz = g(z(t))\frac{dz}{dt}dt$. This kind of description of holomorphic forms is awkward and not particularly elegant. It may be asked whether there is not a description in closed form which more strongly uses the fact that C is given to us not only as an abstract complex manifold, but concretely, with an embedding in 2-dimensional projective space, in which we can use a single coordinate system (x_0, x_1, x_2) when we operate with homogeneous coordinates. This is in fact possible — indeed, we have given such a description of the differential forms in the Corollary to Theorem 1. We now want to understand this description better. The essential new idea will be that there is a close connection between the 1-forms on C and the 2-forms on the complement $P_2(\mathbb{C}) - C$.

We first want to investigate this connection in a somewhat more qualitative topological way. To do this we begin with some quite general remarks on the description of cohomology by differential forms. Differential forms are there to be integrated over cycles. If one integrates a p-form ω on a manifold X over a p-cycle, then one obtains a complex number which, by Stokes' theorem, depends only on the homology class of the cycle when ω is closed, i.e. when $d\omega = 0$. Of course, one uses p-cycles with complex coefficients, which are quite analogous to the p-cycles with integer coefficients defined in 6.3. Their homology classes form the pth homology group of X with complex coefficients,

$$H_p(X,\mathbb{C}).$$

Thus a closed p-form yields a linear form on $H_p(X,\mathbb{C})$, and hence an element of the dual vector space. This dual vector space is (by the universal coefficient theorem) just the pth cohomology group of X with complex coefficients :

$$H^p(X,\mathbb{C}) \cong H_p(X,\mathbb{C})^*.$$

Thus a closed p-form ω yields a cohomology class $[\omega] \in H^p(X,\mathbb{C})$, and Stokes' theorem shows that this class does not change when we add to ω an exact form, i.e. a differential of the form $d\psi$, where ψ is a (p-1)-form. It is therefore meaningful to divide the vector space $Z^p(X)$ of closed p-forms on X with complex coefficients by the vector space of exact forms $B^p(X)$. Then one obtains the pth de Rham cohomology group

$$H^p_{DR}(X) = Z^p(X)/B^p(X)$$

and our above discussion can be summarised by saying that the integration of closed p-forms over p-cycles induces a canonical homomorphism

$$H^p_{DR}(X) \to H^p(X,\mathbb{C}).$$

Until now we have left open what differential forms we want to work with.

When X is given as a differentiable manifold, it is natural to work with differentiable forms, and the resulting differentiable de Rham cohomology satisfies the following well-known theorem (cf. e.g. Narasimhan [N6]).

<u>De Rham Theorem (1931)</u>

The canonical homomorphism for differentiable de Rham cohomology,

$$H^p_{DR}(X) \to H^p(X,\mathbb{C}),$$

is an isomorphism.

When X is a complex manifold, it is natural to work with holomorphic differential forms. Under certain assumptions the resulting analytic de Rham cohomology, despite the fact that we are working with far fewer differential forms, is still isomorphic to the cohomology $H^*(X,\mathbb{C})$. This holds, in particular, when there are a lot of holomorphic functions on X — more precisely : when X is a <u>Stein manifold</u> (these manifolds may be characterised as those which admit embeddings as closed submanifolds in an affine space $\mathbb{C}^N$). But in general the analytic de Rham cohomology is no longer isomorphic to cohomology. This holds especially in the case of interest to us, where X is a compact 1-dimensional complex manifold. Such an X admits very few holomorphic functions, namely, just the constant functions. For that reason, the first analytic de Rham cohomology group $H^1_{DR}(X)$ is isomorphic to the vector space of holomorphic 1-forms. (Each holomorphic 1-form ω is closed, because $d\omega$ is a holomorphic 2-form, but X is 1-dimensional, hence $d\omega = 0$.) Thus $H^1_{DR}(X)$ is a complex vector space of dimension p, whereas it is well known that $H^1(X,\mathbb{C})$ has dimension 2p, where p is the genus of X. Thus in this case the holomorphic forms yield only half the cohomology.

However, when one wants to describe the rest of the cohomology by means of differential forms related to the complex analytic structure, one has several options.

First option : One replaces the definition of the analytic de Rham cohomology, which in a certain sense is addressed very much to a local situation, namely the Stein or affine case, by a more global definition of de Rham cohomology. One defines the analytic de Rham hypercohomology $\mathbb{H}^*_{DR}(X)$, and then one again obtains an isomorphism $\mathbb{H}^*_{DR}(X) \cong H^*(X,\mathbb{C})$. These homological methods are important standard techniques in modern algebraic geometry, but they lead beyond the scope of this course.

Second option : One considers not only the cohomology classes which come from holomorphic forms, but also others, e.g. those which come from antiholomorphic forms. We want to explain this a little more precisely. Let X be a complex manifold. A differentiable complex-valued differential form ω is a (p,q)-form (or "form of type (p,q)") when it has a representation of the following form in local complex coordinates $z_1,\ldots,z_n$:

$$\omega = \Sigma a_{i_1,\ldots,i_p,j_1,\ldots,j_q} dz_{i_1} \wedge \ldots \wedge dz_{i_p} \wedge d\bar{z}_{j_1} \wedge \ldots \wedge d\bar{z}_{j_q} .$$

Each m-form on X may be expressed uniquely as a sum of (p,q)-forms with $p + q = m$. When X is not only a complex manifold, but also a projective algebraic manifold (or more generally : a compact Kähler manifold), then this decomposition leads to a corresponding decomposition of the cohomology, the Hodge decomposition.

Hodge decomposition :

Let X be a projective algebraic manifold, and let $H^{p,q}(X)$ be the subspace of $H^{p+q}(X,\mathbb{C})$ generated by the closed (p,q)-forms. Then one has the direct sum decomposition

$$H^m(X,\mathbb{C}) = \bigoplus_{p+q=m} H^{p,q}(X).$$

The closed forms of type $(p,0)$ are just the holomorphic p-forms, and $H^{p,0}(X)$ is canonically isomorphic to the vector space of holomorphic p-forms.

This fundamental theorem was proved by Hodge by tying together the ideas of Riemann with his own theory of harmonic forms in a series of works (from 1931 onwards), cf. Hodge [H8]. A concise introduction to these ideas may be found e.g. in [D3] VII.1, and a good summary and overview is in [G8] §3.

As an example, let us consider the case of a nonsingular algebraic curve C of genus p. In this case we obtain the Hodge decomposition:

$$H^1(X,\mathbb{C}) = H^{1,0}(X) \oplus H^{0,1}(X).$$

The half $H^{1,0}(X)$ is — as we established above — the p-dimensional vector space of holomorphic 1-forms. If ω is a holomorphic 1-form, then $\bar{\omega}$ is a closed (0,1)-form, because $d\bar{\omega} = \overline{d\omega} = 0$. Thus $H^{0,1}(X)$ is generated by the forms which result from the holomorphic 1-forms by conjugation, the "antiholomorphic" forms. There is a conjugation on the complex vector space $H^1(X,\mathbb{C})$, because $H^1(X,\mathbb{C})$ is the complexification of $H^1(X, \mathbb{R})$, and this conjugation corresponds precisely to the conjugation $\omega \mapsto \bar{\omega}$, so that : $H^{1,0}(X) = \overline{H^{0,1}(X)}$. The closed (p,q)-forms we are working with here are differentiable, even real analytic, but neither complex analytic nor meromorphic.

Third option : One can allow the forms used to describe the cohomology classes to be meromorphic and to have poles. We confine ourselves to the case of a nonsingular curve C. It follows immediately from the residue theorem that the integral of a meromorphic differential form of the second kind over a 1-cycle of C is always a well defined number (see above), which depends only on the homology class of the cycle. (For differential forms of the 3rd kind this is not so). For that reason one can also associate a cohomology class $[\omega] \in H^1(C,\mathbb{C})$ with ω of the 2nd kind. One can show — and we shall carry this out in detail later — that one in fact obtains all cohomology classes in $H^1(C,\mathbb{C})$ in this way.

We have now indicated three different, but related, methods for describing the cohomology of a complex manifold X by differential forms. Now when X is not only a complex manifold, but also a projective algebraic manifold, then it is natural to work only with rational differential forms, i.e. those which have local representations, in Zariski open sets, of the form

$$\Sigma a_{i_1 \dots i_p} dz_{i_1} \dots dz_{i_p}$$

with entire rational functions $a_{i_1 \dots i_p}$. When we consider differential forms defined on the entire projective algebraic manifold (i.e. without poles) then these are only a priori fewer in number than the holomorphic forms. For curves, we have already seen this, because Theorem 1 in fact gives a description of <u>all</u> holomorphic 1-forms by rational forms. However, this also holds generally : holomorphic forms on a projective algebraic manifold X and rational forms without poles are the same, and so too are meromorphic forms and rational forms with poles (see Serre [S14] GAGA).

For forms on a non-projective algebraic manifold, e.g. on an affine algebraic manifold, it is quite different. One already sees this in the simplest case of 0-forms, i.e. functions on $X = \mathbb{C}$: the entire rational functions are a much smaller class than the entire functions, which also include transcendental functions, such as the exponential function.

Interestingly enough, it nevertheless holds for non-projective algebraic manifolds that rational forms suffice for computing the cohomology. More precisely, one has the following important theorem of Grothendieck :

Algebraic de Rham Theorem :

For a nonsingular algebraic manifold X the algebraic de Rham hypercohomology is isomorphic to the analytic de Rham hypercohomology, and hence to $H^*(X,\mathbb{C})$.

For affine X, such as the complement of a hypersurface in $P_n(\mathbb{C})$, the de Rham hypercohomology coincides with the usual de Rham cohomology. For that reason, it follows that :

Theorem :

For a nonsingular affine X the algebraic de Rham cohomology $\mathcal{H}^*(X)$ is isomorphic to the cohomology with complex coefficients :

$$\mathcal{H}^*(X) \cong H^*(X,\mathbb{C}).$$

This theorem allows us to work with just rational differential forms in what follows.

With this, we wish to conclude our general remarks on the connection between cohomology and differential forms, and return to the problem of concretely describing the forms on a plane curve. The general remarks were intended only as background information, not necessary to the understanding of what follows, but perhaps useful.

We want to study the relation between the differential forms on a plane algebraic curve C and the differential forms on its complement. If we go over to the cohomology classes defined by the differential forms, then this means investigating the relations between the cohomology of C and P_2-C. Dual to this, we can also study the relations between the homology of C and P_2-C, and this is what we want to do first.

For this purpose we first choose a tubular neighbourhood T of C in P_2. (When we provide $P_2(\mathbb{C})$ with a Riemannian metric, then for sufficiently small $\varepsilon > 0$ we can take T to be the set of all points at distance $< \varepsilon$ from C. Details may be found in all textbooks on differentiable manifolds, e.g. [B8].) The boundary ∂T of the tubular neighbourhood is a 3-dimensional orientable manifold and can be viewed as an oriented fibre bundle $\partial T \to C$ with fibre S^1. The orientation of the fibres comes from the complex structure of C and $P_2(\mathbb{C})$: T looks like a neighbourhood of the zero section in the complex normal bundle, and in the lines of this complex normal bundle we have an orientation in the sense of a positive circuit round the origin.

Now let $[\gamma]$ be any homology class in $H_1(C)$. We choose a cycle γ representing $[\gamma]$ which consists of disjoint oriented circles – one easily sees that this is always possible.

The preimage $\tau(\gamma)$ of γ in ∂T is a two-dimensional manifold – one can even say that it is a disjoint union of two-dimensional tori, because each component is an oriented S^1-bundle over a circle. In addition, the orientation of the cycle γ and the orientation of the fibre S^1 give an orientation of $\tau(\gamma)$ in a natural way. The oriented submanifold $\tau(\gamma)$ of P_2-C defines a homology class $[\tau(\gamma)]$ in $H_2(P_2-C)$ in the sense of 6.3, and one can prove that it depends only on the homology class of $[\gamma]$. Thus

$$[\gamma] \mapsto [\tau(\gamma)]$$

defines a homomorphism

$$\tau : H_1(C) \to H_2(P_2-C).$$

In a similar way, one can define a homomorphism

$$\tau : H_p(V) \to H_{p+1}(V-W)$$

for each complex 1-codimensional submanifold W of a complex manifold V. When a homology class $[\gamma]$ is represented by an oriented submanifold, the definition of $\tau(\gamma)$ goes exactly as above. In general the idea is the same, but not quite so easy to carry out technically.

A technically more elegant way to define τ, though not so geometrically intuitive, is, e.g., by means of the following commutative diagram

$$\begin{array}{ccc} H_p(W) & \xrightarrow{\tau} & H_{p+1}(V-W) \\ \phi \| \wr & & \| \wr \ \psi \\ H^{2n-p}(W) & \xrightarrow{\partial} & H^{2n-p+1}(V,W). \end{array}$$

Here the vertical isomorphisms ϕ and ψ are those from Poincaré duality and generalised Alexander duality respectively, and ∂ is the boundary homomorphism in the exact cohomology sequence. However, we do not want to make further use of this technical definition, and for details we refer e.g. to [S15].

The mapping τ is also called the <u>Leray coboundary</u>. The simplest conceivable situation in which the Leray coboundary appears is of course the situation of the classical residue theorem : $V = \mathbb{C}$ and $W = \{0\}$.

$H_0(W)$ is generated by the 0-cycle γ which consists of the point 0, and $\tau(\gamma)$ is represented by a circle around the origin with orientation in the positive direction

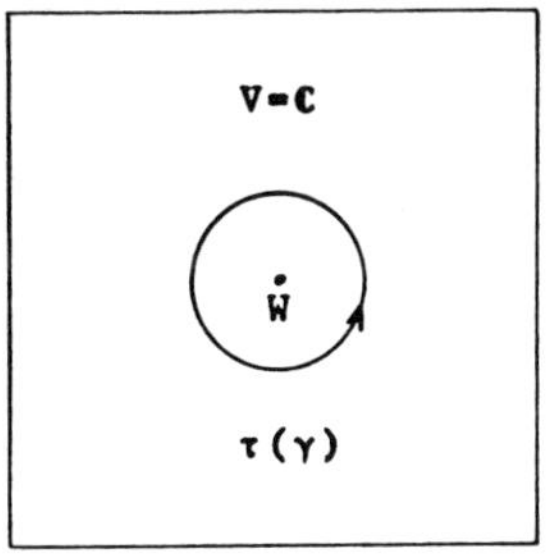

The Leray coboundary

$$\tau : H_p(W) \to H_{p+1}(V-W)$$

has been defined in a geometric way. By passing to homology with complex coefficients and finally dualising we also obtain a dual mapping

$$\tau^* : H^{p+1}(V-W,\mathbb{C}) \to H^p(W,\mathbb{C}).$$

Now when we recall that cohomology classes can be described with the help of differential forms, the question arises whether there is an operation on differential forms corresponding to the mapping τ^*. This is in fact the case, and the operation in question is a construction of the <u>residue</u>. We shall define the residue only for those forms on $V-W$

which have no worse singularities than poles along W. A differentiable $(p+1)$-form ω on $V-W$ has at most a <u>pole</u> of order k along W when the following holds : if, in any open subset U of V, the submanifold $W \cap U$ is described by the defining equation $f(z) = 0$, then the restriction of $f^k\omega$ to U is a differential form on U.

Let ω be a $(p+1)$-form on $V-W$ with a pole of order 1 along W, and suppose in addition that $d\omega$ has at most a pole of order 1 along W. Such forms are also called <u>forms with a logarithmic pole</u> along W. One sees immediately that ω has a logarithmic pole just in case the following holds : if U is a coordinate neighbourhood and f is a complex coordinate function in U for which W is the zero set of f, then the restriction of ω to U has a representation

$$\omega = \alpha \wedge \frac{df}{f} + \beta \qquad (*)$$

where α and β are differentiable forms without poles on U. If $\omega = \alpha' \wedge \frac{df}{f} + \beta'$ is any other such representation in U, then $\alpha-\alpha'$ is a form divisible by f, hence $\alpha|W = \alpha'|W$. If, in addition, $f' = uf$ with $u \neq 0$ is another defining equation for W and if $\omega = \alpha' \wedge \frac{df'}{f'} + \beta'$, then it follows that $\omega = \alpha' \wedge \frac{df}{f} + (\alpha' \wedge \frac{du}{u} + \beta')$ and hence $\alpha|W = \alpha'|W$ again. Thus the restriction $\alpha|W \cap U$ depends only on ω, and not on the choice of representation (*), or on the choice of f. If one now chooses a covering of a neighbourhood of W in V by coordinate neighbourhoods U of the above kind, and for each U chooses a representation

$$\omega|_U = \alpha_U \wedge \frac{df}{f} + \beta$$

then the restriction of α_U to $W \cap U$ defines a well-determined differential form on W. This form is called the <u>residue</u> of ω and is denoted by $\mathrm{Res}(\omega)$. Thus $\mathrm{Res}(\omega)$ is defined by

$$\mathrm{Res}(\omega)\big|_{W\cap U} = \alpha_U\big|_{W\cap U}.$$

The residue construction for forms with logarithmic pole is compatible with the exterior derivative construction, because the local representation (*) of such a form ω implies the corresponding representation

$$d\omega = d\alpha \wedge \frac{df}{f} + d\beta$$

for the derivative $d\omega$.

Thus in particular we have associated each closed $(p+1)$-form ω on $V-W$ with a pole of order 1 of the closed p-form $\mathrm{Res}(\omega)$ on W.

If in addition ω is holomorphic on V-W, then $\mathrm{Res}(\omega)$ is also a holomorphic form on W. Residues were introduced by Poincaré.

For forms with a pole of higher order the definition of residue is not quite so simple. Let ω be a $(p+1)$-form with a pole of order k at most, and suppose in addition that $d\omega$ has a pole of order k at most. This is the same as saying that ω has a local representation of the following form :

$$\omega = \frac{\alpha \wedge df}{f^k} + \frac{\beta}{f^{k-1}}$$

where α and β are differentiable forms without poles, and $k > 1$.

We set

$$\omega' = \frac{(-1)^p}{k-1} \frac{d\alpha}{f^{k-1}} + \frac{\beta}{f^{k-1}}$$

$$\psi = \frac{(-1)^{p+1}}{k-1} \frac{\alpha}{f^{k-1}} .$$

Then an easy calculation shows :

$$\omega = \omega' + d\psi.$$

Thus, modulo the exact form $d\psi$, we can replace the form ω with a pole of order k at most by the form ω' with a pole of order $k-1$ at most. For a closed form ω we can iterate this argument and hence replace ω locally, up to an exact form, by a form $\tilde{\omega}$ with a pole of order 1. If ω is meromorphic, then $\tilde{\omega}$ can also be chosen meromorphic. But this representation only holds locally! Using a differentiable partition of unity one then obtains a global representation on the whole of V-W,

$$\omega = \tilde{\omega} + d\tilde{\psi},$$

where $\tilde{\omega}$ and $\tilde{\psi}$ are differentiable forms on V-W and $\tilde{\omega}$ has at most a pole of order 1 along W.

One can define

$$\mathrm{Res}(\omega) = \mathrm{Res}(\tilde{\omega}) .$$

Admittedly, the p-form $\mathrm{Res}(\omega)$ on W is no longer uniquely determined by ω, but depends on the choice of $\tilde{\omega}$. However, an easy calculation shows that $\mathrm{Res}(\omega)$ is determined up to an exact differential form on W, and hence the cohomology class defined by $\mathrm{Res}(\omega)$ is unique. Consequently, the integral of $\mathrm{Res}(\omega)$ over each cycle γ on W has a unique value, depending only on ω and γ. We point out once again that, because of the construction of the global representation $\omega = \tilde{\omega} + d\tilde{\psi}$, the form $\mathrm{Res}(\omega)$ is only a differentiable form, even when ω is a meromorphic form with poles along W.

As an illustration of the above definitions we consider the simplest conceivable case, the classical case in which V is a neighbourhood of the origin in $\mathbb{C}$ and W is the origin. A meromorphic form ω with a pole of order k at 0 is then given by

$$\omega = \Big(\sum_{i=-k}^{\infty} a_i z^i\Big) dz.$$

We set

$$\tilde{\omega} = \frac{a_{-1}}{z} dz$$

$$\tilde{\psi} = \sum_{i\neq -1} \frac{a_i}{i+1} z^{i+1}.$$

Then

$$\omega = \tilde{\omega} + d\tilde{\psi}$$

and, according to our definition, $\mathrm{Res}(\omega) = \mathrm{Res}(\tilde{\omega})$. The residue of $\tilde{\omega}$ is, again by definition, the restriction of the 0-form a_{-1} to the point $0 \in \mathbb{C}$, i.e. simply the complex number a_{-1}. Thus we have, altogether,

$$\mathrm{Res}(\omega) = a_{-1}.$$

Now it is well known that the classical residue theorem, in its simplest form, says that the integral of ω over a small circle δ around the origin is :

$$\int_{\delta} \omega = 2\pi i \cdot a_{-1}.$$

If we regard δ as the Leray coboundary $\tau(\gamma)$ of the cycle γ consisting of the point 0, and if we view the complex number a_{-1} as the integral of the 0-form $\mathrm{Res}(\omega)$ over γ, then we can also formulate the residue theorem as

$$\int_{\tau(\gamma)} \omega = 2\pi i \int_{\gamma} \mathrm{Res}(\omega).$$

This formulation best expresses the essence of the situation, because in this form the theorem generalises to the general case considered above.

Theorem 2 (Residue theorem)

Let V be a complex manifold, let W be a 1-codimensional

complex submanifold, let ω be a closed $(p+1)$-form on $V-W$ with a pole along W, and let $\mathrm{Res}(\omega)$ be – up to an exact form – the residue of ω. Then if γ is a p-cycle on W and $\tau(\gamma)$ is its Leray coboundary we have :

$$\int_{\tau(\gamma)} \omega = 2\pi i \int_{\gamma} \mathrm{Res}(\omega).$$

We do not wish to carry out the proof here, but refer, say, to [S15] or the original work of Leray. One can say that in the final analysis the proof reduces to the classical residue theorem with the help of algebraic topological arguments.

We can also formulate the residue theorem as follows : if ω is a closed $(p+1)$-form on $V-W$ with a pole along W, if $[\omega]$ is the cohomology class in $H^{p+1}(V-W,\mathbb{C})$ defined by ω, and if $[2\pi i\ \mathrm{Res}(\omega)]$ is the cohomology class in $H^{p}(W,\mathbb{C})$ defined by the closed form $2\pi i\ \mathrm{Res}(\omega)$, then

$$\tau^*[\omega] = [2\pi i\ \mathrm{Res}(\omega)].$$

Thus the residue is – up to the inessential factor $2\pi i$ – exactly the operation on the plane of differential forms which is dual to the Leray coboundary on the homological plane.

For this reason, assertions about the relation between the homology of W and $V-W$ given by the Leray coboundary are reflected by corresponding assertions about the relation between the differential forms on W and $V-W$ given by the residue. In what follows, this principle will be applied to the case in which V is the complex projective plane P_2 and W is a nonsingular curve C in P_2.

We first investigate the Leray coboundary homomorphism for this situation.

<u>**Proposition 3**</u> :

The Leray coboundary

$$\tau : H_1(C) \to H_2(P_2-C)$$

is an isomorphism.

<u>**Proof**</u> :

We first show surjectivity. Let δ be a 2-cycle in P_2-C. The homology class $[\delta]$ in P_2 represented by δ and the homology class $[C]$ represented by C have the intersection number

$[C] \cdot [\delta] = 0,$

because δ does not meet the curve C at all. Hence it follows from our computation of the intersection ring of P_2 in 6.3.5, 6.3.6 and 6.3.7 that

$[\delta] = 0.$

Thus we can find a 3-chain α in P_2 with $\partial\alpha = \delta$. We can choose α in such a way that α meets the curve C transversely in a 1-cycle γ (a cycle, because $\partial(\alpha.C) = \partial\alpha \cdot C \pm \alpha \cdot \partial C = 0$).

We claim that the Leray coboundary $\tau(\gamma)$ is just the homology class in P_2 represented by δ. Proof : let T be a small tubular neighbourhood of C in P_2, so small that the boundary ∂T meets the chain α transversely in a 2-cycle δ'. By definition, δ' represents the Leray coboundary of γ. On the other hand, it is clear that δ and δ' are homologous, because $\delta-\delta'$ is the boundary of the chain $\alpha - \alpha \cap T$. Thus we have shown

$\tau(\gamma)$ is homologous to δ,

and hence the surjectivity of τ is proved.

Now we prove the injectivity. Let γ be a 1-cycle in C and let α be the 3-chain which one obtains as the preimage of γ in a small tubular neighbourhood $T \to C$. By definition, the boundary $\partial\alpha$ represents the Leray coboundary of γ. We have to show that when $\partial\alpha$ is null homologous in P_2-C, γ is null homologous in C. Suppose then that $\partial\alpha$ is the boundary of a 3-chain β in P_2-C. Then $\alpha-\beta$ is a 3-cycle in P_2 which meets C transversely in the cycle γ. But, since $H_3(P_2) = 0$, $\alpha-\beta$ is null homologous. It follows that γ is also null homologous, as was to be proved.

The result just proved gives us the opportunity – as we have already remarked in a more general way – to express the differential forms on C as residues of differential forms on P_2-C with poles along C.

To do this, we first consider the general problem of describing all rational differential forms on the n-dimensional projective space $P_n(\mathbb{C})$. One can arrive at quite an explicit description of these forms in a purely algebraic way (cf. [G7] Theorem 2.9). For the sake of simplicity, we want to confine ourselves here to forms of the highest degree, the n-forms.

It is completely clear how one describes the rational n-forms affinely : if $(x_0,\dots,x_n)$ are homogeneous coordinates and $(z_1,\dots,z_n)$ are affine coordinates with $z_i = {}^{x_i}/x_0$, then the rational n-forms are just the forms

$$\frac{g(z)}{f(z)}\, dz_1 \wedge \dots \wedge dz_n \qquad (*)$$

where f and g are entire rational functions. Under the substitution $z_i = {}^{x_i}/x_0$, the form $dz_1 \wedge \dots \wedge dz_n$ goes to the following form :

$$dz_1 \wedge \dots \wedge dz_n = x_0^{-(n+1)} \sum_{i=0}^{n} (-1)^i x_i dx_0 \wedge \dots \wedge \widehat{dx_i} \wedge \dots \wedge dx_n.$$

We set

$$\omega := \sum_{i=0}^{n} (-1)^i x_i dx_0 \wedge \dots \wedge \widehat{dx_i} \wedge \dots \wedge dx_n.$$

In homogeneous form, the rational function ${}^{g(z)}/f(z)$ is written as the quotient of two homogeneous polynomials of the same degree. If one then multiplies the denominator by x_0^{n+1}, then one obtains the differential form (*) as the homogeneous expression

$$\frac{G(x)}{F(x)}\, \omega, \qquad (**)$$

where F and G are homogeneous polynomials and

$$\text{degree } F = \text{degree } G + (n+1).$$

Conversely, each homogeneous form (**) yields an affine rational form (*) by the substitution $x_0 = 1$, $x_i = z_i$. Hence we have proved the following :

Lemma 4 :

Written homogeneously, the rational n-forms on $P_n(\mathbb{C})$ are just the forms

$$\frac{G(x)}{F(x)} \cdot \omega$$

where $\omega = \sum_{i=0}^{n} (-1)^i x_i dx_0 \wedge \dots \wedge \widehat{dx_i} \wedge \dots \wedge dx_n$, and F and G are homogeneous polynomials with

$$\text{degree } G = \text{degree } F - (n+1).$$

We now assume that $F(x) = 0$ is the homogeneous equation of a non-singular curve of order m, and we want to compute the residues of the form

$$\frac{G(x)}{F(x)} \cdot \omega$$

with a pole of order 1 along C. G is then a homogeneous polynomial of degree m-3, by Lemma 4.

Proposition 5 :

Let C be a nonsingular curve of order m in $P_2(\mathbb{C})$, with homogeneous equation $F(x) = 0$. Let

$$\omega = x_0 dx_1 \wedge dx_2 - x_1 dx_0 \wedge dx_2 + x_2 dx_0 \wedge dx_1 .$$

Then the rational forms on $P_2(\mathbb{C})$ with a pole of order 1 along C are just the forms

$$\frac{\Phi(x)}{F(x)} \cdot \omega,$$

where Φ runs through all the homogeneous polynomials of degree m-3. The residue of such a form may be computed as follows. Let $\Sigma a_i \frac{\partial F}{\partial x_i} = 0$ be the equation of an arbitrarily chosen polar of C. Then

$$\operatorname{Res}\left(\frac{\Phi(x)}{F(x)}\right) \cdot \omega = \frac{\Phi(x) \cdot \Sigma(-1)^{\operatorname{sign}(i,j,k)} a_i x_j dx_k}{\Sigma a_i \frac{\partial F}{\partial x_i}} .$$

Proof :

The first assertion follows from Lemma 4. The second assertion, concerning the residue, is proved as follows. Let $L(x) = 0$ be the equation of an arbitrarily chosen line through the pole (a_0, a_1, a_2). Using the Euler identity $\Sigma x_i \frac{\partial F}{\partial x_i} = mF$ (Theorem 4.4.3), one checks the following equation by a simple calculation :

$$\frac{\Phi}{F}\omega = \frac{\Phi \cdot \Sigma \pm a_i x_j dx_k}{\Sigma a_i \frac{\partial F}{\partial x_i}} \wedge \left(\frac{dF}{F} - m\frac{dL}{L}\right) . \qquad (*)$$

But this identity implies the assertion about the residue, because the rational form

$$\frac{dF}{F} - m\frac{dL}{L}$$

yields just the form

$$\frac{df}{f}$$

on restriction to the complement U of the line $L(x) = 0$, where $f = 0$ is the associated affine equation of C. Thus by restriction to U and use of (*) we have written our form $\frac{\Phi}{F} \cdot \omega$ in the form $\alpha \wedge \frac{df}{f}$, and then, by definition, $\alpha|C$ is the residue of $\frac{\Phi}{F} \cdot \omega$

restricted to $C \cap U$. This proves agreement between the residue and the form in question on an open set, and hence they are identical over the whole of C.

Remark :

The preceding proof would perhaps be simpler and more easily remembered if one wrote it in affine coordinates $z_1 = {}^{x_1}/x_0$, $z_2 = {}^{x_2}/x_0$ and made the additional assumption that the coordinates were chosen so that the pole had homogeneous coordinates $(0,0,1)$ or $(0,1,0)$. Then

$$\mathrm{Res}\left(\frac{\phi dz_1 \wedge dz_2}{f}\right) = \frac{\phi dz_1}{\partial f/\partial z_2} = \frac{-\phi dz_2}{\partial f/\partial z_1} .$$

Proof :

$$\frac{\phi dz_1}{\partial f/\partial z_2} \wedge \frac{\frac{\partial f}{\partial z_1}dz_1 + \frac{\partial f}{\partial z_2}dz_2}{f} = \frac{\phi dz_1 \wedge dz_2}{f} ,$$

and one proves the second equation analogously.

Comparison of Theorem 1 with Proposition 5 shows the following :

Proposition 6 :

The differential forms of 1st kind on a nonsingular plane curve C are just the residues of the 2-forms on P_2-C with poles of order 1 along C.

With this, we have already acquired a wide understanding of the dual isomorphism of the Leray coboundary :

$$\tau^* : H^2(P_2-C,\mathbb{C}) \to H^1(C,\mathbb{C}) .$$

If p is the genus of C, then inside the 2p-dimensional complex vector space $H^1(C,\mathbb{C})$ we have the p-dimensional vector space $H^{1,0}(C)$ generated by the differential forms of 1st kind, and inside the 2p-dimensional vector space $H^2(P_2-C,\mathbb{C})$ we have the p-dimensional subspace $\mathcal{H}_1$ generated by the rational forms with a pole of order 1, and Proposition 6 just says :

$$\tau^*(\mathcal{H}_1) = H^{1,0}(C) .$$

Now, how do things stand with the other half of the Hodge decomposition $H^1(C,\mathbb{C}) = H^{1,0}(C) + H^{0,1}(C)$, with the differential forms of 2nd kind on C and with the forms with a pole of higher order along C?

Let $\mathcal{A}^p_k$ be the vector space of rational p-forms on P_2-C with a pole of order $\leq k$ along C, and let $\mathcal{A}^p = \bigcup_k \mathcal{A}^p_k$ be the vector space of all rational p-forms with poles along C. By definition, the second algebraic de Rham cohomology group $\mathcal{H}^2$ of P_2-C is just

$$\mathcal{H}^2 = \mathcal{A}^2/d\mathcal{A}^1.$$

The algebraic de Rham theorem of Grothendieck implies :

$$\mathcal{H}^2 \cong H^2(P_2-C,\mathbb{C}).$$

We now introduce a kind of refinement of the de Rham cohomology groups $\mathcal{H}^2_k$ which takes into account the orders of the poles of the differential forms.

$$\mathcal{H}^2_k := \mathcal{A}^2_k/d\mathcal{A}^1_{k-1}.$$

One can now prove the following by purely algebraic methods (cf. [G7], Theorems 4.2, 4.3).

(i) For each rational 2-form ω on P_2-C with a pole of order $k > 2$ there is a rational 1-form ψ such that $\omega + d\psi$ has a pole of order ≤ 2.

(ii) If ω has a pole of order k along C and if there is a ψ such that $\omega + d\psi$ has a pole of order $< k$ along C, then there is such a ψ with a pole of order $< k$ along C.

With the help of the above refinement of the de Rham groups, this result can also be formulated as follows :

Proposition 7 :

The natural homomorphisms

$$\mathcal{H}^2_k \to \mathcal{H}^2$$

are surjective for $k \geq 2$ and injective for all k. In particular, one has isomorphisms (for $k \geq 2$)

$$\mathcal{H}^2_k \cong \mathcal{H}^2 \cong H^2(P_2-C)$$

and a filtration

$$\mathcal{H}^2_1 \subset \mathcal{H}^2_2 .$$

Thus to describe the cohomology $H^2(P_2-C)$ by means of differential forms, we only need to acquire a grasp of $\mathcal{H}^2_2$. By definition, $\mathcal{H}^2_2 = \mathcal{A}^2_2/d\mathcal{A}^1_1$. By Lemma 4, the forms in $\mathcal{A}^2_2$ are just the

$$\frac{\Phi}{F^2}\,\omega,$$

where Φ is a homogeneous form of degree $2m-3$. The forms in $d\mathcal{A}^1_1$

can be described explicitly as follows. We consider the form

$$\psi = -\frac{(x_0A_1-x_1A_0)dx_2-(x_0A_2-x_2A_0)dx_1+(x_1A_2-x_2A_1)dx_0}{F}$$

where the A_i are any homogeneous polynomials of degree $m-2$. Then ψ is in $\mathcal{A}_1^1$, and each element of $\mathcal{A}_1^1$ is of this form (cf. [G7] 2.9).

One finds $d\psi$ by a small computation, involving multiple use of the Euler identity 4.4.3, to be

$$d\psi = \frac{\Sigma A_i \frac{\partial F}{\partial x_i} - F\Sigma \frac{\partial A_i}{\partial x_i}}{F^2} \cdot \omega \ .$$

This determines the form of the elements of $d\mathcal{A}_1^1$, and we obtain the following result :

<u>**Proposition 8**</u> :

(i) $\mathcal{H}_1^2$ is the vector space of forms $\frac{\Phi}{F} \cdot \omega$, where Φ is homogeneous of degree $m-3$ and $\omega = x_0dx_1 \wedge dx_2 - x_1dx_0 \wedge dx_2 + x_2dx_0 \wedge dx_1$.

(ii) Let $\mathcal{F}_{2m-3}$ be the vector space of homogeneous polynomials of degree $2m-3$ and let $\mathcal{G}$ be the subspace of all polynomials $\Sigma A_i \frac{\partial F}{\partial x_i}$. Let $\Phi_1,\ldots,\Phi_p$ be the basis of $\mathcal{F}_{2m-3}/\mathcal{G}$. Then the forms

$$\frac{\Phi_r}{F^2} \cdot \omega, \quad r = 1,\ldots,p$$

are a basis of $\mathcal{H}_2^2/\mathcal{H}_1^2$.

With this result we have completely computed the cohomology group $H^2(P_2-C,\mathbb{C})$ by means of 2-forms. The residues of these 2-forms then yield a complete description of $H^1(C,\mathbb{C})$ by means of 1-forms, by the residue theorem and Proposition 3. By definition of residue, the rational forms on P_2-C with a pole of order 1 along C yield holomorphic forms, hence differential forms of 1st kind, on C. On the other hand, the forms on P_2-C with poles of order 2 yield only differentiable forms on C, when one applies the definition of residue sketched earlier. Now one can also compute the residues of these forms differently, and this leads to meromorphic forms on C, though admittedly with poles. This goes as follows :

<u>**Proposition 9**</u> :

Let C be a nonsingular plane curve with affine equation $f(x,y) = 0$, and let

$$\alpha = \frac{g}{f^2}\, dx \wedge dy$$

be a rational 2-form on the complement of C with a pole of order 2 along C. Then the cohomology class in $H^1(C,\mathbb{C})$ determined by $\mathrm{Res}(\alpha)$ is the same as that determined by the meromorphic differential $\mathrm{Res}(\beta)$ of 2nd kind, where

$$\beta = \frac{g\frac{\partial^2 f}{\partial y^2} - \frac{\partial g}{\partial y}\frac{\partial f}{\partial y}}{f\left(\frac{\partial f}{\partial y}\right)^2}\, dx \wedge dy.$$

Here the residue of the form β, which is meromorphic in $P_2 - C$ with a pole of order 1 along C, is defined analogously to the residue of a form which is holomorphic in $P_2 - C$ with a pole of order 1.

<u>**Proof**</u> :

Let $\tilde{C}$ be the polar with equation $\frac{\partial f}{\partial y} = 0$ and let ψ be the rational 1-form

$$\psi = \frac{g}{\frac{\partial f}{\partial y} f}\, dx.$$

Then we obviously have

$$d\psi = \alpha - \beta.$$

Now let γ be any cycle on the curve C which does not pass through the intersection points of C and $\tilde{C}$, and let $\tau(\gamma)$ be a cycle in $P_2 - C - \tilde{C}$ which represents the Leray coboundary of $[\gamma]$. By the residue theorem we then have

$$\int_\gamma \mathrm{Res}(\beta) = \frac{1}{2\pi i}\int_{\tau(\gamma)} \beta = \frac{1}{2\pi i}\int_{\tau(\gamma)} \alpha - d\psi = \frac{1}{2\pi i}\int_{\tau(\gamma)} \alpha = \int_\gamma \mathrm{Res}(\alpha).$$

If γ is a small oriented circle in C around a pole p of $\mathrm{Res}(\beta)$, then γ is null homologous in C and hence $\int_\gamma \mathrm{Res}(\alpha) = 0$, since $\mathrm{Res}(\alpha)$ is a differentiable form over the whole of C in the sense of the earlier definition (equivalent argument : $\tau(\gamma)$ is null homologous in $P_2 - C$, and α is differentiable (even holomorphic) over the whole of $P_2 - C$, hence $\int_{\tau(\gamma)} \alpha = 0$.) But $\int_\gamma \mathrm{Res}(\alpha) = 0$ simply means that the classical residue of the meromorphic 1-form $\mathrm{Res}(\alpha)$ on C vanishes at the pole p. $\mathrm{Res}(\alpha)$ is therefore a meromorphic differential form of 1st or 2nd kind on C. Hence $\mathrm{Res}(\beta)$ uniquely defines a cohomology class. This class is the same as the one given by $\mathrm{Res}(\alpha)$, because its application to the homology class $[\gamma]$ of a cycle γ is given by integration of the differential form over the cycle, and since one can always choose γ so that γ avoids the intersection points of C and $\tilde{C}$, the equation

$$\int_\gamma \mathrm{Res}(\beta) = \int_\gamma \mathrm{Res}(\alpha)$$

shows the equality of the two homology classes.

For later calculations it is useful to have an explicit affine description of $\mathrm{Res}(\beta)$. This description follows immediately from the definition. Multiplying numerator and denominator of β by $\frac{\partial f}{\partial y}$ one gets :

$$\beta = \frac{g\cdot\frac{\partial^2 f}{\partial y^2} - \frac{\partial g}{\partial y}\frac{\partial f}{\partial y}}{(\frac{\partial f}{\partial y})^3} \cdot \frac{\frac{\partial f}{\partial y}}{f}\, dx \wedge dy = \frac{g\frac{\partial^2 f}{\partial y^2} - \frac{\partial g}{\partial y}\frac{\partial f}{\partial y}}{(\frac{\partial f}{\partial y})^3}\, dx \wedge \frac{df}{f},$$

whence we have

Corollary to Proposition 9 :

$$\mathrm{Res}(\beta) = \frac{g\frac{\partial^2 f}{\partial y^2} - \frac{\partial g}{\partial y}\cdot\frac{\partial f}{\partial y}}{(\frac{\partial f}{\partial y})^3}\, dx .$$

Remark :

More generally, one can use the relation

$$d\Big(\frac{1}{(k-1)}\frac{g}{\frac{\partial f}{\partial y}f^{k-1}}\,dx\Big) = -\frac{g}{f^k}\,dx \wedge dy + \frac{1}{(k-1)f^{k-1}}\,d\Big(\frac{g}{\partial f/\partial y}\Big) \wedge dx$$

to progressively lower the pole order of a meromorphic form $\frac{g}{f^k}dx \wedge dy$ on the complement with a pole of higher order along C, until one has a form with a pole of order 1.

We summarise our results on the description of the cohomology of a curve by means of differential forms in the following theorem.

Theorem 10 :

Let C be a nonsingular algebraic curve, with the following associated complex vector spaces of differential forms resp. cohomology classes :

A^0	vector space of rational functions on C
A^1	vector space of differentials of 1st or 2nd kind
dA^0	vector space of differentials of rational functions
$H = A^1/dA^0$	meromorphic "de Rham cohomology" of C

$H^1 \subset H$ subspace of differentials of 1st kind

$H^1(C, \mathbb{C})$ 1st cohomology group of C with complex coefficients

$H^{1,0} \subset H^1(C, \mathbb{C})$ the subspace generated by closed $(1,0)$-forms

$H^{0,1} \subset H^1(C, \mathbb{C})$ the subspace generated by closed $(0,1)$-forms

$H^2(P_2 - C, \mathbb{C})$ 2nd cohomology group of the complement

$H^2_1 \subset H^2_2$ the refined de Rham cohomology groups of 2-forms on $P_2 - C$ with poles of order 1 resp. 2 along C.

For these objects one has the following commutative diagram of maps, in which the horizontal arrows are induced by inclusions or coset maps and the vertical arrows are isomorphisms.

$$2\pi i \, \text{Res} \left(\begin{array}{c} H^2_1 \hookrightarrow H^2_2 \longrightarrow H^2_2/H^2_1 \\ \downarrow \alpha \\ H^2(P_2-C, \mathbb{C}) \\ \downarrow \tau^* \\ H^{1,0} \hookrightarrow H^1(C,\mathbb{C}) \longrightarrow H^1(C,\mathbb{C})/H^{1,0} \cong H^{0,1} \\ \uparrow \quad \uparrow \beta \quad \uparrow \\ H^1 \hookrightarrow H \longrightarrow H/H^1 \end{array} \right) 2\pi i \, \text{res}$$

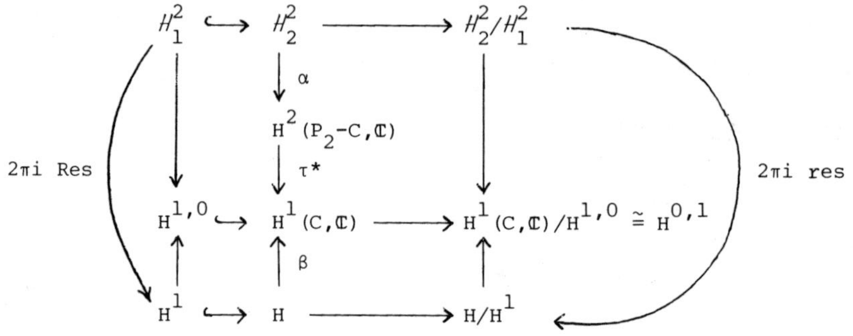

Here Res denotes the operation of constructing the residue of a form with a pole of order 1 along C, and the map res is defined by Proposition 9 ; τ^* is the dual isomorphism to the Leray coboundary. The maps α and β are induced by integration of differential forms over cycles, and the remaining isomorphisms are induced by β and $\tau^* \circ \alpha$.

Proof :

The theorem follows easily from Propositions 3, 6, 7 and 9, together with the residue theorem and the cited assertions about the Hodge decomposition, as soon as it is proved that $\beta : H \to H^1(C, \mathbb{C})$ is injective. But this is clear — a differential of 1st or 2nd kind whose periods all vanish is the differential of a rational function, namely, its own indefinite integral.

Propositions 8 and 9, and Theorem 10, give a complete description of the differentials of 1st and 2nd kind on a nonsingular plane algebraic curve. We want to lay out this description explicitly for two examples. Since the nonsingular curves of order 1 and 2, i.e. the lines and quadrics, are in fact rational, the theorems just mentioned show that they have no differentials of 1st and 2nd kind, so the first interesting case is that of cubics. We therefore choose the cubics in Hesse and Legendre normal form as examples.

Example 1 :

Cubics in Hesse normal form

By 7.3.4, a cubic C_μ in Hesse normal form has homogeneous equation $F(x_0,x_1,x_2) = 0$ and affine equation $f(x,y) = 0$, where

$$F = x_0^3 + x_1^3 + x_2^3 - 3\mu x_0x_1x_2$$

$$f = 1 + x^3 + y^3 - 3\mu xy.$$

Since C_μ has degree $m = 3$, the adjoint polynomials of degree $m-3$ have degree 0, and hence are constant. Hence by Theorem 1 (or Proposition 8), the differentials of 1st kind on C_μ are just $c\cdot\omega_1$, where $c \in \mathbb{C}$ and $\omega_1 = \frac{1}{3}\frac{dx}{\partial f/\partial y}$, so

$$\omega_1 = \frac{dx}{y^2-\mu x}\ .$$

To determine the differentials of 2nd kind we have, by Proposition 8, to find a basis of $\mathcal{F}_3/\mathcal{F}_3 \cap \mathcal{I}$, where $\mathcal{I}$ is the ideal generated by the partial derivatives $\frac{\partial F}{\partial x_0}, \frac{\partial F}{\partial x_1}, \frac{\partial F}{\partial x_2}$. One easily convinces oneself that $x_0x_1x_2$ represents such a basis. In this way one finds a basis of $\mathcal{H}_2^2/\mathcal{H}_1^2$ to be the form

$$\alpha = \frac{xy}{f^2}\, dx \wedge dy.$$

If one now uses Proposition 9 and the Corollary to go from α to the equivalent form β, and constructs $\mathrm{Res}(\beta)$, then one finds the differential of 2nd kind to be, up to a constant,

$$\omega_2 = \frac{x(y^2+\mu x)}{(y^2-\mu x)^3}\, dx\ .$$

Thus we obtain the result : the differentials

$$\omega_1 = \frac{dx}{y^2-\mu x} \quad \text{and} \quad \omega_2 = \frac{x(y^2+\mu x)}{(y^2-\mu x)^3}\, dx$$

of 1st and 2nd kind represent a basis of $H^1(C_\mu,\mathbb{C})$, where C_μ is a cubic in Hesse normal form.

Example 2 :

Cubics in Legendre normal form

It follows from 7.3.11 that every nonsingular cubic has an affine equation of the following form relative to suitable coordinates :

$$y^2 = x(x-1)(x-t) \qquad t \neq 0,1,\infty.$$

We denote the cubic with this equation by C_t. The equation is also called Legendre normal form. We set

$$f(x,y,t) = y^2 - x(x-1)(x-t)$$

$$F(x_0,x_1,x_2,t) = x_0x_2^2 - x_1(x_1-x_0)(x_1-tx_0).$$

By Proposition 8 and Theorem 1 the differentials of 1st kind are the multiples of $(\frac{\partial f}{\partial y})^{-1}dx$, i.e. of

$$\omega_1 = \frac{dx}{y}.$$

We remark that the indefinite integral $\int \omega_1$ has already been considered at the end of 7.4.3. It is the elliptic integral

$$\int \frac{dx}{\sqrt{x(x-1)(x-t)}},$$

which describes the — many-valued — inverse mapping of the universal covering

$$\mathbb{C} \to C_t.$$

(We recall that the elliptic curve C_t is isomorphic to the complex torus

$$\mathbb{C}/a_1\mathbb{Z} + a_2\mathbb{Z},$$

where $a_i = \int_{\gamma_i} \omega_1$ and γ_1, γ_2 are cycles which represent a basis of $H_1(C_t)$.)

To compute the differentials of 2nd kind we again have to find a basis of $\mathcal{F}_3/\mathcal{F}_3 \cap (\frac{\partial F}{\partial x_0}, \frac{\partial F}{\partial x_1}, \frac{\partial F}{\partial x_2})$. One can check that the homogeneous polynomial of degree 3,

$$x_0x_1^2 - x_0^2x_1$$

represents a basis of this vector space. Hence one obtains the 2-form

$$\alpha := \frac{x^2-x}{f^2}dx \wedge dy$$

as basis for $\mathcal{H}_2^2/\mathcal{H}_1^2$. If one uses the Corollary to Proposition 9 to go to an equivalent meromorphic form β with a pole of order 1 along C, and constructs the residue of β, then one finds the differential of 2nd kind to be, up to constant,

$$\omega_2 = \frac{x(x-1)dx}{2y^3} .$$

Thus we have the following result for a cubic C_t in Legendre normal form :

The two differentials

$$\omega_1 = \frac{dx}{y} \quad \text{and} \quad \omega_2 = \frac{x(x-1)dx}{2y^3} ,$$

of 1st and 2nd kind, represent a basis of $H^1(C_t,\mathbb{C})$.

The explicit description of the differentials of 1st and 2nd kind on a curve makes possible a whole series of important further investigations of curves. We cannot go into all these questions here. We shall therefore demonstrate the usefulness of the explicit description of differentials of 1st and 2nd kind in just one problem, by way of example. My reason for choosing this example is that, in itself, it is a simple, beautiful and classical example of the investigation of periods of integrals which – particularly through the work of P.A. Griffiths, W. Schmid, P. Deligne, N. Katz and others – has proved to be fruitful in recent years and also is of relevance to my own work. In order to allow the significance of this example to emerge clearly, I must backtrack again, as this is necessary to give a proper account of the example in the shortest possible form.

The problem we pose is the following : in the family of cubics C_t in Legendre normal form

$$y^2 - x(x-1)(x-t) = 0,$$

we want to investigate the dependence of the periods of differentials of 1st kind on the parameter t. The significance of this problem became apparent in the course of sections 7.3 and 7.4, in which we first investigated cubics. There, among other things, we stated and partly proved the following two facts :

(1.) Associated with each nonsingular cubic is a certain number, its j-invariant, and two cubics are isomorphic just in case they have the same j-invariant. For the family of cubics in Hesse normal form we computed the j-invariant in 7.3.10, with the help of our analysis of the inflection point configuration – it is a certain rational function of the parameter of the family. From this one could also compute the j-invariant of the Legendre normal form C_t as a function of t (cf. remark (iv) after 7.3.11).

(2.) If a_1 and a_2 are the periods $a_i = \int_{\gamma_i} \omega$ of a differential of first kind on a cubic C, then C is isomorphic to the 1-dimensional complex torus $\mathbb{C}/a_1\mathbb{Z}+a_2\mathbb{Z}$, and hence to the 1-dimensional torus

$$\mathbb{C}/\mathbb{Z} + \frac{a_1}{a_2}\mathbb{Z}.$$

One can compute the j-invariant as a certain meromorphic function

$$j = J(\frac{a_1}{a_2})$$

which is invariant under the operations in the group $PSL(2,\mathbb{Z})$ of integral linear fractional transformations (cf. 7.4.4 and the remarks at the end of 7.4).

In our investigations which follow we shall not only compute the j-invariant $j = J(\frac{a_1}{a_2})$, but even express the ratio $\frac{a_1}{a_2}$ of the periods of C_t as a function of t.

The cubics in Legendre normal form constitute a linear system of cubics. By passing to homogeneous coordinates in the (x_0,x_1,x_2)-plane with $x = {}^{x_1}/x_0$, $y = {}^{x_2}/x_0$, and to homogeneous coordinates (t_0,t_1) for the parameter $t = t_1/t_0$, the cubics of this linear system become described by the following equations

$$t_0[x_0x_2^2-x_1^2(x_1-x_0)] + t_1[x_0x_1(x_1-x_0)] = 0.$$

The system is parametrised by the points (t_0,t_1) of a 1-dimensional complex projective space $P_1(\mathbb{C})$. For the three parameter values (1,0), (1,1) and (0,1) the corresponding cubics C_0, C_1, C_∞ are singular. C_0 and C_1 are irreducible cubics with ordinary double points at $p_1 = (1,0,0)$ resp. $p_2 = (1,1,0)$, and C_∞ decomposes into three lines which all meet at the point $p_3 = (0,0,1)$. Any two distinct cubics of the linear system meet at precisely these points p_1, p_2, p_3.

The fact that the different cubics of the system have intersection points in common makes the investigation in a certain sense unclear and complicated. We would prefer to go to a simpler situation, as in section 9.2, where the curves no longer intersect, but instead lie disjoint from each other as fibres of a fibration. With this objective, we blow up the projective plane $P_2(\mathbb{C})$ at the three points p_1, p_2, p_3. In fact we blow up repeatedly, namely, at the points at which the proper preimages of the curves C_t still intersect. After a total of 9 σ-processes, two at p_1, two at p_2 and 5 at p_3, or at infinitely near points of these, one obtains a 2-dimensional complex manifold X in which the proper preimages of the curves C_t are all disjoint.

One can then obtain a well defined mapping $\phi : X \to P_1(\mathbb{C})$ with these curves as fibres, in the following way. Let

$$\pi : X \to P_2(\mathbb{C})$$

be the modification mapping which results from blowing up the 9 points. Also let

$$F : P_2(\mathbb{C}) \to P_1(\mathbb{C})$$

be the rational "mapping" with $F(x_0,x_1,x_2) = (t_0,t_1)$ where

$$t_0 = x_0x_1(x_0-x_1)$$
$$t_1 = x_0x_2^2-x_1^2(x_1-x_0).$$

F is not well defined at the three points p_3,p_1,p_2, but the composition

$$\phi = F \circ \pi$$

is a well defined holomorphic mapping

$$\phi : X \to P_1(\mathbb{C}).$$

What do the fibres $X_t = \phi^{-1}(t)$, $t \in P_1(\mathbb{C})$, look like for this mapping?

It is clear that for $t \neq 0,1,\infty$ the mapping π induces an isomorphism

$$X_t \cong C_t$$

between the fibre X_t and the nonsingular cubic C_t. The singular fibres X_0,X_1,X_∞ consist of the proper preimages of the three singular cubics C_0,C_1,C_∞ together with some of the exceptional curves. It is an easy exercise to determine the precise configuration of these singular fibres, and the result is the following : X_0 consists of two nonsingular rational curves which meet transversely at two points. Schematic picture :

One of the two curves is the preimage of the cubic with the ordinary double point p_1, the other is the exceptional curve introduced at p_1, which separates the two branches of the double point.

The fibre X_1 looks just like X_0. The fibre X_∞ consists of 7 nonsingular rational curves which meet transversely in a configuration which is described schematically by the following picture :

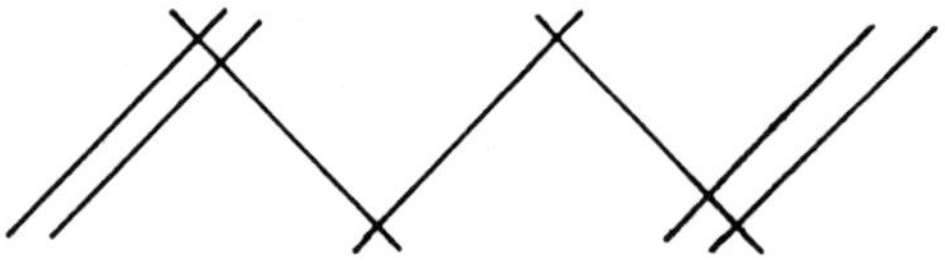

Three of these curves are the preimages of the three components of C_∞, the other four are the first four of the five exceptional curves introduced at p_3. All reduced components of the three singular fibres X_0, X_1, X_2 have self-intersection number -2.

Thus we have constructed a 2-dimensional complex manifold X together with a holomorphic mapping $\phi : X \to S$ onto a nonsingular curve S (namely, $S = P_1(\mathbb{C})$), such that the fibres X_t, with finitely many exceptions, are all elliptic curves. Such a surface is called an <u>elliptic surface</u>. In the sixties, elliptic surfaces were investigated by K. Kodaira in a series of beautiful works. One of the first and simplest problems here is the description of the possible singular fibres, the exceptional fibres. In our case we have described the exceptional fibres X_0, X_1, X_2 quite explicitly. It is useful to represent the associated configuration by a graph similar to the resolution graph of 8.4, whose vertices correspond to curves and whose edges correspond to intersections. Then one obtains the following graphs for X_0, X_1 resp. X_∞ :

A_1

D_6

Those who know the theory of Lie groups and algebras will recognise these graphs as the extended Dynkin diagrams of types A_1 resp. D_6. This coincidence is no accident. Kodaira's classification of the exceptional fibres of elliptic surfaces shows that, apart from a pair of simple special cases, the graphs of exceptional fibres which can appear are just the extended Dynkin diagrams of type A_k , $k = 1,2,\ldots$, or D_k , $k = 4,5,\ldots$ or E_6, E_7, E_8 (Kodaira [K5]).

Our elliptic surface is a very special elliptic surface : it is an <u>elliptic modular surface</u> of level 2 (cf. T. Shioda [S16]). This family has 4 points of order 2, whose values for each t are just the 4 points of order 2 of the abelian variety C_t - in much the same way the family of cubics in Hesse normal form has 9 intersections, whose values are the 9 points of order 3, i.e. the 9 inflection points (cf. 7.4.7).

After this small digression on elliptic surfaces in general, we return to our special elliptic surface $\phi : X \to P_1(\mathbb{C})$ which we have obtained from the cubics in Legendre normal form. We consider the cohomology $H^1(X_t,\mathbb{C})$ and the homology $H_1(X_t, \mathbb{Z})$ of the fibres. As far as the cohomology of the fibres is concerned, in Example 2 we have already described a basis for $H^1(X_t,\mathbb{C})$ with the help of two differentials, of 1st and 2nd kind, namely

$$\omega_1 = \frac{dx}{y} = \frac{dx}{\sqrt{x(x-1)(x-t)}}$$

$$\omega_2 = \frac{x(x-1)dx}{2y^3} = \frac{x(x-1)dx}{2(\sqrt{x(x-1)(x-t)})^3} .$$

The problem we have set ourselves is to investigate the dependence of the periods of ω_1 on the parameter t. This means : we must provide ourselves with two cycles $\gamma_1(t)$ and $\gamma_2(t)$ in C_t or X_t which represent a basis of $H_1(X_t)$, and investigate the dependence of the integrals

$$\int_{\gamma_j(t)} \omega_i$$

on the parameter t. To do this sensibly it is obviously necessary to choose the family of cycles in a sensible way, and we address ourselves to this problem first.

The problem lies in the fact that for different values of t the cycles $\gamma_j(t)$ must lie in different fibres. Naturally, the cycles should depend continuously on the variable t, in a suitable sense. It is a matter of making this condition precise. Now this is quite a general problem which arises with each differentiable family of differentiable manifolds. Let $\phi : X' \to S'$ be a differentiable proper mapping of differentiable manifolds, whose rank at all points equals the dimension of S', so that the fibres $X_t = \phi^{-1}(t)$ are compact differentiable manifolds. How does one then define continuous families of cycles $\gamma(t)$ in X_t? (In our special example $X' = X - (X_0 \cup X_1 \cup X_\infty)$ and $S' = P_1(\mathbb{C}) - \{0,1,\infty\}$.) We shall now give a construction which permits a path $g : [0,1] \to S'$ to be lifted uniquely, for each $x \in X_{g(0)}$, to a "horizontal" path $\tilde{g}_x : [0,1] \to X'$ with $\tilde{g}_x(0) = x$. The continuous family of cycles $\gamma(t)$ then results from a fixed cycle γ in X_s by "parallel displacement" along a path from s to t in S' by means of the horizontal lifting.

In order to be able to lift the path g uniquely, we need a decomposition of the tangent bundle into a "vertical" part and a "horizontal" part. This is certainly possible locally. Because the rank condition on ϕ is equivalent, by the implicit function theorem, to the condition that for each point $x \in X'$, there is a neighbourhood U of x in X', neighbourhoods $V = \phi(U)$ of $s = \phi(x)$ in S' and W of x in X_s, and a diffeomorphism $h : U \to V \times W$ which converts the mapping ϕ into projection onto the first factor, i.e. which makes the following diagram commute.

$$\begin{array}{ccc} U & \xrightarrow{\;h\;} & V \times W \\ & {\scriptstyle\phi}\searrow \quad \swarrow{\scriptstyle p} & \\ & V & \end{array}$$

At each point $x \in U$ the tangent mapping of h^{-1} gives us a direct sum decomposition into the <u>vertical</u> tangent space T^v of the fibre through x and a <u>horizontal</u> tangent space T^h. One obtains the decomposition of the tangent bundle into a direct sum of vertical and horizontal subbundles,

$$TX' = T^vX' \oplus T^hX'$$

by pasting together local decompositions $T^v \oplus T_i^h$ in the neighbourhoods U_i by means of a partition of unity, $\{\psi\}$, where the ψ_i are differentiable functions with support U_i, $\psi_i \geq 0$ and $\Sigma\psi_i = 1$. One obtains the desired subspace T^h as the set of vectors

$$T^h = \{a = \Sigma\psi_i(x)a_i\},$$

where the a_i are the intersections of the horizontal spaces T_i^h with the vertical affine spaces parallel to T^v, so that $\{a_i\} = T_i^h \cap (T^v+c)$.

The vertical subbundle T^vX' is therefore just the kernel of the tangent mapping $T^v : TX' \to TS'$, while the horizontal subbundle T^hX' is a complement of T^vX'. Such a horizontal subbundle T^hX' is also called an <u>Ehresmann connection</u> for $\phi : X' \to S'$. In contrast to T^vX', T^hX' is not unique. When we choose a fixed Ehresmann connection T^hX', this allows us to carry out a kind of "parallel displacement" of the fibres, in fact parallel displacement along any differentiable path in the base space S' of the mapping $X' \to S'$. If $g : [0,1] \to S'$ is any differentiable path and if $x \in X_{g(0)}$, then there is an interval $[0,r(x)] \subset [0,1]$ and a horizontal lifting $\tilde{g}_x : [0,r(x)] \to X'$ of g. I.e. $\tilde{g}_x$ is a differentiable path in X' such that (i) $\tilde{g}_x(0) = x$, (ii) $\phi \circ \tilde{g}_x = g$, and (iii) each tangent vector to $\tilde{g}_x$ is horizontal, i.e. an element of T^hX'.

Condition (iii) means that $\tilde{g}_x$ must satisfy a certain system of ordinary differential equations, and it follows from the existence and uniqueness theorems for ordinary differential equations that the horizontal lifting $\tilde{g}_x$ exists and is unique on a small interval $[0,r(x)]$, where r depends continuously on x. Since ϕ was assumed proper, $X_{g(0)}$ is compact and $r(x) \geq r > 0$ for all x. By repeated local liftings we finally obtain, for each $x \in X_{g(x)}$, a unique horizontal lifting $\tilde{g}_x : [0,1] \to X'$ of g. The endpoint of the path $\tilde{g}_x(s)$ is then a point $h_g(x)$ in the fibre $X_{g(1)}$ over the endpoint $g(1)$ of the path g, and in this way one gets a diffeomorphism

$$h_g : X_{g(0)} \to X_{g(1)}$$

between the fibres over the initial and final points. This is the parallel displacement of $X_{g(0)}$ along the path g.

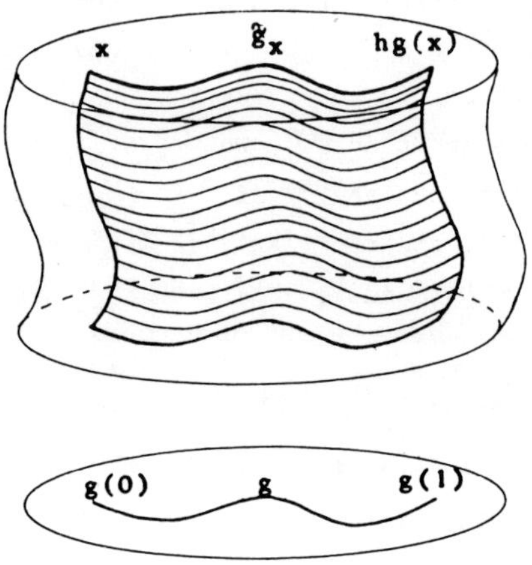

Let us fix a basepoint $t \in S'$. With each closed piecewise differentiable path g in S' with endpoint t we associate the corresponding diffeomorphism $h_g : X_g \to X_t$. The set of diffeomorphisms of X_t obtained in this way forms a group, called the holonomy group of the connection with respect to t. If the closed paths g_0 and g_1 are homotopic, then the corresponding diffeomorphisms h_{g_0} and h_{g_1} are isotopic. In particular, for null homotopic closed paths g the diffeomorphisms h_g are isotopic to the identity. They form a normal subgroup of the holonomy group, the restricted holonomy group. If we divide by this subgroup, $g \mapsto h_g$ induces a homomorphism of the fundamental group $\pi_1(S',t)$ into the quotient group. This may still depend on the connection. To get something independent of the choice of connection, we have to pass from the group $\mathrm{Diff}(X_t)$ of all diffeomorphisms of X_t to its quotient $\mathrm{Diff}(X_t)/\mathrm{Is}(X_t)$, where $\mathrm{Is}(X_t)$ is the normal subgroup of diffeomorphisms isotopic to the identity. Then we get a homomorphism

$$\pi_1(S',t) \to \mathrm{Diff}(X_t)/\mathrm{Is}(X_t)$$

The image of this homomorphism is sometimes called the geometric monodromy group. (The word "monodromic" means "running uniquely" and refers to the unique lifting of paths.

Using the parallel displacement of an Ehresmann connection one can prove that $\phi : X' \to S'$ is a locally trivial differentiable fibre bundle. One gives S' a Riemannian metric. For each point $s \in S'$

one chooses a normal geodesic coordinate neighbourhood U_s in S' and defines a trivialising diffeomorphism $X_s \times U_s \to \phi^{-1}(U_s)$ by $(x,s') \mapsto h_g(x)$, where g is the geodesic arc from s to s'.

The assertion that $\phi : X' \to S'$ is a locally trivial fibre bundle under the given hypotheses is the Ehresmann fibration theorem. The proof sketched here is given by J.A. Wolf [W8]. It is easily generalised to manifolds with boundary. If X' is a differentiable manifold with boundary $\partial X'$, and if $\phi|\partial X'$ as well as ϕ has maximal rank dim S', then $\phi : X' \to S'$ and $\phi|\partial X' : \partial X' \to S$ are locally trivial fibre bundles.

Parallel displacement also allows immediate definition of continuous families of cycles, for whose sake we developed the whole theory above. If γ is any p-cycle in a fibre X_s, then we can transport γ into any fibre X_t by parallel displacement, and thus obtain a family of cycles $\gamma(t)$. Admittedly, the result of the parallel displacement depends on the choice of path along which displacement occurs. Since one can choose different paths from s to t, one obtains different cycles in one and the same fibre X_t, convertible to each other by operations in the corresponding holonomy group. The resulting family of cycles $\gamma(t)$ is therefore a many-valued function which associates each $t \in S$ with all cycles in X_t obtainable from a fixed cycle by parallel displacement. These families of cycles are what we mean by a "continuous family of cycles". Since they come into being through parallel displacement by means of a horizontal lifting of paths, in future we shall prefer to speak of a horizontal family of cycles. Locally, of course, such a horizontal family may be represented by single-valued branches, since locally we may use the same kind of parallel displacement as in the trivialisation of the bundle.

The definition of horizontal families of cycles depends on the choice of Ehresmann connection $\phi : X' \to S'$. However, if one passes from the cycles to their homology classes, then simple algebraic topological arguments show that for homology classes the result of parallel displacement is independent of the choice of connection and depends only on the fibration $X' \to S'$. In this way one obtains the concept of a horizontal family of homology classes. In addition, one obtains isomorphisms of the homology of the fibres, induced by parallel displacement, and in particular a homomorphism

$\pi_1(S',t) \to Aut(H_p(X_t))$.

One calls this the monodromy of the fibration, and its image is called the monodromy group of the fibration.

For practical work with horizontal families it is sufficient, and simpler, to use any families of cycles which induce horizontal families of homology classes. This is how we shall proceed, but we call these more general families "horizontal families of cycles" as well.

Now we want to compute an example of the monodromy group of a fibration quite explicitly, and by elementary methods. As example we choose the elliptic surface which we have already obtained from the family of cubics C_t in Legendre normal form. In order to be able to give the most explicit possible description of parallel displacement for cycles in C_t, we again describe the cubic C_t, as in 3.4, as a branched double covering of the Riemann sphere with four branch points : $0, 1, t$ and ∞.

We briefly recall the construction in 3.4. For a fixed $t = t_0$ we can describe C_t as follows : we triangulate the sphere as a tetrahedron with vertices $0, 1, t, \infty$. We cut the tetrahedron along the three edges going to ∞, and thereby obtain a large "triangle" consisting of four small "triangles", the faces of the tetrahedron. If one cuts the 2-fold covering C_{t_0} of the sphere in the same way, then it falls into two sheets, two "large triangles". Identification along one side yields a parallelogram, further identifications along opposite sides yield a real 2-dimensional torus, homeomorphic to C_{t_0} by construction. The following pictures illustrate the situation :

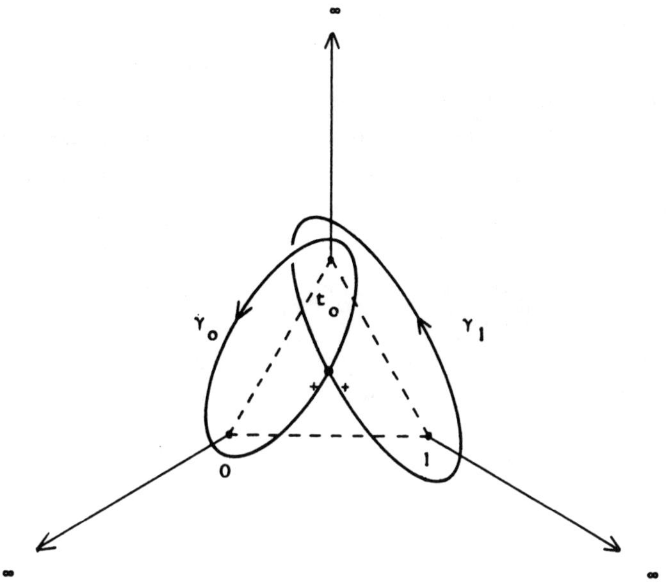
∞
t_o
γ_o
γ_1
+ +
0
1
∞
∞

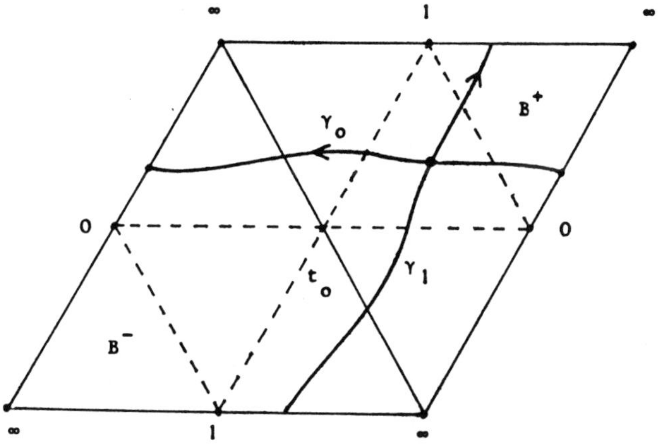
∞
1
∞
B^+
γ_o
0
0
t_o
γ_1
B^-
∞
1
∞

The decomposition into triangles is not essential to describe the covering, but it makes it easier to follow how cycles in C_t are mapped to cycles in the plane, and conversely how cycles in the plane lift to curves in C_t. Each curve in the plane which does not run through a branch point lifts uniquely to a curve in the covering, as soon as one chooses a point q in the covering to correspond to a fixed point p of the curve in the plane. This can be done by specifying in which of the two sheets B^+, B^- the point q is to lie. Thus we can describe a curve in the covering by giving its image in the plane together with a mark of + or - on a suitable point on the image curve, according as its preimage lies in B^+ or B^-. Without this information, a given curve in the plane has two possible preimage curves, convertible into each other by the covering transformation which exchanges the two sheets.

For example, in the above diagram we have described two cycles in the plane, consisting of simple circuits around 0 and t_0, resp. 1 and t_0. With the help of the triangle decomposition one can clearly see that their preimages in the parallelogram run in the way shown in the diagram. In the torus which results from identification of opposite sides the preimage cycles γ_0 and γ_1 of the two cycles in the plane are therefore homologous to a latitude circle and a meridian on the torus. Thus the cycles γ_0 and γ_1 represent a basis of $H_1(C_{t_0})$. We want to parallel displace these cycles in our family of elliptic curves C_t and thus compute the monodromy group of the fibration.

To parallel displace the cycles, it suffices to parallel displace the curves themselves. More precisely, it suffices, for a given path $w : [0,1] \to P_1(\mathbb{C}) - \{0,1,\infty\}$ with initial point $t_0 = w(0)$ and final point $t_1 = w(1)$, to parallel displace the curve C_{t_0} into the curve C_{t_1}, i.e. to construct a continuous family of homeomorphisms

$$\Psi_s : C_{t_0} \to C_{w(s)}, \quad s \in [0,1]$$

with

$$\Psi_0 = \mathrm{id}_{C_{t_0}} .$$

We obtain such a family of homeomorphisms Ψ_s as follows : we choose a continuous family of homeomorphisms

$$\phi_s : P_1(\mathbb{C}) \to P_1(\mathbb{C})$$

with the following properties :

$$\phi_0 = id_{P_1(\mathbb{C})}$$

$$\phi_s(0) = 0,\ \phi_s(1) = 1,\ \phi_s(\infty) = \infty,\ \phi_s(t_0) = w(s).$$

Then one easily convinces oneself, using familiar elementary properties of branched coverings, that there is a unique continuous family of continuous mappings $\Psi_s : C_{t_0} \to C_{w(s)}$ which forms, together with the mappings ϕ_s and the branched covering mappings, a commutative diagram

$$\begin{array}{ccc} C_{t_0} & \xrightarrow{\Psi_s} & C_{w(s)} \\ \downarrow & & \downarrow \\ P_1(\mathbb{C}) & \xrightarrow[\phi_s]{} & P_1(\mathbb{C}) \end{array} ,$$

and for which $\Psi_0 = id$. The ϕ_s are obviously homeomorphisms and hence the desired family is constructed.

There are of course many possible choices for the family ϕ_s. The particular one chosen is quite irrelevant : parallel displacement always leads to the same result on homology classes. To choose ϕ_s one can, e.g., proceed as follows. Assume, for the sake of simplicity, that w is a simple closed path around the point $t = 0$. We identify a neighbourhood of the closed curve in $P_1(\mathbb{C}) - \{0,1,\infty\}$ with the annulus $S^1 \times [-1,1]$, and in fact in such a way that the path w becomes identified with $w(s) = (e^{2\pi is},0)$. Then for $0 \leq s \leq 1$ we define :

$$\phi_s \mid P_1(\mathbb{C}) - S^1 \times [-1,1] = id$$

$$\phi_s(e^{2\pi i\alpha},\rho) = (e^{2\pi i(\alpha+1-|\rho|s)},\rho) \quad \text{for} \quad 0 \leq \alpha \leq 1,\ -1 \leq \rho \leq 1.$$

The following picture illustrates the definition :

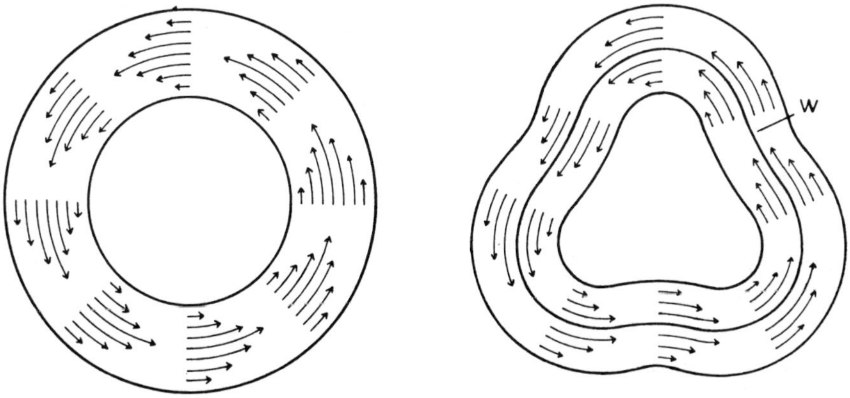

In practice, when one wants to parallel displace a given cycle γ along a given path w, one does not of course explicitly write down the above identification of an annulus with a neighbourhood of w ; one takes it for granted and proceeds intuitively to find what $\Psi_1(\gamma)$ looks like qualitatively ; i.e. one finds a cycle homologous to $\Psi_1(\gamma)$ intuitively.

In this connection it is necessary and useful to make clear – at least in principle – how the homeomorphism ϕ_1 lifts to the homeomorphism Ψ_1 of the double covering $C_{t_0} \to P_1(\mathbb{C})$. It is easy to see the following : the preimage of the spine of the annulus, i.e. the curve w along which one makes the displacement, is a lemniscate-like curve $\tilde{w}$ with a self-intersection, and the preimage of the annulus looks qualitatively like a neighbourhood of the lemniscate bounded by Cassini curves, as shown in the following picture :

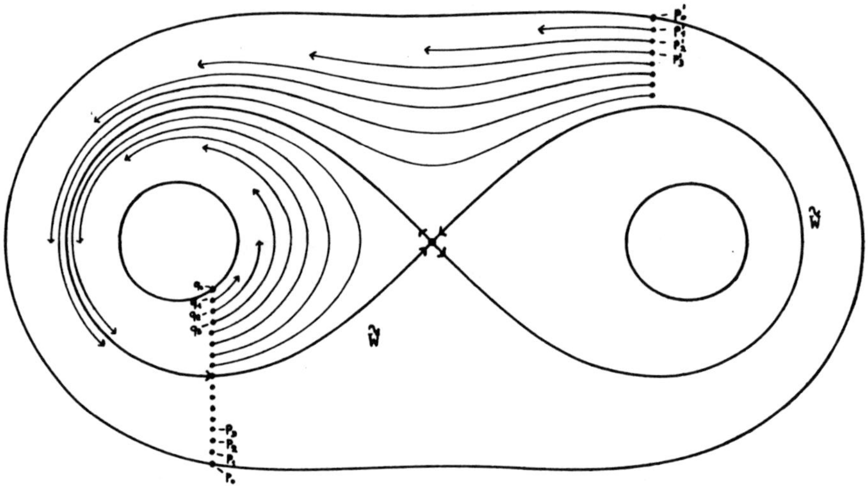

(The arrowheads indicate the positions of the images of the points p_i and q_j under Ψ_1.)

The picture also shows where a few points p_i, q_j are transported by the diffeomorphism Ψ_1. On the lemniscate itself Ψ_1 is the identity, and likewise on the two ovals which bound the region on the inside. On the outer Cassini curve Ψ_1 is the covering transformation which looks qualitatively like reflection in the double point of the lemniscate. The points q_j inside the lemniscate are carried a bit further in the direction of the orientation of the arc $\tilde{W}$ induced by the orientation of W. How far, depends on their distance from the arc. On the other hand, the points p_i outside are first sent to the points p_i' by the covering transformation, and then a bit further in the direction of the lemniscate arc. Outside the preimage of the annulus, Ψ_1 is the identity or the covering transformation. In fact, it is necessarily the covering transformation on the component which meets the outer Cassini curve, and the identity on the other component.

This completely describes the homeomorphism Ψ_1 of the torus and one can immediately deduce, from this description, the cycles γ_0' and γ_1' to which γ_0 resp. γ_1 are sent by Ψ_1. On the homology we obviously have :

$$\gamma_0' = \gamma_0$$

$$\gamma_1' = -2\gamma_0 + \gamma_1 .$$

The following picture shows the mapping of γ_0, γ_1 into γ_0', γ_1' by the geometric monodromy Ψ_1.

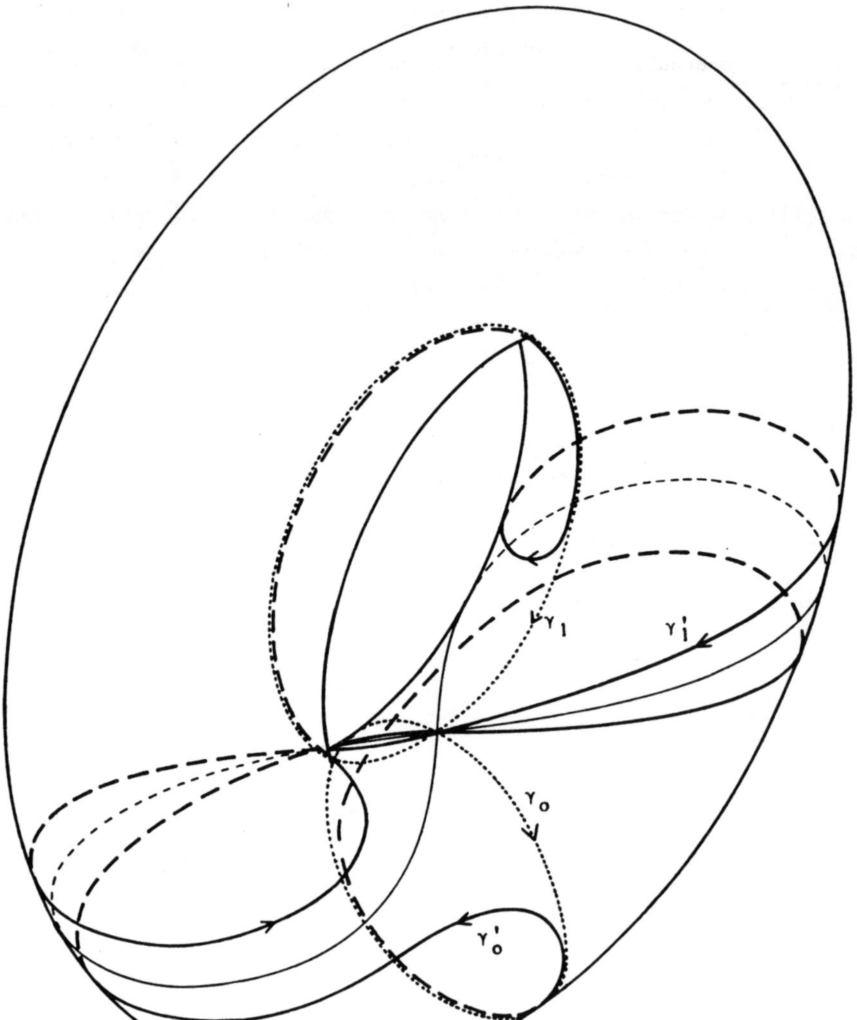
γ_1
γ_1'
γ_o
γ_o'

We shall now replace the geometric monodromy Ψ_1 by a similar, but somewhat simpler, homeomorphism Ψ. One sees very easily that the homeomorphism Ψ_1 of the torus just described is homotopic to the following, more simply described, homeomorphism Ψ. Ψ is the identity outside two annuli over the arcs of the lemniscate, whose centre lines are therefore homologous to γ_0. Inside these annuli Ψ rotates the circles parallel to the centre line through an angle which increases from 0 to 2π across the annulus from one boundary to the other. The accompanying figure shows the cycles γ_0' and γ_1' into which γ_0 and γ_1 are mapped by the homeomorphism.

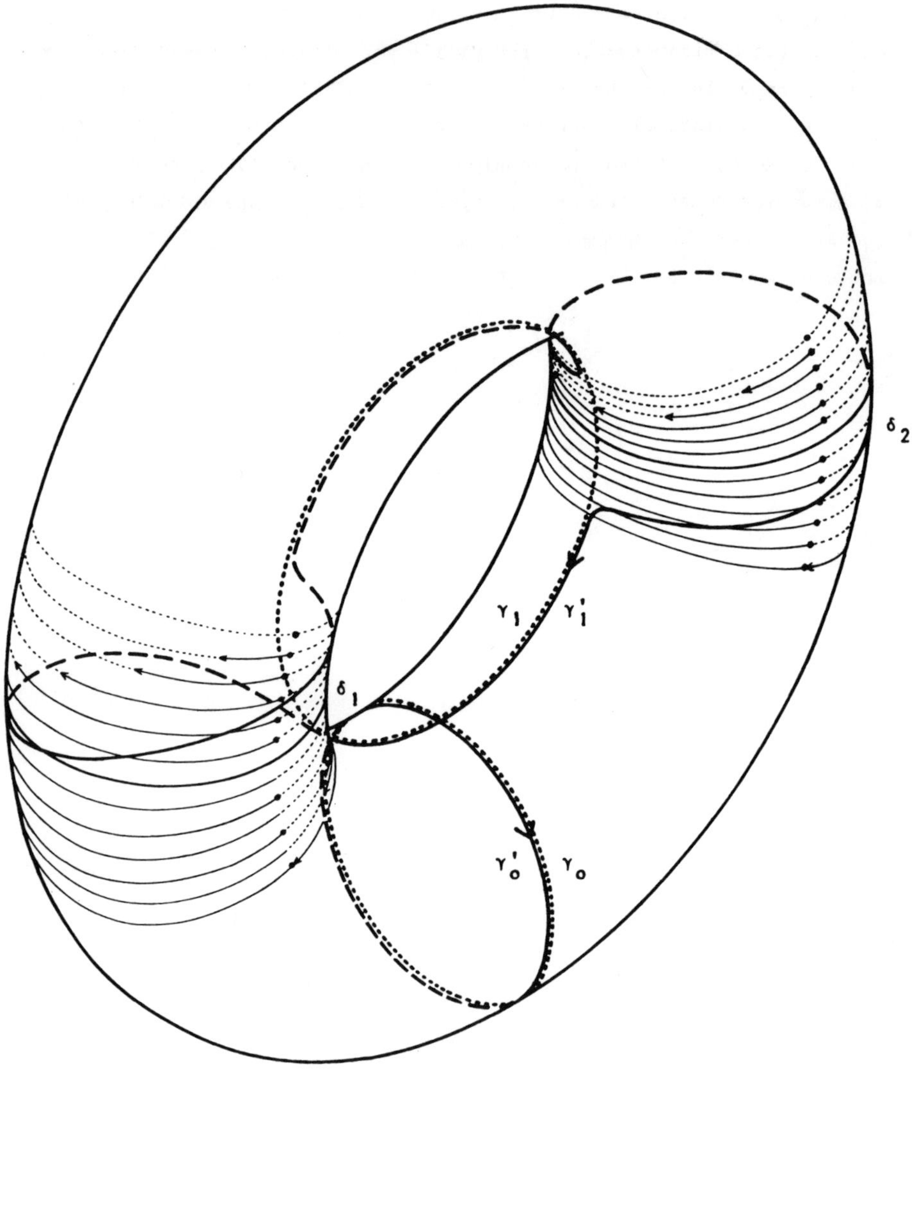
δ_2
γ_1
γ_1'
δ_1
γ_0'
γ_0

The new description of the geometric monodromy by the homeomorphism Ψ is not only simpler than the preceding one, but it better expresses the essence of the situation. We now want to go further into this.

The homeomorphism describes the parallel displacement of the fibre X_{t_0} during a circuit around the singular fibre X_0 along a circle about $t = 0$ in the t-plane. As we have seen previously, the singular fibre consists of two nonsingular rational curves, hence topological spheres, which meet transversely at two points. Here we have a situation similar to that in 3.4, where we first investigated a family of cubics. There we established that the singular cubics resulted from the nonsingular ones by contracting certain 1-cycles on the nonsingular curves to points. We called these cycles the vanishing cycles. The situation here is similar : the singular fibre X_0 results from the nonsingular fibre X_{t_0} by contracting two vanishing cycles δ_1, δ_2 to the two singular points p_1 , p_2 of X_0 :

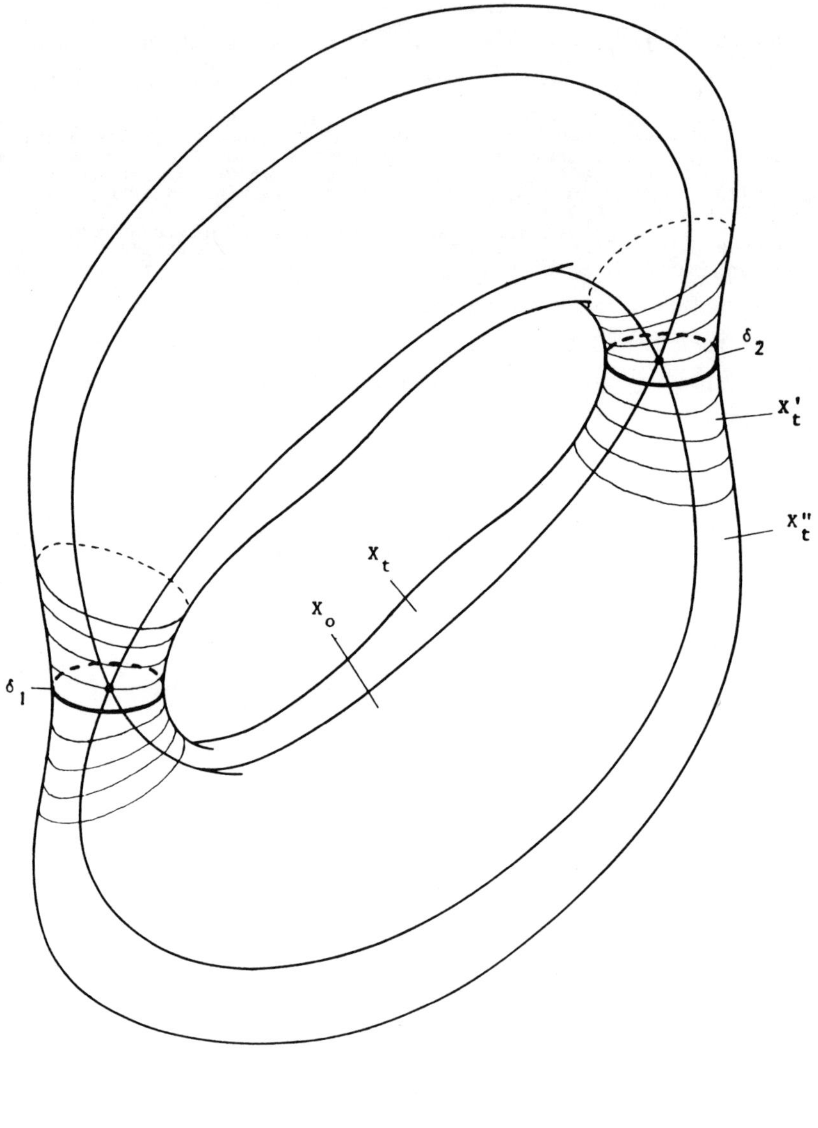

δ_2
X'_t
X''_t
X_t
X_o
δ_1

In section 9.2 we have already analysed such situations in detail. We compared the homology of the singular and nonsingular fibres. To do this we chose small balls B_i around the singular points p_i and decomposed the nonsingular fibre X_t into two parts :

$$X'_t = X_t \cap \bigcup_i B_i$$

$$X''_t = X_t - \bigcup_i B_i .$$

When t traverses a small circle around $t = 0$, the X'_t form a trivial differentiable fibre bundle. The X''_t form the Milnor fibration of the singular points p_i. Since in our case the p_i are ordinary double points, the Milnor fibre is an annulus $S^1 \times [-1,1]$ with the centre line $S^1 \times \{0\}$ as vanishing cycle δ_i.

One can now construct the geometric monodromy $\Psi : X_{t_0} \to X_{t_0}$ to suit this decomposition. On X'_t one chooses Ψ to be the identity, because of the triviality of the fibration :

$$\Psi | X'_{t_0} = id_{X'_{t_0}} .$$

On the annuli one can take the monodromy Ψ to be rotation of the concentric circles through an angle which increases from 0 to 2π from one boundary to the other. The following picture shows how Ψ maps a cycle γ which cuts one vanishing cycle δ into a new cycle γ' :

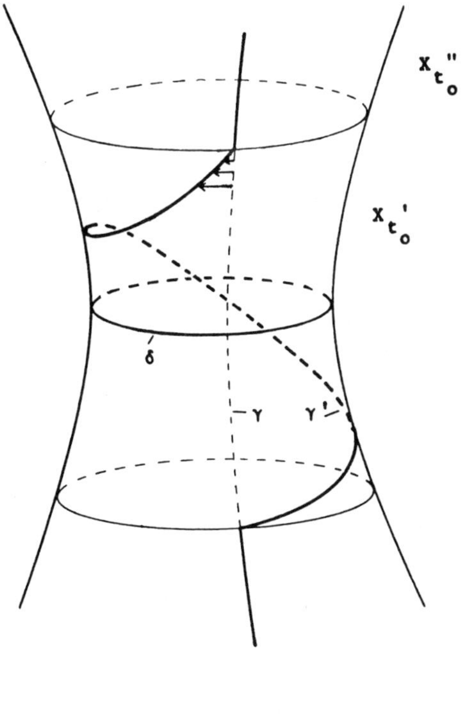

When one takes the sense of rotation carefully into account, this description of the geometric monodromy easily gives the following formula for the image γ' of a cycle γ under parallel displacement along a small positively oriented circle around a singular fibre : up to homology,

$$\gamma' = \gamma - \sum_i \langle\gamma,\delta_i\rangle\delta_i ,$$

where the δ_i are the vanishing cycles relative to the singular fibre being encircled, and $\langle\gamma,\delta_i\rangle$ is the intersection number of γ and δ_i. This is the famous "Picard-Lefschetz formula" (cf. [P4], [L1], [P2]).

With the preceding considerations, which in some respects can be considerably generalised, we have exhaustively described the monodromy for a circuit around the singular fibre X_0 in our special elliptic surface $X \to P_1(\mathbb{C})$. We have

$$\gamma_0' = \gamma_0$$

$$\gamma_1' = \gamma_1 - 2\gamma_0.$$

There is an analogous description of the monodromy for a circuit around the singular fibre X_1, where the circuit is taken in the positive sense along a circle around $t = 1$. It is clear from the symmetry of the situation that γ_0 and γ_1 then exchange rôles. In X_1 two vanishing cycles homologous to γ_1 contract to the two singular points and the Picard-Lefschetz formulae give the cycles γ_0'', γ_1'' to which γ_0, γ_1 are sent, up to homology :

$$\gamma_0'' = \gamma_0 + 2\gamma_1$$

$$\gamma_1'' = \gamma_1.$$

We have now completely determined the monodromy group Γ of our fibration of elliptic curves in Legendre normal form : if we choose γ_0 and γ_1 as basis for the homology group $H_1(X_{t_0})$, then the monodromy for circuits around the fibres X_0 and X_1 is described by the following matrices T_0 and T_1 :

$$T_0 = \begin{pmatrix} 1 & -2 \\ 0 & 1 \end{pmatrix}$$

$$T_1 = \begin{pmatrix} 1 & 0 \\ 2 & 1 \end{pmatrix}.$$

The subgroup Γ of $SL(2,\mathbb{Z})$ generated by these two matrices can also be characterised as follows.

$$\Gamma = \{\begin{pmatrix} a & b \\ c & d \end{pmatrix} \in SL(2,\mathbb{Z}) \mid b \equiv c \equiv 0 \pmod 2,\ a \equiv d \equiv 1 \pmod 4\}.$$

This is a subgroup of index 2 in the second <u>principal congruence subgroup</u>

$$\Gamma(2) = \{ \begin{pmatrix} a & b \\ c & d \end{pmatrix} \in SL(2,\mathbb{Z}) \,|\, b \equiv c \equiv 0 \pmod 2,\ a \equiv d \equiv 1 \pmod 2 \}$$

$$= \mathrm{Kern}(SL(2,\mathbb{Z}) \to SL(2,\mathbb{Z}/2\mathbb{Z})).$$

If one goes over to the modular group $PSL(2,\mathbb{Z})$ by the homomorphism $SL(2,\mathbb{Z}) \to PSL(2,\mathbb{Z})$ then one obtains an isomorphism of Γ onto its image $\bar{\Gamma} \subset PSL(2,\mathbb{Z})$, the "<u>projective monodromy group</u>". In our case $\bar{\Gamma}$ is the free group with generators $\bar{T}_0$, $\bar{T}_1$, and this is just the second <u>principal congruence subgroup</u>

$$\bar{\Gamma}(2) = \mathrm{Kern}(PSL(2,\mathbb{Z}) \to PSL(2,\mathbb{Z}/2\mathbb{Z})).$$

(For the properties of these groups we refer, e.g., to the book of Magnus [M7], Chapter III.2.)

Thus we have proved :

<u>Proposition 11</u> :

For the elliptic surface $X \to P_1(\mathbb{C})$ with singular fibres X_0, X_1, X_∞ which is associated with the family of cubics in Legendre normal form $y^2 = x(x-1)(x-t)$, the monodromy group $\Gamma \subset SL(2,\mathbb{C})$ is generated by the matrices

$$T_0 = \begin{pmatrix} 1 & -2 \\ 0 & 1 \end{pmatrix} \qquad T_1 = \begin{pmatrix} 1 & 0 \\ 2 & 1 \end{pmatrix}.$$

By passing to the projective monodromy group one obtains an isomorphism of free groups

$$\pi_1(P_1(\mathbb{C}) - \{0,1,\infty\}) \cong \bar{\Gamma} = \bar{\Gamma}(2),$$

where $\bar{\Gamma}(2) \subset PSL(2,\mathbb{Z})$ is the second principal congruence subgroup of the modular group. This is a normal subgroup of index 6 in the modular group.

With this theorem we have computed the monodromy group of our family of elliptic curves C_t. In addition, we have given a method for pushing the cycles of C_{t_0} along any given path w with initial point t_0, by means of an explicitly constructed family of homeomorphisms $\psi_s : C_{t_0} \to C_{w(s)}$. Since the homeomorphisms are lifts of corresponding homeomorphisms $\phi_s : P_1(\mathbb{C}) \to P_1(\mathbb{C})$ with $\phi_s(0) = 0$, $\phi_s(1) = 1$, $\phi_s(t_0) = w(s)$, one can obtain the cycles $\gamma(s) = \psi_s(\gamma)$ by lifting the cycles $\phi_s(\bar{\gamma})$, where $\bar{\gamma}$ is the image of γ in $P_1(\mathbb{C})$. In this way one can follow the parallel displacement of cycles in the complex plane, which is both very simple and technically useful. The following sequence of pictures shows, for example, the displacement of γ_1 as t makes a circuit around 0 with initial and final point t_0.

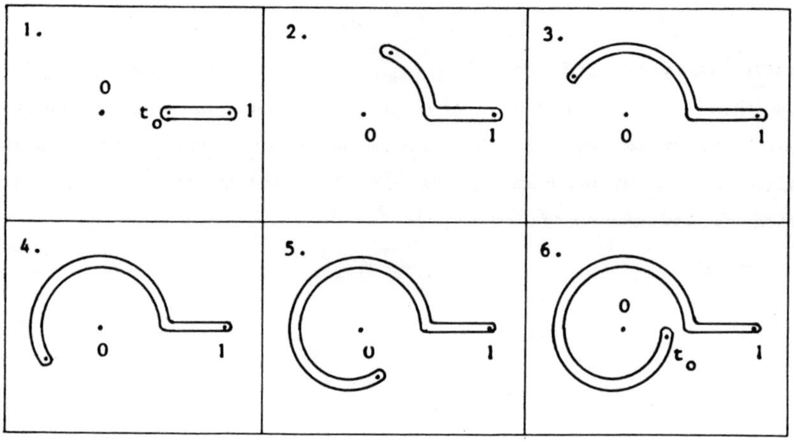

Now we return to the original problem which brought us up against the problem of parallel transport and monodromy. Since the latter problems have now been solved, our original problem can be made precise as follows :

Let $\omega_1(t)$ and $\omega_2(t)$ be the following differential forms on the cubic with equation

$$y^2 = x(x-1)(x-t)$$

$$\omega_1(t) = \frac{dx}{y} = \frac{dx}{\sqrt{x(x-1)(x-t)}}$$

$$\omega_2(t) = \frac{x(x-1)dx}{2y^3} = \frac{x(x-1)dx}{2(\sqrt{x(x-1)(x-t)})^3} .$$

Also, let $\gamma(t)$ be a horizontal family of 1-cycles in this family of cubics. The problem is to investigate the many-valued analytic functions on $P_1(\mathbb{C}) - \{0,1,\infty\}$ given by the periods

$$\int_{\gamma(t)} \omega_i(t) .$$

Our goal is not to calculate any particular one of these integrals, but rather, to characterise the totality of functions

$$\int_{\gamma(t)} \omega_1(t)$$

in a suitable way, for all possible choices of the families $\gamma(t)$. The most important step will be to show that all these functions are solutions of a certain differential equation. In order to work out this differential equation, we must of course compute the derivatives of our period functions. Thus we are faced with the general problem of computing the derivative

$$\frac{d}{dt}\int_{\gamma(t)}\omega(t)$$

for a family of differential forms $\omega(t)$ on a family of curves C_t in the plane, or a family of fibres X_t in a fibration, and a horizontal family of cycles $\gamma(t)$. In our special case of the forms ω_1, ω_2 it is clear – if we proceed naively enough – what we have to do. We simply differentiate under the integral sign and obtain :

$$\frac{d}{dt}\int_{\gamma(t)}\omega_1(t) = \frac{d}{dt}\int_{\gamma(t)}\frac{dx}{\sqrt{x(x-1)(x-t)}}$$

$$= \int_{\gamma(t)}\frac{x(x-1)dx}{2(\sqrt{x(x-1)(x-t)})^3} = \int_{\gamma(t)}\omega_2(t)$$

$$\frac{d}{dt}\int_{\gamma(t)}\omega_2(t) = \frac{d}{dt}\int_{\gamma(t)}\frac{x(x-1)dx}{2(\sqrt{x(x-1)(x-t)})^3}$$

$$= \int_{\gamma(t)}\frac{3}{4}\,\frac{dx}{(x-t)^2\sqrt{x(x-1)(x-t)}}$$

However, we still have to justify this procedure, because the cycle $\gamma(t)$ is also variable. To do this we consider the general problem of computing

$$\frac{d}{dt}\int_{\gamma(t)}\omega(t),$$

which we have just formulated above. Thus we consider a general family of curves C_t, given by the affine equation

$$f(x,y,t) = 0,$$

where $f(x,y,t)$ is a polynomial in x, y and t. In addition, we consider a family of differential forms

$$\omega(t) = P(x,y,t)dx + Q(x,y,t)dy$$

where P and Q are rational functions of x, y and t and where we regard $\omega(t)$ as a meromorphic form on the normalisation X_t of C_t. Since the restriction of the differential df to X_t vanishes identically, we have there

$$\frac{\partial f}{\partial x}dx + \frac{\partial f}{\partial y}dy = 0.$$

Hence we can also write $\omega(t)$ in the following form :

$$\omega(t) = R(x,y,t)dx.$$

Here, $R(x,y,t)$ is a rational function of x, y and t. Now let $\gamma(t)$ be a horizontal family of 1-cycles for the family of curves X_t. (We assume that the X_t, apart from finitely many exceptional fibres, again form a locally trivial differentiable fibre bundle, so that parallel displacement of cycles is defined.) The problem of computing

$$\frac{d}{dt}\int_{\gamma(t)}\omega(t)$$

is solved by the following :

<u>Lemma 12</u> :

$$\frac{d}{dt}\int_{\gamma(t)} R(x,y,t)dx = \int_{\gamma(t)} \left[\frac{\partial R}{\partial t} - \frac{\partial R}{\partial y}\left(\frac{\partial f}{\partial y}\right)^{-1}\frac{\partial f}{\partial t}\right]dx.$$

<u>Proof</u> :

We sketch two proofs, the first of which is tailor-made for the intuitive description of curves as branched coverings of the complex line, while the second is better suited to generalisation to the higher-dimensional case.

<u>1st proof</u> :

We view $y = y(x,t)$ as a (many-valued) algebraic function of x and X_t as the concrete Riemann surface of this function, realised as a branched covering of the sphere by the projection $(x,y) \mapsto x$. We can choose the family of cycles $\gamma(t)$ to avoid the branch points and possible poles of the form $\omega(t)$. Let $\bar{\gamma}(t)$ be the image of $\gamma(t)$ under the projection. Then with proper choice of the branch of $y = y(x,t)$, y becomes a single-valued function on $\bar{\gamma}(t)$, and we can interpret our period as a completely ordinary line integral in the complex plane :

$$\int_{\gamma(t)} R(x,y,t)dx = \int_{\bar{\gamma}(t)} R(x,y(x,t),t)dx.$$

In this integral the path of integration $\bar{\gamma}(t)$ varies with the parameter t. However, for each (nonsingular) t_0 and sufficiently small neighbourhood $U(t_0)$ of t_0 we can replace the variable cycle $\bar{\gamma}(t)$ by the constant cycle $\bar{\gamma}(t_0)$ for all t in $U(t_0)$. Hence

$$\int_{\bar{\gamma}(t)} R(x,y(x,t),t)dx = \int_{\bar{\gamma}(t_0)} R(x,y(x,t),t)dx \quad \text{for} \quad t \in U(t_0).$$

<u>Proof</u> :

There is a neighbourhood V of $\bar{\gamma}(t_0)$ in C such that, for all t in a sufficiently small neighbourhood $U'(t_0)$ the form

$R(x,y(x,t),t)dx$ is a holomorphic 1-form on V. Moreover, it is obvious on continuity grounds that one can choose the parallel displacement of cycles so that $\bar{\gamma}(t) \subset V$ for all t in a sufficiently small neighbourhood $U''(t_0)$. Now if one chooses $U(t_0)$ to be a connected neighbourhood of t_0 in $U_0'(t_0) \cap U_0''(t_0)$, then parallel displacement along a path from t_0 to t in $U(t_0)$ gives

$\bar{\gamma}(t)$ is homologous to $\bar{\gamma}(t_0)$ in V

and hence the equality of integrals claimed above follows from the Cauchy integral theorem.

Thus it suffices to compute

$$\frac{d}{dt}\int_{\bar{\gamma}(t_0)} R(x,y(x,t),t)dx.$$

But this is obtained from well-known theorems on the interchange of differentiation and integration in line integrals whose integrand depends on a parameter :

$$\frac{d}{dt}\int_{\bar{\gamma}(t_0)} R(x,y(x,t),t)dx$$

$$= \int_{\bar{\gamma}(t_0)} \frac{\partial}{\partial t}(R(x,y(x,t),t)dx$$

$$= \int_{\bar{\gamma}(t_0)} [\frac{\partial R}{\partial t} + \frac{\partial R}{\partial y}\frac{\partial y}{\partial t}]dx$$

$$= \int_{\bar{\gamma}(t_0)} [\frac{\partial R}{\partial t} - \frac{\partial R}{\partial y}(\frac{\partial f}{\partial y})^{-1}\frac{\partial f}{\partial t}]dx.$$

The last identity follows from the fact that

$$f(x,y(x,t),t) \equiv 0$$

gives

$$\frac{\partial f}{\partial y}\cdot\frac{\partial y}{\partial t} + \frac{\partial f}{\partial t} = 0$$

on differentiation by t. If we now replace integration over $\bar{\gamma}(t_0)$ by integration over $\gamma(t)$ while keeping the same argument in the integral just computed, then we finally obtain the formula claimed :

$$\frac{d}{dt}\int_{\gamma(t)} R(x,y,t)dx = \int_{\gamma(t)} [\frac{\partial R}{\partial t} - \frac{\partial R}{\partial y}(\frac{\partial f}{\partial y})^{-1}\frac{\partial f}{\partial t}]dx.$$

<u>2nd proof</u> :

We view our forms $\omega(t)$ as the residues of forms on the complement and compute the derivatives as follows :

$$\frac{d}{dt}\int_{\gamma(t)} \operatorname{res} \frac{g(x,y,t)}{f^k} dx\wedge dy = \frac{d}{dt}(2\pi i)^{-1}\int_{\tau(\gamma(t))} \frac{g(x,y,t)}{f^k} dx\wedge dy$$

$$= \frac{d}{dt}(2\pi i)^{-1}\int_{\tau(\gamma(t_0))} \frac{g(x,y,t)}{f^k} dx\wedge dy$$

$$= (2\pi i)^{-1}\int_{\tau(\gamma(t_0))} \frac{\partial}{\partial t} \cdot \frac{g(x,y,t)}{f^k} dx\wedge dy$$

$$= (2\pi i)^{-1}\int_{\tau(\gamma(t_0))} \frac{\frac{\partial g}{\partial t}\cdot f - kg\cdot\frac{\partial f}{\partial t}}{f^{k+1}} dx\wedge dy$$

$$= \int_{\gamma(t)} \operatorname{res} \frac{\frac{\partial g}{\partial t}\cdot f - kg\cdot\frac{\partial f}{\partial t}}{f^{k+1}} dx\wedge dy.$$

If one computes the residue in the sense of the Corollary to Proposition 9 and the subsequent remarks there, and applies the formula from the first proof in differentiating the resulting integral, then an easy calculation shows that the result coincides with the formula just obtained in the first proof.

Lemma 12 justifies, in particular, our earlier computation of the derivative of the integral $\int_{\gamma(t)} \omega_i(t)$. We now restate the results obtained. We set

$$\omega = \frac{dx}{y}$$

$$\omega' = \tfrac{1}{2}\,\frac{dx}{y(x-t)}$$

$$\omega'' = \frac{3}{4}\,\frac{dx}{y(x-t)^2}\,.$$

Earlier, we had

$$\omega = \omega_1$$

$$\omega' = \omega_2,$$

and with the computation of the derivatives we obtained :

$$\frac{d}{dt}\int_{\gamma(t)} \omega = \int_{\gamma(t)} \omega'$$

$$\frac{d}{dt}\int_{\gamma(t)} \omega' = \int_{\gamma(t)} \omega''.$$

The forms ω and ω' represent a basis for the cohomology $H^1(C_t,\mathbb{C})$ of cubics in Legendre normal form – this was just the result of our second example of the computation of cohomology by differentials of first and second kind. For that reason, ω'' must be, up to an exact form, a linear combination of ω and ω', where the coefficients of the linear combination depend on the parameter t :

$$\omega'' = p\omega + q\omega' + d\psi.$$

We can even find the coefficients p and q in this linear combination explicitly. We claim :

$$p = \frac{1}{4t(1-t)}$$

$$q = \frac{-1+2t}{t(1-t)} .$$

__Proof__ :

Let ψ be the meromorphic function

$$\psi = \frac{y}{2(x-t)^2 t(1-t)} .$$

Then a trivial calculation shows that (for constant t)

$$\omega'' = p\omega + q\omega' + d\psi.$$

This gives us the following result :

__Proposition 13__ :

Let C_t be the cubic in Legendre normal form,

$$y^2 = x(x-1)(x-t).$$

Then the family of cubics C_t has the properties :

(i) If ω_1 and ω_2 are the following differential forms of first resp. second kind on C_t :

$$\omega_1 = \frac{dx}{y} \qquad \omega_2 = \frac{dx}{2y(x-t)} ,$$

then ω_1 and ω_2 represent a basis for the cohomology group $H^1(C_t,\mathbb{C})$.

(ii) If $\gamma(t)$ is any horizontal family of cycles, and if $z_1(t)$ resp. $z_2(t)$ are the many-valued analytic functions

$$z_i(t) = \int_{\gamma(t)} \omega_i ,$$

Then these functions are solutions of the first order system of ordinary differential equations

$$\frac{dz_1(t)}{dt} = z_2$$

$$\frac{dz_2(t)}{dt} = pz_1 + qz_2 ,$$

where p and q are the following functions :

$$p = \frac{1}{4t(1-t)} \qquad q = \frac{-1+2t}{t(1-t)} .$$

(iii) In particular, the period of the holomorphic 1-form ω_1, i.e. the function

$$z(t) = \int_{\gamma(t)} \omega_1$$

in the above notation, is a solution of the hypergeometric differential equation

$$z'' - qz' - pz = 0.$$

More generally, the hypergeometric differential equation with parameters $a, b, c \in \mathbb{C}$ is the ordinary homogeneous linear differential equation of second order

$$z'' + \frac{c-(a+b+1)t}{t(1-t)} z' - \frac{ab}{t(1-t)} z = 0.$$

In our case the parameter values are

$$a = \tfrac{1}{2}$$
$$b = \tfrac{1}{2}$$
$$c = 1.$$

The hypergeometric differential equation has three singular points, namely 0, 1 and ∞, and these are regular singular points.

An ordinary homogeneous linear differential equation

$$\frac{d^n z}{dt^n} + a_1(t)\frac{d^{n-1}z}{dt^{n-1}} + \ldots + a_n(t)z = 0$$

with meromorphic coefficients a_i has a regular singular point at t_0 if, when $t \to t_0$ (in any sector with vertex t_0), the solutions of the differential equation increase like powers of t at most. A necessary and sufficient condition for this is that $a_i(t)$ have at most a pole of order i at t_0. Another necessary and sufficient condition is that the solutions in the neighbourhood of t_0 be representable as linear combinations, with complex coefficients, of functions of the form

$$(t-t_0)^{\lambda} \sum_{i=0}^{\infty} c_i (t-t_0)^i (\ln(t-t_0))^k,$$

where λ is a complex number and $k \leq n$ is a natural number.

The solutions of the hypergeometric differential equation have been investigated in depth by such important mathematicians as Euler, Gauss and Riemann, in terms of series expansions such as this, on the one hand, and also in terms of integral representations like those considered above. A solution of the hypergeometric equation with parameters a, b, c where $c \neq 0, -1, -2, \ldots$ is obtained by Gauss' hypergeometric series

$$F(a,b;c;t) = \sum_{n=0}^{\infty} \frac{a(a+1)\dots(a+n-1)b(b+1)\dots(b+n-1)}{c(c+1)\dots(c+n-1)} \cdot \frac{t^n}{n!}$$

For our special hypergeometric equation with parameters $a = \frac{1}{2}$, $b = \frac{1}{2}$, $c = 1$ we obtain from this a solution $z_0 = F(\frac{1}{2},\frac{1}{2};1;t)$. But this linear differential equation, being homogeneous and linear of second order, has two linearly independent solutions. The way to find another solution is explained in the books on special functions, e.g. "Higher Transcendental Functions" [B15] Vol. 1, 2.7.1, p. 95. There one finds :

$$z_0 = F(\tfrac{1}{2},\tfrac{1}{2};1;t)$$

$$z_1 = -iF(\tfrac{1}{2},\tfrac{1}{2};1;1-t)$$

are linearly independent solutions of the hypergeometric differential equation with parameters $a = \frac{1}{2}$, $b = \frac{1}{2}$, $c = 1$. One also finds the monodromy group of the hypergeometric equation computed there. If one analytically continues the solutions z_0 and z_1, which indeed are many-valued analytic functions in $\mathbb{C} - \{0,1\}$, along a path in $\mathbb{C} - \{0,1\}$ with initial and final point t_0, then on returning to t_0 one obtains two new solutions z_0', z_1', and since z_0, z_1 is a basis for the vector space of all solutions, z_0' and z_1' must be linear combinations of z_0 and z_1 :

$$z_0' = \alpha z_0 + \beta z_1$$

$$z_1' = \gamma z_0 + \delta z_1 .$$

In this way one obtains a homomorphism from the fundamental group $\pi_1(P_1(\mathbb{C}) - \{0,1,\infty\})$ into the group of invertible 2×2 matrices, and the image of this homomorphism is the monodromy group of the differential equation. In our case this monodromy can be computed as follows :

One has the relation (loc. cit.)

$$F(\tfrac{1}{2},\tfrac{1}{2};1;t) = \frac{-1}{\pi} \ln(1-t)F(\tfrac{1}{2},\tfrac{1}{2};1;1-t) + \frac{2}{\pi} \sum_{n=0}^{\infty} \left[\frac{\frac{1}{2}\cdot(\frac{1}{2}+1)\dots(\frac{1}{2}+n-1)}{n!}\right]^2 [\psi(n+1)-\psi(n+\tfrac{1}{2})](1-t)^n$$

where $\psi = \Gamma'/\Gamma$ is the logarithmic derivative of the gamma function. The precise form of the second summand on the right-hand side is not essential here. What is essential, is that there is a relation

$$z_0 = \frac{1}{\pi i} \ln(1-t)\, z_1 + \phi(1-t))$$

where $\phi(1-t)$ is single-valued in the neighbourhood of $t = 1$. Since z_1 is also single-valued in the neighbourhood of $t = 1$, and since $\ln(1-t)$ becomes $\ln(1-t) + 2\pi i$ on a small circuit around $t = 1$ in the positive direction, we have :

A small circuit around $t = 1$ changes z_0, z_1 into z_0'', z_1'' where

$$z_0'' = z_0 + 2z_1$$

$$z_1'' = z_1.$$

One shows analogously that a small circuit around $t = 0$ changes z_0 and z_1 into z_0' and z_1' where

$$z_0' = z_0$$

$$z_1' = -2z_0 + z_1.$$

Thus the analytic route has brought us back to the monodromy group which we found topologically in Proposition 11. In addition, we have found that the solutions z_0 , z_1 of the hypergeometric differential equation transform by exactly the same formulae as the cycles γ_0 , γ_1. (Compare the formulae for γ_0' , γ_1' and γ_0'', γ_1'' in combination with the Picard-Lefschetz formulae.) This fact explains the connection between the solutions z_0 , z_1 of our equation we have just described by power series, and the solutions by integrals

$$\int_{\gamma_i(t)} \omega_1 \qquad i = 0,1$$

described earlier in Proposition 13(iii). Because one easily checks that the only invertible 2×2 matrices with

$$\begin{pmatrix} a & b \\ c & d \end{pmatrix} \begin{pmatrix} 1 & 2 \\ 0 & 1 \end{pmatrix} \begin{pmatrix} a & b \\ c & d \end{pmatrix}^{-1} = \begin{pmatrix} 1 & 2 \\ 0 & 1 \end{pmatrix}$$

$$\begin{pmatrix} a & b \\ c & d \end{pmatrix} \begin{pmatrix} 1 & 0 \\ -2 & 1 \end{pmatrix} \begin{pmatrix} a & b \\ c & d \end{pmatrix}^{-1} = \begin{pmatrix} 1 & 0 \\ -2 & 1 \end{pmatrix}$$

are matrices from the centre, and hence matrices of the form

$$\begin{pmatrix} a & 0 \\ 0 & a \end{pmatrix}.$$

It follows that, up to a common constant $c \in \mathbb{C}$, the solutions by series and integrals must coincide :

$$\int_{\gamma_0(t)} \omega_1 = cz_0 \quad \text{and} \quad \int_{\gamma_1(t)} \omega_1 = cz_1.$$

Thus, by a combination of purely topological and purely analytic methods, we have explicitly calculated the periods of the form ω as functions of t, without carrying out the integration directly. We want to put the result on record :

Proposition 14 :

Let C_t be the family of cubics in Legendre normal form

$y^2 = x(x-1)(x-t)$.

Choose a base point t_0 and two cycles γ_0, γ_1 in C_{t_0} in the way described earlier. Thus their images in the x-plane are cycles which run once around t_0 and 0 resp. 1, as shown in the following picture:

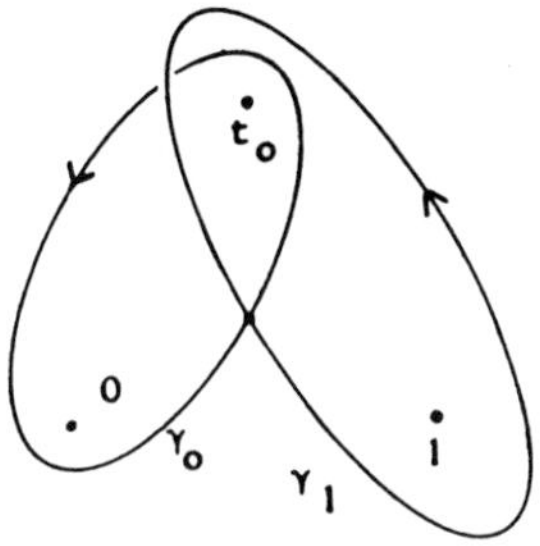

Let $(\gamma_0(t),\gamma_1(t))$ be the horizontal family of pairs of cycles which results from the pair (γ_0,γ_1) by parallel displacement ; $(\gamma_0(t),\gamma_1(t))$ represents a basis of $H_1(C_t)$.

In addition, let $\omega(t)$ be the following family of differentials of first kind on C_t :

$$\omega(t) = \frac{dx}{y} = \frac{dx}{\sqrt{x(x-1)(x-t)}} .$$

By Proposition 13, the periods $\int_{\gamma_i(t)}\omega(t)$ are solutions of the hypergeometric differential equation with parameters $a = \frac{1}{2}$, $b = \frac{1}{2}$, $c = 1$. These elliptic integrals of 1st kind may be identified (up to a constant $c_0 \in \mathbb{C}$ whose determination we omit) with the fundamental system of solutions of the hypergeometric equation described above by hypergeometric series : one has

$$\int_{\gamma_0(t)} \frac{dx}{\sqrt{x(x-1)(x-t)}} = c_0 F(\tfrac{1}{2},\tfrac{1}{2};1;t)$$

$$\int_{\gamma_1(t)} \frac{dx}{\sqrt{x(x-1)(x-t)}} = -ic_0 F(\tfrac{1}{2},\tfrac{1}{2};1;1-t) .$$

Corollary 15 :

For the cubics C_t in Legendre normal form,

$$y^2 = x(x-1)(x-t),$$

and the notation of Proposition 14, one has : the ratio $\int_{\gamma_0(t)} \omega(t) / \int_{\gamma_1(t)} \omega(t)$ of the periods equals

$$a(t) = i \cdot \frac{F(\frac{1}{2},\frac{1}{2};1;t)}{F(\frac{1}{2},\frac{1}{2};1;1-t)} .$$

Thus C_t is isomorphic to the 1-dimensional complex torus

$$\mathbb{C}/\mathbb{Z} + a(t)\mathbb{Z}.$$

Proof :

The first assertion follows immediately from Proposition 14, the second from the first and the facts about the relation between cubic curves and 1-dimensional complex tori (in Theorems 7.4.3 and 7.4.4) which we recalled at the beginning of our present investigation of cubics in Legendre normal form.

Our new results represent definite progress over our previous ones. In 7.4.4 we gave an isomorphism between a fixed 1-dimensional torus $\mathbb{C}/\mathbb{Z} + a\mathbb{Z}$ and a particular cubic in Weierstrass normal form, namely the cubic

$$y^2 = 4x^3 - g_2(a)x - g_3(a).$$

Here, g_2 and g_3 were explicitly described functions of a, and the mapping of the torus was explicitly described by the Weierstrass $\wp$-function, i.e. by $x = \wp_a$ and $y = \wp'_a$, where $\wp_a$ is doubly periodic with periods 1 and a : $\wp$ was the inverse function of the indefinite elliptic integral

$$\int \frac{dx}{\sqrt{4x^3 - g_2x - g_3}} .$$

These results are very satisfactory when one begins with the torus $\mathbb{C}/\mathbb{Z} + a\mathbb{Z}$. But what remains open is the converse question : given a cubic in Weierstrass normal form, $y^2 = 4x^3 - g_2x - g_3$, to which tori is it isomorphic ? In 7.4 we have answered this question only implicitly. It is isomorphic to each torus $\mathbb{C}/\mathbb{Z} + a\mathbb{Z}$ with $g_2 = g_2(a)$, $g_3 = g_3(a)$, and more generally, to each torus $\mathbb{C}/\mathbb{Z} + a\mathbb{Z}$ with the J-invariant

$$J(a) = \frac{g_2^3}{g_2^3-27g_3^2} .$$

However, we did not explain how one computes the number a for a given cubic. This is what we have now achieved for cubics C_t in Legendre normal form

$$y^2 = x(x-1)(x-t),$$

Corollary 15 contains the desired formula.

<u>**Example**</u> :

The cubic with equation

$$y^2 = x(x-1)(x-\tfrac{1}{2})$$

is isomorphic to the complex torus $\mathbb{C}/\mathbb{Z} + i\,\mathbb{Z}$.

These investigations, in combination with the earlier investigations in 7.3 and 7.4, have already given us a very good insight into the nature of cubic curves. Even more interesting things remain to be said, however, we can be quite satisfied with what we have already achieved, and we want to leave it at that.

In the first half of this long section we were concerned with obtaining general statements about the periods of differential forms on curves, and in the second half we analysed quite a special example, the family of cubics in Legendre normal form, from this point of view. In conclusion, we want to make a few sketchy remarks on the aspects of this special example which are typical and which generalise to other situations.

Consider, for example, the following general situation : let X be a nonsingular projective algebraic manifold and let $X \to S$ be a holomorphic mapping onto a nonsingular curve. In our special example X was the elliptic modular surface and $S = P_1(\mathbb{C})$. Suppose that for finitely many $t_1,\dots,t_k \in S$ the fibres X_{t_i} are singular. If one sets

$$X' = X - \bigcup_i X_{t_i}$$

$$S' = S - \{t_1,\dots,t_k\},$$

then $X' \to S'$ is a differentiable locally trivial fibre bundle. One can again define a geometric monodromy, and for each p one obtains a <u>monodromy</u>

$\pi_1(S',t_0) \to \mathrm{Aut}(H_p(X_{t_0}))$.

One has a parallel displacement of homology classes, i.e. one has a connection on the vector bundle over S' whose fibres are the homology groups $H_p(X_t,\mathbb{C})$. Analogously, one has a parallel displacement for cohomology classes, and hence a connection on the $H^p(X_t,\mathbb{C})$. One can describe the cohomology by means of differential forms, as we have indicated at the beginning of this section. This yields a description of the connection at the level of differential forms, called the Gauss-Manin-connection. This connection is described explicitly by differential equations for the periods. If $\omega(t)$ is a holomorphic family of p-forms $\omega(t)$ on X_t, $t \in S'$, and if $\gamma(t)$ is a horizontal family of p-cycles, then the periods

$$\int_{\gamma(t)} \omega(t)$$

satisfy, as functions of t, an ordinary linear differential equation with poles at $t_1,\ldots,t_k$. In our special example we had three singular fibres X_t for $t = 0,1,\infty$, and the differential equation was the hypergeometric equation. The hypergeometric equation may be essentially characterised as the ordinary linear differential equation with three regular singular points. What is typical here is the regularity. More generally, one has the following theorem :

Regularity Theorem :

The differential equations for the periods $\int_{\gamma(t)} \omega(t)$ have regular singularities.

For our hypergeometric differential equation we have determined the monodromy group explicitly and described it by generators which correspond to circuits around the singular fibres. With this description one finds, among other things, that the monodromy group is completely reducible, i.e. that each invariant subspace has an invariant complement. (In our case there are no invariant subspaces at all apart from $\{0\}$ and the whole space.)

We computed the generators explicitly by travelling round the fibres X_0, X_1, X_∞. It follows immediately from this computation that their eigenvalues are 1 for X_0, X_1 and -1 for X_∞. More generally, the following theorem holds :

Monodromy Theorem :

(i) The eigenvalues of the monodromy transformations for

circuits around singular fibres X_{t_i} on small circles round t_i are roots of unity.

(ii) The monodromy group is fully reducible.

There are many very different proofs of the Regularity - and Monodromy Theorem. We refer those who are interested in these results and methods beyond the scope of this course to the survey article of Griffiths [G2]. I hope that this course has helped to awaken interest in these beautiful, deep and far-reaching questions.

Bibliography

[A1] Arnold, V.I. : Critical points of smooth functions. In: Proceedings of the Int. Congress of Mathematicians, Vol. 1, 19-39, Vancouver 1974.

[A2] Artin, M. : On the solution of analytic equations. Invent. math. 5, 277-291 (1968).

[A3] Artin, M. : The implicit function theorem in algebraic geometry. In: Proc. of the Bombay Colloquium on Algebraic Geometry, 13-14.

[A4] A'Campo, N. : Sur la monodromie des singularités isolées d'hypersurfaces complexes. Invent. Math. 20, 147-169 (1973).

[A5] A'Campo, N. : La fonction zêta d'une monodromie. Comment. Math. Helvetici 50, 233-248 (1975).

[A6] A'Campo, N. : Le groupe de monodromie du déploiement des singularités isolées de courbes planes I. Math. Ann. 213, 1-32 (1975).

[A7] A'Campo, N. : Le groupe de monodromie du déploiement des singularités isolées de courbes planes II. In: Proceedings of the Int. Congress of Mathematicians, Vol. 1, 395-404, Vancouver 1974.

[A8] Artin, E. : Theorie der Zöpfe. Hamburger Math. Abhandlungen 4, 47-72 (1925).

[A9] Ashley, C.W. : The Ashley book of knots. Doubleday & Company, Inc., Garden City, New York 1944.

[A10] Atiyah, M. - Macdonald, I. : Introduction to Commutative Algebra. Addison-Wesley, 1969.

[B1] Behnke, H. - Sommer, F. : Theorie der analytischen Funktionen einer komplexen Veränderlichen. Die Grundlehren der math. Wissenschaften in Einzeldarstellungen, Bd. 77, Springer-Verlag, Berlin-Göttingen-Heidelberg 1955.

[B2] Berzolari, L. : Allgemeine Theorie der höheren ebenen algebraischen Kurven. In : Enzyklopädie der math. Wissenschaften, Bd. III 2, 1, 313-455 (1906).

[B3] Borel, A. - Haefliger, A. : La classe d'homologie fondamentale d'un espace analytique. Bull. Soc. math. France 89, 461-513 (1961).

[B4] Brieskorn, E. : Beispiele zur Differentialtopologie von Singularitäten. Invent. Math. 2, 1-14 (1966).

[B5] Brieskorn, E. : Über die Dialektik in der Mathematik. In : Mathematiker über Mathematik, edited by M. Otte, Springer-Verlag, Berlin-Heidelberg-New York 1974.

[B6] Brieskorn, E. : The development of geometry and topology. Notes of introductory lectures given at the University of La Habana in 1973. In : Materialien zur Berufspraxis des Mathematikers 17, 109-204 (1976).

[B7] Brieskorn, E. : Singularitäten. Manuskript eines Vortrages auf der Jahrestagung der DMV, Sept. 1975. Jber. Deutsch. Math.-Verein. 78, H.2, 93-112 (1976).

[B8] Bröcker, Th. - Jänich, K. : Einführung in die Differentialtopologie, Heidelberger Taschenbücher 143, Springer-Verlag, Berlin-Heidelberg-New York 1973. English translation : Introduction to Differentiable Topology, Camb. U.P. 1982.

[B9] Bröcker, Th. : Differentiable germs and catastrophes. London Math. Soc. Lecture Notes Series 17, Cambridge University Press 1975.

[B10] Birman, J. : Braids, links and mapping class groups. Ann. of Math. Studies 82, Princeton University Press, Princeton 1975.

[B11] Brieskorn, E. : Die Monodromie der isolierten Singularitäten von Hyperflächen. Manuskripta Math. 2, 103-161 (1970).

[B12] Brill, A. : Über den Weierstraßschen Vorbereitungssatz. Math. Ann. 64, 538-549 (1910).

[B13] Burde, G. - Zieschang, H. : Neuwirthsche Knoten und Flächenabbildungen. Abh. Math. Seminar Hamburg 31, 239-246 (1967).

[B14] Berzolari, L. : Algebraische Transformationen und Korrespondenzen. In: Enzyklopädie der Mathematischen Wissenschaften III C 1,1, 1781-2218 (1933).

[B15] Bateman, H. : Higher transcendental functions. (Bateman manuscript project), edited by A. Erdélyi. McGraw-Hill Book Company Inc. New York, Toronto, London 1953/55.

[C1] Carmichael, R.D. : Introduction to the theory of groups of finite order. Dover Publications Inc., New York 1956.

[C2] Cartan, H. : Théorie élémentaire des fonctions analytiques d'une ou plusieurs variables complexes. Hermann, Paris 1961. English translation : Elementary theory of analytic functions of one or several complex variables. Hermann, Paris and Addison-Wesley Publ. Comp., London 1963.

[C3] Chern, S.S. : Complex manifolds without potential theory. Second edition. Springer-Verlag, Berlin-Heidelberg-New York, 1979.

[D1] Deligne, P. - Katz, N. : Groupes de monodromie en géometrie algébrique (SGA 7 II). Lecture Notes in Math. 340, Springer-Verlag, New York-Heidelberg-Berlin 1973.

[D2] Dickson, L.E. : Linear Groups. Dover Publications, Inc. New York 1958.

[D3] Dieudonné, F. : Cours de géométrie algébrique. Presses universitaires de France 1974.

[D4] Dold, A. : Lectures on Algebraic Topology. Die Grundlehren der math. Wissenschaften in Einzeldarstellungen, Bd. 200, Springer-Verlag, Berlin-Heidelberg-New York 1972.

[D5] Duistermaat, J.J. : Oscillatory integrals, Lagrange immersions and unfoldings of singularities. In: Commun. in Pure and Appl. Math. Vol. XXVII, 207-281 (1974).

[D6] Dürer, A. : Unterweisung der Messung mit dem Zirkel und Richtscheid. Stocker-Schmid, Zürich 1966. (Reprint of 1525 ed.)

[E1] Ehlers, F. : Newtonpolyeder und die Monodromie von Hyperflächensingularitäten. Bonner math. Schriften 111, 1979.

[E2] Enriques, F. - Chisini, O. : Lezioni sulla teoria geometrica delle equazioni e delle funzioni algebriche. 3 Vols., Bologna 1915, 1918, 1924.

[F1] Fulton, W. : Algebraic curves. W.A. Benjamin Inc., New York-Amsterdam 1969.

[F2] Fox, R.H. : A quick trip through knot theory. In: Topology of 3-manifolds and related topics. Prentice Hall 1962.

[G1] Golubitsky, M. - Guillemin, V. : Stable mappings and their singularities. Springer-Verlag GTM 14, New York-Heidelberg-Berlin 1973.

[G2] Griffiths, Ph.A. : Periods of integrals on algebraic manifolds : Summary of main results and open problems. Bull. AMS 76, 228-296 (1970).

[G3] Grothendieck, A. : Sur quelques propriétés fondamentales en théorie des intersections. In: Seminaire C. Chevalley 1958: "Anneaux de Chow et Applications", E.N.S., Paris.

[G4] Grauert, H. - Fritzsche,K. : Einführung in die Funktionentheorie mehrerer Veränderlicher, Hochschultext. Springer-Verlag, Berlin-Heidelberg-New York 1974.

[G5] Grauert, H. - Remmert, R. : Analytische Stellenalgebren. Die Grundlehren der math. Wissenschaften in Einzeldarstellungen, Bd. 176, Springer-Verlag, Berlin-Heidelberg-New York 1971.

[G6] Gunning, R. - Rossi, H. : Analytic functions of several complex variables. Prentice Hall 1965.

[G7] Griffiths, Ph.A. : On the periods of certain rational integrals I, II. Ann. of Math. 90, 460-495 and 496-541 (1969).

[G8] Griffiths, Ph.A. - Cornalba, M. : Some transcendental aspects of algebraic geometry. In: Algebraic Geometry, Arcata 1974. Proceedings of Symposia in Pure Mathematics, Vol. 29, A.M.S., Providence 1975.

[H1] Hilbert, D. - Cohn-Vossen, S. : Anschauliche Geometrie. Die Grundlehren der math. Wissenschaften in Einzeldarstellungen, Bd. 37 (Reprinted by Wissenschaftlichen Buchgesellschaft, Darmstadt). English translation : Geometry and the Imagination, Chelsea, New York 1952.

[H2] Hironaka, H. : On the equivalence of singularities. In : Arithmetical Algebraic Geometry (Conference at Purdue University 1963). Harper & Row, New York.

[H3] Hironaka, H. : Resolution of singularities of an algebraic variety over a field of characteristic zero. Ann. of Math. 79, 109-326 (1964).

[H4] Hironaka, H. : Introduction to the theory of infinitely near singular points. Memorias de Matematico des Instituto "Jorge Juan" 28, Madrid 1974.

[H5] Hirzebruch, F. : Über vierdimensionale Riemannsche Flächen mehrdeutiger Funktionen von zwei komplexen Veränderlichen. Math. Ann. 126, 1-22 (1953).

[H6] Hirzebruch, F. : Topological Methods in Algebraic Geometry (Third Edition). Die Grundlehren der math. Wissenschaften in Einzeldarstellungen, Bd. 131, Springer-Verlag, New York-Heidelberg-Berlin 1966.

[H7] Hwa, R.C. - Teplitz, V.L. : Homology and Feynman Integrals. Benjamin, New York 1966.

[H8] Hodge, W. : The theory and application of harmonic integrals. Cambridge 1941.

[J1] Lejeune - Jalabert, M. - Teissier, B. : Contributions à l'étude des singularités du point de vue du polygone de Newton, Thèse. Univ. Paris VII, 1973.

[K1] Klein, F. : Vorlesungen über die Entwicklung der Mathematik im 19. Jahrhundert. Die Grundlehren der math. Wissenschaften in Einzeldarstellungen, Bd. 24, Springer-Verlag, Berlin 1926.

[K2] Knutson, D. : Algebraic Spaces. Lecture notes in Math. 203, Springer-Verlag, Berlin-Heidelberg-New York 1971.

[K3] Kohn, G. - Loria, G. : Spezielle ebene algebraische Kurven. In: Enzyklopädie der Math. Wissenschaften, Bd. III 2,1, 457-634 (1914).

[K4] Kurke, H. - Pfister, G. - Roczen, M. : Henselsche Ringe und algebraische Geometrie. Math. Monographien Bd. 11, VEB Deutscher Verlag der Wissenschaften, Berlin 1975.

[K5] Kodaira, K. : On compact analytic surfaces. In: Analytic Functions, Princeton University Press, Princeton, N.Y. 1960.

[L1] Lefschetz, S. : L'analysis situs et la géométrie algébrique. Gauthier -Villars, Paris 1924.

[L2] Lefschetz, S. : Differential equations : Geometric theory. Interscience Publishers Inc., New York 1957.

[L3] Lojasiewicz, S. : Triangulation of semianalytic sets. Ann. Scuola Norm. Sup. Pisa 18, 449-474 (1964).

[L4] Loria, G. : Spezielle algebraische und transzendente ebene Kurven, I, II. Sammlung von Lehrbüchern auf dem Gebiete der math. Wissenschaften, Bd. 5, 1(2). Teubner-Verlag, Leipzig-Berlin 1910.

[L5] Lê Dũng Trang : Sur les nœuds algébriques. Comp. Math. 25, 281-321 (1972).

[L6] Levinson, N. : A canonical form for an analytic function of several variables at a critical point. Bull. Amer. Math. Soc. Vol. 66, 68-69 (1960).

[L7] Levinson, N. : A polynomial canonical form for a critical point. Bull. Amer. Math. Soc. Vol. 66, 366-368 (1960).

[M1] Milnor, J. : Singular points of complex hypersurfaces. Ann. of Math. Studies 61, Princeton University Press, Princeton, N.Y., 1968.

[M2] Milnor, J. - Orlik, P. : Isolated singularities defined by weighted homogeneous polynomials. Topology 9, 385-393 (1970).

[M3] Müller, H.R. : Trochoidenhüllbahnen und Rotationskolbenmaschinen. In: Selecta Mathematika III, Springer-Verlag, Berlin-Heidelberg-New York 1971.

[M4] Mumford, D. : Abelian Varieties. Tata Institute of Fundamental Research studies in Math. 5, Oxford University Press 1970.

[M5] Malgrange, B. : The preparation theorem for differentiable functions. In: Differential Analysis. Bombay Colloquium, 203-208 (1964).

[M6] Mumford, D. : The topology of normal singularities of an algebraic surface and a criterion for simplicity. Publ. Math. IHES 9, 5-22 (1961).

[M7] Magnus, W. : Noneuclidean tesselations and their groups. Academic Press, New York 1974.

[M8] Mumford, D. : Algebraic Geometry I, Complex projective varieties. Die Grundlehren der math. Wissenschaften in Einzeldarstellungen, Bd. 221, Springer-Verlag, Berlin-Heidelberg-New York 1976.

[M9] Mumford, D. : Lectures on curves on an algebraic surface. Ann. of Math. studies 59, Princeton University Press, Princeton, N.Y. 1966.

[M10] Mumford, D. : Curves and their Jacobians. The University of Michigan Press, Ann Arbor 1975.

[N1] Nagata, M. : Local rings. Interscience tracts in pure and applied Math. 13, John Wiley & Sons, New York-London 1962.

[N2] Newton, I. : Curves by Sir Isaak Newton. In: Lexicon Technicum. Or, an Universal Dictionary of Arts and Sciences. By John Harris. Vol. 2, London 1710. Also in : The mathematical works of I. Newton, Vol. 2, 135-161, Johnson Reprint Corporation, New York-London 1967.

[N3] Narasimhan, R. : Introduction to the theory of analytic spaces. Lecture Notes in Math. 25, Springer-Verlag, New York-Berlin-Heidelberg 1966.

[N4] Newton I. : The correspondence of Isaak Newton. Vol. 2 (1676-1687), Cambridge University Press 1960, 20-42 and 110-163.

[N5] Noether, M. : Rationale Ausführungen der Operationen in der Theorie der algebraischen Funktionen. Math. Ann. 23, 311-358 (1883).

[N6] Narasimhan, R. : Analysis on real and complex manifolds. Advanced studies in pure mathematics, Vol. 1. Masson & Cie, Paris, North Holland Publ. Comp., Amsterdam 1968.

[P1] Pham, F. : Singularités des courbes planes : Une introduction à la géométrie analytique complexe. Cours de 3e cycle, Faculté des sciences de Paris, année universitaire 1969-1970.

[P2] Pham, F.: Formules de Picard-Lefschetz généralisées et ramification des intégrales. Bull. Soc. Math. France 93, 333-367 (1965).

[P3] Puiseux, M. : Recherches sur les fonctions algébriques. Journal de Math. (1) 15, 365-480 (1850).

[P4] Picard, E. -Simart, G. : Théorie des fonctions algébriques de deux variables indépendantes. Vol. I, II. Gauthiers-Villars, Paris 1897 and 1905 (Reprinted by Chelsea Publ. Company, New York 1971).

[P5] Plücker, J. : System der analytischen Geometrie. Berlin 1835.

[P6] Plücker, J. : Theorie der algebraischen Kurven. Bonn 1839.

[P7] Pascal, E. : Repertorium der höheren Mathematik II 1, 2nd ed. Teubner-Verlag, Leipzig-Berlin 1910.

[R1] Reiffen,H.J. - Scheja, G. - Vetter, V. : Algebra. Bibliogr. Inst. 110/110a, Mannheim-Wien-Zürich 1969.

[R2] Reeve, J.E. : A summary of results in the topological classification of plane algebroid singularities. Università e Politecnico di Torino, Rendiconti del Seminaro Matematico 1954/55.

[R3] Riemann, B. : Theorie der Abel'schen Functionen. Journal für reine und angew. Math., Bd. 54, Nr. 14, 115-155 (1857). Also in: Gesammelte Werke 88-144, Leipzig 1876, 2nd ed. 1892.

[S1] Schilling, F. : Über neue kinematische Modelle zur Verzahnungstheorie nebst einer geometrischen Einführung in dieses Gebiet. Zeitschr. f. Math. und Phys., Bd. 51, 1-29. Teubner-Verlag 1904.

[S2] Scholz, E. : Geschichte des Mannigfaltigkeitsbegriff von Riemann bis Poincaré. Birkhäuser 1980.

[S3] Seifert, H. - Threlfall, W. : Lehrbuch der Topologie. Chelsea Publishing Company, New York 1945. English translation : A Textbook of Topology. Academic Press 1980.

[S4] Shafarevich, I.R. : Basic algebraic geometry. Die Grundlehren der math. Wissenschaften in Einzeldarstellungen, Bd. 213, Springer-Verlag, Berlin-Heidelberg-New York 1974.

[S5] Siegel, C.L. : Topics in complex function theory. Vols. I-III. John Wiley & Sons, Inc. New York-London-Sydney-Toronto 1960, 1971, 1973.

[S6] Smith, D.E. : History of Mathematics, Vol. II. Dover Publications, Inc., New York 1958.

[S7] Steinitz, B. : Konfigurationen der projektiven Geometrie. In: Enzyklopädie der Math. Wissenschaften, Bd. III 11, 481-516 (1910).

[S8] Struik, D.J. : Abriß der Geschichte der Mathematik. VEB Deutscher Verlag der Wissenschaften, Berlin 1961. English edition : A Concise History of Mathematics, Dover 1948.

[S9] Schulze-Röbbecke, Th. : Algorithmen zur Auflösung und Deformation von Singularitäten ebener Kurven. Bonner mathematische Schriften 96, 1978.

[S10] Serre, J.P. : Algèbre locale: Multiplicités. Lecture Notes in Math. 11, Springer-Verlag, Berlin-Heidelberg-New York 1965.

[S11] Snow, M. - Snow, W. : Tablet Weaving. Golden Press, New York.

[S12] Speltz, A. : Der Ornamentstil zeichnerisch dargestellt. Bruno Hessling GmbH, Berlin-New York 1904.

[S13] Stallings, J. : On fibering certain 3-manifolds. In: Topology of 3-manifolds and related topics. Prentice Hall 1962.

[S14] Serre, J.P. : Géométrie algébrique et géométrie analytique. Ann. Inst. Fourier 6, 1-42 (1956).

[S15] Sebastiani, M. : Un Exposé de la Formule de Leray-Norguet. Bol. Soc. Bras. Math. 1, No. 2, 47-57 (1970).

[S16] Shioda, T. : On elliptic modular surfaces. Journ. of the Math. Soc. Japan 24, 20-59 (1972).

[S17] Seifert, H. : Über das Geschlecht von Knoten. Math. Ann. 110, 570-592 (1934).

[T1] Thom, R. : Stabilité structurelle et morphogénèse. W.A. Benjamin, Inc. Reading, Mass. 1972.

[T2] Teixeira, F.G. : Traité de courbes spéciales remarquables, Paris 1908. Translated and enlarged edition of : Tratado des curvas especiales notables, Madrid 1905.

[T3] Toomer, G.J. : Diocles on Burning Mirrors. Springer-Verlag 1976.

[T4] Tougeron, F.C. : Ideaux des fonctions différentiables. Ergebnisse der Mathematik und ihrer Grenzgebiete 71, Springer-Verlag, Berlin-Heidelberg-New York 1972.

[T5] Trotter, H.F. : Some knots spanned by more than one unknotted surface of minimal genus. In: Knots, groups and 3-manifolds, Ann. of Math. Studies 84, Princeton University Press, Princeton 1975.

[W1] van der Waerden, B.L. : Algebra I and II. Heidelberger Taschenbücher, Bd. 12 & 23, Springer-Verlag, Berlin-Heidelberg-New York 1971 & 1967.

[W2] van der Waerden, B.L. : Einführung in die algebraische Geometrie. Die Grundlehren der math. Wissenschaften in Einzeldarstellungen, Bd. 51, Springer-Verlag, Berlin-Heidelberg-New York 1973.

[W3] Walker, R.J. : Algebraic curves. Springer-Verlag, New York-Heidelberg-Berlin 1978.

[W4] Weber, H. : Lehrbuch der Algebra, Bd. I & II. Friedrich Vieweg und Sohn Verlag, Braunschweig 1898/99.

[W5] Wunderlich, W. : Ebene Kinematik. BI 447/447a. Mannheim 1970.

[W6] Wassermann, G. : Stability of unfoldings. Lecture Notes in Math. 393, Springer-Verlag, Berlin-Heidelberg-New York 1974.

[W7] Weil, A. : Number of solutions of equations in finite fields. Bull. Amer. Math. Soc. 55, 497-508 (1949).

[W8] Wolf, J. : Differentiable fibre spaces and mappings compatible with Riemann metrics. Michigan Math. Journ. 11, 65-70 (1964).

[W9] Whitney, H. : Complex analytic varieties. Addison-Wesley Publication Company, Reading, Mass. 1972.

[Z1] Zeeman, E.C . : Levels of structure in catastrophe theory illustrated by applications in the social and biological sciences. In: Proceedings of the Int. Congress of Math., Vol. 2, 533-546, Vancouver 1974.

[Z2] Zariski, O. : Algebraic Surfaces. Second supplemented edition. Ergebnisse der Mathematik, Bd. 61. Springer-Verlag, Berlin-Heidelberg-New York 1971.

INDEX

A Roman I before the page number indicates that the reference is to the first part of the lecture notes, i.e. to §§1-7.

A

Abelian integral 628
Addition on cubics 306
Adjoint
  curve 598
  polynomial 601
Algebra
  application in geometry I 69
  of holomorphic function germs 354
  reduced 354
Algebraic
  curve I 79, I 202
  cycle I 274
  local ring 328
  manifold I 119
  de Rham theorem 639
  space 337
  structure I 203
Algorithm
  euclidean (for Puiseux expansion) 514
Analysis and synthesis I 66
Analytic
  set 337
  —, at a point 348
  —, locally 349
  set germ 349
  set germ ideal 351
  subset 349
Angle trisection I 3, I 7, I 15
Arithmetic genus 610
Artin approximation theorem 336
Astroid I 32
  equation I 81, I 86
  genus 619
  natural equation I 86
  Plücker formulae 585
  polar 582
  singular points I 81

B

Barycentric
  calculus I 110
  subdivision I 265
Base point (of a linear system) 615
Basis theorem
  Hilbert 345
  Rückert 345

Bertini theorems 617
Betti number I 270
  of a fibre 550, 573
Bézout's theorem I 227, I 232, I 264, I 277
Birational
  isomorphism 468
  transformation 468
  ——————— group 468
Blowing up a point 462
Boundary (of a chain) I 262
Bouquet (of topological spaces) I 143
Brachistochrone I 27
Braid (torus knots) 551
Branch
  of a set germ 361
  of the multiplicity sequence 504
  point (of a covering) I 147
Brianchon's theorem I 256
Bundle
  Hopf 461
  line 460
  of vector spaces 460

C

Canonical
  divisor 622
  line bundle 622
Cardano formula I 68, I 182
Cardioid I 30
Cartier divisor 622
Cassini curves I 18, I 77
Catacaustic I 31
Catastrophe I 53
  theory I 53
Caustic I 31
  of a parabola I 55
  stable I 55
Cell
  chain, of a triangulation I 266
  decomposition, dual I 264
Central projection I 102, I 111
Chain
  cell- I 266
  closed I 261
  p- I 261
Characteristic
  exponent, first 484
  term of the Puiseux expansion 411
Chern class 623
  numbers 611
Chevalley dimension 368
Chow
  ring I 275
  theorem I 140
Circle I 3
  equation I 73, I 117, I 127, I 145

Circle
- evolute I 19
- quadrature I 3, I 7
- Seifert 537, 551

Cissoid (of Diocles) I 9, I 91
- equation I 74
- polar 579

Class
- Chern 623
- intersection (in homology theory) I 268
- of a curve I 254, 576

Classification
- of compact orientable 2-manifolds 324
- of cubics I 87, I 283
- of curves I 87, I 91, 323
- of quadrics I 278
- of singularities 323

Clebsch formula 619

Closed
- chain I 261
- p-form 635

Coboundary, Leray 641

Coherent sheaf 360

Cohomology
- class 653
- de Rham 635
- de Rham, group 635

Collinear points I 243

Collineation I 111, I 115, I 123, I 137
- homogeneity of polynomials preserved under I 204

Complete intersection 369

Completion
- of a curve I 207
- of the affine plane I 113

Complex, simplicial I 260

Conchoid of Nicomedes I 13
- equation I 75

Condition, linear (in a linear system) I 242

Configuration
- Desargues' I 248
- Kummer's 586
- of a quartic 586
- of lines on a cubic surface 586
- Pascal's I 248
- tactical I 296

Conformal mapping, main theorem I 280

Conic section I 4, I 100
- general equation I 74
- organic generation I 6

Conjugation (as a mapping) I 137

Connection
- Ehresmann 663
- Gauss-Manin 692

Connectivity, Riemann 610

Contact
- exponent 484

Contact
    maximal 481
    ———, stability under blowing up 491
Continuity principle (of Poncelet) I 136
Contour (caustic) I 41
Contraction of graphs 529
Coordinates
    affine I 124, I 136
    cartesian I 106
    development of I 66
    exchange I 116, I 137
    homogeneous I 110, I 114, I 123, 456
    local I 119
    neighbourhood I 119
    system, homogeneous I 123
    triangular I 108
Covering
    branched I 147, I 150
    ———, of Riemann sphere I 154
Cremona
    group 468
    transformation 468
Cross ratio I 106
Crossing, normal (of divisors) 498
Cube
    cell decomposition I 266
    duplication of I 3, I 6, I 10
Cubic I 87, I 118, I 207, I 245, 689
    as abelian variety I 308
    as topological space I 163
    classification I 304
    cohomology 684
    genus 619
    Hesse normal form I 293, 655
    Legendre normal form 656
    Plücker formulae 584
Curve
    abstract I 203
    algebraic I 79, I 203
    ———, affine I 172, I 191, I 203
    ———, projective I 203
    ——— ———, complex I 203
    as complex analytic subset I 139
    as topological subset I 140
    Cassini I 18, I 77
    class I 254
    classification I 88
    complete I 203
    complex I 139
    ———, projective I 203
    defined by a differential equation I 79
    dual I 86, I 253, 576
    elliptic I 319, 324
    embedding in a space I 219
    equation I 71, I 86
    exceptional 463, 470

Curve
  genus 324
  geometric (Descartes) I 79
  Hessian I 289
  hyperelliptic 620
  in the complex projective plane I 139
  in the projective plane I 117
  intersection number I 238
  irreducible I 204
  Lissajou I 65
  order I 88, I 207
  pencil I 240
  plane I 202
  ———, algebraic I 203
  ———, complex algebraic I 203
  rational I 280
  transcendental I 79
  Watt I 58
  with multiple components I 239
Cusp (of a curve) I 10
Cycle
  algebraic I 274
  bounding I 260
  fundamental I 269
  horizontal family 665
  intersection I 267, I 274
  p- I 260, I 262
  vanishing I 162
Cycloid I 20
  equation I 78
Cycloidal
  gear I 33
  pendulum I 30

D

Decomposition (of a set germ) 361
Deformation
  of a curve 620
  of a singularity 366
Degree
  n, homogeneous I 194
  of a curve I 207
  of a polynomial I 174, I 194
Delian problem I 3, I 6, I 10
Derivative I 71
Desargues'
  configuration I 248
  theorem I 248
Descartes, folium of I 88, I 91, 589
Dialectic (in mathematics) I 66, I 82
Diffeomorphism I 126
Differentiable structure I 119
Differential equation, hypergeometric 686
  monodromy 687
Differential form
  1st species 628
  ——————— (on a curve) 649

Differential form
    2nd species 629
    3rd species 629
    and cohomology 635
    closed 635
    exact 635
    holomorphic 628
    meromorphic 628
    on a curve 634, 654
    on a Riemann surface 627
    (p,q) 637
    rational 638
Dimension
    Chevalley 368
    Krull 368
    of a set germ 367
    of an analytic subset 367
    Weierstrass 368
Disappearing line (in central projection) I 105
Discriminant I 181, I 196
    of polynomials of degree 3 and 4 I 182
Division algorithm, euclidean I 174
Division theorem (for convergent power series) 339
    special 340
Divisor
    canonical 624
    Cartier 622
    common (of polynomials) I 177
    greatest common I 175
    intersection number of 623
    of a function (principal divisor) I 313
    of a singular curve 529
    on the projective plane I 240
    Weil 622
Dodecahedron I 266
Double point (of a curve) I 213
Dual
    cell decomposition I 264
    curve I 86
    ——— (polar) 576
    graph 509
    projective plane I 251
Dualisation I 251
Duality principle I 252, I 255
Duplication (of the cube) I 3, I 6, I 10

E

Ehresmann connection 663
Elementary
    chain (of a resolution graph) 522
    symmetric functions, main theorem on 344
Elimination (in equations) I 179
Ellipse I 4, I 73
    parallel curves of I 41
Ellipsograph I 8

Elliptic
    curve   I 306, I 319, 324
    function   I 318
    integral   I 318, 656
    modular surface   661
Embedding, Segre   500
Envelope   I 31, I 82
Epicycles   I 19
Epicyclic curves   I 19, I 77
Epicycloid   I 19
Epitrochoid   I 23
Equation
    affine   I 207
    from a parametrisation   I 180
    of a curve   I 206
    of a hypersurface   355
    of a polynomial, implicit   334
Essential term of Puiseux expansion   411
Étale topology   335
Euclidean
    algorithm   I 174
    ———— (for Puiseux pairs)   514
    geometry   I 117, I 120
    ring   I 174
Euler characteristic
    of a curve   614
    of a fibre   550
    of a knot spanning surface   540
    of a singular fibre   603
    of a surface   609
    of the Milnor fibre and resolution graph   568
    of the Milnor fibre and the multiplicity sequence   569
Euler's partial differential equation   I 197
Exact p-form   635
Exceptional
    curve   463, 470
    ——— of 1st kind   463
    hypersurface   470
    line   463
Exponent, first characteristic   484

F

Factorial ring (= UFD)   I 174
Feynman integral   574
Fibration (of a knot)   543
Fibre (see also Milnor fibre)
    1st Betti number   550, 573
    genus   544
Fibre bundle
    locally trivial   I 130
    locally trivial differentiable   542
Folium of Descartes   I 88, I 91, 590
Function(s)
    doubly periodic   I 313
    elliptic   I 318
    holomorphic   329
    homogeneous   I 194

Function(s)
- main theorem on symmetric 344
- ring of holomorphic 329
- transcendental I 79
- unfolding of 573
- Weierstrass $\wp$- I 303, I 315
- ———, differential equation I 317

Function germs
- algebra 354
- holomorphic 330
- ring of holomorphic 330, 354
- theorem on implicit 335, 348

Fundamental
- class I 264, I 269
- cycle I 269

G

Gaussian hypergeometric series 687
Gauss-Manin connection 692
General polynomial 342
Generation, organic (of curves) I 6
Genus
- arithmetic 610
- formula (for curves), generalised 624
- ——— of Clebsch and Noether 601
- of a connected orientable surface 540
- of a knot 541
- of a surface 609
- of curves 324, 576, 596, 603, 610, 614, 619, 621
- Todd 611
- virtual 613

Geometric monodromy group 664
Geometry
- analytic I 66
- ———, development of I 69
- complex-projective I 136
- euclidean I 117, I 120
- of position I 106
- projective I 90, I 116
- ———, development of I 102
- Riemannian I 119

Graph, dual 509
Grothendieck topology 335
Group
- complex-projective-linear I 137
- Cremona 468
- real-projective-linear I 115, I 123

H

Harmonic mean point 591
Henselian local ring 335
Henselisation 335
Hensel's lemma 335
Hessian
- curve I 289

Hessian
    group I 297
    matrix I 289
    normal form of cubics I 293, 655
Hexagon I 246
    and quadric I 246
Hilbert's
    basis theorem 345
    Nullstellensatz I 192, 352
Hodge decomposition 637, 649
Holomorphic function 329
    germs 330
    ——, algebra of 354
    ——, ring of 330, 354
    -s, ring of 329
Holonomy group (of a fibration) 664
Homogeneous
    coordinate system I 123
    coordinates 456
    equation I 117
    function I 194
    of degree n I 194
    weighted I 195
Homogenisation I 117, I 199
Homology
    class I 262, I 264
    group I 260
    ——, p^{th} singular I 263
    —— of a submanifold I 272
    —— of a projective algebraic subset I 273
    of Milnor fibre 573
    theory, singular I 263
Hopf
    fibration I 138
    line bundle 461
    ——, complex, over $P_1(\mathbb{C})$ 531
Horizontal
    family of cycles 665
    —— homology classes 665
    tangent space 663
Hurwitz formulae 613
    generalised 626
Hyperbola I 5, I 72, I 127, I 145, I 157
    projective equation I 117
Hyperboloid I 161
Hypercohomology, de Rham 637
    algebraic 639
    analytic 637
Hyperelliptic curve 620
Hypergeometric differential equation 686
    monodromy group 687
    series 687
Hyperplane I 123
Hypersurface 355
    affine algebraic I 191
    equation 355

Hypersurface
    exceptional 470
    projective-algebraic I 199
Hypocycloid I 19
    star curve, astroid I 32
    three-cusped (Steiner's) I 32
    ————, polar 581
Hypotrochoid I 22

I

Icosahedron I 266
Ideal, maximal 359
    of an analytic set germ 351
Implicit
    equation of a polynomial 334
    function theorem 334, 348
Index theorem of Atiyah-Singer 611
Infinitely near point 475, 504
    ν-tuple 476
Infinitesimal calculus in geometry I 71
Inflection point
    configuration (of cubics) I 295, I 320
    improper I 288
    number of 583
    of cubic I 284, I 305, I 320
    r-tuple I 288
Integral I 71
    abelian 628
    elliptic I 318, 656
    Feynman 574
    of 1st species 628
    of 2nd species 629
    of 3rd species 629
    rectification I 86
Intersection
    class I 264
    complete 369
    cycle I 267, I 274
    multiplicity I 231
    ———— of cycles I 269
    number
    ——, in the multiplicity sequence 512
    —— of curves I 238, I 274
    —— of cycles I 269, I 274
    —— of divisors 623, 632
    of curves I 90, I 208, I 227
    ring I 260, I 270, I 275
    theory I 90, I 227, I 232
Invariant
    j- I302, I322, 657
    of curve, global 576
    order of curves I 88, I 207
Irreducible
    component (of a set germ) 361
    curve I 204
    in integral domain I 173
    set germ 361

Isobaric I 195
Isolated singularity 359
Isomorphic, as algebraic manifolds I 313
Isomorphism
birational 468
(mapping germ) 356
Poincaré 271

J

Jacobian variety 322

K

Kind of a differential form 628
Knot
fibred 542, 543
genus 541
Neuwirth-Stallings 543
of singularity 535
torus I 223
Krull
dimension 368
principal ideal theorem 369
Kummer configuration 586

L

Lattice (in $\mathbb{C}$) I 312
Legendre normal form (of cubics) 656
Lemniscate I 18
Length, proper (of a multiplicity sequence) 507
Leray coboundary 641, 645
Line I 2, I 91, I 207
at infinity I 114
bundle 460
———, canonical 624
———, Hopf 401
———, complex (over $P_1(\mathbb{C})$) 531
complex-projective I 140
equation I 72
exceptional 463
pencil I 240
projective I 116
through a point 459
Linear system (of curves) I 240, 615
dimension I 240
given by a linear condition I 242
of cubics I 281
Linear transformation 455
Lissajou curves I 65
equation I 78
Local ring 328
henselian 335
Localisation 328
Locally analytic 349
Locus I 3
Logarithmic pole (of a differential form) 642
Lojaziewicz trick 343

M

Main theorem
    of conformal mapping I 280
    on elementary symmetric functions 344
Malgrange
    preparation theorem 345
    trick 340
Manifold
    abelian I 313
    classification of compact orientable 2-dim. 324
    complex I 119
    ——— algebraic I 119
    development of concept 120
    differentiable I 119
    Jacobian 322
    n-dimensional topological I 118
    projective plane as I 118
    real algebraic I 119
    —— analytic I 119
    Riemannian I 119
    Stein 636
Mapping
    germ 355
    main theorem of conformal I 280
    rational 468
    regular I 311
Maximal contact 481
    stability under blowing up 491
Maximal ideal 359
Mean point, harmonic 591
Metric, Riemannian I 119
Milnor fibration 544
    and Newton polygon 574
    1st Betti number 550
    homology 573
    in the standard resolution 561
Milnor number 573
Möbius band I 130, 462, 533
Modular surface, elliptic 661
Moduli I 322
    space 322
    ———, dimension 322
Monkey saddle 550
Monodromic 664
Monodromy 691
    geometric 574
    theorem 692
Monodromy group 665
    geometric 664
    of a differential equation 687
    of a fibration 665
    of the hypergeometric equation 687
    projective 679
Multiplicity
    at infinitely near points 598
    of a curve at a point I 212, I 225

Multiplicity
of cycles I 269
of intersection of two curves I 90, I 208
Multiplicity sequence 504
and Puiseux expansion 512
and resolution graph 521
proper length of 507
system 505

N

Neighbourhood
formal, in $\mathbb{C}[[x_1,\ldots,x_n]]$ 336
germ 336
in classical topology 329
small 328
Zariski open 328
Neil's parabola (see Parabola, Neil's)
Nephroid I 31
Neuwirth-Stallings knots 543
Newton polygon 370, 477
and Milnor fibration 574
Nilpotent 354
Noether
formula 614
normalisation lemma 354
Noetherian ring 347
Non-orientability I 130
Normal crossings (of divisors) 498
Normal form, Weierstrass I 302
Nullstellensatz
Hilbert I 192, 352
Rückert 351

O

Octahedron I 266
One-sided surface (in space) I 130
Order (of a plane curve) I 88
Organic generation
of conic sections I 6
of the cissoid I 11
of the conchoid I 15
Orientability I 130
Outline I 41, I 53, I 104, I 184

P

Parabola I 4, I 72, I 127, I 145
diverging I 92, I 163
generalised I 87, I 214, I 223
Neil's (= semicubical) I 87, I 88, I 101, I 118, I 127, I 215, I 223, I 280, 473, 496
———, rectification I 87
projective equation I 118
Parallel
curve I 41
——— of a parabola I 55

Parallel
projection I 104
Parametrisation, rational (of curves) I 77, I 280
Pascal
configuration I 248
theorem I 247
Pasting
of disc bundles 533
topological I 153
p-
boundary I 262
chain I 261
cycle I 262
———, bounding I 260
Pencil, linear (of curves) I 240
Pericycloid I 19
Period
of a differential form 628
relations (Riemann) I 321
residual 629
Peritrochoid I 24
Perspectivity I 111
p-form
closed 635
exact 635
rational on $P_n(\mathbb{C})$ 646
with logarithmic pole 642
Plane
complex projective (see projective n-dim. space)
dual projective I 251
real projective I 114
Platonic solids I 266
Plumbing 533
Plücker formulae 576, 584
Poincaré
duality theorem I 270
isomorphism I 271
Point
at infinity I 90, I 105
infinitely near 475, 504
multiplicity at infinitely near 598
ν-tuple infinitely near 476
ordinary n-tuple I 225
regular I 213, 363, 686
singular I 213, 363, 686
Polar 592
coordinates I 78
description 589
general 594
in a circle 578
of a curve 587
of a curve relative to a point 588
of a family of curves 592
r^{th} 588
Pole (of a differential equation) 642
Pole-polar relationship 578
Polygon, Newton 477

Polyhedra, regular I 266
Polynomial I 172
adjoint 601
as implicit equation 334
associated I 200
division I 174
general 342
homogeneous I 194
————, weighted I 195
primitive I 176
ring I 174
Weierstrass 338
Position
general I 243
special I 243
Power series
ring of convergent 331
ring of formal 331
————, neighbourhood 336
Preimage
proper 470
strict 470
total (of the standard resolution) 529
Preparation theorem
Malgrange 345
Weierstrass 338
Prime element I 173
Prime ideal chain 368
Principal congruence subgroup 679
Principal divisor (of a function) I 313
Principal ideal theorem, Krull 369
Projection
centre of I 102
stereographic I 142
Projective
algebraic hypersurface I 199
———— subset I 139
geometry I 116
line I 116
——, complex I 140
linear group, complex I 137
————, real I 123
linear subspace I 123
monodromy group 679
plane, real I 114
properties (Poncelet) I 106
space 459
——, complex n-dimensional I 136, I 198
————, intersection ring I 260
space, real n-dimensional I 122
subspace, linear I 123, I 137
viewpoint I 102
Projectively related figures I 111
Projectivity I 113
Puiseux
chain 522
pairs I 214

Puiseux
pairs and multiplicity sequence 512
——— and resolution graph 522
Pure-dimensional (set germ) 367

Q

Quadratic transformation 455, 498
Quadrature (of the circle) I 3, I 7
Quadric I 91, I 207, I 243, I 246, I 278, 455, 619
and hexagon I 246, I 247
projective I 142
Quartic I 207, 455
Quasihomogeneous (see Weighted homogeneous)
Quintic I 207

R

Radical (of an ideal) 351, 354
Rational mapping 468
Rectification (of curves) I 86
Reduced algebra 354
Reducible set germ 361
regular I 82
curve I 79
mapping (of algebraic manifolds) I 311
point I 213, 363
polyhedron I 266
singular point 686
Regularity theorem 692
Relatively prime I 175
Remainder (in polynomial division) I 174
Representative (of a set germ) 350
Residual period 629
Residue 641
Residue theorem 644
Resolution
and Puiseux expansion 522
by quadratic transformations 455, 496
graph 503, 506
———, contractible 529
of singularities 323, 325, 366, 503
standard 498
topology of 531
Resultant (of polynomials) 178, 196, 202
computation of intersection of curves 230
de Rham
algebraic theorem 639
cohomology group 635
hypercohomology 637
theorem 636
Riemann
connectivity 610
period relations I 321
sphere I 141
Riemann surface I 119, I 140, 564

Riemann surface
abstract I 156
concrete I 156
covering the sphere I 153
of a function I 155
Riemannian
geometry I 119
manifold I 119
metric I 119
Riemann-Roch theorem 611
Ring
algebraic local 328
local 328
———, henselian 335
noetherian 345
of convergent power series 331
of entire algebraic functions 333
of formal power series 331
of holomorphic function germs 330, 354
of holomorphic functions 329
of rational functions in Zariski neighbourhoods 328, 331
spectrum of 332
Rotation reciprocator I 33, I 38
Rouché's theorem 353
Rückert
basis theorem 345
Nullstellensatz 351

S

Schemes 332
Segre embedding 500
Seifert circles 537, 551
Self-intersection number (of zero section of line bundle) 531
Set
affine-algebraic 191
analytic 337
Set germ
analytic 349
————, ideal 351
decomposition 361
morphisms 357
representative 350
Sextic I 207
Sheets (of a covering) I 152
Shell curve (Dürer) (see also Conchoid) I 47
σ-process 462, 491
n-dimensional 464
Simplex I 260
singular I 263
Singular
homology group I 263
———— theory I 263
point I 81, I 211, I 213, I 258, 363
——— (on outline) I 52
simplex I 263

Singularity I 10, I 82, 350
isolated 359
resolution of 323, 326
set 363, 366
Solid, Platonic I 266
Space
algebraic 337
projective 459
Species (= kind)
of a differential form 628
of integral of a differential form 628
Spectrum (of a ring) 332
Sphere
covering of I 153
Riemann I 141
Spider lines I 24
Spiric sections (of Perseus) I 16, I 76
Stallings, Neuwirth-Stallings knots 543
Standard resolution 498
Star (of a point in a triangulation) I 266
Stein manifold 636
Strata 366
Stratification 366
Structure
algebraic (of a curve) I 203
differentiable I 119
Study's lemma I 192
Subdivision, barycentric I 265
Submanifold, homology class of a I 272
Subset
analytic 349
complex-analytic I 139
projective-algebraic I 139
topological I 140
Subspace, projective linear I 123, I 137
Surface (see projective curve)
elliptic 660
one-sided, in space I 130
Swallowtail I 188
System (of multiplicity sequences) 505

T

Tangent
equation I 80
problem I 71, I 79, I 211
to a curve I 213, I 227
Tangent space
horizontal 663
vertical 663
Tautochrone I 29
Tetrahedron I 266
group I 299
t-general 338
Theorem
Artin approximation 336
Atiyah-Singer 611

Theorem
- Bertini 617
- Bézout I 227, I 232, I 277
- Brianchon I 256
- Chow I 140
- Enriques-Chisini 516
- Liouville I 314
- Milnor 543
- M. Noether 518
- Pascal I 247
- Poincaré duality I 270
- Riemann-Roch 611
- Rolle I 286
- Rouché 353
- Seifert 536
- Torelli I 322

Todd genus 611
Topology
- classical 329
- Étale 335
- Grothendieck 335
- of resolution 531

Toroid I 41
Torus
- complex I 313
- equation I 76
- higher-dimensional I 320
- knot I 223, 544, 550
- ——, as braid 551

Transformation
- birational 468
- Cremona 468
- group, Cremona 468
- linear 455
- quadratic 455, 496
- Tschirnhaus I 182

t-regular 338
Triangular coordinates I 108, 456
Triangulation I 260
Trick
- of Lojaziewicz 343
- of Malgrange 340

Triple point I 213
Trisection of angle I 3, I 7, I 15
Trochoid I 21
- rotation-reciprocator I 33, I 38

Tschirnhaus transformation I 182
Tubular neighbourhood I 132

U

UFD (= factorial ring) I 174
Unfolding (of a holomorphic function) 573

V

Vanishing
- cycle I 162

Vanishing
line I 105
point I 105
Variety I 191
abelian I 306
Vertical tangent space 663
Virtual genus 613

W

Wankel motor I 38
Watt curves I 58
Wave front (of a parabola) I 55
Weierstrass
dimension 368
formula 339
normal form I 302
$\wp$-function I 303, I 316
——————, differential equation I 317
polynomial 338
preparation theorem 338
Weighted homogeneous = quasi-homogeneous = isobaric I 195
Weil divisor 622
Wheel curve I 19

Z

Zariski-open neighbourhood 328
Zero I 190
of a cubic I 68, I 183
set I 190